MECHANICAL DRAWING

CAD-COMMUNICATIONS

TWELFTH EDITION

THOMAS E. FRENCH•
CARL L. SVENSEN•
JAY D. HELSEL•
BYRON URBANICK

GLENCOE
McGraw-Hill

New York, New York   Columbus, Ohio   Mission Hills, California   Peoria, Illinois

Glencoe/McGraw-Hill

A Division of The **McGraw·Hill** *Companies*

Copyright ©1997 by Glencoe/McGraw-Hill. Previous copyrights 1990, 1985, 1980, 1974, 1966, 1957, 1948, 1940, 1934, 1927, and 1919 by McGraw-Hill, Inc. All rights reserved. Except as permitted under the United States Copyright Act, no part of this publication may be reproduced or distributed in any form or by any means, or stored in a database or retrieval system, without the prior written permission of the publisher.

Send all inquiries to:
Glencoe/McGraw-Hill
3008 W. Willow Knolls Drive
Peoria, IL 61614-1083

ISBN 0-02-667958-2 (Student Text)
ISBN 0-02-677959-0 (Teacher's Resource Binder)
ISBN 0-02-667961-2 (Student Workbook)

Printed in the United States of America

2 3 4 5 6 7 8 9 10 VHJ 00 99 98 97

CONTENTS

ABOUT THE AUTHOR

About the Author

Jay D. Helsel is a professor and chairman of the Department of Industry and Technology at California University of Pennsylvania. He completed his undergraduate work at California State College and was awarded a master's degree from Pennsylvania State University. He received his doctoral degree in educational communications and technology from the University of Pittsburgh.

Dr. Helsel has worked in industry and has taught drafting and a variety of laboratory and professional courses at the high school, college, and university levels. During the past 35 years, he has also owned and operated Technical Graphics, a technical art and illustrating company. His work appears in many technical publications. He is a member of many professional organizations, including the American Society for Engineering Education and the International Technology Education Association.

Dr. Helsel is coauthor of *Mechanical Drawing–CAD Communications, Engineering Drawing and Design, Fundamentals of Engineering Drawing, Computer-Aided Engineering Drawing,* and *Engineering Drawing and Design Workbook,* and is the author of many other technical publications.

Acknowledgments

The publisher gratefully acknowledges the cooperation and assistance received from many persons and companies during the development of *Mechanical Drawing.* Special recognition is given to the following persons for their contributions:

for editing the text and creating CAD illustrations,
  Jody James
  James Editorial Consulting
  Oviedo, Florida

for revising Chapter 1,
  David Sinclair
  Instructor, Design Technology/
    Engineering
  American River College
  Los Rios Community College
    District
  Sacramento, California

for contributing to Chapter 6,
  Patrick McCuistion
  Ohio University
  Athens, Ohio

for writing Chapter 13,
  Michael Eleder
  Instructional Design Systems
  Orland Park, Illinois

for revising Chapter 14 and writing Chapters 15, 16, and 24,
  Ron Shea
  Quality Corporation
  Rochester, Washington

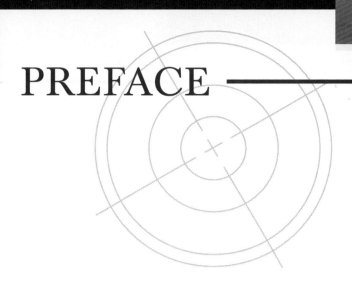

PREFACE

From the very first edition of *Mechanical Drawing*, written by Thomas E. French and Carl L. Svensen in 1919, the authors carefully prepared each chapter to help students learn to visualize in three dimensions, to build imaginations, to think precisely, and to understand the language of industry. Although it was one of the first drafting textbooks written, it was so well oriented to industrial drafting practices that it established the basic concepts for drafting education. Each subsequent edition was updated to conform with changes in standards set by the American National Standards Institute (ANSI) and with advancements in the technologies of industry.

This twelfth edition is no exception. The text and illustrations have been thoroughly updated to conform to the latest ANSI standards. The coverage of CAD has been rewritten and expanded to reflect the current state of the technology, both hardware and software. We have also added a new chapter on the various roles drafters may be asked to play in a company's in-house desktop publishing efforts.

No matter how much technology alters the tools of drafting, knowledge of the concepts remains the foundation for a successful career. Therefore, even though this edition provides information specific to manual and CAD drafting, it greatly emphasizes basic drafting elements and concepts. Regardless of a teacher's available equipment, time, and experience, this book provides the essential skills students will need to pursue a career in drafting.

Chapter 1 begins with a discussion of the variety of careers and positions available to drafters in industry today. "Plans—For Industry, Your Career, Your Future" describes various types of drafting careers and emphasizes the specific preparation needed for each.

Chapters 2 through 12 have been updated but remain conceptually unchanged. Since today's students are uniformly familiar with the decimal system, many of the drawings that formerly appeared with fractional dimensions have been changed to incorporate decimal inches instead. Students are therefore able to concentrate initially on drafting concepts, and later on the incorporation of fractions as needed specifically for architectural and structural drafting.

Chapter 13, "Desktop Publishing with CAD," is an entirely new chapter designed to broaden students' understanding of drafting-related tasks. Many industries now act as their own publishing houses, producing technical manuals, newsletters, and a variety of other documents that incorporate text and graphics. Often, the task of preparing these documents falls to the drafting department, since that department already has the capability of producing technical drawings. Today's drafting students must therefore have knowledge not only of how to create drawings but also of how to prepare text and design pages.

Chapter 14 provides insights in current management techniques for graphic communication of all types. "Managing Graphic Communication" emphasizes that the system is the solution to storage and retrieval of drawings and electronic drawing files. The networking of information from CAD or design studies is essential for creating accessibility, reproducibility, and distribution.

A basic knowledge of computer-aided drafting (CAD) is becoming a job requirement even for entry-level drafters. Therefore, Chapters 15 and 16 are devoted to helping students understand more about the concepts and techniques of CAD and how they differ from manual drafting. Chapter 15 is a conceptual chapter that explains in general how CAD systems work, using examples from two widely used CAD software packages. Chapter 16 is a hands-on tutorial designed to be read and worked through at a CAD station. Written for the Auto-CAD software, this chapter allows students to get the "feel" of creating a finished CAD drawing.

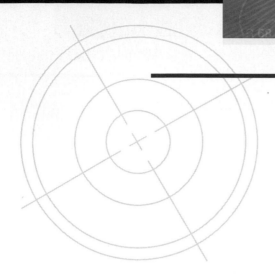

The remaining chapters explore various special topics within the drafting field. Chapters 17, 20, 21, 23, and 24 address drafting specialties such as welding drafting, architectural and structural drafting, map drafting, electrical and electronics drafting, and aerospace drafting. Chapters 18, 19, and 22 concentrate on advanced topics such as surface developments, intersections, cams and gears, and the use of graphs, charts, and diagrams in drafting.

In this edition, we have added elements to the beginning of each chapter to help focus student interest. Objectives help students understand the purpose of the chapter. In addition, we have moved the vocabulary lists to the beginning of each chapter so that students know in advance to look for these terms as they read the chapter.

At the end of the chapters, learning activities, review questions, and problems have been expanded and updated. We have also added a new element, "School-to-Work: Solving Real-World Problems." The purpose of these problems is to give students experience in solving drafting problems such as they might encounter on the job. These problems range from design to problem-solving, and they require the use of higher-level thinking skills.

It is hoped that faithful users of this text will again find continuity that enhances their coursework and builds a new technology base. We thank the many teachers who helped shape this edition. Their continued suggestions and support are gratefully solicited.

TWENTY-FIRST CENTURY DRAFTING

IMAGINATION–COMMUNICATION–INNOVATION

Imagine! What do you think you'll need for a better lifestyle in the twenty-first century? What kind of house will you live in? What kind of car will you drive? What magic will your computer appear to perform? If you stretch your imagination, you can start thinking about how things will be designed in the future. All designers need creative imaginations. Walter Gropius, an influential German architect, once said, "The gift of creative imagination is more important than all technology." Technology is the result of creative researchers, engineers, and designers thinking about what will be needed, tomorrow and beyond (Fig. I-1).

When you have an idea, how do you tell others about it? You could talk, write, or draw a picture. In industry, ideas about a design are often communicated through a special kind of technical drawing process called *drafting*. Drafting is a form of communication that is technical and very exact. Drafters the world over have agreed to use the same lines and symbols and the same methods for drawing (Fig. I-2). For this reason, drafting is called "the universal language of industry." Technical drawings are used for communication in industry because they are the clearest way to tell others what to make and how to make it. Think how hard it would be to tell a friend in words about the shape and size of all the parts in an automobile!

Fig. I-1 *Any product (or building) must first be imagined before it can become a reality.* (Blocks & Materials ™ for AccuRender™/KETIV Technologies, Inc.)

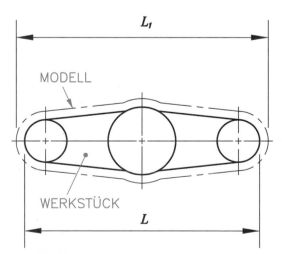

Fig. I-2 *The same lines, symbols, and methods are used by drafters all over the world. What advantages do you think this might provide?* (Courtesy of Cloyd Richardson)

Figure I-3 describes a simple tool, a V-block. By reading this description, can you imagine how the V-block will look? Could you make one from this description? Figure I-4 is a drawing of this same V-block. See how much easier it is to understand the shape and size from the drawing than from the written description. It would be much easier to make the V-block from this drawing.

THE V-BLOCK IS TO BE MADE OF CAST IRON AND MACHINED ON ALL SURFACES. THE OVERALL SIZES ARE TWO AND ONE-HALF INCHES WIDE, AND SIX-INCHES LONG. A V-SHAPED CUT HAVING AN INCLUDED ANGLE OF 90° IS TO BE MADE THROUGH THE ENTIRE LENGTH OF THE BLOCK. THE CUT IS TO BE MADE WITH THE BLOCK RESTING ON THE THE THREE-INCH BY SIX INCH SURFACE. THE V-CUT IS TO BEGIN ONE-QUARTER INCH FROM THE OUTSIDE EDGES. AT THE BOTTOM OF THE V-CUT, THERE IS TO BE A RELIEF SLOT ONE-EIGHTH INCH WIDE BY ONE-EIGHTH INCH DEEP.

Fig. I-3 *Read this description of a V-block. Is it easy to picture how the block should look?*

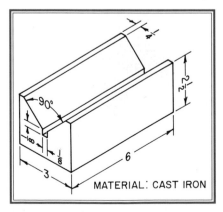

Fig. I-4 *A technical drawing of the V-block shows its size and shape. This drawing communicates the same information as Fig. I-3, but in a more efficient way.*

Imagination and communication lead to *innovation*, the introduction of something new. Often, one new product leads to the development of others. Such products are called *spin-offs*.

Consider the wheel. It was invented during the Bronze Age, a period of human culture that began around 3500 B.C. When fastened to an axle and wagon, the wheel makes it possible for animals to move greater loads. From this use of the wheel came countless other applications, or spin-offs.

Ancient innovations included the grindstone, the pulley, and the spinning wheel. In more recent times gears, cams, rotating shafts, propellers, and turbine engines have evolved. From casters on handcarts or skateboards, to doorknobs and steering wheels, the spin-off is generated by creative imaginations. The people who work on spin-offs are creative, and they are good at communicating their ideas so that the new products can be made (Fig. I-5).

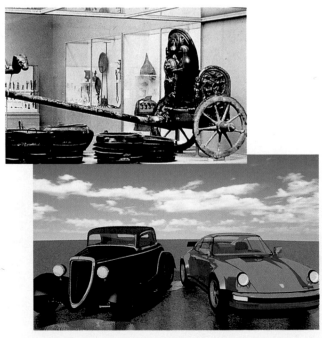

Fig. I-5 *One innovation leads to another. The chariot was one of the first wheeled vehicles. What do you think it has in common with a modern automobile?* (A. Bettmann Archive/B. Blocks & Materials™ for Accurender™/KETIV Technologies, Inc.)

ACHIEVEMENTS THROUGH DESIGN AND DRAFTING

Think of some of the world's most outstanding engineering achievements: the United States' space shuttle, the British-French *Concorde* supersonic airplane, the Japanese high-speed trains. These achievements required careful planning by teams of experts. One of the experts on such teams is the drafting technician (drafter).

If you could examine the technical drawings of all types of products, you would learn about the importance of drafting details. The drawings for space vehicles, automobiles, buildings, electronic equipment, and all other products are essential to manufacturing. Products are made following these drawings.

Behind every major product or outstanding achievement are men and women of exceptional ability, determination, and creative imagination. They serve on engineering teams that are planning for your tomorrow.

For a design project to go right and finish right, it has to start right. Drafting is an important part of starting any project. That's why drafting "know-how" is essential to all who are on the engineering design team.

Without a command of the graphic language, engineers and designers would find it nearly impossible to communicate their ideas effectively to members of their teams. Technical drawing is one of the most important courses for an engineering or design technology career (Fig. I-6).

This textbook provides opportunities to learn drafting standards and techniques through hands-on practice. It assumes from the beginning that you are not experienced or skilled in drafting. It starts with the basics and moves on to more complex techniques.

By using this textbook, you can start to build the best possible portfolio (collection) of traditional (hand-drawn) and CAD drawings for a successful venture into drafting. You have a chance to add professional dimensions to your career plans for tomorrow.

Fig. I-6 *Members of the engineering design team work together to solve problems and create new products.* (Courtesy of General Motors Corp.)

PLANS–FOR INDUSTRY, YOUR CAREER, YOUR FUTURE

OBJECTIVES

Upon completion of Chapter 1, you will be able to:

○ Describe the role and importance of the drafting technician (drafter).
○ Define key terms used in the field of drafting.
○ List and describe the four levels of graphic communication.
○ Describe the importance of drafting as a means of communicating technical ideas.
○ Identify careers available within the engineering design team.

VOCABULARY

In this chapter, you will learn the meanings of the following terms:

- computer-aided design and drafting (CADD)
- computer literacy
- conceptual design
- concurrent engineering
- design drafting technician
- design drawings
- design engineer
- designer
- drafter
- drafting technology
- engineer
- production drawings
- prototype
- rapid prototyping
- specifications
- technical drawings
- technical promotional drawings

Choose a new product you would like to buy. Do you think a ten-speed bike, a dirt bike, a stereo sound system, a home computer game, or a pair of sports shoes could be produced without detailed plans? The people who work on the design for a new product can affect our style of living. **Design drawings** (drawings that describe the design of a new product) and **specifications** (information about size, shape, materials, and surface finishes) are needed for all new products (Fig. 1-1).

THE ENGINEERING DESIGN TEAM

You will find various types of industrial plants in or near every major city in the United States. Each of these plants that make our products has an engineering design team to produce design drawings and specifications that describe their products. The product description is generally defined in a specification, and the product's size and shape are detailed in a set of **technical drawings**. The drawings define the shape of the product by using lines, dimensions, and notes (Fig. 1-2).

Technology has produced useful items that are important parts of our daily lives. Each year, new discoveries and inventions help make our lives more interesting. From ordinary products we use every day to extraordinary ones like the space shuttle, technology will continue to shape our lives and our hopes for the future.

The new products you buy or see advertised on TV or in magazines, catalogs, and newspapers have been designed by men and women on an engineering design team. An understanding of the role of the engineer, designer, and drafter will help you to see how new ideas are created and detailed so that they work. You will realize how hard the design team works to develop designs to meet ever-changing lifestyles.

There is an old saying that when something doesn't work, "it's back to the drawing board." This term is used to mean that the design has to be modified or improved. Sometimes it means the design team has to develop new products for a

Fig. 1-1 *Even a simple device such as a handle cannot be manufactured without technical drawings and specifications.* (Courtesy of Computervision Corp.)

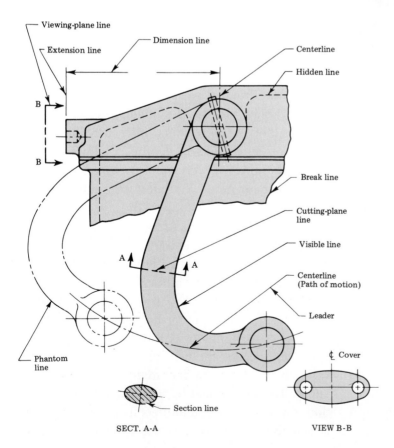

Fig. 1-2 *Various kinds of lines are used on technical drawings. For example, a phantom line (long lines alternating with two short lines) is used to show alternate positions of movable parts.*

Fig. 1-3 *Perhaps you will become a designer or drafter and help create products that many people will use.* (Courtesy of Hewlett Packard)

Tremendous changes in engineering design have occurred recently throughout the United States and the industrialized world. Traditionally, a product is developed and engineered by a system of linear steps from concept to production. This process takes a large amount of time. For example, a simple product such as a pump may take as long as a year; more complex products, such as an automobile, may take five or more years from concept to market. This time is required because the design has to be prototyped and then modified before it is ready to be used by customers. A traditional engineering system is shown in Fig. 1-4. You can see how many steps the design would have to go through. In addition, any changes made can return it to a previous step at any time in the process. The drafting work in a traditional system often is done by hand, creating the need for many designers and drafters. Technical drawings are two-dimensional, and require many drawings to explain every detail.

COMPUTERS AND CONCURRENT ENGINEERING

Since the introduction of the computer for design and drafting, many changes have occurred in the design process. Engineers, designers, and drafters look at the product differently. Since personal computers have become so powerful and reasonably priced, even small engineering companies can incorporate the latest computer tools in the design of products.

Today many companies have changed their engineering system from the traditional system to concurrent system of engineering (Fig. 1-5). **Concurrent engineering** allows everyone in the company to have access to the product information database. For example, at the same time that the designer is working on a product design, the engineer can access the design and analyze it for strength and its component interferences and clearances. The drafter can be working on the production drawings. The manufacturing engineer can access the design to start the production tooling

changing marketplace. Another way of putting it is to say that people want a faster bike, a better sound system, a better camera, a simpler computer language, or fancier and faster sporting shoes.

A new product starts with a **conceptual design** (a rough design idea) and a set of preliminary **design drawings**. This is followed by a product **prototype**: a functional model that allows engineers to test and analyze the product. When the product is ready for the marketplace, a set of **production drawings**, specifications, and **technical promotional drawings** are completed. The production drawings and specifications are used to manufacture the product, and the technical promotional drawings (illustrations) are used to promote the product in the market. They are considered graphic communication, or the language of industry. The engineering design teams are responsible for all new achievements. Everyone, in every part of society, is a consumer of new products. No matter which direction your life takes, the engineer and design drafter will influence your future—at work, at play, in worship, and at home (Fig. 1-3).

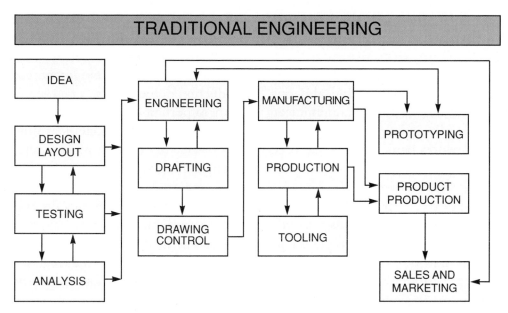

Fig. 1-4 *A traditional engineering system is organized as shown here.* (Courtesy of Circle Design)

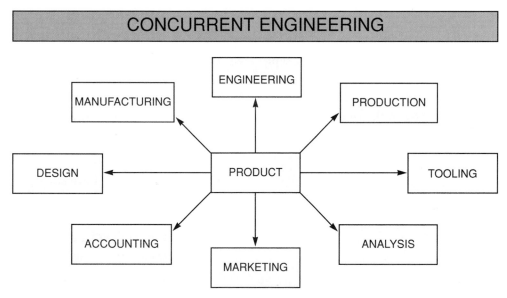

Fig. 1-5 *Compare the concurrent engineering system, shown here, with the traditional system in Fig. 1-4.* (Courtesy of Circle Design)

and the manufacturing machine programming. The accounting department can access cost information, and the production control department can look at shop routing, materials requirement planning, and shop capacity planning. Furthermore, the sales promotion department can insert information into their word processing and promotional programs from the product database. Finally, products can be sent electronically to company subsidiaries throughout the world. This system for designing a product has revolutionized the way we conduct business. We can expect even greater advances in the future.

Concurrent engineering has significantly reduced the amount of time to engineer and manufacture a product. In some cases, it has cut product engineering time in half, or even more. It has also placed higher standards on the engineer, designer, and drafter. Technical people must now be competent in visualizing designs in both two and three dimensions. They must be technically capable in the preparation and interpretation of technical drawings. **Computer literacy** (knowing how to use computers effectively) has become a necessary job skill. In addition, technical people must be team players and be able to communicate effectively both verbally and in writing.

Many products are now being designed as three-dimensional (3D) solid models. Because these models are actually three-dimensional representations, the viewer can rotate or "turn" the model and view it from any angle. The engineer uses the solid model to analyze the product design for strength, tolerances, and types of material. Other team members use it to analyze product performance and its operating behavior. The drafter uses the model to generate two-dimensional (2D) production drawings by setting up the necessary views and sections that will be required in production. A computer-generated solid model can even be used in a process known as **rapid prototyping** to create a polymer (plastic) prototype of the product.

This design approach has reduced the number of technical people required in the development of a product. It has forced colleges and universities to modify engineering curriculums so that they better prepare students to work in, and with, the latest technical tools. High-school students who are considering a career in design must plan on completing at least a two-year degree at a community or technical college in a design or drafting technology program that will provide instruction in the fundamentals of design and drafting as well as advanced classes in computer-aided design. Industries need people who are willing to commit to lifelong learning and contributions to the advancement of technical knowledge.

FROM SCHOOL TO WORK: MAKING CAREER PLANS

You can make plans to serve on an engineering design team or to have the team serve you. Many careers are available at several technical levels within a company. Businesses need employees that can understand and relate to technical design

requirements—not only the engineers that design products, but also the people that sell and maintain them. All companies need technical services for growth in their development, construction, or production departments.

An important position on any design team is that of the drafter. You can learn the career skills necessary for career placement and growth through hands-on experiences in high-school beginning drafting courses and later in a two-year design or drafting technology program at a technical or community college.

A career in the design world may begin with high school courses in drafting, science, mathematics, and English. These should be followed by coursework in a community or technical college's design or drafting technology program. People with a two-year degree in design or drafting technology can find good employment opportunities when they have an aptitude for problem-solving along with design and drafting skills. This requires an understanding of the design process, manual drafting and CAD, manufacturing, materials, communication, technical writing, applied science, and mathematics (Fig. 1-6).

JOINING THE TEAM

Employment opportunities for technical people can be found in advertisements or through employment consultants. Industry generally posts the following entry-level positions:

1. **Junior Drafter** Must have good manual drafting and CAD skills. Requires at least one year of high-school drafting and an associate degree in drafting technology from a technical or community college.
2. **Drafting Technician** Must have an associate degree in drafting technology from a technical or community college and at least one year of drafting experience.
3. **Designer** A **designer** is a person who has completed a two-year degree from a technical or community college and who has had at least five years industrial experience. A knowledge of the design process and drawing requirements is essential.

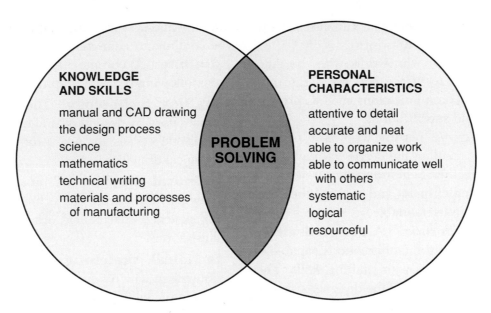

Fig. 1-6 *Becoming successful in design/drafting requires knowledge, skills, and the right attitudes.*
(Courtesy of Circle Design)

4. **Engineer** Must have four years of college with a degree in a specialized area; for example: aerospace, architectural, mechanical, biomedical, electrical, or chemical engineering.

There are so many technical advances that no one person or team can know all the areas of the new technology well. Therefore, most engineers, designers, and drafters specialize in one aspect of the technology. However, they must also have a broad knowledge of design processes, CAD, manufacturing, and testing in order to understand the total product picture.

THE ENGINEERING DESIGN TEAM

All design-oriented jobs require creative communication skills in concept design, manual and CAD technical drawing, technical writing, and presentation of ideas. A person working in this field must be able to communicate ideas and be able to interpret ideas or information obtained from others.

Major Positions and Their Creative Challenges

The engineering design team is a part of most major corporations and is considered essential in preparing legal documents for contracting services. Large offices may assign more than 100 people to a team; many small offices have 5 to 10 on the team; and medium-size offices have from 10 to 100 team members. Many specialized design offices have 4 to 12 people on a team. The design team consists of people in various positions working together to design and create a new product. The basic positions are described below.

Research and Development People who are responsible for creating experiments that will explain new theories. They propose concepts for using new technologies through creative sketches and the use of science and mathematics.

Development Engineer An engineer who designs research projects and collects data so that the data can be applied to the development of new products.

Project Engineer A person who coordinates all the specialized areas of engineering and design for production or construction projects.

Design Engineer A **design engineer** is a person, usually with an engineering degree, who applies math, science, and technology principles to solve problems for production and construction. Design engineers are team members who work with project engineers, designers, and drafters.

Technical Illustrator A specialized drafter who can create a three-dimensional pictorial or a 3D CAD model from the details of an engineering drawing.

Three-dimensional CAD models can also serve as the basis for production drawings.

Designer   A person who works with the engineers and drafters to turn a concept design into usable technical communications such as production drawings and specifications.

Senior Detailer   A person who is especially skilled in understanding the details of how things work and go together. Senior detailers are capable of detailing complex parts and making the complicated details understandable.

Design Drafting Technician   A **design drafting technician** is a team member who is capable of combining design skills with drafting skills. This requires the interpretation of the designer's sketches and the engineer's details so that the design drawings can be prepared accurately for the senior detailer.

Drafter   A **drafter** is a junior technician who uses drafting skills to prepare single technical drawings under the direction of a senior detailer or designer.

CAD Operator   A drafting specialist capable of preparing design and production drawings from a CAD workstation. In many companies, the concept design work is done using CAD. Production and design drawings are often a by-product of a single CAD database.

Reproduction Specialist   A junior drafter or clerk who serves the engineering design team by developing prints and properly storing engineering drawings. The reproduction machines provide hard copies of the original drawings. Drawings are also stored on a computer database where any design team member can view and retrieve them.

Understanding Engineering Design Terms

The universal graphic language is used by engineering design teams all over the world. This text will emphasize an introduction to freehand and mechanical drawing skills, as well as computer-aided design and drafting concepts.

To understand engineering design, you must be familiar with basic terms used to describe tasks and concepts associated with it. These terms and their definitions are listed below. Review the terms carefully and be sure you understand them before you continue to later chapters of this text.

drafting   A common term that refers to all the following types of graphic communication.

drafting technology   The tools and techniques used by designers, drafters, and engineers to develop ideas for new products into usable technical drawings.

technical drawing   A broad term for any drawing which expresses technical ideas, including sketches, mechanical drawings, charts, and illustrations.

technical sketching   Freehand technical drawing in which you use a pencil and paper to communicate the shape of ideas to others.

mechanical drawing   A technical drawing made with drafting instruments.

engineering drawing   A mechanical drawing used by engineers and other members of the engineering design team to describe the production of a part, its shape and size, and the materials used to produce it.

engineering graphic   A graphic illustration or drawing that represents physical objects used in engineering and science.

descriptive geometry   A method for solving spacial and graphical problems using precise geometric descriptions.

computer-aided design and drafting (CADD)   An interactive design tool that is used to produce design ideas and technical drawings using computer hardware and software.

CAD   Most commonly used to mean "computer-aided drafting." However, the meaning of this acronym is based on the context in which it is used. It can also mean "computer-aided design." In other cases, CAD is used interchangeably with CADD.

basic drafting   A beginning course in the basic language of graphics which acquaints the student with basic processes, activities, and skills for advanced work.

Before beginning your experience with the graphic language, let's examine the role of the drafter on the engineering design team. The engineer, designer, and drafter form the hub of the engineering process (Fig. 1-7). This means everything revolves around the engineer and his or her team. Engineering is the reason for the existence of drafting and the graphic language.

THE MANY BRANCHES OF ENGINEERING

An **engineer** is a person who has at least a four-year degree in an engineering specialty. Engineering has evolved into many specialized branches, including aerospace, agricultural, architectural, chemical, civil, electrical, industrial, mechanical, mining and metallurgical, nuclear, petroleum, plastics, and safety. Each branch uses technical drawings to communicate ideas and products for manufacturing or construction. The science and mathematics common to each branch help convert natural resources into processes that provide useful new material for products and machines.

The chart in Fig. 1-8 illustrates five common branches of engineering. Notice the different types of activities for the engineering design team. The list of products they work with could be increased. Can you think of a few more activities to add to the list? Can you name more products? For example, could helicopters be added to aerospace engineering? And could houses be added to architectural engineering? How about civil, electrical, and

Fig. 1-7 *The engineering design team forms the hub of the engineering process. The drafter is an important member of the team.* (Courtesy of Circle Design)

mechanical engineering products? What is the most common activity that engineers engage in? You should have identified designing as one of the most common activities.

Even though there are many branches of engineering with areas of specialization, some activities are common to all engineers. Other activities may be limited to just a few engineers. Let's examine some of these activities, also referred to as the "levels of engineering functions or services."

LEVELS OF ENGINEERING FUNCTIONS

In each branch of engineering, the engineer performs services at one or more of the levels of engineering functions discussed below. The chart in Fig. 1-9 shows the levels in order of decreasing emphasis on mathematics and science.

Look at the working processes that the engineer engages in at each level. As less mathematics and science are required, the working process becomes less theoretical and more applied. That is to say, research and development generally involves high-level conceptual thinking about things that haven't been achieved–in many cases, those that don't work. Production, construction, and operation levels are developed around things we know will work.

THE IMPORTANCE OF GRAPHIC COMMUNICATION

Graphic communication is communication that is written or drawn. It has four separate levels: (1) creative communication, (2) technical communication, (3) market communication, and (4) manufacturing or construction communication. Each of these levels has its place in the field of drafting (Fig. 1-10).

Notice how the forms of communication, or types of drawings, change from level to level as you move from the unknown in research to the more concrete in manufacturing and construction. At the highest level, in which mathematics and science are emphasized, the drawings have to be creative–they are often unrefined and lack a final finished form. Ideas are in the formative stage.

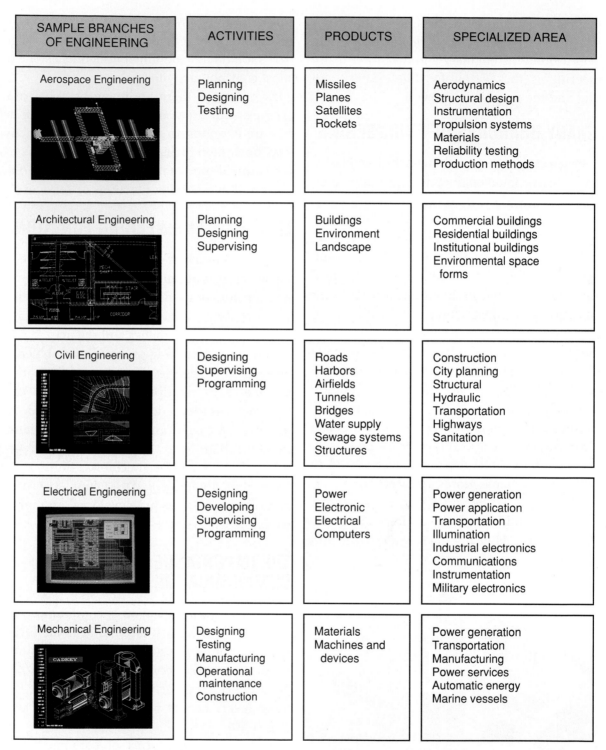

SAMPLE BRANCHES OF ENGINEERING	ACTIVITIES	PRODUCTS	SPECIALIZED AREA
Aerospace Engineering	Planning Designing Testing	Missiles Planes Satellites Rockets	Aerodynamics Structural design Instrumentation Propulsion systems Materials Reliability testing Production methods
Architectural Engineering	Planning Designing Supervising	Buildings Environment Landscape	Commercial buildings Residential buildings Institutional buildings Environmental space forms
Civil Engineering	Designing Supervising Programming	Roads Harbors Airfields Tunnels Bridges Water supply Sewage systems Structures	Construction City planning Structural Hydraulic Transportation Highways Sanitation
Electrical Engineering	Designing Developing Supervising Programming	Power Electronic Electrical Computers	Power generation Power application Transportation Illumination Industrial electronics Communications Instrumentation Military electronics

Mechanical Engineering	Designing
	Testing
	Manufacturing
	Operational
	  maintenance
	Construction
		Materials
		Machines and
		  devices
			Power generation
			Transportation
			Manufacturing
			Power services
			Automatic energy
			Marine vessels

Fig. 1-8 *The five most common branches of engineering can be examined in this chart. Designing is the engineer's most common activity.* (Courtesy of Circle Design)

ENGINEERING FUNCTION	WORKING PROCESSES	COMMUNICATION FORMS
RESEARCH. The research engineer through experimentation and inductive reasoning employs the theories of science and mathematics in his or her attempts to find new processes.	Imagination Inductive reasoning Ideation	**Brainstorming -** Creating and developing ideas; conceptual sketching **Sketching -** creating and refining
DEVELOPMENT. The development engineer uses the results of research by creatively applying new knowledge to produce ingenious products.	Creative Imagination Clever Models	**Sketching -** refining **Delineation -** a graphic outline **Drawing -** construction
DESIGN. The design engineer creates products by determining construction or manufacturing methods, materials, and physical shapes to satisfy technical requirements, performance specifications, economic constraints, and marketing esthetics.	Creative Clever	**Sketching -** refining **Designing -** charts reports delineation
PRODUCTION. The production engineer selects equipment and plans plant facilities to accommodate human and economic factors and allow for efficient material flow, testing, and inspection.	Plans Supervises Organizes	**Reports Flow diagrams Architectural and engineering diagrams**
CONSTRUCTION. The construction engineer prepares the building environment to yield the safest, most economical, highest quality product possible.	Supervises Organizes	**Reports Charts Flow diagrams Engineering blueprints**
OPERATION. The operating engineer supervises the day-to-day control of machines and facilities providing power, transportation, and communication.	Supervises Operates Maintains	**Flow diagrams Charts Architectural and engineering drawings**

Fig. 1-9 *When you examine the various services performed by the engineer in this chart, you will find that the drafter or CADD technician serves at all levels.* (Courtesy of Circle Design)

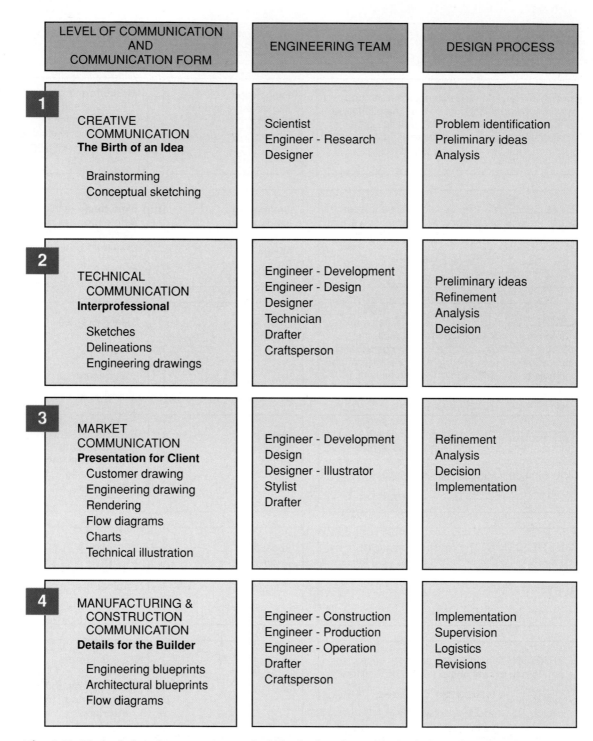

LEVEL OF COMMUNICATION AND COMMUNICATION FORM	ENGINEERING TEAM	DESIGN PROCESS
1 CREATIVE COMMUNICATION **The Birth of an Idea** Brainstorming Conceptual sketching	Scientist Engineer - Research Designer	Problem identification Preliminary ideas Analysis
2 TECHNICAL COMMUNICATION **Interprofessional** Sketches Delineations Engineering drawings	Engineer - Development Engineer - Design Designer Technician Drafter Craftsperson	Preliminary ideas Refinement Analysis Decision
3 MARKET COMMUNICATION **Presentation for Client** Customer drawing Engineering drawing Rendering Flow diagrams Charts Technical illustration	Engineer - Development Design Designer - Illustrator Stylist Drafter	Refinement Analysis Decision Implementation
4 MANUFACTURING & CONSTRUCTION COMMUNICATION **Details for the Builder** Engineering blueprints Architectural blueprints Flow diagrams	Engineer - Construction Engineer - Production Engineer - Operation Drafter Craftsperson	Implementation Supervision Logistics Revisions

Fig. 1-10 *The level of graphic communication should be clearly understood by the drafter and CADD technician.* (Courtesy of Circle Design)

Level One: Creative Communication

Graphic communication usually begins as an idea in the mind of an engineer or a designer. The designer's first concept sketches are the beginning of a product. The first work may be thought of as the birth of an idea. The creative communication of sketching is important to people who need to express ideas quickly. Sketching can capture ideas for further study.

Level Two: Technical Communication

When the engineer or designer gives a sketch to the other members of the design team, it is ready for level two. Engineers, designers, and technicians or architects and their assistants study and change the original design to make it more practical. The design is refined, or improved, through the ideas of several people. Two (or more) heads are better than one! The result is a set of technical drawings that fully explain the design.

Level Three: Market Communication

This level includes the evaluation of a design by or for a client or customer. This is especially useful in architectural design. Before designs are made final, customers look at them to see if they like them. Clients evaluate designs for style, form, and function. Approving the designs of the floor plan and the outside of a house before actual construction work on the house begins is an example.

Level Four: Construction Communication

This level includes all the details needed for manufacturing or construction. These drawings must be complete so that estimators can figure the exact costs of a project and factory superintendents can know exactly how a product is to be made. People who use the drawings should not have to guess about details or ask the designer exactly what was meant.

SYSTEM OF GRAPHIC COMMUNICATION

Various types of drawings are used to communicate technical information in the engineer's office. Look back at the chart in Fig. 1-10 showing the four levels of graphic communication. Notice that the column to the right of each level shows who might be working at each level. The Design Process column shows the design stages related to the four levels.

The successful drafter and design engineer must be able to communicate ideas accurately. Whether the ideas are for new structures, vehicles, or household products, the engineer must understand how to prepare and interpret technical drawings drawn up by team members. The engineer must know the standards set up for his or her specialized area of design.

The penmanship, so to speak, of the universal language is a very necessary skill to learn. The standards are established by the American National Standards Institute (ANSI) in New York. Large companies also maintain a set of inter-company standards for drafting and design. The standards help large design teams accomplish major achievements that would otherwise be difficult to organize.

Society needs students who can continue technical progress through drafting. If you develop a thorough understanding of the first 12 chapters of this book, you will have the basic skills to accept the challenge of the advanced chapters.

Computers help organize the data for an emerging twenty-first century technology. Students today often work at personal computer workstations in the drafting classroom as they become prepared for professional roles on the design team. Computer literacy is essential as you begin to learn the universal language called Mechanical Drawing.

CHAPTER 1

REVIEW

Learning Activities

1. Check employment advertisements in local newspapers for listings of positions related to the fields of engineering and drafting. Prepare a chart showing job titles, experience and educational background required, and starting salaries.

2. Collect technical drawings or prints from industries in your area. Sort them into several categories as described in the chapter. Produce a bulletin board display of the best examples from each group.

3. Visit an industrial drafting room. Ask questions about topics such as job opportunities, organizational structure, and the educational background of drafters, designers, and engineers.

4. Invite engineers and drafters to visit your classes. Ask them to describe the nature of their work, their educational background, and the salary potential in various aspects of drafting and engineering.

5. Survey several industries in your area to determine the use of manual drafting vs. computer-aided drafting. Make a pie chart showing the results.

Questions

Write your answers on a separate sheet of paper.

1. Design drawings and _____ are needed before any new product is manufactured in an industrial plant.

2. Describe the role of the drafter on an engineering design team.

3. What is concurrent engineering, and how does it differ from traditional engineering methods?

4. Why is computer literacy now considered necessary for members of the engineering design team?

5. Freehand technical drawing is also called _____.

6. What do the letters CAD stand for?

7. Name at least three of the six most common branches of engineering.

8. List the four levels of graphic communication.

9. What do the letters ANSI stand for?

10. What information do detailed technical drawings provide about a part or product to be manufactured?

11. An illustrator is a specialized drafter who prepares _____ drawings.

12. The letters CADD stand for _____.

13. Name three entry-level positions on the engineering design team.

14. Professionals who apply math, science, and technical drawing to solve problems for production and construction are called _____.

SKETCHING AND LETTERING

OJECTIVES

Upon completion of Chapter 2, you will be able to:

○ Describe the importance of freehand sketching for communicating technical ideas.
○ Letter clear, neat freehand notes and directions on a technical drawing or sketch.
○ Communicate technical ideas through freehand sketching.
○ Develop design ideas through multiview or pictorial sketches.
○ Explain the advantages of using a CAD program to create notes and other text on a drawing.

VOCABULARY

In this chapter, you will learn the meanings of the following terms:

• **arcs**	• **isometric lines**	• **overlay**
• **axis (axes)**	• **isometric sketch**	• **plane**
• **composition**	• **lettering**	• **point**
• **concentric circles**	• **line**	• **proportion**
• **ellipses**	• **nonisometric**	• **radius (radii)**
• **Gothic lettering**	**lines**	• **tangent arcs**
• **guidelines**	• **oblique sketch**	• **texture**

When you see sketches bring new ideas to life, you know that the saying *progress begins on paper* is true. When words alone cannot describe new or futuristic forms, sketches are needed to show the thoughts that cannot be said (Fig. 2-1).

Freehand sketching is the simplest form of drawing. It is one of the quickest ways to express ideas. For example, a sketch can help simplify a technical discussion. Designers, drafters, technicians, engineers, and architects often explain complicated or unclear thoughts with a freehand sketch. Ideas imagined in the mind can be caught in sketches and thus held in simple lines for further

study. Figure 2-2 shows how sketching can become an important part of technical discussions and design decisions.

The language of sketching has four basic *visual symbols* (things that can be seen). These are a point, a line, a plane, and a **texture,** or surface quality (Fig. 2-3). A **point** is a symbol that describes a location in space. The path between two points is called a **line.** Two *nonparallel lines* (lines that cross) can define a **plane,** which is a flat surface. Any idea, no matter how simple or complicated or how plain or spectacular, can be sketched using these four visual symbols.

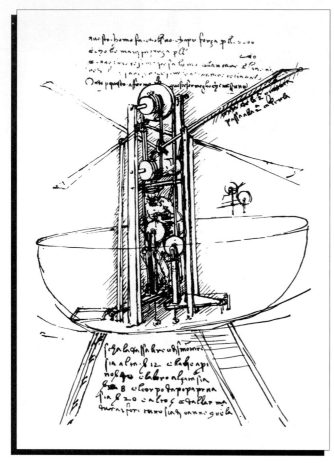

Fig. 2-1 *During Leonardo da Vinci's time (1452-1519), no one had yet seen a helicopter. His sketch helped convey the idea. The person in the center powers the rotating wings.* (Courtesy of the Bettmann Archive)

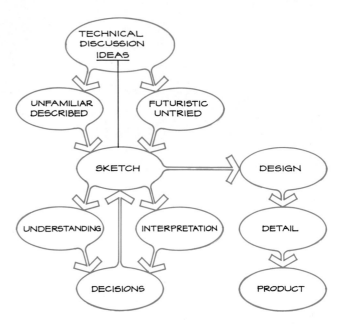

Fig. 2-2 *Sketches can assist technical discussions by showing the relationships among ideas. How else might a sketch be of use in a technical discussion?*

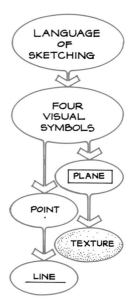

Fig. 2-3 *Simple visual symbols are basic elements of sketching.*

This chapter emphasizes the need for freehand sketching and lettering. It also introduces lettering (called *text*) and sketching that can be developed with CAD systems. As you may remember from Chapter 1, a CAD (computer-aided drafting) system is a computer system that allows drafters to create accurate drawings using a computer.

REASONS FOR SKETCHING

People create technical sketches for many reasons. The following are the nine most important.

1. To persuade people who make decisions about a project that an idea is good.

2. To develop a refined sketch of a proposed solution to a problem so that a client can respond to it (Fig. 2-4).

3. To clarify a complicated detail of a drawing that has more than one view by enlarging it or by creating a simple *pictorial* (picturelike) *sketch.*

4. To give design ideas to drafters so that they can do the detail drawings.

5. To develop a series of ideas for refining a new product or machine part.

Fig. 2-4 *A refined sketch for the client.*
(Courtesy of A. W. Wendell and Sons, Architects)

6. To develop and analyze the best methods and materials for making a product.

7. To record permanently a design improvement on a project that already exists. The change may result from the need to repair a part that breaks over and over again. It may result from the discovery of an easier and less expensive way to make a part.

8. To show that there are many ways to look at or solve a problem.

9. To spend less time in drawing. It is quicker to make a sketch, which takes only a pencil and a sheet of paper, than to create a mechanical drawing.

SKETCHING IN CAD

Beyond the reasons described in the previous section, one of the most important reasons for sketching has become to communicate quickly with the drafter at the CAD station. CAD technicians are doing more and more mechanical drafting in companies all over the world.

Sketching skills are a means of preparing drawings for CAD communications. The ability to prepare, read, and relate to sketches enables CAD technicians to translate the sketches into two- and three-dimensional mechanical drawings stored in computer-created drawing files.

LETTERING

Freehand drawings generally need some freehand lettering to explain features of a new idea or product. **Lettering** is the practice of printing clear, concise words on a drawing to help people understand the drawing. A well-planned drawing may be worth a thousand words, as an old saying goes, but a few choice words well organized can explain some important details. What kind of notes could be used to help explain the sketch of the robot in Fig. 2-5?

The notes lettered on the rough sketch in Fig. 2-6 describe features that are functional and important to operation. On sketches you have prepared or studied in the past, can you recall a need for notes to explain an idea? Simple freehand lettering complements an idea that is captured in a sketch, especially if the lettering is neat and carefully placed on the drawing.

Composition

In lettering, **composition** means arranging words and lines with letters of the right style and size. Letters in words are not placed at equal distances from each other. They are placed so that

Fig. 2-5 *A sketch can shape a futuristic design. This robot lives inside a pet door and goes out daily to fetch the newspaper.*
(Courtesy of Cloyd Richardson)

the spaces between the letters *look equal*. The distance between words, called *word spacing,* should be about equal to the height of the letters. Figure 2-7 shows examples of proper and improper letter and word spacing. The clear distance (open space) between lines of letters is from ½ to 1½ times the height of the letters.

Tools such as lettering triangles and the Ames lettering instrument are available to help drafters create neat, uniform lettering using the proper spacing (Fig. 2-8). On mechanical drawings, drafters create ruled **guidelines** spaced .12 in. apart to help keep their lettering uniform. When you are sketching, however, you will need to estimate the appropriate distances.

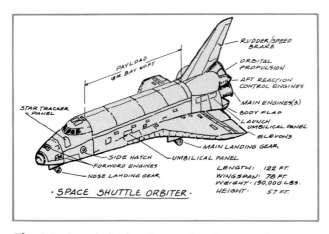

Fig. 2-6 *A rough sketch with notes about important features.*

Drafters need good lettering skills throughout their careers. Freehand lettering is one of the first things students of drafting study. This is because you do not learn how to do good lettering all at once. You learn it by practicing it little by little for a long time. Your sketching should get better with every sketch or drawing you do.

Fractions

Fractions are always made with a division line that is *horizontal* (side-to-side). The whole fraction is usually twice the height of regular numbers. The numerals (numbers) in a fraction must never touch the division line (Fig. 2-9).

CAD LETTERING

CAD drawings maintain uniform lettering and dimensioning (Fig. 2-10). Each CAD software program has text and dimension commands that allow the drafter to prepare neat, precise notes. Because the software automatically creates the text and dimensions in a uniform size, it is possible to describe a simple or complex object without first creating guidelines. See Fig. 2-11 for an example of AutoCAD text options.

Even drafters who spend most of their time using CAD to create drawings should understand the basic principles of lettering. As you become good at freehand lettering, you will be better able to place text in computer-aided drafting.

Notes can appear to be hand-lettered even on drawings (Fig. 2-11). Examine the text options available to the CAD technician within the software. Note that lettering in a CAD drawing is generally referred to as *text.*

LE TT ERIN G CO MPOSITI O N
INV OL VE S THE   SP A CIN G OF   LETTERS,
W ORD S,   A ND LIN ES AND   THE CH OI CE
OF A PPROPRI A TE STYL ES AND SIZ E S.

INCORRECT LETTER, WORD, AND LINE SPACING

LETTERING COMPOSITION
INVOLVES THE SPACING OF LETTERS,
WORDS, AND LINES AND THE CHOICE
OF APPROPRIATE STYLES AND SIZES.

CORRECT LETTER, WORD, AND LINE SPACING

Fig. 2-7 *Single-stroke lettering. Study the word spacing.*

TYPES OF LETTERING

The choice of lettering style should be made carefully. For example, it is usually not practical to use fancy roman-style lettering for notes and dimensions on technical drawings. Such lettering requires more pencil strokes, so it is more costly in time and money than the single-stroke styles. Yet it would serve the purpose of the drawing no better.

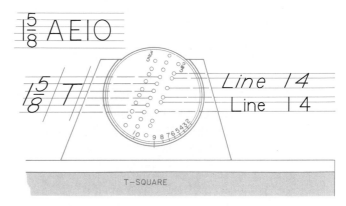

Fig. 2-8 *Lettering guidelines should be evenly spaced. Use a lettering triangle or an Ames lettering guide.*

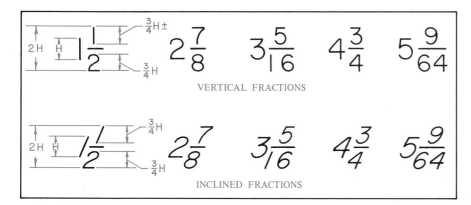

Fig. 2-9 *Position guidelines for fractions as shown here.*

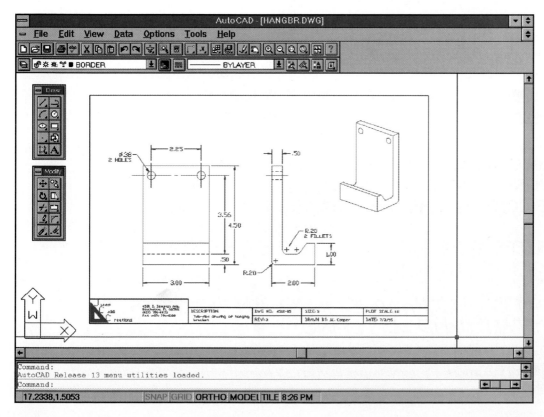

Fig. 2-10 *A typical CAD drawing with lettering and dimensioning.*

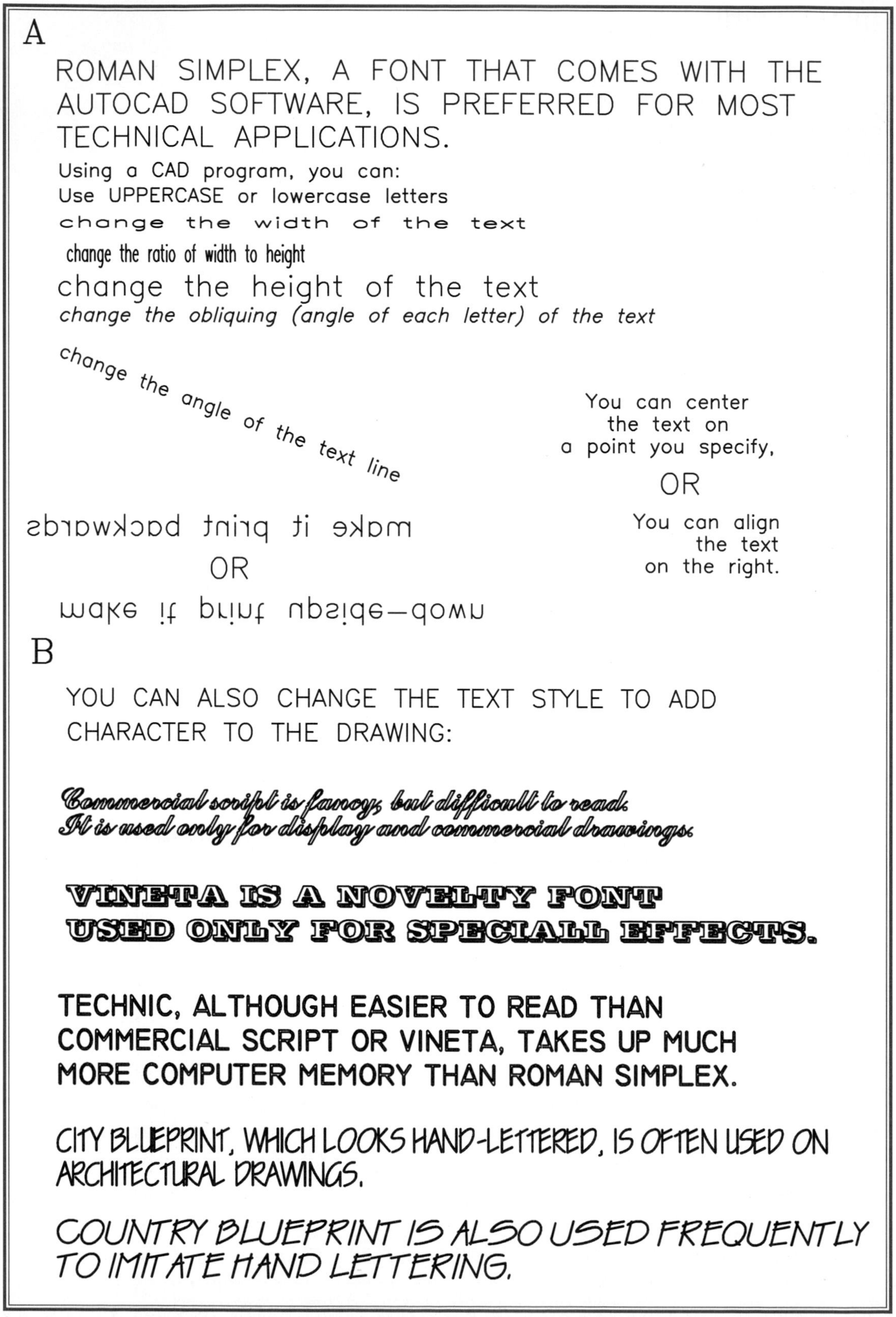

A

ROMAN SIMPLEX, A FONT THAT COMES WITH THE AUTOCAD SOFTWARE, IS PREFERRED FOR MOST TECHNICAL APPLICATIONS.

Using a CAD program, you can:

Use UPPERCASE or lowercase letters

change the width of the text

change the ratio of width to height

change the height of the text

change the obliquing (angle of each letter) of the text

change the angle of the text line

You can center
the text on
a point you specify,

OR

make it print backwards

OR

make it print upside-down

You can align
the text
on the right.

B

YOU CAN ALSO CHANGE THE TEXT STYLE TO ADD CHARACTER TO THE DRAWING:

Commercial script is fancy, but difficult to read. It is used only for display and commercial drawings.

VINETA IS A NOVELTY FONT USED ONLY FOR SPECIALL EFFECTS.

TECHNIC, ALTHOUGH EASIER TO READ THAN COMMERCIAL SCRIPT OR VINETA, TAKES UP MUCH MORE COMPUTER MEMORY THAN ROMAN SIMPLEX.

CITY BLUEPRINT, WHICH LOOKS HAND-LETTERED, IS OFTEN USED ON ARCHITECTURAL DRAWINGS.

COUNTRY BLUEPRINT IS ALSO USED FREQUENTLY TO IMITATE HAND LETTERING.

Fig. 2-11 *In addition to ensuring evenly-spaced, uniform lettering, CAD programs provide several fonts (type styles) and other options.*

The lettering style most commonly used on working drawings is single-stroke **Gothic lettering** (Fig. 2-12). This style is best because it is easy to read and easy to hand letter. It is made up of *uppercase* (capital) letters, *lowercase* (small) letters, and numerals. Nearly all companies now use only uppercase lettering. As a result, this book stresses uppercase lettering. Letters and numerals may be either vertical or inclined. You can vary your lettering to make it more individual. Figure 2-13 shows some of the possible variations. They are common styles for designers and architects. Whichever style you choose, remember that the same style should be followed throughout a set of drawings.

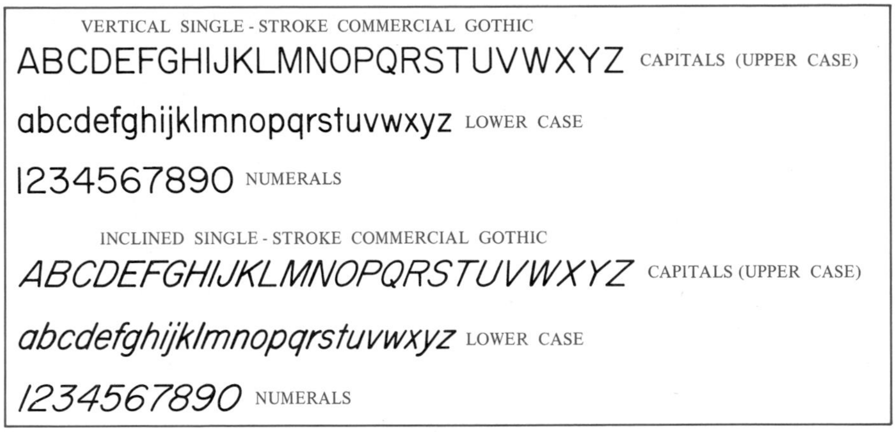

Fig. 2-12 *Single-stroke Gothic letters, vertical and inclined.*

Fig. 2-13 *Variations in lettering styles for the designer.*

Single-Stroke Vertical Lettering

The shapes and proportions of *vertical* (straight) letters and numerals are shown in Fig. 2-14. The figure also demonstrates the pencil strokes needed to create each letter and the order in which they should be made. Study these characters carefully until you understand completely how each is made. Each character is shown in a square 6 units high. The squares are divided into unit squares. By following these, you can easily learn the right shapes, proportions, and strokes. (Proportion is discussed in greater detail later in this chapter.)

Single-Stroke Inclined Lettering

Inclined (slanted) letters and numerals should slope at an angle of about $67\frac{1}{2}°$ from the horizontal. Figure 2-15 shows single-stroke inclined letters and numerals on an inclined grid. Notice that the only difference between vertical (straight) and inclined letters is the slant.

Display Lettering

Display lettering is often used for title sheets, posters, and other places where large, easy-to-read lettering is needed. Although any style may be used for the display, the drafter most often uses Gothic, which is considered the most legible lettering style. Figure 2-16 shows a typical block style.

The block style is the easiest display style to make. It has no curved strokes and only takes a T-square and triangle. A light-line grid may be drawn on the sheet and later erased. If you are using a tracing medium (drawing material), you can put a grid sheet under the drawing sheet. Block in the letters in pencil first. Then, if desired, trace them in ink. Letters can be either outlined or filled in, as shown in Fig. 2-16. The thickness of the strokes can vary from one fifth to one tenth the height of the letter. In Fig. 2-16, one seventh is used.

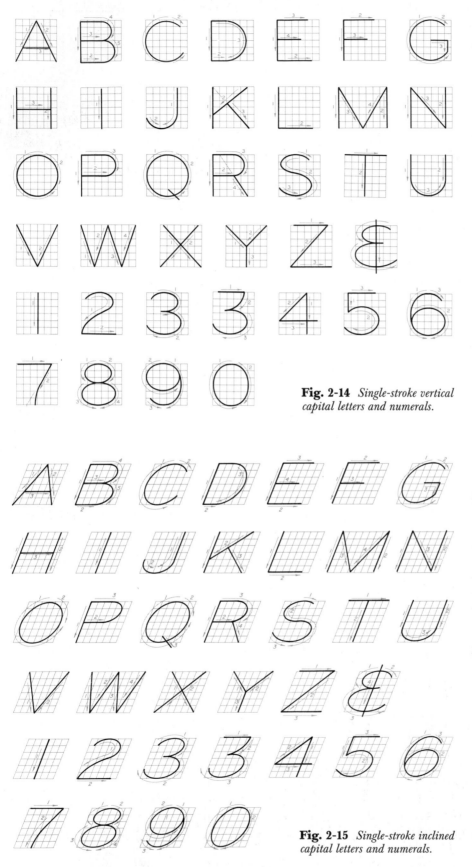

Fig. 2-14 *Single-stroke vertical capital letters and numerals.*

Fig. 2-15 *Single-stroke inclined capital letters and numerals.*

Regular Gothic capital letters are made in much the same way. The only difference is that you have to make curved strokes. Do this with circle templates or a compass, or draw the curves freehand.

CAD LETTERING STYLES

Most CAD programs provide various lettering styles. Each style is right for a particular use. It is up to the drafter to choose the correct style and size for a particular drawing. As in freehand lettering, you should be careful to choose a style that is appropriate for the drawing.

Lettering styles in CAD and other computer programs are known as *fonts*. AutoCAD provides several types of fonts, including native Auto-CAD ("shape compiled") fonts as well as TrueType® and PostScript® fonts. A small sample of these are shown in Fig. 2-11B. The TrueType and PostScript formats are used by other types of computer programs, such as word processing applications. Using these fonts, a drafter can create the text in a word processor and then import the text into the CAD drawing. This is useful when a drawing will contain a large amount of text because word processors are, in general, more powerful text editors than those included in CAD programs.

Lettering in CAD is much simpler and faster than lettering by hand because the drafter doesn't have to be concerned about letter spacing, word spacing, or uniform height. The software manages those issues automatically. However, CAD lettering introduces new pitfalls for the beginning drafter. The availability of many different fonts may tempt the drafter to use fancier lettering or to use more than one type or style of lettering on a drawing. Complex fonts (such as roman) take no more time to create using a CAD program than single-stroke fonts.

It is therefore especially important to keep good drafting practices in mind when adding notes to a CAD drawing. The same rules that apply to hand lettering also apply to CAD lettering. Remember that there are good reasons for these rules. For example, complex fonts are difficult to read when you use all upper-case letters.

CAD programs provide one or more commands that allow you to place text on a drawing. In AutoCAD, for example, you could use the TEXT, DTEXT, or MTEXT command to place text on a drawing. To change the font, you would use the STYLE command (Fig. 2-17). You can also edit and change the style of text that is already present in a CAD drawing by using a command such as CHANGE or DDMODIFY (Fig. 2-18).

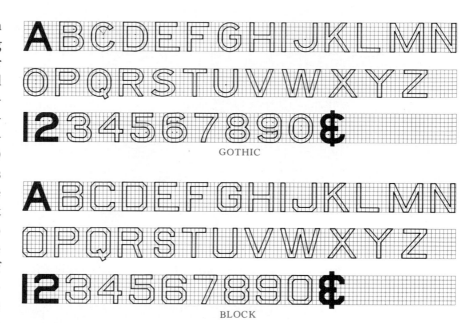

Fig. 2-16 *Regular-style Gothic and Block-style lettering.*

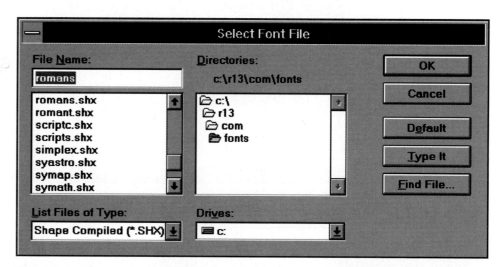

Fig. 2-17 *When you choose the STYLE command in AutoCAD, the Select Font File dialog box appears to help you choose a font (text style). The font files appear in the list box on the left.*

TYPES OF SKETCHES

Any image drawn on paper freehand (without a straightedge or other tools) may be called a *sketch*. Most drafters use several types of sketches. The type of sketch used depends on the purpose of the sketch and its intended life span (Figs. 2-19 and 2-20).

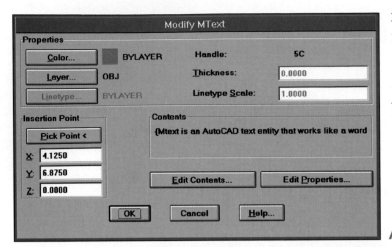

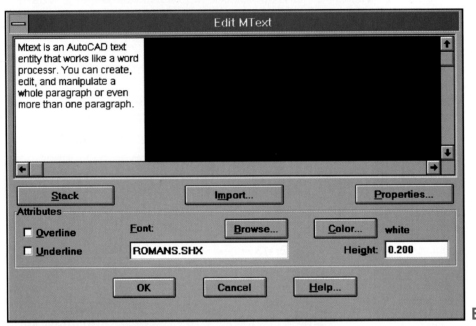

Fig. 2-18 *To edit AutoCAD's mtext, use the DDMODIFY command. The Modify Mtext dialog box appears (A), allowing you to change the color and layer of an mtext entity. You can also pick Edit Contents... to edit the wording and attributes of the text (B) or Edit Properties... to edit the text style, height, and other physical characteristics (C).*

Rough Sketches

Rough sketches are usually drawn quickly with jagged lines. Their primary purpose is to express thoughts quickly. Figure 2-21 shows rough sketches that were used to develop preliminary (early) designs of a two-position automobile mirror. The sketches show several design choices.

Never use instruments or straightedges to prepare a rough sketch. Instruments tend to restrict the creative expressions developed with good pencil techniques. Avoid a mechanical, hard-line look. Concentrate on using good proportions, and add a few choice notes if necessary to clarify the drawing.

Refined Sketches

Refined sketches are drawn more carefully than rough sketches. They show good proportion and excellent line values. They may be more persuasive than an unrefined sketch. Many refined sketches are based on a rough sketch that has captured the general idea.

You may use a straightedge to control long lines on a refined sketch. However, never allow the line to look mechanically drawn. Sketched lines should have some irregular character.

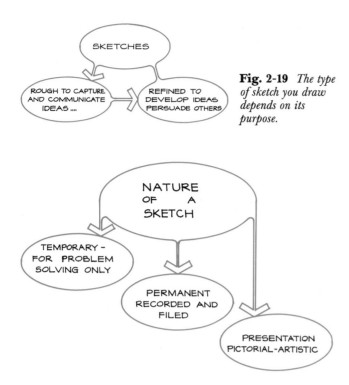

Fig. 2-19 *The type of sketch you draw depends on its purpose.*

Fig. 2-20 *The life span of a sketch depends on its nature or purpose.*

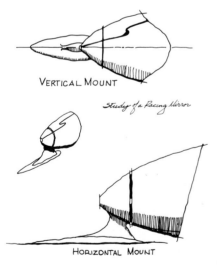

Fig. 2-21 *Study of two-position mirror for a racing car.*

Presentation Sketches

Pictorial (picturelike) sketches that have been greatly refined are known as *presentation sketches*. These sketches are used to convince a client or management to accept and approve the ideas presented. Pictorial sketches have a three-dimensional

view that can be understood easily by nontechnical people. Such sketches are generally drawn so that they look glamorous, artistic, or eye-appealing.

Temporary Sketches

Many technical sketches have short lives. Some are done merely to solve an immediate problem. Then they are thrown away. Other technical sketches are kept longer. It may take weeks or even months to study some sketches and make mechanical drawings from them. However, these sketches too may be thrown away some day.

Permanent Sketches

Sometimes the engineering department or the management of a company will include a sketch in a notice to other employees. Such a sketch is an important record and should be kept. Therefore, some sketches are filed as part of a company's permanent records.

THE OVERLAY

A very good way to refine or improve a sketch is to use an overlay. An **overlay** is a piece of translucent (tracing) paper that is placed on top of a sketch or drawing. Because you can see through the paper, you can quickly trace the best parts of the sketch or drawing underneath. Refining ideas often means sketching over and over again on tracing paper, changing the drawing until the design is right.

Overlays are used in three important ways. The first is reshaping an idea. This might include refining the proportions of the parts of an object or changing its shape entirely. Second, an overlay can be used to refine the drawing itself without really changing the design. Although these two uses may seem very similar, they are both important to sketch development (Fig. 2-22). Finally, an overlay can be used to add various options to a basic drawing. In Fig. 2-23, for example, separate overlays could be used to demonstrate various seating and steering wheel designs. This type of overlay is often used to help clients visualize various possibilities.

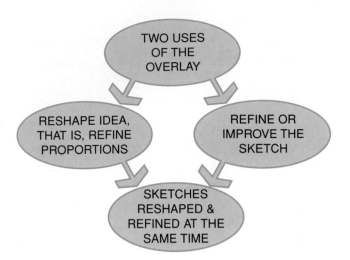

Fig. 2-22 *An overlay can help the drafter reshape an idea as well as refine the sketch in which the idea is presented.*

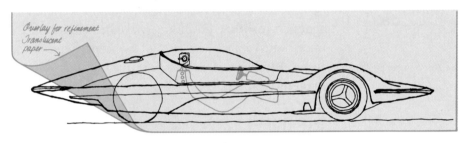

Fig. 2-23 *The overlay can speed up the design process.*

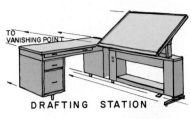

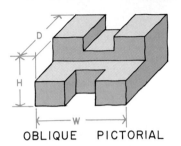

Fig. 2-24 *Pictorial drawings show all three dimensions in one view. The two main types of pictorial drawings are the perspective pictorial, in which objects that are farther away become smaller as they approach a **vanishing point**, and the oblique pictorial, in which perspective is not used.*

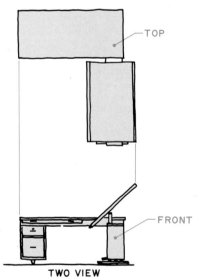

VIEWS NEEDED FOR A SKETCH

In Chapter 5, multiview projection will be discussed in full. However, you need to know some basic things about views and how they are placed in order to create good sketches.

There are two types of drawings that you can sketch easily. One is called a *pictorial drawing*. In this picturelike type of drawing, the width, height, and depth of an object are shown in one view (Fig. 2-24).

In the other type of drawing, an object is usually shown in more than one view. You do this by drawing sides of the object and relating them to each other, as shown in Fig. 2-25. The system by which the views are arranged in relation to each other is known as *multiview projection,* or *orthographic projection.*

Fig. 2-25 *Typical multiview drawings (two-view and three-view).*

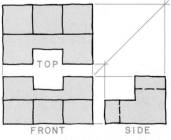

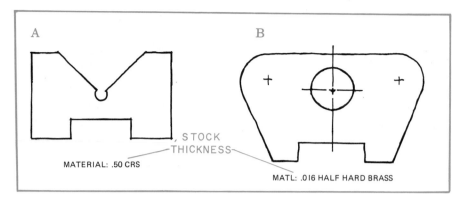

Fig. 2-26 *Typical one-view drawings. Notice that in each case the thickness is shown as a note.*

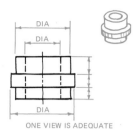

Fig. 2-27 *A cylindrical object may require only one view.*

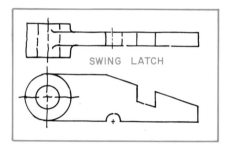

Fig. 2-28 *A two-view sketch.*

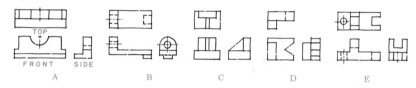

Fig. 2-29 *Select the two required views for the objects shown in A, B, C, D, and E.*

One-View Sketches

If an object can be described in two dimensions (height and width, for example), a one-view drawing is generally sufficient. Objects shown in one-view drawings generally have a depth or thickness that is uniform (the same throughout). In these cases, drafters may give the depth in a note rather than drawing an extra view. Typical one-view drawings are shown in Fig. 2-26. The thickness of the stamping is shown by a note on the sketch. Many objects that are shaped like cylinders can also be shown in single views if the diameter of the cylindrical part is noted, as in Fig. 2-27.

Two-View Sketches

Many objects, such as the one shown in Fig. 2-28, can be described in two views. If you are careful to select two views that describe the object well, this can help simplify the drawing. Figure 2-29 shows five three-view drawings in which the objects could have been described well in only two views. Which view is not needed in each case? In drawing A, is it the top view?

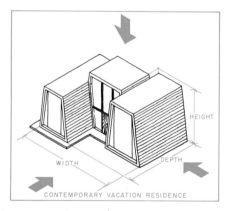

Fig. 2-30 *An A-frame residence in pictorial, showing three dimensions.*

Multiview Sketches

A pictorial drawing shows how the object looks in three-dimensional form. Three directions are suggested for viewing the residence in Fig. 2-30. From the front, you can see the width and height. This is the front view. From the side, you can see the depth and height. This is the right-side view. From above, the depth and width show. This is the top view. However, a three-view pictorial does not fully describe this residence. Some of the lines and details are not entirely visible (Fig. 2-31).

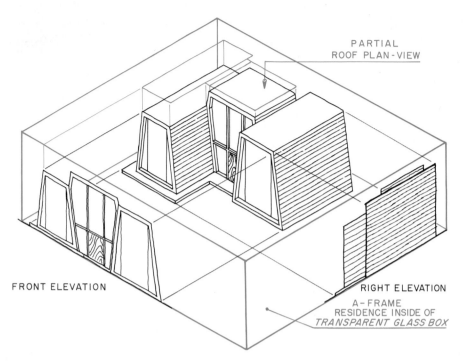

Fig. 2-31 *Two elevations can be projected from the pictorial drawing to the transparent glass box.*

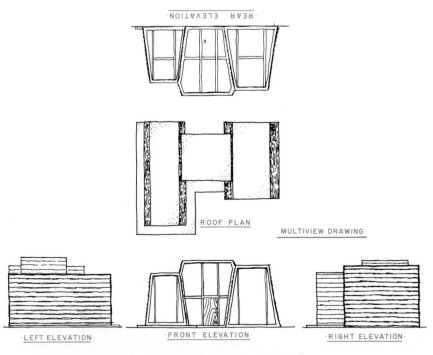

Fig. 2-32 *When the glass box is opened, five views are projected.*

The Glass Box

To get an idea of how many views are required to describe an object, many drafters use an imaginary glass box. For example, the residence in Fig. 2-31 can be thought of as being inside a transparent (clear) glass box. By looking at each side of the building through the glass box, you can see the five views of the building. Fig. 2-32 shows the glass box opened up into one plane as the views would be drawn on paper.

MATERIALS FOR SKETCHING

Sketching has two major advantages over formal drawings. First, only a few materials are required to create a sketch. Second, you can create a sketch anywhere. You are ready to sketch with a pencil, an eraser, and a pad of paper. If you need more equipment than that, you are probably not as good a drafter as you could be.

Paper

You can use plain paper for sketching. If you need to refine the sketch, use tracing paper. You may also use graph paper to control proportions while sketching. The most common type of graph paper has heavily ruled 1.00-inch (in.) squares. The 1.00-in. squares are then subdivided into lightly ruled .10-, .12-, .25-, or .50-in. squares. This paper is called 10 to the inch, 8 to the inch, and so on.

Graph paper ruled in millimeters (mm) is also available. In addition, there are many specially ruled types of graph paper for particular kinds of drawing. For example, you can use special graph paper for isometric or perspective drawings. These kinds of drawings are explained further in Chapter 12.

You can sketch on any convenient size of paper. However, standard 8.50 × 11.00 in. (216 × 279 mm) letter paper is the best for making small sketches quickly. You can hold the paper on stiff cardboard or on a clipboard while working on it. If you use graph paper, put it under tracing paper to help guide line spacing.

Pencils and Erasers

Most drafters like to use soft lead pencils (grades F, H, or HB), properly sharpened. They also use an eraser that is good for soft leads, such as a plastic eraser or a kneaded-rubber eraser.

Use a drafter's pencil sharpener to remove the wood from the *plain* end of a pencil so that you don't remove the grade mark (F, H, or HB) on the other end. Sharpen the lead to a point on a sandpaper block or on a file. If you are using a lead holder, use a lead pointer. Do not forget to adjust the grade mark in the window of the lead holder if it has one. Be careful to remove the needle point that the sharpener leaves by touching it gently on a piece of scrap paper. Then the pencil will not groove or tear the drawing paper. For more detailed information about pencils, lead holders, sharpeners, and erasers, refer to Chapter 3.

Four types of points are used for sketching: sharp, near-sharp, near-dull, and dull (Fig. 2-33A). The points should make lines of the following kinds (Fig. 2-33B):

- Sharp point—a thin black line for center, dimension, and extension lines

- Near-sharp point—visible or object lines
- Near-dull point—cutting plane and border lines
- Dull point-construction lines

AUTOCAD SKETCH AND PLINE COMMANDS

Sketching in CAD is different from freehand sketching. Although most CAD programs offer a method of freehand "sketching," this is usually not the most efficient way to create a CAD sketch. For example, the AutoCAD software has a SKETCH command that allows freehand drawing (Fig. 2-34). With this command, you can capture the movement of the cursor on a digitizer or the movement of a mouse to record a series of small lines. However, in spite of its name and function, the most common use of the SKETCH command is to convert hand-drawn documents to CAD drawings.

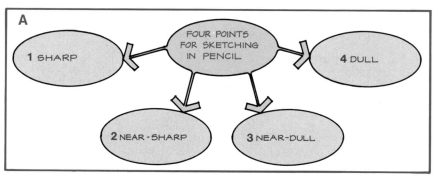

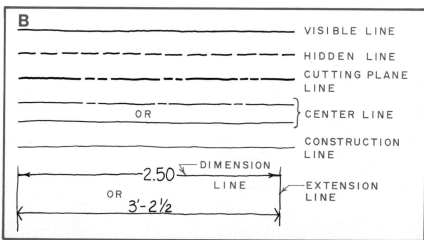

Fig. 2-33 *(A) Four convenient pencil points for sketching. (B) Types of lines used in sketching.*

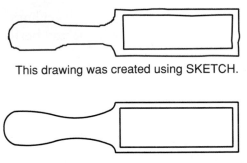

This drawing was created using SKETCH.

This drawing was created using PLINE and other AutoCAD commands.

Fig. 2-34 *The SKETCH command allows the drafter complete freedom to draw freehand on the computer.*

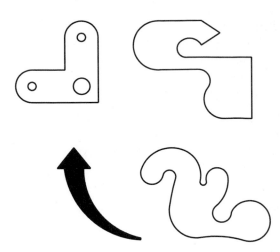

Fig. 2-35 *A polyline is a connected series of lines and arcs. Because the CAD programs consider it a single entity, an object drawn as a polyline is easier to move and edit than the same object drawn as a collection of individual lines and arcs.*

Because it is difficult to use accurately, the SKETCH command is generally not used to create sketches. Why create an inaccurate sketch when the CAD software allows you to create a much more refined sketch in the same amount of time? In fact, freehand sketching can actually take more time, and the results are generally not impressive. Therefore, "sketching" in CAD is much more likely to be done using the standard CAD commands.

Drafters often use the PLINE command to create "freeform" lines and arcs. PLINE allows you to create a series of connected lines and arcs at any size (Fig. 2-35). It produces a much cleaner appearance than the SKETCH command. although it is more difficult to control the exact placement of an arc using PLINE.

DRAWING LINES

Lines drawn freehand have a natural look (Fig. 2-36). Their slight changes in direction show freedom of movement. To draw a line, hold the pencil far enough from the point that you can move your fingers easily and yet can put enough pressure on the point to make dense, black lines when necessary. Draw light construction lines with very little pressure on the point. They should be light enough that they need not be erased.

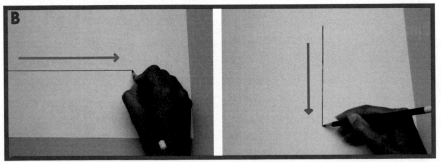

Fig. 2-36 *Some ways of sketching straight lines.* (Courtsy of Ann Garvin and Circle Design)

Straight Lines

You can sketch lines in the following ways: (1) Draw one long, continuous line. (2) Draw short dashes where the line should start and end. Then place the pencil point on the starting dash. Keeping your eye on the end dash, draw toward it. (3) Draw a series of strokes that touch each other or are separated by very small spaces. (4) Draw a series of overlapping strokes (Fig. 2-36A).

Before you try to draw objects, practice sketching straight lines to improve your line technique. Draw vertical lines from the top down (Fig. 2-36B). Draw horizontal lines from left to right (if you are right-handed).

Slanted Lines and Specific Angles

Sketch slanted, or inclined, lines from left to right. It might be easiest to turn the paper and draw an inclined line the same way as a horizontal one. When trying to sketch a specific angle, first draw a vertical line and a horizontal line to form a right (90°) angle. Divide the right angle in half to form two 45° angles (Fig. 2-37). Or divide it in thirds to form three 30° angles. By starting with these simple angles, you can estimate (guess) other angles more exactly. Note the direction of the inclined lines drawn to form a desired angle in Fig. 2-38.

DRAWING CIRCLES AND ARCS

There are several ways to sketch a circle. One way is to draw very light horizontal and vertical lines. Estimate the length of the **radius** (the distance from the center of the circle to its edge; plural, *radii*) and mark it off. Using the marks as guides, draw a square in which you can sketch the circle (Fig. 2-39).

The second way to draw a circle is to draw very light centerlines. Then draw bisecting (halving) lines through the center at convenient angles (Fig. 2-40). Next, estimate the length of the radius and mark off this distance on all the lines (Fig. 2-41). The bottom of the curve is generally easier to form, so draw it first. Then turn the paper so that the rest of the circle is on the bottom. Finish drawing the circle. Draw a curved line that runs through all the radius marks (Fig. 2-42).

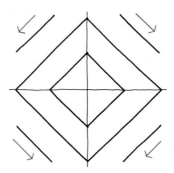

Fig. 2-37 *Draw horizontal and vertical lines before sketching slanted lines.*

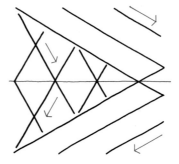

Fig. 2-38 *Sketching slanted lines and angles.*

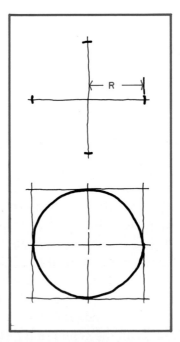

Fig. 2-39 *Mark off the radii and draw a square in which to sketch a circle.*

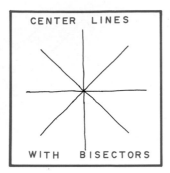

Fig. 2-40 *For a circle, draw centerlines with bisecting lines.*

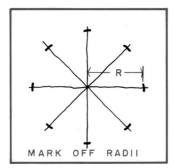

Fig. 2-41 *Mark off the estimated radii on all lines before sketching the circle.*

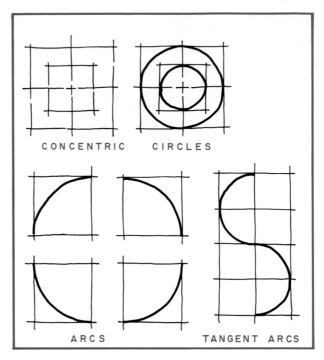

Fig. 2-43 *Control the shape of arcs and concentric circles by first sketching squares.*

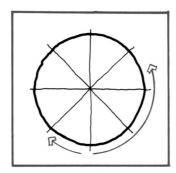

Fig. 2-42 *The bottom side of a curve is the easiest side to form.*

You can use these same methods to sketch **arcs** (parts of a circle); **tangent arcs** (parts of two circles that touch); and **concentric circles** (circles of different diameters that have the same center), as shown in Fig. 2-43. Use light, straight construction lines to block in the area of the figure.

For large circles and arcs and for **ellipses** (ovals), use a scrap of paper with the radius marked off along one edge. Put one end of the marked-off radius on the center of the circle. Draw the arc by placing a pencil at the other end of the radius and turning the scrap paper (Fig. 2-44).

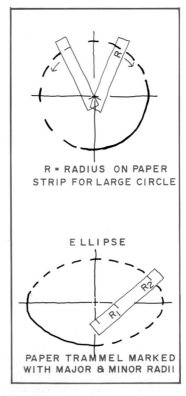

Fig. 2-44 *Large circles, large arcs, and ellipses can be sketched easily with the aid of a strip of paper.*

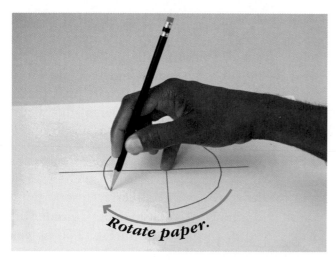

Fig. 2-45 *To use your hand as a compass, use your little finger as a pivot point.* (Courtesy of Ann Garvin and Circle Design)

Fig. 2-46 *Two pencils can serve as a compass.* (Courtesy of Ann Garvin and Circle Design)

Note in Fig. 2-44 that two radii are needed for an ellipse. To draw the ellipse, sketch both center-lines. Keep both radius points on the centerlines as you draw with the pencil at the other end.

You can also use your hand as a compass. To do this, use your little finger as a pivot (turning point) at the center of the circle. Use your thumb and forefinger to hold the pencil rigidly at the radius you want. Turn the paper carefully under your hand, thereby drawing the circle (Fig. 2-45).

Another way to draw circles is to use two crossed pencils. Hold them rigidly with the two points as far apart as the length of the desired radius. Put one pencil point at the center. Hold it there firmly and turn the paper, drawing the circle with the other point (Fig. 2-46).

PROPORTIONS FOR SKETCHING

Sketches are not usually made to scale (exact measure). Nonetheless, it is important to keep sketches in **proportion** so that each part of the drawing is approximately the right size in relation to other parts of the drawing.

Estimating Proportions

In order to sketch well, the you must be able to *eyeball* (estimate by eye) an object's proportions. In preparing the layout, look at the largest overall dimension, usually width, and estimate the size.

Next, determine the proportion of the height to the width. Then as the front view with the width and height takes shape, compare the smaller details with the larger ones and fill them in, too (Fig. 2-47).

It is important that the design drafter, in sketching an object, have a good sense of how distances relate to each other. This allows the drafter to show the width, height, and depth of an object in the right proportions.

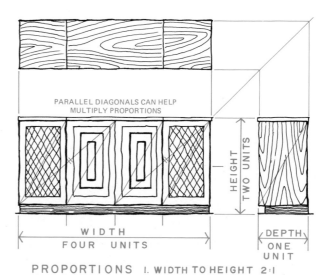

Fig. 2-47 *Sketching a contemporary cabinet with proportional units.*

For example, suppose that the design drafter plans a cabinet to be 4 units wide. The height of the cabinet is 2 units. The depth of the cabinet is 1 unit. This is a proportion of 2 to 1. If you were designing this cabinet in customary units, you might define a unit as 15 inches. The overall width of the cabinet would then be 60 in. (4 × 15 in.). The height would be 30 in. (2 × 15 in.), and the depth 15 in. (1 × 15 in.). The proportions for the cabinet in Fig. 2-47 are developed in 2-to-1 units.

Technique in Developing Proportion

Through practice, you can train your eye to work in two directions so that you can both divide and extend lines accurately. For example, you should be able to divide a line in half by estimating. You can divide the halves again to give fourths, and so on. Using a similar technique, you can expand lines one unit at a time. Start by drawing a line of 1 unit. Increase it by one equal unit so that it is twice as long as at first. Practice adding an equal unit and dividing a unit equally in half. Practice developing units on parallel horizontal lines. Then develop them vertically. By learning to compare distances, you can get better and better at estimating (Fig. 2-48).

Design drafters can use scrap paper or a rigid card as a straightedge when they do not have scales (rulers) at hand. Fold the paper in half to find the length of units on the marked edge, as shown in Fig. 2-49. In this way, you can draw lines of the same length several times in different directions. This will help you maintain the right proportions.

Making a Proportional Sketch

To create drawings in the correct proportions, you should follow seven basic steps. Consider the chair shown in picture form in Fig. 2-50 as you review the steps listed below.

1. Observe the chair shown in Fig. 2-50A.
2. Select the views needed to show all shapes. In this case, you can describe the chair fully using three views: top, front, and right-side.
3. Estimate the proportions carefully. On your drawing paper, mark off major distances for width, height, and depth in all three views (Fig. 2-50B). Use lines that are light enough that you do not have to erase them later.
4. Block in the enclosing rectangles (Fig. 2-50C).
5. Locate the details in each of the views. Block them in (Fig. 2-50D).
6. Finish the sketch by darkening the object lines.
7. Add any dimensions and notes that you need.

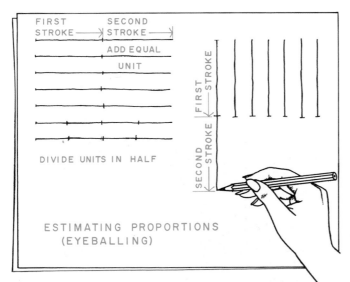

Fig. 2-48 *Practice estimating (eyeballing) proportional units.*

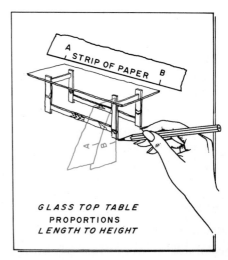

Fig. 2-49 *Drafters develop techniques for blocking out sketches.*

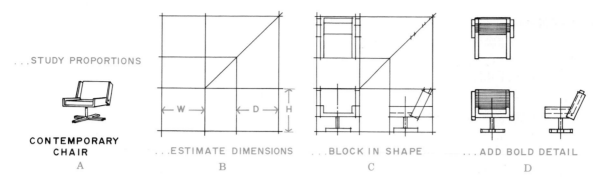

...STUDY PROPORTIONS

CONTEMPORARY CHAIR
A

...ESTIMATE DIMENSIONS
B

...BLOCK IN SHAPE
C

...ADD BOLD DETAIL
D

Fig. 2-50 *The development of a multiview drawing.*

PICTORIAL DRAWING

There are several kinds of pictorial drawings. All types will be discussed in Chapter 12. For sketching, we will consider only two kinds of pictorial drawings: oblique and isometric. Making oblique and isometric sketches will help you learn how to visualize or "see" objects in your mind. You must be able to do this in order to draw multiview projections. Pictorial drawings help people who are not trained to read multiview drawings understand basic shapes.

Oblique Sketching

One of the easiest pictorial drawings to make is the **oblique** (inclined) **sketch.** In an oblique sketch, the depth of an object is drawn at any angle. Every object has three dimensions: width, height, and depth. Each of these dimensions is called an **axis** (plural, *axes*). In oblique drawings, two of the axes are at right (90°) angles to each other. The third axis is drawn at any angle to the other two (Fig. 2-51). You may make any side of an object the front view.

As you may remember, the front view shows the width and height of the object. You usually make the side with the most detail the front view. In Fig. 2-52, a digital clock radio is used as an example of an oblique pictorial drawing. The dial side has been made the front view. This is because it has the most detail and shows the width and height of the radio. The front view is sketched just as the clock radio would appear when you look at it directly from the front.

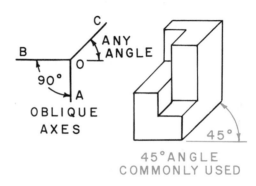

Fig. 2-51 *Oblique drawings always have one right-angle corner.*

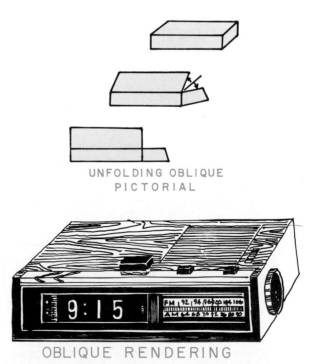

UNFOLDING OBLIQUE PICTORIAL

OBLIQUE RENDERING

Fig. 2-52 *An oblique pictorial drawing.*

Oblique Layout

To make an oblique sketch, as shown in Fig. 2-53, always follow a good layout procedure. If oblique graph paper is available, you may wish to use it as a guide as you follow these steps:

1. Estimate the proportions of the object (Fig. 2-54).
2. Block in lightly the front face of the object with the estimated units for the width and height.
3. Sketch in lightly the receding (going-away) lines at any angle. For example, you might choose a 45° angle from the horizontal. You can choose an angle that shows as much as desired of the top and side. If you draw the third axis at a small angle, such as 30°, the side shows more clearly (Fig. 2-55A). On the other hand, if you choose 60°, the top will show more clearly (Fig. 2-55B).
4. The receding axis may have the same proportioned units as the front axis. Or you can reduce its size by up to one half. When the depth dimension of a drawing is exactly one half of the true dimension, it is called a *cabinet* sketch. Using a full-depth dimension produces a *cavalier* sketch (Fig. 2-53).
5. Darken the final object lines.

Oblique Sketching on Graph Paper

Graph paper is useful for oblique sketching because the front view of the oblique sketch is like the front view of a multiview sketch (Fig. 2-56A and B). If you develop the oblique pictorial drawing on graph paper from a multiview drawing on graph paper, simply transfer the dimensions from one to the other by counting the graph-paper squares:

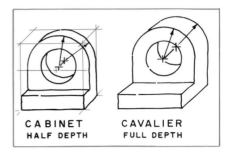

Fig. 2-53 *Oblique drawings can vary in depth.*

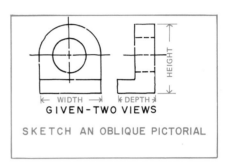

Fig. 2-54 *Check the width, height, and depth on the views given.*

1. Block in lightly the front face of the object by counting squares.
2. Sketch lightly the receding axis by drawing a line diagonally through the squares. You find the depth by using half as many squares as on the side view. Since the side view of Fig. 2-56A shows a depth of four squares, you should use two squares for the depth in the oblique sketch.
3. Sketch in any arcs and circles (Fig. 2-56C).
4. Darken the final object lines (Fig. 2-56D).

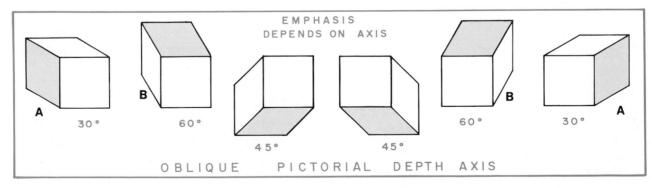

Fig. 2-55 *The many angles of oblique drawing.*

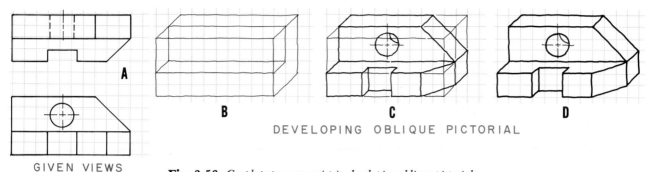

DEVELOPING OBLIQUE PICTORIAL

GIVEN VIEWS

Fig. 2-56 *Graph paper can assist in developing oblique pictorials.*

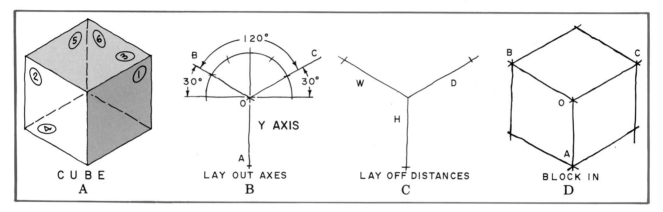

Fig. 2-57 *Sketching isometric angles and layout of Y axis.*

Oblique Circles

In oblique sketching, circles in the front view can be drawn in their true shape. However, circles drawn in the top or side views appear *distorted* (not in their true shape). Indeed, you must draw an ellipse to show such circles. Ellipses are more difficult to represent accurately, and in some cases they may confuse the viewer. Therefore, it is better to show the circular shapes of important parts in the front view (Fig. 2-53).

Isometric Sketches

An **isometric** (equal-measure) **sketch** is based on three lines called *axes*. These are used to show the direction of the three basic dimensions: width, height, and depth. A cube is an object that has six equal square sides. As you can see in Fig. 2-57A, an isometric cube has three equal sides. Thus, there are three equal angles in Fig. 2-57B. The height *OA* is laid off on the vertical leg of the Y axis. The width *OB* is laid off to the left on a line

30° above the horizontal. The depth *OC* is laid off to the right on a line 30° above the horizontal. The 30° lines receding to the left and right can be located by estimating one third of a right angle, as shown in Fig. 2-57B. Lines parallel to the axes are called **isometric lines.** The estimated distances are laid off on them only as shown for the cube at Fig. 2-57C.

The sketched lines for isometric axes tend to become steeper than 30° if you do not prepare the layout carefully. A better pictorial sketch results when the angle is at 30° or a little less. Using isometric graph paper with 30° ruling lets you make sketches quickly and easily (Fig. 2-58). Figure 2-59 shows the steps in making an isometric sketch.

Nonisometric Lines

As you have seen, lines parallel to isometric axes are called *isometric lines*. Therefore, lines that are not parallel to the isometric axes are labeled **nonisometric lines.** Examples of nonisometric lines are shown in Figs. 2-60 and 2-61.

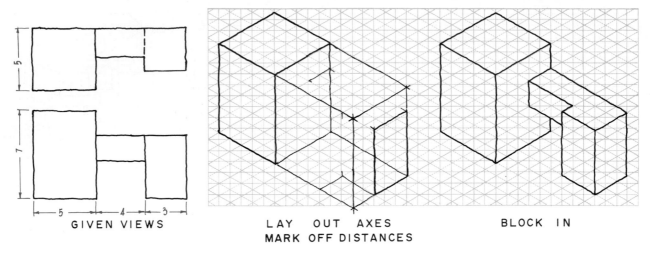

GIVEN VIEWS

LAY OUT AXES
MARK OFF DISTANCES

BLOCK IN

Fig. 2-58 *Isometric grid paper can be used for quick sketches.*

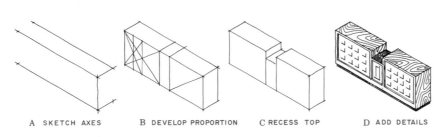

A SKETCH AXES   B DEVELOP PROPORTION   C RECESS TOP   D ADD DETAILS

Fig. 2-59 *The steps in making an isometric sketch.*

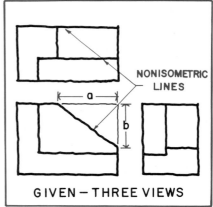

NONISOMETRIC LINES

GIVEN — THREE VIEWS

Fig. 2-60 *Identifying the nonisometric lines on a three-view drawing.*

Any object may be sketched in a box, as suggested in Fig. 2-62. Note, however, that the objects in Figs. 2-63 and 2-64 have some nonisometric lines. You can draw these lines by extending their ends to touch the blocked-in box. Locate points at the ends of the lines by estimating measurements parallel to isometric lines. Having located both ends of the nonisometric lines, you can sketch the lines from point to point. Nonisometric lines that are parallel to each other also appear parallel on the sketch (Fig. 2-64). Note how the ends have been located on lines 1-2 and 1-3 in Fig. 2-64. Distances *a* and *b* are estimated and transferred from the figure at (A) to (B). Any inclined line, plane, or specific angle must be found by locating two points of intersection on isometric lines.

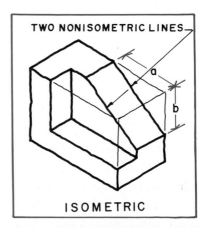

TWO NONISOMETRIC LINES

ISOMETRIC

Fig. 2-61 *The nonisometric lines form an inclined plane on the isometric drawing.*

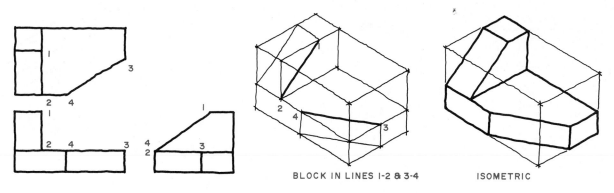

Fig. 2-62 *Development of two inclined planes in an isometric drawing.*

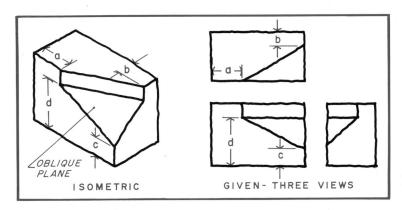

Fig. 2-63 *Developing an oblique plane in an isometric drawing.*

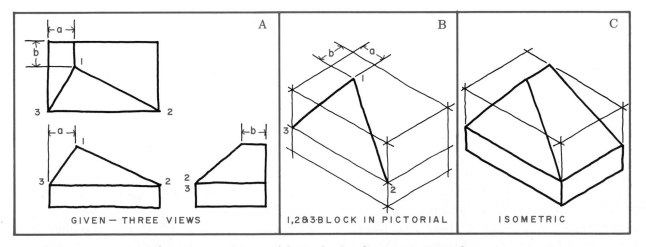

Fig. 2-64 *Development of three inclined surfaces in an isometric drawing.*

Isometric Circles and Arcs

All circles in an isometric view are drawn as ellipses. To sketch a circle in an isometric view (Fig. 2-65), sketch an isometric square first. Sketch the small-end arcs tangent to (touching) the square. Then sketch the larger arcs tangent at points *T* to finish the ellipse. Note that the *major diameter* (long axis) of the ellipse is no longer than the true diameter of the circle. The *minor diameter* (short axis) is shorter. This difference is caused by the isometric angle. Figure 2-65 shows an ellipse for a top view only. Circles on the three faces of an isometric cube are sketched in Fig. 2-66.

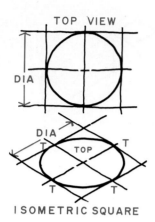

Fig. 2-65 *The isometric square with arcs tangent to form an ellipse (isometric circle).*

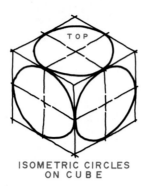

Fig. 2-66 *Isometric circles (ellipses) sketched on the front, top, and side of a cube.*

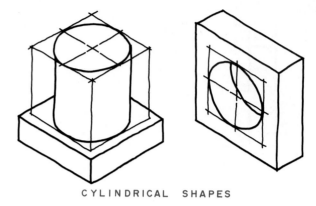

Fig. 2-67 *Isometric circles assist in describing cylindrical forms.*

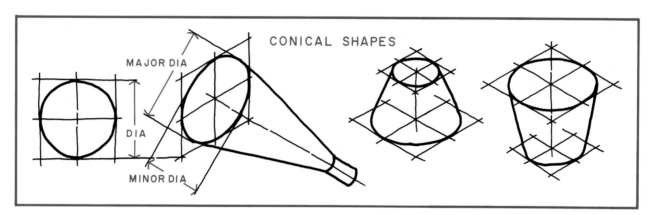

Fig. 2-68 *Isometric circles help form conical shapes.*

You will need isometric circles to develop cylindrical and conical (conelike) shapes in a sketch. Some ways to block in cylindrical shapes are shown in Fig. 2-67. Methods of blocking in conical shapes are shown in Fig. 2-68.

Arcs developed in an isometric view are shown in Fig. 2-69. A semi-circular opening and rounded corners appear in the front view. The object is blocked in at Fig. 2-69A. The outline of the object is darkened in at Fig. 2-69B. Note that only partial circles are needed here. The rounded corners take up only a quarter of the full isometric circle that was plotted.

Irregular Curves

Draw irregular (noncircular) curves in pictorial sketches by plotting coordinates (points). To locate the curve, you generally transfer the coordinates from a multiview sketch. In Fig. 2-70A, coordinates are plotted on the front view that stand for the width from the left edge and the height from the top edge. The series of lines in Fig. 2-70B are drawn parallel to the vertical and horizontal axes to locate points of intersection as shown. Similar coordinates are plotted on the pictorial in Fig. 2-70C. The intersections serve as points for sketching the pictorial curve.

INTRODUCTION TO DIMENSIONING

An effective sketch must fully describe the object. Generally the sketch is made before the measurements of an object have been decided. After the needed dimensions are determined, they can be recorded on the sketch.

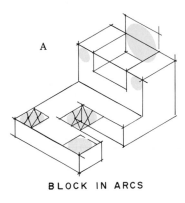

BLOCK IN ARCS

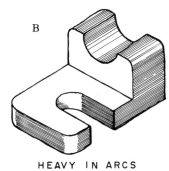

HEAVY IN ARCS

Fig. 2-69 *To block in arcs on an isometric sketch, use a method similar to the one you use for circles.*

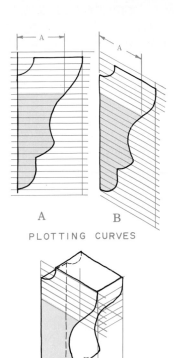

A   B

PLOTTING CURVES

C

Fig. 2-70 *Plot irregular curves using a coordinated grid.*

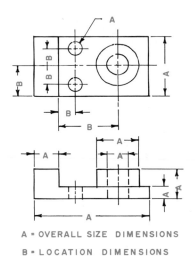

A = OVERALL SIZE DIMENSIONS

B = LOCATION DIMENSIONS

Fig. 2-71 *(A) Size dimensions; (B) location dimensions.*

Types of Dimensions

Two types of dimensions are used on sketches. Size dimensions (Fig. 2-71A) describe the overall geometric elements that give an object form. Location dimensions (Fig. 2-71B) relate these geometric elements to each other. Together, the two types of dimensions accurately describe the size and shape of the object.

Definitions

Several special types of lines are used to dimension a drawing. Refer to Fig. 2-72 as you read the following definitions.

- A *dimension line* is used to show the direction of a dimension (Fig. 2-72). Dimension lines have an arrowhead at each end to show where the dimension begins and ends.
- *Extension lines* are thin lines used to extend the shape of the object to the dimension line (Fig. 2-72A).

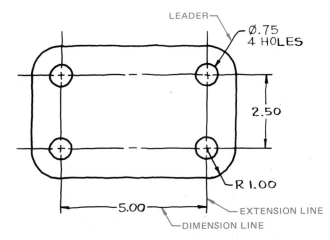

Fig. 2-72 *Drafters use special kinds of lines to clarify dimensioned drawings.*

- A *leader* is a thin line drawn from a note or dimension to the place where it applies. A leader starts with a horizontal dash and angles off to the part featured, usually at 30°, 45°, or 60°. It ends with an arrowhead (Fig. 2-72A).

CHAPTER 2
REVIEW

Learning Activities

1. Collect six sketches and six mechanical drawings that show the various types of lettering used by the professional drafting technician. Compare and evaluate the quality of the work with your teacher. Rate the items as excellent, good, and fair. Use magazines and prints where possible.

2. Bring prints for architectural studies to class and to review the architectural lettering styles on several residential or commercial prints from local architects. Are there freehand and mechanically drawn letters and numbers? Which best represents the professional designer?

3. Compare the styles of lettering used on working drawings in Chapter 11 with the styles used on architectural drawings in Chapter 20. Which style has the more creative form? Is it easy to tell the mechanical lettering from the freehand lettering? Explain.

4. Collect six sketches from your favorite subject, such as racing cars or bicycles, that do not have notes. Prepare at least four lines of lettering on each sketch to describe special, important features. Try to make the lettering look like it was done by the designer.

5. Examine current CAD and design journals and list the CAD software available to the drawing technician. Note on the list the major features advertised for each program.

Questions

Write your answers on a separate sheet of paper.

1. What kind of paper is used in the overlay method?

2. Why do the drafters, designers, and engineers who must take part in highly technical discussions consider sketching a valuable skill?

3. Of all the reasons for sketching, which three seem most important to you? Explain your answer.

4. Name the three terms that describe the nature of a sketch.

5. Why are the four visual symbols in the language of sketching essential in graphic communication?

6. Describe the qualities needed for paper that will be used for refining a sketch.

7. What is the term applied to estimating proportions? Why is estimating important to the drafter and designer?

8. List the uses for the four kinds of pencil points used in freehand sketching.

9. Explain the proper methods for sketching horizontal and vertical lines.

10. Name two advantages of pictorial sketching.

11. When preparing sketches of objects, what is the drafter's first important decision?

12. What is another name for multiview drawing?

13. Name two basic types of pictorial sketching.

14. What are the three principal planes of projection? What view appears on each plane?

15. Describe the advantages and disadvantages of using a CAD system to create notes on a drawing.

CHAPTER 2
PROBLEMS

Beginning with these problems, you will find a small CAD symbol. This symbol indicates that a problem is appropriate for assignment as a computer-aided drafting problem and suggests a level of difficulty (one, two, or three) for the problem.

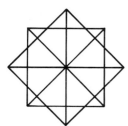

Fig. 2-73

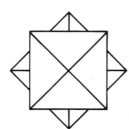

Fig. 2-74

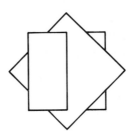

Fig. 2-75

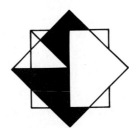

Fig. 2-76

Figs. 2-73 through 2-76 *CAD1 Sketch in 2.00" overlapping squares as creative visual studies. Create two of your own designs.*

Fig. 2-77

Fig. 2-78

Fig. 2-79

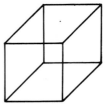

Fig. 2-80

Figs. 2-77 through 2-80 *CAD1 Sketch the squares overlapping, diminishing, and as a transparent cube. Sizes are about 1.50 in., 1.12 in., and .75 in.*

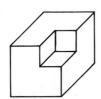

Fig. 2-81 *CAD1*
Sketch a cube with the rectangular shape as shown. Observe the optical illusion.

Fig. 2-82 *CAD1*
Sketch a rectangular solid with 2-to-1 proportions. Use .50", 1.00", and 2.00".

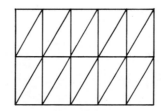

Fig. 2-83 *CAD1*
Sketch the apparent two-dimensional form using six diagonals.

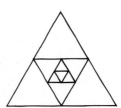

Fig. 2-84 *CAD1*
Sketch a 3.00" equilateral triangle with diminishing triangles at midpoints. Note the proportions. How many triangles can you make diminish inside?

PROBLEMS

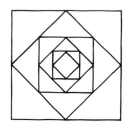

Fig. 2-85 *CAD1*
Sketch a 3.00" square with diminishing squares at midpoints. Note the proportions.

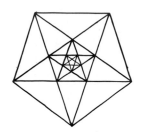

Fig. 2-86 *CAD1*
Sketch a pentagon using 2.00" sides with diminishing five-pointed stars. Note the proportions.

TETRAHEDRON

OCTAHEDRON

HEXAHEDRON

ICASAHEDRON

DODECAHEDRON

Fig. 2-87 **Fig. 2-88** **Fig. 2-89** **Fig. 2-90** **Fig. 2-91**

Figs. 2-87 through 2-91 *CAD2* *Sketch the five basic solids and label each.*

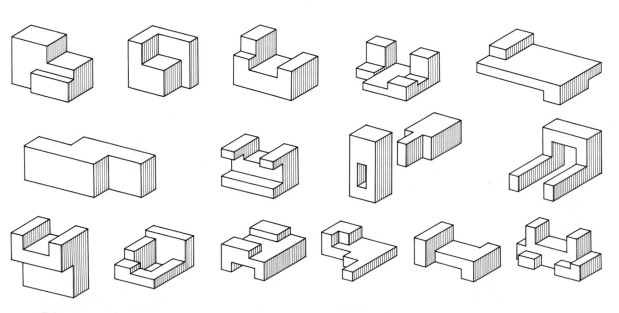

Fig. 2-92 *CAD2* *Sketch the three views needed for a multiview drawing. Select approximate proportions and use selected methods as shown in Figs. 2-35 through 2-65 (straight lines).*

PROBLEMS

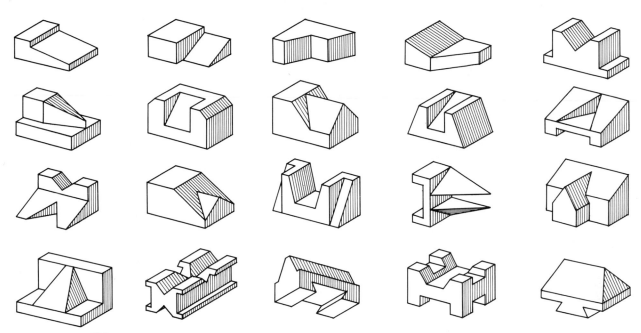

Fig. 2-93 *CAD2 Sketch the three views needed to define the machine part assigned. Select approximate sizes. Use selected methods as shown in Figs. 2-35 through 2-65 (inclined lines).*

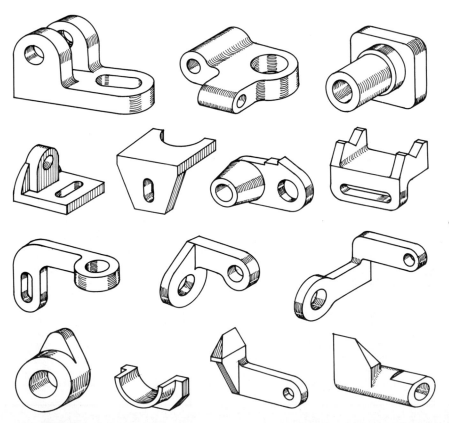

Fig. 2-94 *CAD2 Sketch the three views required to define the problems assigned with curved lines.*

PROBLEMS

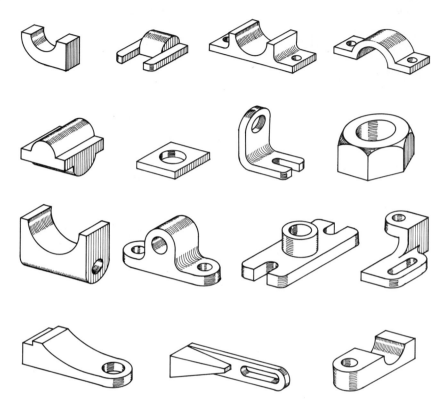

Fig. 2-95 *CAD2 Sketch the three views required to define the problems assigned.*

Figs. 2-96 through 2-99 *CAD1 Lettering is best learned in short, unhurried periods. The problems are planned for a working space of 7.00 in. × 5.00 in. Each problem may be worked on an 8.50 × 11.00 in. sheet. Use an H or a 2H pencil with a conical point. The lettering should be black (2H or H pencil) in order to reproduce, as most drawings are made on a translucent medium. Use a sharp 4H pencil to rule very light guidelines for the heights of the letters. Rotate the pencil in the fingers after each few strokes to keep the point in shape; sharpen the pencil as often as necessary to keep a proper point.*

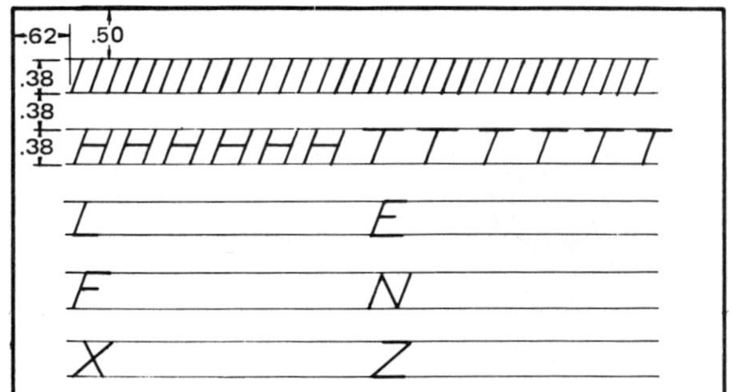

.62   .50

.38 |||||||||||||||||||||||||||||||||||||

.38

.38 HHHHHH TTTTTT

L   E

F   N

X   Z

INLET FILLET HELIX

Fig. 2-96 *CAD1 Letter .38" inclined capitals in pencil. Draw a few very light guidelines.*

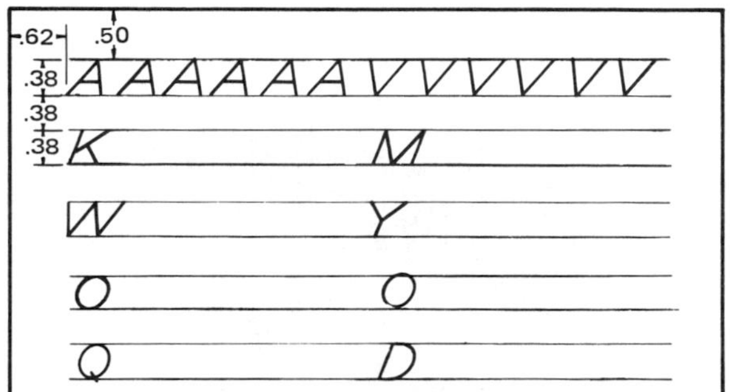

.62   .50

.38 AAAAAA VVVVVV

.38 K   M

W   Y

O   O

Q   D

LAMINATED WOOD

Fig. 2-97 *CAD1 Letter .38" capitals in pencil. Complete each line.*

PROBLEMS

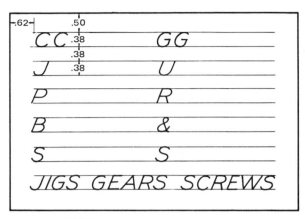

Fig. 2-98 *CAD1 Letter .38" inclined capitals in pencil. Complete each line.*

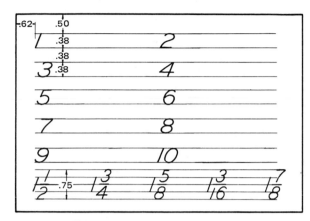

Fig. 2-99 *CAD1 Make the total height of the fractions two times the height of the numerals.*

Figs. 2-100 through 2-103 *CAD1 Prepare a lettering sheet as you did for the four inclined lettering assignments and practice the same letters with vertical capitals for the four assignments.*

── SCHOOL-TO-WORK ──

Solving Real-World Problems

Figs. 2-104 through 2-110 *SCHOOL-TO-WORK problems are designed to challenge a student's ability to apply skills learned within this text. Be creative and have fun!*

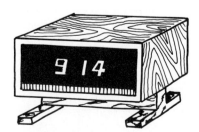

Fig. 2-104 *CAD2 Make a multiview sketch of the digital clock. Use your imagination and redesign the base using your choice of materials, shapes, and sizes. Prepare an oblique sketch of your finished design. Add dimensions to your sketches if instructed to do so.*

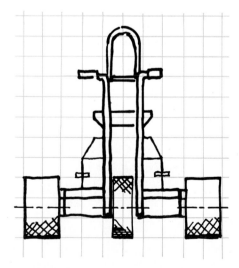

Fig. 2-105 *CAD2 Make a three-view sketch of the three-wheel vehicle. Be creative in designing and redesigning all parts. Use the front view shown as the basis for the size and general shape of your design. Use overlays to aid in refining your design ideas (see Fig. 2-23).*

PROBLEMS

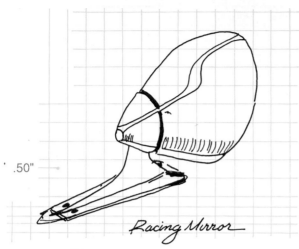

Fig. 2-106 *CAD2 BE CREATIVE: Prepare a two- or three-view sketch of the race car mirror. Add a pictorial sketch if your design is somewhat different from the sketch shown.*

Fig. 2-108 *CAD2 Prepare a sketch of the space shuttle from Figs. 2-6, and 24-25. Two views are required, with notes.*

Fig. 2-110 *CAD2 Prepare your name and your school name on a cover sheet for your drawings. Use block lettering as shown in Fig. 2-16.*

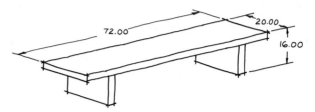

Fig. 2-107 *CAD1 Design a park bench using the overall sizes shown in the sketch. Be as creative as you wish in the design and the selection of materials and fasteners. Sketch two or three views as required.*

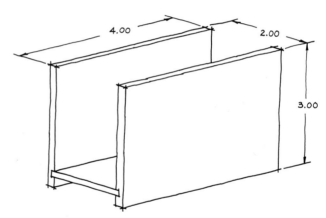

Fig. 2-109 *CAD1 Redesign the letter holder shown in the sketch. The finished product should hold approximately 15 letters in either vertical or horizontal position. Initials or other decorative overlays can be added as desired. Sketch two or three views as required. Add a pictorial sketch if instructed to do so.*

3

THE USE AND CARE OF DRAFTING EQUIPMENT

OBJECTIVES

Upon completion of Chapter 3, you will be able to:

○ Describe the components of a CAD system.
○ Properly and efficiently use basic drafting tools and equipment to produce technical drawings.
○ Identify and use the lines and line symbols recommended by the American National Standards Institute (ANSI).
○ Produce a finished ink tracing.

VOCABULARY

In this chapter, you will learn the meanings of the following terms:

- acute angle
- alphabet of lines
- angle
- circumference
- compass
- drafting film
- erasing shields
- inclined
- india ink
- irregular curves
- obtuse angle
- opaque
- protractor
- right angle
- scales
- symmetrical
- T-square
- transparent
- vellum
- vertex

You have seen why drawings are used in industrial, engineering, and scientific work. Designers and scientists often make freehand sketches to help them study new ideas or to show their ideas to other people. Usually, though, technical drawings are made with drafting *instruments* (tools). The plans and directions that are needed for doing engineering work of all kinds are prepared in the drafting room (Fig. 3-1). There, the designs are worked out and checked.

Technical drawing is really a language. In learning to read and write the drafting language, you must learn which instruments to use on a particular problem. You need to know how to use drafting equipment skillfully, accurately, and quickly.

Fig. 3-1 *An industrial drafting room.* (Courtesy of Brent Phelps and University of North Texas/Benedict Wong, Ph.D.)

Fig. 3-2 *Drafting equipment.* (Courtesy of Teledyne Post)

BASIC MANUAL DRAFTING EQUIPMENT

Figure 3-2 shows a variety of basic drafting equipment. Many of these items are included in basic student drafting kits. Other common drafting equipment is listed below.

- drafting board
- T-square, or parallel-ruling straightedge, or drafting machine
- drawing sheets (paper, cloth, or film)
- drafting tape
- drafting pencils
- pencil sharpener
- erasing shield
- triangle, 45° and 30°-60° (not required with drafting machine)
- architect's, engineer's, or metric scale
- irregular curve
- drawing instrument set

- lettering instruments
- black drawing ink
- technical pens
- brush or dust cloth
- protractor
- cleaning powder

Your teacher can tell you exactly what equipment will be needed for your course.

Drawing Tables and Desks

Drawing tables and desks come in many different sizes and types (Fig. 3-3). One kind of table has a fixed top and a separate drawing board that can be moved around. Another has a fixed top that is made to be used as a drawing board itself. A third has an adjustable top that holds a separate drawing board at the slope you want. Still another has an adjustable top that is made to be used as a drawing board.

Some drawing tables are made to be used while you are standing up or sitting on a high stool. Other tables are the same height as regular desks. There are also combined tables and desks that you can use either standing up or sitting down. The type that combines a drafting table, desk, and regular office chair is the most comfortable and efficient. This kind is replacing the high drawing table in many drafting rooms. The cost of drawing tables and desks varies according to their size, style, and number of components.

Fig. 3-3 *Drafting tables are available in a variety of sizes and styles.* (Courtesy of Mayline Co.)

Drawing Boards

The drawing sheet is attached to a drawing board. Drawing boards used in school or at home usually measure 9.00 × 12.00 in. (230 × 300 mm), 16.00 × 21.00 in. (400 × 530 mm), or 18.00 × 24.00 in. (460 × 600 mm). Boards used to make engineering or architectural drawings are typically larger and may be any size needed. Boards are generally made of soft pine or basswood. They are made so that they will stay flat and so that the guiding edge (or edges) will remain straight. Hardwood or metal (steel or aluminum) strips are used on some boards to provide truer and more durable guiding edges.

T-Squares

A **T-square** (Fig. 3-4) is a drafting instrument that consists of a head that lines up with a true edge of the drafting board and a blade, or straightedge, that provides a true edge. T-squares are made of various materials. However, most have plastic-edged wood or clear plastic blades, and heads of wood or plastic. Where extreme accuracy is needed, stainless steel or hard aluminum blades with metal heads are used. The blade must be very straight. It must be attached securely to the top surface of the T-square head.

You can easily find how accurate your T-square is, as shown in Fig. 3-5. First, on a clean sheet of paper, draw a sharp, thin line along the drawing edge of the T-square. Second, turn the drawing sheet around and line up the drawing edge of the T-square with the other side of the line. If the drawing edge and the pencil line do not match, the T-square is not accurate. When you find this condition, you should replace the T-square.

T-squares come in several sizes. The size and type of T-square you choose should be appropriate for the type of drawing you will be doing. It should extend most of the way across the drafting board you will be using. The size of a T-square is found by measuring along the blade from the contact surface of the head to the end of the blade.

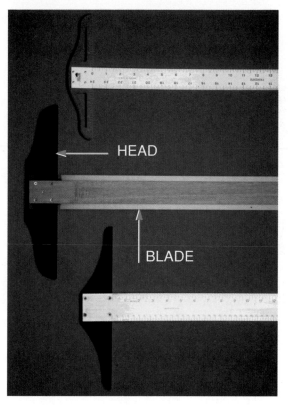

Fig. 3-4 *T-squares are available in various styles and materials.* (Courtesy of Mishima and Circle Design)

Drafting Machines

Two kinds of drafting machines are currently in wide use. The arm, or elbow, type is shown in Fig. 3-6. It uses an anchor and two arms to hold a movable protractor head with two scales. The scales are ordinarily at right angles to each other. The arms allow the scales to be moved to any place on the drawing that is parallel to the starting position. (The two are *parallel* when their edges are exactly the same distance apart at all points. Parallelism will be discussed further in Chapter 4.)

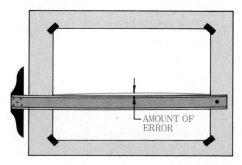

Fig. 3-5 *Check to see that the T-square is accurate.*

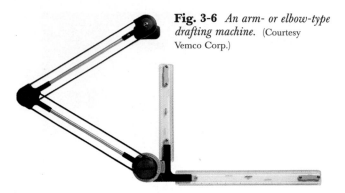

Fig. 3-6 *An arm- or elbow-type drafting machine.* (Courtesy Vemco Corp.)

Fig. 3-7 *The track-type drafting machine is especially adapted for wide drawings. It may also be used for drawings of regular sizes.* (Courtesy Vemco Corp.)

The track type of drafting machine, shown in Fig. 3-7, uses a horizontal guide rail at the top of the board and a moving arm rail at right angles to the top rail. An adjustable protractor head and two scales, usually at right angles, move up and down on the arm. The scales may be moved to any place on the drawing that is parallel to the starting position. This type of drafting machine is easy to use on large boards or on boards placed vertically or at a steep angle.

Most industrial drafting departments and many schools now use drafting machines. Drafting machines combine the functions of the T-square, triangles, scales, and protractor. You can draw lines to any desired length, at any location, and at any angle just by moving the scale ruling edge to the desired location. This lets you draw faster and with less work. You should learn how to use the drafting machine in all the possible ways and how to take care of it. If you do so, you will soon see how valuable a tool the drafting machine is.

Other Basic Tools

In addition to the equipment described thus far in this chapter, drafters use a variety of other tools, instruments, and equipment. Triangles, protractors, erasers and eraser shields, and parallel-ruling straightedges, among others, form part of a drafter's everyday toolkit.

Triangles

Drafters use two types of triangles in combination with the T-square to draw lines at various angles. The 45° triangle has one 90° angle and two 45° angles. The 30°-60° triangle has 30°, 60°, and 90°

angles. Techniques for using triangles are presented later in this chapter and in Chapter 4.

Protractors

A **protractor** is an instrument that is used to measure, or lay out, angles. A semicircular protractor is shown in Fig. 3-8, where an angle of 43° is measured.

Parallel-Ruling Straightedges

Many drafters prefer to use parallel-ruling straightedges when working on large boards placed vertically or nearly so. A guide cord is clamped to the ends of the straightedge. The cord runs through a series of pulleys on the back of the board. This allows the straightedge to slide up and down on the board in parallel positions (Fig. 3-9).

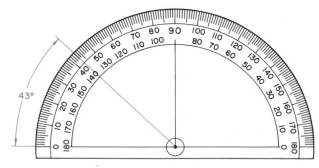

Fig. 3-8 *A protractor is used to lay out, or measure, angles.*

Fig. 3-9 *A parallel-ruling straightedge is another convenient instrument used to save time.* (Courtesy of Staedtler)

Irregular Curves

Irregular curves, or *French curves* (Fig. 3-10), are used to draw noncircular curves (involutes, spirals, ellipses, and so forth). Irregular curves are also used to draw curves on graphic charts. In addition, they can be used to plot motions and forces and to make some engineering and scientific graphs.

Irregular curves are made of sheet plastic. They come in many different forms, some of which are shown in Fig. 3-10. Sets are made for ellipses, parabolas, hyperbolas, and many special purposes. Many drafters also use *flexible curves,* which can be adjusted to complex curves that may be difficult to draw using other types of irregular curves (Fig. 3-11).

To use an irregular curve (Fig. 3-12), find the points through which a curved line is to pass. Then set the path of the curve by drawing a light line, freehand, through the points. Adjust it as needed to make the curve smooth. Next, match the irregular curve against a part of the curved line and draw part of the line. Move the irregular curve to match the next part, and so on. Each new position should fit enough of the part just drawn to make the line smooth. Note whether the radius of the curved line is increasing or decreasing and place the irregular curve in the same way. Do not try to draw too much of the curve with one position. If the curved line is **symmetrical** (mirrored around an axis), mark the position of the axis or centerline on the irregular curve on one side. Then turn the irregular curve around to match and draw the other side.

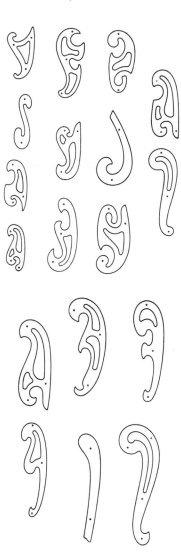

Fig. 3-10 *Some irregular, or French, curves. They are made in a great variety of forms.* (Teledyne Post)

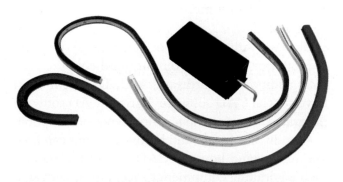

Fig. 3-11 *Flexible curves for plotting smooth curves. Some drafters use "ducks" such as the one shown here (the rectangular object) to position flexible curves more accurately and to hold the curves in place.* (Courtesy of Ann Garvin and Vanderberg Drafting Supply, Inc.)

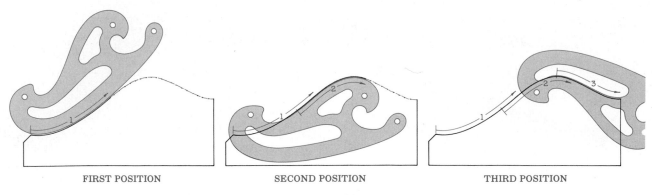

FIRST POSITION   SECOND POSITION   THIRD POSITION

Fig. 3-12 *Steps in drawing a smooth curve.*

Templates

Templates (Fig. 3-13) are an important part of the equipment of engineers and professional drafters. They save a great deal of time in drawing shapes of details. These include bolt heads, nuts, and electrical, architectural, and plumbing symbols.

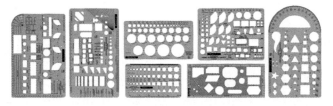

Fig. 3-13 *Templates are made for all possible uses and save a good deal of time.* (Courtesy of Staedtler)

CAD EQUIPMENT AND SUPPLIES

In schools and engineering offices around the world, computer-aided drafting (CAD) systems are beginning to replace manual drafting equipment. You will find the newest CAD equipment in all contemporary design studios and in major corporate design offices. A typical CAD system is shown in Fig. 3-14.

A CAD system consists of a combination of *hardware* (equipment) and *software* (computer programs). In general, a CAD system requires the following hardware components.

- central processing unit (CPU)
- high-resolution monitor (CRT)
- keyboard
- digitizer (with pen stylus or button puck) or mouse
- plotter or printer for output (hard copy)

In addition to hardware, a CAD system must have special CAD software. Various CAD software programs are available for most types of computer.

A study of software and hardware terms can help you understand how the CAD system can replace manual drafting equipment to accomplish the same end result: a technical drawing. Table 3-1 compares CAD drafting elements with their corresponding manual drafting tools, techniques, and procedures.

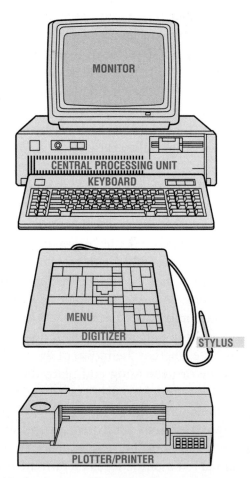

Fig. 3-14 *A computer-aided drafting system.* (Courtesy of Jeff Stoecker)

CAD System	Manual Drafting
Drafter sets the drawing area on the screen to represent any standard sheet size defined by ANSI or ISO.	Drafter selects the appropriate size sheet and attaches it to the drawing board.
Drafter selects a standard unit of measurement as the default for the drawing.	Drafter uses the appropriate architect's, engineer's, or metric scale.
Most CAD software provides a user-definable nonprinting grid that can be spaced as the drafter requires, as well as an increment snap that makes it easy to create drawings with exact measurements.	Drafter may use ruled (lined) media or grid sheets.
CAD drawings are stored on the computer's hard disk drive or on removable 3½ in. or 5¼ in. floppy disks.	Drawing sheets are stored in large files that require a large amount of space.
CAD software provides various techniques for drawing horizontal, vertical, and angular lines accurately.	Drafter uses the track-drafter, T-square, and triangles to draw horizontal, vertical, and angular lines.
To create basic geometry such as circles, polygons, ellipses, arcs, and irregular curves, the drafter enters commands at the keyboard or by selecting them from easy-to-use menus.	Drafter uses templates, compasses, curves, and dividers to lay out basic geometry.
To modify, delete, erase, or unerase anything on the drawing, the CAD drafter enters the appropriate commands.	Drafter uses an eraser and erasing shield or uses scissors and paste-up methods to revise a drawing.
When the drawing is finished, the drafter tells the computer to send as many copies as necessary to the plotter or printer. Each "copy" is therefore an original.	When the drawing is finished, the drafter copies the original using a reproduction machine.

Table 3-1 *Comparison of CAD and Manual Drafting Techniques.*

DRAWING SHEETS

Drawings are made on many different materials. Papers may be white, tinted cream, or pale green. They are made in many thicknesses and qualities. Most drawings are made directly in pencil or ink on tracing paper, vellum, tracing cloth, glass cloth, or plastic film. When such materials are used, copies can be made by printing or other reproduction methods (Chapter 14).

Vellum is tracing paper that has been treated to make it more **transparent** (clear). Tracing cloth is a finely woven cloth. It is treated to provide a good working surface and good transparency. Polyester (plastic) films are widely used in industrial drafting rooms. They are very transparent, strong, and lasting. Drawing films are made with a *matte* (dull and rough) surface. They are suitable for both pencil and ink work.

Drawing Sheet Sizes

Trimmed sizes of drawing sheets follow standards set by two organizations: the American National Standard Institute (ANSI) and the International Standards Organization (ISO). Table 3-2 lists the specifications by both organizations.

ANSI provides two series of standards, which are commonly called the *U.S. customary series.* Both are developed upward in size, from smallest to largest. The first is based on standard-size 8.50 × 11.00 in. letter paper, and the second is based on a 9.00 × 12.00 in. sheet.

The ISO standard is developed downward in size from a base sheet with an area of about 1 square meter (m²). Sheet sizes are based on a length-to-width ratio of one to the square root of two $(1 : \sqrt{2})$. Each smaller size has an area equal to half the preceding size. Multiples of these sizes are used for larger sheets also.

U.S. Customary Series			ISO Standard	
Size	First series	Second series	Size	Third series
A	8.50 x 11.00 in.	9.00 x 12.00 in.	A0	841 x 1189 mm
B	11.00 x 17.00 in.	12.00 x 18.00 in.	A1	594 x 841 mm
C	17.00 x 22.00 in.	18.00 x 24.00 in.	A2	420 x 594 mm
D	22.00 x 34.00 in.	24.00 x 36.00 in.	A3	297 x 420 mm
E	34.00 x 44.00 in.	36.00 x 48.00 in.	A4	210 x 297 mm

Table 3-2 *Standard drawing sheet sizes.*

Fastening the Drawing Sheet to the Board

By attaching the drawing sheet to the board, you have the freedom to move the T-square and triangles freely over the whole sheet. The sheet may be held in place on the board in several ways. Some drafters put drafting tape across the corners of the sheet and, if needed, at other places. Others use small, precut circular pieces of tape, called dot tape. Neither of these two methods will damage the corners or the edges of the sheet. They also can be used on composition boards or other boards with hard surfaces. As a result, most drafters prefer to use one of these methods.

To fasten the paper or other drawing sheet, place it on the drawing board with the left edge 1.00 in. (25 mm) or so away from the left edge of the board (Fig. 3-15). (Left-handed students should work from the right edge.) Put the lower edge of the sheet at least 4.00 in. (100 mm) up from the bottom of the board so you can work on it comfortably. Then line up the sheet with the T-square blade, as shown in Fig. 3-15A. Hold the sheet in position. Move the T-square down, as shown in Fig. 3-15B, keeping the head of the T-square against the edge of the board. Then fasten each corner of the sheet with drafting tape.

PENCILS AND ERASERS

Both regular wooden pencils and mechanical pencils are used for technical drawing. Four kinds of lead are now used in drawing pencils. One pencil lead is made from *graphite,* a form of the element carbon. It also contains clay and resins (sticky substances). Graphite pencils have been used for more than 200 years, and they are still the most important kind. They are used mostly on paper or vellum, but they may also be used on cloth. Graphite drafting pencils are usually made in 17 degrees of hardness, or grades, as shown in Fig. 3-16.

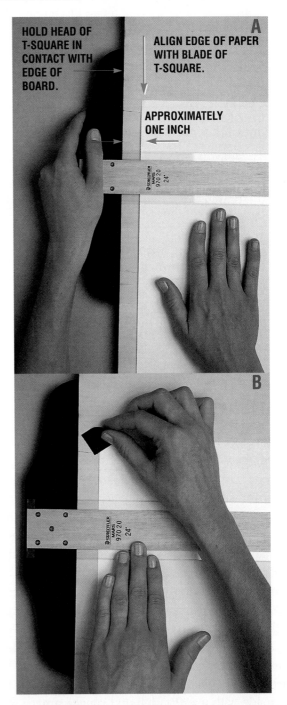

HOLD HEAD OF T-SQUARE IN CONTACT WITH EDGE OF BOARD.

A

ALIGN EDGE OF PAPER WITH BLADE OF T-SQUARE.

APPROXIMATELY ONE INCH

B

Fig. 3-15 *Fastening the drawing sheet to the board.* (Courtesy of Mishima and Circle Design)

6B	softest and blackest
5B	extremely soft
4B	extra soft
3B	very soft
2B	soft, plus
B	soft
HB	medium soft
F	intermediate, between soft and hard
H	medium hard
2H	hard
3H	hard, plus
4H	very hard
5H	extra hard
6H	extra hard, plus
7H	extremely hard
8H	extremely hard, plus
9H	hardest

Fig. 3-16 *Standard grades of graphite drawing pencils.*
(Courtesy of Circle Design)

The grade of pencil you use depends on the kind of surface on which you are drawing. It also depends on how **opaque** (dark) and how thick you want the finished line to be. To lay out views on fairly hard-surfaced drawing paper, use grades 4H and 6H. When you use tracing paper or cloth and draw finished views that are to be reproduced (copied by a machine), use an H or 2H pencil. Grades HB, F, H, and 2H are sometimes used for sketching and lettering and for drawing arrowheads, symbols, border lines, and so on. The exact grade you use depends on the drawing and the surface. Very hard and very soft leads are seldom used in ordinary drafting.

Since film has come into use for drawings, new kinds of pencil lead have been developed. Three types are described here, based on information furnished by the Joseph Dixon Crucible Company. The first is a *plastic pencil*. This type is a black crayon. Its lead is extruded (squeezed out) in a "plasticizing" process. Drawings made with this lead reproduce well on microfilm. The second type of lead is a combination of plastic and graphite and is made by heating. This kind stays sharp, draws a good opaque line, does not smear easily, erases well, and microfilms well. It can be used on paper or cloth as well as on film. The third type is used mostly on film. It does not remain as sharp as the other types. However, it draws a fairly opaque line, erases well, does not smear easily, and microfilms well.

These three types of lead are made in only five or six grades. Their grades are not the same as those used for regular graphite leads. The companies that make these pencils use different systems of letters and numbers to tell what kind of lead is in each pencil and how hard it is. The drafter must experiment with various grades to determine which best suits specific needs.

Sharpening the Pencil

To sharpen a wooden pencil, cut away the wood at a long slope, as shown in Fig. 3-17A. Always sharpen the end opposite the grade mark, being careful not to cut the lead. Leave about .38 to .50 in. (10 to 13 mm) exposed. Then shape the lead to a long conical (cone-shaped) point. Do this by rubbing the lead back and forth on a sandpaper pad (or on a fine file), as shown in Fig. 3-17B, while turning it slowly to form the point, as in (C) or (D). Some drafters prefer the flat point (or chisel point) shown in (E). Keep the sandpaper pad at hand so that you can sharpen the point often.

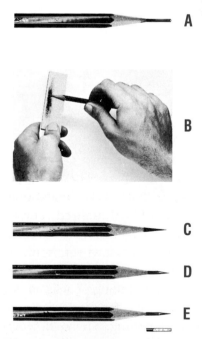

Fig. 3-17 *Sharpening the pencil properly is important.*
(Courtesy of Mishima and Circle Design)

Fig. 3-18 *A drafter's pencil sharpener cuts the wood, not the lead.*
(Courtesy of Mishima)

Fig. 3-19 *This lead pointer allows a choice of point shapes.*
(Courtesy of Staedtler and Circle Design)

Many drafters like to shape and smooth the point further by "burnishing" it on a piece of rough paper such as drafting paper.

Mechanical sharpeners have special drafter's cutters that remove the wood as shown in Fig. 3-18. Special pointers are made for shaping the lead, as in Fig. 3-19. Such devices may be hand-operated or electrically powered.

Mechanical pencils, also called *lead holders,* are widely used by drafters. They hold plain sticks of lead in a chuck that allows the exposed lead to be extended to any length desired. The lead for most lead holders is shaped in the same way as the lead in wooden pencils. However, some refill pencils have a built-in sharpener that shapes the lead. Still other mechanical drafting pencils use leads made to specific thicknesses for desired line widths and require no sharpening. These, however, have not generally be found to be very practical for the beginning drafting student.

Techniques for Using a Drawing Pencil

Pencil lines must be clean and sharp. They must be dark enough for the views to be seen when standard line widths are used. If you use too much pressure, you will groove the drawing surface. You can avoid this by using the correct grade of lead.

Develop the habit of turning the pencil between your thumb and forefinger as you draw a line. This will help make the line uniform and keep the point from wearing down unevenly. *Never sharpen a pencil over the drawing board.* After you sharpen a pencil, wipe the lead with a cloth to remove the dust. Being careful in these ways will help keep the drawing clean and bright. This is important when you plan to use the original pencil drawing to make copies.

Erasers and Erasing Shields

Use soft erasers, such as the vinyl type, the Pink Pearl, or the Artgum, to clean soiled spots or light pencil marks from drawings. Rubkleen, Ruby, or Emerald erasers are generally good for removing pencil or ink. On film, use a vinyl eraser made especially for erasing on film. Electric erasing machines similar to the one shown in Fig. 3-20 are also in common use in drafting rooms.

When working on paper or cloth, erase lines *along* the direction of the work. On film, erase *across* the direction of the work. Always erase carefully to avoid marring the finish on the drawing sheet. Keep in mind that regular ink erasers often contain grit. If you use them at all, use them very carefully to keep from damaging the drawing surface.

To avoid erasing nearby lines accidentally, most drafters use an erasing shield. **Erasing shields** are made of metal or plastic and have openings of different sizes and shapes. By positioning the shield so that the part to be erased shows through one of the openings, you can protect lines and areas that you do not want to erase (Fig. 3-21).

Fig. 3-20 *An electric erasing machine saves time.* (Courtesy of Staedtler)

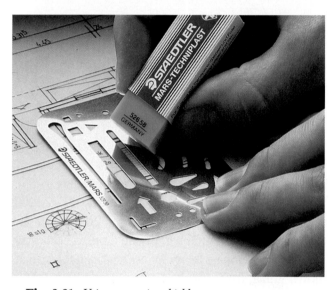

Fig. 3-21 *Using an erasing shield.* (Courtesy of Staedtler)

CASE INSTRUMENTS

A full set of instruments usually includes compasses with pen part, pencil part, lengthening bar, dividers, bow pen, bow pencil, bow dividers, and one or two ruling pens (optional).

Large-bow sets (Fig. 3-22) are favored by most drafters. They are known as master, or giant, bows and are made in several patterns. With large bows, 6.00 in. (152 mm) or longer, circles can be drawn up to 13.00 in. (330 mm) in diameter or, with lengthening bars, up to 40.00 in. (1016 mm) in diameter. Large-bow sets let you use one instrument in place of the regular compasses, dividers, and small-bow instruments. Large-bow instruments let you hold the radius securely at any desired distance, up to their largest possible radius.

The Dividers

Lines can be divided and distances transferred (moved from one place to another) with dividers. Adjust the dividers as shown in Fig. 3-23A.

To transfer a distance using the dividers, adjust the dividers to exactly the length to be transferred (the radius of a circle or the length of a line, for example). Transfer the length by positioning the dividers at the new location.

You can also use the dividers to divide a line, arc, or circle into equal parts. To divide a line into three equal parts, adjust the points of the dividers until they seem to be about one third the length of the line.

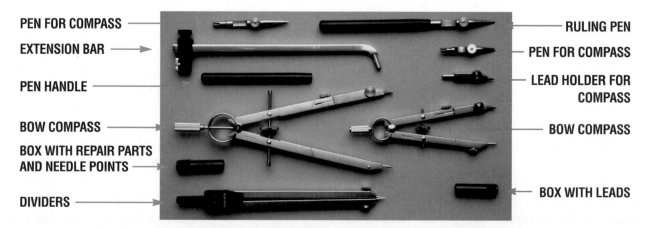

PEN FOR COMPASS

EXTENSION BAR

PEN HANDLE

BOW COMPASS

BOX WITH REPAIR PARTS AND NEEDLE POINTS

DIVIDERS

RULING PEN

PEN FOR COMPASS

LEAD HOLDER FOR COMPASS

BOW COMPASS

BOX WITH LEADS

Fig. 3-22 *A large-bow set of drawing instruments.* (Courtesy of Mishima and Circle Design)

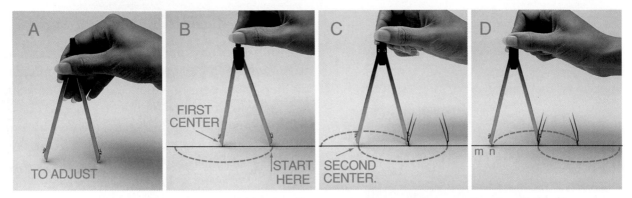

Fig. 3-23 *The dividers are used to divide and transfer distances.* (Courtesy of Ann Garvin and Circle Design)

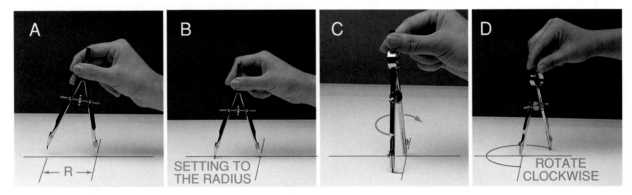

Fig. 3-24 *The compass is used to draw circles and arcs.* (Courtesy of Mishima and Circle Design)

Put one point on one end of the line and the other point on the line (Fig. 3-23B). Turn the dividers about the point that rests on the line, as in Fig. 3-23C. Then turn it in the alternate direction, as in Fig. 3-23D. If the last point falls short of the end of the line, increase the distance between the points of the dividers by an amount about one third the distance *mn*. Then start at the beginning of the line again. You may have to do this several times. If the last point overruns the end of the line, decrease the distance between the points by one third the extra distance. For four, five, or more spaces, follow the same rules, but correct by one fourth, one fifth, or so forth, of the overrun or underrun. You can divide an arc or the **circumference** (perimeter) of a circle in the same way.

The Compass

Regular curves (those that are true circles or arcs) can be drawn with a **compass** (Fig. 3-24). Leave the legs of the compass straight for radii under 2.00 in. (50 mm). For larger radii, make the legs *per-*

pendicular (at a 90° angle to) to the paper (Fig. 3-25). When you need a radius of more than 8.00 in. (200 mm), put in a lengthening bar (Fig. 3-26) to increase the length of the pencil leg, or use a beam compass.

To get the compass ready for use, sharpen the lead as shown in Fig. 3-27, allowing it to extend about .38 in. (10 mm). Using a long *bevel* (slant) on the outside of the lead will keep the edge sharp when you increase the radius. Then adjust the shouldered end of the needle point until it extends slightly beyond the lead point, as shown in Fig. 3-27. You cannot use as much pressure on the lead in the compass as you can on a pencil. Therefore, use lead one or two degrees softer in the compass to get the same *line weight* (thickness and darkness).

To draw a circle or an arc with the compass, locate the center of the arc or circle by two *intersecting* (crossing) lines. Lay off the radius by a short, light dash (Fig. 3-24A). Open the compass by pinching it between your thumb and second finger. Set it to the desired radius by putting the needle point at the center and moving the pencil leg with your first and second fingers (Fig. 3-24B).

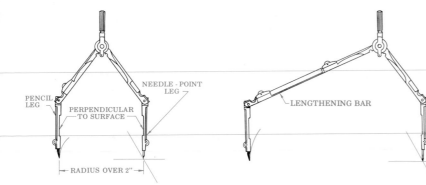

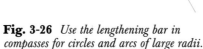

Fig. 3-25 *Adjusting the compass for large circles.*

Fig. 3-26 *Use the lengthening bar in compasses for circles and arcs of large radii.*

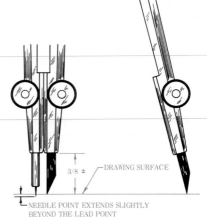

Fig. 3-27 *Adjusting the point and shaping the lead of the compass.*

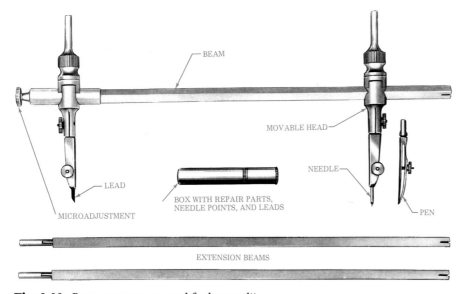

Fig. 3-28 *Beam compasses are used for large radii.*

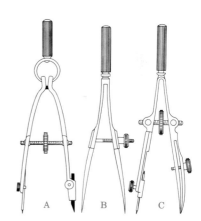

Fig. 3-29 *Bow instruments are used for drawing small circles and arcs and for stepping off short distances.*

When the radius is set, raise your fingers to the handle (Fig. 3-24C). Turn the compass by twirling the handle between your thumb and finger. Start the arc near the lower side and turn clockwise (Fig. 3-24D). As you draw the curve, slant the compass a little in the direction of the line. *Do not force the needle point into the paper.* Use only enough pressure to hold the point in place.

The Beam Compass

Beam compasses (Fig. 3-28) are used to draw arcs or circles with large radii. The beam compass is made up of a bar (beam) on which movable holders for a pencil part and a needle part can be put and fixed as far apart as desired. A pen part may be used in place of the pencil point. If a needle point is put in both holders, a beam compass can be used as a divider.

The usual bar is about 13.00 in. (330 mm) long. However, by using a coupling to add extra length, you can draw circles of almost any size.

Bow Instruments

A set of bow instruments (Fig. 3-29) is made up of the bow pencil (A), the bow dividers (B), and the bow pen (C). They can be adjusted with either a center wheel, as in (A) and (C), or a side wheel, as at (B). They may be of the hook-spring type, as at (A), or the fork-spring type, as at (B) and (C). They are usually about 4.00 in. (102 mm) long.

The bow instruments are easy to use and accurate for small distances or radii (less than 1.25 in. or 32 mm). They hold small distances better than the large instruments.

Bow Dividers

Use the bow dividers for transferring small distances or for marking off a series of small distances. You may also use them to divide a line into small spaces.

Bow Pencil

The bow pencil is used to draw small circles. Whether you use instruments with center wheels or with side wheels is up to you. Sharpen and adjust the lead for the bow pencil as shown in Fig. 3-30A. The inside bevel holds an edge for small circles and arcs, as shown at (B). For larger radii, the outside bevel at (C) is better. Some drafters prefer a conical center point or an off-center point, as at (D), (E), and (F).

Use the bow pencil (Fig. 3-31) with one hand. Set the radius as in (A). Start the circle near the lower part of the vertical centerline (B). Turn clockwise. (Left-handers will need to reverse the above procedure.)

Bow Pen

The bow pen is included in a set of bow instruments to offer drafters a choice of marking points (pen or pencil). Use the bow pen in the same manner as the bow pencil.

Drop-Spring Bow Compass

The drop-spring bow compass (Fig. 3-32) is designed for drawing very small circles. It is especially useful for drawing many small circles of the same size, such as when drawing rivets. Attach the marking point to a tube that slides on a pin. Set the radius with the spring screw.

To use the drop-spring bow compass, keep the pin still and turn the marking point around it. Hold the marking point up while putting the pin on the center. Then drop the marking point and turn it. The circles drawn will all be the same size.

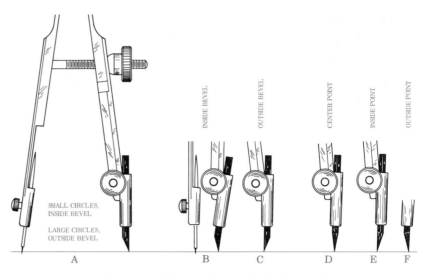

Fig. 3-30 *Adjusting the lead for the bow pencil.*

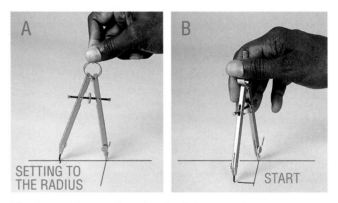

Fig. 3-31 *Adjusting the radius for the bow pencil compass.*
(Courtesy of Ann Garvin/Vanderberg Drafting Supply Inc. and Circle Design)

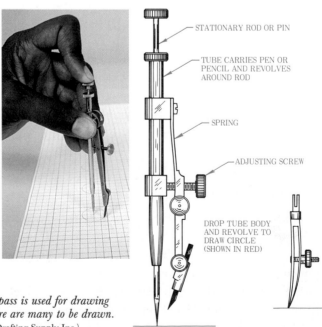

Fig. 3-32 *The drop-spring bow compass is used for drawing very small circles, especially where there are many to be drawn.*
(Courtesy of Ann Garvin and Vanderberg Drafting Supply Inc.)

Adjusting Bow Instruments

You can make large adjustments quickly with the side-wheel bows by pressing the fork and spinning the adjusting nut. Some center-wheel bows are also built for making large, rapid adjustments. To do this, hold one leg in each hand and either push to close or pull to open. Make small adjustments with the adjusting nut on both the side-wheel and the center-wheel bows.

SCALES

Scales are used for laying off distances and for making measurements. Measurements can be full size or in some exact proportion to full size. Scales are made in various shapes (Fig. 3-33A). They are usually made of boxwood, plastic, plastic on boxwood, or metal. Some scales are open-divided (Figs. 3-33B and 3-33C), with only the end units subdivided. Others are full-divided (Figs. 3-33D and 3-33E), with subdivisions over their entire length.

Types of Scales

Different types of scales are used to make different kinds of drawings. Commonly used customary-inch scales include the architect's scale (Fig. 3-33B), the mechanical engineer's scale (C), and the civil engineer's scale (D). By contrast, the proportional metric scale can be used on all types of engineering drawings when metric units are used. Both customary and metric scales are made in all of the shapes shown in Fig. 3-33A.

The Architect's Scale

The architect's scale (Fig. 3-33B) is divided into proportional feet and inches. The triangular form shown is used in many schools and in some drafting offices because it has many scales on a single stick. However, many drafters prefer flat scales, especially when they do not have to change scales often. The usual proportional scales are listed in Table 3-3.

The symbol ′ is used for feet and ″ for inches. Thus, three feet four and one-half inches is written 3′-4½″. When all dimensions are in inches, the symbol is usually left out. Also, on architectural and structural drawings, inch marks (″) are not shown and the dimension is given as 3′-4½ (see Chapter 20).

Proportional scales are used in drawing buildings and in making mechanical, electrical, and other engineering drawings. They are also used in drafting in general. The proportional scale to which the views are drawn should be given on the drawing. This is done in the title block if only one scale is used. If different parts of a drawing are in different scales, the scales are given near the views in this way:

- Scale: 6″ = 1′-0
- Scale: 3″ = 1′-0
- Scale: 1½″ = 1′-0

The Mechanical Engineer's Scale

The mechanical engineer's scale (Fig. 3-33C) has inches and fractions of an inch divided to represent inches. The usual divisions are the following:

- *Full size*-1 in. divided into 32nds
- *Half size*-½ in. divided into 16ths
- *Quarter size*-¼ in. divided into 8ths
- *Eighth size*-⅛ in. divided into 4ths

These scales are used for drawing parts of machines or where larger reductions in scale are not needed. The proportional scale to which the views are drawn should be given on the drawing. This is done in the title block.

Proportion	Gradations	Ratio
Full size	12" = 1'-0"	1:1
¼ size	3" = 1'-0"	1:4
⅛ size	1½" = 1'-0"	1:8
1/12 size	1" = 1'-0"	1:12
1/16 size	¾" = 1'-0"	1:16
1/24 size	½" = 1'-0"	1:24
1/32 size	⅜" = 1'-0"	1:32
1/48 size	¼" = 1'-0"	1:48
1/64 size	3/16" = 1'-0"	1:64
1/96 size	⅛" = 1'-0"	1:96
1/128 size	3/32" = 1'-0"	1:128

Table 3-3 *Proportional scales.* (Courtesy of Circle Design)

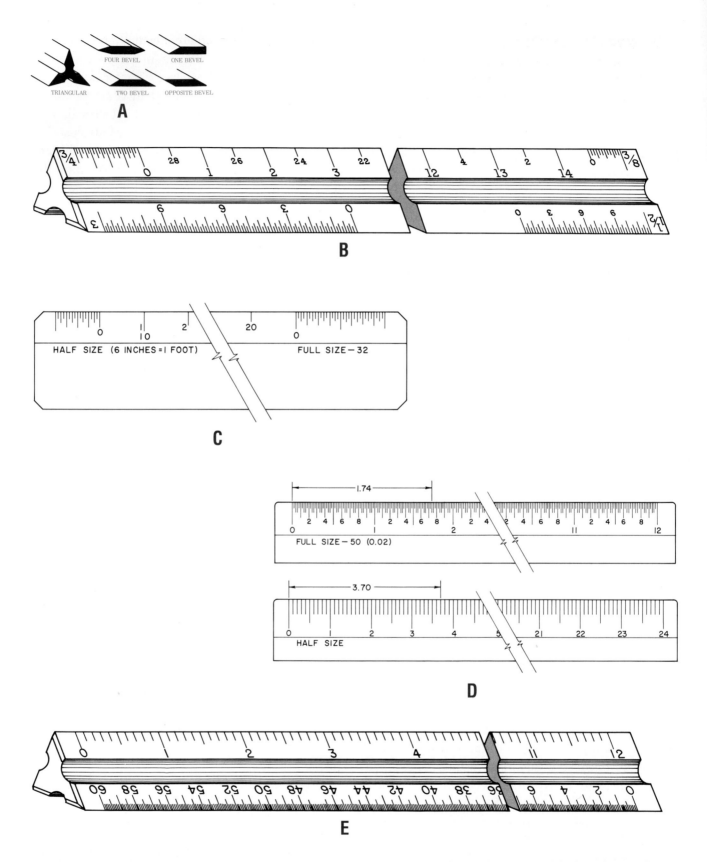

Fig. 3-33 *Scales. (A) Examples of the various scale shapes. (B) Architect's scale, open divided. The triangular form has many proportional scales. (C) Mechanical engineer's scale, open divided. (D) Civil engineer's scale, divided into decimals. (E) Decimal-inch scales are often used in drawing machine parts.*

The Civil Engineer's Scale

The civil engineer's scale and decimal-inch scale (Fig. 3-33D and E) have inches divided into decimals. The usual divisions follow:

- 10 parts to the inch
- 20 parts to the inch
- 30 parts to the inch
- 40 parts to the inch
- 50 parts to the inch
- 60 parts to the inch

With the civil engineer's scale, 1 inch may stand for feet, rods, miles, and so forth. It may also stand for quantities, time, or other units. The divisions may be single units or multiples of 10, 100, and so on. Thus, the 20-parts-to-an-inch scale may stand for 20, 200, or 2,000 units.

This scale is used for civil engineering work. This includes maps and drawings of roads and other public projects. It is also used where decimal-inch divisions are needed. These uses include plotting data and drawing graphic charts.

Fig. 3-34 *Civil engineers place graphic scales on maps to show people how to interpret them.*

Enlarged	Same size	Reduced
2000:1		
100:1	1:1	
500:1		1:2
200:1		1:5
100:1		1:10
50:1		1:20
20:1		1:50
10:1		1:100
5:1		1:200
2:1		1:500
		1:1000
		1:2000

Table 3-4 *Metric proportional scales.*

The scale used should be given on the drawing or work as follows:

- Scale: 1″ = 500 pounds
- Scale: 1″ = 100 feet
- Scale: 1″ = 500 miles
- Scale: 1″ = 200 pounds

For some uses, a graphic scale is put on a map, drawing, or chart, as shown in Fig. 3-34.

The Metric Scale

Metric scales (Fig. 3-35) are divided into millimeters. The usual proportional scales in the metric system are listed as a ratio in Table 3-4.

Metric architectural proportions are used in drawing buildings and in making many mechanical, electrical, and other engineering drawings. They are also often used in drafting in general. The proportional scale to which the views are drawn should be given on the drawing. This is done in the title block if only one scale is used. If different parts on the same drawing are in different scales, the scales are given near the views in this way:

- Scale: 1:2
- Scale: 1:5
- Scale: 1:10
- Scale: 1:50
- Scale: 1:100

To reduce the size of an object being drawn, a drafter uses one of the scales shown in Fig. 3-36A. The proportional scales used to enlarge drawings of machine parts are shown in Fig. 3-36B.

Using a Scale

Scales are used for full-size drawings as well as for scaled drawings. A drawing is created "to scale" when the object being drawn is too large for the drawing sheet or too small to be read easily.

Fig. 3-35 *Metric scales are divided into millimeters.*

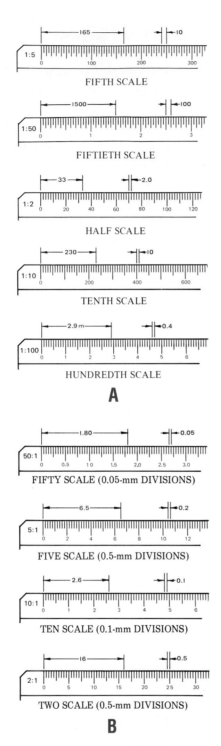

A

B

Fig. 3-36 *Metric scales for reduction (A) and enlargement (B)*

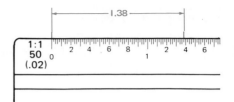

Fig. 3-37 *Making a measurement with the full-size scale.*

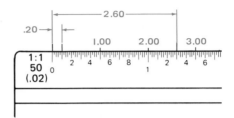

Fig. 3-38 *Measuring to half size on a full-size scale.*

Full-Size Drawings

When an object is not too large for the paper on which you are drawing it, draw it full size or full scale (1:1). To make a measurement, put the scale on the paper in the direction in which you are measuring. Make a short, light dash next to the zero on the scale and another dash next to the division at the distance you want. Do not make a dot or punch a hole in the drawing sheet. Figure 3-37 shows a distance of 1.38″ laid off on a scale.

Drawing to Scale

Views of large parts and projects have to been drawn to a very small scale, in a *reduced proportion*. In the customary system, the first reduction is to the scale of .50″ = 1.00″ (1:2), commonly called *half size*. You can use a full-size scale to draw half size by letting each half-inch unit on the scale (.50 in.) stand for 1.00 in. and each 12.00-in. unit stand for 24.00 in. This is shown in Fig. 3-38, where 3.60 is laid off by two half inches and .30 of the next inch. Always think full size. A scale divided and marked for half size (Fig. 3-33D) is easier to use.

If smaller views are needed, the next reduction that can be used is the scale of 3″ = 1′-0″, called *quarter size* (1:4). Find this scale on the architect's scale. The actual length of 3.00 in. represents 1 foot (ft.) divided into 12 parts. Each part stands for 1.00 in. and is further divided into eighths. Learn to think of the 12 parts as standing for real inches. Figure 3-39 shows how to lay off the distance of 1′-3½″. Notice where the zero mark is. It is placed so that inches can be measured in one direction from it and feet in the other direction.

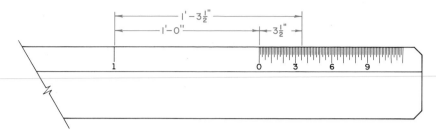

Fig. 3-39 *Reading the scale of 3" = 1'-0, called quarter size.*

For 1:4 reductions, you can use a regular mechanical engineer's scale with a scale of $\frac{1}{4}'' = 1''$. For other reductions, use the proportional scales listed in the sections on the architect's scale and the mechanical engineer's scale.

For small parts, you can use *enlarged proportions.* For example, for double-size views, use a scale of $24'' = 1'$-0 (2:1). For very small parts. draw at a scale of 4:1 or 8:1 or, for some purposes, 10:1, 20:1, or larger.

When drawing to scale using the metric system, you can get reduced or enlarged proportions with the scales shown in Fig. 3-36.

ALPHABET OF LINES

The different lines, or line symbols, used on drawings form a kind of graphical alphabet commonly known as the **alphabet of lines.** The line symbols recommended by ANSI are shown in Fig. 3-40. Two line widths–thick and thin–are generally used. (Formerly, a medium thick line was used as a hidden line.) Drawings are easier to read when there is good contrast between different kinds of lines. All lines must be uniformly sharp and black.

TECHNIQUES FOR DRAWING LINES

Now that you are familiar with the basic drafting tools, you need to know how to use the tools correctly. The sections that follow discuss basic drawing techniques. Additional, more complex techniques will be presented in Chapter 4.

Horizontal Lines

To draw a horizontal line, use the upper edge of the T-square blade as a guide. With your left hand, place the head of the T-square in contact with the left edge of the board. Keeping the head in contact,

move the T-square to the place you want to draw the line. Slide your left hand along the blade to hold it firmly against the drawing sheet. Hold the pencil about 1.00 in. (25 mm) from its point. Slant it in the direction in which you are drawing the line. (This direction should be left to right for right-handers and right to left for left-handers.) While drawing the line, rotate the pencil slowly and slide your little finger along the blade of the T-square (Fig. 3-41A). This will give you more control over the pencil.

On film, keep the pencil at the same angle (55° to 65°) all along the line. You must also use less pressure on film than on paper or other material. Always keep the point of the lead a little distance away from the corner between the guiding edge and the drawing surface, as shown in Fig. 3-41A. This will let you see where you are drawing the line. It will also help you avoid a poor or smudged line. *Be careful to keep the line parallel to the guiding edge.*

Vertical Lines

Use a triangle and a T-square to draw vertical lines (Fig. 3-41B). Place the head of the T-square in contact with the left edge of the board. Keeping the T-square in contact, move it to a position below the start of the vertical line. Place a triangle against the T-square blade. Move the triangle to where you want to begin the line. Keeping the vertical edge of the triangle toward the left, draw upward. Slant the pencil in the direction in which you are drawing the line. Be sure to keep this angle the same when drawing on film. Keep the point of the lead far enough out from the guiding edge that you can see where you are drawing the line. Be careful to keep the line parallel to the guiding edge.

Inclined Lines

Inclined (slanted) lines are drawn using triangles, a protractor, or a drafting machine. Before you can lay out inclined lines accurately, however, you must understand the terms used to describe and measure angles. An **angle** is formed when

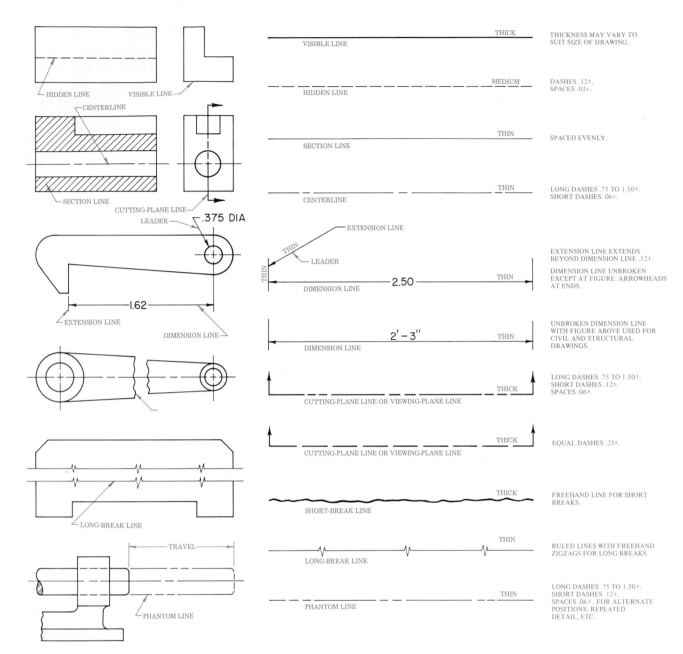

Fig. 3-40 *Alphabet of lines for pencil drawing and inking. (Inking is discussed later in this chapter.)*

two straight lines meet at a point. The point where the lines meet is called the **vertex** of the angle. The lines are the sides of the angle. Any nonhorizontal line must be drawn at a specified angle from the T-square.

Angles are measured in a unit called a *degree*. Degrees are marked by the symbol °. If the **circumference** (rim) of a circle is divided into 360 parts, each part is an angle of 1°. One fourth of a circle is $^{360}/_4 = 90°$, as shown in Fig. 3-42A. As you may recall, a part of a circle is called an *arc*. Two

lines drawn to the center of a circle from the ends of a 90° arc on that circle form a **right angle.** An **acute angle** (B) is less than 90°. An **obtuse angle** (C) is greater than 90°. For closer measurements, a degree is divided into 60 equal parts called *minutes* (′). Minutes, in turn, are divided into 60 equal parts called *seconds* (″). Notice that the size of an angle does not depend on the length of its sides. The number of degrees in a given angle are the same no matter what the size of the circle and arc that define it.

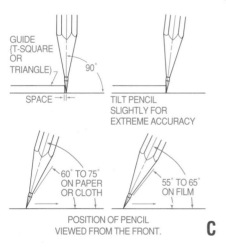

Fig. 3-41 *(A) Drawing a horizontal line. (B) Drawing a vertical line. (C) To ensure accurate drawing, position the pencil as shown here.* (Courtesy of Mishima and Circle Design)

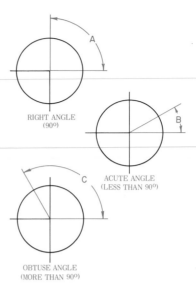

Fig. 3-42 *Angles are measured in degrees, minutes, and seconds.*

Decimal degree	Degrees, minutes, and seconds
10° ± 0.5°	10° ± 0°30'
0.75°	0°45'
0.004°	0°0'14"
90° ± 1.0°	90° ± 1°
25.6° ± 0.2°	25°36' ± 0°12'
25.51°	25°30'36"

Table 3-5 *Decimal-degree equivalents of degrees, minutes, and seconds.*

While angles are commonly given in degrees, minutes, and seconds, the use of *decimal degrees* is now growing in popularity. For example, 50.5° means the same as 50°30'. Table 3-5 shows some additional examples.

30°, 45°, and 60° Lines

You can draw lines at 30°, 45°, or 60° from the horizontal or vertical by using the triangles. Lines inclined at 30° and 60° from horizontal or vertical lines are drawn with the 30°-60° triangle held against the T-square blade, as shown in Fig. 3-43, or against a horizontal straightedge. The 30°-60° triangle can also be used to lay off equal angles, 6 at 60° or 12 at 30°, about a center point.

To draw lines inclined at 45° from horizontal or vertical lines, hold the triangle against the T-square blade, as shown in Fig. 3-44, or against a horizontal straightedge. The 45° triangle can also be used to lay off eight equal angles of 45° about a center point.

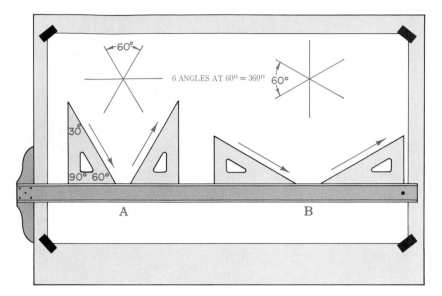

Lines Inclined at 15° Increments

The 45° and 30°-60° triangles, alone or together and combined with the T-square, can be used to draw angles increasing by 15° from the horizontal or vertical line. Some ways of placing the triangle to draw angles of 15° and 75° are shown in Fig. 3-45. Ways of achieving the angles are shown in Fig. 3-46. Fig. 3-46O shows lines drawn for all the positions possible.

INKING

Most technical drawings are made with pencil on a good-quality tracing paper (vellum) or on a special film medium known as **drafting film.** However, inking has become a major trend in the past few decades. Ink is often used to make high-quality tracings that can be blueprinted, photographed, or microfilmed. As this trend continues, better inking techniques and equipment have been developed.

Fig. 3-43 *The 30°-60° triangle has angles of 30°, 60°, and 90°.*

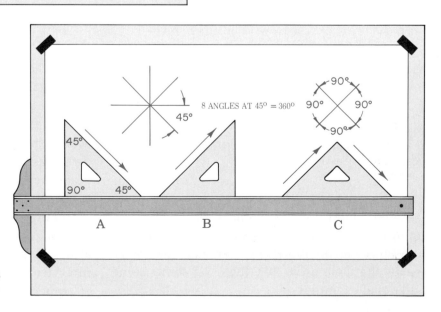

Fig. 3-44 *The 45° triangle has angles of 45° and 90°.*

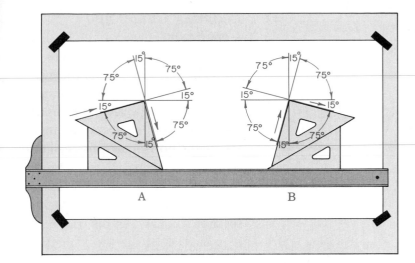

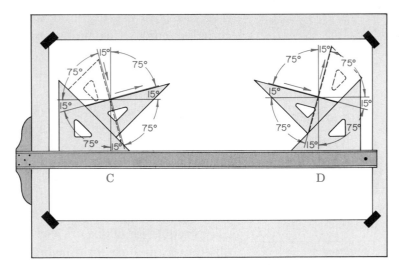

Fig. 3-45 *Drawing lines at 15° and 75° using the two triangles.*

Technical pens are available with points of various sizes to draw specific line widths (Fig. 3-49). Points for technical pens are made of different materials for use on different types of media. There are three main kinds of points:

- Hard-chrome stainless steel for paper or vellum.
- Tungsten-carbide for long wear on film, vellum, and paper (most commonly used in plotters).
- Jewel for long continuous use on film.

PENS FOR CAD PLOTTERS

Technical pens are also available for plotters used in computer-aided drafting. These pens provide several reliable, uniform line widths, and the ink is durable. The special inks for plotters come in various colors and types. They include waterproof inks for ball-point pens, as well as fiber-tip colored-ink cartridges for paper media.

Another trend in CAD is the use of *inkjet* plotters (Fig. 3-50). These plotters spray the ink directly onto the media from tiny nozzles, avoiding the need for pens entirely. The ink used in inkjet plotters has a special consistency that allows it to flow through the "jets," or nozzles, without clogging. Because this technology is accurate and fast, it has replaced pen plotters in many larger companies. However, both types are still used extensively.

Electrostatic plotters provide an alternative to pen and inkjet plotters. In these plotters, a dry ink, or *toner,* is used. The toner is transferred to the paper in the form of electrically charged particles. This is the same principle used by laser printers and copiers to transfer an image to paper.

Drawing Ink

Ink used for technical drawings is called **india ink.** It must be completely opaque in order to produce good, uniform line tone and yet be erasable on all drafting media.

Equipment for Inking

Inking requires a few tools in addition to those needed for drawing in pencil. As with other drafting equipment, several options are available for each tool.

Technical Pens

Drafting instrument sets are available with technical fountain pens and attachments (Fig. 3-47). *Technical pens* (Fig. 3-48) are precise instruments that provide known line widths. Over the years, technical pens have made inking more efficient. Some technical pens have a refillable cartridge for storing ink. Others have a cartridge that is used once and then replaced. The disposable (stainless steel point) technical pen, now commonly used, performs the same tasks as the refillable type and eliminates many maintenance procedures.

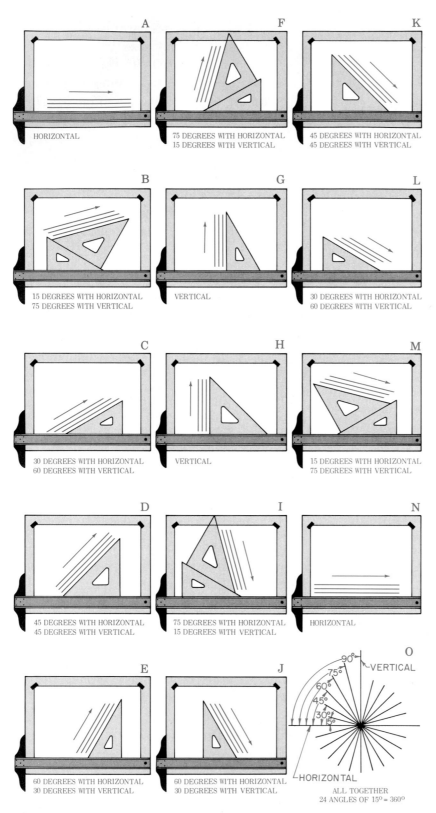

Fig. 3-46 *Drawing lines with the T-square and triangles.*

Lettering Guides and Equipment

The lettering set shown in Fig. 3-51 has three basic tools for inking: a scriber, a set of lettering templates, and a set of technical pens. The scriber has a tailpin to follow the horizontal groove in the bottom of the template, a tracer pin to follow the engraved letter, and a barrel slot. A technical pen is set in the slot. When the tracer pin is moved along the letter, the pen copies the letter above the template (Fig. 3-52). Lettering templates come in many styles and lettering sizes, from about .06 to 2.00 in. (1.5 to 50 mm) high. Other templates are used to draw symbols and other shapes (Fig. 3-53). The width of the pen point used depends on the height of the letters and size of the symbols.

Inking Techniques

Techniques for inking are slightly different from those for drawing in pencil. Hand position and the order in which items are drawn are affected by the fact that ink, unlike pencil, must be allowed to dry to help avoid smudges.

Inking Straight Lines

Figure 3-54 shows the correct position for drawing lines with a technical pen. Note the direction of the stroke and the angle of the pen. Hold the technical pen in a nearly vertical position (perpendicular to the media) to get the most uniform line.

Inking Circles, Arcs, and Irregular Curves

Templates, compasses, and French curves can be used with technical pens for inking circles, arcs,

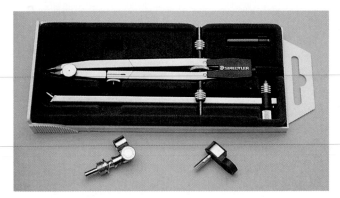

Fig. 3-47 *Instruments with technical pen attachments.*
(Courtesy of Ann Garvin and Vanderberg Drafting Supply, Inc.)

Fig. 3-48 *A nine-pen and a three-pen technical inking set.*
(Courtesy of Staedtler)

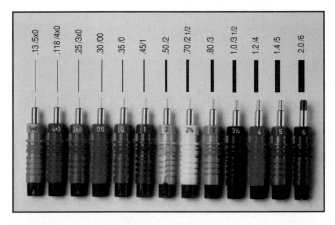

Fig. 3-49 *The range of lines and point sizes available in technical pens.* (Courtesy of Staedtler and Circle Design)

Fig. 3-50 *Plotters produce accurate inked reproductions of CAD drawings. Line color and line width are controlled by the pens used in the plotter. The CAD file tells the plotter which lines to produce with which pens.* (Courtesy of Hewlett Packard)

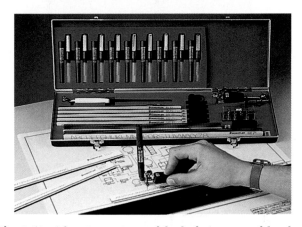

Fig. 3-51 *A lettering set is one of the drafter's most useful tools.*
(Courtesy of Staedtler)

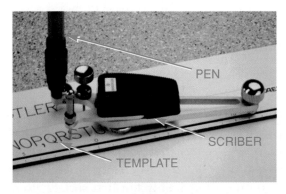

Fig. 3-52 *Three basic parts of a lettering set are the pen, the template, and the scriber.* (Courtesy of Staedtler and Circle Design)

and irregular curves. (Fig. 3-55). Some compasses are designed specifically to hold the technical pen while others require the use of an adapter, as shown in Fig. 3-55. The lengthening bar, or beam compass, can be used for large circles.

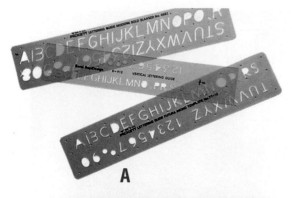

Fig. 3-53 *(A) Templates are available in many styles of lettering and kinds of symbols. (B) Technical pens are easy to use with templates.* (Courtesy of Mishima)

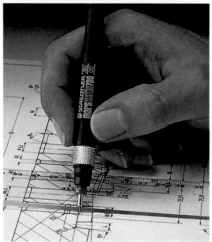

Fig. 3-54 *The position of the technical pen is important when drawing lines.* (Courtesy of Staedtler)

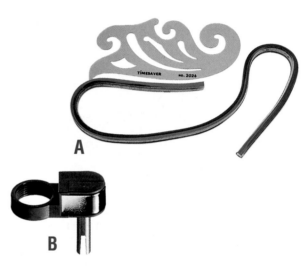

Fig. 3-55 *(A) The technical pen follows various types of curves easily. (B) Adapter for use of technical pens in compasses.* (A–Courtesy of Ann Garvin and Vanderberg Drafting Supply, Inc. ; B–Courtesy of Staedtler)

Irregular (French) curves can be used to guide the pen when curves that are not true circular arcs are being inked. Use either fixed curve templates or irregular adjustable curves (Fig. 3-55).

Erasing Techniques

The ink used on polyester drafting film is waterproof. However, you can easily remove ink from the film by rubbing it with a moistened plastic eraser (Fig. 3-56A) or by using an electric erasing machine (B). Do not use any pressure in rubbing. The polyester film does not absorb ink, and therefore all ink dries on top of its highly finished surface. Remove ink from other surfaces, such as tracing vellum or illustration board, with regular ink erasers or chemically imbibed ink erasers that absorb ink. But be very careful. Press lightly with strokes in the direction of the line to remove ink caked on the surface. Too much pressure damages the surface and makes it hard to revise the drawing.

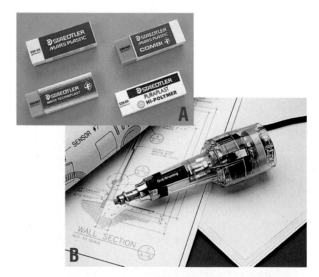

Fig. 3-56 *(A) Several types of ink erasers are available. Some absorb ink; others remove it from the drawing. The type you choose depends on the medium used for the drawing. (B) You can also use an electric erasing machine to remove ink from a drawing.* (A–Courtesy of Staedtler; B–Courtesy of OCÉ Bruning)

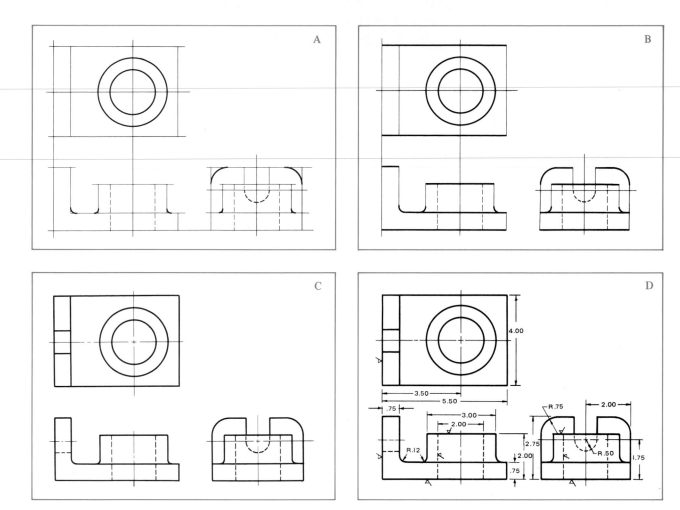

Fig. 3-57 *Order of inking, or steps, in developing a finished drawing.*

Order of Inking

Smooth joints and tangents, sharp corners, and neat fillets (rounded corners) make a drawing look better and make it easier to read. *Good inking requires careful practice and a definite order of working procedures.*

The usual order of inking or tracing is shown in Fig. 3-57. First, ink the arcs, centered over the pencil lines, as in (A). Ink the horizontal lines next (B). Complete the drawing with the vertical lines (C). Then add the dimension lines, arrowheads, and so on, and fill in the dimensions (D). The order of inking is:

1. Ink main centerlines.
2. Ink small circles and arcs.
3. Ink large circles and arcs.
4. Ink hidden circles and arcs.
5. Ink irregular curves
6. Ink horizontal full lines
7. Ink vertical full lines.
8. Ink slanted full lines.
9. Ink hidden lines.
10. Ink centerlines.
11. Ink extension and dimension lines.
12. Ink arrowheads and figures.
13. Ink section lines.
14. Letter notes and titles.
15. Ink border lines.
16. Check drawing carefully.

Chapter 6 describes many of these items in greater detail.

CHAPTER 3

Learning Activities

1. Visit an industrial drafting room and pay special attention to the types of drafting tables, equipment, and instruments being used.

2. Collect technical drawings done on various types of drafting sheets and compare them for qualities such as transparency and durability. Make prints of several and compare the quality of reproduction.

3. Draw sample lines on different types of tracing sheets using as many different grades of leads as are available, and make prints. Compare the line qualities.

4. Draw sample lines on different types of tracing sheets using india ink, and make prints. Compare the quality of prints made from ink drawings with those made from pencil drawings in Activity 3.

5. Test different types of erasers on various types of drafting sheets to determine which works best on each. Test both pencil and ink lines.

6. If available, try using a T-square and triangles, a parallel-ruling straightedge and triangles, and the two basic types of drafting machines (elbow- and track-type). Which do you find most convenient? Which do you prefer to use? Why?

7. Try to obtain basic drafting equipment of your own to use at home. The more you practice, the more quickly your drafting skills will improve.

8. Check the accuracy of some of the dimensions on industrial drawings. This will give you additional practice in using drafting scales.

Questions

Write your answers on a separate sheet of paper.

1. Explain how you can check the accuracy of a T-square.

2. Name three kinds of drawing sheets.

3. Drafting pencils are made in 17 degrees of hardness, from _____ (softest and blackest) to _____ (hardest).

4. The shape of a drafting pencil point may be _____ or _____.

5. How many widths, or thicknesses, of lines are generally used in drafting?

6. You can draw angles in _____° intervals by using the 45° and 30°-60° triangles with the T-square.

7. What name is given to the different lines, or line symbols, used on drawings?

8. An instrument used in measuring, or laying out, angles is called a(n) _____.

9. Name two types of drafting machines.

10. What are the five major components of a CAD system?

11. What is the best position in which to hold a technical pen?

12. Name two ways to ink circles and arcs.

13. Name the instruments used to guide the pen when you are drawing curved lines that are not true circles or arcs.

14. How do you erase india ink from drafting film?

15. How many widths of lines are recommended for use on ink drawings? Name them.

16. Why are circles and arcs inked before straight lines?

17. List three kinds of points for technical pens. Why are there different types?

18. Name the three basic parts of a lettering set.

19. Ink curves to the points of _____ first.

CHAPTER 3

PROBLEMS

Beginning with these problems, you will find a small CAD symbol. This symbol indicates that a problem is appropriate for assignment as a computer-aided drafting problem and suggests a level of difficulty (one, two, or three) for the problem.

GENERAL INSTRUCTIONS

The trim sizes of sheets recommended by the American National Standards Institute (ANSI) and the International Standards Organization (ISO) are in almost universal use in industry and are, therefore, desirable for use in drawing courses. Standard drawing sheet sizes are given on page 66 in Table 3-2. Most of the drawing problems throughout this book are planned for working on A or B (ANSI) or A4 or A3 (ISO) sheets. It is possible, of course, to use other sizes and arrangements where necessary or where desired by the instructor. To assist in such cases, Fig. 3-58 is given with letters to indicate the dimensions. The desired numerical values can be filled in after the equal signs (=) on the figure. When this special layout is used, your instructor will supply the sizes.

Problems designed for B-size (11.00 × 17.00 in.) or A3-size (297 × 420 mm) sheets are designed for the layouts shown in Fig. 3-59. Slight adjustments and conversions are necessary when the metric sheets are used. Other layouts for B- and A3-size sheets are suggested in Fig. 3-60.

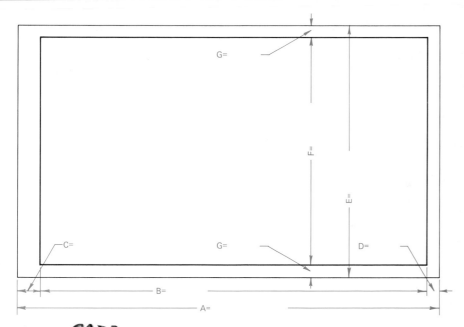

Fig. 3-58 *CAD1* *Adjustable layout for any size sheet.*

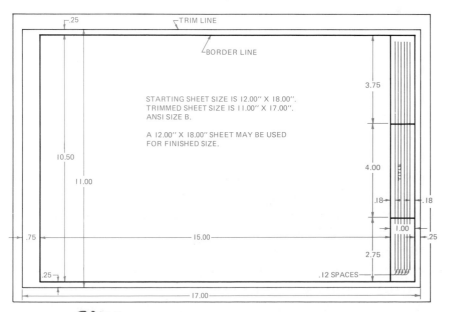

Fig. 3-59A *CAD1* *Standard layout of a B-size sheet.*

PROBLEMS

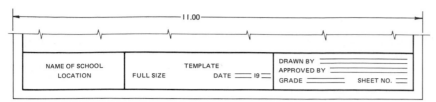

Fig. 3-59B *CAD1* *Suggested title block for a B-size sheet. Slight adjustments are needed for A3-size (metric) sheets.*

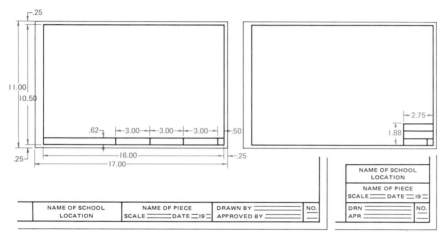

Fig. 3-60 *CAD1* *Alternate layouts.*

Sheet Layout: Two-Minute Method

If you follow the method illustrated in progressive steps in Fig. 3-61, the layout of a sheet should not take more than two minutes.

Begin with a 12.00 × 18.00 in. sheet and tape it to the board as described earlier in this chapter. With the scale, measure 17.00″ near the bottom of the sheet (or on the bottom of the sheet) making short vertical marks, not dots. Measure and mark .75″ in from the left-hand mark and .25″ in from the right-hand mark (Fig. 3-61A). From this last mark, measure 1.00″ toward the left and mark. Lay the scale vertically near the left of the paper (or on the left edge) and make short horizontal marks 11.00″ apart. Make short marks .25″ up from the bottom mark and .25″ down from the top mark. With the T-square, draw horizontal lines through the four marks last made, as in (B).

Next, draw vertical lines through the five vertical marks (C). Then darken the border lines and the sheet will appear as in (D). It is now ready to be used for drawing. Lettering in the title block should be done after the assigned drawing problem is completed. This will help to avoid smudging the lettering.

The order of work for an A- or A4-size sheet (8.50 × 11.00 in. or 210 × 297 mm, respectively) is given in Fig. 3-62. The method is the same except for the dimensions. If sheets measuring 8.50 × 11.00 in. or 210 × 297 mm trimmed size are used, the method is the same except that measurements are made from the edges of the sheet. Also, since the sizes given are in inches, slight adjustments will be needed for metric-size sheets.

Begin with a 9.00 × 12.00 in. sheet and tape it to the board. Lay the scale horizontally near the bottom of the sheet, and make short vertical marks 11.00″ apart (Fig. 3-62A). Hold the scale in position and make short vertical marks .25″ from the left-hand and right-hand marks.

Lay the scale vertically near the left-hand edge of the sheet, and make short horizontal marks 8.50″ apart (B). Hold the scale in position and make short marks .75″ down from the top mark and .25″ up from the bottom mark.

Place the T-square in position and draw light horizontal lines through the four horizontal marks (C). Place the T-square and triangle in position and draw light vertical lines through the four vertical marks (D). Go over the border lines to make them clear and black. Use an F or HB pencil. The sheet is now ready to be used for making a drawing.

PROBLEMS

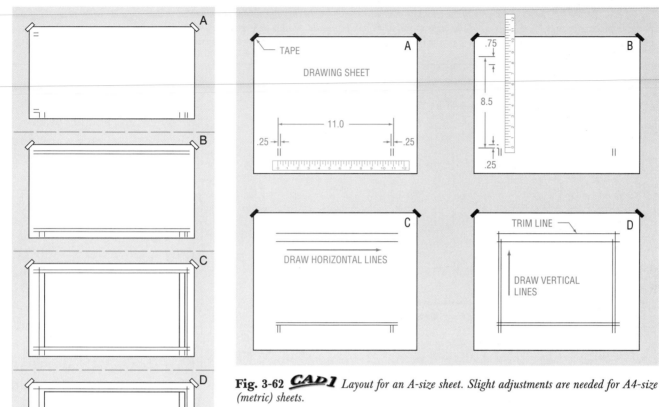

Fig. 3-61 *CAD1 Two-minute method.*

Fig. 3-62 *CAD1 Layout for an A-size sheet. Slight adjustments are needed for A4-size (metric) sheets.*

The sheet may be used with the long measurement horizontal, as in Fig. 3-63A, or vertical, as in (B). A title strip may be used, as suggested. The height of the title strip may be .62″ or more, to suit the information required by the instructor. Four problems may be worked on a B- or an A3-size sheet, as shown in (C).

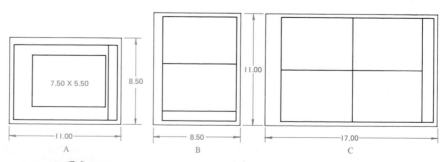

Fig. 3-63 *CAD1 Three arrangements of sheets.*

PROBLEMS

Fig. 3-64. *CAD1 Make a drawing of the template shown in Fig. 3-64. In this and several of the one-view drawings, the order of working is shown in progressive steps (A, B, C, etc.). These steps should be followed carefully, because they represent the drafter's procedure in making drawings. You should not consider the explanations as applying only to the particular problem. Instead, try to understand the system and apply it to all drawings.*

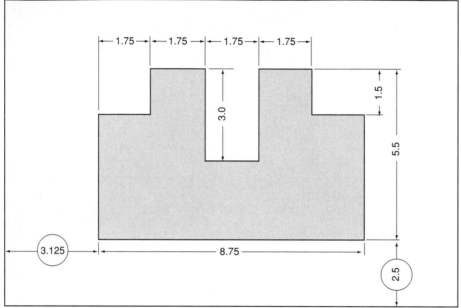

Fig. 3-64 *CAD1 Template.*

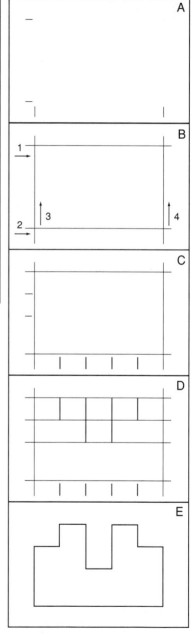

1. On an 11.00 × 17.00 in. drawing sheet, lay out a working space 15.00″ wide and 10.50″ high.

2. Measure 3.12″ from the left border line, and from this mark measure 8.75″ toward the right.

3. Lay the scale on the paper vertically near (or on) the left edge, make a mark 2.50″ up, and from this measure 5.50″ more. The sheet will appear as in (A).

4. Draw horizontal lines 1 and 2 with the T-square and triangle (B).

5. Lay the scale along the bottom line of the figure with the measuring edge on the upper side and make marks 1.75″ apart. Then with the scale on line 3, with its measuring edge to the left, measure from the bottom line two vertical distances, 2.50″ and 1.50″ (C).

6. Through the two marks, draw light horizontal lines.

7. Draw the vertical lines with T-square and triangle by setting the pencil on the marks on

the bottom line and starting and stopping the lines on the proper horizontal lines (D).

8. Erase the lines not wanted (if necessary) and darken the lines of the figure to finish the drawing (E). Do not add dimensions unless instructed to do so.

PROBLEMS

Fig. 3-65. *CAD 1 Make a drawing of the stencil shown in Fig. 3-65. This drawing gives practice in accurate measuring with the scale and making neat corners with short lines. The construction shown in the order of working should be drawn very lightly with a well-sharpened 3H or 4H pencil.*

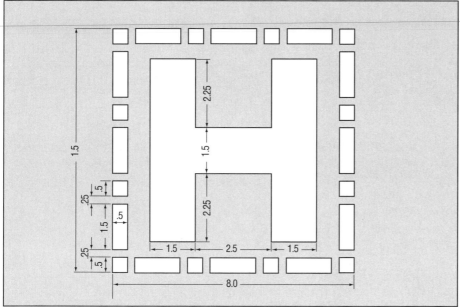

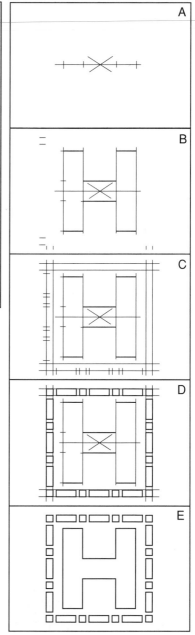

Fig. 3-65 *CAD 1 Order of working for drawing the stencil.*

1. On an 11.00 × 17.00 in. (B-size) drawing sheet, lay out a working space 15.00″ wide and 10.50″ high. Find the center of this space by laying the T-square blade face-down across the opposite corners and drawing short lines where the diagonals intersect (A).

2. Draw a horizontal centerline through the center, and on it, measure and mark off points for the four vertical lines. The drawing will appear as in (A).

3. Draw the vertical lines lightly with T-square and triangle. On the first vertical line, at the extreme left, measure and mark off points for all horizontal lines (B).

4. Draw the horizontal lines as finished lines. Measure points for the stencil border lines on the left side and bottom (C).

5. Draw the border lines. On the lower and left-hand border lines, measure the points for the ties.

6. Complete the border by drawing the cross lines as finished lines and darkening the other lines as in (D).

7. Darken the vertical lines and finish as in (E).

8. Fill in the title block.

PROBLEMS

Figs. 3-66 through 3-74. *CAD1* *Make an instrument drawing of the figure assigned by your instructor. Include all centerlines. Do not add dimensions unless instructed to do so. Scale and sheet size are suggested. Your instructor may wish to change these.*

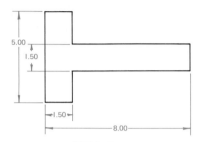

Fig. 3-66 *CAD1 Sheet-metal pattern. A- or A4-size sheet. Scale: Full size.*

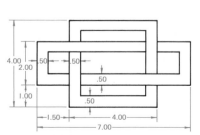

Fig. 3-67 *CAD1 Template for letter T. A- or A4-size sheet. Scale: Full size.*

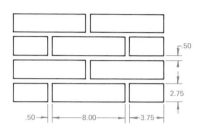

Fig. 3-68 *CAD1 Brick pattern. B- or A3-size sheet. Scale: ³/₄ size.*

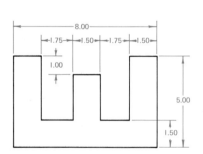

Fig. 3-69 *CAD1 Template for letter E. A- or A4-size sheet. Scale: Full size.*

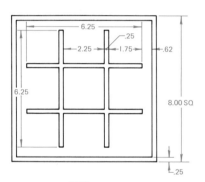

Fig. 3-70 *CAD1 Tic-tac-toe board. A- or A4-size sheet. Scale: ³/₄ size.*

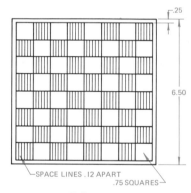

Fig. 3-71 *CAD1 Checker board. A- or A4-size sheet. Scale: Full size.*

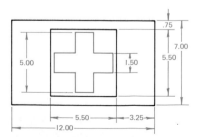

Fig. 3-72 *CAD1 First-aid station sign. B- or A3-size sheet. Scale: Full size.*

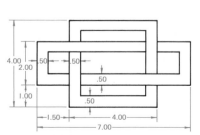

Fig. 3-73 *CAD1 Inlay. A- or A4-size sheet. Scale: Full size.*

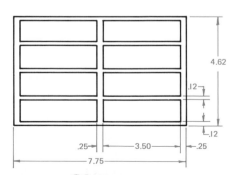

Fig. 3-74 *CAD1 Grill. A- or A4-size sheet. Scale: Full size.*

PROBLEMS

Fig. 3-75. *CAD2 Make a drawing of the shearing blank shown in Fig. 3-75. When a view has inclined lines, it should first be blocked in with square corners. Draw angles of 15°, 30°, 45°, 60°, and 75° with the triangles after locating one end of the line.*

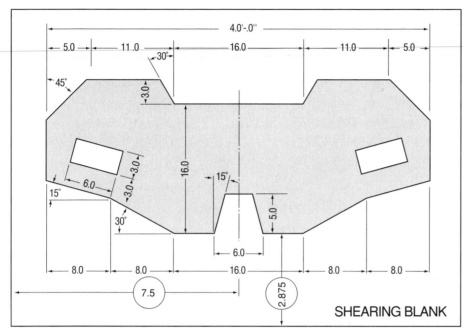

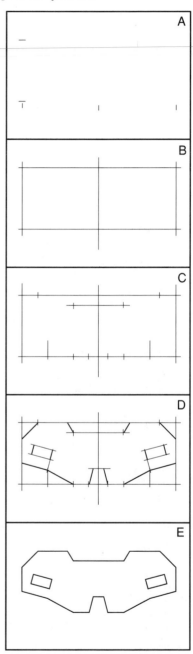

SHEARING BLANK

Fig. 3-75 *CAD2 Order of working for drawing the shearing blank.*

1. Locate vertical centerline and measure 2′ on each side (A). Note that this drawing must be made to the scale of 3″ = 1′ (1:4).

2. Locate vertical distances for top and bottom lines.

3. Draw main blocking-in lines as in (B).

4. Make measurements for starting points of inclined lines (C).

5. Draw inclined lines with T-square and triangles (D).

6. Finish as in (E).

7. Fill in the title block.

PROBLEMS

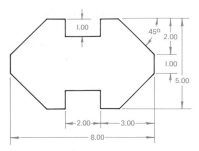

Fig. 3-76 *CAD2 Template. A- or A4-size sheet. Scale: Full size.*

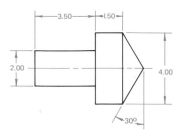

Fig. 3-77 *CAD2 Pivot. A- or A4-size sheet. Scale: Full size.*

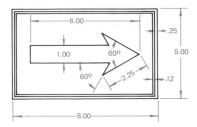

Fig. 3-78 *CAD2 Direction sign. A- or A4-size sheet. Scale: Full size.*

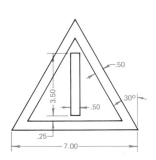

Fig. 3-79 *CAD2 International danger road sign. A- or A4-size sheet. Scale: Full size.*

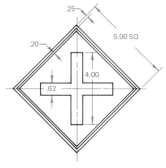

Fig. 3-80 *CAD2 Highway warning sign. A- or A4-size sheet. Scale: ³/₄ size.*

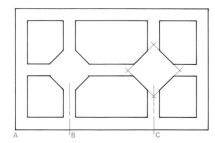

Fig. 3-81 *CAD2 Grill plate. Make all ribs .50″ wide. The distance AB is 2.25″; BC is 3.50″; AD is 2.50″. The diamond shapes are 1.50″ square. Scale: 1:1.*

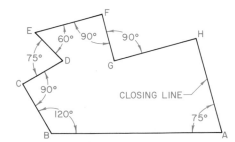

Fig. 3-82 *CAD2 Metric measurement. Draw horizontal line AB 180 mm long. Work clockwise around the layout. Remember! Angular dimensions are the same in the customary and metric systems. BC = 60 mm; CD = 48 mm; DE = 42 mm; EF = 74 mm; FG = 50 mm; GH = 90 mm. Measure the closing line and measure and label the angle at H. A- or A4-size sheet. Scale: 1:1.*

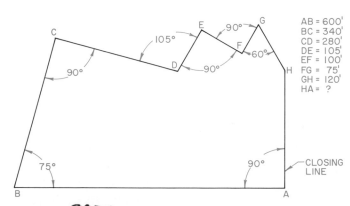

AB = 600′
BC = 340′
CD = 280′
DE = 105′
EF = 100′
FG = 75′
GH = 120′
HA = ?

Fig. 3-83 *CAD2 Land parcel. Practice using a civil engineer's scale by drawing the land parcel shown above. Use a scale of 1″ = 40′-0. Measure the length of the closing line to the nearest tenth of a foot and note it on your drawing. B- or A3-size sheet.*

PROBLEMS

Fig. 3-84. *CAD2 Make a drawing of the cushioning base in Fig. 3-84 using triangles, compasses, and scale. Centers of arcs and tangent points must be carefully located. Order of working for drawing the cushioning base:*

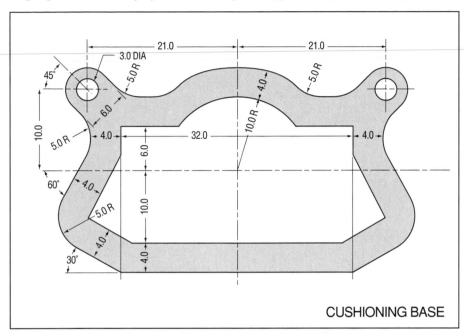

CUSHIONING BASE

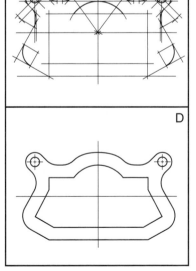

Fig. 3-84 *CAD2 Order of working for drawing the cushioning base.*

1. Through the center of the working space, draw horizontal and vertical distances (A). This drawing must be made to the scale of $3'' = 1'$ (1:4). Then draw horizontal and vertical lines (B).

2. Draw inclined lines with 45° and 30°-60° triangles. Then draw large arcs and two semi-circles with tangents at 45° (B).

3. Locate centers and tangent points for the 5.00″-radius tangent arcs. To do this, measure 5.00″ perpendicularly from each tangent line and draw lines parallel to the tangent lines. The intersection of these lines will be the required centers. To find the points of tangency, draw lines from the centers perpendicular to the tangent lines. To find the centers for the 5.00″ arcs tangent to the middle arc, proceed as follows: Increase the radius of the larger arc by 5.00″ and draw two short arcs cutting lines parallel to and 5.00″ above the top horizontal tangent line. These points will be the centers. Lines joining these centers with the center of the large arc will locate the points of tangency of the arcs (C).

4. Complete the view by drawing the lines for the opening. Darken the lines and finish the drawing as shown in (D).

PROBLEMS

Figs. 3-85 through 3-96. *Make an instrument drawing of the figure assigned. Include all centerlines. Do not dimension unless instructed to do so. Determine an appropriate drawing sheet size and scale before you begin.*

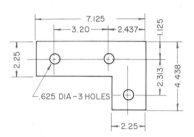

Fig. 3-85 *CAD1* *Angle bracket.*

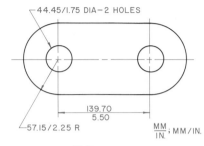

Fig. 3-86 *CAD1* *Link plate.*

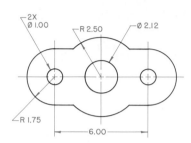

Fig. 3-87 *CAD1* *Armature support.*

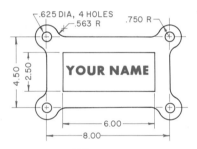

Fig. 3-88 *CAD1* *Identification plate.*

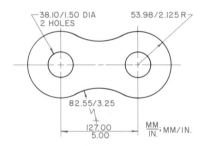

Fig. 3-89 *CAD1* *Bicycle chain link.*

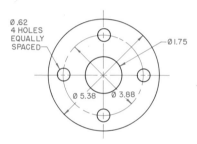

Fig. 3-90 *CAD1* *Round flange.*

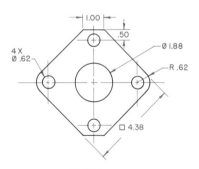

Fig. 3-91 *CAD2* *Bronze shim.*

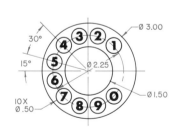

Fig. 3-92 *CAD2* *Telephone dial.*

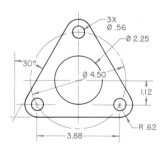

Fig. 3-93 *CAD2* *Base plate.*

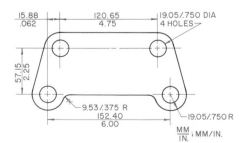

Fig. 3-94 *CAD2* *Cover plate.*

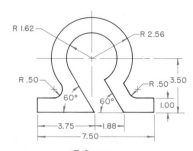

Fig. 3-95 *CAD2* *Housing.*

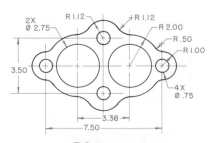

Fig. 3-96 *CAD2* *Carburetor gasket.*

PROBLEMS

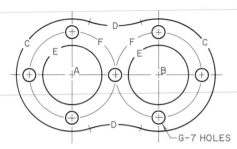

Fig. 3-97 *CAD2 Head gasket. Use metric scale. The distance between center points A and B is 45.50 mm. Radius of arc C is 30 mm. Radius of arc D is 43 mm. Diameter of hole E is 32 mm. Diameter of circle F is 45.50 mm. Holes labeled G have a diameter of 7 mm. Scale: as assigned.*

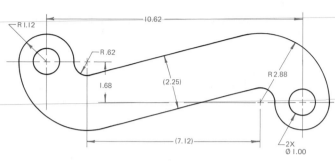

Fig. 3-98 *CAD2 Offset bracket. Locate all center points before beginning to draw circles and arcs.*

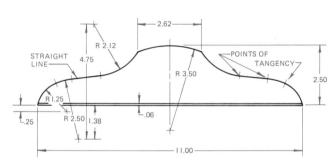

Fig. 3-99 *CAD2 T-square head. Draw to a scale of ¾″ = 1″. Be sure to locate points of tangency.*

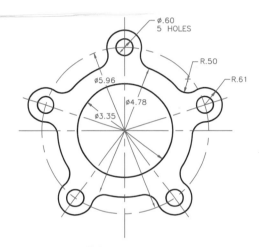

Fig. 3-100 *CAD2 Draw the gasket, being sure to make neat tangent joints. Scale: Full size.*

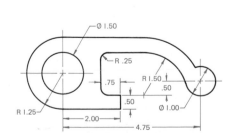

Fig. 3-101 *CAD2 Draw the adjustable sector. Make neat tangent joints. Use compass and bow pencil. Scale: Full size.*

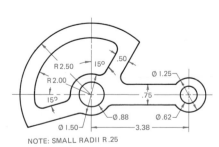

NOTE: SMALL RADII R.25

Fig. 3-102 *CAD2 Adjusting ring.*

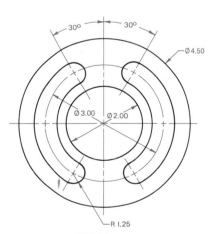

Fig. 3-103 *CAD2 Anchor plate.*

PROBLEMS

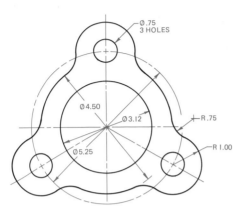

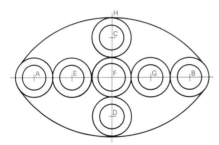

Fig. 3-104 *CAD2* Gasket.

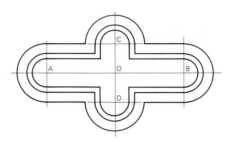

Fig. 3-105 *CAD2* Make a drawing of the frame using T-square, triangle, scale, and bow pencil. Draw intersecting lines at right angles through the center of the working space. Lay off AO = OB = 5.25″ and OC = OD = 2.25″ with the scale. With centers at A, B, C, and D, draw semicircles with radii of 1.12″, 1.50″, and 2.25″. Draw horizontal and vertical tangent lines. Brighten lines necessary to show the view. Scale: 6″ = 1′-0 (1:2).

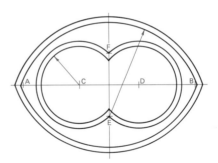

Fig. 3-106 *CAD2* Make a drawing of the multiple dial plate using T-square, triangle, scale, compasses, and bow pencil. Draw center lines at right angles. Lay off FC = FD = FG = FE = EA = GB = 6.00″. With centers at A, B, C, D, E, F, and G, draw circles with a diameter of 6.00″. With center at F, draw a circle with a diameter of 4.50″. With centers at A, B, C, D, E, and G, draw circles with a diameter of 4.00″ With centers at H and I, draw tangent arcs with a radius of 18.00″. Brighten lines necessary to show the view. Scale: 3″ = 1′-0 (1:4).

Fig. 3-107 *CAD2* Make a drawing of the double dial plate using T-square, triangle, scale, dividers, and compasses. Draw line AB = 7.00″ and divide it into three equal parts with the dividers. With centers at C and D, draw arcs with radii of 1.50″ and 1.75″. With centers at E and F, draw arcs with radii of 3.75″ and 4.00″ to complete the view. Scale: Full size.

Fig. 3-108 *CAD2* Measuring practice. Measure the lengths of lines A through N at full-size, ¾″ = 1″, ½″ = 1″, 1″ = 40′-0, 1″ = 1′-0, etc., as assigned by your instructor.

PROBLEMS

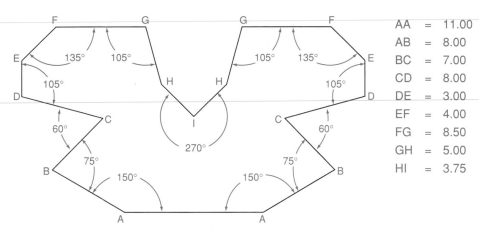

AA	= 11.00
AB	= 8.00
BC	= 7.00
CD	= 8.00
DE	= 3.00
EF	= 4.00
FG	= 8.50
GH	= 5.00
HI	= 3.75

Fig. 3-109 *CAD2 Irregular polygon. Construct as shown. Begin by drawing line AA centered horizontally near the bottom of the sheet. The length of each line is given at the right of the figure. All angles may be drawn with the T-square and a combination of triangles.*

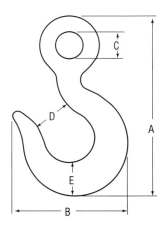

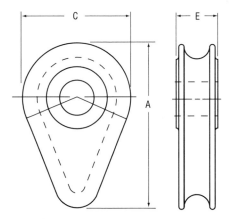

WIRE ROPE HOOK				
A	B	C	D	E
4.94	3.20	.88	1.06	.84
5.44	3.50	1.00	1.12	.90
6.25	4.10	1.12	1.25	1.12
6.88	4.54	1.25	1.38	1.30
7.62	4.88	1.38	1.50	1.38
8.60	5.75	1.50	1.70	1.56
9.50	6.38	1.16	1.88	1.70

CAST-STEEL THIMBLE			
WIRE ROPE Ø	A	C	E
.50	3.06	2.12	.88
.62	5.06	3.38	1.25
.75	5.06	3.38	1.25
.88	6.56	4.50	1.50
1.00	6.56	4.50	1.50

Fig. 3-110 *CAD2 Prepare an ink drawing of the wire rope hook using the dimensions selected by your instructor. Determine the radii necessary for smooth tangencies.*

Fig. 3-111 *CAD2 Prepare a two-view ink drawing of the solid cast-steel thimble with dimensions selected by your instructor. Select suitable radii for all curves. Note the points of tangencies to determine the tapered shape.*

PROBLEMS

Figs. 3-112 through 3-114. *Make an instrument drawing of the figure assigned. Include all centerlines. Do not dimension unless instructed to do so.*

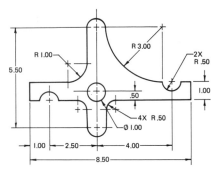

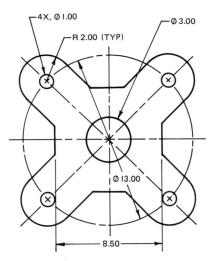

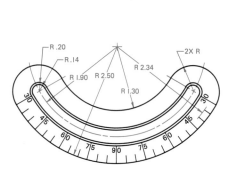

Fig. 3-112 *Paul. Scale: 1:1.*

Fig. 3-113 *Gasket. Scale: 1:2.*

Fig. 3-114 *Tilt scale. Scale: 1:1.*

SCHOOL-TO-WORK

Solving Real-World Problems

Fig. 3-115 *SCHOOL-TO-WORK problems are designed to challenge a student's ability to apply skills learned within this text. Be creative and have fun!*

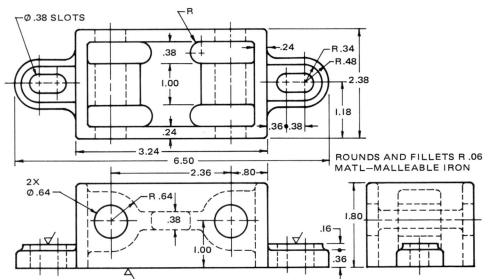

Fig. 3-115 *Draw a gasket for the bottom of the guide block. It should be shaped so that when cut out, it will touch only the metal surface on the bottom. Scale: 1:1.*

GEOMETRY FOR TECHNICAL DRAWING

OBJECTIVES

Upon completion of Chapter 4, you will be able to:

○ Identify and describe various geometric shapes and constructions used by drafters.
○ Construct various geometric shapes using points, lines, and planes from technical specifications using drafting instruments.
○ Apply geometric construction as a problem-solving tool in technical drawing.
○ Use geometry to reduce or enlarge a drawing or to change the proportions of a drawing.

VOCABULARY

In this chapter, you will learn the meanings of the following terms:

- **bisect**
- **chord**
- **circumscribed**
- **diameter**
- **entities**
- **geometric constructions**
- **geometry**
- **hypotenuse**
- **inscribed**
- **intersect**
- **object snaps**
- **ogee curve**
- **parallel**
- **perpendicular**
- **polygon**
- **regular polygon**
- **right angle**
- **tangent**

Geometry has always been important to people. It was used in ancient times for measuring land and making right-angle corners for buildings and other kinds of construction. (A **right angle** is a "square" angle that measures 90°.) The Egyptian *rope stretchers* used rope with marks or knots at 12 equal spaces. The rope was divided into 3-, 4-, and 5-space sections, as shown in Fig. 4-1. A square corner was made by stretching the rope and driving pegs into the ground at the 3-, 4-, and 5-space marks. This was one way an ancient people used geometry.

Fig. 4-1 *Egyptian rope stretchers.* (Courtesy of Circle Design)

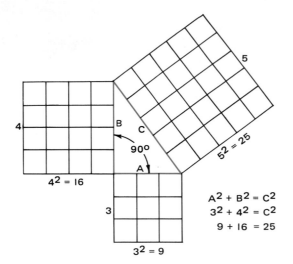

Fig. 4-2 *Pythagorean theorem shown graphically and mathematically.*

The use of the 3-4-5 triangle for making a right angle was proved by the mathematician Pythagoras in the sixth century B.C. This proof is called the Pythagorean theorem. The theorem is shown graphically and mathematically in Fig. 4-2.

This method also works well for triangles that have the same proportions, such as 6, 8, and 10 units.

$$6^2 + 8^2 = 10^2$$
$$36 + 64 = 100$$
$$100 = 100$$

The units may be millimeters, meters, inches, fractions of an inch, or any other units of measure.

Geometry is the study of the size and shape of things. The relationship of straight and curved lines in drawing shapes is also a part of geometry. Geometric figures used in drafting include circles, squares, triangles, hexagons, and octagons. Many other shapes and relationships are shown in Fig. 4-3.

Drawings made of individual lines and points drawn in proper relationship to one another are known as **geometric constructions.** Geometric constructions are used by drafters, surveyors, engineers, architects, scientists, mathematicians, and designers. Therefore, nearly everyone in all technical fields needs to know the constructions explained in this chapter. Study the "dictionary of drafting geometry" in Fig. 4-3 before beginning the geometric constructions on the following pages.

GEOMETRY IN COMPUTER-AIDED DRAFTING

Computer-aided drafting (CAD) software combines the basic elements of geometric construction with an electronic medium (computer) that makes certain drafting tasks easier and faster to perform. Lines, circles, arcs, ellipses, and many other two-dimensional forms, as well as cubes, cylinders, spheres, pyramids, prisms, and other three-dimensional forms, can be combined using CAD software to develop a technical drawing of any object. A clear understanding of these elements and how they are combined to develop more complex shapes makes it easier and faster for beginning CAD students to learn the system.

Designers and drafters can also use CAD to solve design problems geometrically. Using CAD, it is possible to create a quick, but accurate, sketch and then check the angles, fit, tolerances, and other geometric values before committing to a particular design. If the design doesn't seem to work, the designer hasn't invested much time in it; if it does work, the sketch can be modified for further use.

SOLVING PROBLEMS USING GEOMETRY

One of the purposes for geometric construction is to help drafters solve design problems. The constructions you will practice in the next section may seem at first to have little application to everyday problems. However, drafters frequently use these techniques and others to help solve practical design problems.

For example, suppose a designer is designing a complicated mechanical assembly that has moving parts. The designer must determine whether the current design allows enough clearance for the parts to move freely. Depending on the actual design, the drafter may need to construct accurate circles and various tangent, parallel, and perpendicular lines to measure clearances.

In other cases, designers are given very specific *parameters* (guidelines) within which they must work. For example, a designer may be told that the part must fit within a predefined area. Again, the drafter uses principles of geometric construction to meet the specifications.

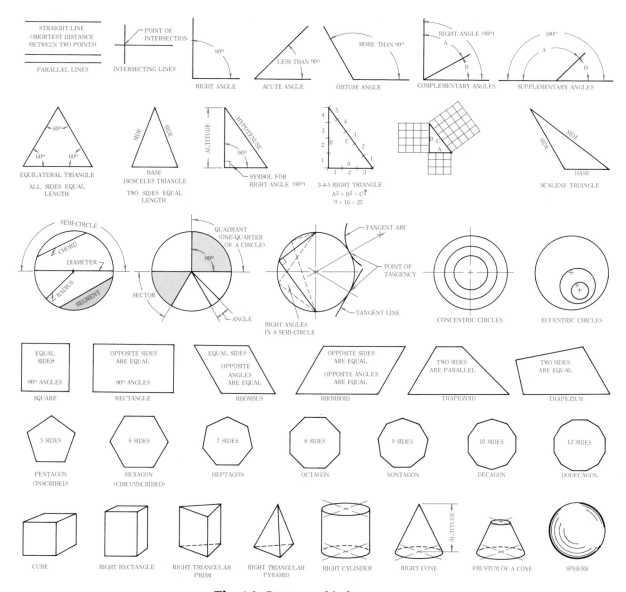

Fig. 4-3 *Dictionary of drafting geometry.*

What other practical design and drafting problems can be solved using geometric constructions? As you complete the constructions in the next section, think about practical drawing problems that could be solved using each technique.

CONSTRUCTIONS

This section consists of a series of example exercises. By working through these constructions, you will begin to understand how to draw the basic geometry used in drafting.

As you work, remember that one of the most important concepts in drafting is accuracy. Other key concepts include clarity (legibility) and neatness. One reason these concepts are important is that sloppy work may give the designer the wrong answers, which may lead to a part being manufactured with the wrong dimensions or specifications. Lack of clarity may cause people to misread the information, again causing errors in the manufactured product.

Therefore, it is a good idea to practice these concepts even when the work is not critical. Practice working accurately, legibly, and neatly as you perform the following constructions and throughout the rest of this course.

Bisect a Line or Arc

This construction demonstrates how to bisect a line or arc. **Bisect** means to divide into two equal parts. Follow the instructions in Fig. 4-4 to bisect line *AB* and arc *AB*.

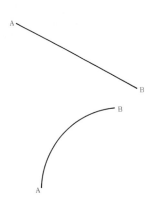

Given line AB or arc AB.

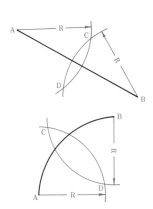

*With A and B as centers and any radius R greater than one half of AB, draw arcs to **intersect** (cut across) at C and D. The **radius (pl. radii)** is the distance from the center of an arc or circle to any point on the arc or circle.*

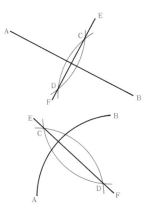

Draw line EF through intersections C and D.

Fig. 4-4 *Bisect a straight line or arc.*

Divide a Line into Equal Parts

Two methods of dividing a line into equal parts are described below. Try both methods. Can you think of situations in which you would need to use one method instead of the other?

Method 1

In this construction, you will divide a straight line into eight equal parts. This method can be applied to create any number of equal divisions. Follow the instructions in Fig. 4-5.

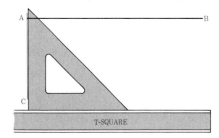

*Draw a line of any length at A perpendicular to line AB. Lines are **perpendicular** when they cross at right (90°) angles.*

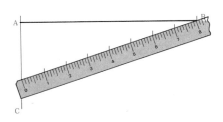

Place the scale with zero on line AC at an angle so that the scale touches point B. Keeping zero on line AC, adjust the angle of the scale until any eight equal divisions are included between line AC and point B (in this case, 8 in.). Mark the divisions.

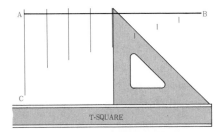

*Draw lines parallel to AC through the division marks to intersect line AB. Two lines are **parallel** when they are always the same distance apart.*

Fig. 4-5 *Divide a straight line into any number of equal parts (first method).*

Method 2

Follow the instructions in Fig. 4-6 to divide a line into five equal parts.

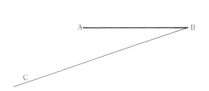

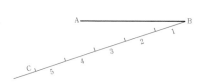

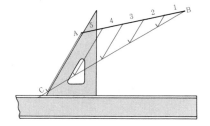

Draw line BC from point B at any convenient angle and of any length.

Use dividers or a scale to step off five equal spaces on line BC beginning at point B.

Draw a line connecting points A and C. Draw lines through each point on BC parallel to line AC to intersect line AB.

Fig. 4-6 *Divide a straight line into any number of equal parts (second method).*

Draw a Perpendicular Line

Many procedures exist for constructing a line that is perpendicular to another line. Each method is useful in certain drafting situations. Examples of several methods are shown below.

Method 1

Follow the steps in Fig. 4-7 to draw a line at a given point on another line so that the two lines are perpendicular. A line is **perpendicular** to another line when the lines cross at right angles (90°).

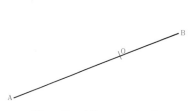

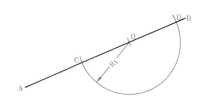

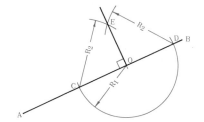

Given line AB and point O.

With O as the center and any convenient radius R_1, construct an arc intersecting line AB, locating points C and D.

With C and D as centers and any radius R_2 greater than OC, draw arcs intersecting at E. Draw a line connecting points E and O to form the perpendicular.

Fig. 4-7 *Construct a line perpendicular to a given line through a given point on the line (first method).*

Method 2

This construction presents another method of drawing a line perpendicular to a given line at a given point on that line. Use this method when the given point lies near one end of the line. Follow the instructions in Fig. 4-8.

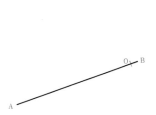

Given line AB and point O.

From any point C above line AB, construct an arc using CO as the radius and passing through line AB to locate point D.

Draw a line through points D and C, extending it through the arc to locate point E. Connect points E and O to form the perpendicular line.

Fig. 4-8 *Construct a line perpendicular to a given line through a given point on the line (second method).*

Method 3

This construction demonstrates another way to draw a line perpendicular to a given line through a given point on the line. Follow the steps in Fig. 4-9.

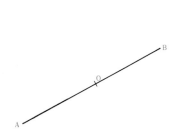

Given line AB and point O.

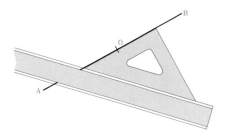

Place the T-square and triangle as shown.

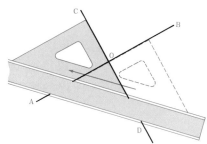

Slide the triangle along the T-square until the edge aligns with point O on line AB. Draw the perpendicular CD.

Fig. 4-9 *Construct a line perpendicular to a given line through a given point on the line (third method).*

Method 4

This exercise demonstrates a method of constructing a line perpendicular to a given line through a point that does *not* lie on the given line. Follow the instructions in Fig. 4-10.

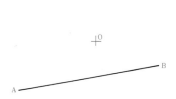

Given line AB and point O.

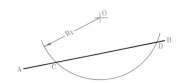

With O as the center, draw an arc with radius R_1 long enough to intersect line AB to locate points C and D.

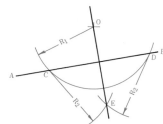

With C and D as centers and radius R_2 greater than one half of CD, draw intersecting arcs to locate point E. A line drawn through points O and E is the perpendicular line.

Fig. 4-10 *Construct a line perpendicular to a given line through a point that is not on the given line (first method).*

Method 5

Follow the steps in Fig. 4-11 to practice another way to draw a line perpendicular to a given line through a point that is not on the line.

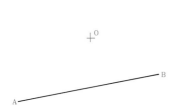

Given line AB and point O.

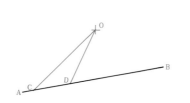

Draw lines from point O to any two points on line AB, locating points C and D.

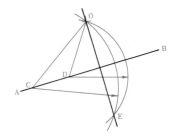

With C and D as centers and CO and DO as radii, draw arcs to intersect, locating point E. Connect points O and E to form the perpendicular line.

Fig. 4-11 *Construct a line perpendicular to another line through a point that is not on the given line (second method).*

Draw a Parallel Line

The following construction methods create a line that is parallel to another line. Recall that lines are parallel when they are always the same distance apart.

Method 1

This construction allows you to place a line parallel to another line through a given point. Follow the instructions in Fig. 4-12.

Given line AB and point P.

Fig. 4-12 *Use a compass to construct a line parallel to a given line through a given point.*

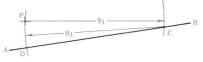

With point P as the center and any convenient radius R_1, draw an arc intersecting line AB to locate point C. With point C as the center and the same radius R_1, draw an arc through point P and line AB to locate point D.

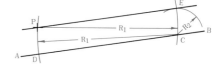

*With C as the center and radius R_2 equal to chord PD, draw an arc to locate point E. (A **chord** is a straight line between two points on a circle.) Draw a line through points P and E. Line PE is parallel to AB.*

Method 2

The steps in Fig. 4-13 demonstrate another way to construct a line parallel to another line through a given point.

Given line AB and point P.

Place the T-square and triangle as shown.

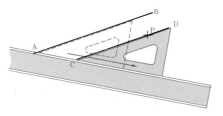

Slide the triangle until the edge aligns with point P. Draw the parallel line CD.

Fig. 4-13 *Use a triangle and T-square to construct a line parallel to a given line through a given point.*

Method 3

Use this method to construct a line parallel to a given line at a specified distance from the given line. Follow the instructions in Fig. 4-14. *NOTE:* See "Construct a Tangent Line" later in this chapter for instructions to create a tangent line.

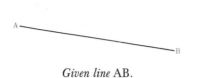

Given line AB.

Draw two arcs with centers anywhere along line AB. The arcs should have a radius R equal to the specified distance between the two parallel lines.

*Draw a parallel line CD tangent to the arcs. A line is **tangent** to an arc or circle when it touches the arc or circle at one point only.*

Fig. 4-14 *Construct a line parallel to a given line at a given distance from the line.*

Bisect an Angle

This construction demonstrates a method of bisecting a given angle. Follow the steps in Fig. 4-15.

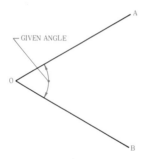

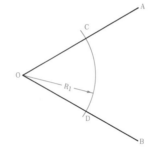

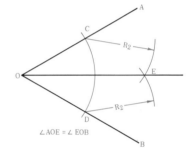

Given angle AOB.

With point O as the center and any convenient radius R_1, draw an arc to intersect AO and OB at C and D.

With C and D as centers and any radius R_2 greater than one half the radius of arc CD, draw arcs to intersect, locating point E. Draw a line through points O and E to bisect angle AOB.

Fig. 4-15 *Bisect an angle.*

Construct an Angle

The construction in Fig. 4-16 demonstrates a method of copying a given angle to a new location and orientation. Follow the steps in Fig. 4-16.

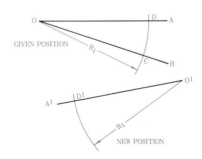

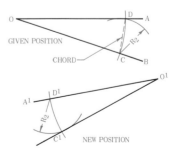

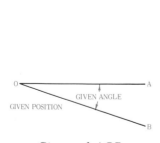

Given angle AOB.

Draw one side O^1A^1 in the new position. With O and O^1 as centers and any convenient radius R_1, construct arcs to intersect BO and AO at C and D and A^1O^1 at D^1.

With D^1 as the center and radius R_2 equal to chord DC, draw an arc to locate point C^1 at the intersection of the two arcs. Draw a line through points O^1 and C^1 to complete the angle.

Fig. 4-16 *Construct an angle.*

Construct a Triangle

A *triangle* is a **polygon** (closed figure) that contains three sides. The following constructions show methods for drawing various types of triangles.

Method 1

This method constructs an isosceles triangle. An *isosceles triangle* is one in which two sides are of equal length. Follow the steps in Fig. 4-17.

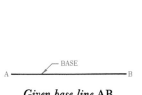

Given base line AB.

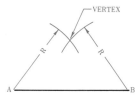

*With points A and B as centers and a radius R equal to the length of the sides you want, draw intersecting arcs to locate the third vertex of the triangle. (A **vertex** is the point at which two lines meet.) The other two vertices (plural of vertex) are at the endpoints of the base line.*

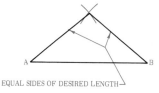

Draw lines through point A and the vertex and through point B and the vertex to complete the triangle.

Fig. 4-17   *Construct an isosceles triangle.*

Method 2

This method constructs an equilateral triangle. An *equilateral triangle* is one in which all three sides are of equal length and all three angles are equal. Follow the steps in Fig. 4-18 to construct an equilateral triangle.

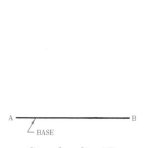

Given base line AB.

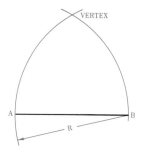

With points A and B as centers and a radius R equal to the length of line AB, draw intersecting arcs to locate the third vertex.

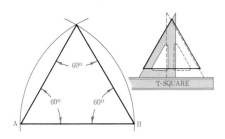

Draw lines through point A and the vertex and through point B and the vertex to complete the triangle. NOTE: An equilateral triangle may also be constructed by drawing 60° lines through the ends of the base line with the 30°-60° triangle, as shown at right.

Fig. 4-18   *Construct an equilateral triangle.*

Method 3

Construct a right triangle using this method when you know the length of two sides of the triangle. A *right triangle* is one that has a right (90°) angle at one of its vertices. Follow the steps in Fig. 4-19.

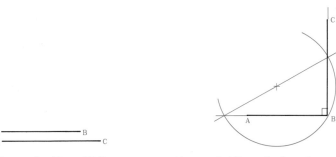
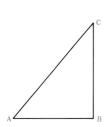

Given sides AB and BC.

Draw side AB in the desired position. Draw a line perpendicular to AB at B equal to BC. NOTE: Use the method of Fig. 4-7 or the method of Fig. 4-11 to construct the perpendicular line.

Draw a line connecting points A and C to complete the right triangle.

Fig. 4-19 *Construct a right triangle given the lengths of two sides.*

Method 4

Use this method to construct a right triangle when you know the length of one side and the length of the hypotenuse. The **hypotenuse** of a right triangle is the side opposite the 90° angle. Follow the steps in Fig. 4-20.

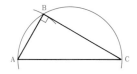

Given hypotenuse AC and side AB.

Draw the hypotenuse in the desired location. Draw a semicircle on AC using ½ AC as the radius.

With point A as the center and a radius equal to side AB, draw an arc to intersect the semicircle at B. Draw AB and then draw a line to connect B and C to complete the triangle.

Fig. 4-20 *Construct a right triangle given the length of one side and the length of the hypotenuse.*

Method 5

This construction illustrates the 3-4-5 method of drawing a right triangle. Follow the steps in Fig. 4-21.

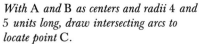

Given base line AB 3 units long.

With A and B as centers and radii 4 and 5 units long, draw intersecting arcs to locate point C.

Draw lines AC and BC to complete the triangle.

Fig. 4-21 *Construct a right triangle using the 3-4-5 method, given a 3-unit base.*

Method 6

Use this method to construct a triangle when you know the lengths of all three sides. This construction is useful for *scalene* triangles (those that include three different angles and sides of three different lengths). Follow the steps in Fig. 4-22.

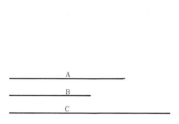

Given triangle sides A, B, and C.

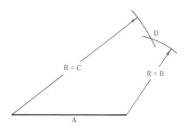

Draw base line A in the desired location. Construct arcs from the ends of line A with radii equal to lines B and C to locate point D.

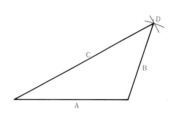

Connect both ends of line A with point D to complete the triangle.

Fig. 4-22 *Construct a triangle given the lengths of all three sides.*

Construct a Circle

This construction describes a method of creating a circle given three points that lie on the circle. Follow the steps in Fig. 4-23.

Given points A, B, and C, draw lines AB and BC.

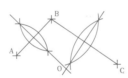

Draw perpendicular bisectors of AB and BC to intersect at point O.

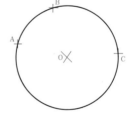

Draw the required circle with point O as the center, and radius R = OA = OB = OC.

Fig. 4-23 *Construct a circle, given three points that lie on the circle.*

Construct a Tangent Line

The constructions that follow present methods of creating lines tangent to a circle.

Method 1

Use this method to construct a line tangent to a given point on a circle without using a triangle or T-square. Follow the steps in Fig. 4-24.

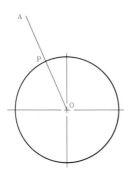 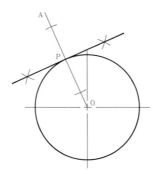

Given circle with center point O and tangent point P, draw line OA from the center of the circle to extend beyond the circle through point P.

Draw a line perpendicular to OA at P. The perpendicular line is the tangent line.

Fig. 4-24 *Construct a line tangent to a circle through a given point on the circle (first method).*

Method 2

Use this method to construct a line tangent to a given point on a circle using a 30°-60° triangle and a T-square. Follow the steps in Fig. 4-25.

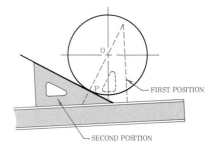

Given a circle with center point O and tangent point P, place a T-square and triangle so that the hypotenuse of the triangle passes through points P and O.

Hold the T-square, turn the triangle to the second position at point P, and draw the tangent line.

Fig. 4-25 *Construct a line tangent to a given point on a circle (second method).*

Method 3

This construction creates lines tangent to a circle from a given point outside the circle. Follow the steps in Fig. 4-26.

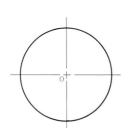 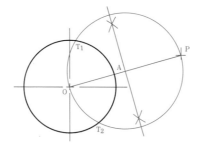 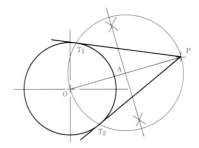

Given a circle with center point O and point P outside the circle.

Draw line OP and bisect it to locate point A. Draw a circle with center A and radius R = AP = AO to locate tangent points T_1 and T_2.

Draw PT_1 and PT_2. These lines are tangent to the circle.

Fig. 4-26 *Construct a line tangent to a circle from a given point outside the circle.*

Method 4

Use this method to construct a line tangent to the exterior of two circles. Follow the steps in Fig. 4-27.

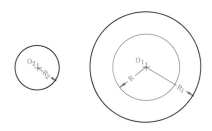
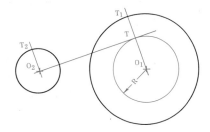
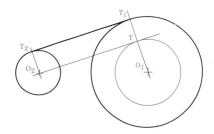

Given two circles with centers O_1 and O_2 and radii R_1 and R_2. $R = R_1 - R_2$. Using radius R and point O_1 as the center, draw a circle.

From center point O_2 draw a tangent O_2T to the circle of radius R. Draw radius O_1T. Extend it to locate tangent point T_1. Draw O_2T_2 parallel to O_1T_1.

Draw the needed tangent T_1T_2 parallel to TO_2.

Fig. 4-27 *Draw an exterior common tangent to two circles of unequal radii.*

Method 5

Use this method to construct a line tangent to the interior of two circles. Follow the steps in Fig. 4-28.

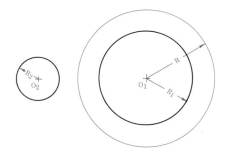
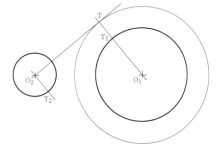
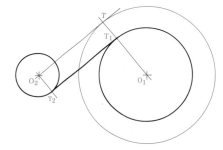

Given two circles with centers O_1 and O_2, and radii R_1 and R_2. $R = R_1 + R_2$. Using radius R and point O_1 as the center, draw a circle.

From center point O_2, draw a tangent O_2T to the circle of radius R. Draw radius O_1T to locate tangent point T_1. Draw O_2T_2 parallel to O_1T.

Draw the needed tangent T_1T_2 parallel to TO_2.

Fig. 4-28 *Draw an interior common tangent to two circles of unequal radii.*

Construct a Tangent Arc

The following constructions demonstrate methods to draw arcs tangent to other geometry, such as straight lines and other arcs.

Method 1

Use this method to construct an arc tangent to two straight lines. The technique is shown for two lines at an *acute* (less than 90°) angle, at an *obtuse* (greater than 90°) angle, and at a right angle (exactly 90°). Follow the steps in Fig. 4-29.

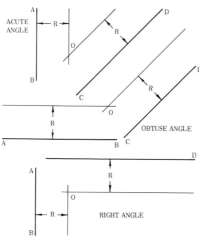
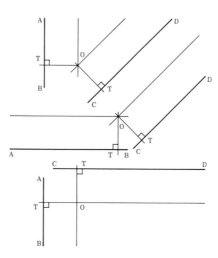
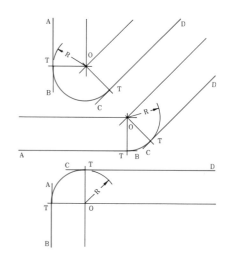

Given lines AB and CD, draw lines parallel to AB and CD at a distance R from them on the inside of the angle. The intersection O will be the center of the arc you need.

Draw perpendicular lines from O to AB and CD to locate the points of tangency T.

With O as the center and radius R, draw the needed arc.

Fig. 4-29 *Construct an arc tangent to two straight lines at an acute angle, an obtuse angle, and a right angle.*

Method 2

Use this method to construct an arc tangent to two given arcs. Follow the steps in Fig. 4-30.

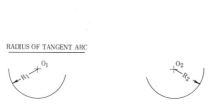
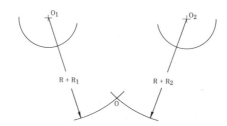
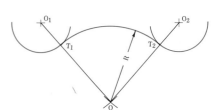

Given two arcs having radii R_1 and R_2 (radii may be equal or unequal) and radius R of the tangent arc.

Draw an arc with center O_1 and radius = $R + R_1$. Draw an arc with center O_2 and radius = $R + R_2$. The intersection at O is the center of the tangent arc.

Draw lines O_1O and O_2O to locate tangent points T_1 and T_2. With point O as the center and radius R, draw the tangent arc needed.

Fig. 4-30 *Construct an arc of a given radius tangent to two given arcs.*

Method 3

Use this method to construct an arc tangent to a line and an arc. Follow the steps in Fig. 4-31.

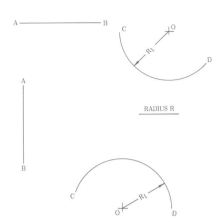

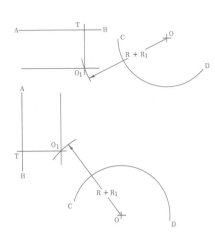

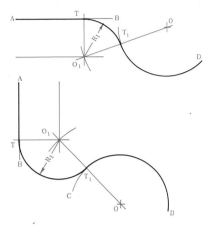

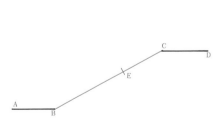

Given line AB, arc CD, and radius R.

Draw a line parallel to AB, at distance R, toward arc CD. Use radius $R_1 + R$ to locate point O_1. Draw a line from O_1 perpendicular to AB to locate tangent point T.

Draw a line from O to O_1 to locate tangent point T_1 on CD. With point O_1 as the center and radius R, draw the tangent arc.

Fig. 4-31 *Construct an arc of given radius tangent to an arc and a straight line.*

Construct an Ogee Curve

An **ogee curve** is a reverse curve that looks something like an S. To construct an ogee curve, follow the steps in Fig. 4-32.

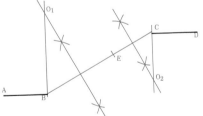

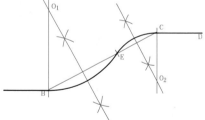

Given lines AB and CD, draw line BC. Select a point E on line BC through which the curve is to pass.

Draw perpendicular bisectors of BE and EC. Draw lines perpendicular to AB at B and to CD at C. They must cross the bisectors of BE and EC at O_1 and O_2, respectively.

Draw one arc with center O_1 and radius O_1E and the other with center O_2 and radius O_2E to complete the required curve.

Fig. 4-32 *Draw a reverse, or ogee, curve.*

Construct a Square

There are several ways to construct a square. The method you choose depends on the rest of the geometry in the drawing.

Method 1

Use this method to construct a square when you know the length of a side. Follow the steps in Fig. 4-33.

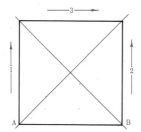

Given the length of the side AB, construct 45° diagonals from ends of line AB. Complete the square by drawing perpendicular lines at each end of line AB to intersect the diagonals. Draw the last line from the intersection of the diagonal and vertical lines. Draw the sides in the order shown by the numbered arrows.

Fig. 4-33 *Construct a square given the length of a side.*

Method 2

Use this method to construct a square inscribed in a circle. A square (or other polygon) is **inscribed** in a circle when its four corners are tangent to (touch but do not cross) the circle. Follow the steps in Fig. 4-34.

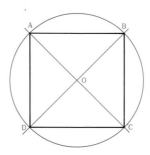

Given a circle with center point O, draw 45° diagonals through the center point O to locate points A, B, C, and D. Connect A and B, B and C, C and D, and D and A to complete the square.

Fig. 4-34 *Construct a square inscribed within a circle.*

Method 3

Use this method to construct a square circumscribed about a circle. A square (or other polygon) is **circumscribed** about a circle when the square fully encloses the circle and the circle is tangent to the square on all four sides. Follow the steps in Fig. 4-35.

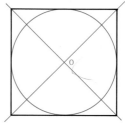

Given a circle with center point O, draw 45° diagonals through the center point O. Draw sides tangent to the circle, intersecting at the 45° diagonals to complete the square.

Fig. 4-35 *Construct a square circumscribed about a circle.*

Construct a Pentagon

A *pentagon* is a five-sided polygon. When all five sides of a polygon are exactly the same length and all its angles are equal, it is called a **regular polygon.** The following constructions demonstrate methods for constructing pentagons.

Method 1

Use this method to construct a regular pentagon when you know the length of one side. Follow the steps in Fig. 4-36 to create a regular pentagon.

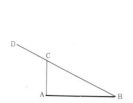

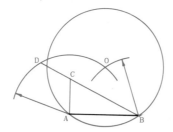

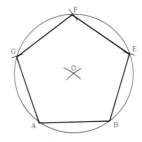

Given line AB, construct a perpendicular line AC equal to one half the length of AB. Draw line BC and extend it to make the line CD equal to AC.

With radius AD and points A and B as centers, draw intersecting arcs to locate point O. With the same radius and O as the center, draw a circle.

Step off AB as a chord to locate points E, F, and G. Connect the points to complete the pentagon.

Fig. 4-36 *Construct a regular pentagon given the length of one side.*

Method 2

This construction demonstrates a method of inscribing a pentagon within a circle. Follow the steps in Fig. 4-37. *NOTE:* The **diameter** of a circle is the distance across the circle through its center point. The symbol for diameter is Ø.

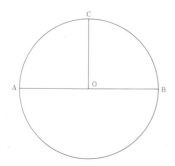

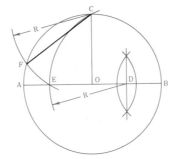

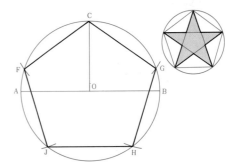

Given a circle with diameter AB and radius OC.

Bisect radius OB to locate point D. With D as center and radius DC, draw an arc to locate point E. With C as center and radius CE, draw an arc to locate point F. Chord CF is one side of the pentagon.

Step off chord CF around the circle to locate points G, H, and J. Draw the chords to complete the pentagon.

Fig. 4-37 *Construct a regular pentagon inscribed within a circle, given the circle.*

Construct a Hexagon

A *hexagon* is a six-sided polygon. The following constructions demonstrate methods for constructing regular hexagons.

Method 1

Use this method to construct a regular hexagon when you know the distance across the flats (sides). The distance across the flats is the distance from the *midpoint* (exact middle) of one side of the polygon through the center point to the midpoint of the opposite side of the polygon. Follow the steps in Fig. 4-38.

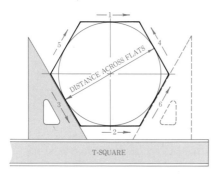

Given the distance across the flats of a regular hexagon, draw centerlines and a circle with a diameter equal to the distance across the flats. With the T-square and 30°-60° triangle, draw the tangents in the order shown.

Fig. 4-38 *Construct a regular hexagon, given the distance across the flats.*

Method 2

Use this method to construct a regular hexagon when you know the distance across the corners. The distance across the corners is the distance from one vertex of a polygon through the center point to the opposite vertex. Follow the steps in Fig. 4-39.

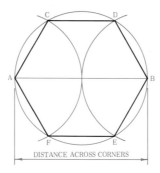

Given the distance AB across the corners, draw a circle with AB as the diameter. With A and B as centers and the same radius, draw arcs to intersect the circle at points C, D, E, and F. Connect the points to complete the hexagon.

Fig. 4-39 *Construct a regular hexagon, given the distance across the corners (first method).*

Method 3

This construction demonstrates another method of constructing a regular hexagon, given the distance across the corners. Follow the steps in Fig. 4-40.

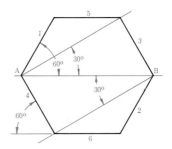

Given the distance AB across the corners, draw lines from points A and B at 30° to line AB. The lines can be any convenient length. With the T-square and 30°-60° triangle, draw the sides of the hexagon in the order shown.

Fig. 4-40 *Construct a regular hexagon, given the distance across the corners (second method).*

Construct an Octagon

An *octagon* is an eight-sided polygon. The following constructions demonstrate methods of drawing regular octagons.

Method 1

Use this method to construct an octagon circumscribed about a circle. Follow the steps in Fig. 4-41.

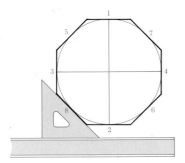

Given the distance across the flats, draw centerlines and a circle with a diameter equal to the distance across the flats. With the T-square and 45° triangle, draw lines tangent to the circle in the order shown to complete the octagon.

Fig. 4-41 *Construct a regular octagon circumscribed about a circle, given the distance across the flats.*

Method 2

Use this method to construct an octagon inscribed within a circle. Follow the steps in Fig. 4-42.

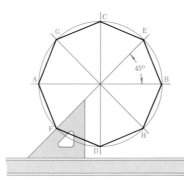

Given the distance across the corners, draw centerlines AB and CD and a circle with a diameter equal to the distance across the corners. With the T-square and 45° triangle, draw diagonals EF and GH. Connect the points to complete the octagon.

Fig. 4-42 *Construct a regular octagon inscribed within a circle, given the distance across the corners.*

Method 3

Use this method to construct an octagon inscribed within a square. Follow the steps in Fig. 4-43.

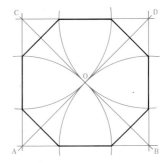

Given the distance across the flats, construct a square having sides equal to AB. Draw diagonals AD and BC with their intersection at O. With A, B, C, and D as centers and radius R = AO, draw arcs to intersect the sides of the square. Connect the points to complete the octagon.

Fig. 4-43 *Construct a regular octagon inscribed within a square, given the distance across the flats.*

Construct an Ellipse

The constructions in this section demonstrate methods to draw an *ellipse,* or regular oval.

Method 1

This construction demonstrates the use of the pin-and-string method to draw an ellipse. Follow the steps in Fig. 4-44.

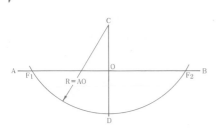

Given major axis AB and minor axis CD intersecting at O. With C as center and radius R = AO, an arc locates F_1 and F_2.

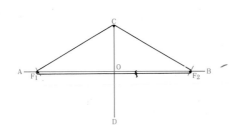

Place pins at points F_1, C, and F_2. Tie a string around the three pins and remove pin C.

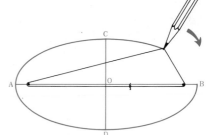

Put the point of a pencil in the loop and draw the ellipse. Keep the string taut (tight) when moving the pencil.

Fig. 4-44 *Construct an ellipse by the pin-and-string method.*

Method 2

This construction demonstrates the use of the trammel method to draw an ellipse. A *trammel* is a piece of paper or plastic on which specific distances have been marked off. Follow the steps in Fig. 4-45.

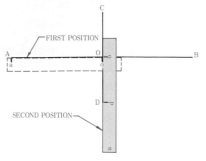

Given major axis AB *and minor axis* CD *intersecting at point* O. *Cut a strip of paper or plastic (trammel). Mark off distances* AO *and* OD *on the trammel.*

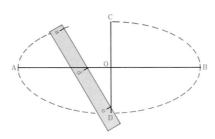

On the trammel, move point O *along line* CD *(minor axis) and point* D *along line* AB *(major axis) and mark points at* A.

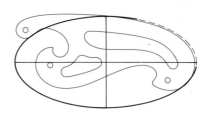

Use a French curve or flexible curve to connect the points to draw the ellipse.

Fig. 4-45 *Draw an ellipse by the trammel method.*

Method 3

Use this construction method to draw an approximate ellipse using the major and minor axes of the ellipse. This method works when the *minor axis* (the shorter axis) is at least two-thirds the size of the *major axis* (the longer axis). Follow the steps in Fig. 4-46.

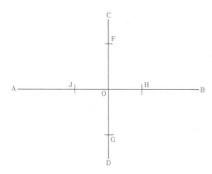

Given major axis AB *and minor axis* CD *intersecting at point* O. *Lay off* OF *and* OG, *each equal to* AB − CD. *Lay off* OJ *and* OH, *each equal to three fourths of* OF.

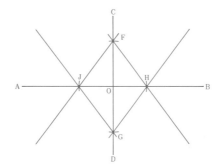

Draw and extend lines GJ, GH, FJ, *and* FH.

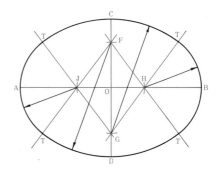

Draw arcs with centers F *and* G *and radii* FD *and* GC *to the points of tangency. Draw arcs with centers* J *and* H *and radii* JA *and* HB *to complete the ellipse. The points of tangency are marked* T.

Fig. 4-46 *Draw an approximate ellipse when the minor axis is at least two-thirds the size of the major axis.*

Method 4

Use this construction method to draw an approximate ellipse when the minor axis is less than two-thirds the size of the major axis. Follow the steps in Fig. 4-47.

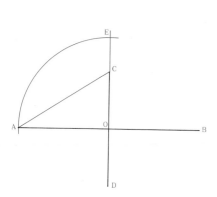

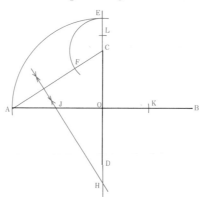

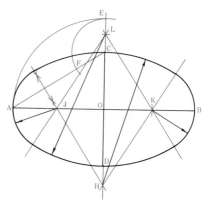

Given major axis AB and minor axis CD intersecting at point O. Draw line AC. Draw an arc with point O as the center and radius OA and extend line CD to locate point E.

Draw an arc with point C as the center and radius CE to locate point F. Draw the perpendicular bisector of AF to locate points J and H. Locate points L and K. OL = OH and OK = OJ.

Draw arcs with J and K as centers and radii JA and KB. Draw arcs with H and L as centers and radii HC and LD to complete the ellipse.

Fig. 4-47 *Draw an approximate ellipse when the minor axis is less than two-thirds the size of the major axis.*

Reduce or Enlarge a Drawing

The following techniques reduce or enlarge an existing drawing.

Method 1

If the drawing is square or rectangular, use a diagonal line to enlarge or reduce the drawing. Follow the directions in Fig. 4-48.

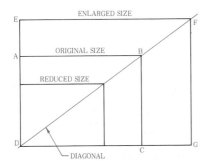

Draw a diagonal through corners D and B. Measure the width or height you need along line DC or DA (example: DG). Draw a perpendicular line from that point (G) to the diagonal. Draw a line perpendicular to DE intersecting at point F. Reductions are made in the same way.

Fig. 4-48 *Reduce or enlarge a square or rectangular area.*

Method 2

If the drawing is not square or rectangular, use the method described in Fig. 4-49.

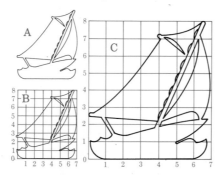

Lay a grid over the drawing. Use squares of an approximate size. Draw a larger or smaller grid on a separate sheet of paper. The size of the grid depends upon the amount of enlargement or reduction needed. Use dots to mark key points on the new grid corresponding to points on the original drawing. Connect the points to complete the new drawing.

Fig. 4-49 *Reduce or enlarge the drawing of a sailboat shown at A.*

Change the Proportion of a Drawing

Occasionally, you may need to change the proportion of a drawing. Use the technique shown in Fig. 4-50.

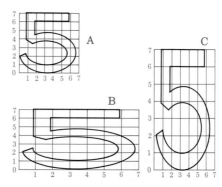

Draw a grid over the original drawing. Draw a grid on a separate sheet of paper in the needed proportion, as shown at B or C. Use dots to mark key points from the original drawing. Connect the points to complete the new drawing.

Fig. 4-50 *Change the proportion of the drawing shown at A.*

GEOMETRIC CONSTRUCTION USING CAD

In computer-aided drafting, **entities** are individual geometric elements such as lines and circles from which you develop a drawing. (Some CAD programs refer to these elements as *objects*.) The CAD system stores mathematical information about these entities. It remembers, for example, where you selected the beginning (start) point of a line, where you selected the end point, and then automatically draws the line. Because AutoCAD provides a good example of construction methods, this section focuses on AutoCAD. However, the techniques and definitions are similar in almost all CAD programs.

Basic geometric entities include points, lines, circles, arcs, ellipses, text, 3D solids, dimensions, polylines, and so on. Many CAD programs also have more advanced entities. For example, AutoCAD recognizes attributes, polygon meshes, splines, regions, viewports, and block references as individual entities.

Figure 4-51 shows examples of a simple 2D geometric construction generated using a CAD system. The more powerful CAD systems can also create 3D objects such as spheres, cylinders, and cubes (Fig. 4-52), as well as intricate *composite solids* made up of many individual 3D entities (Fig. 4-53).

To create a drawing using a CAD program, you use *commands* to tell the software which entities you want to use and where you want to place them. The method of entering commands varies among CAD software from simply entering the command at the keyboard to selecting *icons* (small pictures associated electronically with the command) to using menus. Some examples of icons are shown in Fig. 4-54.

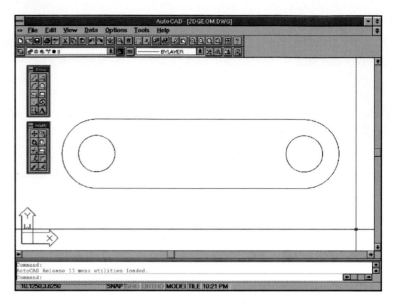

Fig. 4-51 *CAD programs allow you to create various two-dimensional entities. The object shown here was created using the LINE, ARC, and CIRCLE commands.*

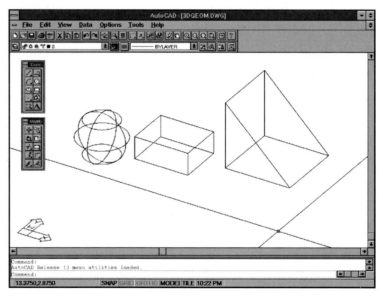

Fig. 4-52 *AutoCAD provides several 3D "primitives" from which a drafter can build a three-dimensional drawing. The shapes shown here were created using the SPHERE, BOX, and WEDGE commands. Although they may look like isometrics, they exist in the CAD drawing file in three dimensions.*

CAD commands for basic entities such as arcs and circles offer many options. The CAD technician needs these options to create complex drawings accurately. For example, you can draw a circle (using the CIRCLE command) by specifying its center point and its radius, or by specifying its center point and its diameter. But what if you don't know where the center point is supposed to be?

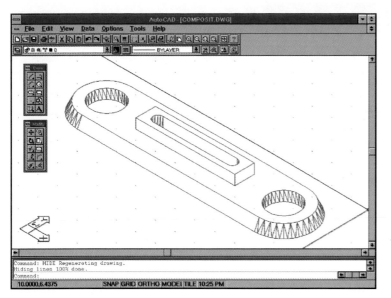

Fig. 4-53 *CAD programs that offer 3D drafting allow you to create complex objects in three dimensions.*

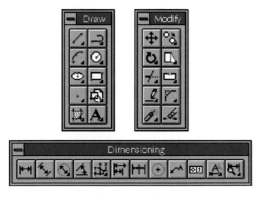

Fig. 4-54 *In AutoCAD for Windows®, you can issue commands by picking icons. The icons are arranged into several "toolbars" for convenience. The Draw, Modify, and Dimensioning toolbars are shown here.*

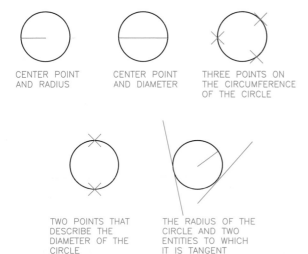

CENTER POINT AND RADIUS

CENTER POINT AND DIAMETER

THREE POINTS ON THE CIRCUMFERENCE OF THE CIRCLE

TWO POINTS THAT DESCRIBE THE DIAMETER OF THE CIRCLE

THE RADIUS OF THE CIRCLE AND TWO ENTITIES TO WHICH IT IS TANGENT

Fig. 4-55 *AutoCAD offers several ways to create a circle. In what circumstances might you use each of these options?*

Other circle-drawing options allow you to create the circle. You can specify three points on the circle, or even two points (Fig. 4-55).

You can create squares, rectangles, triangles, hexagons, and other polygons either by using individual lines or by entering the POLYGON command. This command is an example of a CAD command that is flexible enough to draw many different shapes, depending on your specifications. For example, to draw a triangle, you could enter the POLYGON command and specify three sides. To draw a hexagon, specify six sides.

One very important aspect of CAD drawing is the relationship of the individual entities to each other. Suppose the center of a circle needs to be at the exact midpoint of a line. In manual drafting, you would use instruments to find the midpoint of the line. To accomplish the same thing in AutoCAD, you can use **object snaps.** In this example, when the computer asks where you want the center of the circle, you could enter the Midpoint object snap and specify the line (Fig. 4-56). AutoCAD places the center of the circle exactly at the midpoint of the line. Other object snaps include Endpoint, Quadrant, Intersection, Center, Perpendicular, Tangent, and several more. Using the object snaps, you can create an accurate CAD drawing quickly and easily.

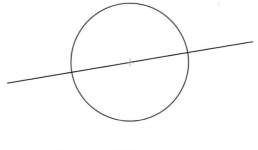

Command: CIRCLE
3P/2P/TTR/<Center point>: MIDPOINT
of (pick the line)
Diameter/<Radius>: 1.5

Fig. 4-56 *To snap to the midpoint of a line in AutoCAD, enter the Midpoint object snap. The command sequence to perform this operation in AutoCAD is shown below the circle.*

CHAPTER 4
REVIEW

Learning Activities

1. Vocabulary is an important part of a complete understanding of drafting geometry. Study Fig. 4-3 carefully in an attempt to learn the vocabulary of geometry and to associate terms with their respective graphic symbols (shapes).

2. Describe the Pythagorean theorem using Fig. 4-2. Calculate side *C* when side *A* is 10 and side *B* is 15. Is *C* an even number?

3. Collect pictures and drawings of objects that contain various geometric shapes. Can you name a well-known building that has five sides?

4. How many geometric shapes can you find that were used in the design of your school building? List or sketch as many as you can find.

5. At your school or public library, find books or other materials that describe the design process. Look for examples of how geometric construction is used in the design process.

6. Talk to drafters or designers in your area to find out how they use geometry in their work. What types of math courses are required for their jobs? How often do they use the information they learned in their math courses?

Questions

Write your answers on a separate sheet of paper.

1. What kind of triangle has one right (90°) angle?

2. What is a geometric construction? How can geometric constructions be helpful to designers and engineers?

3. Name three of the most important concepts in drafting. Why are they important?

4. A hexagon has how many sides?

5. An octagon has how many sides?

6. What term describes a closed figure that has equal sides and equal opposite angles?

7. Describe at least one method of creating an ellipse.

8. What is an obtuse angle?

9. Explain the 3-4-5 method of creating a triangle. What type of triangle does this method create?

10. What name is given to a triangle in which all three sides are of equal length?

11. What is the name of the point at which a line touches an arc or circle?

12. What is the difference between inscribing a polygon in a circle and circumscribing it around a circle?

13. In CAD, what are entities?

14. What is the purpose of using object snaps in a CAD program?

15. A plane figure in which opposite sides are equal and all angles are 90° is a _____.

16. A triangle in which two sides are equal is an _____ triangle.

17. How many sides does a heptagon have?

18. Concentric circles have a common _____.

19. Where two lines intersect, they form a _____.

20. The longest side of a right triangle is called the _____.

CHAPTER 4

PROBLEMS

The small CAD symbol next to each problem indicates that a problem is appropriate for assignment as a computer-aided drafting problem and suggests a level of difficulty (one, two, or three) for the problem.

Draw each problem three times the size shown below. Use dividers to pick up the dimensions from Figs. 4-58 through 4-77 as assigned, and step off each measurement three times.

Problems 4-58 through 4-77 are designed for working four problems on an A- or A4-size sheet, laid out as shown in Fig. 4-57.

Problems 4-78 through 4-90 are single-view drawings. These are designed to give additional practice in geometric constructions. Draw each problem using the assigned sheet size and scale. You may omit the dimensions unless your instructor directs you to include them. Nearly all the problems for Chapter 3 may also be used as problems for Chapter 4 and vice versa.

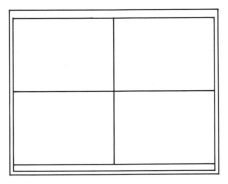

Fig. 4-57 *Layout for Problems 4-58 through 4-77.*

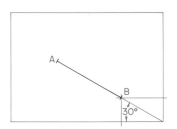

Fig. 4-58 *CAD1* *Bisect line AB.*

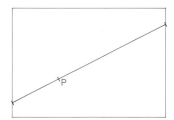

Fig. 4-59 *CAD1* *Construct a perpendicular at point* P.

Fig. 4-60 *CAD1* *Divide line AB into five equal parts.*

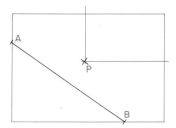

Fig. 4-61 *CAD1* *Construct line* CD *through point* P *so that* CD *is parallel to* AB *and equal in length to* AB.

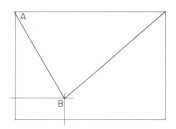

Fig. 4-62 *CAD1* *Bisect angle* ABC.

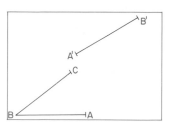

Fig. 4-63 *CAD1* *Copy angle* ABC *in a new location, beginning with* A′B′.

PROBLEMS

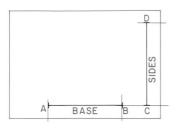

Fig. 4-64 *CAD1 Construct an isosceles triangle on base AB with sides equal to CD.*

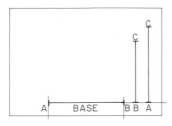

Fig. 4-65 *CAD1 Construct a triangle on base AB with sides equal to BC and AC.*

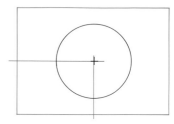

Fig. 4-66 *CAD1 Construct a square in the Ø3.00-inch circle with corners touching the circle.*

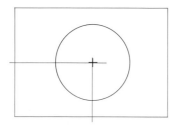

Fig. 4-67 *CAD1 Construct a regular pentagon in the Ø3.00-inch circle.*

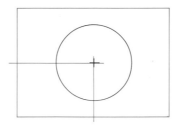

Fig. 4-68 *CAD1 Construct a regular hexagon around the Ø3.00-inch circle.*

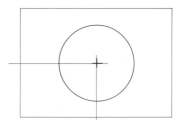

Fig. 4-69 *CAD1 Construct a regular octagon around the Ø3.00-inch circle.*

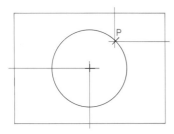

Fig. 4-70 *CAD1 Construct a tangent line through point P on the Ø3.00-inch circle.*

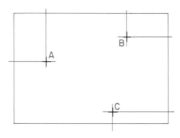

Fig. 4-71 *CAD1 Construct a circle through points A, B, and C.*

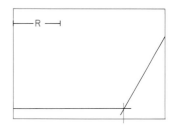

Fig. 4-72 *CAD1 Construct an arc having a radius R tangent to the two lines.*

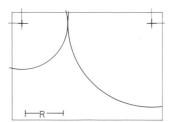

Fig. 4-73 *CAD1 Construct an arc having a radius R tangent to two given arcs.*

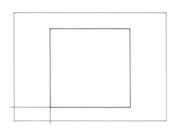

Fig. 4-74 *CAD1 Construct a regular octagon within the 3.00-inch square.*

Fig. 4-75 *CAD1 Construct an ellipse on the 4.00-inch major and 2.50-inch minor axes.*

PROBLEMS

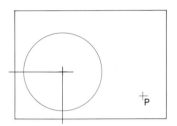

Fig. 4-76 *CAD1 Construct two lines from point P tangent to the Ø2.50-inch circle.*

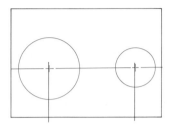

Fig. 4-77 *CAD1 Construct two lines tangent to the exterior of the two circles.*

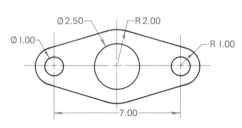

Fig. 4-78 *CAD1 Gasket. Draw the view of the gasket full-size or as assigned. Mark all points of tangency. Do not dimension.*

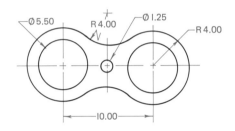

Fig. 4-79 *CAD1 Pipe support. Scale: as assigned. Locate and mark all centers and all points of tangency. Do not dimension.*

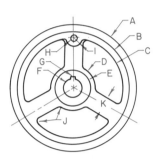

Fig. 4-80 *CAD2 Handwheel. Scale: as assigned. A = Ø7.00″, B = Ø6.12″, C = 2.75″R, D = 1.25″R, E = Ø2.00″, F = Ø1.00″, G (keyway) = .20″wide × .10″ deep, H = 0.38″, I = .38″R, J = .20″ R, K = 1.00″.*

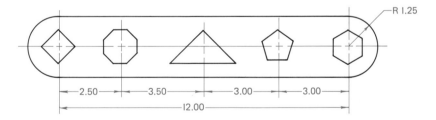

Fig. 4-81 *CAD1 Combination wrench. Scale: as assigned. Square: 1.00″; octagon: 1.38″ across flats; isosceles triangle: 2.75″ base, 2.00″ sides; pentagon: inscribed within Ø1.38″ circle; hexagon: 1.25″ across flats. Use the method of your choice for constructing geometric shapes. Do not erase construction lines.*

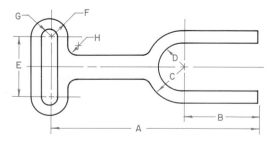

Fig. 4-82 *CAD1 Adjustable fork. Scale: as assigned. A = 220 mm, B = 80 mm, F = 20 mm, G = 8 mm, H = 10 mm.*

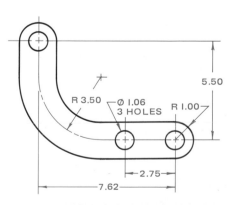

Fig. 4-83 *CAD1 Rod support. Scale: as assigned.*

PROBLEMS

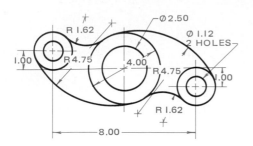

Fig. 4-84 **CAD1** *Rocker arm. Scale: full-size or as assigned.*

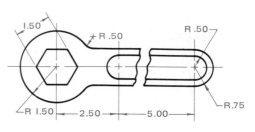

Fig. 4-85 **CAD1** *Hex wrench. Scale: as assigned. Mark all tangent points. Do not erase construction lines.*

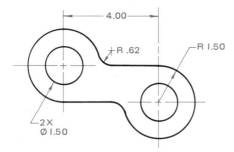

Fig. 4-86 **CAD1** *Offset link. Scale: full size or as assigned. Locate and mark all points of tangency.*

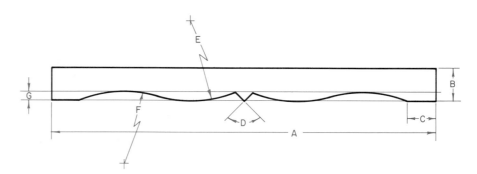

Fig. 4-87 **CAD1** *Valance board. Scale: 1" = 1'-0 or as assigned. A = 8'-0, B = 0'-8, C = 0'-7, D = 90°, E = 2'-6, F = 2'-6, G = 0'-2. Locate and mark points of tangency. Do not erase construction lines.*

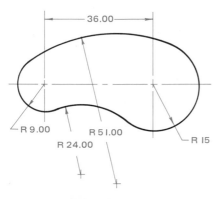

Fig. 4-88 **CAD2** *Kidney-shaped table top. Scale: full-size or as assigned.*

PROBLEMS

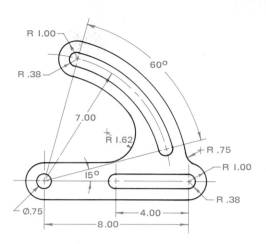

Fig. 4-89 *CAD2 Adjustable table support.*
Scale: as assigned.

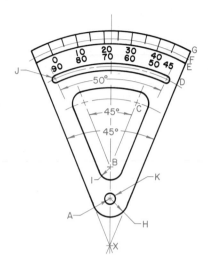

Fig. 4-90 *CAD2 Tilt scale. Scale:
as assigned. AB = 1.75, AX = 2.62,
AC = 5.50, AD = 7.25, AE = 8.50,
AF = 8.75, AG = 9.25, H = 1.00" R,
I = R.62, J = R.20, K = R.50.*

SCHOOL-TO-WORK

Solving Real-World Problems

Fig. 4-91 *SCHOOL-TO-WORK problems are designed to challenge
a student's ability to apply skills learned within this text. Be creative
and have fun!*

Fig. 4-91 *Design and draw a cover for your 8.50 × 11.00 inch or 11.00 × 17.00
inch set of technical drawings. Use various geometric shapes in the design. Use block
lettering similar to that shown in Fig. 2-16. Geometric shapes such as circles, squares,
hexagons, octagons, ellipses, etc., can be used in the background. A transparent or
translucent overlay with only the lettering can be used to enhance the design.
Information on the cover should include your name, your school name, the name of the
course, your instructor's name, and the year.*

MULTIVIEW DRAWING

OBJECTIVES

Upon completion of Chapter 5, you will be able to:

○ Identify and select the various views of an object.
○ Determine the number of views needed to describe fully the shape and size of an object.
○ Define the term *orthographic projection*.
○ Describe the difference between first- and third-angle projection.
○ Visualize the "glass box" concept and apply it to the process of selecting and locating views on a drawing.
○ Develop a multiview drawing, following a prescribed step-by-step process, from the initial idea to a finished drawing.
○ Describe the procedures used to create orthographic projections using CAD.

VOCABULARY

In this chapter, you will learn the meanings of the following terms:

• first-angle projection	• normal views	• solid model
• front view	• orthographic projection	• spherical
• horizontal plane	• pictorial drawing	• third-angle projection
• implementation	• profile plane	• top view
• multiview drawing	• quadrant	• vertical plane
• negative cylinder	• right-side view	• visualization

People communicate ideas by verbal and written language and by *graphic* (pictorial) means. One of the graphic means is technical drawing. It is a language used and understood in all countries. When accurate visual (sight) understanding is necessary, technical drawing is the most exact method that can be used.

Technical drawing involves two things: (1) visualization and (2) implementation. **Visualization** is the ability to see clearly in the mind's eye what a machine, device, or other object looks like. **Implementation** is the process of drawing the object that has been visualized. In other words, the designer, engineer, or drafter first visualizes the object and then explains it pictorially using a technical drawing. Thus, the idea is recorded in a form that can be used as a means of communication.

A technical drawing, properly made, gives a clearer, more accurate description of an object than a photograph or written explanation. Technical drawings made according to standard *principles* (rules) result in views that give an exact visual description of an object (Fig. 5-1).

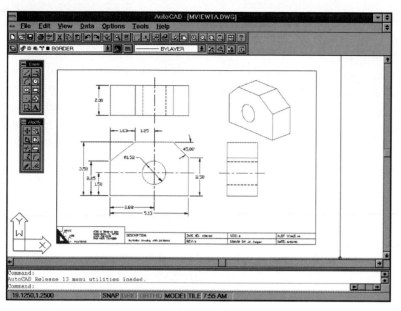

Fig. 5-1 *This three-view drawing gives an accurate description of the object.*

Fig. 5-2 *Photograph of a V-block.*

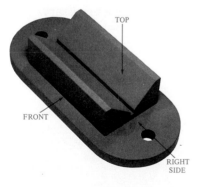

TOP

FRONT

RIGHT
SIDE

Fig. 5-3 *Photograph of a V-block with front, top, and side views labeled.*

MULTIVIEW DRAWING

A photograph of a V-block is shown in Fig. 5-2. It shows the object as it appears to the eye. Notice that three sides of the V-block are shown in a single view. In Fig. 5-3, the same photograph is shown with the three sides labeled according to their *relative* (related) positions. If all sides could be shown in a single photograph, it would also include a left-side view, a rear view, and a bottom view. Nearly all objects have six sides.

Obviously, an object cannot be photographed before it has been built. Therefore, it is necessary to use another kind of graphic representation. One possibility is to make a pictorial drawing, as shown in Fig. 5-4. A **pictorial drawing** is a drawing that shows an object as it appears to the human eye— as it would appear in a photograph. It shows, just as a photograph, the way the object looks in general. However, it does not show the exact forms and relationships of the parts that make up the object. It shows the V-block as it appears, not as it really is. For example, the holes in the base appear as ellipses, not as true circles.

The problem, then, is to represent an object on a sheet of paper in a way that will describe its exact shape and proportions. This can be done by drawing views of the object as it is seen from different positions. These views are then arranged in a standard order so that anyone familiar with drafting practices can understand them immediately.

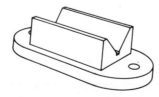

Fig. 5-4 *Pictorial drawing of a V-block.*

In order to describe accurately the shape of each view, imagine a position directly in front of the object, then above it, and finally at the right side of it. This is where the ability to visualize is important. Figure 5-5 shows the **front view** of the V-block, which reveals the exact shape of the V-block when viewed from the front. The dashed lines are used to show the outline of details behind the front surface, called *hidden details.* Notice that this view shows the width and the height of the object.

Figure 5-6 is a **top view** of the V-block. It shows the width and the depth. Since the view is taken directly from above, the exact shape of the top is shown. Notice that the holes are true circles and that the rounded ends of the base are true radii. In the photograph and in the pictorial drawing, these appeared as elliptical shapes.

Figure 5-7 is a **right-side view** of the V-block. It shows the depth and the height. Notice that the shape of the V appears to be symmetrical in the drawing. It appears distorted or misshapen in the photo and in the pictorial drawing.

The front, top, and right-side views are the ones most often used to describe an object in a technical drawing. They are therefore known as the **normal views.**

The Relationship of Views

Views must be placed in proper relationship to one another. Only in this way can technical drawings be read and understood properly. Figure 5-8 shows the V-block and how its three normal views have been *revolved* (turned) into their proper places. Notice that the top view is directly above the front view. The right-side view is directly to the right of the front view. Each of the normal views is where it logically belongs. When the normal views are placed in proper relationship to one another, the result is a **multiview drawing.** Multiview drawing is the exact representation of an object on one plane. These three views usually give a complete description of the object. Figure 5-8 is an example of a multiview drawing.

Other Views

Most objects have six sides, or six views. In most cases, two or three views can be used to describe completely the shape and size of all parts of an object. However, in some cases it may be necessary to show views other than the front, top, and right side. In Fig. 5-9, dice are shown in the upper left-hand corner. Since the detail is different on each of the six sides, six views are needed for a complete graphic description. The six views are shown in their proper locations in the lower part of Fig. 5-9. Only in unusual cases are six views necessary.

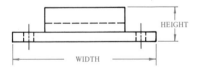

Fig. 5-5 *Front view of V-block.*

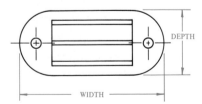

Fig. 5-6 *Top view of V-block.*

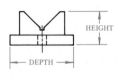

Fig. 5-7 *Right-side view of V-block.*

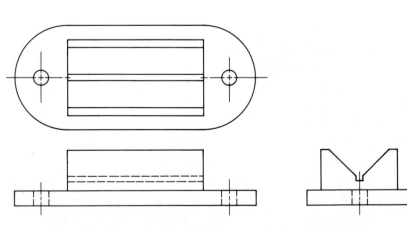

Fig. 5-8 *The relationship of the three normal views of the V-block.*

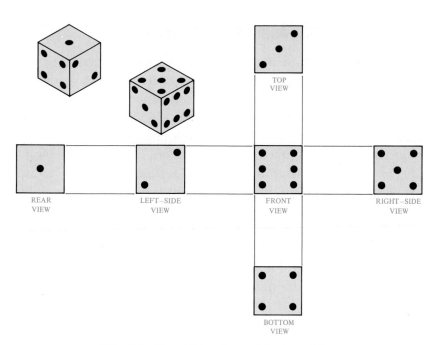

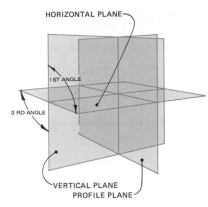

Fig. 5-10 *The three planes used in orthographic projection.*

Fig. 5-9 *Pictorial drawing and six views of dice.*

CAD-GENERATED MULTIVIEW DRAWINGS

Notice that the drawing in Fig. 5-1 is a multiview drawing with dimensions and a pictorial view. It was developed on a CAD system using the Auto-CAD software. Chapters 15 and 16 will show you in detail how this is done.

CAD systems have several drawing aids that help make multiview drawings easier to accomplish. Lettering and line technique, for example, require much less attention than they do on manual drawings. The views can be placed accurately with little trouble. However, it is still important to develop a strong basic understanding of multiview drawing because the basic principles apply to both manual and CAD drawings. Because multiview drawing is the basic building block of detailed working drawings, it is considered by many to be the single most important aspect of learning technical drawing. What you learn about multiview drawing through hands-on experiences in both manual drafting and CAD drafting can be instrumental in preparing you for a drafting career.

ORTHOGRAPHIC PROJECTION

Earlier we said that multiview drawing is the exact representation of two or more views of an object on a plane. These views are developed through the principles (rules) of orthographic projection. *Ortho* means "straight or at right angles." *Graphic* means "written or drawn." *Projection* comes from two Latin words: *pro,* meaning "forward," and *jacere,* meaning "to throw." Thus, orthographic projection literally means "thrown forward, drawn at right angles." **Orthographic projection** is the method of representing the exact form of an object in two or more views on planes, usually at right angles to each other, by lines drawn perpendicular from the object to the planes.

Angles of Projection

On a technical drawing, a *plane* is an imaginary flat surface that has no thickness. Orthographic projection involves the use of three planes. They are the **vertical plane** (up-and-down), the **horizontal plane** (side-to-side), and the **profile plane** (side view). These are shown in Fig. 5-10. A view of an object is projected and drawn upon each plane. Notice that the vertical and horizontal planes divide space into four **quadrants** (quarters of a circle). In orthographic projection, quadrants are usually called *angles*. Thus, we get the names **first-angle projection** and **third-angle projection.** First-angle projection is used in European countries. Third-angle projection is used in the United States and Canada. Second- and fourth-angle projection are not used.

First-Angle Projection

Figure 5-11 shows an object within the three planes of the first quadrant for developing a three-view drawing in first-angle projection. The front view is projected to the vertical plane. The top view is projected to the horizontal plane. The left-side view is projected to the profile plane. *In first-angle projection, the front view is located above the top view.* The left-side view is to the right of the front view (Fig. 5-12).

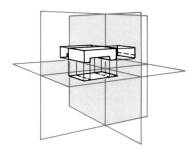

Fig. 5-11 *The position of the three planes used in first-angle projection.*

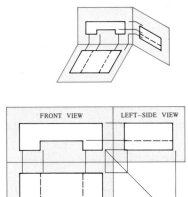

Fig. 5-12 *Three views in first-angle projection.*

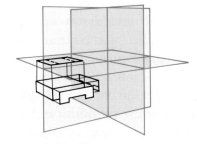

Fig. 5-13 *The position of the three planes used in third-angle projection.*

Third-Angle Projection

Third-angle projection uses the same basic principles as first-angle projection. The main difference is in the relative positions of the three planes. Figure 5-13 shows the same object placed within the third quadrant for developing a three-view drawing in third-angle projection. In this case the front view is projected to the vertical plane. The top view is projected to the horizontal plane. The right-side view is projected to the profile plane. The horizontal and profile planes are rotated so that all the views lie in a single plane. *In third-angle projection, the top view is above the front view.* The right-side view is to the right side of the front view (Fig. 5-14).

THE GLASS BOX

In each case, the three views of an object have been developed by using imaginary transparent (see-through) planes. The views are projected onto these planes. We mentioned earlier that most objects have six sides. Therefore, six views may result. To explain the theory of projecting all six views, we will use an imaginary glass box.

Imagine a transparent glass box around the bookend shown in Fig. 5-15A. Figure 5-15B shows the glass box partially opened with the six views labeled. When the box is fully opened up into one plane (Fig. 5-16), the views are in their relative positions as if they had been drawn on paper.

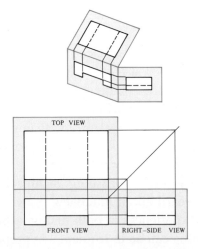

Fig. 5-14 *Three views in third-angle projection.*

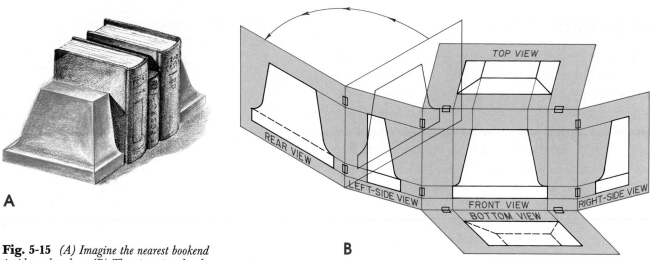

Fig. 5-15 *(A) Imagine the nearest bookend inside a glass box. (B) Then imagine the glass box opening as shown here.*

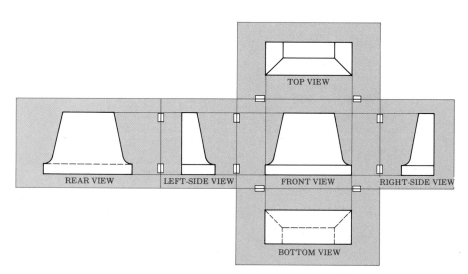

Fig. 5-16 *When the glass box has been completely opened, you can see all six sides of the object clearly.*

These views are arranged according to proper order for the six views. Notice that the rear view is located to the left of the left-side view.

Also notice that some views give the same information found in other views. The views may also be mirror images of one another. Thus, it is not necessary to show all six views for a complete description of the object. The three normal views, as ordinarily drawn, are shown in Fig. 5-17.

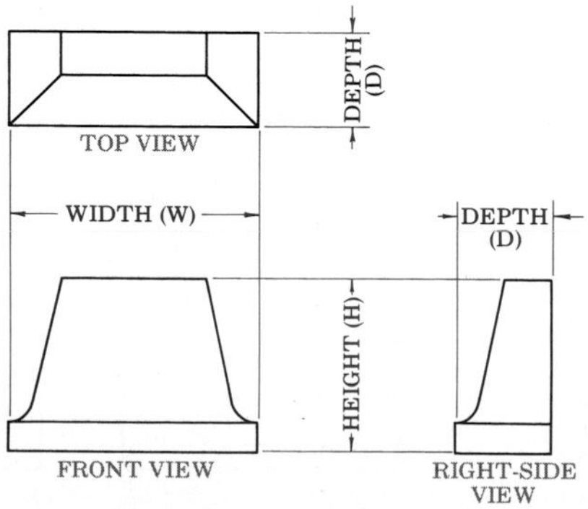

TECHNIQUES FOR SPECIAL LINES AND SURFACES

To describe an object fully, you must show every feature in each view, whether or not it can ordinarily be seen. You must also include other lines that are not actually part of the object to clarify relationships and positions in the drawing. To reduce confusion, special line symbols (linetypes) are used to differentiate between object lines and lines that have other special meanings.

TOP VIEW

DEPTH (D)

WIDTH (W)

DEPTH (D)

FRONT VIEW

HEIGHT (H)

RIGHT-SIDE VIEW

Fig. 5-17 *The front, top, and right-side views.*

Hidden Lines

It is necessary to describe every part of an object. Therefore, everything must be represented in each view, whether or not it can be seen. Both interior (inside) and exterior (outside) features are projected in the same way. Parts that cannot be seen in the views are drawn with *hidden lines* that are made up of short dashes (Fig. 5-18). Notice that the first dash of a hidden line touches the line where it starts (Fig. 5-18A). If a hidden line is a continuation of a visible line, space is left between the visible line and the first dash of the hidden line (B). If the hidden lines show corners, the dashes touch the corners (C).

Dashes for hidden arcs (Fig. 5-19A) start and end at the tangent points. When a hidden arc is tangent to a visible line, leave a space (B). When a hidden line and a visible line project at the same place, show the visible line (C). When a centerline and a hidden line project at the same place (Fig. 5-20A), draw the hidden line. When a hidden line crosses a visible line (B), do not cross the visible line with a dash. When hidden lines cross (C), the nearest hidden line has the "right of way." Draw the nearest hidden line through a space in the farther hidden line.

Centerlines

Centerlines are special lines used to locate views and dimensions. (See the alphabet of lines, Fig. 3-40.) Primary centerlines, marked *P* in Fig. 5-21, locate the center on symmetrical views in which one part is a mirror image of another. Primary centerlines are used as major locating lines to help in making the views. They are also used as base lines for dimensioning. Secondary centerlines, marked *S* in Fig. 5-21, are used for drawing details of a part.

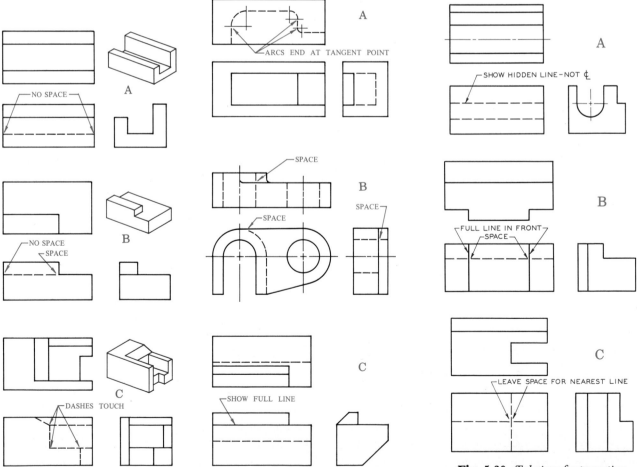

Fig. 5-18 *Hidden lines.*   **Fig. 5-19** *Hidden arcs.*   **Fig. 5-20** *Technique for presenting hidden and visible lines.*

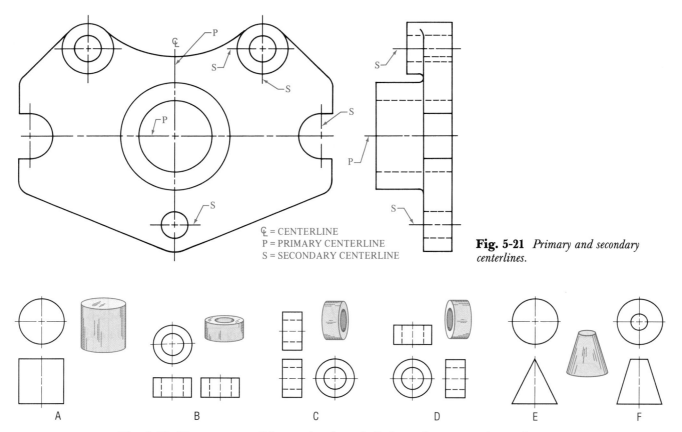

Fig. 5-21 *Primary and secondary centerlines.*

Ⴛ = CENTERLINE
P = PRIMARY CENTERLINE
S = SECONDARY CENTERLINE

Fig. 5-22 *The appearance of the curved surfaces of cylinders and cones in multiview drawings.*

Primary centerlines are the first lines to be drawn. The views are developed from them. Note that centerlines represent the axes of cylinders in the side view. The centers of circles and arcs are located first so that measurements can be made from them to locate the lines on the various views. As you may recall from the previous section, when a hidden line falls on a centerline, the hidden line is drawn (Fig. 5-20A). When a hidden line falls on a visible line, draw the visible line.

Curved Surfaces

Some curved surfaces, such as cylinders and cones, do not show as curves in all views. This is illustrated in Fig. 5-22. A cylinder with its axis (centerline) perpendicular to a plane will show as a circle on that plane. It will show as a rectangle on the other two planes. Three views of a cylinder in different positions are shown in Fig. 5-22B, C, and D. The holes may be thought of as **negative cylinders.** (In mathematics, *negative* means an

amount less than zero. A hole is a "nothing" cylinder. However, it has size. Thus, in a sense, it is negative.)

A cone appears as a circle in one view. It appears as a triangle in the other, as shown in Fig. 5-22E. One view of a frustum of a cone appears as two circles (F). In the top view, the conical surface is represented by the space between the two circles.

Cylinders, cones, and frustums of cones have single curved surfaces. They are represented by circles in one view and straight lines in the others. The handles in Fig. 5-23A have double curved surfaces that are represented by curves in both views. The ball handle has **spherical** (ball-shaped) ends. Thus, both views of the ends are circles because a sphere appears as a circle when viewed from any direction. The slotted link in Fig. 5-23B and C is an example of tangent plane and curved surfaces. The rounded ends are tangent to the sides of the link, and the ends of the slot are tangent to the sides. Therefore, the surfaces are smooth. There is no line of separation.

CHAPTER 5
PROBLEMS

The small CAD symbol next to each problem indicates that a problem is appropriate for assignment as a computer-aided drafting problem and suggests a level of difficulty (one, two, or three) for the problem.

The following problems provide practice in representing objects by views to help you gain a thorough understanding of the theory of shape description. Study chapters 3 and 5 carefully before you attempt these problems. Views and pictures are given for some of the problems to assist you in visualizing the object. In others, two views are given, from which you must work out the third view. In still others, pictures are given, from which you must determine which views are necessary, plan and arrange the views in the working space, and then work out the views.

Sheet layouts and title blocks are covered at the end of Chapter 3. Chapter 5 describes a systematic method for placing the views on a drawing sheet. Review this material carefully before beginning to solve the drawing problems.

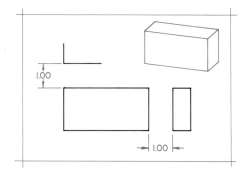

Fig. 5-37 *CAD1* *Sanding block. Draw the two views shown and complete the third (top) view. Do not draw the pictorial view. The block is .75″ × 1.75″ × 3.50″. Scale: Full size.*

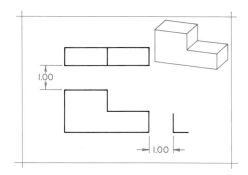

Fig. 5-38 *CAD1* *Step block. Draw the front and top views, but not the pictorial view. Complete the right-side view in its proper location. The step block is .75″ × 1.75″ × 3.50″. The notch is .88″ × 1.75″. Scale: Full size.*

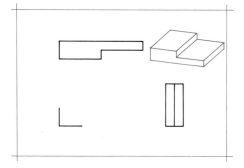

Fig. 5-39 *CAD1* *Half lap. Draw the top and right-side views, but not the pictorial view. Complete the front view in its proper shape and location. The half lap is .75″ × 1.75″ × 3.50″. The notch is .38″ × 1.75″. Scale: Full size.*

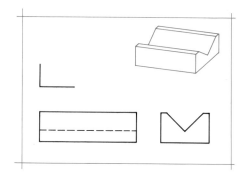

Fig. 5-40 *CAD1* *V-block. Draw the front and right-side views as shown. Complete the top view in its proper location. Do not draw the pictorial view. The overall sizes are 1.25″ × 2.00″ × 4.00″. Scale: Full size.*

PROBLEMS

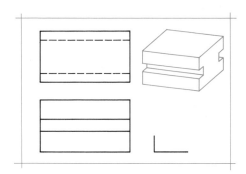

Fig. 5-41 *CAD1 Slide. Draw the front and top views as shown. Complete the right-side view in its proper location. Do not draw the pictorial view. The overall sizes are 2.12″ square × 3.75″. The slots are .38″ deep and .50″ wide. Scale: Full size.*

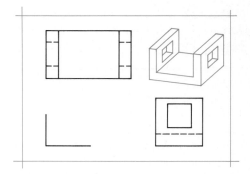

Fig. 5-42 *CAD1 Rod support. Draw the top and right-side views, but not the pictorial. Complete the front view. The overall sizes are 2.00″ square × 3.50″. Bottom and ends are .50″ thick. The holes are 1.00″ square and are centered on the upper portions. Scale: Full size.*

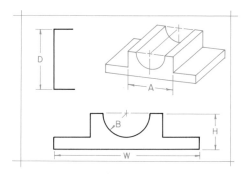

Fig. 5-43 *CAD1 Cradle. Draw the front view, but not the pictorial. Complete the top view in the proper shape and location. Height = 2.00″, width = 6.00″, depth = 2.25″. Base = .50″ thick. A = 3.00″, B = 1.00″. Scale: Full size.*

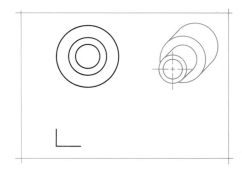

Fig. 5-44 *CAD1 Spacer. Draw the top view, but not the pictorial. Complete the front view. Base = 2.50″ × 1″. Top = Ø1.50″ × .75″ Hole = Ø1.00″. A vertical sheet will permit a larger scale. Scale: As assigned.*

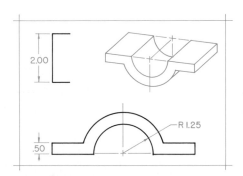

Fig. 5-45 *CAD1 Strap. Draw the front view as shown. Complete the top view in the proper shape and location. Do not draw the pictorial view. Overall width is 6.00″. Scale: Full size.*

PROBLEMS

Figs. 5-46 through 5-57 Two- and three-view problems. Each problem has one view missing. Draw the view or views given and complete the remaining view in the proper shape and location. Scale: Full size or as assigned. Do not dimension unless instructed to do so.

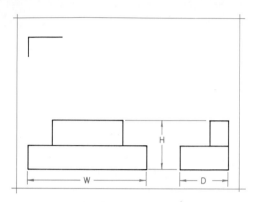

Fig. 5-46 **CAD1** *Stop. W = 5.00″, H = 2.00″, D = 2.00″, base = 1.00″ × 2.00″ × 5.00″, top = .75″ × 1.00″ × 3.00″.*

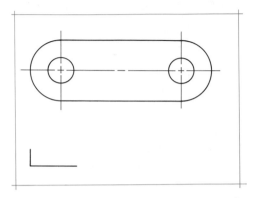

Fig. 5-47 **CAD1** *Link. W = 7.50″, D = 2.50″, H = 1.25″, holes = Ø1.06″.*

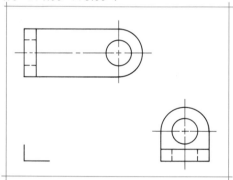

Fig. 5-48 **CAD1** *Angle bracket. W = 5.00″, D = 2.00″, H = 2.25″, material thickness = .50″, holes = Ø1.00″.*

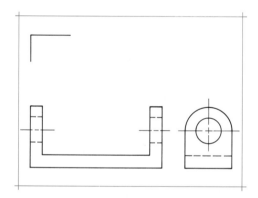

Fig. 5-49 **CAD1** *Saddle. W = 5.50″, D = 2.00″, H = 2.25″, material thickness = .50″, hole = Ø1.00.*

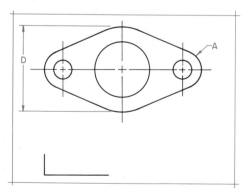

Fig. 5-50 **CAD1** *Spacer. W = 6.50″, D = 3.25″, thickness = 1.00″, holes = Ø2.38″, 0.75″, A = .75″*

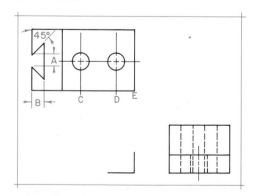

Fig. 5-51 **CAD1** *Dovetail slide. W = 4.25″, D = 2.50″, H = 2.00″, base thickness = .75″, upright thickness = 1.25″, holes = Ø.62″, A = .50″, B = .50″, CD = 1.50″, DE = .75″.*

PROBLEMS

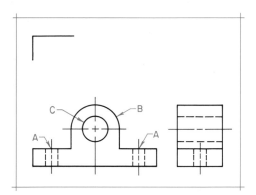

Fig. 5-52 *CAD1* *Rod guide. W = 5.12″, D = 1.88″, H = 2.50″, C = Ø1.00″, A = Ø.50″ 3.62″ apart, base thickness = .75″, B = R1.00″.*

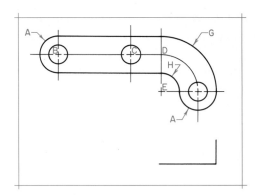

Fig. 5-53 *CAD1* *Hinge plate. A = R.75″, BC = 3.00″, CD = 1.25″, DE = 1.50″, EF = 1.50″, G = R2.25″, H = R.75″, holes = Ø.75″, thickness = 1.00″.*

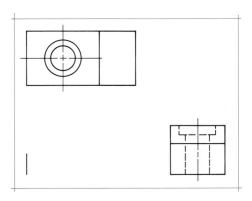

Fig. 5-54 *CAD1* *Offset lug. W = 4.50″, D = 2.25″, H = 2.00″, notch = .75″ × 1.50″, hole = Ø1.00″, counterbore = Ø1.50″ × .38″ deep.*

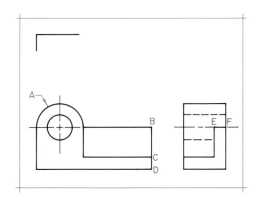

Fig. 5-55 *CAD1* *Pin holder. W = 4.75″, D = 1.75″, H = 2.75″, hole = Ø1.00″, A = R1.00″, BC = 1.25″, BD = 1.75″, EF = .50″.*

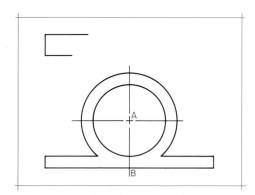

Fig. 5-56 *CAD1* *Ring. Base = .50″ × .88″ × 7.00″ outer diameter, 3.00″ inner diameter, AB = 2.00″.*

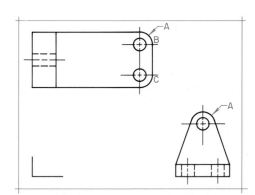

Fig. 5-57 *CAD1* *Bracket. L = 5.00″, W = 2.25″, H = 2.75″, base thickness = .50″, upright = 1.00″, A = R.50″, BC = 1.25″, holes = Ø.50″.*

PROBLEMS

Figs. 5-58 through 5-66 Use instruments to create two- or three-view drawings of these objects. Scale: As assigned. Do not dimension unless instructed to do so. Include all centerlines.

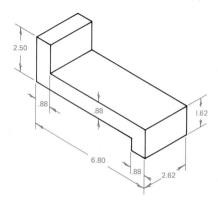

Fig. 5-58 _CAD1_ _Stop._

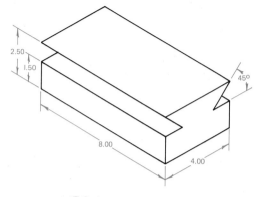

Fig. 5-59 _CAD1_ _Dovetail slide._

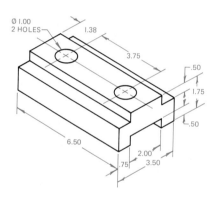

Fig. 5-60 _CAD1_ _Slide._

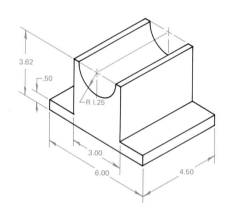

Fig. 5-61 _CAD1_ _Cradle block._

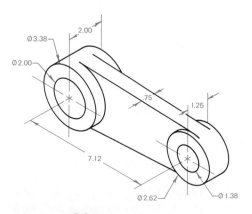

Fig. 5-62 _CAD1_ _Pivot arm._

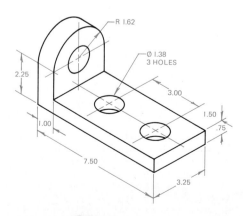

Fig. 5-63 _CAD1_ _Base._

PROBLEMS

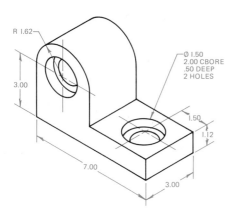

Fig. 5-64 *CAD2* Shaft support.

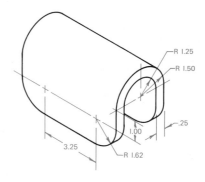

Fig. 5-65 *CAD2* Edge protector.

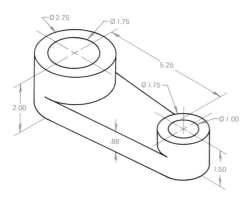

Fig. 5-66 *CAD2* Swivel arm.

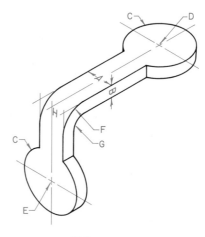

Fig. 5-67 *CAD2* Offset ring. A = 1.50″, B = .62″, C = R1.62″, D = Ø1.00″, E = Ø1.12″, F = R1.50″, G = R.88″, EH = 4.00″, HD = 6.75″.

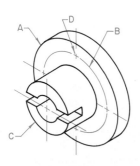

Fig. 5-68 *CAD2* Socket. All dimensions are in millimeters. A = Ø50.5 mm × 7 mm thick, B = 38 mm, C = 25.2 × 17 mm long with Ø13 mm hole through, slots = 4.5 mm wide × 8 mm deep, D = Ø6 mm, 4 holes equally spaced. Scale: As assigned.

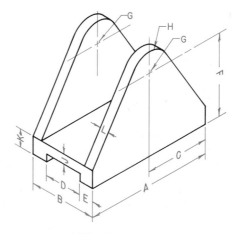

Fig. 5-69 *CAD2* Shaft guide. A = 7.25″, B = 3.88″, C = 3.62″, D = 2.25″, E = .80″, F = 4.50″, G = Ø1.00″, 2 holes, H = R1.25″, J = .50″, K = 1.00″, L = .50″. Scale: As assigned.

PROBLEMS

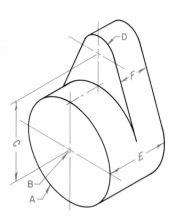

Fig. 5-70 *CAD2* *Cam. A = Ø2.62″, B = Ø1.25″, 1.75″ counterbore .25″ deep on both ends, C = 2.12″, D = R.56″, E = 1.75″, F = .88″. Scale: As assigned.*

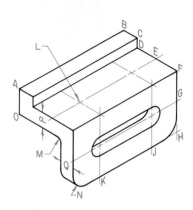

Fig. 5-71 *CAD2* *Adjustable stop. AB = 10.00″, BC = 1.38″, CD = .75″, DE = 1.88″, EF = 2.00″, FG = 2.50″, GH = 2.50″, FH = 5.00″, HJ = 2.50″, JK = 5.00″, L = Ø1.00″, 2 holes, M = R.50″, N = R1.25″, AO = 2.00″, P = 1.25″, Q = 1.25″, slot = 1.50″ wide. Scale: As assigned.*

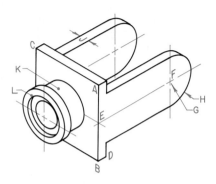

Fig. 5-72 *CAD2* *Camera swivel base. AB = 40 mm, AC = 38 mm, BD = 5.5 mm BE = 20 mm EF = 45 mm, H = R12 mm, G = Ø10 mm, J = 6 mm. Boss: K = Ø24 mm × 9 mm long, L = Ø28 mm × 4 mm long, hole = Ø12 mm × 14 mm deep, counterbore = 18 mm × 3 mm deep. Scale: As assigned.*

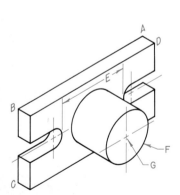

Fig. 5-73 *CAD2* *Pipe support. AB = 8.00″, BC = 4.00″, AD = .75″, E = 3.88″, F = Ø2.75″ × 2.25″ long, G = Ø1.12″ hole through, slots = 1.00″ wide. Scale: As assigned.*

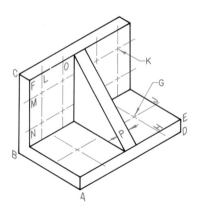

Fig. 5-74 *CAD2* *Angle plate. AB = 6.00″, BC = 6.50″, AD = 9.75″, DE = 1.00″, CF = 1.00″, G = Ø.75″, 2 holes, EH = 2.00″, EJ = 2.50″, FL = 1.12″, LO = 2.00″, FM = 1.50″, MN = 2.50″, P = 1.12″, K = Ø.50″, 8 holes. Scale: As assigned.*

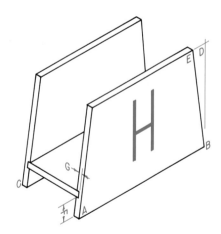

Fig. 5-75 *CAD2* *Letter holder. Draw all necessary views. All material is .20″ thick (plastic or wood). AB = 4.00″, AC = 2.00″, BD = 3.00″, DE = .38″, F = .38″, G = .10″. Add a design to the front view. See Chapter 2 for layout of block-style lettering. Dimension only if instructed to do so. Scale: Full size or as assigned.*

PROBLEMS

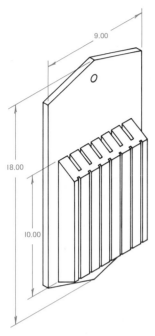

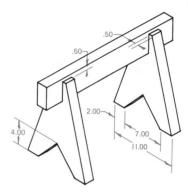

Fig. 5-78 *CAD2* *Mini sawhorse.
Top rail is 2.00″ × 4.00″ × 24.00″.
Legs are cut from 2.00″ × 12.00″
stock, 14.50″ long. Dimension only if
instructed to do so. Scale: ¹/₄″ = 1″ or
as assigned.*

Fig. 5-77 *CAD2* *Desktop book rack. Draw
all necessary views. Material is laminated
wood, plastic, or aluminum. AB = 8.00″, BC
= 9.50″, CD = 2.00″, DE = .38″, EF =
6.00″, G = R1.25″. All bends are 90°. Scale:
As assigned.*

Fig. 5-76 *CAD2* *Knife rack. Draw
all necessary views. Back is .50″ ×
9.00″ × 18.00″. Front is 1.50″ ×
7.00″ × 10.00″ with 30°-angle bevels
on each end. Slots for knife blades are
.12″ wide × 1.00″ deep. Grooves on
front are .12″ wide × .12″ deep.
Estimate all sizes not given. Redesign as
desired. Scale: Half size or as assigned.*

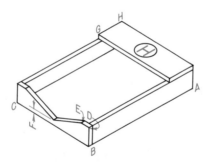

Fig. 5-79 *CAD2* *Note-paper box.
All stock is .25″ thick. AB = 6.62″,
BC = 4.62″, BD = 1.00″, DE =
.50″, F = .25″, GH = 1.50″. Draw
all necessary views. Do not dimension
unless instructed to do so. Initial inlay is
optional. Redesign as desired. Scale: Full
size or as assigned.*

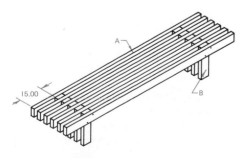

Fig. 5-80 *CAD2* *Garden bench. A =
2.00″ × 4.00″ × 6′-0, B = 2.00″ × 4.00″
× 1′-4. Draw front, top, and right-side views
of the bench. Use six or more top rails. Scale:
1″ = 1′-0 or as assigned.*

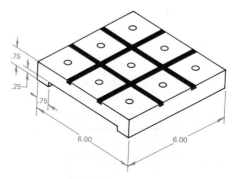

Fig. 5-81 *CAD2* *Tic-tac-toe board.
Material: hardwood with black Plexiglas®
inlays. Inlays are .12″ × .25″ and are
located 2.00″ on center. Holes are Ø.31″ ×
.25″ deep and are centered within squares.
Game board is designed to use marbles. Scale:
Full size or as assigned.*

PROBLEMS

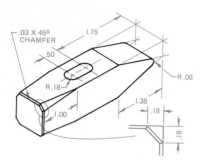

Fig. 5-82 **CAD2** *Hammer head. Draw all necessary views. Overall sizes are .88″ square × 3.50″ long. Do not dimension unless instructed to do so. Scale: 2.00″ = 1.00″ (2:1) or as assigned.*

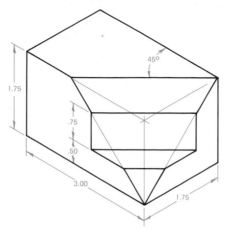

Fig. 5-83 **CAD2** *Draw three views of the corner lock.*

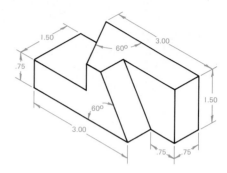

Fig. 5-84 **CAD2** *Draw three views of the wedge spacer.*

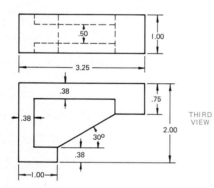

Fig. 5-85 **CAD2** *Draw three views of the bracket.*

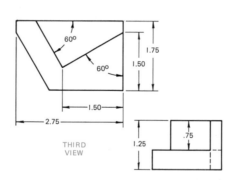

Fig. 5-86 **CAD2** *Draw three views of the locator.*

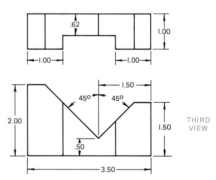

Fig. 5-87 **CAD2** *Draw three views of the V-slide.*

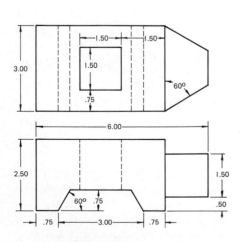

Fig. 5-88 **CAD2** *Draw three views of the locating support.*

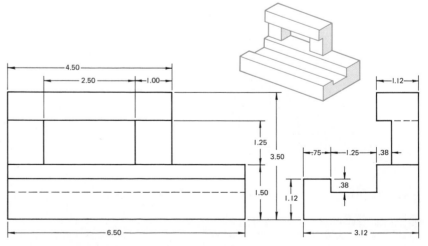

Fig. 5-89 **CAD3** *Draw three views of the support guide.*

PROBLEMS

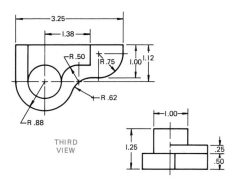

THIRD
VIEW

Fig. 5-90 *CAD3* *Draw three views of the pivot.*

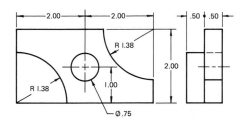

Fig. 5-92 *CAD2* *Draw three views of the locating plate.*

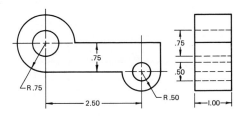

Fig. 5-91 *CAD3* *Draw three views of the link.*

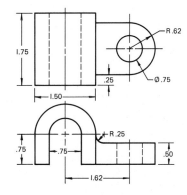

Fig. 5-93 *CAD3* *Draw three views of the holder.*

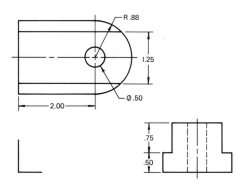

Fig. 5-94 *CAD3* *Draw three views of the slide.*

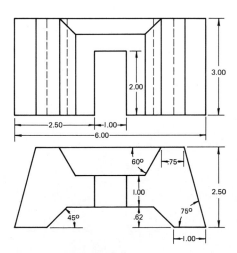

Fig. 5-95 *CAD3* *Draw three views of the base.*

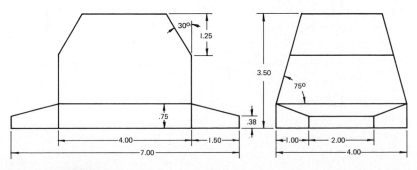

Fig. 5-96 *CAD3* *Draw three views of the base.*

PROBLEMS

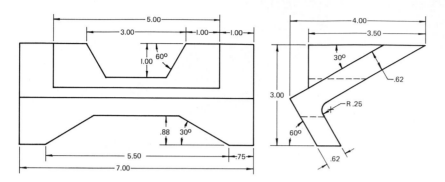

Fig. 5-97 *CAD3 Draw three views of the separator.*

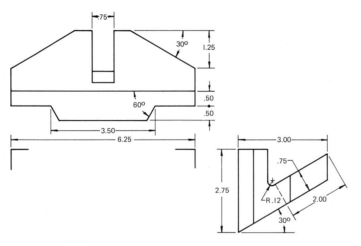

Fig. 5-98 *CAD3 Draw three views of the secondary guide lug.*

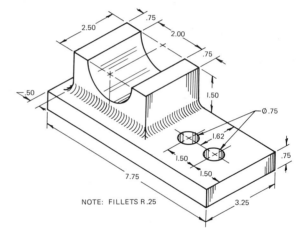

Fig. 5-99 *CAD3 Draw three views of the horizontal guide.*

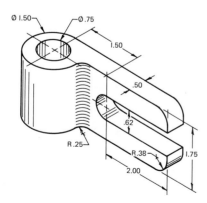

Fig. 5-100 *CAD3 Draw two views of the vertical stop.*

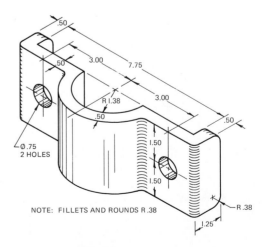

Fig. 5-101 *CAD3 Draw three views of the side cap.*

PROBLEMS

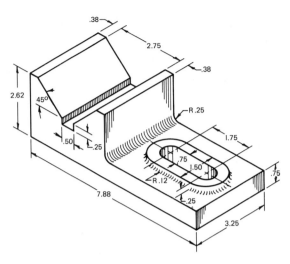

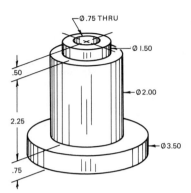

Fig. 5-103 *CAD3 Draw two complete views of the cast-iron collar.*

Fig. 5-102 *CAD3 Draw three views of the V-block base.*

Figs. 5-104 through 5-107 *Draw two or three views of Figs. 5-104 through 5-107 as assigned. Scale: optional. Do not dimension unless instructed to do so.*

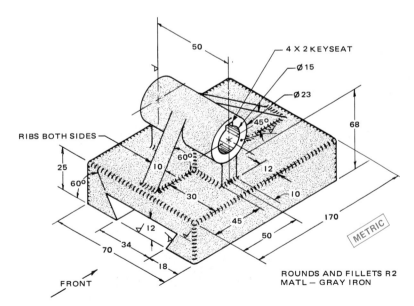

Fig. 5-104 *CAD3 Cross slide.*

PROBLEMS

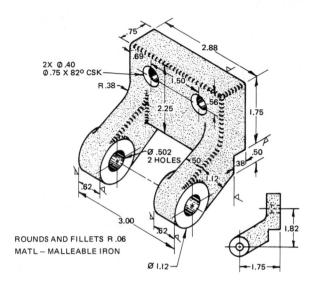

.75
2.88
2X Ø.40
Ø.75 X 82° CSK
.69
R.38
1.50
.56
2.25
1.75
.50
Ø.502
2 HOLES
.50
.38
.62
1.12
3.00
.62
Ø 1.12
1.82
1.75

ROUNDS AND FILLETS R .06
MATL — MALLEABLE IRON

Fig. 5-105 CAD3 *Shaft support.*

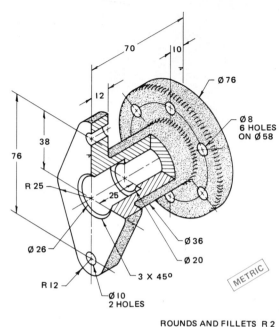

70
10
Ø 76
12
Ø 8
6 HOLES
ON Ø 58
38
76
R 25
25
Ø 36
Ø 26
Ø 20
3 X 45°
R 12
METRIC
Ø 10
2 HOLES

ROUNDS AND FILLETS R 2
MATL — CI

Fig. 5-106 CAD3 *Flanged coupling.*

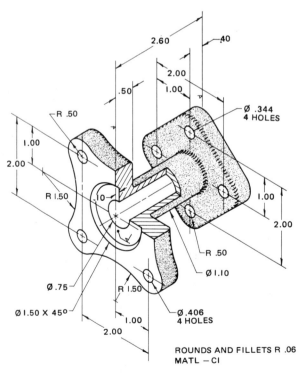

2.60
.40
2.00
.50
1.00
R .50
Ø .344
4 HOLES
1.00
2.00
R 1.50
.10
1.00
2.00
R .50
Ø.75
R 1.50
Ø 1.10
Ø 1.50 X 45°
Ø.406
4 HOLES
1.00
2.00

ROUNDS AND FILLETS R .06
MATL — CI

Fig. 5-107 CAD3 *Adapter.*

SCHOOL-TO-WORK

Solving Real-World Problems

Figs. 5-108 through 5-109. *SCHOOL-TO-WORK problems are designed to challenge a student's ability to apply skills learned within this text. Be creative and have fun!*

Fig. 5-108 *Design a tic-tac-toe board similar to the one shown in Fig. 5-81 to use golf tees rather than marbles. Devise a convenient way to store the golf tees with the board. Draw all necessary views.*

Fig. 5-109 *Design a knife rack similar to the one shown in Fig. 5-76. Design it to set on a counter or table rather than hang on the wall. Draw all necessary views and specify the materials from which it is to be made.* **Remember! Simplicity is the key to good design.**

DIMENSIONING

OBJECTIVES

Upon completion of Chapter 6, you will be able to:

○ Apply measurements, notes, and symbols to orthographic views on a technical drawing.
○ Use ANSI standards for dimensioning and notes.
○ Differentiate between size dimensions and location dimensions.
○ Dimension a technical drawing using either SI (metric) units or U.S. customary units.
○ Determine appropriate sizes for precision fits between matching parts.
○ Specify geometric tolerances using symbols and notes.
○ Designate appropriate surface textures.

VOCABULARY

In this chapter, you will learn the meanings of the following terms:

- aligned system
- assembly drawing
- basic hole system
- basic shaft system
- bilateral tolerances
- datums
- detail drawing
- dimension line
- dimensioning
- dual dimensioning system
- extension lines
- finish mark
- geometric dimensioning and tolerancing
- leader
- outline view
- tolerance
- unidirectional system
- unilateral tolerances

To describe an object completely, a drafter needs to define both the shape and the size of the object. Most of this book deals with the ways of describing shape. This chapter, however, discusses how to show the size of the objects that you draw. It is very important to understand clearly the rules and principles of size description. After all, a machinist cannot make parts correctly unless all the sizes on the drawing are accurate.

Another name for size description is **dimensioning.** *Dimensions* (sizes) are measured in either U.S. customary or metric (SI) units. In the customary system, measurements are given in feet and inches, in inches and fractions, or in decimal divisions of inches. Decimal divisions and metric units are now most commonly used throughout industry and are used exclusively in ANSI Y14.5M (the drafting standard on dimensioning). Refer to Chapter 3 for more information about U.S. customary and metric units.

Metric units, such as meters and millimeters, are often used for engineering drawings. For example, civil engineering drawings may be dimensioned in meters. Architects may use both meters and millimeters, depending on the size of the item they are drawing.

Notes and symbols that show the kind of finish, materials, and other information needed to make a part are also part of dimensioning. A complete *working drawing* (the drawing or set of drawings from which the part is manufactured) includes shape description, measurements, notes, and symbols (Fig. 6-1).

Dimensions on working drawings must be as precise as necessary to allow the manufacturer to create the part or object. When dimensions must be precise, they are given in hundredths, thousandths, or ten-thousandths of an inch. If the metric system is being used, the measurements may be in tenths, hundredths, or even thousandths of a millimeter.

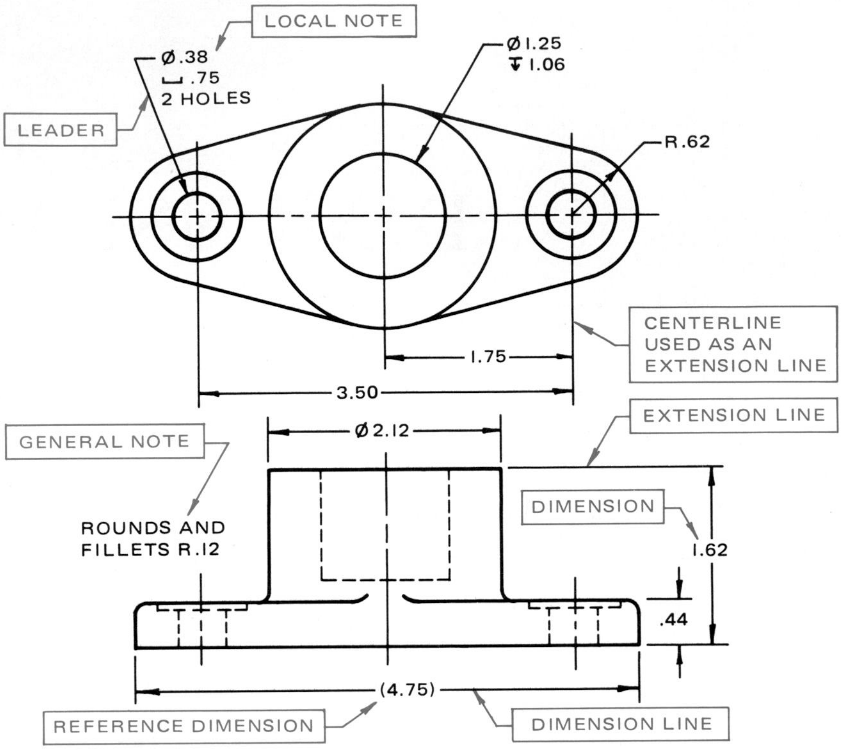

Fig. 6-1 *Dimensioning includes measurements, notes, and symbols.*

LINES AND SYMBOLS FOR DIMENSIONING

The views on drawings describe the shape of an object. In theory, size could be found by measuring the drawing and applying a scale. In reality, though, this is not practical, even when the views are drawn to actual or full size. The measuring simply takes too much time. More important, it is impossible to measure a drawing accurately enough for many interchangeable parts that must fit closely together. To ensure accuracy and efficiency, the drafter adds size information to the drawing using a system of lines, symbols, and numerical values.

Lines and symbols are used on drawings to show where the dimensions apply (Fig. 6-2). Professional and trade associations, engineering societies, and certain industries have agreed upon the symbols so that the people who use the drawings can recognize their meaning. The latest standards information on drawings and symbols can be found in publications from the American National Standards Institute (ANSI), the Society of Automotive Engineers (SAE), the Military Standards, and the International Standards Organization (ISO).

Dimension Lines

A **dimension line** is a thin line that shows where a measurement begins and where it ends (Fig. 6-3). It is also used to show the size of an angle. The dimension line should have a break in it for the dimension numbers. To keep the numbers from getting crowded, draw dimension lines at least .38 in. from the lines of the drawing and at least .25 in. from each other. On metric drawings, dimension lines should be at least 10 mm from the lines of the drawing and 6 mm from each other.

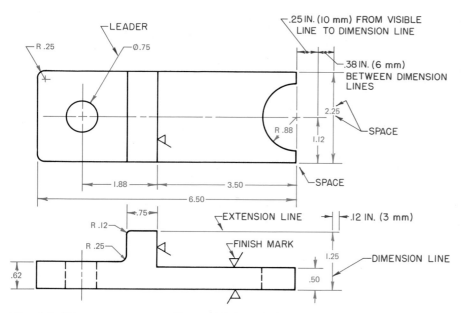

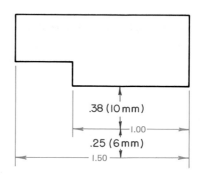

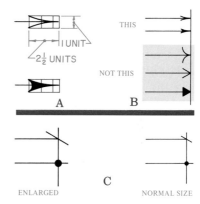

Fig. 6-2 *Dimensioning consists of lines, symbols, and placement techniques.*

Fig. 6-3 *Dimension lines must be spaced to provide clarity.*

Fig. 6-4 *Arrowheads.*

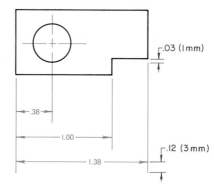

Fig. 6-5 *Extension lines. A centerline may be used as an extension line.*

Arrowheads

Arrowheads are placed at the ends of dimension lines to show where a dimension begins and ends (Fig. 6-3). They are also used at the end of a leader (Figs. 6-2 and 6-8) to show where a note or dimension applies to a drawing.

Arrowheads can be open or solid. Their shapes are shown enlarged in Fig. 6-4A and reduced to actual size in (B). Draw arrowheads carefully. In any one drawing, they should all be the same size and shape. However, in a small space, you may have to vary the size somewhat.

Some industries use other means to point out the endpoint of a dimension line or leader. Figure 6-4C shows some examples of these. These methods do the same job as arrowheads. For example, slash marks are often used instead of arrowheads in architectural drafting. For most working drawings, however, the arrowheads shown in (A) and (B) are preferred.

Extension Lines

Extension lines are thin lines that extend the lines or edges of the views. They are used to locate center points and to provide space for dimension lines (Fig. 6-5). Since extension lines are not part of the views, they should not touch the outline. Start the extension line about .03 to .06 in. (1 to 1.5 mm) from the part, and extend it about .12 in. beyond the last dimension line.

Numerals and Notes

Numerals and notes have to be easy to read, so you must create them carefully. Do not make them unnecessarily large, however. Capital letters are preferred on most drawings. As you discovered in Chapter 2, letters and numerals can be either vertical or inclined. However, vertical letters are becoming more commonly used.

In general, make numerals about .12 in. (3 mm) high. When you use fractions, make the fractions about .25 in. (6 mm) and the fraction numerals about .09 in. (2.5 mm) high. Always draw the fraction bar (division line) in line with the dimension, not at an angle. Also, allow some space between the fraction numerals and the fraction bar. Refer to Fig. 2-9 for specific heights and spaces. Draw light guidelines for figures and fractions using a lettering triangle or an Ames lettering instrument. Figure 6-6 shows the procedure for using guidelines for numerals and notes.

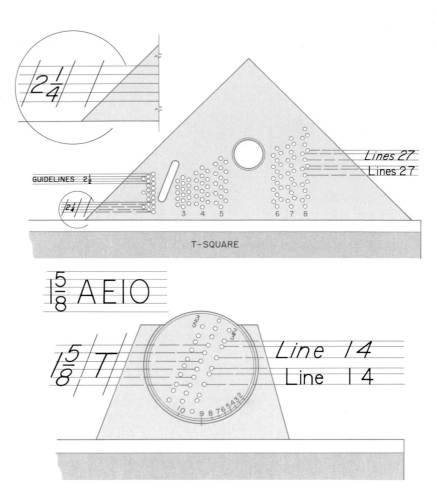

Fig. 6-6 *Guidelines for letters, whole numbers, and fractions.*

Sometimes drawings are made to be microfilmed or to be reduced photographically and used at a smaller size. When this is the case, make the numerals larger and with heavier strokes so that they will be clear and easy to read when reduced.

The Finish Mark

To dimension a drawing correctly, drafters must know the correct symbols to use as well as the principles of dimensioning. In most cases, drafters must also know the shop processes that are used to build or make the products they draw. Sometimes drafters include symbols on the drawing to show which processes are needed.

The **finish mark,** or surface-texture symbol (Fig. 6-7), shows that a surface is to be machined (finished). The old symbol form $\vee$ is still used

occasionally, but it is being replaced. An even older symbol form, f may be found on some very old drawings. The standard symbol now in general use is shown in Fig. 6-7. The point of the symbol should touch the edge view of the surface to be finished. Modified forms of this symbol can be used to show that allowance for machining is needed, that a certain surface condition is needed, and other information. These symbols are described later in this chapter and in ANSI B46.1, *Surface Texture.*

Leaders

A **leader** is a thin line drawn from a note or dimension to the place where it applies (Fig. 6-8). Always draw leaders at an angle to the horizontal. An angle of 60° is preferred, but 45°, 30°, or other

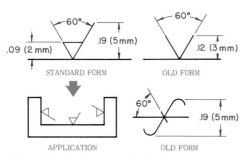

Fig. 6-7 *The finish mark tells which surfaces are to be machined.*

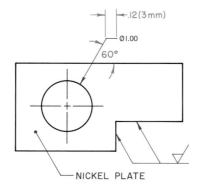

Fig. 6-8 *Leaders point to the place where a note or dimension applies.*

angles may be used. A leader starts with a dash, or short horizontal line. This line should be about .12 in. (3 mm) long, but it may be longer if needed. It generally ends with an arrowhead. However, a dot is used if the leader is pointing to a surface rather than to an edge (Fig. 6-8).

Other rules for drawing leaders include:

- When a number of leaders must be drawn close together, drawing them parallel to each other.
- Draw a leader to a circle or arc so that the arrowhead points to its center.
- Do not draw long leaders.
- Avoid drawing leaders horizontally, vertically, or at a small angle.
- Do not draw leaders parallel to dimension, extension, or section lines.

SCALE OF A DRAWING

Scales used in making drawings are described in Chapter 3. The scale used should be given in or near the title. If a drawing has views of more than

one part and different scales are used, the scale should be given close to the views. Scales are stated as full or full-size, 1:1; half-size, 1:2; and so forth. If enlarged views are used, the scale is shown as 2 times full size, 2:1; 5 times full size, 5:1; and so forth.

The scales used on metric drawings are based on divisions of 10. Scales such as 100:1, 1:50, and 1:100 are examples.

UNITS AND PARTS OF UNITS

When you use the customary system, give the measurements in feet and decimals of a foot, feet and inches, inches and fractions of an inch, or inches and decimals of an inch. When customary dimension are in inches, omit the inch symbol("). When feet and inches are used, standard practice is to show the symbol for feet but *not* for inches: 7′-5, 7′-0, and so forth.

Feet and inches are used with common fractions, such as ½, ¼, ⅛, and so on, when particular accuracy is not necessary. They are used, for example, when measurements do not have to be closer than ±¹⁄₆₄ in.

Sometimes, parts must fit together with extreme accuracy. In that case, the machinist must work within specified limits. If the measurements are customary, the decimal inch is used. This is called *decimal dimensioning.* Such dimensions are used between finished surfaces, center distances, and pieces that must be held in a definite, accurate relationship to each other.

With customary measures, you may use decimals to two places where limits of ±.01 in. are close enough (Fig. 6-9A, B, and C). Use decimals to three or more places where limits smaller than ±.01 in. are required, as in Fig. 6-10A and B. For two-place decimals, fiftieths, such as .02, .04, .24 (even numbers), are preferred over decimals such as .03 and .05 (odd numbers). Fiftieths can be divided by two and still end up as two-place decimals. This is useful when, for example, a diameter is divided by 2 to get a radius. The decimal point should be clear and placed on the bottom guideline in a space about the width of a 0.

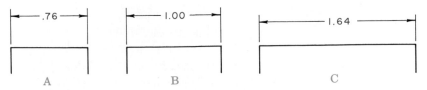

Fig. 6-9 *Decimal dimensions: two places.*

Fig. 6-10 *Decimal dimensions: three places.*

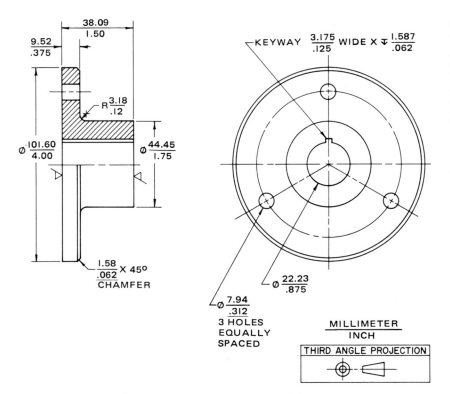

Fig. 6-11 *Typical dual-dimensioned drawing.*

Decimal dimensioning is used in most industries. It is the preferred method of drafting with customary measure. A **dual dimensioning system** is sometimes used in industries involved in international trade. This system uses both the decimal inch and the millimeter (Fig. 6-11). However, in many industries it is becoming more common to use the metric system alone.

When you use the metric system, give the dimensions in millimeters, meters and, for special applications, micrometers. With metric dimensions, use decimals to one place where limits of ±0.1 mm are close enough. Use decimals to two places or more where limits smaller than ±0.1 mm are required.

SPECIFYING UNITS IN CAD DRAWINGS

The appearance of units is controlled in most CAD programs by setting up a *dimensioning style*. Figure 6-12 shows the dialog boxes that appear in AutoCAD Release 13 when you enter the DDIM (Dimensioning Styles) command. You can set the height of the units and the gap between the dimension line and the units from the Annotation dialog box (A). When the check box next to Enable Units is checked, AutoCAD automatically switches to the dual dimensioning system. By default, the primary units are customary, and the secondary units are metric. Note also that you can control primary and alternate (secondary) units separately.

When you pick the Units... button in the Primary Units section, the Primary Units dialog box appears. You can control the type of units (B) and the number of decimal places (C) from the Primary Units dialog box.

After you have set and saved the configuration for the units for a drawing, all the dimensions you create automatically have that style. For further flexibility, you can have as many different dimensioning styles as necessary in a drawing file. Remember, however, that the dimensions should normally be consistent within a drawing.

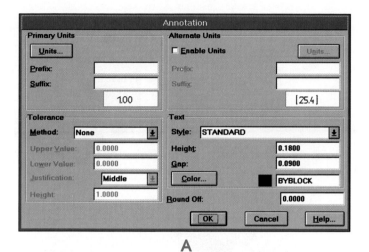

A

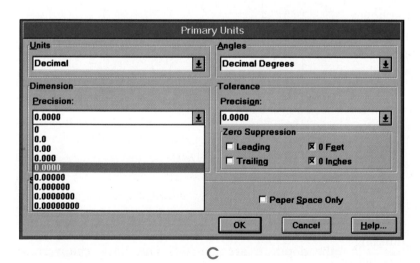

B

C

Fig. 6-12 *(A) From the Annotation dialog box in AutoCAD, you can set up basic unit parameters. (B) A drop-down box in the Primary Units dialog box allows you to choose the type of units to use in the drawing. (C) Another drop-down box allows you to set the number of decimal places. Notice that you can specify angles and their precision (decimal places) separately.*

PLACING DIMENSIONS

Two methods of placing dimensions are currently in use: the *aligned system* and the *unidirectional system*. In the **aligned system** of dimensioning (Fig. 6-13), the dimensions are placed in line with the dimension lines. Horizontal dimensions always read from the bottom of the sheet. Inclined dimensions read in line with the inclined dimension line. If possible, all dimensions should be kept outside the shaded area in Fig. 6-14. The aligned system was once the only system in use.

In the **unidirectional system** of dimensioning (Fig. 6-15), all the dimensions read from the bottom of the sheet, no matter where they appear. In both systems, notes and dimensions with leaders should read from the bottom of the drawing. The unidirectional system has now replaced the aligned system in most industries. ANSI Y14.5M uses this system exclusively.

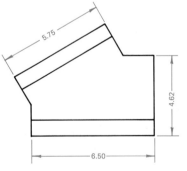

Fig. 6-13 *The aligned system of placing dimensions.*

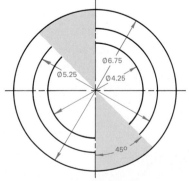

Fig. 6-14 *Avoid placing dimensions in the shaded area.*

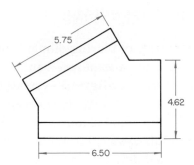

Fig. 6-15 *The unidirectional system of placing dimensions.*

THEORY OF DIMENSIONING

There are two basic kinds of dimensions: size dimensions and location dimensions. *Size dimensions* define each piece. Giving size dimensions is really a matter of giving the dimensions of a number of simple shapes. Every object is broken down into its geometric forms, such as prisms, cylinders, pyramids, cones, and so forth, or into parts of such shapes. This is shown in Fig. 6-16, where the bearing is separated into simple parts. A hole or hollow part has the same outlines as one of the geometric shapes. Think of such open spaces in an object as *negative* shapes.

The idea of open spaces is especially valuable to certain industries. Drafters in the aircraft industry need to know the weights of parts. These weights are worked out from the volumes of the parts as solids. From these solids, the volumes of the holes and hollow or open spaces (negative or minus shapes) are subtracted. To get the total weight, the result is then multiplied by the weight per cubic inch or per cubic millimeter of the material.

When the object being dimensioned has a number of pieces, the positions of each piece must also be given. These are given by *location dimensions*. Each piece is first considered separately and then in relation to the other pieces. When the size and location dimensions of each piece are given, the size description is complete. Dimensioning a whole machine, a piece of furniture, or a building is just a matter of following the same orderly pattern that is used for a single part.

Size Dimensions

The first shape is the *prism*. For a rectangular prism (Fig. 6-17), the width *W,* the height *H,* and

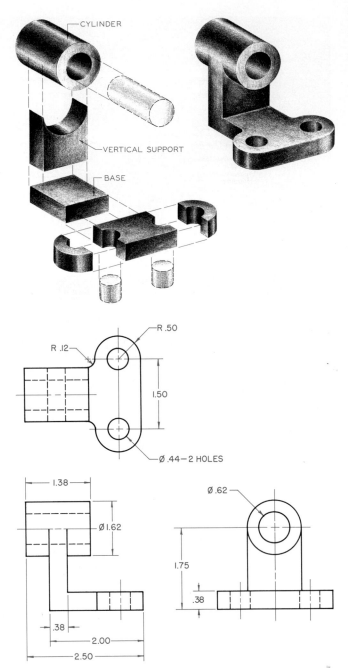

Fig. 6-16 *Parts can usually be broken down into basic geometric shapes for dimensioning.*

the depth *D* are needed. This basic shape may appear in a great many ways. A few of these are shown in Fig. 6-18. Flat pieces of irregular shape are dimensioned in a similar way (Fig. 6-19). The rule for dimensioning prisms is as follows.

For any flat piece, give the thickness in the edge view and all other dimensions in the outline view.

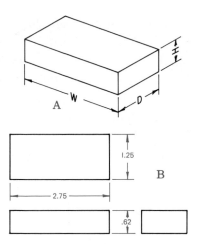

Fig. 6-17 *A rectangular prism.*

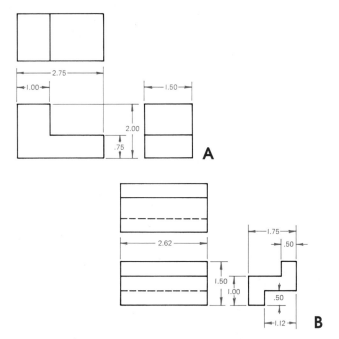

Fig. 6-18 *Dimensioning a rectangular prism.*

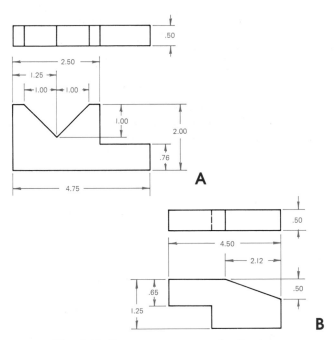

Fig. 6-19 *Dimensioning an irregular flat shape.*

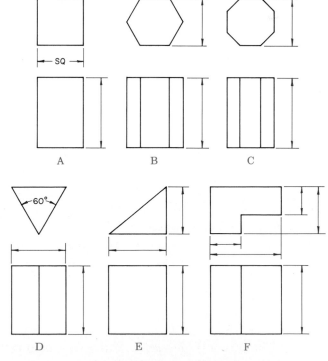

Fig. 6-20 *Dimensioning prisms.*

The **outline view** is the one that shows the shape of the flat surface or surfaces. The front views in Fig. 6-19 are the outline views.

Methods for dimensioning other prisms are shown in Fig. 6-20. A square prism needs two dimensions. A hexagonal or an octagonal prism may also use two dimensions. A triangular prism may need three dimensions, and so on for other regular or irregular prisms.

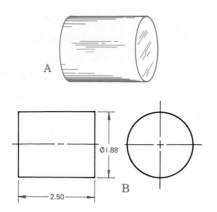

A

Ø1.88

2.50

B

Fig. 6-21 *Dimensioning a cylinder.*

The second shape is the *cylinder* (tube shape). The cylinder needs two dimensions: the diameter and the length (Fig. 6-21). Three cylinders are dimensioned in Fig. 6-22A. One of these is the hole. This is because a hollow cylinder can be thought of as two cylinders of the same length, as shown by the washer in Fig. 6-22B. The rule for dimensioning cylinders is as follows.

For cylindrical pieces, give the diameter and the length on the same view.

The symbol Ø is always placed with the diameter dimension. On simple drawings, the circular view may be eliminated by using the Ø symbol as shown in Fig. 6-22C. The abbreviation DIA may be found on older drawings instead of the symbol.

Notes are generally used to give the sizes of holes. Such a note is usually placed on the outline view (Fig. 6-23A). These notes are sometimes used to show what operations are needed to form or finish the hole. For example, drilling, punching, reaming, lapping, tapping, countersinking, spotfacing, and so forth, may be specified (Figs. 6-24 and 6-25). The symbols used in these figures are defined in Table 6-1.

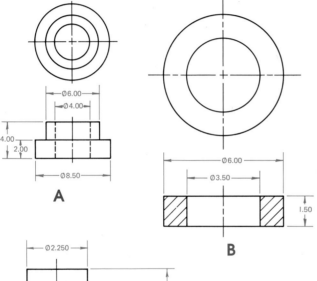

Ø6.00
Ø4.00
4.00
2.00
Ø8.50

A

Ø6.00
Ø3.50
1.50

B

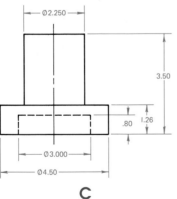

Ø2.250
3.50
.80   1.26
Ø3.000
Ø4.50

C

Fig. 6-22 *A hole in a cylinder is dimensioned in the same way as the inner and outer surfaces of the cylinder.*

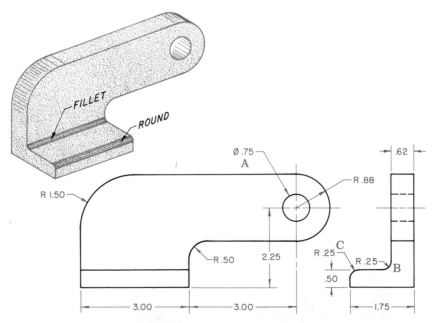

Fig. 6-23 *Fillets, rounds, and radii.*

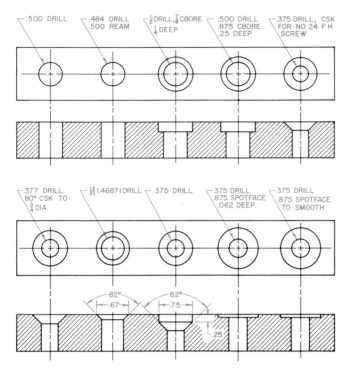

Fig. 6-24 *Older methods use notes to specify the information needed for dimensions and some common operations.*

However, the trend in industry is away from specifying machining operations. Sizes are given and the machining methods are determined by the manufacturer or machinists. For example, the note .500 DRILL would now be Ø.500 in most industries.

When parts of cylinders occur, such as fillets (Fig. 6-23B) and rounds (Fig. 6-23C), they are dimensioned in the views in which the curves show. The radius dimension is given and is preceded by the abbreviation R.

Some of the other shapes are the cone, the pyramid, and the sphere. The cone, the frustum, the square pyramid, and the sphere can be dimensioned in one view (Fig. 6-26). In these instances, the second view may not even be needed. To dimension rectangular or other pyramids and parts of pyramids, two views are needed.

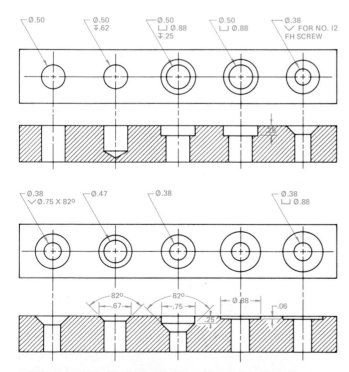

Fig. 6-25 *More current (and preferred) methods for specifying dimensions and operations.*

TABLE 6-1	Notes and Symbols for Machining Operations	
Traditional Method	**Preferred Method**	
½ Drill or .50 Drill	Ø	.50
.48 Drill .500 Ream	Ø	.500
½ Drill, 7/8 Cbore ¼ Deep	Ø ⊔ ▽	.500 .875 .25
.38 Drill 82° CSK To .75 DIA	Ø Ø	.38 .75 X ∨82°
.38 Drill .88 Spotface .06 Deep	Ø ⊔ ▽	.38 .88 .06

Table 6-1 *Notes and symbols for machining operations.*

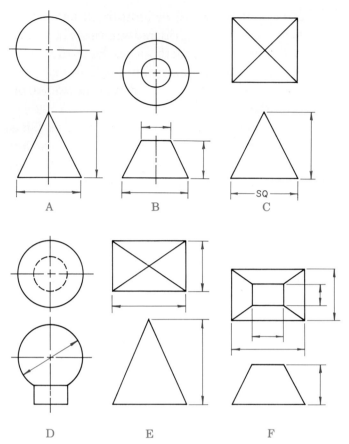

Fig. 6-26 *Dimensioning some elementary shapes.*

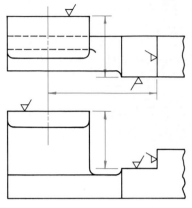

Fig. 6-27 *Locating dimensions for prisms.*

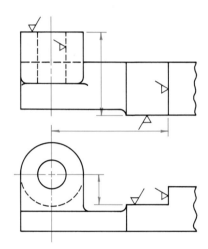

Fig. 6-28 *Locating dimensions for prisms and cylinders.*

Location Dimensions

Location dimensions are used to show the relative positions of the basic shapes. They are also used to locate holes, surfaces, and other features. In general, location dimensions are needed in three mutually perpendicular directions: up and down, crossways, and forward and backward.

Finished surfaces and centerlines, or axes, are important for fixing the positions of parts by location dimensions. In fact, finished surfaces and axes are used to define positions. Prisms are located by surfaces, surfaces and axes, or axes. Cylinders are located by axes and bases. Three location dimensions are needed for both forms.

All the surfaces and axes must be studied together so that the parts will go together as accurately as necessary. That means that in order to include the right location dimensions and notes on a drawing, drafters must be familiar with the engineering practices needed for the manufacture, assembly, and use of a product. There are two general rules for showing location dimensions. The first rule is:

Prism forms are located by the axes and the surfaces (Fig. 6-27). Three dimensions are needed.

The second rule is:

Cylinder forms are located by the axis and the base (Fig. 6-28). Three dimensions are needed.

Combinations of prisms and cylinders are shown in Fig. 6-29. The dimensions marked *L* in Fig. 6-29B are location dimensions.

Datum Dimensioning

Datums are points, lines, and surfaces that are assumed to be exact (Fig. 6-30). Such datums are used to compute or locate other dimensions. Location dimensions are given from them. When positions are

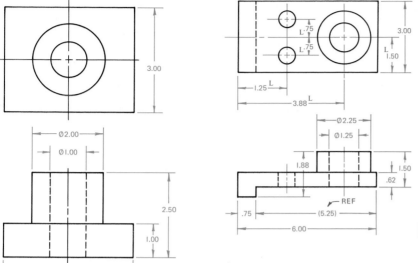

Fig. 6-29 *Examples of dimensioning prisms and cylinders.*

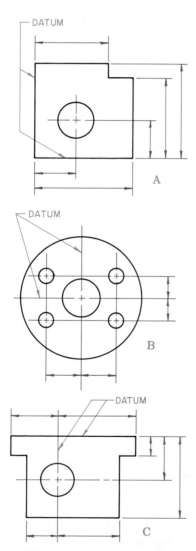

Fig. 6-30 *Datum dimensioning.*

are located from datums, the different features of a part are all located from the same datum.

Two surfaces, two centerlines, or a surface and a centerline are typical datums. In Fig. 6-30A, two surface datums are used. In (B), two centerlines are used. In (C), a surface and a centerline are used. A datum must be clear and visible while the part is being made. Mating parts should have the same datums because they fit together.

GENERAL RULES FOR DIMENSIONING

The following are general rules for dimensioning:

1. Dimension lines should be spaced about .25 in. (6 mm) apart and about .38 in. (10 mm) from the view outline (Fig. 6-3).

2. If the aligned system is used, dimensions must read in line with the dimension line and from the lower or right-hand side of the sheet (Fig. 6-12). If the unidirectional system is used, all dimensions must read from the bottom of the sheet (Fig. 6-14).

3. On machine drawings, dimensions should be given in decimal inches or millimeters, even if the drawings are of large objects such as airplanes or automobiles. In customary dimensioning, values are given to two digits, except when greater accuracy is required. Whole numbers use a decimal marker followed by two zeros to the right of the decimal point; for example, 3.00. In metric dimensioning, values are given to one digit, except when greater accuracy is required. Whole numbers do not need a decimal point or zeros. A millimeter value of less than 1 is shown with a zero to the left of the decimal point; for example, 0.2.

4. When all dimensions are in inches or millimeters, the symbol is generally omitted. Add a note to the drawing such as ALL DIMENSIONS ARE IN MILLIMETERS, or use the metric symbol.

5. You may dimension very large areas on architectural and structural drawings in meters. In that case, whole numbers stand for millimeters and decimalized dimensions stand for meters. However, this is true *only* on architectural and structural drawings. When customary measure is used on architectural and structural drawings, any measurement over 12 in. is given in feet and inches.

6. In general, dimension sheet-metal drawings in inches or in millimeters.

7. Dimension furniture and cabinet drawings in millimeters and inches.

8. When customary measure is used, feet and inches are shown as 7′-3. Where the dimension is in even feet, it is written 7′-0.

9. Position all dimensions clearly. Do not repeat the same dimension on different views.

10. Do not give dimensions that are not needed. This is especially important in interchangeable manufacturing processes in which limits are used. Figure 6-31A shows unnecessary dimensions. They have been omitted in Fig. 6-31B.

11. Place overall dimensions outside the smaller dimensions (Figs. 6-31 and 6-32). When you give the overall dimension, leave out the dimension of one of the smaller distances (Fig. 6-31B) unless it is needed for reference. If a dimension is needed for reference, put parentheses around it to show that it is for reference only, as in Fig. 6-32.

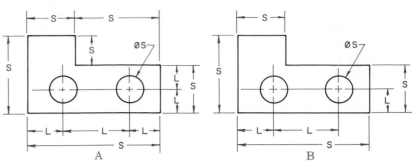

Fig. 6-31 *Omit unnecessary dimensions.*

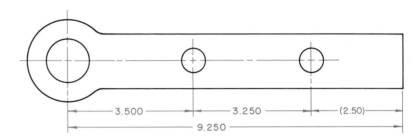

Fig. 6-32 *Enclose reference dimensions in parentheses.*

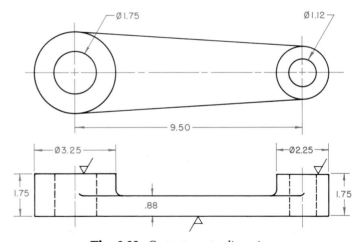

Fig. 6-33 *Center-to-center dimensions.*

12. On circular end parts, give the center-to-center dimension instead of an overall dimension (Fig. 6-33).

13. When a dimension must be placed within a sectioned area, leave a clear space for the number (Fig. 6-34).

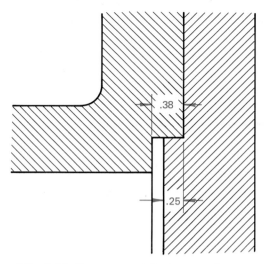

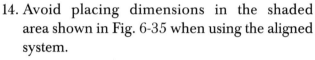

Fig. 6-34 *Dimensions within a sectioned area.*

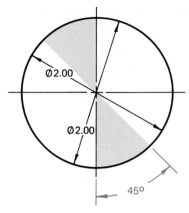

Fig. 6-35 *Avoid the shaded area when using aligned dimensions.*

14. Avoid placing dimensions in the shaded area shown in Fig. 6-35 when using the aligned system.
15. Give dimensions from centerlines, finished surfaces, or datums where needed.
16. Do not use a centerline or a line of a view as a dimension line.
17. Do not draw a dimension line that extends from a line of a view.
18. Avoid crossing a dimension line with another line.
19. Give the diameter of a circle, not the radius. Use the symbol Ø before the dimension.
20. Place the abbreviation R before the dimension when giving the radius of an arc.
21. In general, dimensions should be placed outside the view outlines.
22. Do not draw extension lines that cross each other or cross dimension lines if you can avoid doing so without making the drawing more complicated.
23. Avoid dimensioning to hidden lines if possible.

Remember that there are no hard-and-fast rules or practices that are not subject to change under the special conditions or needs of a particular industry. However, when there is a variation of any rule, there must be a reason to justify it.

STANDARD DETAILS

The shape, methods of manufacture, and use of a part generally tell you which dimensions must be given and how accurate they must be. A knowledge of manufacturing methods, pattern-making, foundry and machine-shop procedures, forging, welding, and so on, is very useful when you are choosing and placing dimensions. The quantity of parts to be made must also be considered. If many copies are to be made, quantity-production methods have to be used. In addition, some items may incorporate purchased parts, identified by name or brand, that call for few, if any, dimensions.

Some companies have their own standard parts for use in different machines or constructions. The dimensioning of these parts depends on how they are used and produced. There are, however, certain more-or-less standard details or conditions. For these, there are suggested ways of dimensioning.

Angles and Chamfers

In the past, angles have generally been dimensioned in degrees (°), minutes (′), and seconds (″). However, decimalized angles are now preferred (Fig. 6-38A). To convert from minutes to decimal degrees, divide minutes by 60, because 60′ = 1°.

Example: $30' = 30 \div 60$ or $.5°$. To convert seconds $(")$ to decimal degrees, divide seconds by 3600 $(60' \times 60'' = 3600)$. Example: $45'' = 45 \div 3600 = .01°$. Therefore, $15°30'45'' = 15.51°$.

$$\begin{array}{r} 15.00 \\ + \quad .50 \\ + \quad .01 \\ \hline 15.51° \end{array}$$

Angular tolerance is usually bilateral (shown on both sides of the angle as plus or minus); for example, the tolerance might be $±.08$. If decimalized angles are not being used, the dimensions might be written $±5'$ for minutes (Fig. 6-36B). Angular tolerance is stated either on the drawing or in the title block. Angular measurements on structural drawings are given by run and rise (Fig. 6-37A).

A similar method is used for slopes, as in (B) and (C). Here one side of the triangle is made equal to 1. Two usual methods of dimensioning chamfers (bevels) are shown in Fig. 6-38.

Tapers

Tapers can be dimensioned by giving the length, one diameter, and the taper as a ratio, as shown in Fig. 6-41A. Another method is shown in (B). In this method, one diameter or width, the length, and either the American National Standard or another standard taper number are given. For a close fit, the taper is dimensioned as shown in (C). The diameter is given at a located gage line. In (D), one diameter and the angle are given.

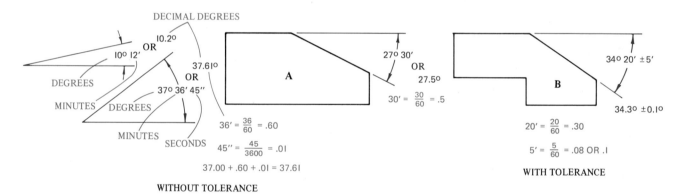

Fig. 6-36 *Dimensioning an angle.*

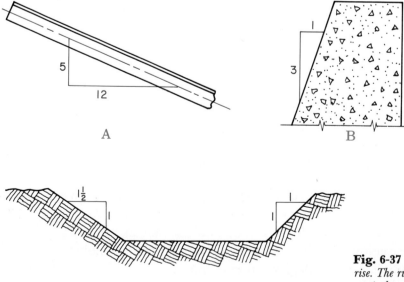

Fig. 6-37 *Angles of slopes are dimensioned using run and rise. The run is the horizontal measurement, and the rise is the vertical measurement.*

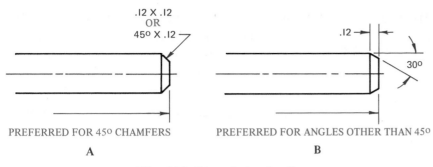

PREFERRED FOR 45° CHAMFERS   PREFERRED FOR ANGLES OTHER THAN 45°

A   B

Fig. 6-38 *Dimensioning chamfers.*

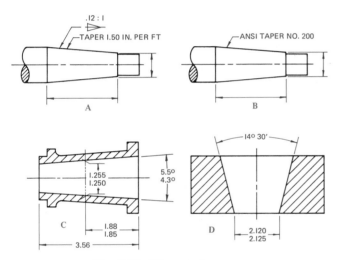

Fig. 6-39 *Dimensioning tapers.*

Curves

A curve made up of arcs is dimensioned by the radii that have centers located by points of tangency (Fig. 6-40). Noncircular or irregular curves can be dimensioned as shown in (Fig. 6-41A). They can also be dimensioned from datum lines, as in (B). A regular curve can be described and dimensioned by showing the construction or naming the curve, as in (C). The basic dimensions must also be given.

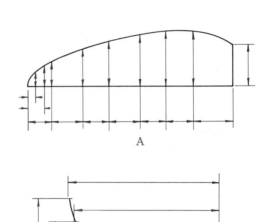

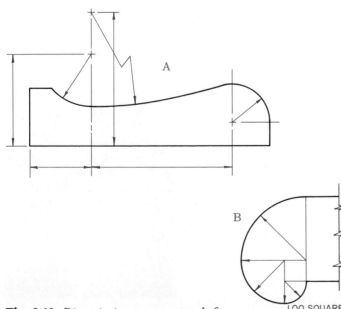

Fig. 6-40 *Dimensioning curves composed of circular arcs.*

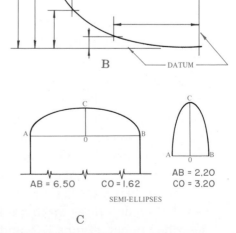

Fig. 6-41 *Dimensioning noncircular curves.*

DIMENSIONING A DETAIL DRAWING

A drawing for a single part that includes all the dimensions, notes, and information needed to make that part is called a **detail drawing.** The dimensioning should be done in the following order:

1. Complete all the views of a drawing before adding any dimensions or notes.
2. Think about the actual shape of the part and its characteristic views. With this in mind, draw all of the extension lines and lengthen of any centerlines that may be needed.
3. Think about the size dimensions and the related location dimensions. Draw the dimension lines, leaders, and arrowheads.
4. After considering any changes, put in the dimensions and add any notes that may be needed.

DIMENSIONING AN ASSEMBLY DRAWING

When the parts of a machine are shown together in their relative positions, the drawing is called an **assembly drawing.** If an assembly drawing needs a complete description of size, the rules and methods of dimensioning apply.

Drawings of complete machines, constructions, and so on, are made for different uses. The dimensioning must show the information that the drawing is designed to supply.

1. If the drawing is only to show the appearance or arrangement of parts, the dimensions can be left off.
2. If a drawing is needed to tell the space a product requires, give overall dimensions.
3. If parts have to be located in relation to each other without giving all the detail dimensions, center-to-center distances are usually given. Dimensions needed for putting the machine together or erecting it in position may also be given. *Photo-drawings* (photographs of products with dimensions, notes, and other details drawn on them), instead of regular drawings, can be made of a machine for the uses described in paragraphs 1, 2, and 3.

4. In some industries, assembly drawings are completely dimensioned (Chapter 11). These *composite drawings* are used as both detail and assembly drawings.

For furniture and cabinet work, sometimes only the major dimensions are given. For example, length, height, and sizes of stock may be given. The details of joints are left to the cabinet maker or to the standard practice of the company. This is especially true if construction details are standard.

NOTES FOR DIMENSIONS

In some industries, notes are used to supply the information needed for some operations. Among these are the operations for making drilled holes (Fig. 6-42), reamed holes (Fig. 6-43), and counterboring or spotfacing (Fig. 6-44). For specific notations, review Figs. 6-24 and 6-25.

Fig. 6-42 *Drilling a hole.*

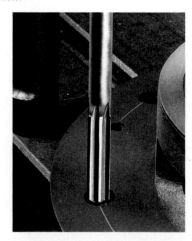

Fig. 6-43 *Reaming a hole.*

Sometimes a hole is to be made in a piece after assembly with its mating piece. In that case, the words AT ASSEMBLY should be added to the note. Because such a hole is located when it is made during assembly with its mating part, no dimensions are needed for its location. Some other dimensions with machining operations are suggested in Fig. 6-45.

The trend, however, in industry is to avoid the use of operational names in dimensions and notes. Terms such as turn, bore, grind, drill, ream, and other similar operational terms are no longer used in dimensioning in many industries. If the drawing is adequately dimensioned, it remains a shop problem to meet the drawing specifications.

DIMENSIONING A CAD DRAWING

As you read earlier in this chapter, you can define dimensions in a CAD drawing by setting up dimension styles. After you have set up the style you need, you create the dimensions by issuing a series of dimensioning commands. Table 6-2 lists and describes common AutoCAD dimensioning commands. AutoCAD Release 13 for Windows offers a Dimensioning toolbar, as shown in Fig. 6-46A. To use the toolbar, use the cursor control to pick the appropriate icon, or tool, and then click on the part of the object you want to dimension. For example, to dimension the angle in the object shown in Fig. 6-46B, pick the Angular Dimension icon to select the DIMANGULAR command, and then select the two sides of the angle.

Fig. 6-44 *Counterboring to a specified depth. Spotfacing is generally used to provide smooth spots.*

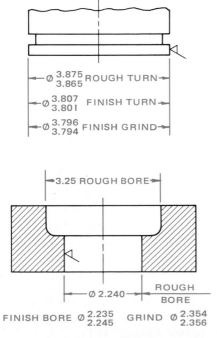

Fig. 6-45 *Operations with limits specified.*

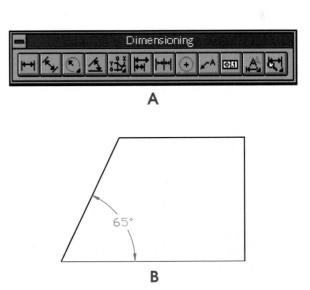

Fig. 6-46 *(A) The dimensioning toolbar contains many of the most commonly used dimensioning commands. To dimension an angle, for example, enter the DIMANGULAR command by picking the fourth icon from the left. Then pick the two lines that make up the angle, as shown in (B).*

TABLE 6-2	Dimensioning Commands in AutoCAD Release 13	
Command	Associated Icon (Windows only)	Action
DIMLINEAR	[icon]	Creates both vertical and horizontal linear dimensions.
DIMALIGNED	[icon]	Creates a dimension in which the dimension line is aligned with the object line.
DIMRADIUS	[icon]	Creates a radius dimension.
DIMDIAMETER	[icon]	Creates a diameter dimension.
DIMANGULAR	[icon]	Creates an angular dimension.
DIMORDINATE	[icon]	Creates an ordinate dimension.
DIMBASELINE	[icon]	Creates a series of baseline dimensions.
DIMCONTINUE	[icon]	Continues another dimension from the last extension line entered.
DIMCENTER	[icon]	Creates a center mark or center line.
DIMLEADER	[icon]	Creates a leader and allows you to attach one or more lines of text.
TOLERANCE	[icon]	Used in geometric dimensioning and tolerancing (See "Geometric Dimensioning and Tolerancing" later in this chapter).
DDIM	[icon]	Allows you to set up and save various dimension styles.

Table 6-2 *Dimensioning commands in AutoCAD Release 13.*

ABBREVIATIONS

You already know many of the abbreviations used in dimensioning. Examples from American National Standards are listed in Appendix C. Other abbreviations may be found in the latest edition of *American National Standard Abbreviations for Use on Drawings,* ANSI Y1.1.

INTERCHANGEABLE PARTS

When one part is to be assembled with other parts, it must be made to fit into place without any more machining or handwork. These parts are called *mating parts,* or *interchangeable parts.* For mating parts to fit together, specified size allowances are necessary. For example, suppose two mating parts are a rod or shaft and the hole in which it fits or turns. For these parts to fit together, variation in the diameter of the rod and the hole must be limited. If the rod is too large in diameter, it will not turn. If it is too small, the rod will be too loose and will not work properly.

LIMIT DIMENSIONING

Absolute accuracy cannot be expected. Instead, workers have to keep within a fixed limit of accuracy. They are given a number of tenths, hundredths, thousandths, or ten-thousandths of an inch or millimeter that the part is allowed to vary from the absolute measurements. This permitted variation is called the **tolerance.** The tolerance may be stated in a note on the drawing or written in a space in the title block. An example would be DIMENSION TOLERANCE ±.01 UNLESS OTHERWISE SPECIFIED.

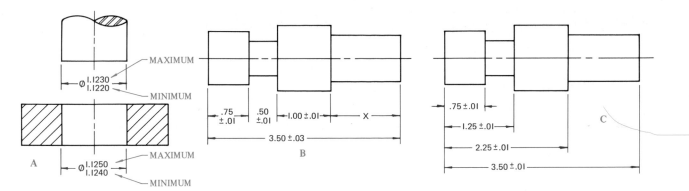

Fig. 6-47 *Limit dimensions.*

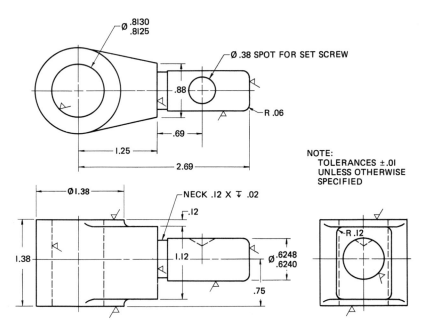

Fig. 6-48 *A detail drawing with limits.*

NOTE:
TOLERANCES ±.01
UNLESS OTHERWISE
SPECIFIED

(B). In this case, the dimension *X* could have some variation. This dimension would not be given unless it was needed for reference. If it is given, it would be enclosed in parentheses.

Progressive dimensions (each starting at the same place) are shown in (C). Here they are all given from a single surface. This kind of dimensioning is called *baseline dimensioning*.

Very accurate or limiting dimensions should not be called for unless they are truly needed, because they greatly increase the cost of making a part. The detail drawing in Fig. 6-48 has limits for only two dimensions. All the others are *nominal* dimensions. The amount of variation in these parts depends on their use. In this case, the general note calls for a tolerance of ±0.1 in.

PRECISION OR EXACTNESS

The latest edition of ANSI Y14.5 gives precise information on accurate measurement and position dimensioning. The following paragraphs are adapted from *American National Standard Drafting Practices* with the permission of the publisher, The American Society of Mechanical Engineers, 345 East 47th St., New York, NY 10017.

Limit dimensions, or limits that give the maximum and minimum dimensions allowed, are also used to show the needed degree of accuracy. This is illustrated in Fig. 6-47A. Note that the maximum limiting dimension is placed above the minimum dimension for both the shaft (external dimension) and the hole in the ring (internal dimension).

In Fig. 6-47B and C, the basic sizes are given, and the plus-or-minus tolerance is shown. *Consecutive dimensions* (one after the other) are shown in

Expressing Size

Size is a designation of magnitude. When a value is given to a dimension, it is called the *size* of that dimension. *NOTE:* The words *dimensions* and *size* are both used to convey the meaning of magnitude. Several different size descriptions can be used to describe a part. Study the following definitions.

nominal size The nominal size is used for general identification. Example: ½-in. (13-mm) pipe.

basic size The basic size is the size to which allowances and tolerances are added to get the limits of size.

design size The design size is the size to which tolerances are added to get the limits of size. When there is no allowance, the design size equals the basic size.

actual size An actual size is a measured size.

limits of size The limits of size (usually called *limits*) are maximum and minimum sizes.

Expressing Position

Dimensions that fix position usually call for more analysis than size dimensions. Linear and angular sizes locate features in relation to one another (point-to-point) or from a datum. Point-to-point distances may be enough to describe simple parts. If a part with more than one critical dimension must mate with another part, dimensions from a datum may be needed.

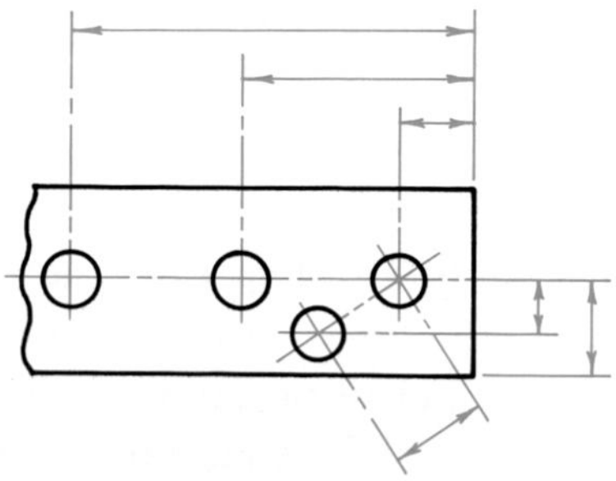

Locating Round Holes

Figures 6-49 through 6-54 show how to position round holes by giving distances, or distances and directions, to the hole centers. These methods can also be used to locate round pins and other features. Allowable variations for the positioning dimensions are shown by stating limits of dimensions or angles, or by true position expressions.

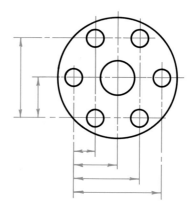

Fig. 6-50 *Locating holes by rectangular coordinates. (ANSI)*

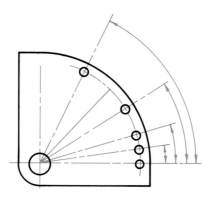

Fig. 6-51 *Locating holes on a circle by polar coordinates. (ANSI)*

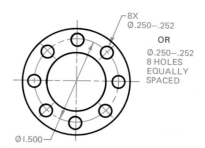

Fig. 6-49 *Locating holes by linear distances.*

Fig. 6-52 *Locating holes on a circle by radius or diameter and the words "equally spaced."*

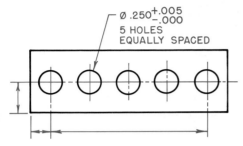

Fig. 6-53 *"Equally spaced" holes in a line. (ANSI)*

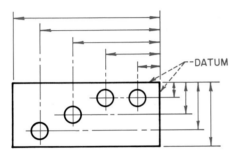

Fig. 6-54 *Dimensions for datum lines. (ANSI)*

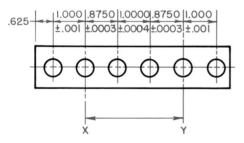

Fig. 6-55 *Point-to-point, or chain, dimensioning. (ANSI)*

TOLERANCE

A tolerance is the total amount a given dimension may vary. A tolerance should be expressed in the same form as its dimension. The tolerance of a decimal dimension should be expressed by a decimal to the same number of places. The tolerance of a dimension written as a common fraction should be expressed as a fraction. An exception is a close tolerance on an angle that can be expressed by a decimal representing a linear distance.

In a "chain" of dimensions with tolerances, the last dimension may have a tolerance equal to the sum of the tolerances between it and the first dimension. In other words, tolerances accumulate; they are added together. The datum dimensioning method of Fig. 6-54 avoids overall accumulations. The tolerance on the distance between two features (first and second hole, for example) is equal to the tolerances on two dimensions from the datum added together. Where the distance between the two points must be controlled closely, the distance between the two points should be dimensioned directly, with a tolerance. Figure 6-55 illustrates a series of chain dimensions where tolerances accumulate between points *X* and *Y.* Datum dimensions in Fig. 6-56 show the same accumulation with larger tolerances. Figure 6-57 shows how to avoid the accumulation without the use of extremely small tolerances.

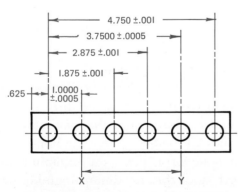

Fig. 6-56 *Datum dimensioning. (ANSI)*

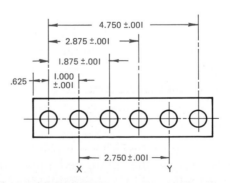

Fig. 6-57 *Dimensioning to prevent tolerance accumulation between X and Y. (ANSI)*

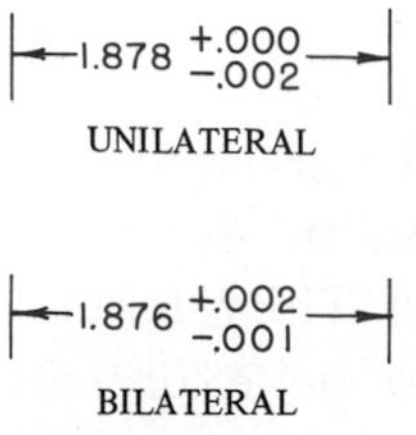

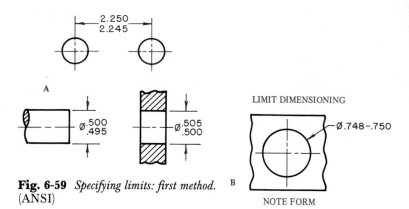

UNILATERAL

BILATERAL

Fig. 6-58 *(A) A unilateral tolerance allows deviation in one direction only. (B) A bilateral tolerance allows deviation on both sides of the design size.* (ANSI)

Fig. 6-59 *Specifying limits: first method.* (ANSI)

Unilateral Tolerance System

Unilateral tolerances allow variations in only one direction from a design size. This way of stating a tolerance is often helpful where a critical size is approached as material is removed during manufacture (Fig. 6-58A). For example, close-fitting holes and shafts are often given unilateral tolerances.

Bilateral Tolerance System

Bilateral tolerances allow variations in both directions from a design size. Bilateral variations are usually given with locating dimensions. They are also used with any dimensions that can be allowed to vary in either direction (Fig. 6-58B).

LIMIT SYSTEM

A limit system shows only the largest and smallest dimensions allowed (Figs. 6-59 and 6-60). The tolerance is the difference between the limits.

The amount of variation permitted when dimensioning a drawing can be given in several ways. The ways recommended in this book are as follows:

1. If the plus tolerance is different from the minus tolerance, two tolerance numbers are used, one plus and one minus (Fig. 6-58). *NOTE:* Two tolerances in the same direction should not be called for.

2. When the plus tolerance is equal to the minus tolerance, use the combined plus-and-minus sign ($\pm$) followed by a single tolerance number (Fig. 6-61).

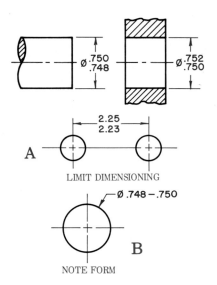

Fig. 6-60 *Specifying limits: second method.* (ANSI)

3. Show the maximum and minimum limits of size. Place the numerals in one of the following two ways. Do not use both on the same drawing.

   a. Place the high limit above the low limit where dimensions are given directly. Place the low limit before the high limit where dimensions are given in note form (Fig. 6-61).

   b. For location dimensions given directly (not by note), place the high-limit number (maximum dimension) above. Place the low-limit number (minimum dimension) below. For size dimensions given directly, place the number representing the maximum material condition below. Where the limits are given in note form, place the minimum number first and the maximum number second. See Fig. 6-60 for an example.

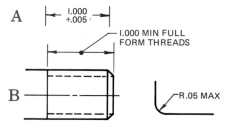

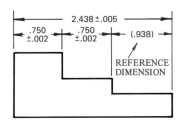

Fig. 6-61 *Using a combined plus-and-minus sign. (ANSI)*

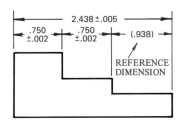

Fig. 6-62 *Expressing a single tolerance or limit. (ANSI)*

Fig. 6-63 *Placing tolerance and limit numerals. Note the reference dimensions. (ANSI)*

4. You do not always have to give both limits.

    a. A unilateral tolerance is sometimes given without stating that the tolerance in the other direction is zero (Fig. 6-62A).

    b. MIN or MAX is often placed after a number when the other limit is not important. Depths of holes, lengths of threads, chamfers, etc., are often limited in this way (Fig. 6-62B).

5. The above recommendations are for *linear* (in-line) tolerances, but the same forms are also used for angular tolerances. Angular tolerances may be given in degrees, minutes and seconds, or decimals. Decimals are now preferred.

PLACING TOLERANCE AND LIMIT NUMERALS

A tolerance numeral is placed to the right of the dimension numeral and in line with it. It may also be placed below the dimension numeral with the dimension line between them. Figure 6-63 shows both arrangements.

DIMENSIONING FOR FITS

The tolerances on the dimensions of interchangeable parts must allow these parts to fit together at assembly (Fig. 6-64). Figure 6-65 shows a way to dimension mating parts that do not need to be interchangeable. The size of one part does not need to be held to a close tolerance. It is to be made the proper size at assembly for the desired fit. *NOTE:* For further information on limits and fits, see ANSI B4.1.

To calculate dimensions and tolerances of cylindrical parts that must fit well together, you must first decide which dimension you will use for the basic size. You can use either the minimum hole size (*hole basis*) or the maximum shaft size (*shaft basis*) as the basic size.

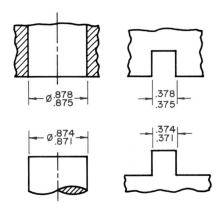

Fig. 6-64 *Indicating dimensions or surfaces that must fit closely. (ANSI)*

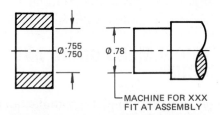

Fig. 6-65 *Dimensioning noninterchangeable parts that must fit closely. (ANSI)*

Basic Hole System

A **basic hole system** is one in which the design size of the hole is the basic size and the allowance is applied to the shaft. To determine the limits for a fit in the basic hole system, follow these steps:

1. Give the minimum hole size.

2. For a clearance fit, find the maximum shaft size by subtracting the desired allowance (minimum clearance) from the minimum hole size. For an interference fit, add the desired allowance (maximum interference).

3. Adjust the hole and shaft tolerances to get the desired maximum clearance or minimum interference (Fig. 6-66).

By using the basic hold system, you can often keep tooling costs down. This is possible because standard tools such as a reamer or broach can be used for machining.

Basic Shaft System

A **basic shaft system** is one in which the design size of the shaft is the basic size and the allowance is applied to the hole. To figure out the limits for a fit in the basic shaft system, follow these steps:

1. Give the maximum shaft size.

2. For a clearance fit, find the minimum hole size by adding the desired allowance (minimum clearance) to the maximum shaft size. Subtract for an interference fit.

3. Adjust the hole and shaft tolerances to get the desired maximum clearance or minimum interference. See Fig. 6-67.

Use the basic shaft method only if there is a good reason for it, such as when a standard-size shaft is necessary or desired. For additional information on American National Standard limits and fits, see Appendix Tables A-26 through A-30.

GEOMETRIC DIMENSIONING AND TOLERANCING

An engineering drawing of a manufactured part is intended to convey information from the designer to the manufacturer and inspector. It must contain all

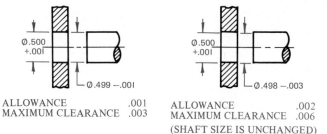

Fig. 6-66 *Fits in the basic hole system.* (ANSI)

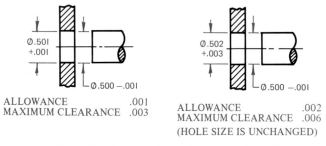

Fig. 6-67 *Fits in the basic shaft system.* (ANSI)

information necessary for the part to be correctly manufactured. It must also enable an inspector to make a precise determination of whether the parts are acceptable.

Therefore, each drawing must convey three essential types of information:

- the material to be used
- the size or dimensions of the part
- the shape or geometric characteristics

The drawing must also specify permissible variations for each of these aspects, in the form of tolerance or limits. The addition of this material, size, and shape information to an engineering drawing is known as **geometric dimensioning and tolerancing.**

Materials are usually covered by separate specifications or supplementary documents, and the drawings need only make reference to these. *Size* is specified by linear and angular dimensions. Tolerances may be applied directly to these dimensions or may be specified by means of a general tolerance note. *Shape* and geometric characteristics, such as orientation and position, are described by views on the drawing, supplemented to some extent by dimensions.

Geometric dimensioning and tolerancing can be one of the most important subjects learned by those who will be entering the manufacturing workplace. As a communication system, it incorporates knowledge of design, tooling, production, and inspection. It also touches on many other aspects of industry that affect these areas. To fully understand and appreciate geometric dimensioning and tolerancing requires a concerted effort over a long period of daily use. It is like a language that uses graphic symbols instead of letters.

In the past, tolerances were often shown for which no precise interpretation existed, such as on dimensions which originated at nonexistent center lines. The specification of datum features was often omitted, resulting in measurements being made from actual surfaces when, in fact, datums were intended. There was confusion concerning the precise effect of various methods of expressing tolerances and of the number of decimal places used. While tolerancing of geometric characteristics was sometimes specified in the form of notes, no precise methods or interpretations were established. Straight or circular lines were drawn, without specifying how straight or round they should be. Square corners were drawn without specifying by how much the 90° angle could vary.

Modern systems of tolerancing, which include geometric and positional tolerancing, use of datum and datum targets, and more precise interpretation of linear and angular tolerances, provide designers and drafters with a means of expressing permissible variations in a very precise manner. Furthermore, the methods and symbols are international in scope and therefore help break down language barriers. This section covers the application of modern geometric dimensioning and tolerancing methods used on technical drawings.

It is not necessary to use geometric tolerances for every feature on a part drawing. In most cases, it is to be expected that if each feature meets all dimensional tolerances, form variations will be adequately controlled by the accuracy of the manufacturing process and equipment used.

Datums

Prior to a discussion of the geometric dimensioning language, you need to understand how parts are positioned for machining, inspection, and assembly. Parts are positioned on datums. A *datum* is a theoretically exact point, line, axis, or area from which geometric measurements are taken. A theoretical datum is established by the contact of a datum feature and a simulated datum (Fig. 6-68).

The *datum feature* is any physical portion of a part. The *simulated datum* is what the datum feature contacts and should imitate the mating part in the assembly. A simulated datum may be a mounting surface of a machine tool, a surface of an assembly fixture, or a surface of an inspection holding fixture.

The role of an engineering drawing is to specify what the part should be like after machining or assembly. Therefore, finished surfaces are generally selected as datum features. However, this is not always possible. In many cases specific points, lines, or areas of a surface are identified as *datum targets*. Datum targets are specified where datum features are rough, uneven, or on different levels, such as on castings, forgings, or weldments. It is very common for a part to be supported by one or more machined surfaces and one or more datum targets.

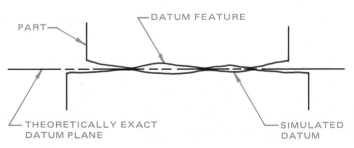

Fig. 6-68 *A theoretical datum.*

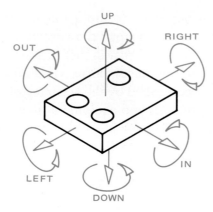

Fig. 6-69 *The six degrees of freedom. Limiting a part's movement in one or more of these directions during manufacture improves the accuracy of the part.*

Datum Reference Frame

It may not be possible to make parts exactly the same, but it is possible to design a reliable and repeatable support structure while they are being machined. This is called restricting the *degrees of freedom*. Parts must be restrained from moving in the directions shown in Fig. 6-69.

Parts may move up and down, in and out, and from side to side. They may also rotate. The restrictive environment created to hold the parts is called a *datum reference frame*. One of the major aspects of designing and machining parts is figuring out what the datum reference frame should be.

Specifying Datum Features on Drawings

Datum features may be identified by the methods shown in Fig. 6-70. This symbol may be attached to a visible line representing the datum feature (A) or to an extension line (B). If the datum feature is a rectangular size feature, the straight line connecting the square and the triangle must be in line with the dimension line that states the size of the feature (C).

The letters used in the square box do not have to be

in alphabetical order. The important thing is how each letter is used on the rest of the drawing. Choose letters that will not be misunderstood due to their appearance elsewhere in a different context.

Specifying Datum Targets on Drawings

If specific portions of a feature will be used to establish the theoretical datums, they are identified with datum target symbols. Datum targets are of three types: points, lines, and areas. Figure 6-71 illustrates how each of the types are shown on a drawing.

Figure 6-71A shows an example of a datum target point. A large X is placed where the part will rest on the tooling. Figure 6-71B shows an example of a datum target line. A phantom line is used to show where the line of contact will be. Figure 6-71C shows an example of a datum target area. The area is shown with a phantom line that has been crosshatched. The area may be any shape.

In all cases, the datum target is identified with a letter and number placed in the bottom half of a circle. The letter identifies the datum, while the number identifies the specific target. The datum target area size is placed in the upper half of the datum target symbol.

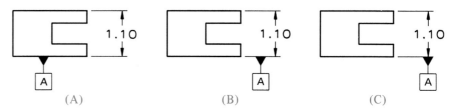

Fig. 6-70 *Placement of the datum feature symbol.*

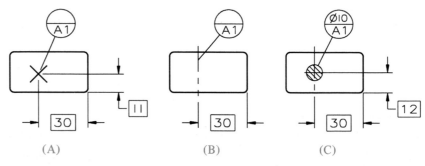

Fig. 6-71 *The datum target may be a point (A), a line (B), or an area (C).*

The dimensions for datum targets may be basic dimensions as shown or general toleranced dimensions. If basic dimensions are used, the actual location tolerances for the datum targets are determined by the employees who make the tooling.

Geometric Dimensioning Sentence Structure

The ASME Y14.5M-1994 standard defines fourteen main geometric symbols used to describe geometric conditions. In addition to these, several other symbols may also be used. The feature control symbols and their names are shown in Fig. 6-72. The sizes of these symbols are shown in Appendix A-27.

Symbol	Symbol Name	Symbol	Symbol Name
—	Straightness	◎	Concentricity
▱	Flatness	⊕	Position
○	Circularity	＝	Symmetry
⌀	Cylindricity	∅	Diameter
⌒	Profile of a Line	Ⓜ	Maximum Material Condition
⌓	Profile of a Surface	Ⓣ	Tangent Plane
//	Parallelism	Ⓛ	Least Material Condition
∠	Angularity	Ⓟ	Projected Tolerance Zone
⊥	Perpendicularity	Ⓕ	Free State
↗	Circular Runout	⑤Ⓣ	Statistical Tolerance
↗↗	Total Runout	◁—▷	Between

Fig. 6-72 *Geometric characteristic symbols.*

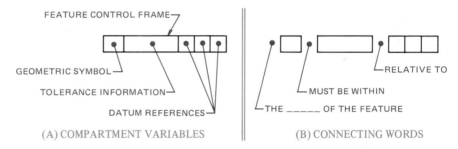

Fig. 6-73 *The structure and use of a feature control frame.*

These symbols, along with numbers, are placed in a rectangular box called a *feature control frame,* which is divided into two or more compartments. Figure 6-73A shows that the first compartment contains the geometric characteristic symbol. The second compartment contains the tolerance information. Additional compartments can be added to contain datum references. These are the variables within the basic sentence structure.

The information contained in the feature control frame may be read like a sentence. The first words spoken include the geometric characteristic name. For example, an introductory phrase for Position would be "The Position of the feature." The term *axis, axes,* or *center plane* is added when the control is related to the size features.

The lines dividing the compartments are where the connecting phrases are spoken. The first connecting phrase is "must be within" (Fig 6-73B). The

second connecting phrase is "relative to." These connecting phrases can remain the same for all geometric specifications. English translations are provided for each specification that follows.

Tolerance Zones

This section will show logical relationships between the characteristics by examining their common attributes. The most common attribute is the *type of tolerance zone* they use. The characteristics may be divided into three different tolerance zone types: parallel lines, parallel planes, and cylinders (Fig. 6-74).

Parallel Lines

The parallel lines tolerance zone type is a two-dimensional area. The distance between the parallel lines is the tolerance zone. It is specified by the

Parallel Lines	Parallel Planes	Cylinders
Straightness	Flatness	Straightness
Circularity	Parallelism	Parallelism
Circular Runout	Perpendicularity	Perpendicularity
Profile of a Line	Angularity	Angularity
Parallelism	Cylindricity	Position
Perpendicularity	Total Runout	Concentricity
Angularity	Profile of a Surface	
	Straightness	
	Position	
	Symmetry	

Fig. 6-74 *Types of tolerance zones.*

geometric tolerance in the feature control frame. In each of the parallel lines cases presented below, the tolerance zone may be at any or all positions on the surface. (The number of inspections is determined by company policy.) Each individual trace of the surface is separate from all others.

Figure 6-75 illustrates four examples of how a parallel lines tolerance zone may be applied to a plane surface. Example A shows the geometric characteristic Straightness. Because Straightness is not related to a datum, it is considered a refinement of the size dimension. The tolerance zone will always remain within the size zone. It is placed in the view where the inspection will take place.

Other geometric controls shown in Fig. 6-75 that use two parallel lines as a tolerance zone are (B) Parallelism, (C) Perpendicularity, and (D) Angularity. Notice that each feature control frame has a note below it that reads, "EACH ELEMENT." This note is added to specify a two-dimensional inspection.

The definition of parallel lines used in this section is, "A line extending in the same direction with and equidistant at all points from another line." This definition may include circles or cylinders that are concentric. The Circularity and Circular Runout examples shown in Fig. 6-76 display two parallel lines about a common center point.

The main difference between Circularity and Circular Runout is that Circularity is considered a refinement of the size dimension, so it requires no datum. The Circular Runout of a surface is controlled relative to an axis derived from a diameter different from the one being controlled. It is considered a surface-to-axis control. The tolerance zone for Circularity must remain within the size tolerance, but in Circular Runout the tolerance may exceed the size tolerance if required.

The example shown in Fig. 6-77 is of Profile of a Line. Profile uses basic dimensions to define a true profile. The area between two parallel splines defines the tolerance zone. In the example, the splines are an equal distance from and on either side of the true profile. Profile is the only parallel lines specification that may control size as well as form.

Parallel Planes

The parallel planes tolerance zone types are very similar to the parallel lines examples, except they are three-dimensional volumes instead of two-dimensional areas. The parallel planes tolerance zone is the space between two parallel surfaces. The distance between the surfaces is specified on a drawing by the geometric tolerance in the feature control frame.

Figure 6-78 illustrates four different geometric characteristics that use the distance between two parallel flat planes as their tolerance zone. Flatness (A) and Parallelism (B) are the same except that Parallelism is related to another surface, while Flatness is not. Because Flatness is not related to a datum, it is considered a refinement of the size dimension. Even though Parallelism is related to a datum, it is still a refinement of the size dimension because it is a control of opposing surfaces.

Perpendicularity (Fig. 6-78C) and Angularity (D) are the same except the tolerance zone in Perpendicularity is always oriented at a basic 90° angle to the datum surface, while the basic angle in Angularity must be specified.

Figure 6-79 shows examples of Cylindricity (A) and Total Runout (B). Cylindricity may be thought of as a combination of Straightness and Circularity. When combined, they form two concentric cylinders around a common axis. Because Cylindricity is not related to a datum, it is considered a

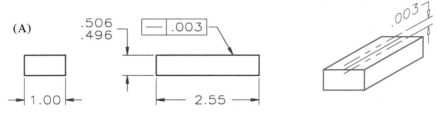

(A)

TRANSLATION: THE STRAIGHTNESS OF THE FEATURE MUST BE WITHIN
THREE THOUSANDTHS.

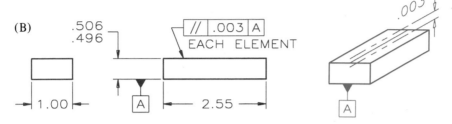

(B)

TRANSLATION: THE PARALLELISM OF EACH FEATURE ELEMENT MUST BE
WITHIN THREE THOUSANDTHS RELATIVE TO DATUM FEATURE A.

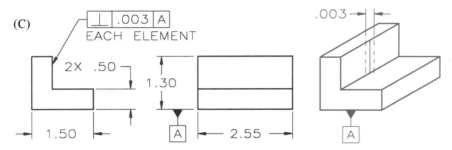

(C)

TRANSLATION: THE PERPENDICULARITY OF EACH FEATURE ELEMENT MUST BE
WITHIN THREE THOUSANDTHS RELATIVE TO DATUM FEATURE A.

(D)

TRANSLATION: THE ANGULARITY OF EACH FEATURE ELEMENT MUST BE
WITHIN THREE THOUSANDTHS RELATIVE TO DATUM FEATURE A.

Fig. 6-75 *Examples of the parallel lines type of tolerance zone.*

refinement of the size dimension. Total Runout uses the same tolerance zone type but, like Circular Runout, the tolerance zone is relative to a datum axis derived from a different diameter than the one being controlled.

The true profile in Profile of a Surface must be specified with basic dimensions. The tolerance zone in the example shown in Fig. 6-80 is equally distributed on either side of the true profile. If Profile of a Surface is applied with datums, it may control size, form, orientation, and position.

Figure 6-81 illustrates three examples of how to control the center plane of a size feature. The Straightness example (A) allows the form of the

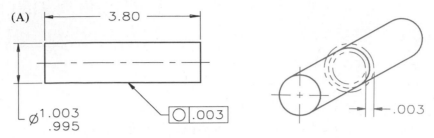

(A)

TRANSLATION: THE CIRCULARITY OF THE FEATURE MUST BE
WITHIN THREE THOUSANDTHS.

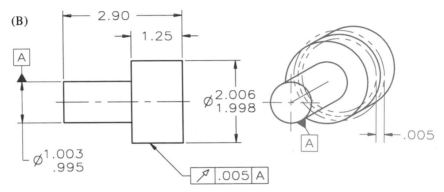

(B)

TRANSLATION: THE CIRCULAR RUNOUT OF THE FEATURE MUST BE
WITHIN FIVE THOUSANDTHS RELATIVE TO DATUM FEATURE A.

Fig. 6-76 *Two parallel lines can define the tolerance zone for concentric cylinders.*

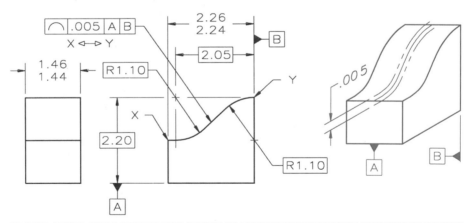

TRANSLATION: THE LINE PROFILE OF THE FEATURE MUST BE WITHIN FIVE THOUSANDTHS
RELATIVE TO DATUM FEATURES A AND B BETWEEN POINTS X AND Y.

Fig. 6-77 *Two parallel splines (curved lines) can define the tolerance zone for the true profile of a line.*

part to bow or warp outside the maximum size dimension. Straightness is the only geometric characteristic that will allow this to happen. The tolerance zone controls the center plane of the part. Because of this, the tolerance is applied to the size of the part and not to a surface.

The Position example (Fig. 6-81B) uses the same tolerance zone type as Straightness, but it is a con-

trol that specifies a centering of the slot relative to the outside surfaces of the part. Position is intended for interchangeable fits.

The Symmetry example in Fig. 6-81C is similar to Position. Their difference lies in how the tolerance is applied. Symmetry can only be applied on a "regardless of feature size" basis. It may be applied in non-interchangeable situations.

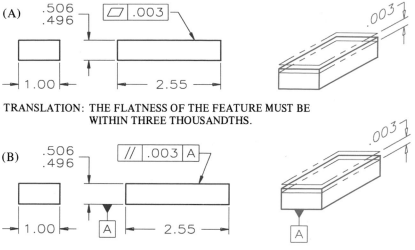

(A)

TRANSLATION: THE FLATNESS OF THE FEATURE MUST BE
WITHIN THREE THOUSANDTHS.

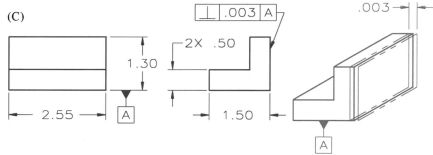

(B)

TRANSLATION: THE PARALLELISM OF THE FEATURE MUST BE WITHIN
THREE THOUSANDTHS RELATIVE TO DATUM FEATURE A.

(C)

TRANSLATION: THE PERPENDICULARITY OF THE FEATURE MUST BE WITHIN
THREE THOUSANDTHS RELATIVE TO DATUM FEATURE A.

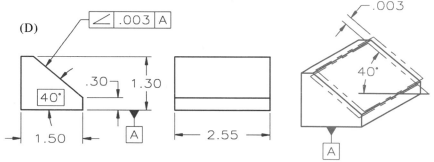

(D)

TRANSLATION: THE ANGULARITY OF THE FEATURE MUST BE WITHIN
THREE THOUSANDTHS RELATIVE TO DATUM FEATURE A.

Fig. 6-78 *Examples of the tolerance zones defined by parallel lines.*

Cylinders

The cylindrical tolerance zone is the most used of the three tolerance zone types. It is a control of the axis of a hole or a cylinder. All geometric characteristics that use this type of zone must have a diameter symbol placed in front of the tolerance value. The tolerance value specifies the diameter of the cylinder.

The following examples may use letters enclosed in circles after the tolerance value or any size datum references. This indicates that the tolerance applies at a specified size condition. If there are no size condition symbols, it means that the tolerance applies regardless of feature size.

The Straightness control in Fig. 6-82 is similar to the example in Fig. 6-81 except for the shape

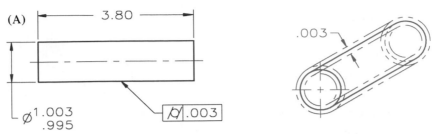

(A)

TRANSLATION: THE CYLINDRICITY OF THE FEATURE MUST BE WITHIN
THREE THOUSANDTHS.

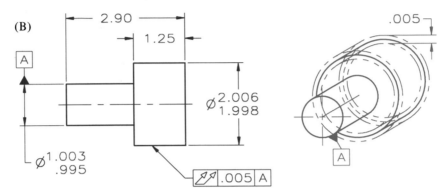

(B)

TRANSLATION: THE TOTAL RUNOUT OF THE FEATURE MUST BE WITHIN
FIVE THOUSANDTHS RELATIVE TO DATUM FEATURE A.

Fig. 6-79 *Cylindricity and Total Runout tolerance zones may be defined using two parallel planes*

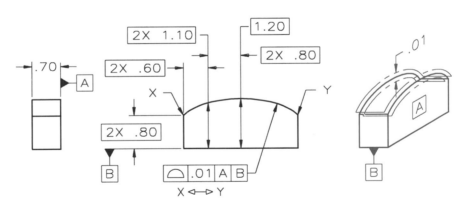

TRANSLATION: THE SURFACE PROFILE OF THE FEATURE MUST BE WITHIN
ONE ONE–HUNDREDTH INCH RELATIVE TO DATUM FEATURES
A AND B BETWEEN LINES X AND Y.

Fig. 6-80 *Specifying the tolerance zone for the profile of a surface using two parallel planes.*

of the tolerance zone. This tolerance also allows the form of the part to bow or warp outside the maximum size dimension. Straightness is the only geometric characteristic that will allow this condition.

The example in Fig. 6-83 illustrates how the parallelism of one hole may be controlled to another using a Parallelism control. The established cylinder for the controlled hole must be parallel to the axis defined by the datum hole.

A Perpendicularity example is shown in Fig. 6-84. The controlling cylinder is oriented at a basic 90° angle relative to the datum surface.

The Position example shown in Fig. 6-85 can be thought of as a combination of Parallelism and Perpendicularity with location. The centers of the tolerance cylinders are located with basic dimensions from the datum surfaces.

Figure 6-86 shows a Concentricity example. The cylinder tolerance zone is aligned with the axis of the datum diameter. It is referred to as an axis-axis control.

Tolerance Zone Combinations

The previous examples have illustrated single tolerance zones only. It is not unusual for different geometric characteristics to be used in combination. Usually, the lower segment of the geometric control is considered a refinement of the upper segment. Two examples are presented to show this concept.

Figure 6-87 depicts Parallelism and Flatness used together. The Parallelism control is a refinement of the size dimension, and the Flatness control is a refinement of the Parallelism.

A very common combination is Position and Perpendicularity. Depending on the datum arrangement, Position may include Perpendicularity. The Perpendicularity control shown in Fig. 6-88 further refines the Position control.

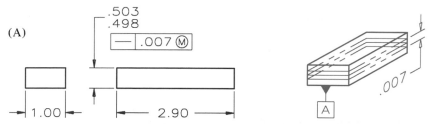

(A)

TRANSLATION: THE STRAIGHTNESS OF THE FEATURE CENTER PLANE MUST BE WITHIN SEVEN THOUSANDTHS AT MAXIMUM MATERIAL CONDITION.

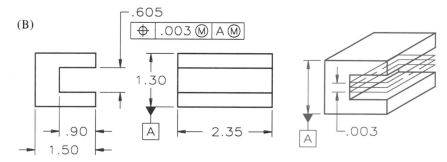

(B)

TRANSLATION: THE POSITION OF THE FEATURE CENTER PLANE MUST BE WITHIN THREE THOUSANDTHS AT MAXIMUM MATERIAL CONDITION RELATIVE TO DATUM FEATURE A AT MAXIMUM MATERIAL CONDITION.

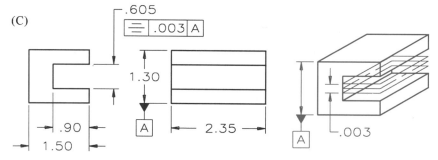

(C)

TRANSLATION: THE SYMMETRY OF THE FEATURE CENTER PLANE MUST BE WITHIN THREE THOUSANDTHS RELATIVE TO DATUM FEATURE A.

Fig. 6-81 *Controlling the center plane of a size feature.*

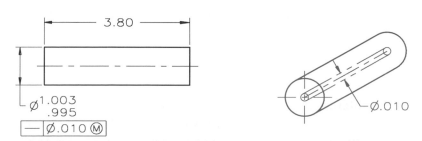

TRANSLATION: THE STRAIGHTNESS OF THE FEATURE AXIS MUST BE WITHIN TEN THOUSANDTHS AT MAXIMUM MATERIAL CONDITION.

Fig. 6-82 *The tolerance zone for Straightness can be defined using a cylinder.*

Uses of Geometric Dimensioning and Tolerancing

Geometric dimensioning and tolerancing is a very flexible communication system that can help designers specify the intent of the design throughout the entire manufacturing process. Geometric dimensioning and tolerancing is used on a daily basis by engineers, tool makers, manufacturers, inspectors, assemblers, and others in many different manufacturing industries. If geometric dimensioning and tolerancing is applied properly, and employees actually follow the geometric specifications on the drawing, the probability of making better parts increases significantly.

SURFACE TEXTURE

There is no such thing as a perfectly smooth surface. All surfaces have irregularities. Sometimes a drafter has to tell how much roughness and waviness the surface of a material can have and the lay direction of both. There are standards for these characteristics. There are also symbols that represent them.

Surface texture is discussed fully in ANSI B46.1. Use that text for study and as a reference. The following paragraphs about surface texture are adapted from *Surface Texture* (ANSI B46.1). They are included here with the permission of the publisher, The American Society of Mechanical Engineers, 345 East 47th St., New York, NY 10017.

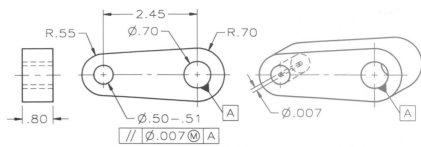

TRANSLATION: THE PARALLELISM OF THE FEATURE AXIS MUST BE WITHIN SEVEN THOUSANDTHS AT MAXIMUM MATERIAL CONDITION RELATIVE TO DATUM FEATURE A.

Fig. 6-83 *The Parallelism control feature can be used to control the parallelism of one hole to another.*

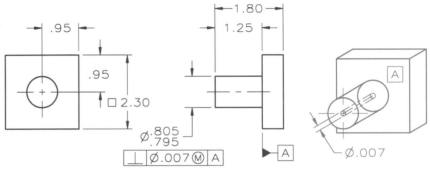

TRANSLATION: THE PERPENDICULARITY OF THE FEATURE AXIS MUST BE WITHIN SEVEN THOUSANDTHS AT MAXIMUM MATERIAL CONDITION RELATIVE TO DATUM FEATURE A.

Fig. 6-84 *Using Perpendicularity to control a cylinder at right angles to the datum surface.*

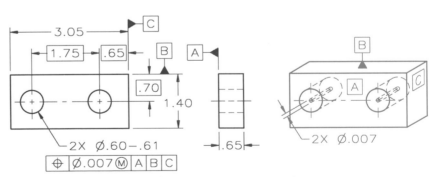

TRANSLATION: THE POSITION OF THE FEATURES AXES MUST BE WITHIN SEVEN THOUSANDTHS AT MAXIMUM MATERIAL CONDITION RELATIVE TO DATUM FEATURES A, B, AND C.

Fig. 6-85 *An example of the Position control feature.*

Surfaces, in general, are very complex in character. This standard deals only with the height, width, and direction of the surface irregularities. These are of practical importance in specific applications.

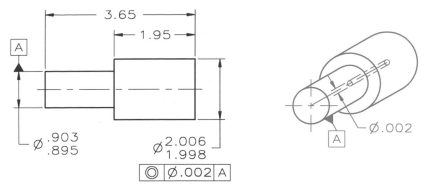

TRANSLATION: THE CONCENTRICITY OF THE FEATURE AXIS MUST BE WITHIN TWO THOUSANDTHS RELATIVE TO DATUM FEATURE A.

Fig. 6-86 *An example of the Concentricity control feature.*

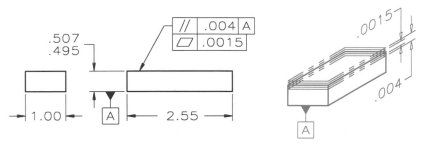

TRANSLATION: THE PARALLELISM OF THE FEATURE MUST BE WITHIN FOUR THOUSANDTHS RELATIVE TO DATUM FEATURE A AND FLAT WITHIN FIFTEEN TEN–THOUSNDTHS.

Fig. 6-87 *Using more than one geometric characteristic.*

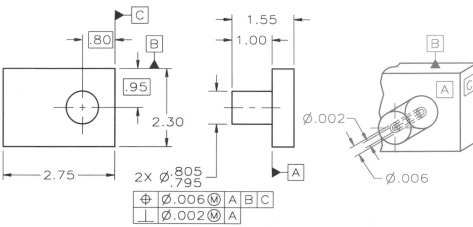

TRANSLATION: THE POSITION OF THE FEATURES AXES MUST BE WITHIN SIX THOUSANDTHS AT MAXIMUM MATERIAL CONDITION RELATIVE TO DATUM FEATURES A, B, AND C AND PERPENDICULAR WITHIN TWO THOUSANDTHS TO DATUM FEATURE A.

Fig. 6-88 *Position and Perpendicularity are often used together.*

Classification of Terms and Ratings Related to Surfaces

The terms and ratings in this book have to do with surfaces made by various means. Among these are machining, abrading, extruding, casting, molding, forging, rolling, coating, plating, blasting, burnishing, and others.

surface texture Surface texture includes roughness, waviness, lay, and flaws. It includes repetitive or random differences from the nominal surface that forms the pattern of the surface.

profile The profile is the contour (shape) of a surface in a plane perpendicular to it. Sometimes an angle other than a perpendicular one is specified.

measured profile The measured profile is a representation of the profile obtained by instruments or other means (Fig. 6-89).

microinch A microinch is one millionth of an inch (.000 0001 in.). Microinches may be abbreviated μin.

micrometer A micrometer is one millionth of a meter (.000 0001 m). Micrometers may be abbreviated μm.

roughness Roughness is the finer irregularities in the surface texture. Roughness usually includes irregularities caused by the production process. Among these are traverse feed marks and other irregularities within the limits of the roughness-width cutoff. See Fig. 6-90.

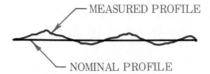

Fig. 6-89 *An enlarged profile shows that a surface is not as it appears. (ANSI)*

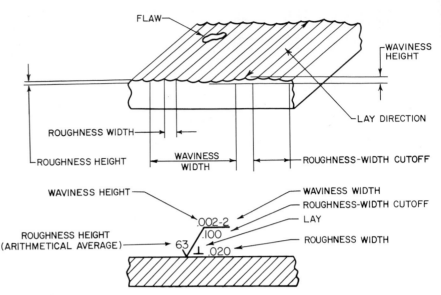

Fig. 6-90 *Relation of symbols to surface characteristics. Refer to Fig. 6-92.*

roughness height For the purpose of this book, roughness height is the arithmetical average deviation. It is expressed in microinches or in micrometers measured normal to the centerline. The preferred series of roughness-height values is shown in Table 6-3.

roughness width Roughness width is the distance between two peaks or ridges that make up the pattern of the roughness. Roughness width is given in inches or in millimeters.

roughness-width cutoff This is the greatest spacing of repetitive surface irregularities to be included in the measurement of average roughness height. Roughness-width cutoff is rated in inches or in millimeters. Standard values are shown in Table 6-4. Roughness-width cutoff must always be greater than the roughness width in order to obtain the total roughness-height rating.

waviness Waviness is covered by surface-texture standards. Geometric tolerancing now covers this surface condition under flatness. Flatness is a condition in which all surface elements are in a single plane. Flatness tolerances are applied to surfaces to control variations in surface texture (Fig. 6-92). In the past, waviness was defined as the usually widely spaced component of surface texture. It generally was of wider spacing than the roughness-width cutoff. Waviness results from factors such as machine or work deflections, vibration, chatter, heat treatment, or warping strains. Roughness may be thought of as superimposed on a "wavy" surface.

TABLE 6-3	Preferred Series Roughness	
Roughness	**Values**	**Grade**
50	2000	12
25	1000	11
12.5	500	10
6.3	250	9
3.2	125	8
1.6	63	7
0.8	32	6
0.4	16	5
0.2	8	4
0.1	4	3
0.05	2	2
0.025	1	1

Table 6-3. *Preferred series roughness.*

waviness height Waviness height is rated in inches as the peak-to-valley distance.

waviness width Waviness width is rated in inches as the spacing of successive wave peaks or successive wave valleys. When specified, the values are the maximum amounts permissible.

TABLE 6-4	Standard Roughness-Width Cutoff Values*					
mm	0.075	0.250	0.750	2.500	7.500	25.000
in	0.003	0.010	0.030	0.100	0.300	1.000

*When no value is specified, the value .030 in. (0.750 mm) is assumed.

Table 6-4. *Standard roughness-width cutoff values.*

lay Lay is the direction of predominant surface pattern. Ordinarily, it is determined by the production method used. Lay symbols are shown in Fig. 6-91.

flaws Flaws are irregularities that occur at one place or at relatively infrequent or widely varying intervals in a surface. Flaws include defects such as cracks, blowholes, checks, ridges, scratches, etc. The effect of flaws is not included in the roughness-height measurements unless otherwise specified.

contact area Contact area is the amount of area of the surface required to be in contact with its mating surface. Contact area should be distributed over the surface with approximate uniformity. Contact is specified as shown in Fig. 6-92.

Designation of Surface Characteristics

Where no surface control is specified, you can assume that the surface produced by the operation will be satisfactory. If the surface is critical, the quality of surface needed should be shown.

Surface Symbol

The symbol used to designate surface irregularities is the check mark with horizontal extension, as shown in Fig. 6-93A. The point of the symbol must touch the line indicating which surface is meant. It may also touch the extension line, or a leader pointing to the surface. The long leg and extension is drawn to the right as the drawing is read. Where only roughness height is shown, the horizontal extension may be left off. Figure 6-93B shows the typical use of the symbol on a drawing.

Where the symbol is used with a dimension, it affects all surfaces defined by the dimension.

Symbol	Designation	Example
||	Lay parallel to the line representing the surface to which the symbol is applied.	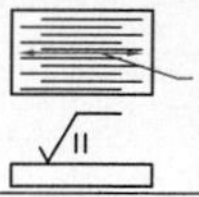 DIRECTION OF TOOL MARKS
⊥	Lay perpendicular to the line representing the surface to which the symbol is applied.	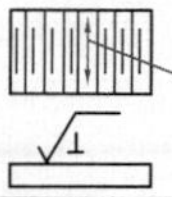 DIRECTION OF TOOL MARKS
X	Lay angular in both directions to line representing the surface to which the symbol is applied.	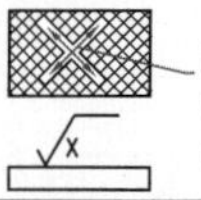 DIRECTION OF TOOL MARKS
M	Lay multidirectional.	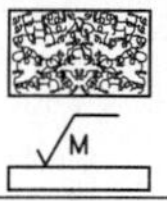
C	Lay approximately circular relative to the center of the surface to which the symbol is applied.	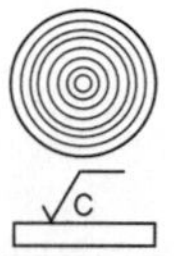
R	Lay approximately radial relative to the center of the surface to which the symbol is applied.	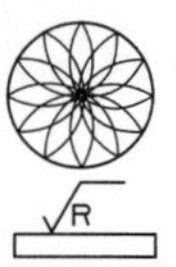

Fig. 6-91 *Lay symbols. (ANSI)*

Areas of transition, such as chamfers and fillets, should usually be the same as the roughest finished area next to them. Surface-roughness symbols always apply to the completed surface unless otherwise indicated. Drawings or specifications for plated or coated parts must tell whether the surface-roughness symbols apply before plating, after plating, or both before and after plating.

Application of Symbols and Ratings

Figure 6-92 shows the way roughness, waviness, and lay are called for on the surface symbol. Only those rating necessary to specify the desired surface need to be shown in the symbol.

Symbols Indicating Direction of Lay

Symbols for lay are shown in Fig. 6-91. Roughness ratings usually apply in a direction that gives the maximum reading. This is normally across the lay.

This is the end of the material extracted and adjusted from *Surface Texture*, ANSI B46.1. For more information, use the complete ISO and ANSI standard.

A	Roughness height rating is centered above and between the two legs. The specification of only one rating shall indicate the maximum value and any lesser value shall be acceptable. A value is here applied to the symbol variations.	E	Minimum requirements for contact or bearing area with a mating part or reference surface shall be indicated by a percentage value placed above the extension line as shown. Further requirements may be controlled by notes.
B	The specification of maximum value and minimum value roughness height ratings indicates the permissible range of value rating.	F	Lay designation is indicated by the lay symbol placed at the right of the long leg.
C	If a final surface texture must be produced by a special production method, it is placed above the horizontal extension.	G	If it is necessary to indicate a sampling length, it is placed below the horizontal extension.
D	Any indication as to treatment or coating is also placed above the horizontal extension. The numerical value of roughness applies to the surface texture after treatment, unless stated otherwise.	H	Where required, maximum roughness width rating shall be placed at the right of the lay symbol. Any lesser rating shall be acceptable.

Fig. 6-92 *Applications of symbols and ratings. (ANSI)*

THE ROLE OF CAD IN QUALITY ASSURANCE

Keep in mind that the reason for drawing accuracy and detail is to allow the product to be manufactured correctly. Many companies have Quality Control or Quality Assurance procedures to verify the accuracy of the final part. The company may use measuring devices, such as a micrometer, caliper, or specialized instruments to check diameters, lengths, or roundness of the product.

The process of checking product part quality can also be performed by a computerized measuring device such as a Computer Coordinate Measuring (CCM) machine. This machine has a small arm device that gathers data by touching the part at preprogrammed critical areas. A computer then compares the findings with the dimensions, tolerances, and feature control frame information. Some CCM machines now gather data by using laser optics. These machines generate a three-dimensional digital wireframe image of the part being measured. If the product was designed using

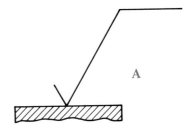

Fig. 6-93A *The surface symbol. (ANSI)*

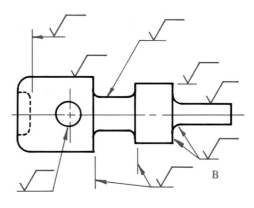

Fig. 6-93B *The surface symbol on a drawing.*

a CAD system, the information gathered can be compared to the digital data already contained in the CAD system's database.

CHAPTER 6

Learning Activities

1. Obtain several objects similar to the one shown in Fig. 6-16 and practice visualizing the basic geometric shapes from which they are made. Draw objects assigned and fully dimension them.

2. Obtain technical drawings from industry. Mark size dimensions *S* and location dimensions *L*.

3. Visit the Quality Control or Quality Assurance department of a local manufacturing or assembly plant. Find out what measures the company uses to ensure that products meet company or industry standards.

4. On drawings obtained from industry, try to determine if more careful dimensioning techniques could result in reducing the number of views required. Use a translucent overlay to sketch the changes.

5. On drawings obtained from industry, study and interpret geometric tolerancing symbols and surface texture data.

Questions

Write your answers on a separate sheet of paper.

1. Name the line used to show where a measurement begins and ends.

2. Another name for size description is _____.

3. The dimensioning system in which the numerals are placed in line with the dimension lines is called the _____ system.

4. There are two basic kinds of dimensions: (1) size dimensions, and (2) _____ dimensions.

5. Angles may be dimensioned by degrees, minutes, and seconds or by _____ angles.

6. Points, lines, and surfaces assumed to be exact are _____.

7. When all dimensions and notes read from the bottom of the sheet, the _____ dimensioning system is being used.

8. Name or sketch the abbreviations or symbols for the following terms: diameter, inch, centerline, millimeter, radius, tolerance, outside diameter, and countersink.

9. The amount by which the accuracy of a part may vary from the absolute measurement is called _____.

10. Explain the purpose of geometric dimensioning and tolerancing.

11. Name the three types of tolerance zones used in geometric dimensioning and tolerancing.

12. What is a feature control frame?

13. Explain the difference between a datum feature and a simulated datum.

14. You can control the appearance of dimensions in a CAD program by setting up one or more _____.

15. A cylinder can be fully described in a single view if you give both the length and the _____ on the same view.

16. The dimension ".500 drill" would be more accurately shown as _____.

17. On machine drawings, dimensions should be given in _____ inches or in _____ (metric sizes).

18. When a hole is to be made in a piece after assembly with its mating piece, the words _____ are added to the note.

CHAPTER 6
PROBLEMS

The small CAD symbol next to each problem indicates that a problem is appropriate for assignment as a computer-aided drafting problem and suggests a level of difficulty (one, two, or three) for the problem.

Figures 6-94 through 6-112 offer a total of 60 dimensioning problems. Additional problems may be chosen from other chapters.

The problems in Figs. 6-94 through 6-97 are to be done as follows:

1. Take dimensions from the printed scale at the bottom of the page, using dividers.

2. Draw the complete views.
3. Add all necessary extension and dimension lines for size and location dimensions.
4. Add arrowheads, dimensions, and notes.

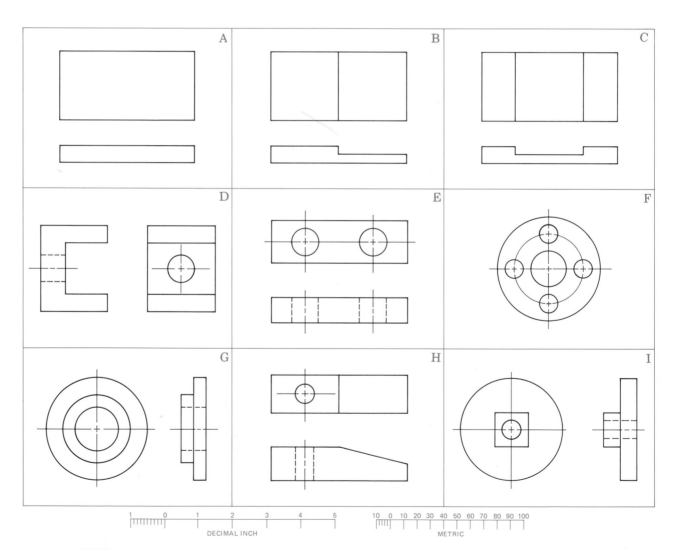

Fig. 6-94 *CAD1* *Problems for dimensioning practice. Take dimensions from the printed scale, using dividers. Draw the views as shown and add all necessary size and location dimensions.*

PROBLEMS

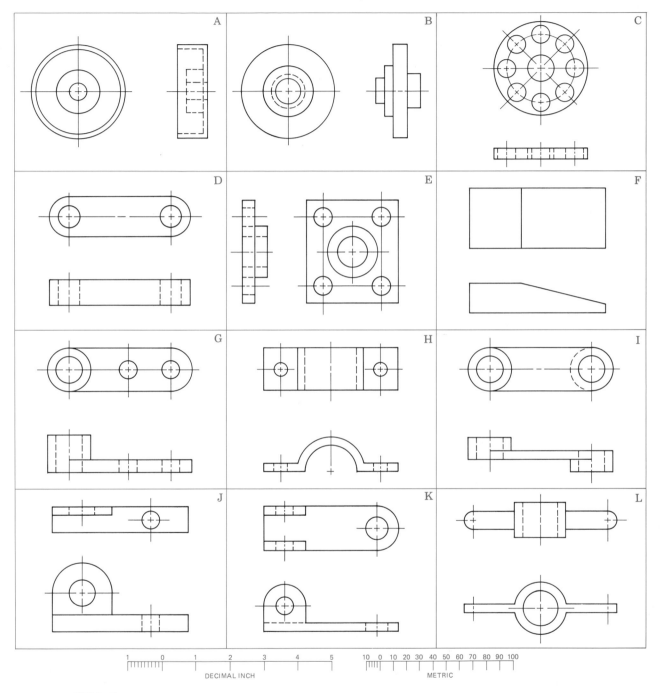

1 0 1 2 3 4 5   10 0 10 20 30 40 50 60 70 80 90 100
DECIMAL INCH   METRIC

Fig. 6-95 *CAD1 Problems for dimensioning practice. Take dimensions from the printed scale, using dividers. Draw the views as shown and add all necessary size and location dimensions.*

PROBLEMS

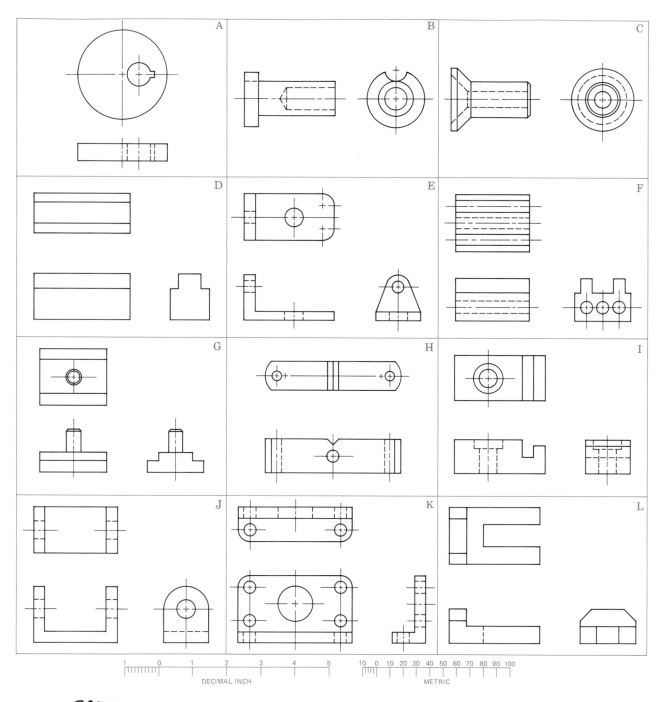

Fig. 6-96 *CAD2* *Problems for dimensioning practice. Take dimensions from the printed scale, using dividers. Draw the views as shown and add all necessary size and location dimensions.*

PROBLEMS

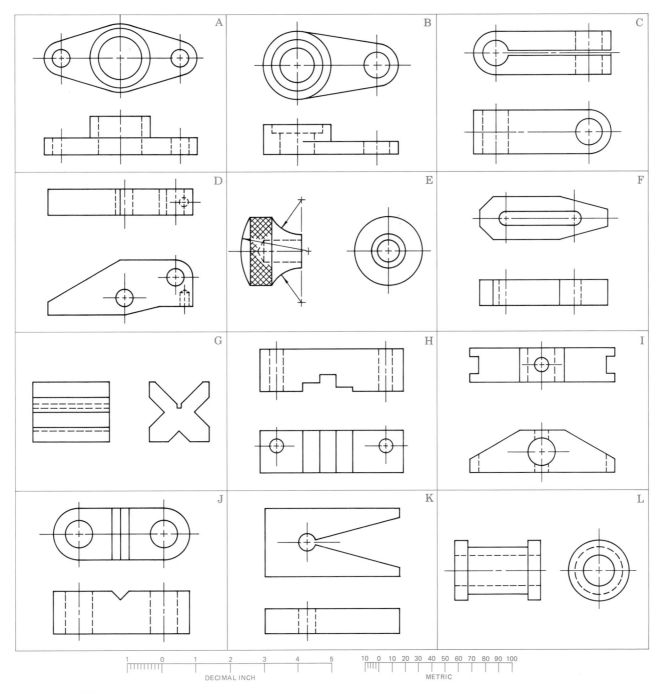

Fig. 6-97 *CAD2 Problems for dimensioning practice. Take dimensions from the printed scale, using dividers. Draw the views as shown and add all necessary size and location dimensions.*

PROBLEMS

Figures 6-98 through 6-112 are more advanced. For these problems, use the following procedure:

1. Determine the necessary views and prepare a freehand sketch.
2. Dimension the sketch.
3. Decide on a scale and draw the views mechanically.
4. Add all necessary dimensions and notes.

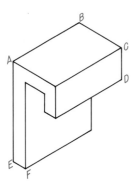

Fig. 6-98 *CAD2 Cutoff stop. Draw all necessary views and dimension. Scale: double size or as assigned.* AB = 40 mm, BC = 26 mm, CD = 17 mm, AE = 53 mm, EF = 7.50 mm.

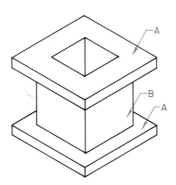

Fig. 6-99 *CAD2 Square guide. Draw all necessary views and dimension. Scale: Full size or as assigned.* A = 5 mm thick × 44 mm square, B = 30 mm square × 30 mm high, hole = 20 mm square.

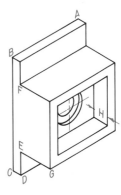

Fig. 6-100 *CAD2 Locator. Draw all necessary views and dimension. Scale: Full size or as assigned.* AB = 40 mm, BC = 60 mm, CD = 5 mm, DE = 12 mm, EF = 36 mm, EG = 18 mm, H = 8 mm, hole = Ø10 mm through, 18 mm Cbore, 2 mm deep.

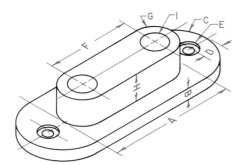

Fig. 6-101 *CAD2 Double-shaft support. Draw all necessary views and dimension. Scale: Full size or as assigned.* A = 67 mm, B = 7 mm, C = R21 mm, D = 10 mm, E = Ø6 mm through, 10 mm Cbore, 2 mm deep, 2 holes, F = 43 mm, G = R12 mm, H = 14 mm.

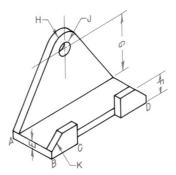

Fig. 6-102 *CAD2 Hanger. Draw all necessary views and dimension. Scale: three-quarter size or as assigned.* AB = 1.25, BC = .80, BD = 3.00, E = .19, F = .50, G = 1.56, H = R.44, J = Ø.38, K = 45°.

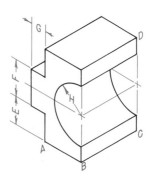

Fig. 6-103 *CAD2 Cradle slide. Draw all necessary views and dimension. Scale: Full size or as assigned.* AB = 2.38, BC = 3.56, CD = 5.12, E = 1.50, F = 2.12, G = .88, H = R1.62.

PROBLEMS

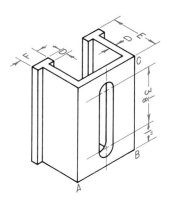

Fig. 6-104 *CAD2 Adjustable stop. Draw all necessary views and dimension. Scale: three-quarter size or as assigned. AB = 3.62, BC = 5.12, D = .30, E = 2.62, F = 1.00, slot = .88 wide.*

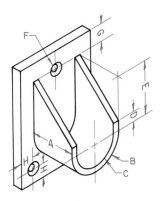

Fig. 6-105 *CAD2 Pipe support. Draw all necessary views and dimension. Scale: half size or as assigned. Base plate = .50 thick × 4.50 wide × 6.50 long, A = 2.38, B = R1.50, C = R1.12, D = .50, E = 3.00, F = Ø.38 hole through, CSK to Ø.75, 3 holes, G = 1.00, H = .75.*

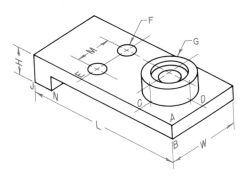

Fig. 6-106 *CAD2 Stop plate. Draw all necessary views and dimension. Scale: Full size or as assigned. Overall sizes: L = 4.25, W = 2.00, H = .75. AB = .38, AC = 1.00, AE = 2.75, AD = 1.00, JN = .50, M = 1.00, F = Ø.44, 2 holes, G = Boss: Ø1.25 × .50 high, Ø.50 through, .88 Cbore = .12 deep.*

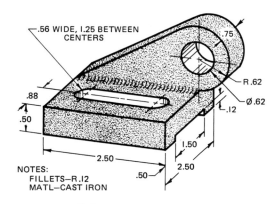

Fig. 6-107 *CAD3 Cast-iron hinge.*

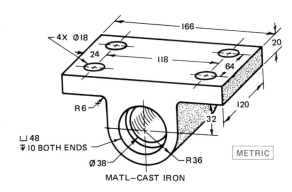

Fig. 6-108 *CAD3 Bearing housing.*

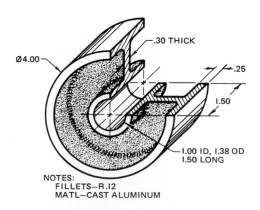

Fig. 6-109 *CAD3 Idler pulley.*

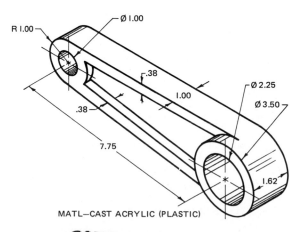

Fig. 6-110 *CAD3 Connecting rod.*

PROBLEMS

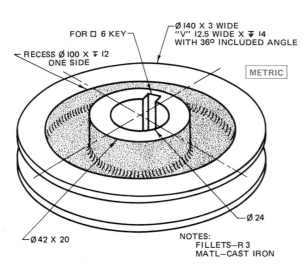

FOR □ 6 KEY

RECESS Ø 100 X ↧ 12
ONE SIDE

Ø 140 X 3 WIDE
"V" 12.5 WIDE X ↧ 14
WITH 36° INCLUDED ANGLE

METRIC

Ø 24

Ø 42 X 20

NOTES:
FILLETS—R 3
MATL—CAST IRON

Fig. 6-111 *CAD3* *Single V-pulley.*

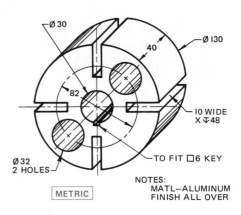

Ø 30

40

Ø 130

82

10 WIDE
X ↧ 48

Ø 32
2 HOLES

TO FIT □ 6 KEY

NOTES:
MATL—ALUMINUM
FINISH ALL OVER

METRIC

Fig. 6-112 *CAD3* *Rotor.*

SCHOOL-TO-WORK

Solving Real-World Problems

Figs. 6-113 through 6-115. *SCHOOL-TO-WORK problems are designed to challenge a student's ability to apply skills learned within this text. Be creative and have fun!*

Fig. 6-113 *Pin and Ring (geometric dimensioning and tolerancing). The pin and ring shown in the illustration are mating parts. For these drawings to be complete, the diameter features need to be related. The diameter features appear to be drawn relative to the center lines. However, there are no dimensional requirements for the position or orientation of the diameter features.*

The dimensional objectives are to establish a geometric relationship between the:
1. Ø1.613 – 1.616 and Ø2.70
2. small diameters and 1.6xx diameters
3. larger diameters and two flat mating surfaces

The datums for the following assignments are:

PIN
Ø1.607 – 1.610 = datum A
Ø.905 – .907 = datum B

RING
2.70 = datum A
Ø1.613 – 1.616 = datum B
Ø.909 – .913 = datum C

Draw and dimension the pin and ring as shown, and add the geometric dimensions that will translate the following information into geometric symbols and numbers:
1. Ring: The position of the feature axis (Ø1.613 – 1.616) must be within seven thousandths at maximum material condition relative to datum feature A.
2. Pin: The position of the feature axis (Ø.905 – .907) must be within two thousandths at maximum material condition relative to the datum feature A at maximum material condition.
   Ring: The position of the feature axis (Ø.909 – .913) must be within three thousandths at maximum material condition relative to the datum feature B at maximum material condition.

PROBLEMS

3. Pin: The perpendicularity of the feature (left face) of the 1.50 dimension must be within .001 relative to datum feature B.

Ring: The perpendicularity of the feature (left face) of the 1.35 dimension must be within .001 relative to datum feature C.

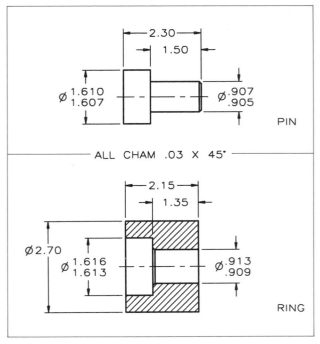

Fig. 6-113 *Pin and ring.*

Fig. 6-114 *Shim and Block (geometric dimensioning and tolerancing). The shim and block shown in the illustration are mating parts. For these drawings to be complete, the width features need to be related. The width features appear to be drawn relative to the center planes. However, there are no dimensional requirements for the position or orientation of the width features. Also, the shim is a stamped part. It is unlikely that the form of the part will fit within the boundary of perfect form at maximum material condition (.055).*

The dimensional objectives are to establish a geometric relationship between the:
1. reality that the shim will be warped and the MMC size of .055
2. slot in the block and the outside surface of the part
3. small width features and larger width features

The datums for the following assignments are:

SHIM
1.392 – 1.395 = datum A
bottom of 2.40 = datum B
left side of 3.30 = datum C
1.398 – 1.403 = datum D

BLOCK
left side of 3.40 = datum A

Draw and dimension the shim and block as shown, and add the geometric dimension that will translate the following information into symbols and numbers:

1. Shim: The Straightness of the feature center plane (.051 – .055) must be within ten thousandths at maximum material condition.
2. Block: The position of the feature center plane (1.398 – 1.403) must be within five thousandths at maximum material condition relative to the datum features A, B, and C.
3. Shim: The position of the feature center plane (.607 – .614) must be within two thousandths at maximum material condition relative to datum feature A at maximum material condition.

   Block: The position of the feature center plane (.602 – .605) must be within three thousandths at maximum material condition relative to datum feature D at maximum material condition.

PROBLEMS

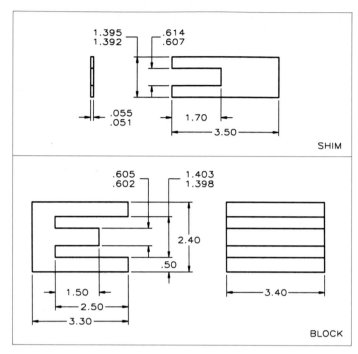

Fig. 6-114 *Shim and block.*

Fig. 6-115 *A designer has submitted this first draft CAD drawing of a roller way. Certain features of the drawing do not meet current ANSI standards. Using a CAD system, create a second draft of the drawing to meet ANSI standards.*

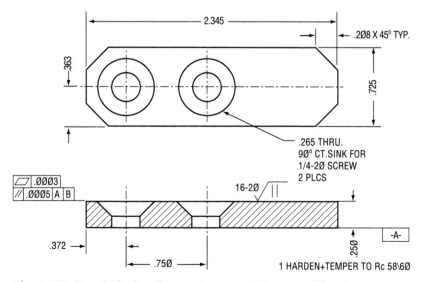

Fig. 6-115 *First draft of a roller way done on a CAD system. (Ektron).*

7

AUXILIARY VIEWS AND REVOLUTIONS

OBJECTIVES

Upon completion of Chapter 7, you will be able to:

○ Develop a primary or secondary auxiliary view of an inclined surface.
○ Determine when a partial auxiliary view is acceptable and when a complete auxiliary view is required.
○ Project and draw an auxiliary sectional view.
○ Use the concept of revolutions to establish the true size and shape of inclined surfaces.

VOCABULARY			
In this chapter, you will learn the meanings of the following terms:	•**auxiliary view** •**auxiliary plane** •**edge view** •**primary auxiliary views**	•**partial auxiliary view** •**auxiliary section** •**secondary auxiliary view** •**oblique plane**	•**reference plane** •**revolution** •**axis of revolution** •**successive revolutions**

In Chapter 5, "Multiview Drawings," you discovered how to describe an object with views on the three regular planes of projection: the top (horizontal), front (vertical), and side (profile) planes. With these planes, you can solve many graphic problems. However, to solve problems involving inclined surfaces, you must draw views on *auxiliary* (additional) planes of projection. This chapter explains how to draw these views on planes that are parallel to the inclined surfaces (Fig. 7-1).

AUXILIARY VIEWS

When an object has an inclined surface, none of the regular views show the inclined part in its true size and shape (Fig. 7-2A). However, a view

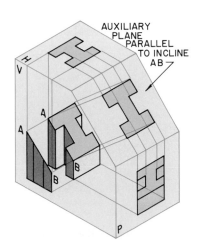

Fig. 7-1 *A primary auxiliary view is parallel to the inclined surface it defines.*

211

on a plane parallel to the inclined surface does show its true size and shape (B). An **auxiliary view** is a projection on an **auxiliary plane** that is parallel to an inclined surface (Fig. 7-3). It is a view that looks directly at the inclined surface in a direction perpendicular to it. Auxiliary views provide a clear, undistorted image of the inclined surfaces on an object.

An anchor with a slanting surface is pictured in Fig. 7-4A. Figure 7-4B shows the three normal views of the same anchor. Not only are these views hard to draw and to understand, but they also show three circular features of the anchor as ellipses. In (C), by contrast, the anchor is described completely in two views, one of which is an auxiliary view.

In Fig. 7-5A, a simple inclined wedge block is shown in the regular views. In none of these views does the slanted surface (surface *A*) appear in its true shape. In the front view, all that shows is its edge line *MN*. In the side view, surface *A* appears, but it is foreshortened. Surface *A* is also foreshortened in the top view. Line *MN* also appears in both views, but looking shorter than its true length, which shows only in the front view. To show surface *A* in its true size and shape, you need to imagine an auxiliary plane parallel to it, as shown in Fig. 7-5B. Figure 7-5C shows the auxiliary view revolved (turned) to align with the plane of the paper. By following this method, you can show the true size and shape of any inclined surface.

Primary Auxiliary Views

Auxiliary views are classified according to which of the three regular planes they are developed from. There are three **primary auxiliary views** (views developed directly from the normal views). Each is developed by projecting as a *primary reference* the height, width, or depth obtained from a normal view. Figure 7-6 shows the three primary auxiliary views.

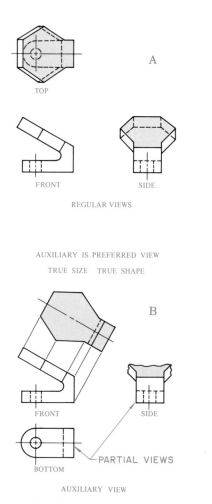

Fig. 7-2 *Compare the regular views (A) with the auxiliary views (B).*

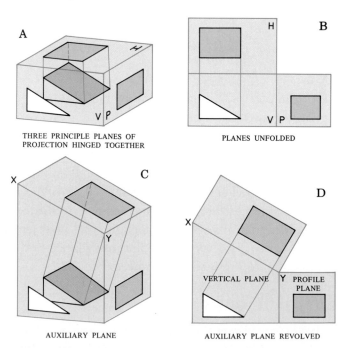

Fig. 7-3 *Basic relationship of the auxiliary plane to the regular planes.*

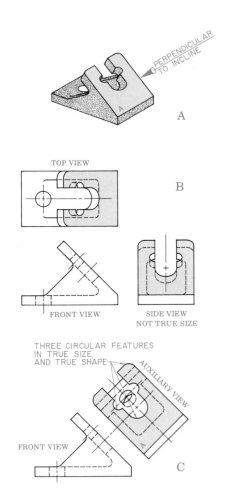

Fig. 7-4 *The pictorial view (A) and the three-view drawing (B) are difficult to draw. The auxiliary view (C) is easier to draw and describes the inclined surface completely.*

Front Auxiliary View

When an auxiliary view is hinged on the front view, the view is known as the *front auxiliary view.* The primary reference of the front auxiliary view is depth. An example is shown in Fig. 7-6A.

Top Auxiliary View

An auxiliary view that is hinged on the top view is known as the *top auxiliary view,* as shown in Fig. 7-6B. The primary reference of the top auxiliary view is the height of the object.

Right-Side Auxiliary View

Finally, a view hinged on the right-side view is the *right-side auxiliary view.* Its primary reference is the width of the object, as shown in Fig. 7-6C.

Constructing an Auxiliary View

To construct any primary auxiliary view, use the following steps, as illustrated in Fig. 7-7. *NOTE:* The method shown in Fig. 7-7 is for a front auxiliary view.

1. Examine the normal views given for an inclined surface.

2. Find the view that shows the **edge view** of the inclined plane. (The inclined plane appears as a line in this view.) The plane associated with this view is the **reference plane,** from which the auxiliary plane will be developed.

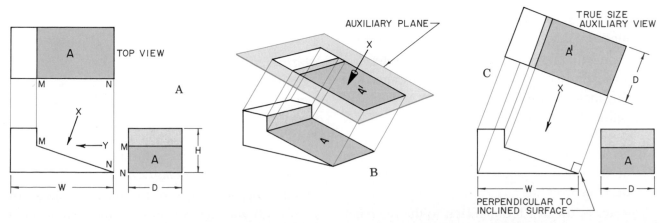

Fig. 7-5 *Basic relationship of the auxiliary view to the three-view drawing.*

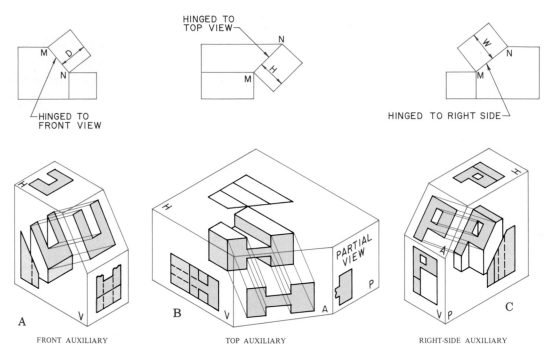

Fig. 7-6 *Three kinds of auxiliary views, showing how the auxiliary plane is hinged.*

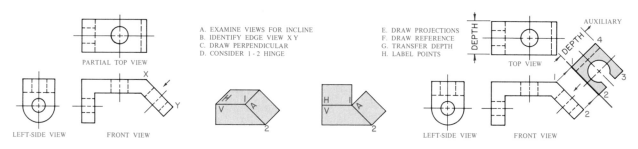

Fig. 7-7 *Steps to construct an auxiliary view. The eight steps apply to three kinds of auxiliary views.*

3. In this view, draw a light construction line at right angles to the inclined surface. This is the line of sight.

4. Think of the auxiliary plane as being attached by hinges to the view from which it is developed.

5. From all important points on the reference view, draw projection lines at right angles to the inclined surface (parallel to the line of sight). In Fig. 7-7, the important points are labeled *1* and *2*.

6. Draw a reference line parallel to the edge view of the inclined surface and at a convenient distance from it.

7. Transfer the depth dimension to the reference line.

8. Project the important points and connect them in sequence to form the auxiliary view. If you labeled the points for reference, do not leave the labels on the final drawing.

Symmetrical Objects

Figure 7-8 shows how to make an auxiliary view of a symmetrical object. In Fig. 7-8A, the object is shown in a pictorial view. Follow these steps, referring to Fig. 7-8 as necessary.

1. Use a center plane as a reference plane (B). This is *center-plane construction*.

2. Find the edge view of the inclined plane. In Fig. 7-8, the edge view of this plane appears as a centerline, line *XY,* on the top view.

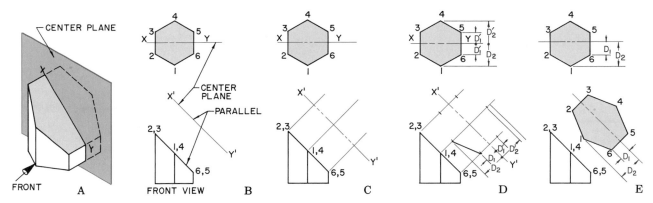

Fig. 7-8 *Steps to draw an auxiliary view using the center–plane reference method.*

3. Label the points on the top view for reference.

4. Transfer these points to the edge view of the inclined surface on the front view, as shown in (B).

5. Parallel to this edge view and at a convenient distance from it, draw the line $X'Y'$, as shown in (C).

6. In the top view, find the distances from the numbered points to the centerline. These are the depth measurement. Transfer them onto the corresponding construction lines you just drew, measuring them off on either side of line $X'Y'$, as shown in (D). The result will be a set of points on the construction lines.

7. Connect and number the points on the construction lines, as shown in (E) to finish the front auxiliary view of the inclined surface.

You could also, if desired, project the rest of the object from the center reference plane.

Using a Vertical Reference Plane

Figure 7-9A shows how to draw a front auxiliary view of a nonsymmetrical object by using a vertical reference plane. First, place the object on reference planes, as shown. These planes are located strictly for convenience in taking reference measurements. The vertical plane can be in front or in back of the object. In this case, it is in back. The construction is similar to that in Fig. 7-8, except that the depth mea-

surements D_1, D_2, and D_3 are laid off in front of the vertical plane. The drawing shows the entire object projected onto the front auxiliary plane.

Using a Horizontal Reference Plane

Figure 7-9B shows how to draw a top auxiliary reference plane. The object is a molding cut at a 30° angle. First, imagine a reference plane XY under the molding, as shown. Then find points *1* through *6* in the top and left-side views. In the top view, find the edge line of the slanted surface. Draw reference line $X'Y'$ parallel to it and at a convenient distance away. Then, from every point in the top view, project a line out to line $X'Y'$ and at right angles to it. Now, in the side view, find height measurements for the various numbered points by measuring up from XY. Lay off these same measurements up from $X'Y'$ along the lines leading to the corresponding points in the top view. Locate more points on the curve as needed in order to draw it accurately. (See the following section, "Curves on Auxiliary Views.") The result is a top auxiliary view, with its base on line $X'Y'$.

Curves on Auxiliary Views

To draw an auxiliary view of a curved line, locate a number of points along that line. Figure 7-10 shows how to make an auxiliary view of the curved cut surface of a cylinder. The cylinder is shown in a horizontal position. It has been cut at an angle, so the true shape of the slanting cut surface is an ellipse.

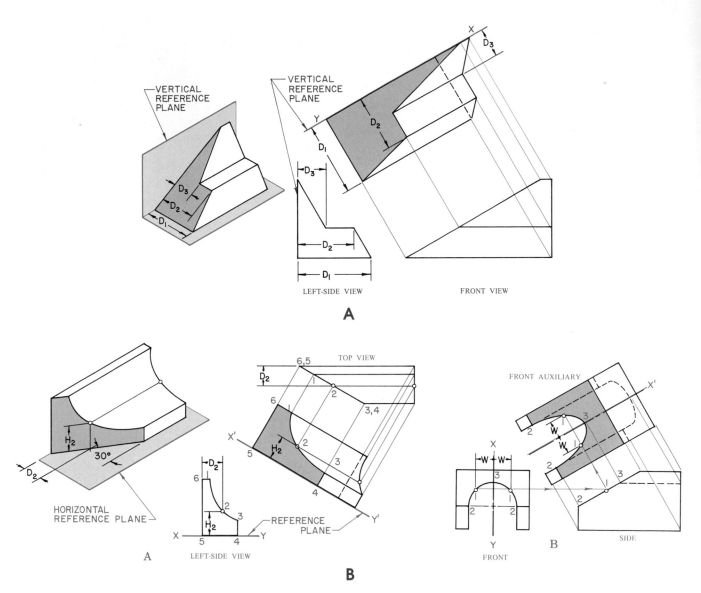

Fig. 7-9 *(A) Drawing a front auxiliary view using a vertical reference plane. (B) Drawing a top auxiliary view with a horizontal reference plane.*

This auxiliary view is a front auxiliary view with the depth as its primary reference. To draw it, follow the steps below.

1. Begin by locating the vertical centerline XY in the side view. This line represents the edge of a center reference plane.

2. Locate a number of points along the rim of the side view. The more points you locate, the more accurate your curve will be.

3. Project lines from these points over to the edge view of the cut surface in the front view.

4. Parallel to this edge view and at a convenient distance from it, draw the new centerline $X'Y'$.

5. From the points you have located on the edge view, project lines out to line $X'Y'$ and perpendicular to it. Continue these lines beyond $X'Y'$.

6. Find the depth measurements in the side view by measuring off the distances D_1, D_2, etc., between the centerline XY and the points located along the rim. Take these distances and measure them off on either side of $X'Y'$.

7. Draw a smooth curve through the points marked to form the ellipse, as shown.

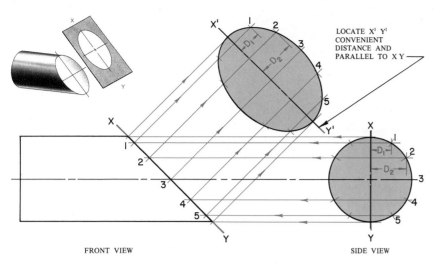

Fig. 7-10 *Drawing an auxiliary-view curve (in this case, an ellipse) about the centerline of the cut surface of a cylinder.*

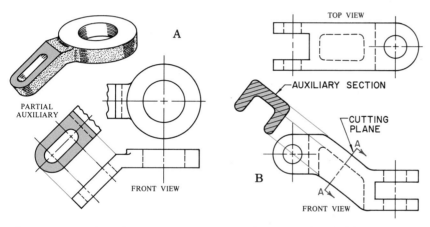

Fig. 7-11 *Partial auxiliary views or sections provide a practical method for explaining details.*

Partial Auxiliary Views

If you use break lines and centerlines properly, you can leave out complex curves while still describing the object completely, as shown in Fig. 7-11A. An auxiliary view in which some elements have been left out is known as a **partial auxiliary view.** In Fig. 7-11A, a half view is sufficient because the symmetrical object is presented in a way that is easy to understand.

Auxiliary Sections

Sometimes it is useful to show a piece of an object in *cross section,* or as it would appear on a *cutting plane* sliced through the object at that point.

When the cutting plane is not parallel to any of the normal views, the resulting cross section is known as an **auxiliary section.** In Fig. 7-11B, the auxiliary section was located by using the cutting plane represented (in edge view) by line *AA.* See Chapter 9, "Sectional Views," for more information about sections.

Secondary Auxiliary Views

A view projected from a primary auxiliary view is called a **secondary auxiliary view.** Secondary auxiliary views are used to find the true size and shape of a surface that lies along an **oblique plane** (one that is inclined to all three of the regular planes).

In Fig. 7-12, surface *1-2-3-4* is inclined to the three regular planes. In Fig. 7-12A, a first auxiliary view has been drawn. It is on a plane perpendicular to the inclined surface. Note that, in this view, points *1, 2, 3,* and *4* appear as a line or edge view of the plane. In (B), a secondary auxiliary view has been drawn from the first. It is on a plane parallel to surface *1-2-3-4.* This view shows the true shape of the surface.

Figure 7-13 shows another example. In this case, an octahedron (eight triangles making a regular solid) is shown in three views. Triangle surface *0-1-2* is inclined to all three. In Fig. 7-13A, a first auxiliary view has been drawn. It is on a plane perpendicular to triangle surface *0-1-2.* Note that line *1-2* in the top view appears as point *1-2* in this auxiliary view and that the triangle now appears as an edge line *0'-1-2.* In (B), a secondary auxiliary view has been drawn. It is on a plane parallel to the edge view of triangle surface *0'-1-2* in the first auxiliary view. This secondary auxiliary view shows the true shape of triangle *0-1-2.*

DEVELOPING AUXILIARY VIEWS WITH CAD

Using CAD, auxiliary views for most two-dimensional drawings are generally created in a manner similar to that used for manual drawings. However, they can usually be drawn in less time because the CAD software provides commands that automate many time-consuming tasks. Figure 7-14 shows the procedure for creating an auxiliary view using AutoCAD.

Three-dimensional drawings save even more time and effort because, as mentioned in previous chapters, you can view a three-dimensional object from any point in space. In other words, you can define a view in which the inclined plane of an object is parallel to the screen (Fig. 7-15).

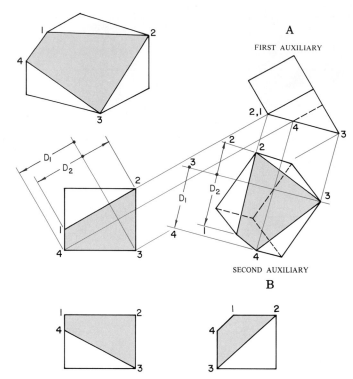

Fig. 7-12 *Secondary auxiliary views assist in finding the true shape of surface 1-2-3-4.*

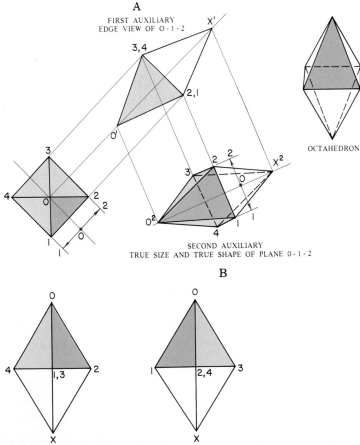

Fig. 7-13 *A secondary auxiliary is needed to develop the true shape of a triangular surface on this octahedron.*

REVOLUTIONS

When the true size and shape of an inclined surface do not show on a drawing, one solution is, as we have seen, to make an auxiliary view. Remember, in auxiliary views, you set up new reference planes to look at objects from new directions. Another solution is to *revolve* (turn) the object. The resulting drawing is called a **revolution.** In a revolution, you use the regular reference planes while imagining that the object has been revolved to an angle that places its primary features parallel to the reference planes.

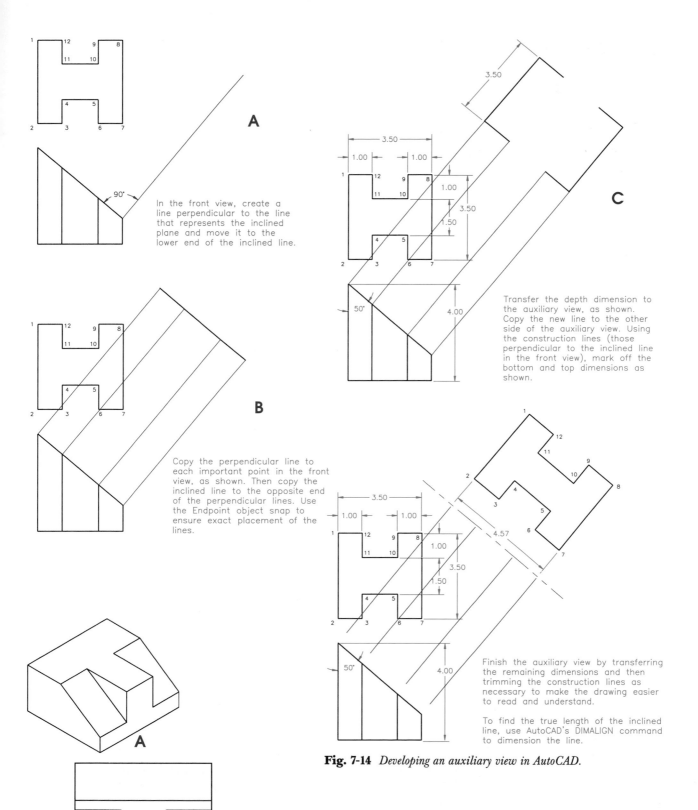

In the front view, create a line perpendicular to the line that represents the inclined plane and move it to the lower end of the inclined line.

A

Copy the perpendicular line to each important point in the front view, as shown. Then copy the inclined line to the opposite end of the perpendicular lines. Use the Endpoint object snap to ensure exact placement of the lines.

B

Transfer the depth dimension to the auxiliary view, as shown. Copy the new line to the other side of the auxiliary view. Using the construction lines (those perpendicular to the inclined line in the front view), mark off the bottom and top dimensions as shown.

C

Finish the auxiliary view by transferring the remaining dimensions and then trimming the construction lines as necessary to make the drawing easier to read and understand.

To find the true length of the inclined line, use AutoCAD's DIMALIGN command to dimension the line.

Fig. 7-14 *Developing an auxiliary view in AutoCAD.*

Fig. 7-15 *When the inclined surfaces of the three-dimensional drawing shown in (A) are parallel to the viewing screen, the resulting view (B) shows their true size and shape.*

The Axis of Revolution

An easy way to picture an object being revolved is to imagine that a shaft or an axis has been passed through it. This imaginary axis is perpendicular to one of the three principal planes. In Fig. 7-16, the three principal planes are shown with an axis passing through each one and through the object beyond. When the object is revolved about one of these axes, the axis is called the **axis of revolution.**

An object can be revolved to the right (clockwise) or to the left (counterclockwise) about an axis perpendicular to either the vertical or the horizontal plane. The object can be revolved backward (clockwise) or forward (counterclockwise) about an axis perpendicular to the profile plane.

The Rule of Revolution

The rule of revolution has two parts.
1. *The view that is perpendicular to the axis of revolution stays the same except in position.* (This is true because the axis is perpendicular to the plane on which it is projected.)
2. *Distances parallel to the axis of revolution stay the same.* (This is true because they are parallel to the plane or planes on which they are projected.)

Figure 7-17 illustrates the two parts of the rule of revolution.

Single Revolution

As you have seen, an axis of revolution can be perpendicular to the vertical, horizontal, or profile plane. This section describes the characteristics of each type of revolution and techniques for creating them.

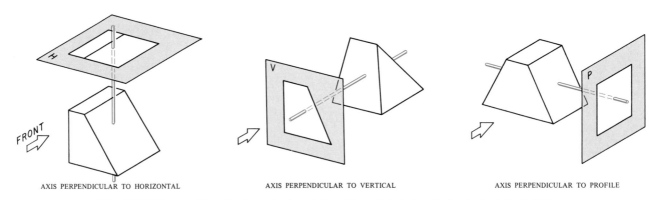

AXIS PERPENDICULAR TO HORIZONTAL   AXIS PERPENDICULAR TO VERTICAL   AXIS PERPENDICULAR TO PROFILE

Fig. 7-16 *Three positions for the axis of revolution. The axis is perpendicular to the principal planes.*

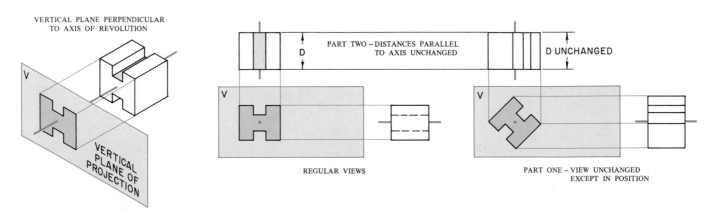

Fig. 7-17 *The rule of revolution. Note that the H-shape in the front view has changed only position, not shape.*

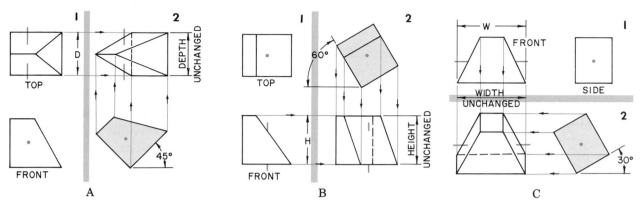

Fig. 7-18 *Single revolution about the three axes.*

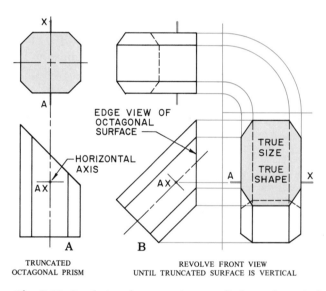

Fig. 7-19 *Revolution about an axis perpendicular to the vertical plane.*

Revolution About an Axis Perpendicular to the Vertical Plane

In Fig. 7-18A, the usual front and top views of an object are shown on the left. On the right, similar views of the same object are shown after the object has been revolved 45° counterclockwise about an axis perpendicular to the vertical plane. Notice that the front view is the same in size and shape, except that it has a new position. The new top view has been made by projecting up from the new front view and across from the old top view. Note that the depth remains the same from one top view to the other.

Figure 7-19 shows how to draw a primary revolution perpendicular to the vertical plane. In (A), an imaginary axis *AX* is passed horizontally through a truncated right octagonal prism. In the front view, it shows as a point. In the top view, and later in the side view, it shows as a line. In (B), the prism has been revolved clockwise about the axis into a new position. You can see that the new front view has the same size and shape as the old. Just its position has changed. However, the side view now shows the true size and shape of the truncated surface. This side view is made by projecting across from the new front view and by transferring the depth from the top view.

Revolution About an Axis Perpendicular to the Horizontal Plane

In Fig. 7-18B, an object is shown on the left in the usual top and front views. The views on the right were drawn after the object had been rotated 60° clockwise about an axis perpendicular to the horizontal plane. The new top view is the same in size and shape as the old top view. The new front view has been made by projecting down from the new top view and across from the old front view. Note that the height remains the same from the original front view to the revolved front view.

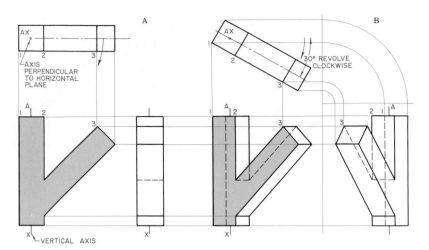

Fig. 7-20 *Revolution about an axis perpendicular to the horizontal plane (clockwise).*

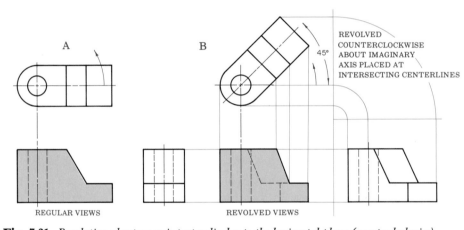

Fig. 7-21 *Revolution about an axis perpendicular to the horizontal plane (counterclockwise).*

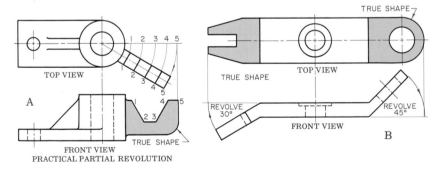

Fig. 7-22 *The inclined surfaces on these objects can be shown clearly by revolving only the inclined parts.*

Figure 7-20 shows how to draw an object that is revolved clockwise about an imaginary vertical axis *AX*. In (A), the three regular views are given. In (B), the top view has been revolved 30° clockwise about the axis *AX*. Since this is a vertical axis, revolution does not change the height of the object. Therefore, points from the old vertical plane

shown in (A) can be projected to make the new one in (B). The new side view is made from the front and top views in the usual way.

Figure 7-21 shows how to draw an object revolved counterclockwise through 45°. Notice that the procedure is the same regardless of the direction or degree of rotation.

Revolution About an Axis Perpendicular to the Profile Plane

At the top of Fig. 7-18C, a third object is shown in the usual front and side views. Below it, the same views show the object after it has been revolved forward (counterclockwise) 30° about an axis perpendicular to the profile plane. The new front view has been made by projecting across from the new side view and down from the old front view. Note that the width remains the same from one front view to the other.

Partial Revolved Views

In a working drawing, you can show an inclined surface by drawing a full or partial view with the object in a revolved position. In Fig. 7-22A, the top view shows the angle of a V-shaped part. In the front view, the part is revolved to show its true shape.

In Fig. 7-22B, the front view shows the angles at which surfaces of a part are inclined. The inclined surfaces are then revolved in the front view. Next, their dimensions (sizes) are transferred to the top view. There, the surfaces appear in true size and shape.

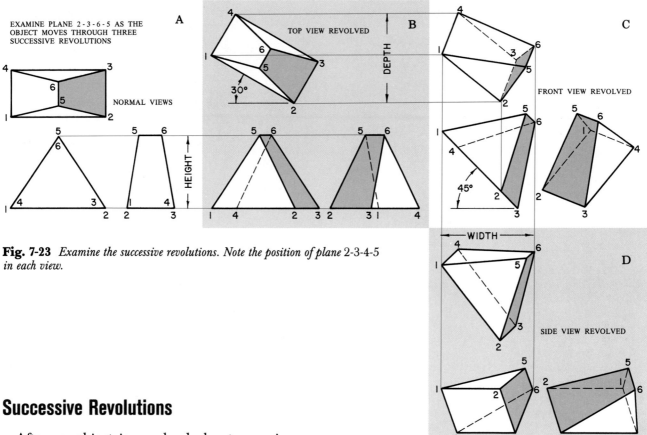

EXAMINE PLANE 2 - 3 - 6 - 5 AS THE OBJECT MOVES THROUGH THREE SUCCESSIVE REVOLUTIONS

NORMAL VIEWS

TOP VIEW REVOLVED

FRONT VIEW REVOLVED

SIDE VIEW REVOLVED

Fig. 7-23 *Examine the successive revolutions. Note the position of plane 2-3-4-5 in each view.*

Successive Revolutions

After an object is revolved about an axis perpendicular to one plane, it can be revolved again about an axis perpendicular to another plane. This process, known as **successive revolutions,** is shown in Fig. 7-23. In (A), an object is shown in the normal views. In (B), the object has been revolved 30° clockwise about an axis perpendicular to the horizontal plane. In (C), the front view from (B) has been revolved 45° clockwise about an axis perpendicular to the vertical plane. In (D), the side view from (C) has been revolved about an axis perpendicular to the profile plane until line *3-4* appears as a horizontal line. In all, three successive revolutions occurred, one in each of the three principal planes of projection.

Industrial Applications

The following illustrations show how a drafter uses revolution. Figure 7-24 shows a tractor with various lift positions. Figure 7-25 shows a product designer's use of a revolved position. Figures 7-26A and B illustrate the *transit* (a measurement device) and the axes of revolution used by the civil engineer.

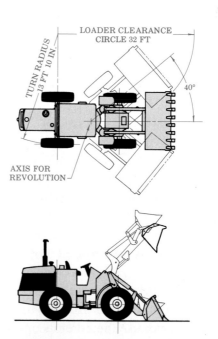

Fig. 7-24 *The profile of a tractor shows several positions of the loading bucket. The plane view shows horizontal rotation.*

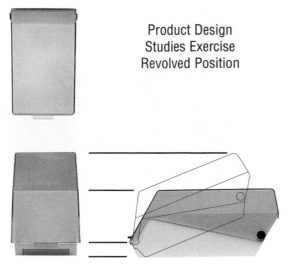

Product Design
Studies Exercise
Revolved Position

Fig. 7-25 *Note that the profile of this disk holder has been revolved to show its true shape.* (Courtesy of Keith Berry and Circle Design)

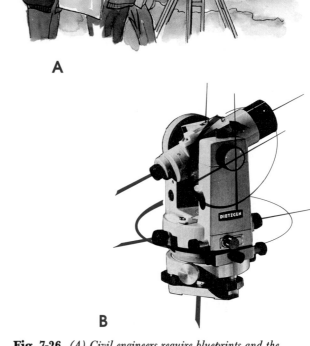

A

B

Fig. 7-26 *(A) Civil engineers require blueprints and the knowledge of revolutions when using the automatic level for examining topography. (B) Note the horizontal and vertical planes of revolution on this transit.* (A- Courtesy of Circle Design.; B- Courtesy of The Dietzgen Co.)

CAD

CAD REVOLUTIONS

The ability of CAD software to rotate objects allows CAD drafters to perform revolutions in CAD. For example, AutoCAD's ROTATE command allows you to rotate any object or group of objects in two or three dimensions (Fig. 7-27).

The ROTATE3D command is a more advanced command meant for use on 3D drawings only. This command contains several options that make it easier to work with an object in three dimensions. You can choose the axis of rotation in many ways. For example, you can align the axis of rotation with an existing object, or you can align it with the current view of the object. You can also use the X, Y, or Z axis as the axis of rotation. If none of these options will produce the revolution you need, you can use the 2 Points option to place the axis at any angle anywhere on the screen.

TRUE SIZE AND SHAPE

You can show the true size of an inclined surface by either an auxiliary view (Fig 7-28A) or a revolved view (Fig. 7-28B and C). In a revolved view, the inclined surface is turned until it is parallel to one of the principal planes. The revolved view in (B) and (C) is similar to the auxiliary in (A).

In the auxiliary view, it is as if the observer has changed position to look at the object from a new direction. Conversely, in the revolved view, it is as if the object has changed position. Both revolutions and auxiliaries help you visualize things better. They also work equally well in solving problems.

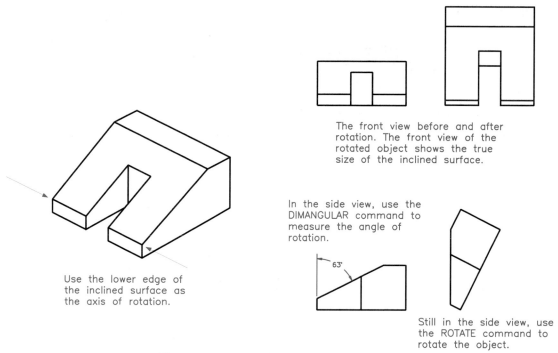

The front view before and after rotation. The front view of the rotated object shows the true size of the inclined surface.

In the side view, use the DIMANGULAR command to measure the angle of rotation.

Use the lower edge of the inclined surface as the axis of rotation.

Still in the side view, use the ROTATE command to rotate the object.

Fig. 7-27 *Rotating a three-dimensional CAD object.*

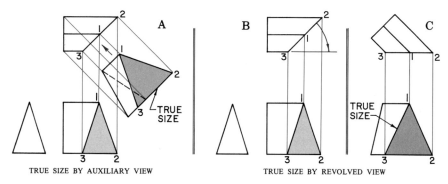

TRUE SIZE BY AUXILIARY VIEW

TRUE SIZE BY REVOLVED VIEW

Fig. 7-28 *You can determine the true size and shape of an object by creating an auxiliary view (A) or by revolving the object (B).*

True Shape of an Oblique Plane

A surface shows its true shape when it is parallel to a plane. Figure 7-29 shows how successive revolutions can be used to find the true shape of a surface. In (D), an object is pictured on which surface *1-2-3-4* is an oblique plane. It is oblique because it is inclined to all three of the normal planes. In (A), the object is drawn in its normal position. In (B), it has been revolved about an axis perpendicular to the horizontal plane until surface

1-2-3-4 is perpendicular to the vertical plane. Now, in the front view, all you see of this surface is its edge line. In (C), the object has been revolved about an axis perpendicular to the horizontal plane until surface *1-2-3-4* is perpendicular to the vertical plane. Now, in the front view, all you see of this surface is its edge line. In (C), the object has been revolved about an axis perpendicular to the front plane until surface *1-2-3-4* is parallel to the profile plane. There it shows its true shape.

True Length of a Line

Since an auxiliary view shows the true size and shape of an inclined surface, it can also be used to find the true length of a line. In Fig. 7-30A, the true length of line *OA* is not apparent in the top, front, or right-side view because it is inclined to all three normal planes. The auxiliary plane shown in (B) does show the true length (TL). This is because the auxiliary plane is parallel to the surface *OAB*.

Figure 7-30C shows another way to show the true length of *OA*. Revolve the object about an axis perpendicular to the vertical plane until surface *OAB* is parallel to the profile plane. The side view then shows the true size of *OAB* and also the true length of *OA*. A shorter method of showing the true length of *OA* is to revolve only the surface *OAB*, as shown in (D).

Figure 7-30E shows the object revolved in the top view until line *OA* is horizontal in that view. The front view now shows *OA* in its true length because this line is now parallel to the vertical plane.

In (F) and (G), still another method is shown. In this case, instead of the whole object being revolved, just line *OA* is turned in the top view until it is horizontal at *OA'*. Point *A'* then can be projected to the front view. There, *OA'* will be shown at its true length.

You can revolve a line in any view to make it parallel to any one of the three principal planes. Projecting the line on the plane to which it is parallel shows its true length. In Fig. 7-30H, the line has been revolved parallel to the horizontal plane. The true length then shows the top view.

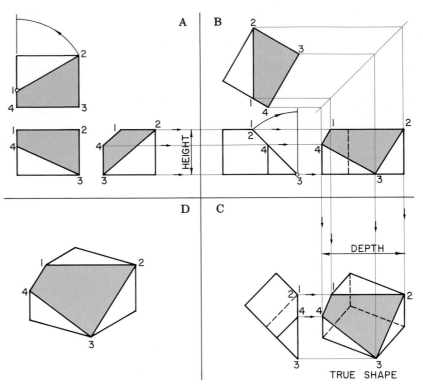

Fig. 7-29 *To find the true size of oblique plane 1-2-3-4, you must perform successive revolutions.*

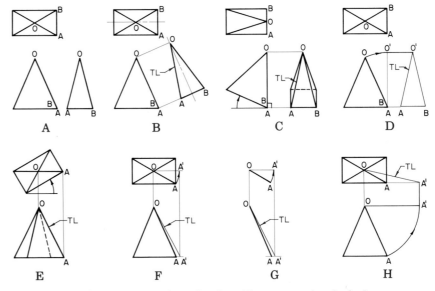

Fig. 7-30 *Typical true-length problems examined and solved.*

CHAPTER 7

Learning Activities

1. Obtain drawings from industry that show auxiliary views. Study the drawings to see if auxiliary views were really necessary for a clear understanding of the part.

2. Use Plexiglas® to construct an auxiliary projection box to show an inclined surface on a part.

3. Collect objects with inclined surfaces. Determine if they require primary or secondary auxiliary views. Do they need three normal views in addition to the auxiliary views? How many normal views are needed? Can partial views be used in some cases?

4. Use a CAD system to compare the complexity of developing an auxiliary view using traditional drafting methods vs. CAD methods. Which is faster? Which is easier?

Questions

Write your answers on a separate sheet of paper.

1. Under what circumstances does a drafter need to draw auxiliary views of an object?

2. How should you place an auxiliary plane in relation to the inclined surface it describes?

3. What is the difference between a primary auxiliary view and a secondary auxiliary view?

4. Name the three primary auxiliary views.

5. List the steps in drawing a primary auxiliary view.

6. What is center-plane construction?

7. Explain the procedure for plotting curved lines on an auxiliary view.

8. What is a partial auxiliary view? Why is it sometimes preferred over a full auxiliary view?

9. Describe how to find the true length of a line using an auxiliary view.

10. When you use a CAD program, what is the advantage of creating a three-dimensional drawing of an object that has an oblique surface?

11. What is the basic reason for revolving the view of an object?

12. What is an axis of revolution?

13. Describe the rule of revolution.

14. Name the three basic single revolutions.

15. What are successive revolutions? Why are they sometimes needed?

16. What two commands in AutoCAD allow you to create revolutions? What is the difference between the two commands?

17. Can you use both auxiliary views and revolved views to find true lengths of inclined and oblique lines? Explain.

CHAPTER 7
PROBLEMS

The small CAD symbol next to each problem indicates that a problem is appropriate for assignment as a computer-aided drafting problem and suggests a level of difficulty (one, two, or three) for the problem.

Figs. 7-31 through 7-50 *In Figs. 7-32 through 7-50, only the top view is given. For each figure, draw the top and front views and either the complete auxiliary view or just the inclined surface, as directed by the instructor. Figure 7-31 has been worked as an example. Dimension the problems assigned by the instructor. The angle X may be 45° or 60° as assigned. The total height of the front view is 3.75"≤ for Figs. 7-31 through 7-50.*

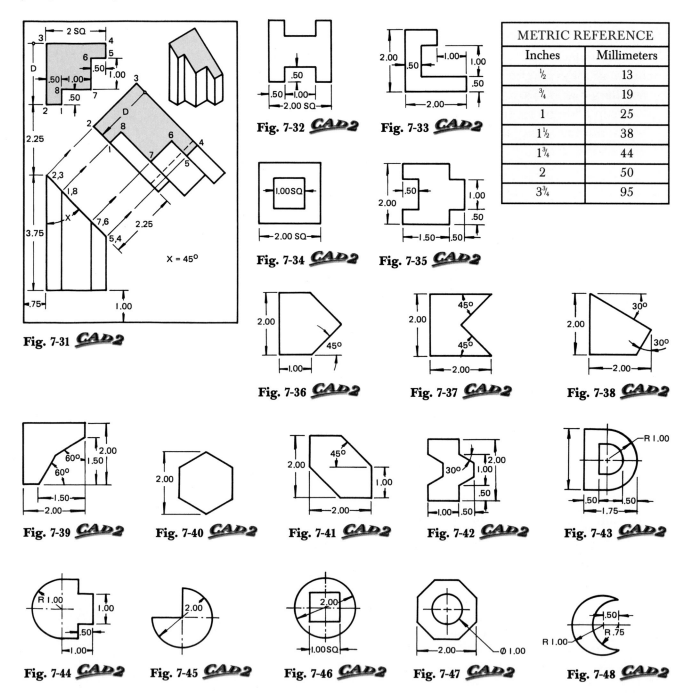

Fig. 7-31 *CAD2*

Fig. 7-32 *CAD2*

Fig. 7-33 *CAD2*

Fig. 7-34 *CAD2*

Fig. 7-35 *CAD2*

Fig. 7-36 *CAD2*

Fig. 7-37 *CAD2*

Fig. 7-38 *CAD2*

Fig. 7-39 *CAD2*

Fig. 7-40 *CAD2*

Fig. 7-41 *CAD2*

Fig. 7-42 *CAD2*

Fig. 7-43 *CAD2*

Fig. 7-44 *CAD2*

Fig. 7-45 *CAD2*

Fig. 7-46 *CAD2*

Fig. 7-47 *CAD2*

Fig. 7-48 *CAD2*

METRIC REFERENCE	
Inches	Millimeters
½	13
¾	19
1	25
1½	38
1¾	44
2	50
3¾	95

CHAPTER 7
PROBLEMS

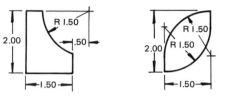

Fig. 7-49 *CAD2*

Fig. 7-50 *CAD2*

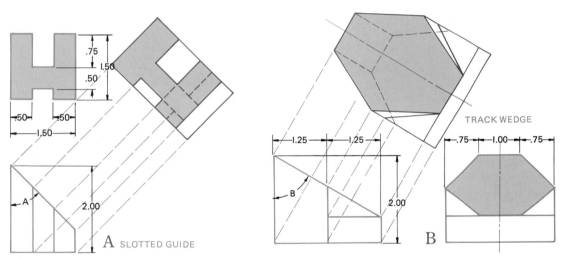

A SLOTTED GUIDE

TRACK WEDGE

B

Fig. 7-51 *CAD3* Part 1: Draw the front, top, and side views of each figure, A and B. Complete the front auxiliary projection. Part 2: Change the angles of the inclined surface in (A) to 30° and in (B) to 45°.

Figs. 7-52 through 7-54 *CAD3* Draw the front, top, and complete auxiliary views. The solutions are given for Figs. 7-52 and 7-53. Change the angle for Fig. 7-52 to 30° and the angle for Fig. 7-53 to 45°. Is there any difference in the solution?

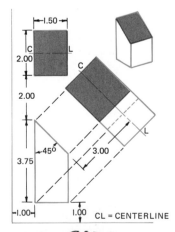

CL = CENTERLINE

Fig. 7-52 *CAD3*

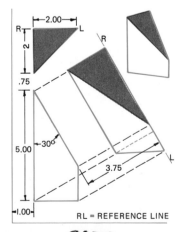

RL = REFERENCE LINE

Fig. 7-53 *CAD3*

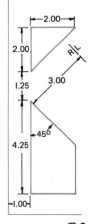

Fig. 7-54 *CAD3*

PROBLEMS

Figs. 7-55 through 7-58 *CAD3* *Draw the front, top, and auxiliary views.*

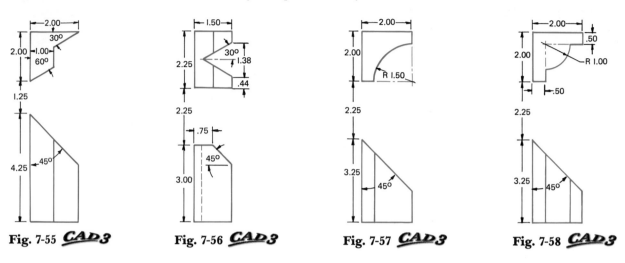

Fig. 7-55 *CAD3* **Fig. 7-56** *CAD3* **Fig. 7-57** *CAD3* **Fig. 7-58** *CAD3*

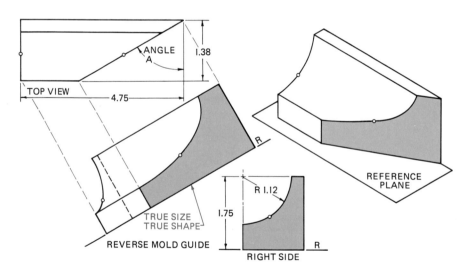

Fig. 7-59 *CAD3* *Develop the three views. Change angle A to 45° and develop the top auxiliary projection.*

PROBLEMS

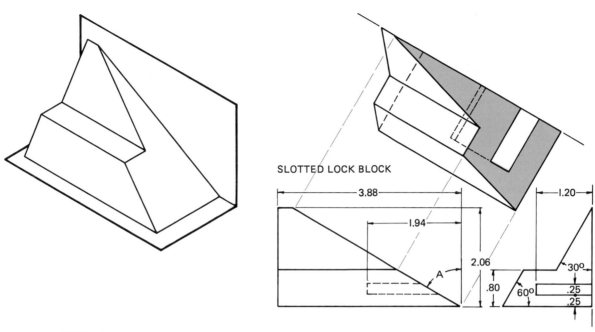

SLOTTED LOCK BLOCK

Fig. 7-60 _CAD3_ _Determine the views necessary to complete the front auxiliary view. Develop views with angle A = 60°. Alternate problems may be assigned with angle A at 45° or 75°._

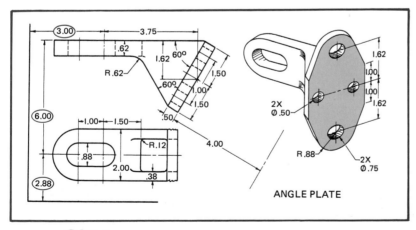

ANGLE PLATE

Fig. 7-61 _CAD3_ _A layout and pictorial for an angle plate are given. Draw the top view and the partial front view as shown. Draw a partial auxiliary view where indicated on the layout. Note that this is an auxiliary elevation. It is made on a plane perpendicular to the horizontal plane._

PROBLEMS

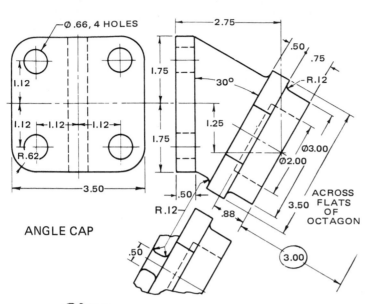

ANGLE CAP

Fig. 7-62 *CAD3 A part front view, a right-side view, and a part auxiliary view of the angle cap are shown on the layout. Draw the views given and another auxiliary view where indicated on the layout. This will be a rear auxiliary view. Dimensioning is required.*

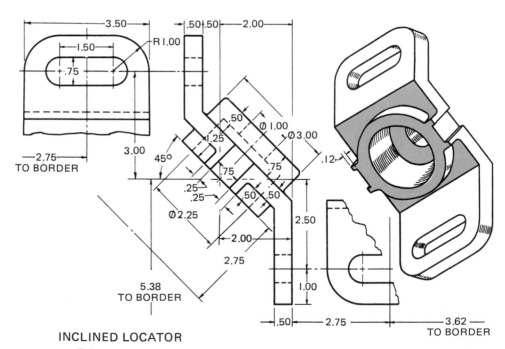

INCLINED LOCATOR

Fig. 7-63 *CAD3 A pictorial and layout of an inclined stop. The complete view in the middle is the right-side view. Draw the complete view and the partial views as necessary. Draw an auxiliary view of the inclined middle, as indicated in the layout.*

PROBLEMS

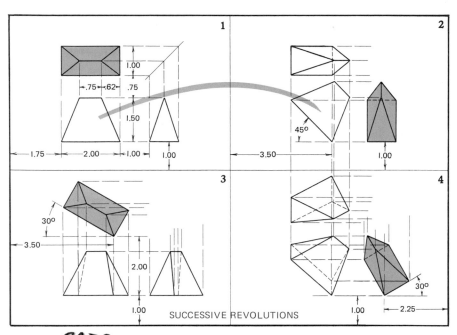

Fig. 7-64 *CAD3* *This figure shows the complete problem. It is given for comparison and is not to be copied. Layout 1 (upper left) contains a three-view drawing of a block in its simplest position. Layout 2 (upper right) shows the block after it has been revolved through 45° about an axis perpendicular to the frontal plane. The front view was drawn first, copying the unrevolved front view. The top view was obtained by projecting up from the front view and across from the unrevolved top view. In layout 3 (lower left), the block has been revolved through 30° about an axis perpendicular to the horizontal plane. The top view was drawn first, copied from the unrevolved top view. In layout 4 (lower right), the block has been rotated 30° from its position in layout 2 about an axis perpendicular to the side plane. The side view was drawn first, copied from the side view of layout 2. The widths of the front and top views were projected from the front view of layout 2.*

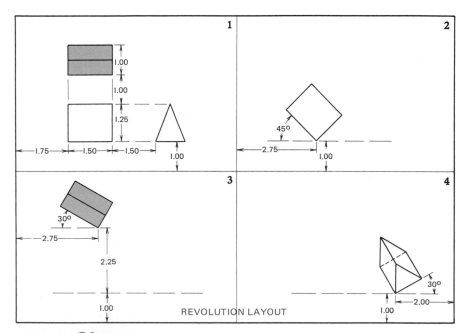

Fig. 7-65 *CAD3* *Draw the revolved views of the wedge shown in layout 1. In layout 2, revolve 45° clockwise about an axis perpendicular to the frontal plane and draw three views. In layout 3, revolve 30° counterclockwise about an axis perpendicular to the horizontal plane.*

PROBLEMS

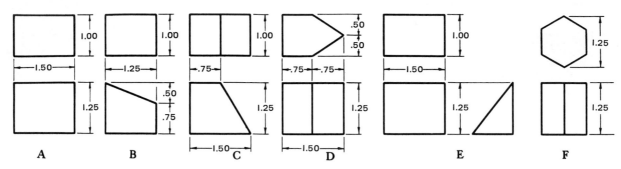

Fig. 7-66 CAD3 *Follow the directions for Fig. 7-65 for the objects in A through F, as assigned.*

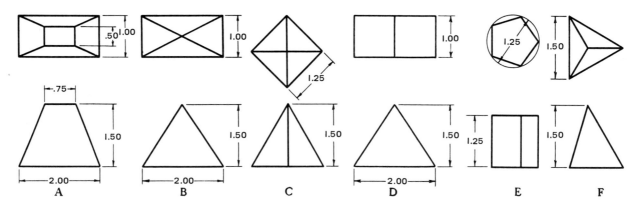

Fig. 7-67 CAD3 *Follow the directions for Fig. 7-65. Optional assignments may change revolutions to determine true sizes of inclined surfaces.*

PROBLEMS

SCHOOL-TO-WORK

Solving Real-World Problems

Fig. 7-68 *SCHOOL-TO-WORK problems are designed to challenge a student's ability to apply skills learned within this text. Be creative and have fun!*

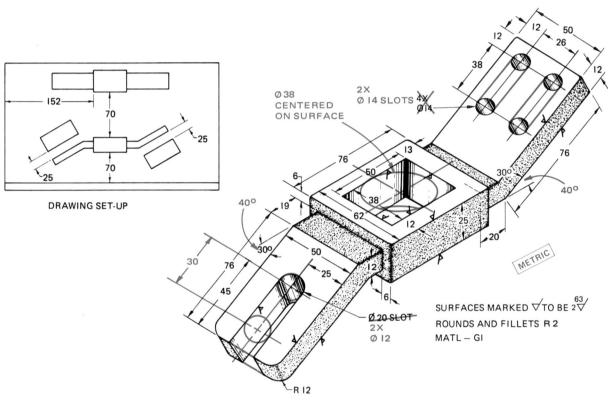

DRAWING SET-UP

Ø 38
CENTERED
ON SURFACE

2X
Ø 14 SLOTS

4X
Ø 14

SURFACES MARKED ∇ TO BE 2∇ 63/

ROUNDS AND FILLETS R 2

MATL — GI

METRIC

Ø 20 SLOT
2X
Ø 12

R 12

Fig. 7-68 *CAD3 The engineers in your company have redesigned this connecting bar to be used on a new tractor hitch. Use the drawing setup shown and make all design changes marked in color by the engineer on the pictorial drawing. Draw front, top, and two auxiliary views. Include all shape and size information necessary for the manufacture of the part. Scale: 1:1.*

BASIC DESCRIPTIVE GEOMETRY

OBJECTIVES

Upon completion of Chapter 8, you will be able to:

○ Manipulate points, lines, and planes in space for the purpose of establishing true positions, true sizes, and true shapes of features.

○ Define terms such as *azimuth, reference plane, slope, bearing,* and *oblique plane.*

○ Establish the true length of an oblique line.

○ Determine the shortest distance between a point and a line or between two lines.

VOCABULARY

In this chapter, you will learn the meanings of the following terms:

- azimuth
- bearing
- coincide
- descriptive geometry
- dihedral angle
- foreshortened
- frontal line
- grade
- inclined line
- inclined plane
- level line
- normal line
- normal plane
- oblique line
- oblique plane
- piercing point
- profile line
- skew lines
- slope
- user coordinate system (UCS)

The designer who works with an engineering team can help solve problems by producing drawings made up of geometric elements. *Geometric elements* are points, lines, and planes defined according to the rules of geometry. Every structure has a three-dimensional form made up of geometric elements (Fig. 8-1). In order to draw three-dimensional forms, you must understand how points, lines, and planes relate to each other in space to form a certain shape. Problems that you might think need mathematical solutions can often be solved through drawings that make manufacturing and construction possible. **Descriptive**

geometry is one of the methods a designer uses to solve problems. In the eighteenth century a French mathematician, Gaspard Monge, developed a system of descriptive geometry (called the *Mongean method*) for solving *spacial* (space) problems related to military structures. Claude Crozet brought descriptive geometry to the U.S. Military Academy at West Point in 1816. While the Mongean method has changed over the years, its basic principles are still taught in engineering schools throughout the world. By studying descriptive geometry, you develop a reasoning ability to solve problems through drawing.

Most structures designed by people are shaped like a rectangle. This is because it is easy to plan and build a structure with this shape. This chapter presents a way of drawing that lets you analyze *all* geometric elements in three-dimensional space. Learning to see geometric elements makes it possible for you to describe a structure of any shape. Figure 8-2 shows the basic geometric elements. It also shows some of the common geometric features found in engineering designs.

Fig. 8-1 *This bridge shows geometric elements.* (Courtesy of Ken Trevarthan)

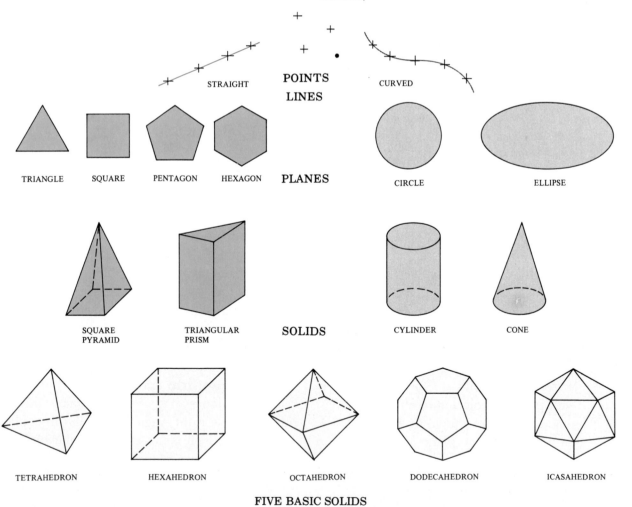

FIVE BASIC SOLIDS

Fig. 8-2 *Basic geometric elements and shapes.*

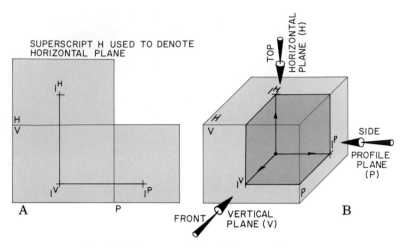

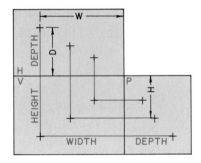

Fig. 8-3 *Locating and identifying a point in space.*

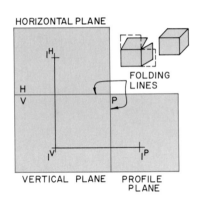

Fig. 8-4 *Points identified on unfolded reference planes.*

Fig. 8-5 *Points are related to coordinated reference planes.*

POINTS

A point can be thought of as having an actual physical existence. On a drawing, you can locate a point with a small dot or a small cross. Normally, a point is identified using two or more projections. In Fig. 8-3A, the small cross for point number 1 is shown in the front, top, and right-side views. In (B), the regular reference planes are shown in a pictorial view with point *1* projected to all three planes. The reference planes are shown again in Fig. 8-4. Notice that when the three planes are unfolded, a flat two-dimensional surface is formed. The fold lines are labeled as shown. *V* stands for the vertical view, *H* stands for the horizontal, or top, view, and *P* stands for the profile, or right-side, view. Points

are also used to identify the intersection of two lines or the corners on an object.

Figure 8-5 shows a group of points. Points are related to each other by distance and direction. These are measured on the coordinated reference planes. You can see the relative vertical height dimensions in the front and side views, the width dimensions in the front and top views, and the depth dimensions in the top and side views. Note that the three basic dimensions are indicated by *H, W,* and *D* (Fig. 8-5).

LINES

If a point moves away from a fixed place, its path forms a line. A line has location, direction, and length. It is easy to draw circular and straight lines. However, plotting irregular curves is somewhat more difficult and must be done very carefully. You can determine a straight line by specifying two points or by specifying one point and a fixed direction.

The Basic Lines

Lines are classified according to how they relate to the coordinating reference planes. The three basic types are described in the following paragraphs.

A **normal line** is one that is perpendicular to one of the three reference planes. It projects onto that plane as a point (Fig. 8-6A, B, and C). If a normal line is parallel to the other two reference planes, as shown in Fig. 8-6D, E, and F, it is shown at its true length (TL), as noted.

An **oblique line** appears inclined in all three reference planes (Fig. 8-8). It makes an angle other than a right angle with all three planes. In other words, it is not perpendicular or parallel to any of the three planes. The true length is not shown in any of these views. Also, the angles of direction cannot be measured on the main reference planes.

True Length of Oblique Lines

Normal lines and inclined lines project parallel to at least one of the main planes of projection. A line parallel to a plane of projection shows true length in that projection. Since an oblique line is not parallel to any of the three main reference planes, you must use an auxiliary reference plane parallel to the oblique line to show its true length (Fig. 8-9A, B, and C). The auxiliary and regular planes of projection have the same relationship as any two main planes. First, they must always be perpendicular to each other (Fig. 8-9D). Second, they must be measured in relation to the plane to which they are related. Obtain the true length as shown. Refer to Chapter 7, "Auxiliary Views and Revolutions," for more information about auxiliary planes.

Line Terminology

It may seem that lines drawn on paper mean little and are worth little. However, they do reflect real things. Therefore, they are used in all aspects of industry every day. The following terms are used in technical areas such as mining, geology, engineering, and navigation.

Slope

A line that makes an angle with the horizontal plane has a **slope** that is measured in degrees. In Fig. 8-10A, the true slope is shown in the front view when the line is in true length. Use an auxiliary projection perpendicular to a horizontal reference to find the slope of an oblique line in true length (B).

Bearing

The angle a line makes in the top view with a north-south line is called its **bearing** (Fig. 8-11A).

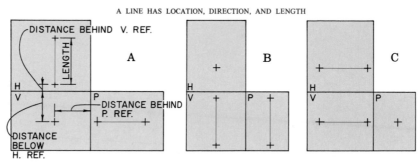

A LINE HAS LOCATION, DIRECTION, AND LENGTH

NORMAL LINES - PERPENDICULAR TO A PRINCIPAL REFERENCE PLANE

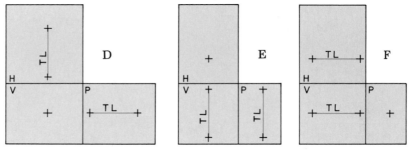

NORMAL LINES - PERPENDICULAR AND PARALLEL WILL ALWAYS BE TRUE LENGTH (T. L.) AS SHOWN.

Fig. 8-6 *Normal lines in true length are parallel to two reference planes.*

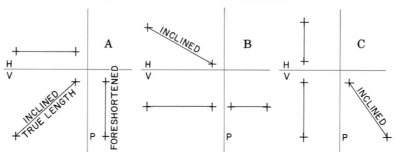

TRUE LENGTH SHOWS ON INCLINED PROJECTION

LINES THAT APPEAR PARALLEL TO REFERENCE PLANES ARE FORESHORTENED

Fig. 8-7 *Inclined lines are parallel to one reference plane.*

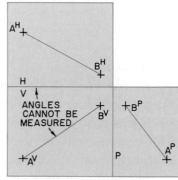

OBLIQUE LINE - THIS LINE IS NOT PARALLEL OR PERPENDICULAR TO ANY ONE OF THE THREE PRINCIPAL REFERENCE PLANES

Fig. 8-8 *Oblique lines appear inclined in all projections.*

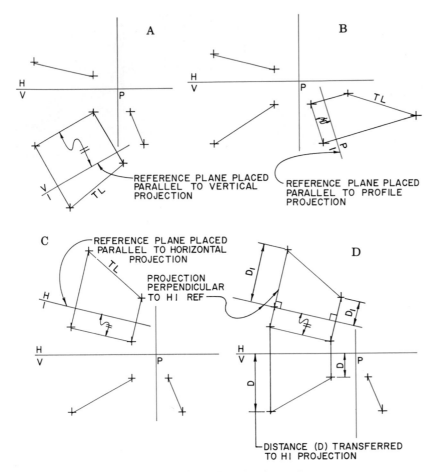

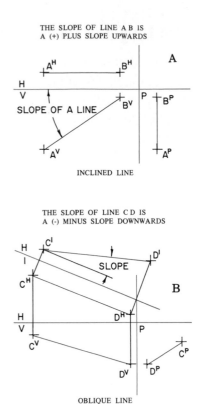

Fig. 8-10 *Slope of a line in the vertical (elevation) projection.*

Fig. 8-9 *True length of an oblique line by auxiliary projection.*

The north-south line is generally vertical, with north at the top. Therefore, right is east and left is west. Make the measurement in the horizontal projection. Dimension it in degrees, as shown in Fig. 8-11B.

Azimuth

A measurement that defines the direction of a line off due north is the **azimuth.** The azimuth is always measured off the north-south line in the horizontal plane. It is dimensioned in a clockwise direction, as shown in Fig. 8-11C.

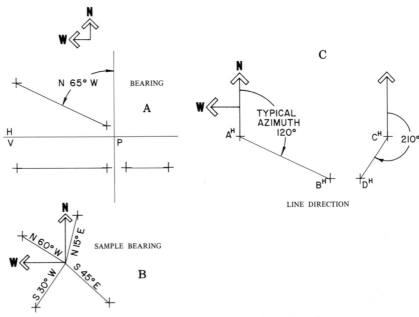

Fig. 8-11 *(A) Bearing is identified. (B) Typical readings. (C) Azimuth readings related to due north.*

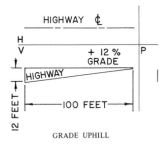

Fig. 8-12 *Grade is measured in the vertical projection.*

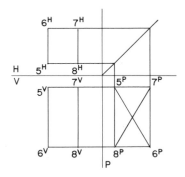

ALL LINES ARE PARALLEL

Fig. 8-13 *Lines are parallel when all three of the reference projections are parallel.*

LINES MUST APPEAR PARALLEL IN ALL THREE VIEWS TO BE PARALLEL

Fig. 8-14 *Lines in 3D space examined for parallel relationship.*

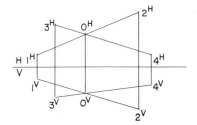

POINT 0 INDICATES ALIGNED INTERSECTION

Fig. 8-15 *Intersecting lines with aligned points.*

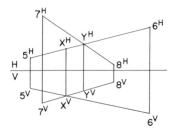

APPARENT POINTS OF INTERSECTION ARE NOT ALIGNED IN TWO PROJECTIONS

Fig. 8-16 *Lines in space examined for intersection.*

Figure 8-15 shows the alignment needed to check the point of intersection of two straight lines. In Fig. 8-16, the points of intersection in the H and V projection are not aligned. Thus, the intersection is incomplete. How would the intersection look if it were completed?

Grade

Incline, or **grade,** is measured as a percentage. Figure 8-12 shows the scale for constructing a highway with a +12 percent grade. The grade rises 12 ft. (3.6 m) in every 100 ft. (30 m) of horizontal distance.

NOTE: In the rest of this chapter, the cross symbol for locating points will no longer be used in the figures.

Parallel Lines

Figure 8-13 shows the relationship of parallel lines in a three-view study. Line projections are parallel if they appear parallel in all three reference planes. Note that the lines in Fig. 8-14 seem parallel in the front and top views, but are not parallel in the side view.

Intersecting Lines

If two lines intersect, they have at least one point in common.

Perpendicular Lines

Perpendicular lines are examined in Fig. 8-17. To find out if two lines are perpendicular, and thus have a right angle between them, first find the true length of one line. This lets you know if the angle between the lines is actually a right angle. Note that the true length in Fig. 8-17A shows that one line is parallel to

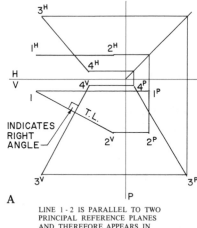

A

LINE 1-2 IS PARALLEL TO TWO PRINCIPAL REFERENCE PLANES AND THEREFORE APPEARS IN TRUE LENGTH IN THE VERTICAL PROJECTION

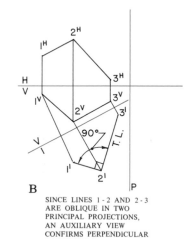

B

SINCE LINES 1-2 AND 2-3 ARE OBLIQUE IN TWO PRINCIPAL PROJECTIONS, AN AUXILIARY VIEW CONFIRMS PERPENDICULAR CONDITION

Fig. 8-17 *The lines shown in (A) are perpendicular in the vertical projection. Those shown in (B) are perpendicular in an auxiliary projection.*

a main plane of projection. Examine the oblique lines in an auxiliary projection as shown in (B) to find true length and the right angle.

PLANES

As a line moves away from a fixed place, its path forms a plane. In drawings, planes are thought of as having no thickness. They are also *infinite*–they can be extended as far as desired. A plane can be determined by intersecting lines, two parallel lines, a line and a point, three points, a triangle, or any other *planar* (flat) surface, such as a two-dimensional polygon.

The Basic Planes

A plane can be classified according to its relation to the three main reference planes. The three basic types of planes are normal, inclined, and oblique. As you read the descriptions below, notice that they closely parallel the descriptions of normal, inclined, and oblique lines.

Normal Plane

A **normal plane** is perpendicular to two of the reference planes and parallel to the third. Figure 8-18 shows three examples of normal planes. Two of the main reference planes in each example show the edge view of the plane. (Recall that the edge view of a plane appears as a line.) In Fig. 8-18A, the plane is parallel to the vertical reference plane and perpendicular to the horizontal and profile planes. In (B), the plane is parallel to the horizontal reference plane and perpendicular to the vertical and profile planes. In (C), the plane is parallel to the profile reference plane and perpendicular to the vertical and horizontal planes.

Inclined Plane

An **inclined plane** is perpendicular to one reference plane and inclined to the other two.

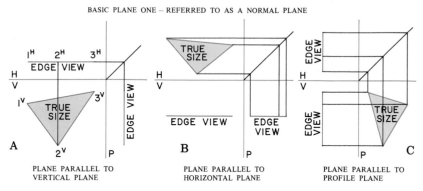

BASIC PLANE ONE – REFERRED TO AS A NORMAL PLANE

Fig. 8-18 *Normal planes shown in three projections.*

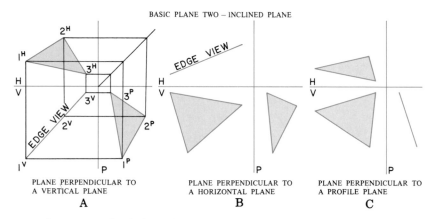

BASIC PLANE TWO – INCLINED PLANE

Fig. 8-19 *Inclined planes shown in the three principal planes of projection.*

Figure 8-19 shows examples of inclined planes. In one of the main reference planes, the inclined plane shows as an edge view. Thus, it is perpendicular to that plane. The other two reference planes show the plane as a foreshortened surface. In Fig. 8-19A, the inclined plane is perpendicular to the vertical reference plane. It is inclined to the horizontal and profile planes, where it is foreshortened. In (B), the inclined plane is perpendicular to the horizontal reference plane, where it shows as a line. The other two reference planes show the plane foreshortened. In (C), the inclined plane is perpendicular to the profile reference plane, where it shows as a line. The plane shows as a foreshortened surface in the other two reference planes.

Oblique Plane

An **oblique plane** is inclined to all three reference planes. An example is shown in Fig. 8-20A. Since the oblique plane is not perpendicular to any of the three main reference planes, by definition it cannot be parallel to any of the three planes. Thus, it shows as a foreshortened plane in each of the three regular views. Fig. 8-20B shows the same oblique plane in a pictorial rendering.

SOLVING DESCRIPTIVE GEOMETRY PROBLEMS

Now that the basic geometric constructions have been described, you may concentrate on how to use the geometry to solve problems. This section begins with rather simple operations and proceeds to describe the solutions to more complex problems.

It should become clear as you work through the problems that Chapters 7 and 8 are closely related. Almost all problems in descriptive geometry can be worked out by using auxiliary planes. You can solve problems by knowing how to find the:

- true length of a line
- point projection of a line
- edge view of a plane
- true size of a plane figure

The ability to understand and solve these problems will build the visual powers necessary for moving on to design problems.

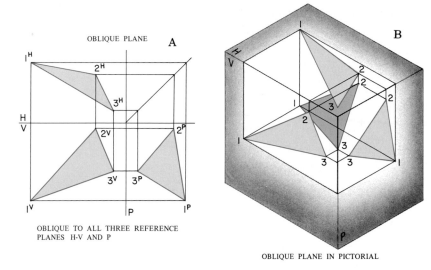

OBLIQUE TO ALL THREE REFERENCE PLANES H-V AND P

OBLIQUE PLANE IN PICTORIAL

Fig. 8-20 *An oblique plane in the three-view projection (A) and pictorial (B).*

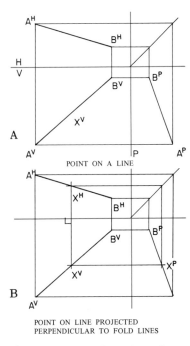

POINT ON A LINE

POINT ON LINE PROJECTED PERPENDICULAR TO FOLD LINES

Fig. 8-21 *A point located on a line.*

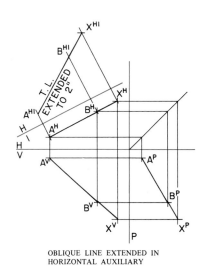

OBLIQUE LINE EXTENDED IN HORIZONTAL AUXILIARY

Fig. 8-22 *Straight lines may be extended.*

Point on a Line

In Fig. 8-21A, line *AB* on the vertical plane has a point *X*. To place the point on the line in the other two reference planes, project construction lines perpendicular to the folding lines, as in (B). Project the construction lines across to $A^H B^H$ and $A^P B^P$ to locate point *X* in the horizontal and profile projections.

Extend straight lines to new points on either end, as needed, to solve a problem (Fig. 8-22).

Note that you cannot tell whether a point is located on a line using just one view. It may seem to be on a line in one view, but another view may show that it is really in front, on top, or in back of the line, as shown in Fig. 8-23.

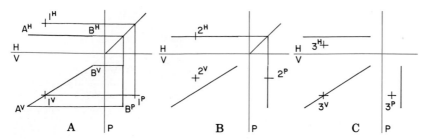

A POINT MAY APPEAR ON A LINE BECAUSE OF POSITION

Fig. 8-23 *Defining a point-line relationship.*

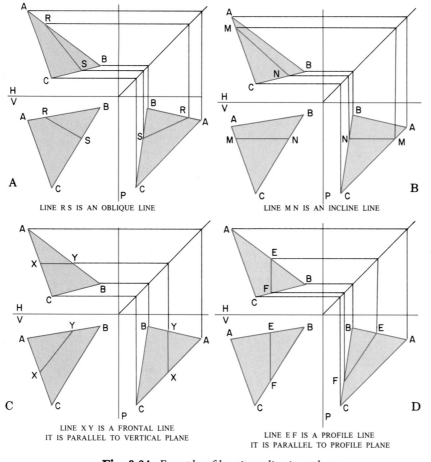

LINE R S IS AN OBLIQUE LINE

LINE M N IS AN INCLINE LINE

LINE X Y IS A FRONTAL LINE
IT IS PARALLEL TO VERTICAL PLANE

LINE E F IS A PROFILE LINE
IT IS PARALLEL TO PROFILE PLANE

Fig. 8-24 *Examples of locating a line in a plane.*

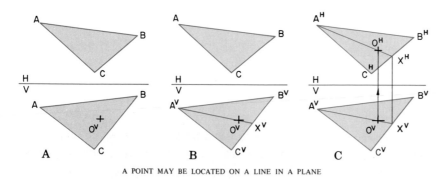

A POINT MAY BE LOCATED ON A LINE IN A PLANE

Fig. 8-25 *Locating a point in a plane.*

Line in a Plane

A line lies in a plane if it intersects two lines of the plane. It also lies in a plane if it intersects one line and is parallel to another line of that plane. Figure 8-24 shows how to add lines to planes. In (A), line *RS* must be a part of plane *ABC,* since *R* is on line *AB* and *S* is on *BC.* You know that line *RS* is an oblique line because it is not parallel to a main plane of projection and is clearly not perpendicular to the reference planes.

In Fig. 8-24B, a horizontal line *MN* is constructed in the vertical projection of plane *ABC.* This line is called a **level line.** Projecting *MN* to the right lines in the horizontal projection shows that it is an inclined line. The top view shows the true length.

In Fig. 8-24C, a line *XY* is constructed parallel to the HV reference line in the horizontal reference plane. Projected into the vertical plane, it shows as an inclined line in true length. This line is called a **frontal line,** since it is parallel to the vertical plane.

Figure 8-24D shows a vertical line *EF* constructed within plane *ABC.* It is parallel to the profile reference plane. Projecting line *EF* to the profile reference shows the line in true length. Line *EF* is called a profile line.

Point in a Plane

To locate a point in a plane, draw a line containing the point to the plane. Figure 8-25A shows that point *O* appears within plane *ABC.* Project line *AX,* which contains point *O,* as shown in (B).

Then project line *AX* to *ABC* in the horizontal reference plane, as shown in (C). Locate point *O* on the line by drawing a vertical projection to line *AX* in the horizontal reference plane.

Point View of a Line

A normal line projects as a point on the plane to which it is perpendicular. In Fig. 8-26A, line *AB* is a normal line—it is parallel to two main reference planes. It therefore shows as a point in the vertical reference plane. In Fig. 8-26B and C, the same conditions exist. The line projects as a point in the horizontal plane (B) and in the profile plane (C).

An inclined line projects as a point to an auxiliary plane (Fig. 8-27). Place a reference plane perpendicular to the inclined line at a chosen distance, and label it H/1 (Fig. 8-28). Transfer distance *D* as shown for a vertical or a horizontal auxiliary projection.

To project an oblique line as a point, use two auxiliary projections. Set up the first auxiliary reference plane parallel to the oblique inclined line, as shown in Fig. 8-29A. Then find the true length. Place the second auxiliary reference plane perpendicular to the true-length line of the first auxiliary. Locate the point projection by transferring distance *X* (Fig. 8-29B).

Distance Between Parallel Lines

Point projection is one way to show the true distance between two parallel lines. In Fig. 8-30, the parallel lines *MN* and *RS* are oblique. Two auxiliary projections are needed to find the point projections. The first auxiliary reference plane V/1 is parallel to *MN* and *RS*. In it, lines *MN* and *RS* are shown true length. The second

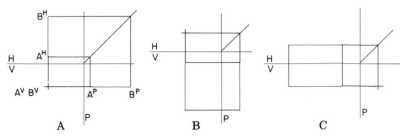

A NORMAL LINE HAS A POINT PROJECTION WHEN PERPENDICULAR TO REFERENCE PLANE

Fig. 8-26 *A point view of a normal line in three positions.*

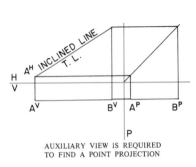

AUXILIARY VIEW IS REQUIRED
TO FIND A POINT PROJECTION

Fig. 8-27 *A point projection of an inclined line can be found in an auxiliary projection.*

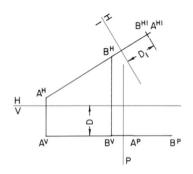

Fig. 8-28 *Transferring a point projection of an inclined line.*

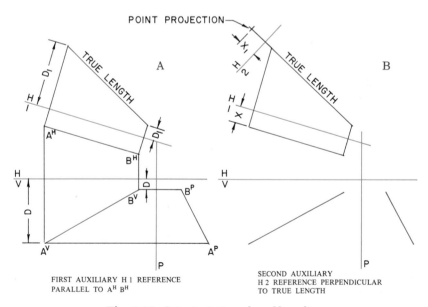

FIRST AUXILIARY H 1 REFERENCE
PARALLEL TO A^H B^H

SECOND AUXILIARY
H 2 REFERENCE PERPENDICULAR
TO TRUE LENGTH

Fig. 8-29 *Point projection of an oblique line.*

auxiliary reference plane V/2 is perpendicular to the true-length lines in the first auxiliary. The distance between the point projections of the lines is a true distance.

A second way of finding the shortest distance between two parallel lines is shown in Fig. 8-31. Think of lines *AB* and *CD* as parts of a plane. Connect points *A, B, C,* and *D* to form the plane. Draw a horizontal line *DX* in the top view and project point *X* into the vertical view. Then draw line *DX* in the vertical plane. Draw the first reference plane V/1 perpendicular to *DX* in the vertical view. Find the edge view of plane *ABCD* by transferring distances 1, 2, 3, and 4, as shown.

The second auxiliary V/2 shows the true lengths of *AB* and *CD*. The plane formed is in true size. Measure the true distance between the lines perpendicularly from *AB* to *CD*, as shown.

Distance Between a Point and a Line

To find the shortest distance from a point to a line, project the line as a point. In Fig. 8-32, project point *A* and oblique line *CD* into the first auxiliary projection H/1. In H/1, label the true length of *CD*. Place the second auxiliary reference plane H/2 perpendicular to line *CD*, and project line *CD* as a point in this plane. As shown, the distance between points in this projection is true length.

Shortest Distance Between Skew Lines

In Fig. 8-33, lines *AB* and *CD* are **skew lines.** That is, the two lines are not parallel and do not intersect. They are both oblique. The shortest distance between these two lines is a perpendicular line between the point view of one line and the other line, as shown in Fig. 8-33B. First, find the true length of one of the lines, *CD*, in the first auxiliary. Do this by placing a V/1 reference line parallel to line *CD*. Place the second auxiliary reference 1/2 perpendicular to the true length of line *CD*. Find the point projection of line *CD* and extend line *AB* as shown. Then construct a perpendicular line from the point projection of *CD* to line *AB*. The perpendicular line intersects *AB* extended at point *X*. Then transfer the intersecting projection back to the first auxiliary, as shown on the extension of line *AB*.

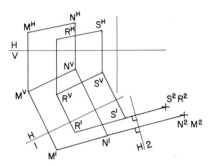

DOUBLE AUXILIARY
POINT - PROJECTION OF PARALLEL
OBLIQUE LINES TO FIND SHORTEST
DISTANCE BETWEEN LINES

Fig. 8-30 *Distance between parallel lines through point projection.*

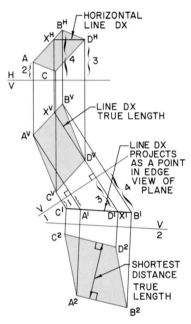

PARALLEL LINES AB - CD FORM A PLANE

Fig. 8-31 *Distance between lines forming a plane.*

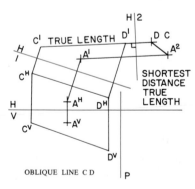

OBLIQUE LINE C D

Fig. 8-32 *Distance from a point to a line.*

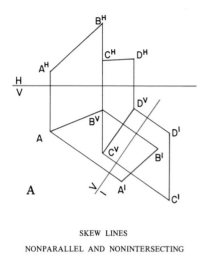

SKEW LINES
NONPARALLEL AND NONINTERSECTING

Fig. 8-33 *Distance between skew lines.*

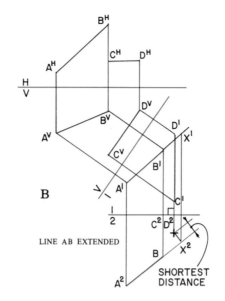

LINE AB EXTENDED

SHORTEST
DISTANCE

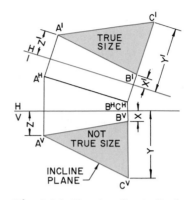

INCLINE
PLANE

Fig. 8-34 *True size of an inclined plane.*

True Size of an Inclined Plane

In Fig. 8-34, plane *ABC* shows as an edge view in the top view. Place the auxiliary reference plane H/1 parallel to the edge view and make perpendicular projections. Transfer the distances *X, Y,* and *Z* as shown to find the true size of the plane in the first auxiliary projection.

True Size of an Oblique Plane

When plane *ABC* in Fig. 8-35A is projected on a plane perpendicular to any line in the figure, it shows an edge view in the first auxiliary. In the top view, draw a line *BX* parallel to the reference plane. Place reference line V/1 perpendicular to the front view of *BX*. Project the front view of *BX* into a point projection in the first auxiliary. The point projection is in the edge view of plane *ABC,* as shown. Place the second auxiliary reference line V/2 parallel to the edge view, as shown in Fig. 8-35B. The projection of plane *ABC* in the secondary auxiliary shows the true size.

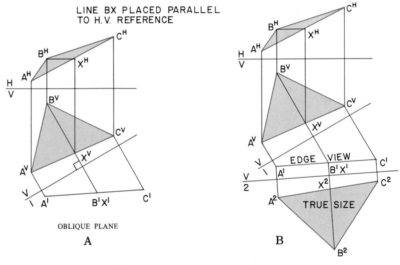

LINE BX PLACED PARALLEL
TO H.V. REFERENCE

OBLIQUE PLANE
A

EDGE VIEW

TRUE SIZE

B

Fig 8-35 *True size of an oblique plane.*

True Angles Between Lines

When two lines show in true length, the angle between them appears in its true value. In Fig. 8-36A, the two lines show as an inclined plane. This is because the vertical view shows that lines *AB* and *AC* **coincide** (lie in a single line). Place the V/1 auxiliary reference parallel to the two lines in the vertical view. The auxiliary view shows the two lines in true length, so it also shows the true angle between the lines.

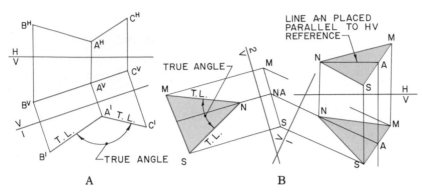

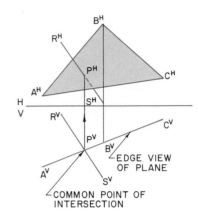

Fig. 8-36 *Finding the true angle between lines using two auxiliary projections.*

Fig. 8-37 *The edge-view system of locating the point at which a line pierces a plane.*

The oblique condition of lines *MN* and *NS* does not show in an edge view. Figure 8-36B shows how to solve this problem using two auxiliary planes. The first reference plane is perpendicular to the plane formed by lines *NA*. The second reference plane is parallel to the first auxiliary view. That is, it is parallel to the edge view of lines *MN* and *NS*. The second auxiliary view shows *MN* and *NS* in true length and the true size of the angle between them.

Piercing Points

If a line does not lie in a plane and is not parallel to it, the line must intersect the plane. The point of intersection is called a **piercing point.** The line can be thought of as piercing the plane.

Edge-View System

A line crossing the edge view of a plane shows the point at which the line pierces the plane. In Fig. 8-37, the straight line is neither in the plane nor parallel to it. It intersects the plane at a point that is common to both. The edge view of plane *ABC* is shown in the vertical plane. Line *RS* in the horizontal plane pierces the plane at point *P* when projected to the vertical plane. If you look at line *RS* closely in the vertical projection, you will see that element *A* of the triangle is lower than point *R* of the piercing line. Therefore, the dashed portion of line *RS* is invisible.

You can find the point at which line *MN* intersects oblique plane *ABC* by using the edge-view system, as shown in Fig. 8-38. The first auxiliary

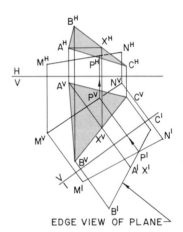

Fig. 8-38 *Finding the piercing point of a line that intersects an oblique plane.*

determines the edge view of plane *ABC*. Carry the piercing point *P* of line *MN* back to the vertical and horizontal using the projections shown by arrows.

Cutting-Plane System

When line *RS* intersects oblique plane *ABC*, you can use a cutting plane containing the line to determine other important points, as shown in Fig. 8-39. The cutting plane seems to intersect the triangle in the front view at points *3* and *4*. Project points *3* and *4* to the top view on lines *AC* and *BC*. Connect the points across the plane to find the piercing point *P*. Then project point *P* to the front view.

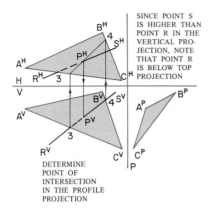

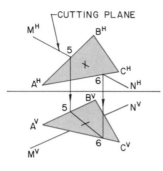

Fig. 8-39 *The cutting-plane system of locating the point at which a line pierces a plane.*

Fig. 8-40 *Intersection of the cutting plane and line AB in the horizontal projection.*

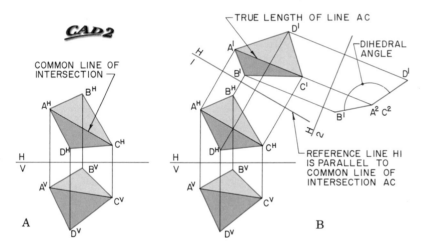

Fig. 8-41 *Finding the angle formed between intersecting planes.*

Since you now know what parts of line *RS* are not visible, you can add hidden lines. Figure 8-40 shows the cutting plane in the horizontal projection intersecting line *AB* at point *5* and line *AC* at point *6*. Can the visibility of line *MN* be determined by inspection?

Angle Between Intersecting Planes

When two planes intersect, they form a **dihedral angle.** The dihedral angle can only be measured perpendicular to the line of intersection.

As you know, the intersection of two planes is a straight line. When you find the point view of the line of intersection, the planes in that view are shown as edges, and the angle between the planes is true in size. In Fig. 8-41A, planes *ABC* and *ACD*

intersect at line *AC*. Take the first auxiliary projection H/1 in Fig. 8-41B in the horizontal plane of the line *AC*. The H/1 auxiliary lets you draw *AC* in true length. In the second auxiliary, the point projection of true-length line *AC* also shows the two given planes as edges. Measure the true angle in the second auxiliary, as shown in Fig. 8-41B.

Angle Between a Line and a Plane

You can use a similar concept to find the angle between a line and a plane. If a view shows a plane in the edge view and a true-length line, then it also shows the true angle between the plane and the line.

Plane Method

In Fig. 8-42, oblique line *XY* intersects oblique plane *ABC*. First find the edge view of oblique plane *ABC*. Place the first auxiliary H/1 perpendicular to a true-length line in the horizontal projection. Note that line *XY* is not shown true length in this first auxiliary view. After projecting the edge view in H/1, place auxiliary plane H/2 parallel to plane *ABC*. In the second auxiliary, plot the true size of plane *ABC*. In the third auxiliary, 2/3, place the reference line parallel to line *XY*. The new edge view of the plane and the true-length line form the true angle.

Line Method

In Fig. 8-43, oblique plane *ABC* and oblique line *XY* intersect in the top view. Place first auxiliary V/1 parallel to line *XY* to find the true length of *XY*. Place second auxiliary V/2 perpendicular to the true length of line *XY* to find the point projection of *XY*. In the third auxiliary, set reference line 2/3 perpendicular to the true length of B^2O^2 so that plane *ABC* is shown as an edge view. The intersection of the line and plane shows the true size of the angle.

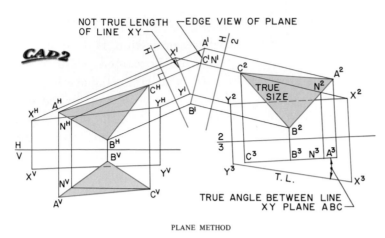

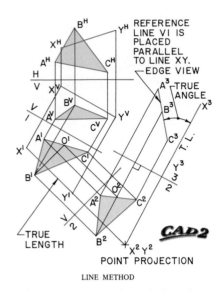

Fig. 8-42 *Finding the angle formed between a line and a plane (plane method).*

Fig. 8-43 *Finding the angle formed between a line and a plane (line method).*

True Size of an Oblique Plane by Revolution

Descriptive geometry problems can be solved by revolving an object. Chapter 7, "Auxiliary Views and Revolutions," discussed the rule of revolution. You can use this rule to solve basic problems by changing the position of an object so that the new view shows needed information.

In Fig. 8-44, it is clear that plane *ABC* is inclined to all the main reference planes. Place line *AX* in plane *ABC* parallel to the horizontal reference line. When you project line *AX* to the top view, it appears as an inclined line in true length. Project the first auxiliary H/1 in the horizontal reference plane, perpendicular to line *AX* to find the point view of *AX*. Plane *ABC* appears as an edge view, with *X* within it. At point *AX,* edge view *B'A'C'* is revolved (dashed line) so that it appears parallel to reference line H/1. Projecting points *B* and *C* to the horizontal projection allows you to draw the true size of plane *ABC* in the top view, as shown.

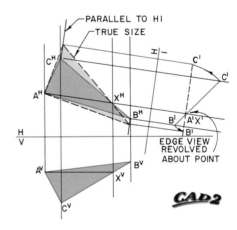

Fig. 8-44 *Finding the true size of an oblique plane by revolution.*

SOLVING PROBLEMS IN CAD DRAWINGS

The Cartesian coordinate system used in CAD corresponds directly to the coordinate system you use to solve descriptive geometry problems. Therefore, you can solve typical descriptive geometry problems, such as those presented in Figs. 8-47 through 8-78, using CAD. In general, the procedure is similar to that used in manual drafting. You simply use the CAD commands to create the geometry instead of doing them by hand. This method can save time if you are familiar with the basic CAD commands.

When the object already exists as a CAD drawing, CAD programs have several commands that allow you to perform certain location and identification procedures without building elaborate geometric constructions. For example, you can identify the location of points, true length of lines, and the true size of angles directly, as shown in Fig. 8-45.

When you solve problems in CAD drawings, it is important to make sure that you are working in the correct coordinate system. Most CAD programs allow you to use several predefined coordinate systems and even define new ones as necessary. In AutoCAD, for example, a **user coordinate system (UCS)** is a user-defined version of the Cartesian coordinate system.

The default coordinate system in AutoCAD is the *world coordinate system (WCS),* which consists of the top right quadrant (quadrant I) of the Cartesian coordinate system. Using the UCS command, you can align a new UCS with any planar object, which means that you can create a special UCS to use with any auxiliary plane you may need. Figure 8-46 shows how to define a custom UCS to find the true size of an angle.

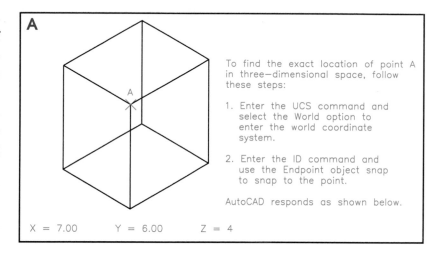

To find the exact location of point A in three—dimensional space, follow these steps:

1. Enter the UCS command and select the World option to enter the world coordinate system.

2. Enter the ID command and use the Endpoint object snap to snap to the point.

AutoCAD responds as shown below.

X = 7.00   Y = 6.00   Z = 4

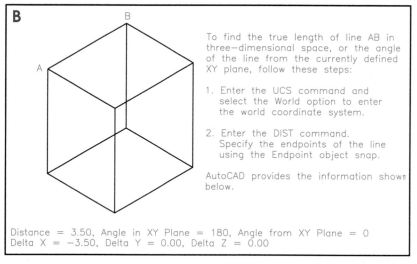

To find the true length of line AB in three—dimensional space, or the angle of the line from the currently defined XY plane, follow these steps:

1. Enter the UCS command and select the World option to enter the world coordinate system.

2. Enter the DIST command. Specify the endpoints of the line using the Endpoint object snap.

AutoCAD provides the information shown below.

Distance = 3.50, Angle in XY Plane = 180, Angle from XY Plane = 0
Delta X = −3.50, Delta Y = 0.00, Delta Z = 0.00

Fig. 8-45 *AutoCAD allows you to find the location of a point (A) and the true length of a line (B) in an existing CAD object. Note that you must be in AutoCAD's world coordinate system.*

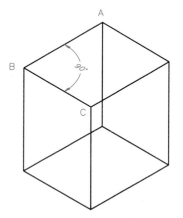

To find the true size of angle ABC, follow these steps.

1. Enter the UCS command and select the 3 Points option to define a user coordinate system along the surface that contains the angle. Select point B for the origin, C for a point on the positive X axis, and A as a point on the positive Y axis.

2. Use the DIMANGULAR command to find the true angle.

Fig. 8-46 *To find the true size of an angle, create your own user coordinate system (UCS) for the plane in which the angle lies.*

CHAPTER 8

REVIEW

Questions

Write your answers on a separate sheet of paper.

1. What are the three basic geometric elements that make up all geometric elements?
2. What is the basic shape of most structures designed by people? Why?
3. Explain the purpose of descriptive geometry.
4. How do you locate a point using descriptive geometry?
5. What are the three characteristics of a line?
6. Name the three basic lines used in descriptive geometry.
7. How does a normal line relate to the three main planes of projection?
8. How does an inclined line relate to the three main planes of projection?
9. In how many of the main planes of projection does an inclined line show in true length?
10. How does an oblique line relate to the three main planes of projection?
11. How many auxiliary projections are needed to find the point projection of an oblique line?
12. Name the three basic planes.
13. What are the major characteristics of a plane?
14. How can you find out whether two lines really intersect?
15. What is the difference between an inclined and an oblique plane?
16. How can you determine the distance between two parallel inclined lines?
17. How should you place the auxiliary reference plane to find the true size of an inclined line?
18. How can you find the true angle between intersecting lines on a manual drawing?
19. In what way is Chapter 7 closely related to Chapter 8?
20. In a three-dimensional CAD drawing, how can you find the exact location (coordinates) of a point?
21. In a CAD drawing, what command should you use to find the exact length of a line?
22. Describe how to find the true size of an angle in a 3D CAD drawing.

PROBLEMS

The small CAD symbol next to each problem indicates that a problem is appropriate for assignment as a computer-aided drafting problem and suggests a level of difficulty (one, two, or three) for the problem.

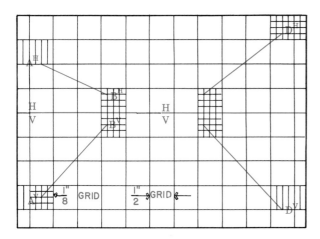

Fig. 8-47 *CAD2 Find the true length of line AB. Determine the true length and slope of line CD.*

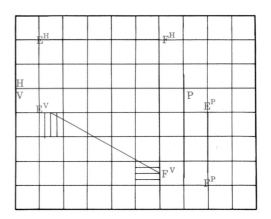

Fig. 8-48 *CAD2 Line EF is the centerline of a pipeline. Scale: 1:500 (1″ = 40′-0). Locate a line X (20′-0) below E on all principal planes of projection. Determine the grade of line EF. Determine the true distance from point E to line X.*

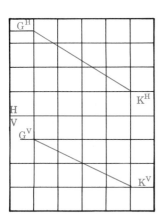

Fig. 8-49 *CAD2 Determine the point projection of line GK.*

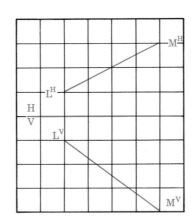

Fig. 8-50 *CAD2 Determine the angle LM makes with the vertical plane. What is the bearing?*

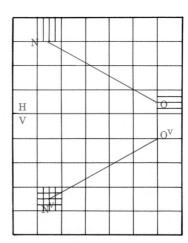

Fig. 8-51 *CAD3 Determine the slope of line NO. Extend NO to measure 2.25″ (56 mm) long. Draw all three views.*

PROBLEMS

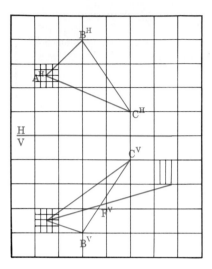

Fig. 8-52 *CAD3 Locate point D in the plan (horizontal) projection. Determine the length of line AD.*

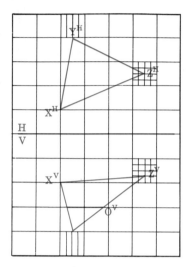

Fig. 8-53 *CAD3 What is the bearing of line NO located on plane XYZ? Determine the true size of plane XYZ.*

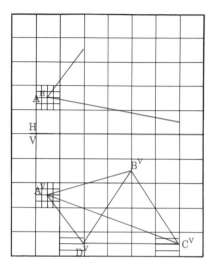

Fig. 8-54 *CAD3 Complete the plan view of plane ABCD and develop a side view.*

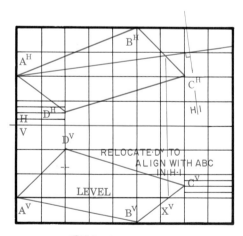

Fig. 8-55 *CAD3 Find the edge view of plane ABCD and determine the angle it makes with the horizontal plane.*

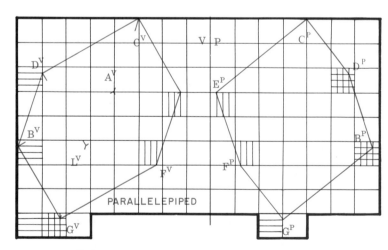

Fig. 8-56 *CAD3 Form the incomplete parallelepiped using points A and L. Determine the proper visibility in all three views.*

PROBLEMS

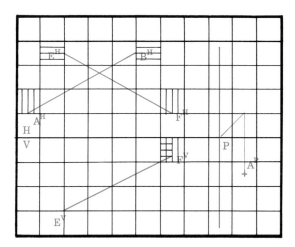

Fig. 8-57 *CAD3 Complete three views showing the intersection of AB and EF.*

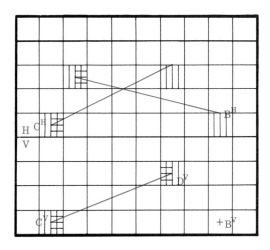

Fig. 8-58 *CAD3 Draw the front view of line AB which intersects line CD. What is the distance from C to A?*

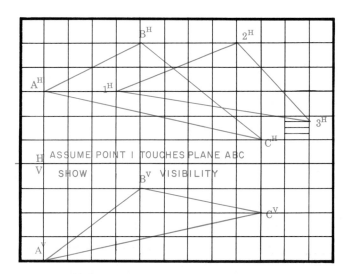

Fig. 8-59 *CAD3 Create a location for plane 1-2-3 in the vertical plane.*

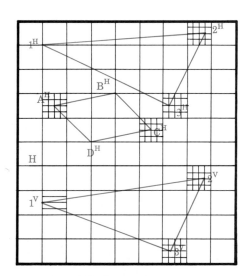

Fig. 8-60 *CAD3 Construct a plane ABCD parallel to plane 1-2-3.*

PROBLEMS

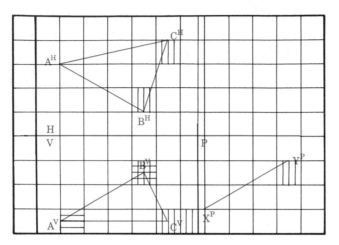

Fig. 8-61 *CAD3 Determine the true size of oblique plane ABC. Draw line XY parallel to plane ABC in the plan view.*

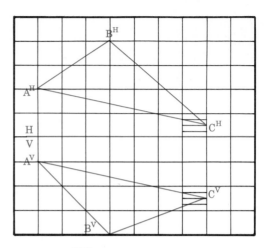

Fig. 8-62 *CAD3 Determine the true size of plane ABC and label its slope.*

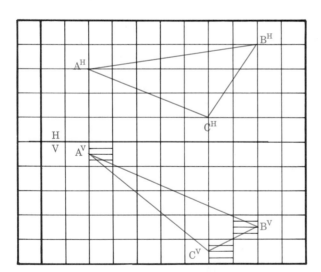

Fig. 8-63 *CAD3 Draw the true size of plane ABC and dimension the three angles of the plane.*

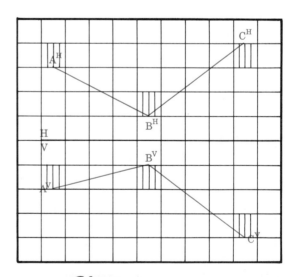

Fig. 8-64 *CAD3 Find the true angle between lines AB and BC.*

PROBLEMS

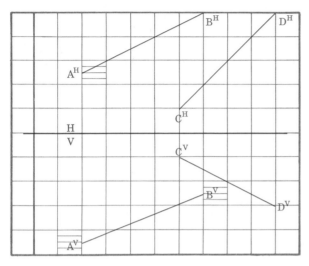

Fig. 8-65 *CAD3 Determine and locate the shortest distance between skew lines AB and CD.*

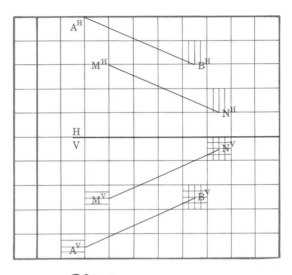

Fig. 8-66 *CAD3 Determine and label the shortest distance between parallel lines AB and MN.*

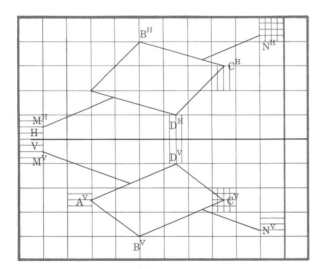

Fig. 8-67 *CAD3 Draw three views of the line and plane and determine whether line MN pierces the plane.*

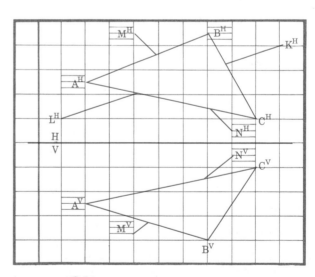

Fig. 8-68 *CAD3 Determine whether line MN pierces plane ABC. Locate line KL so that it pierces the center of plane ABC.*

PROBLEMS

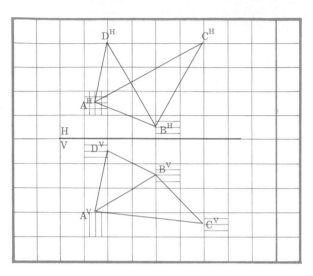

Fig. 8-69 *CAD3 Determine the visibility and angle formed between planes ABC and ABD.*

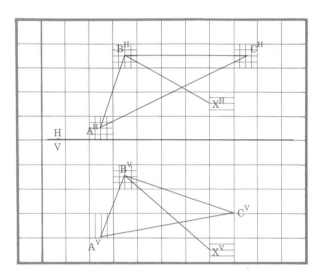

Fig. 8-70 *CAD3 Determine the angle between line BX and plane ABC.*

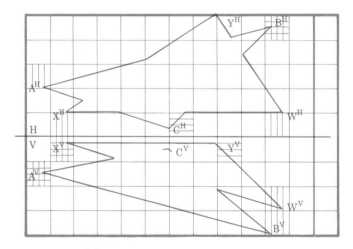

Fig. 8-71 *CAD3 Determine how the two planes intersect and show proper visibility.*

SCHOOL-TO-WORK

Solving Real-World Problems

Fig. 8-72 *SCHOOL-TO-WORK problems are designed to challenge a student's ability to apply skills learned within this text. Be creative and have fun!*

Fig. 8-72 *Study Figs. 8-41 through 8-43 carefully. Create two intersecting oblique planes of your own design and determine the angle formed. Develop your own sizes through experimentation.*

SECTIONAL VIEWS AND CONVENTIONS

OBJECTIVES

Upon completion of Chapter 9, you will be able to:

○ Describe the purpose of a sectional view.
○ Select the appropriate type of sectional view to show the hidden feature.
○ Show ribs, webs, and fasteners, and similar features properly when the cutting plane passes through them.
○ Rotate certain features into the cutting plane.
○ Describe and use conventional breaks and symbols.

VOCABULARY

In this chapter, you will learn the meanings of the following terms:

- auxiliary section
- broken-out section
- conventional break
- crosshatching
- cutting plane
- cutting-plane line
- full section
- half section
- offset section
- phantom section
- removed section
- revolved section
- ribs
- section lining
- webs

Technical drawings must show all parts of an object, including the insides and other parts not easily seen. These hidden details can be drawn with hidden lines made of short dashes. But this method works well only if the hidden part has a rather simple shape. If the shape is complicated, dashed lines may show it poorly. They can also be confusing, as shown in Fig. 9-1A. Instead of dashed lines, a special view called a *section* or **sectional view** should be drawn in these cases. A sectional view shows an object as if part of it were cut away to expose the insides (Fig. 9-1B).

DRAWING SECTIONAL VIEWS

To draw a sectional view, imagine that a wide-blade knife has cut through the object. Call this knife a **cutting plane.** Then imagine that everything in front of the plane has been taken away, so that the cut surface and whatever is inside can be seen (Fig. 9-1B). On a normal view, show where the cutting plane passes through the object by drawing a special line called a **cutting-plane line** (Fig. 9-1C). On the sectional view, show the cut surface by marking it with evenly spaced thin lines.

This is called **section lining** or **crosshatching.** These concepts will be discussed in greater detail later in this chapter.

The basic section lining pattern described above is not the only pattern, but it is the one that is used in most cases. You can use it for objects made of any material, so it is called the *general-purpose symbol.* It is used especially when more than one kind of material need not be shown, such as on a drawing of a single part.

However, special section lining patterns (also called *symbols*) can be used to show what materials are to be used. The American National Standards provide for many symbols to stand for different materials (Fig. 9-2). Under this system, the general-purpose symbol can also mean that an object is made of cast iron. These special symbols are most useful on a drawing that shows several objects made of different materials. These would be used, for example, in an *assembly drawing* (a drawing that shows how different parts fit together). However, do not depend on these symbols alone to describe the materials to be used. Specify the exact materials needed in a note or in a list of materials.

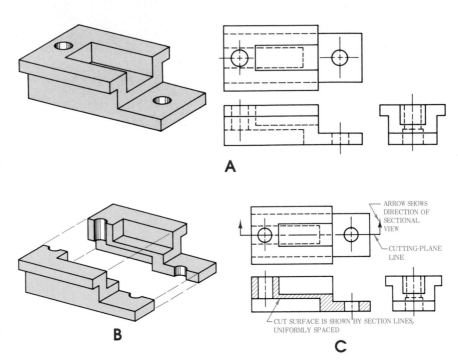

Fig. 9-1 *A) When an object's internal structure is complex, hidden lines become confusing or hard to read. (B) A sectional view provides a much clearer description of the inside of the object.*

Spacing of Section Lines

Section lines are drawn close together or far apart, depending on how much space must be filled (Fig. 9-3). According to the American National Standards, section lines can be spaced from about .03 in. (1.0 mm) to .12 in. (3.0 mm) apart. However, they must be evenly spaced, and they are usually slanted at a 45° angle. The drawing will be neater if you do not space section lines extremely close together. This will also save time. In most cases, the lines will look best spaced about .09 in. (2 mm) apart. The distance between section lines need not be measured; you may space them by eye. If the area to be covered is large, space the lines farther apart, up to .12 in. (3 mm) or more. If the area is small, space the lines closer together, down to .06 in. (1.5 mm) or less apart. If the area is very small, as for thin plates, sheets, and structural shapes, *blacked-in* (solid black) sections may be used, as shown in Fig. 9-4. Note the white space between the parts.

When you are drawing a large sectioned area, one way to save time is to use outline sectioning. This method is shown in Fig. 9-5. Drafters who use it often draw the section lines freehand and spaced widely apart. You can also gray the sectioned area (Fig. 9-6), gray only along its outline (Fig. 9-7), or rub pencil dust over it. Apply a *fixative* (sealing substance) to prevent smudging.

Do not draw section lines parallel to or at right angles to an important visible line (Fig. 9-8). You may, however, draw them at any other suitable angle and space them at any width. Section lines at different spacing and angles are commonly used to identify different sectioned parts.

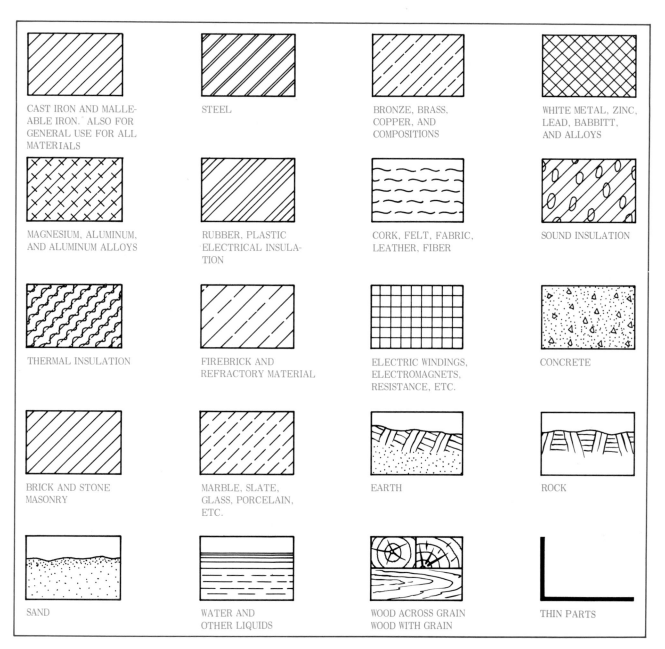

CAST IRON AND MALLE-ABLE IRON. ALSO FOR GENERAL USE FOR ALL MATERIALS	STEEL	BRONZE, BRASS, COPPER, AND COMPOSITIONS	WHITE METAL, ZINC, LEAD, BABBITT, AND ALLOYS
MAGNESIUM, ALUMINUM, AND ALUMINUM ALLOYS	RUBBER, PLASTIC ELECTRICAL INSULA-TION	CORK, FELT, FABRIC, LEATHER, FIBER	SOUND INSULATION
THERMAL INSULATION	FIREBRICK AND REFRACTORY MATERIAL	ELECTRIC WINDINGS, ELECTROMAGNETS, RESISTANCE, ETC.	CONCRETE
BRICK AND STONE MASONRY	MARBLE, SLATE, GLASS, PORCELAIN, ETC.	EARTH	ROCK
SAND	WATER AND OTHER LIQUIDS	WOOD ACROSS GRAIN WOOD WITH GRAIN	THIN PARTS

Fig. 9-2 *American National Standard symbols for section lining.* (ANSI)

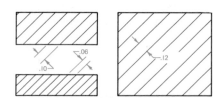

Fig. 9-3 *Space section lines by eye. The distance between section lines varies according to the size of the space to be sectioned.*

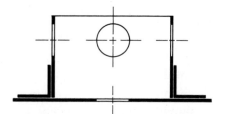

Fig. 9-4 *You may blacken in the entire sectioned area instead of using section lines when areas are very small.*

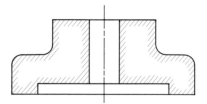

Fig. 9-5 *Outline sectioning.*

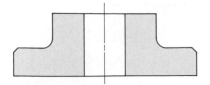

Fig. 9-6 *A cut surface may be grayed.*

Fig. 9-7 *A cut surface may have a grayed outline.*

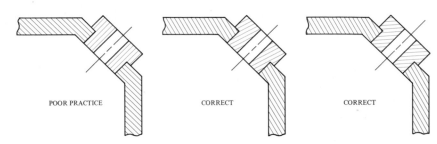

POOR PRACTICE   CORRECT   CORRECT

Fig. 9-8 *Do not draw section lines parallel or perpendicular to a main line of the view.*

The Cutting-Plane Line

The cutting-plane line represents the cutting plane as viewed from an edge (Fig. 9-9). You may draw this line in either of two ways approved by the American National Standards (Fig. 9-10). The first form is more commonly used. The second shows up well on complicated drawings. At each end of the line, draw a short arrow to show the direction for looking at the section. Make the arrows at right angles to the line. Place bold capital letters at the corners as shown, if needed for reference to the section.

A cutting-plane line is not needed when it is clear that the section is taken along an object's main centerline or at some other obvious place. Figure 9-11 shows an example of a symmetrical object in which the centerline serves as the cutting-plane line.

Sections Through Assembled Pieces

If a drawing shows more than one piece in section, draw the section lines in a different direction on each piece (Figs. 9-12 and 9-13). Remember, however, that any piece can, in turn, show several cut surfaces. Make sure that all the cut surfaces of any one piece have section lines in the same direction, as shown in Fig. 9-13.

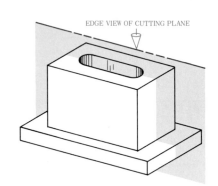

EDGE VIEW OF CUTTING PLANE

Fig. 9-9 *The cutting-plane line represents the edge view of the cutting plane.*

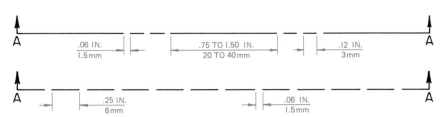

A ——— .06 IN. / 1.5 mm —— .75 TO 1.50 IN. / 20 TO 40 mm —— .12 IN. / 3 mm —— A

A —— .25 IN. / 6 mm —— .06 IN. / 1.5 mm —— A

NOTE: ALL SIZES ARE ESTIMATED, NOT MEASURED.

Fig. 9-10 *Cutting-plane lines.*

CAD SECTIONING

CAD software allows you to show a cut surface by marking it with evenly spaced lines or patterns just as you would in traditional manual drafting. In CAD, these lines are often called *hatch lines, crosshatching,* or *fill.* The CAD software provides a variety of standard patterns, including those recommended by ANSI and ISO. The patterns include symbols for steel, alloys, copper, aluminum, or cast iron, to name a few (Fig. 9-14).

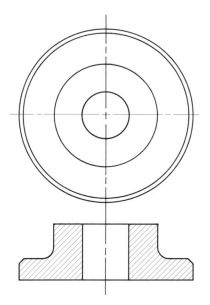

Fig. 9-11 *A centerline may be used to represent a cutting-plane line.*

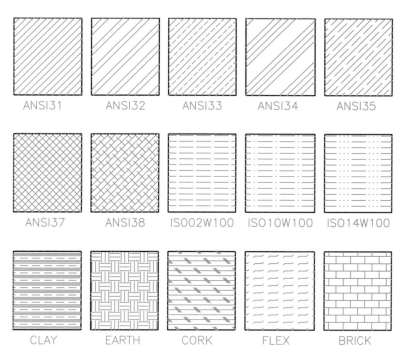

Fig. 9-14 *A sampling of AutoCAD's hatch patterns. Notice that both ANSI and ISO patterns are available, along with other common symbols.*

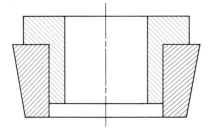

Fig. 9-12 *Because this sectional view contains two different pieces in section, each piece is sectioned using lines at a different angle.*

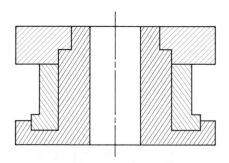

Fig. 9-13 *This sectional view contains three assembled pieces in section, so the section lines are drawn in three different directions.*

The difference in using a CAD system to hatch a section is that the software creates the hatch automatically. You do not have to draw each line individually or worry about spacing the lines evenly. You can control the spacing between the lines by adjusting the scale of the hatch pattern or by specifying the space between lines (depending on the CAD program). You can also control the angle at which the lines are drawn.

Most CAD software today allows you to pick a point in the enclosed area that you want to hatch. The software defines the boundary of the hatch and proceeds to fill the area with the pattern you choose. In AutoCAD, for example, you can use the BHATCH (Boundary Hatch) command to set all the parameters (settings) for the hatch pattern and then determine the hatch boundary (Fig. 9-15). The BHATCH command also provides a preview feature so that you can see what the hatch will look like before you actually apply it to the drawing. You can even specify *islands* (enclosed places within the hatch boundary that should not be hatched).

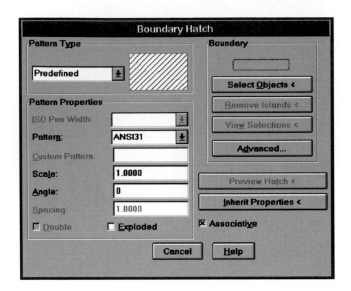

Fig. 9-15 *When you enter the BHATCH command in AutoCAD, the Boundary Hatch dialog box appears. This dialog box allows you to select the hatch pattern scale (or spacing, for user-defined patterns). It also allows you to pick points to define a hatch boundary or to select objects to be hatched.*

TYPES OF SECTIONAL VIEWS

Sectional views can be drawn in many different ways to make the internal features as clear as possible, while keeping the drawing as simple as possible. The various types of sectional views are described in this section.

Full Sections

A **full section** is a sectional view that shows an object as if it were cut completely apart from one end or side to the other, as shown in Fig. 9-16. Such views are usually just called *sections*. The two most common types of full sections are *vertical* and *profile* sections (Figs. 9-17 and 9-18).

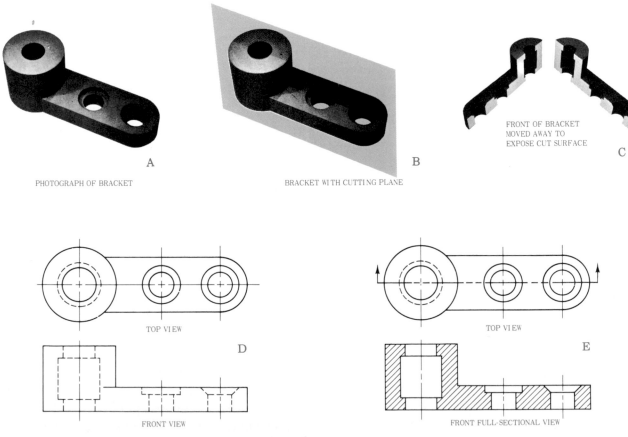

Fig. 9-16 *Full section.*

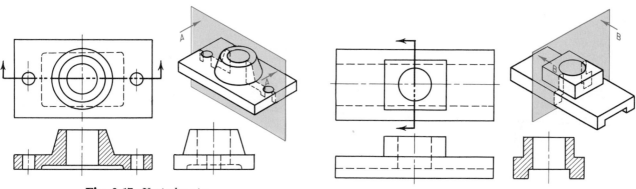

Fig. 9-17 *Vertical section.*

Fig. 9-18 *Profile section.*

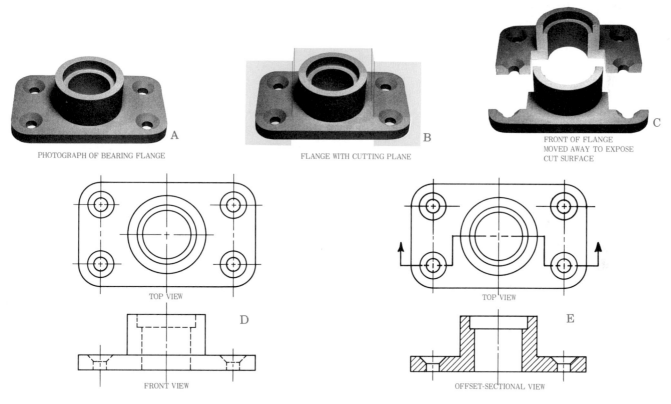

PHOTOGRAPH OF BEARING FLANGE

FLANGE WITH CUTTING PLANE

FRONT OF FLANGE
MOVED AWAY TO EXPOSE
CUT SURFACE

TOP VIEW

FRONT VIEW

TOP VIEW

OFFSET-SECTIONAL VIEW

Fig. 9-19 *Offset section.*

Offset Sections

In sections, the cutting plane is usually taken straight through the object. But it can also be *offset* (shifted) at one or more places to show a detail or to miss a part. This type of section, known as an **offset section,** is shown in Fig. 9-19. In this figure, a cutting plane is offset to pass through the two bolt holes. If the plane were not offset, the bolt holes would not show in the sectional view. Show an offset section by drawing it on the cutting-plane line. Do not show the offset on the sectional view.

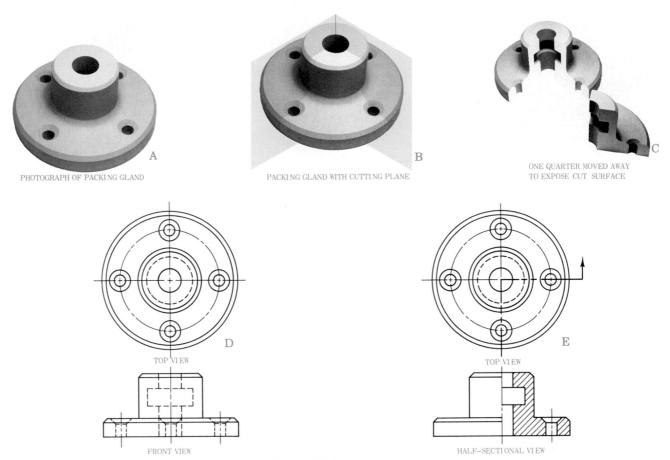

PHOTOGRAPH OF PACKING GLAND

PACKING GLAND WITH CUTTING PLANE

ONE QUARTER MOVED AWAY
TO EXPOSE CUT SURFACE

TOP VIEW

TOP VIEW

FRONT VIEW

HALF–SECTIONAL VIEW

Fig. 9-20 *Half section.*

Half Sections

A **half section** is one half of a full section. Remember, a full section makes an object look as if half of it has been cut away. A half section looks as if one quarter of the original object has been cut away. Imagine that two cutting planes at right angles to each other slice through the object to cut away one quarter of it, as shown in Fig. 9-20. The half section shows one half of the front view in section (Fig. 9-20E). Figure 9-20D shows the exterior of the object.

Half sections are useful when you are drawing a symmetrical object. Both the inside and the outside can be shown in one view. Use a centerline where the exterior and half-sectional views meet,

since the object is not actually cut. In the top view, show the complete object, since no part is actually removed. If the direction of viewing is needed, use only one arrow, as shown in Fig. 9-20E. In the top view (E) the cutting-plane line could have been left out, because there is no doubt where the section is taken.

Broken-Out Sections

A view with a **broken-out section** shows an object as it would look if a portion of it were cut partly away from the rest by a cutting plane and then "broken off" to reveal the cut surface and insides (Fig. 9-21). This view shows some inside detail without drawing a full or half section.

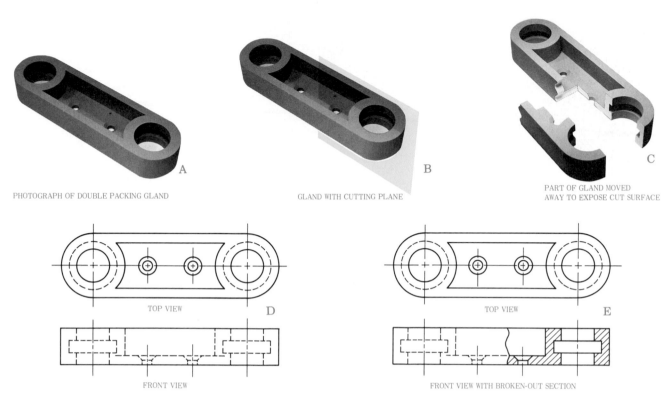

PHOTOGRAPH OF DOUBLE PACKING GLAND

GLAND WITH CUTTING PLANE

PART OF GLAND MOVED
AWAY TO EXPOSE CUT SURFACE

TOP VIEW   D

TOP VIEW   E

FRONT VIEW

FRONT VIEW WITH BROKEN-OUT SECTION

Fig. 9-21 *Broken-out section.*

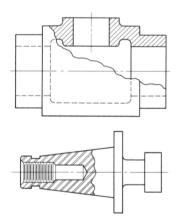

Fig. 9-22 *Two additional examples
of broken-out sections.*

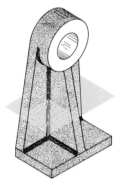

Fig. 9-23 *Cutting plane in position
for revolved section.*

Note that a broken-out section is bounded by a *short-break line* drawn freehand the same thickness as visible lines. Figure 9-22 shows two more examples of broken-out sections.

Revolved Sections

Think of a cutting plane passing through a part of an object, as in Fig. 9-23. Now think of that cut surface as *revolved* or turned 90°, so that its shape can be seen clearly (Fig. 9-24). The result is a **revolved section** (also called a *rotated section*).

Use a revolved section when the part is long and thin and when its shape in cross section is the same throughout (Fig. 9-25). In such cases, the view may be shortened but the full length of the part must be given by a dimension. This lets you draw a large part with a revolved section in a short space.

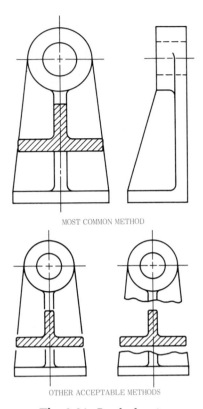

MOST COMMON METHOD

OTHER ACCEPTABLE METHODS

Fig. 9-24 *Revolved section.*

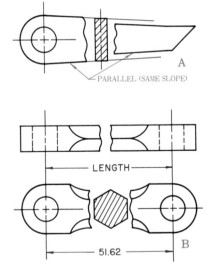

PARALLEL (SAME SLOPE)

A

LENGTH

51.62

B

Fig. 9-25 *Revolved sections in long parts.*

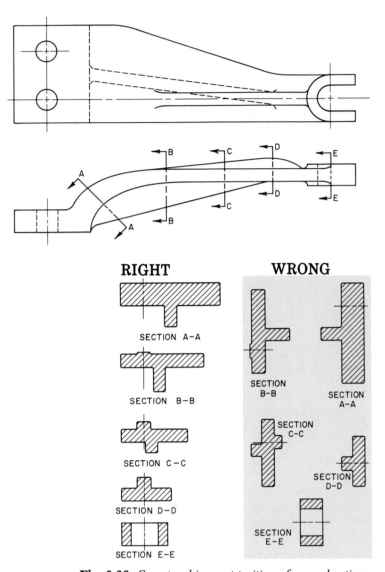

RIGHT

SECTION A-A

SECTION B-B

SECTION C-C

SECTION D-D

SECTION E-E

WRONG

SECTION B-B

SECTION A-A

SECTION C-C

SECTION D-D

SECTION E-E

Fig. 9-26 *Correct and incorrect positions of removed sections.*

Removed Sections

When a sectional view is taken from its normal place on the view and moved somewhere else on the drawing sheet, the result is a **removed section.** Remember, however, that the removed section will be easier to understand if it is positioned to look just as it would if it were in its normal place on the view. In other words, do not rotate it in just any direction. Figure 9-26 shows right and wrong ways to position removed sections. Use bold letters to identify a removed section and its corresponding cutting plane on the regular view (Fig. 9-26).

A removed section can be a *sliced section* (the same as a revolved section), or it can show additional detail visible beyond the cutting plane. You can draw it to fit in dimensions or to a larger scale to show details clearly.

Besides removed sections, you can also draw removed views of the exterior of an object. These too can be made to the same scale or to a larger one. They can also be complete views or partial views.

Auxiliary Sections

When a cutting plane is passed through an object at an angle, as in Fig. 9-27A, the resulting sectional view, taken at the angle of that plane, is called an **auxiliary section.** It is drawn like any other auxiliary view (see Chapter 7).

Usually, on working drawings, only the auxiliary section is shown on the cut surface. However, if needed, any or all parts beyond the auxiliary cutting plane may be shown. In Fig. 9-27B, notice that the auxiliary section contains hidden lines. It also contains three incomplete views.

Phantom (Hidden) Sections

Use a **phantom section** (hidden section) to show in one view both the inside and the outside of an object that is not completely symmetrical. Figure 9-28 shows an object with a circular boss on one side. Since the object is not symmetrical, the inside cannot be shown with a half section. A phantom section is used instead. A partial phantom section can sometimes be better than a broken-out section to show interior detail on an exterior view.

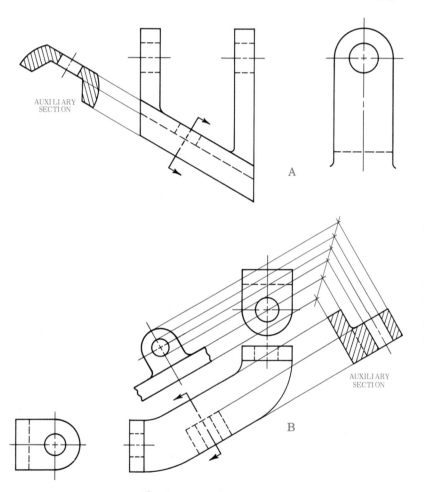

Fig. 9-27 *Auxiliary section.*

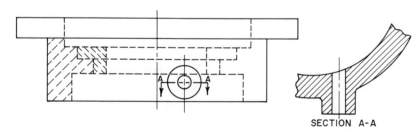

SECTION A-A   **Fig. 9-28** *Phantom section.*

SPECIAL CASES

In practice, drafters make exceptions to the general sectioning rules in certain situations. These exceptions have become standard practice in industry.

Ribs and Webs in Section

Ribs and **webs** are thin, flat parts of an object that are used to brace or strengthen another part of the object. Often, a true section of an object that contains ribs or a web structure does not appear to show a true description of the part. For example, the section shown in Fig. 9-29A would give the idea of a very heavy, solid piece. This would not be a true description of the part. Therefore, when a cutting plane passes through a rib or web parallel to the flat side, do not draw section lining for that part. Instead, draw the part as shown in Fig. 9-29B. Think of the plane passing just in front of the rib.

If a cutting plane passes through a rib, a web, or any other thin, flat part at right angles to the flat side, draw in the section lines for that part. Figure 9-30 shows an example.

Hidden and Visible Lines

Do not draw hidden lines on sectional views unless they are needed for dimensioning or for clearly describing the shape. In Fig. 9-31A, a hub is described clearly using no hidden lines. Compare it with the incorrectly drawn section in (B). On *sectional assembly drawings* (sectional views of how parts fit together), hidden lines are generally omitted. This keeps the drawing from becoming cluttered and hard to read (Fig. 9-32). Sometimes a good way to avoid using hidden lines is to draw a half section or part section.

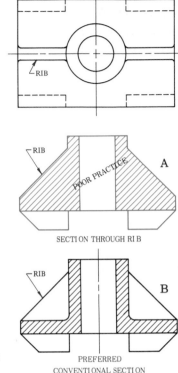

Fig. 9-29 *Ribs in section.*

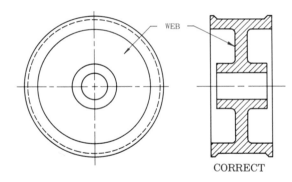

Fig. 9-30 *Web in section.*

Fig. 9-31 *Omit hidden lines when not needed for clarity.*

Normally, in a sectional view, include all the lines that would be visible on or beyond the plane of the section. In Fig. 9-33, for example, the section drawing in (A) correctly includes the numbered lines, which match the lines on the drawing in (B). A drawing without these lines, as shown in (C), would have little value. Do not draw sectional views in this manner. Figure 9-34 provides another example of how to draw visible lines beyond the plane of the section.

Alternate Section Lining

Alternate (or wide) *section lining* is a pattern made by leaving out every other section line. It can be used to show a rib or another flat part in a sectional view when that part otherwise would not show clearly. In Fig. 9-35A, an eccentric piece (two or more circular shapes not using the same centerlines) is drawn in section. A rib is visible in the top view, but in the sectional view, it is not shown by any section lining. As described above, this is standard practice with a flat part like a rib. But there are no visible lines to represent the rib either, because its top and bottom are both even with the surfaces they join. In fact, without the top view, you might not know that the rib was there. A drawing of an eccentric piece without a rib would look exactly the same. The problem is solved, however, in Fig. 9-35B. Here, alternate section lining is used with hidden lines to show the extent of the rib. Alternate section lines are useful to show ribs and other thin, flat pieces in one-view drawings of parts or in assembly drawings.

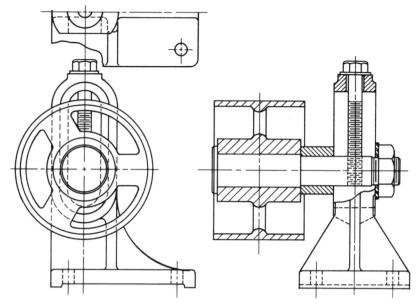

Fig. 9-32 *Omit hidden lines to keep the drawing from becoming confusing.*

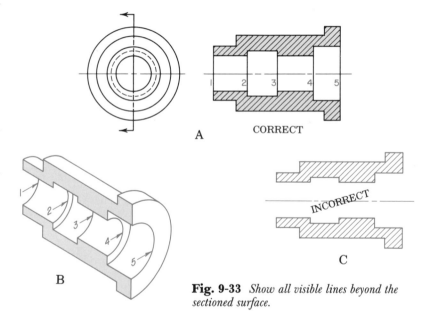

Fig. 9-33 *Show all visible lines beyond the sectioned surface.*

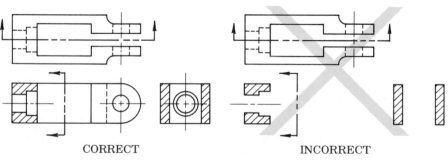

Fig. 9-34 *Correct and incorrect uses of visible lines beyond the plane of the section.*

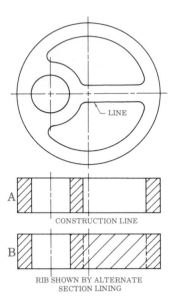

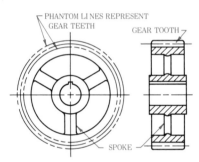

Fig. 9-35 *Alternate, or wide, section lining.*

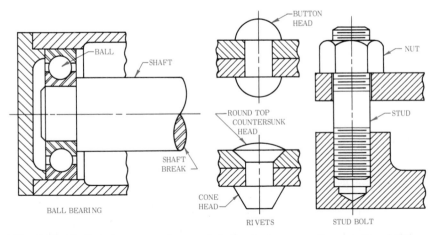

Fig. 9-36 *Spokes and gear teeth should not be sectioned.*

Other Parts Usually Not Sectioned

Do not draw section lines on spokes and gear teeth when the cutting plane passes through them. Leave them as shown in Fig. 9-36. Do not draw section lines either on shafts, bolts, pins, rivets, or similar items when the cutting plane passes through them *lengthwise* (through the axis), as shown in Fig. 9-37. These objects are not sectioned because they have no inside details. Also, sectioning might give a wrong idea of the part. A drawing showing them in full is easier to read. It also takes less time to draw. However, when such parts are cut across the axis, they should be sectioned (Fig. 9-38). See the sectional assembly in Fig. 9-39 for names and drawings of a number of other items that should not be sectioned.

Rotated Features in Section

A section or an *elevation* (side, front, or rear view) of a symmetrical piece can sometimes be hard to read if drawn in true projection. It can also be hard to draw. When drawing such a view, follow the example in Fig. 9-40. In this example, a symmetrical piece with ribs and lugs is drawn twice in section. The first section drawing is a true projection. But it does not show the true shape of the ribs and lugs.

Fig. 9-37 *Shafts, bolts, screws, rivets, and similar parts are usually not sectioned if the cutting plane passes through them lengthwise (along the axis).*

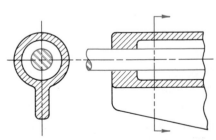

Fig. 9-38 *Section bolts, screws, rivets, and similar parts when the cutting plane cuts the across the axis.*

In the second section, the ribs and lugs have been rotated on the vertical axis until they appear as mirror images of each other on either side of the centerline. Their true shape can now be shown. This is the correct way to draw this kind of object. Note that only the parts that extend all the way around the vertical axis are drawn with section lining. In Fig. 9-41, the lugs are rotated to show true shape. Note that they are not drawn with section lines.

When a section passes through spokes, do not draw section lines on the spokes. Leave them as in the section drawing in Fig. 9-42A. Compare this drawing with the section drawing for a solid web (B). It is the section lining that shows that the web is solid rather than made with spokes.

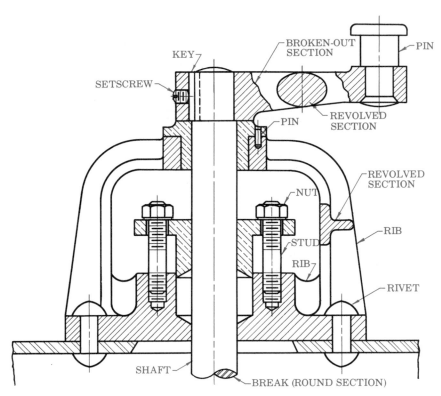

Fig. 9-39 *Examples of features not sectioned.*

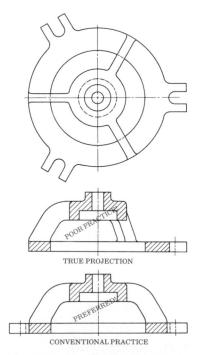

Fig. 9-40 *Some features should be rotated to show true shape.*

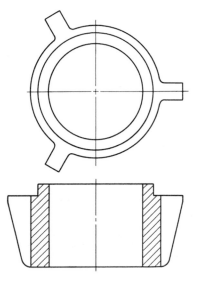

Fig. 9-41 *Do not section lugs.*

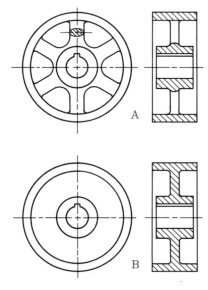

Fig. 9-42 *A section through spokes.*

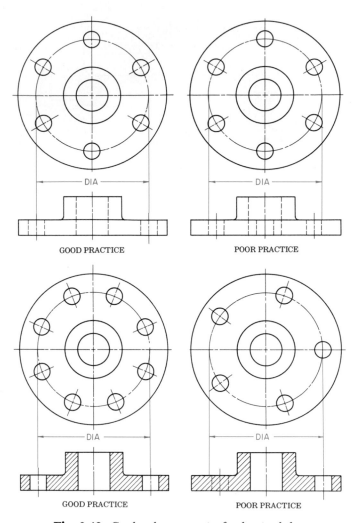

Fig. 9-43 *Good and poor practice for showing holes.*

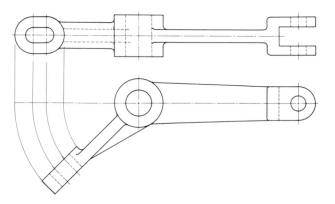

Fig. 9-44 *Rotation of part of a view to show true shape.*

When drawing a section or elevation of a part with holes arranged in a circle, follow the *good practice* examples in Fig. 9-43. In these examples, the holes have been rotated for the section drawing until two of them lie squarely on the cutting plane. These views then show the true distance of the holes from the center, whereas a true projection would not.

Rotating features in drawings is very useful when you want to show true conditions or distances that would not show in a true projection. Moreover, for some objects, only part of the view should be rotated, as shown by the bent lever in Fig. 9-44.

CONVENTIONAL BREAKS AND SYMBOLS

Conventional breaks and symbols are used to show that a uniform part of a very long object has been cut out of the drawing. This makes some details on the drawing easier to draw and easier to understand. Figure 9-45 shows the methods used to draw long, evenly shaped parts and to *break out* (shorten) the drawing of parts. Using a break lets you draw a view to a larger scale. Since the break shows how the part looks in cross section, an end view usually need not be drawn. Give the length by a dimension. The symbols for conventional breaks are usually drawn freehand. However, on larger drawings, conventional breaks are often drawn with instruments to give a neat appearance. Figure 9-46 shows how to draw the break for cylinders and pipes.

INTERSECTIONS IN SECTION

An *intersection* is a point where two parts join (Fig. 9-47). Drawing a true projection of an intersection is difficult and takes too much time. Also, such accuracy of detail is of little or no use to a print reader. Therefore, approximated and preferred sections, such as those shown in Fig. 9-48, are generally drawn.

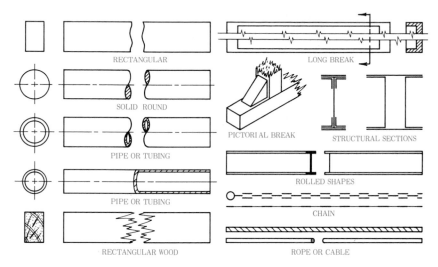

Fig. 9-45 *Conventional breaks and symbols.*

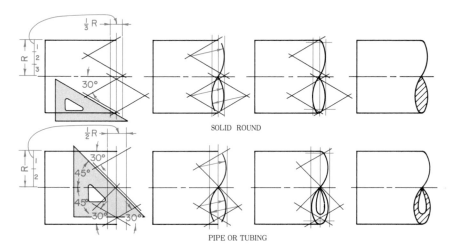

Fig. 9-46 *Drawing the break symbols for cylinders and pipes.*

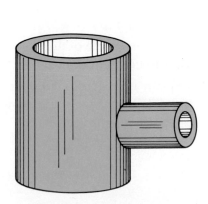

Fig. 9-47 *Intersecting parts.*

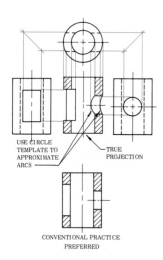

Fig. 9-48 *Approximated and preferred sections.*

CHAPTER 9

Learning Activities

1. Obtain sectional-view drawings from industry. Study them carefully to determine if the appropriate sections were selected and if they were drawn properly. See if any alternative methods would have worked as well or perhaps better.

2. Obtain several objects that have interior detail, such as those shown throughout Chapter 9. Decide which type of section would be most practical, and make complete working drawings with sectional views and dimensions.

3. Examine the next two chapters to see how sections have helped in clarifying drawing details. List each figure that uses a section drawing and label the type of section for class discussion.

4. Locate five sections in the advanced chapters of this book. List each figure number and explain how the section helps clarify the drawing.

Questions

Write your answers on a separate sheet of paper.

1. General-purpose section lining may also be used to designate what material?

2. What type of line is used to show where a section is to be taken?

3. When a cutting-plane line passes through an entire view, a _____ section results.

4. How are thin sections usually shown on a drawing?

5. If you were drawing a sectional view of a symmetrical object, what type of section would you use? Explain.

6. Sketch two types of cutting-plane lines.

7. For drawings that contain several sectioned areas, why might it be faster to use a CAD program than to create the same drawing manually?

8. When one quarter of the object is cut away, a _____ section results.

9. A revolved section is rotated _____°.

10. Name four items that are generally not sectioned even when the cutting plane passes through their axis.

11. A _____ section is used to show both inside and outside details of a nonsymmetrical object.

12. What is the proper position for a removed section on a drawing?

13. Give an example of the use of alternate section lines to clarify a drawing.

14. Explain how a conventional break can make the detail easier to understand in the drawing of a very long object.

CHAPTER 9

PROBLEMS

The small CAD symbol next to each problem indicates that a problem is appropriate for assignment as a computer-aided drafting problem and suggests a level of difficulty (one, two, or three) for the problem.

In Fig. 9-49, problems A and B show examples of full and half sections. In the half sections, the hidden line is optional. Study these examples carefully before attempting any of the drawing assignments for this chapter.

For problems C through L in Fig. 9-49, use dividers to take dimensions from the printed scales below. Draw both views using the scale assigned. Make a full or half section as assigned. Add dimensions if required. Estimate the sizes of fillets and rounds (small radii).

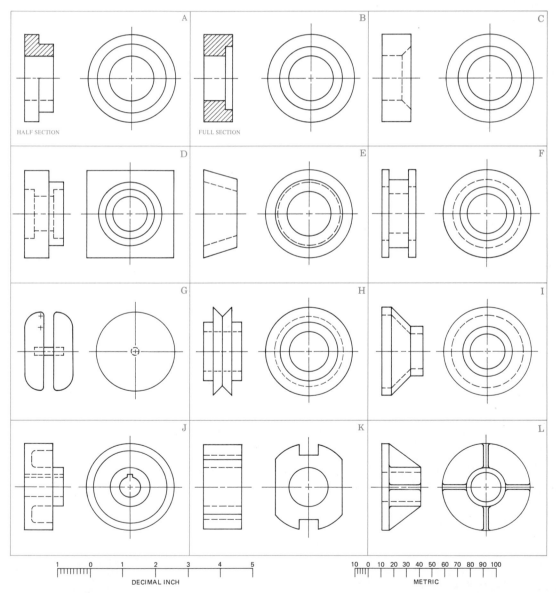

Fig. 9-49 *CAD2* *Problems for practice in sectioning.*

PROBLEMS

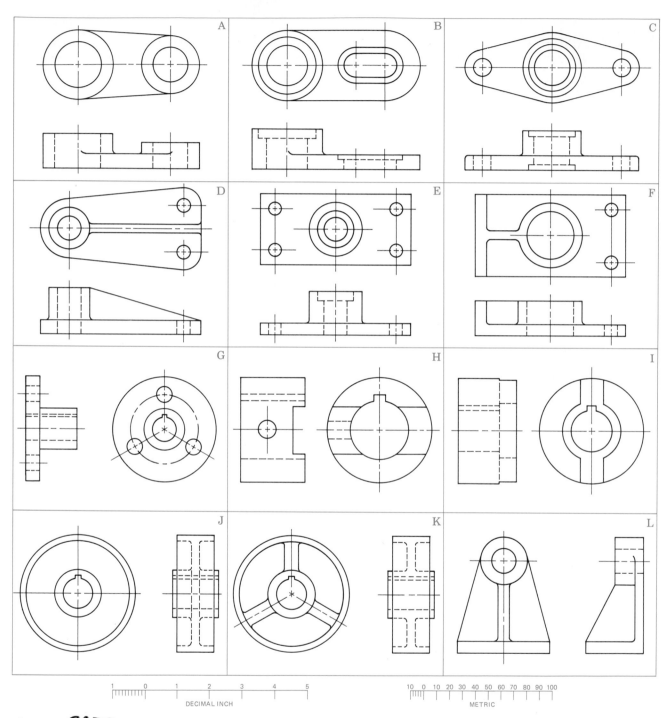

DECIMAL INCH

METRIC

Fig. 9-50 *CAD2 Problems for practice in sectioning. Take dimensions from the printed scale, using dividers. Draw both views. Make a sectional view as assigned. Add dimensions if required. Estimate sizes of fillets and rounds.*

PROBLEMS

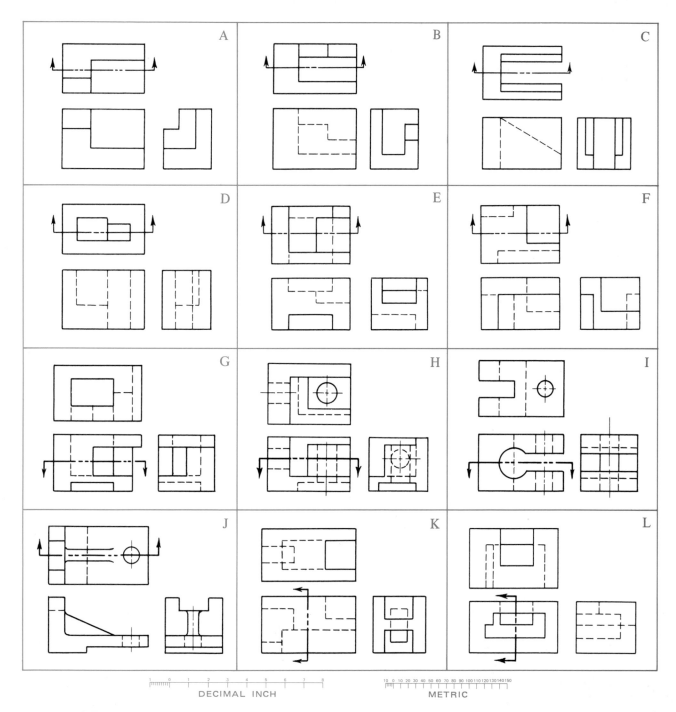

DECIMAL INCH

METRIC

Fig. 9-51 *CAD2* *Problems for practice in sectioning. Take dimensions from the assigned scale, using dividers. Draw all views and section the view indicated by the cutting-plane line. Add dimensions if required.*

PROBLEMS

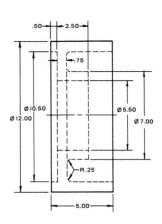

Fig. 9-52 *CAD3 Make a two-view drawing of the collar showing a full or half section as assigned.*

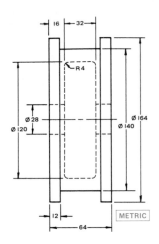

Fig. 9-53 *CAD3 Make a two-view drawing of the steam piston showing a full or half section as assigned.*

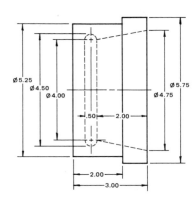

Fig. 9-54 *CAD3 Make a two-view drawing of the shaft cap showing a full or half section as assigned.*

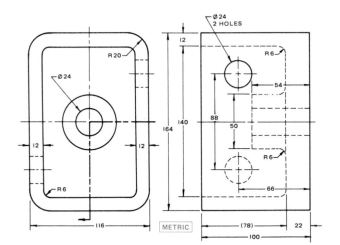

Fig. 9-55 *CAD3 Make a two-view drawing of the protected bearing showing the right-hand view as a half section.*

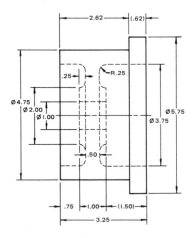

Fig. 9-56 *CAD3 Make a two-view drawing of the water-piston body showing a full or half section as assigned.*

PROBLEMS

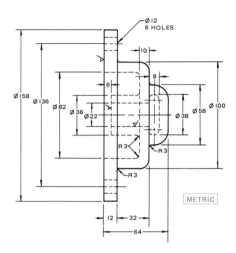

Fig. 9-57 *CAD3 Make a two-view drawing of the cylinder head. Show a full or half section.*

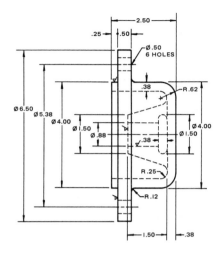

Fig. 9-58 *CAD3 Make a two-view drawing of the cylinder cap. Show a full or half section.*

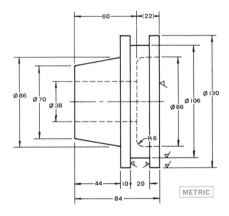

Fig. 9-59 *CAD3 Make a two-view drawing of the cone spacer showing a full or half section.*

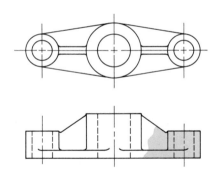

Fig. 9-60 *CAD2 Draw the rod guide, using the assigned scale at the bottom of the page. Make top and front views. Show a broken-out section as indicated by the colored screen.*

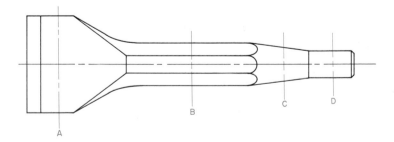

Fig. 9-61 *CAD2 Draw the chisel, using the assigned scale at the bottom of the page. Make revolved or removed sections on the colored centerlines. A is a .25" × 3.00" (6.3 × 76 mm) rectangle, B is a 1.25" (32 mm) (across flats) octagon, and C and D are circular cross sections. The chisel may be drawn full size or to a reduced scale.*

DECIMAL INCH

METRIC

PROBLEMS

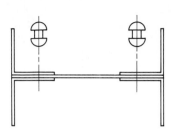

Fig. 9-62 *CAD2 Draw the structural joint, using the assigned scale at the bottom of the page. Make a full-sectional view of the joint with rivets moved into their proper position on the centerlines.*

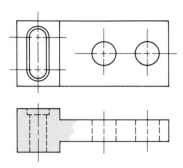

Fig. 9-63 *CAD2 Draw the adjusting plate, using the assigned scale at the bottom of the page. Draw front and top views. Make the broken-out section as indicated by the colored screen.*

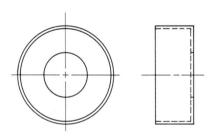

Fig. 9-64 *CAD2 Draw the grease cap using the assigned scale at the bottom of the page. Make front and right full or half sections as assigned.*

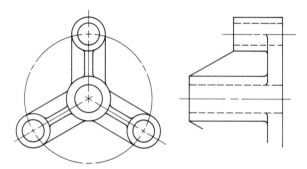

Fig. 9-65 *CAD2 Draw the rotator using the assigned scale at the bottom of the page. Complete the right-side view and make a full or half section.*

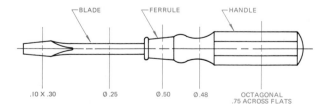

BLADE   FERRULE   HANDLE

.10 X .30   Ø .25   Ø .50   Ø .48   OCTAGONAL
.75 ACROSS FLATS

Fig. 9-66 *CAD2 Draw the screwdriver twice the size shown. Add removed or revolved sections on the colored centerlines. Use dividers to transfer sizes. The overall length is 6.60 in.*

DECIMAL INCH

METRIC

PROBLEMS

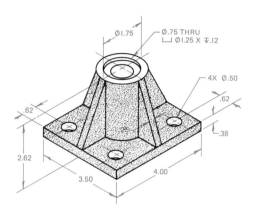

Fig. 9-67 *CAD3 Base plate. Scale: Full size or as assigned. Material: cast iron.*

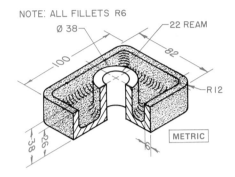

Fig. 9-68 *CAD3 Shaft base. Scale: 1:1 or as assigned. Material: cast iron.*

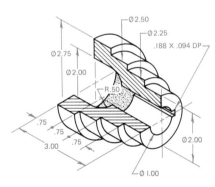

Fig. 9-69 *CAD3 Step pulley. Scale: Full size or as assigned. Material: cast iron.*

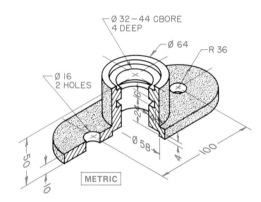

Fig. 9-70 *CAD3 Lever bracket. Scale: 1:1 or as assigned. Material: cast iron.*

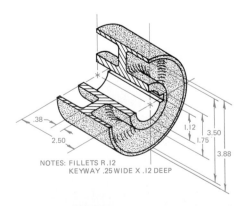

Fig. 9-71 *CAD3 Idler pulley. Scale: Full size or as assigned. Material: cast iron.*

PROBLEMS

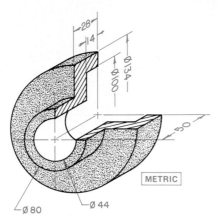

Fig. 9-72 *CAD3* *Retainer. Scale: 1:1 or as assigned. Material: cast aluminum.*

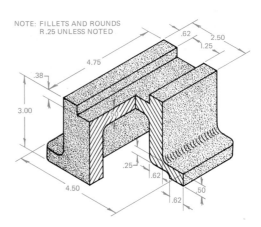

NOTE: FILLETS AND ROUNDS R.25 UNLESS NOTED

Fig. 9-73 *CAD3* *Rest. Scale: three-quarter size or as assigned. Material: cast aluminum.*

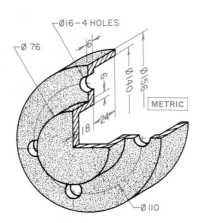

Fig. 9-74 *CAD3* *End cap. Scale: 1:1 or as assigned. Material: cast iron.*

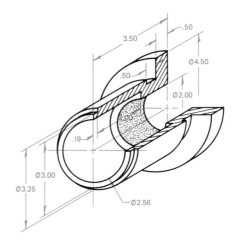

Fig. 9-75 *CAD3* *Flange. Scale: Full size or as assigned. Material: cast aluminum.*

PROBLEMS

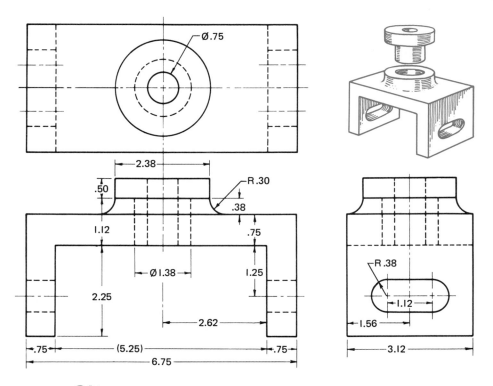

Fig. 9-76 *CAD3 Draw three views of the yoke with the front view in section. There are two pieces: the yoke and the bushing. Do not copy the picture.*

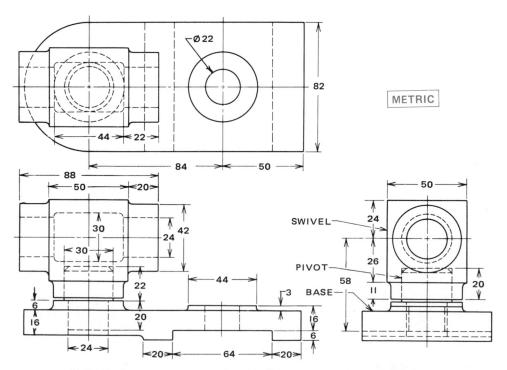

METRIC

Fig. 9-77 *CAD3 Draw three views of the swivel base with the front view in section.*

PROBLEMS

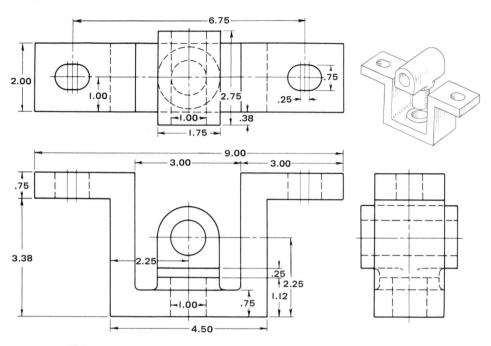

Fig. 9-78 *CAD3 Draw three views of the swivel hanger with the right-side view in section. There are two pieces: the hanger and the bearing.*

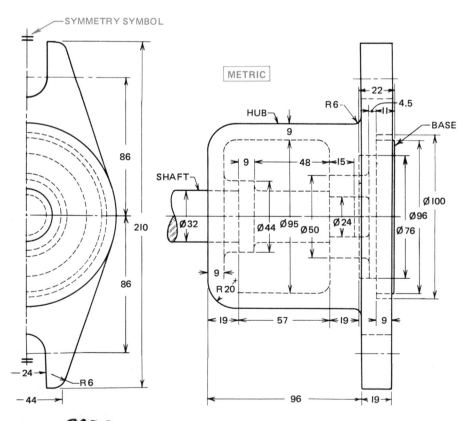

Fig. 9-79 *CAD3 Draw three views of the thrust bearing with the right-hand view in section. There are three parts: the shaft, the hub, and the base.*

PROBLEMS

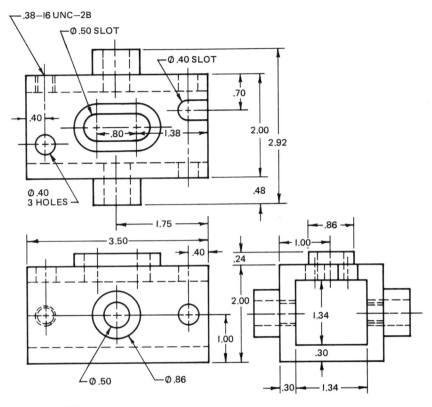

Fig. 9-80 *CAD3 On a B (A3) size sheet, make a three-view drawing of the jacket. Show the front and right-side views in section to improve clarity. Do not dimension unless instructed to do so. Scale: 1:1.*

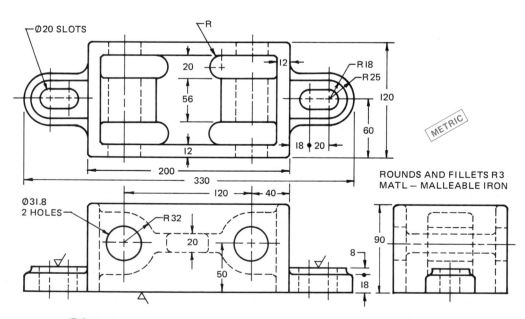

ROUNDS AND FILLETS R3
MATL – MALLEABLE IRON

METRIC

Fig. 9-81 *CAD3 On a B (A3) size sheet, make a three-view drawing of the guide block. Show the front and right-side views in section to improve clarity. Do not dimension unless instructed to do so. Scale: 1:1.*

PROBLEMS

SCHOOL-TO-WORK

Solving Real-World Problems

Fig. 9-82 *SCHOOL-TO-WORK problems are designed to challenge a student's ability to apply skills learned within this text. Be creative and have fun!*

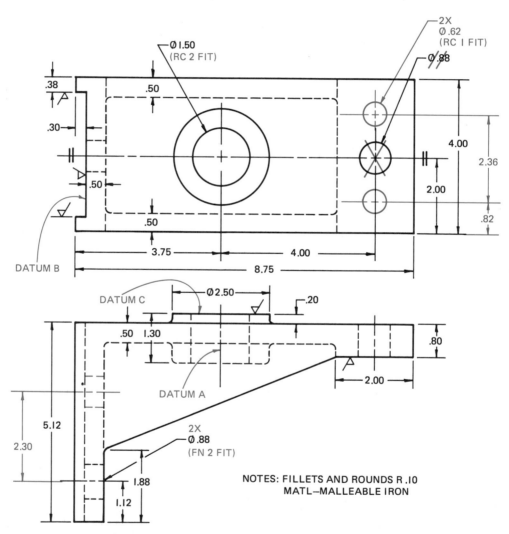

Fig. 9-82 *Prepare a working drawing of the bearing bracket. Show three views, one in section. Make all changes as specified by the design engineer (shown in color). Refer to the appendix tables on limits and fits and dimension the precision holes accordingly. Scale is optional. Add geometric dimensioning and tolerancing symbols to specify the following:*

1. *Datum A to be parallel to datum B to within .003 in. at MMC.*
2. *Datum C to be perpendicular to datum A to within .002 in. at MMC.*
3. *Datum C to be perpendicular to datum B to within .002 in. at MMC.*
4. *Datum C to be flat to within .001 in.*

10

FASTENERS

OBJECTIVES

Upon completion of Chapter 10, you will be able to:

○ Identify and describe various types of fasteners.
○ Define common screw thread terms.
○ Specify threads and fasteners on a technical drawing.
○ Draw detailed, schematic, and simplified thread representations.
○ Name and describe common thread series.
○ Describe and specify classes of thread fits.
○ Draw various types of threaded fasteners.

VOCABULARY

In this chapter, you will learn the meanings of the following terms:

- acme thread
- allowances
- detailed representation
- double thread
- helix
- left-hand thread
- major diameter
- minor diameter
- pitch
- prototype drawing
- right-hand thread
- schematic representation
- screw-thread series
- simplified representation
- single thread
- symbol library
- tolerances
- triple thread

A *fastener* is any kind of device or method for holding parts together. Screws, bolts and nuts, rivets, welding, brazing, soldering, adhesives, collars, clutches, and keys are all fasteners. Each of these fastens in a different way. Some fasten parts permanently; others can later be taken apart again or adjusted.

SCREWS AND SCREW THREADS

The principle of the screw thread has been known for so long that no one knows who discovered it.

Archimedes (287-212 B.C.) a Greek mathematician, put the screw thread to practical use. He used it in designing a screw conveyor to raise water. Similar devices are still used today to move flour and sugar in commercial bakeries, to raise wheat in grain elevators, to move coal in stokers, and for many other purposes.

Screws and other fasteners have so many uses and have become so important that engineers, drafters, and technicians must become familiar with their different forms (Fig. 10-1). They must also be able to draw and specify each type correctly.

The True Shape of a Screw Thread

A *screw thread* can be a helical ridge on the external or internal surface of a cylinder, or it can be in the form of a conical spiral on the external or internal surface of a cone or frustum of a cone. All screw threads are shaped basically as a **helix,** or *helical curve.* Technically, a helix is the curving path that a point would follow if it were to travel in an even spiral around a cylinder and parallel to (in line with) the axis of that cylinder. In simpler terms, if a wire is wrapped around a cylinder in evenly spaced coils, it forms helical curves.

Another way to visualize the shape of screw threads is to cut out a right triangle in paper and wrap it around a cylinder, as shown in Fig. 10-2A. If the triangle's base is the same length as the cylinder's circumference, its hypotenuse, wrapped around the cylinder, forms one turn of a helix. The triangle's altitude is the pitch of the helix. A right triangle and the projections of the corresponding helix are shown in Fig. 10-2B.

To draw the projections of a helix, follow the method shown in Fig. 10-3A and B. First, draw two projections of a cylinder (A). Lay off the pitch. (The *pitch* is the distance from a point on the thread form to the corresponding point on the next form. It will be described in greater detail later in this chapter.) Divide the circumference into a number of equal parts. Divide the pitch into the same number of equal parts. From each division point on the circumference, draw lines parallel to the axis. From each division point on the pitch, draw lines at right angles to the axis. Then draw a smooth curve through the points where these lines meet (B). This gives the projection of the helix.

The application of the helix is shown in Fig. 10-3C, the actual projection of a square thread. However, such drawings are seldom made, since

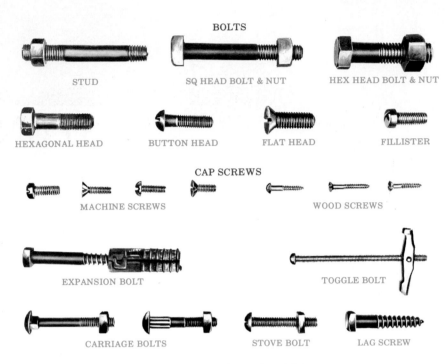

Fig. 10-1 *Examples of some threaded fasteners.*

they take too much time. Also, they are no more useful than conventional representations.

Screw-Thread Standards

The first screws were made for one purpose. They were made without any thought of how anyone else might make one of the same diameter. Later, as industry and mass production developed, goods were produced in quantity by using interchangeable parts. The need then arose for standards for screws and screw threads.

Screw-thread standards in the United States were developed from a system that William Sellers presented to the Franklin Institute in Philadelphia in 1864. Screw-thread standards in England came from a paper presented to the Institution of Civil Engineers in 1841 by Sir Joseph Whitworth. These two standards were not interchangeable.

More and better screw-thread standards have been developed as industrial production has grown more complex. In 1948, standardization committees of Canada, Great Britain, and the United States agreed on the Unified Thread Standards. These standards are now the basic American National Standards. They are listed in *American National*

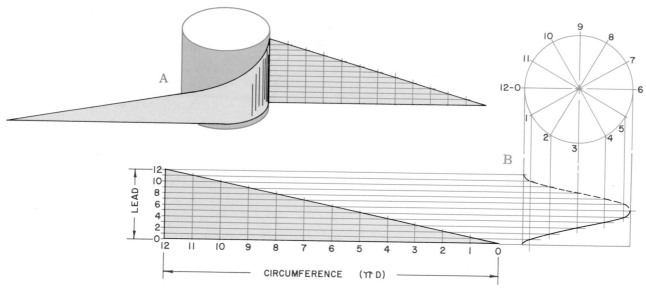

Fig. 10-2 *Picture of a helix (A) and a projection of a helix (B).*

Standard Unified Screw Threads for Screws, Bolts, Nuts and Other Threaded Parts (ANSI B1.1) and in Handbook H-28, *Federal Screw Thread Specifications.*

Since 1948, the different thread systems have been brought increasingly into line with each other. In 1968, the International Standards Organization (ISO) adopted the Unified system as its inch screw-thread system standard. The two systems are alike in some ways.

These similarities are very important as the metric system of measurement comes into worldwide use. To help bring other screw-thread standards into line with this system, the American National Standards Institute has published the *American National Standard Unified Screw Threads-Metric Translation* (ANSI B1.1a).

Screw-Thread Terms

Figure 10-4 shows the main terms used to describe screw threads. The Unified and American (National) screw-thread profile shown in Fig. 10-5 is the form used for fastening in general. Other forms of threads are used to meet special fastening needs. Some of these threads are shown in Fig. 10-6. The sharp V is seldom used today. The square thread and similar forms (worm thread and acme thread) are made especially to transmit motion or power along the line of the screw's axis.

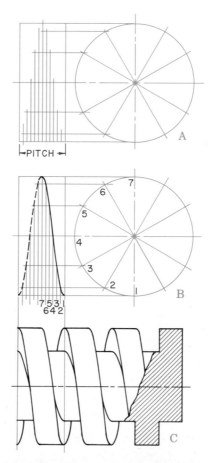

Fig. 10-3 *Drawing the projections of a helix (A and B) and applying the helix to the actual projection of a square thread (C).*

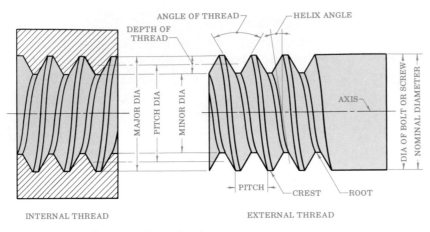

Fig. 10-4 *Screw-thread terms.* (Courtesy of Circle Design)

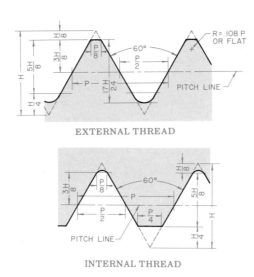

Fig. 10-5 *Unified screw-thread terms.*

The knuckle thread is the thread used in most electric-light sockets. It may also be a "cast" thread. The Dardelet thread automatically locks a screw in place. It was designed by a French military officer but is seldom used today. The former British Standard (Whitworth) has rounded crests and roots. Its profile forms 55° angles. The former United States Standard had flat crests and roots and 60° angles. The buttress thread takes pressure in one direction only: against the surface at an angle of 7°.

Pitch of a Screw Thread

The **pitch** of a thread is the distance from one point on the thread form to the corresponding point on the next form. It may be defined as a formula as follows:

$$\text{Pitch} = \frac{1}{\text{no. of threads per inch}}$$

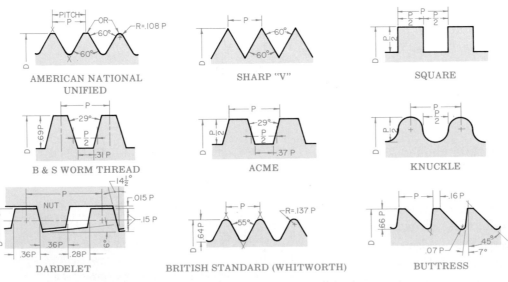

Fig. 10-6 *Some of the various screw-thread profiles.*

Pitch is measured parallel to the thread's axis (Fig. 10-7). The *lead* (pronounced "leed") L is the distance along this same axis that the threaded part moves against a fixed mating part when given one full turn. It is the distance a screw enters a threaded hole in one turn.

Single and Multiple Threads

Most screws have single threads (Fig. 10-7A). A screw has a single thread unless it is marked otherwise. A **single thread** is a single ridge in the form of a helix. If you give a single-thread screw one full turn, the distance it advances into the nut (lead) equals the pitch of the thread.

A **double thread** (Fig. 10-7B) is two helical ridges side by side. The lead, in this case, is twice the pitch. A **triple thread** is three ridges side by side. The lead for this thread is 3 times the pitch. Double and triple threads (also called *multiple threads*) are used where parts must screw together quickly. For example, technical pen caps and toothpaste tube caps have multiple threads.

Right- and Left-Hand Threads

A **right-hand thread** screws in when turned clockwise as you view it from the outside end (Fig. 10-8A). A **left-hand thread** screws in when turned counterclockwise (B). Threads are always right-hand threads unless marked with the initials *LH,* meaning "left-hand." Some devices, such as the turnbuckle (Fig. 10-9), have both right- and left-hand threads. Others, such as bicycle pedals, can have either. A left-hand pedal has left-hand threads; a right-hand pedal has right-hand threads.

DRAWING SCREW THREADS

When drawing screw threads, you use special symbols. These are the same whether the threads are coarse or fine, right hand or left hand. Use notes to give the necessary information.

Under ANSI rules, screw threads can be drawn in three ways, or representations. These are:

- detailed representation
- schematic representation
- simplified representation

The following sections describe these representations and explain how to draw them.

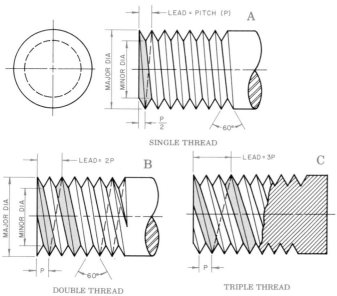

Fig. 10-7 *Single, double, and triple threads.*

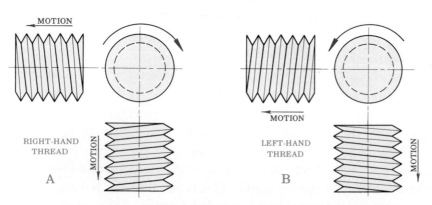

Fig. 10-8 *A right-hand screw thread (A), and a left-hand screw thread (B).*

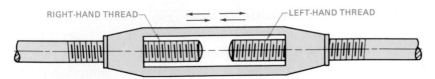

Fig. 10-9 *A turnbuckle uses right- and left-hand screw threads.*

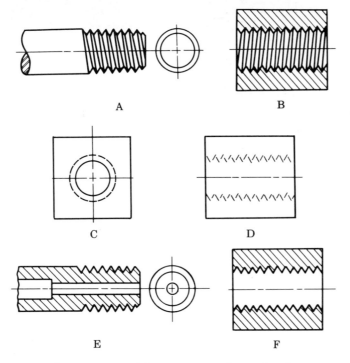

Fig. 10-10 *Detailed representation of screw threads.*

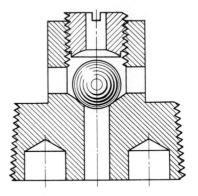

Fig. 10-11 *Threads in section on assembled pieces.*

Detailed Representation

A **detailed representation** approximates the real look of threads (Fig. 10-10). For this kind of drawing, it is not necessary to draw the pitch exactly to scale. Instead, estimate the right size for the drawing. In general, detailed representation is not used in working drawings (see Chapter 11), except where needed for clarity. It is also not usually used if the screw is less than 1.00 in. (25 mm) in diameter. The realistic, or V-form, thread representation can make a drawing clearer where two or more threaded pieces are shown in section (Fig. 10-11).

To draw a detailed representation, draw the screw threads with the sharp-V profile. Use straight lines to represent the helixes of the crest and root lines. To draw the V-form thread, follow the steps in Fig. 10-12, as follows.

1. Lay off the pitch *P,* as shown in (A). The pitch need not be drawn to scale. Make it about the right size for the drawing.
2. Lay off the half pitch $P/_2$ at the end of the thread, as shown.
3. Construct a right triangle with a base of $P/_2$ and an altitude equal to the outside diameter of the screw. Adjust your triangle, or the ruling arm of your drafting machine, to the slope of this right triangle, and draw all the crest lines with this slope.
4. Use the 30°-60° triangle (or the drafting machine ruling arm set at a 30° angle) to draw one side of the V for the threads, as shown in (B).
5. Reverse the triangle, or ruling arm, and complete the Vs.
6. Set the triangle, or the ruling arm of the drafting machine, to the slope of the root lines. Draw them as shown in (C). Notice that the root lines do not parallel the crest lines. This is because the **minor diameter** (root diameter) is less than the **major diameter** (crest diameter).
7. Draw 45° chamfer lines using the construction shown in (D).

A hole with an internal right-hand thread is shown as a sectional view in Fig. 10-13. Notice that the thread-line slope is in the opposite direction from the external right-hand thread lines on a mating screw. This is because the internal thread lines must match the far side of the screw.

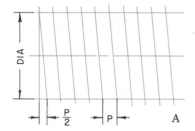

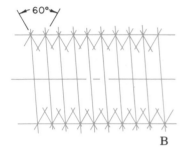

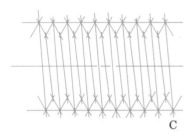

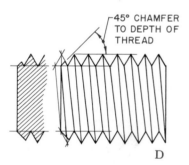

Fig. 10-12 *Drawing the detailed representation of screw threads.*

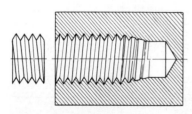

Fig. 10-13 *Internal threads in section (threaded hole).*

Schematic Representation

A **schematic representation** shows the threads using symbols, rather than as they really look. For this kind of drawing, leave out the Vs (Fig. 10-14). Also, the pitch need not be drawn to scale. Make it about the right size for the drawing, and then draw the crest and root lines accordingly. Space them by eye to make them look good.

To draw a schematic representation of screw threads, follow the steps in Fig. 10-15, as follows.

1. Lay off the outside diameter of the screw thread (A).
2. Lay off the thread depth and the chamfer (B).
3. Draw thin crest lines at right angles to the axis (C).
4. Draw thick root lines parallel to the crest lines (D).

The crest and root lines can be at right angles to the axis or slanted to show the helix angle (Fig. 10-16A). To draw slanted threads, give each thread a slope of half the pitch. Notice that the crest lines are thin and root lines are thick. However, you may draw all the lines the same thickness to save time, especially on pencil drawings (Fig. 10-16B).

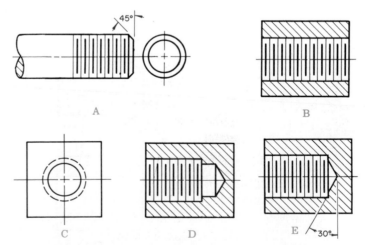

Fig. 10-14 *Schematic representations of screw threads.*

Simplified Representation

A **simplified representation** is much like a schematic representation. In this case, however, draw the crest and root lines as dashed lines (Fig. 10-17B, D, and E), except where either of them would normally show as a visible solid line (Fig. 10-17A, C, F, G, H, I, and J). The drafter saves time by using the simplified representation because it leaves out useless details. Therefore, simplified representation has become the most commonly used in industry.

To draw a simplified representation, follow the steps in Fig. 10-18, as follows.

1. Lay off the outside diameter of the screw (A).
2. Lay off the screw-thread depth (estimate) and the chamfer (B).
3. Draw the chamfer and a line to show the length of the thread (C).
4. Draw dashed lines for the threads (D).

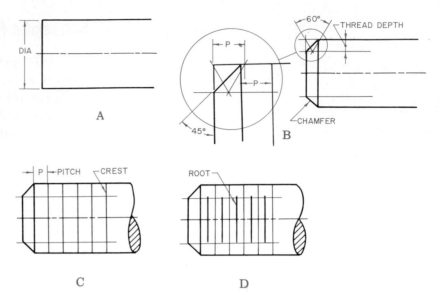

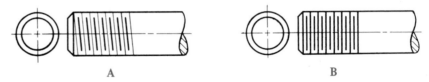

Fig. 10-15 *Drawing the schematic representation of screw threads.*

Fig. 10-16 *(A) Slope-line representation. (B) Uniform-width lines.*

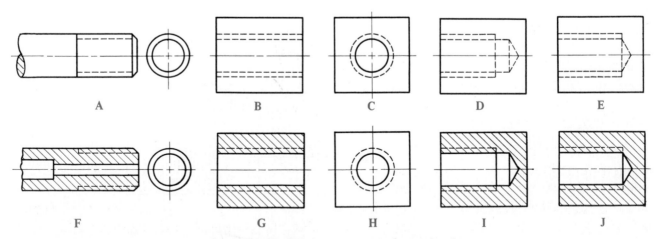

Fig. 10-17 *Simplified representation of screw threads.*

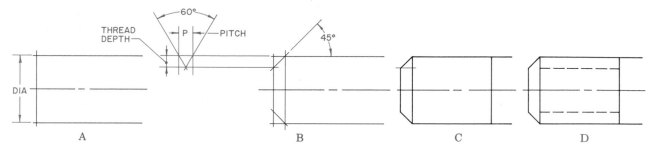

Fig. 10-18 *Drawing the simplified representation of screw threads.*

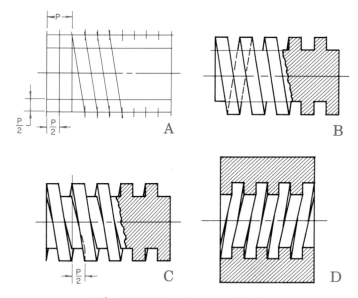

Fig. 10-19 *Drawing square threads.*

Square Screw Threads

The square thread has a depth that is one half its pitch. To draw square threads, perform the steps in Fig. 10-19. First, lay off the diameter, the pitch *P*, one-half-pitch spaces, and the depth of the thread (A). Next, draw the crest lines (B). Then draw the root lines (C). Fig. 10-19(D) shows an internal square thread drawn in section.

Acme Screw Threads

The **acme thread** has a depth that is one half its pitch. To draw acme threads, follow the method shown in Fig. 10-20A. First, lay off the outside diameter, the pitch, and the depth of the thread. Midway between the outside diameter and the depth of the thread is the *pitch diameter*. Along it,

draw the *pitch line*. On the pitch line, lay off one-half-pitch spaces. Use these to draw the thread profile. Now draw the crest lines and root lines to complete the view. Fig. 10-20B shows an enlarged view.

Internal acme threads in section are drawn as shown in Fig. 10-20C. They can also be drawn in other ways, including with dashes for the hidden lines or with the outline in section, as shown in (D).

CLASSIFICATION OF SCREW THREADS

Screws are classified according to several different characteristics. For example, they can be divided according to combinations of diameter and pitch, tolerances and allowances, and various other specifications.

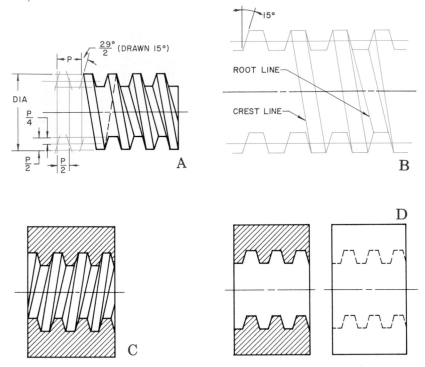

Fig. 10-20 *Drawing acme threads.*

Thread Series

Screws of the same diameter are made with different pitches (numbers of threads per inch) for different uses. In the Unified screw-thread system, the various combinations of diameter and pitch have been grouped in **screw-thread series.** These series are listed in ANSI B1.1. Each is denoted by letter symbols, as follows:

Coarse-Thread Series (UNC or NC) In this series, the pitch for each diameter is relatively large. This series is used for engineering in general.

Fine-Thread Series (UNF or NF) In this series, the pitch for each diameter is smaller (there are more threads per inch) than in the coarse-thread series. This series is used where a finer thread is needed, as in making automobiles and airplanes.

Extra-Fine-Thread Series (EUNF or NEF) In this series, the pitch is even smaller than the fine-thread series. This series is used where the thread depth must be very small, as on aircraft parts or thin-walled tubes.

The Unified system also has several constant-pitch-thread (UN) series. These series have 4, 6, 8, 12, 16, 20, 28, or 32 threads per inch. They offer a variety of pitch-diameter combinations that can be used where the coarse, fine, and extra-fine series are not suitable. However, when selecting a constant-pitch series, the first choices should be the 8-, 12-, or 16-thread series. Constant-pitch threads are generally used as a continuation of coarse-, fine-, and extra-fine-thread series in larger diameters.

Eight-Thread Series (8UN or 8N) This series uses 8 threads per inch for all diameters.

Twelve-Thread Series (12UN or 12N) This series uses 12 threads per inch for all diameters.

Sixteen-Thread Series (16UN or 16N) This series uses 16 threads per inch for all diameters.

Special Threads (UNS, UN, or NS) These are nonstandard, or special, combinations of diameter and pitch.

The American National Standard thread series have been largely replaced by the Unified series described above. However, you should still know what they and their letter symbols are. These series and their symbols are listed below.

Coarse-Thread Series (NC) This series is used for screws, bolts, and nuts produced in quantity, and also for fastening in general.

Fine-Thread Series (NF) This series has a smaller pitch for each diameter than NC. It is used where coarse series is not suitable.

Extra-Fine-Thread Series (NEF) This series is used when an even smaller pitch than NF is needed, as on thin-walled tubes.

Eight-Thread Series (8N) This series is a constant-pitch series for large diameters.

Twelve-Thread Series (12N) This series is a constant-pitch series that has a medium-fine pitch for large diameters.

Sixteen-Thread Series (16N) This series is a constant-pitch series that has an even smaller pitch per diameter.

NOTE: The symbol NS denotes special threads in the American National Standards thread series.

Notice that the American system has only three constant-pitch series, while the Unified system has eight, including all three from the American system. The three in the American system are used most.

Classes of Fits

Screw threads are also divided into *screw-thread classes* based on their **tolerances** (amount of size difference from exact size) and **allowances** (how loosely or tightly they fit their mating parts). The exact screw thread needed can be obtained by choosing both a series and a class. In brief, the classes for Unified threads are Classes 1A, 2A, and 3A for external threads only and Classes 1B, 2B, and 3B for internal threads only.

Classes 1A and 1B These have a large allowance (loose fit). They are used on parts that must be put together quickly and easily.

Classes 2A and 2B These are the thread standards most used for general purposes, such as for bolts, screws, nuts, and similar threaded items.

Classes 3A and 3B These are stricter standards for fit and tolerance than the others. They are used where thread size must be more exact.

The American National Standard also names two other classes: Classes 2 and 3. These classes are described, with tables of dimensions, in Appendix 1 of ANSI B1.1.

Thread Specifications

A screw thread is specified by giving its nominal (major) diameter, number of threads per inch, length of thread, initial letters of the series, class of fit, and external (A) or internal (B), as shown in Fig. 10-21. Any thread you specify is assumed to be both single and right hand unless stated otherwise. If the thread is to be left hand, include the letters LH after the class symbol. If it is to be a double or triple thread, include DOUBLE or TRIPLE.

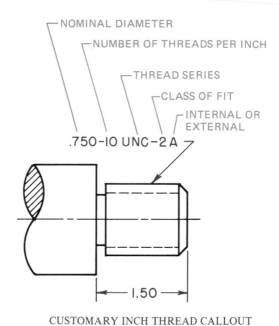

CUSTOMARY INCH THREAD CALLOUT

Fig. 10-21 *Customary thread specifications.*

Some examples of thread specifications are:

- 1.25-7UNC-1A (1.25-in. diameter, 7 threads per inch, Unified threads, coarse threads, Class 1, external)
- .75-10UNC-2A (.75-in. diameter, 10 threads per inch, Unified threads, coarse threads, Class 2, external) (See Fig. 10-21.)
- .88-14UNF-2B (.88-in. diameter, 14 threads per inch, Unified threads, fine threads, Class 2, internal)
- 1.62-18UNEF-3B-LH (1.62-in. diameter, 18 threads per inch, Unified threads, extra-fine threads, Class 3, internal, left hand)

Sizes of threads and fasteners may be given on a drawing by using either fractional-inch sizes or decimal-inch sizes. Since manufacturers of fasteners often specify them in fractions, it is a generally accepted practice to specify them on the drawing using fractions. However, since many industries have switched entirely to decimal-inch dimensioning, the practice of dimensioning and specifying fasteners in decimal-inch sizes is also now a common practice. Either is correct. Drafters use the method required by the standards and practices of individual companies.

If you are using the traditional method, specify tapped (threaded) holes by a note giving the diameter of the tap drill (.42″), depth of hole (1.38″), thread information (.50″ diameter, Unified National Coarse threads, Class 2), internal (B) or external (A), and depth of thread (1.00″). For example:

.42 DRILL × 1.38 DEEP

.50-13UNC-2B × 1 DEEP

In the more current system, tapped (threaded) holes are specified only by giving the thread specification and depth. The tap drill diameter and depth of the drilled hole are not given. For example, in the modern system, the specifications given in the preceding paragraph would be reduced to giving only the thread information (.50), the thread series (UNC), class of fit (2), internal (B) or external (A), and depth of thread (1.00″):

.50-13UNC-2B × 1.00

Metric Threads

Specify an ISO metric screw thread by giving its nominal size (basic major diameter) and pitch, both expressed in millimeters. Include an M to denote that the thread is an ISO metric screw thread. Place the M before the nominal size. Use × to separate the nominal size from the pitch. For the coarse thread series only, do not show the pitch unless you are also giving the length of the thread. If the length is given, separate it from the other designations with an ×. For external threads, the length may be given as a dimension.

For example, designate (specify) a 10-mm diameter, 1.25 pitch, fine-thread series as M10 × 1.25. Specify a 10-mm diameter, 1.5 pitch, coarse-thread series, however, as M10. Remember, for coarse threads, do not give the pitch unless you also give the length of the thread. If in this example the length were 25 mm, the thread would be designated as M10 × 1.5 × 25.

In addition to the basic designation, the designation for the thread's tolerance class must be given. This designation is separated from the basic designation by a dash. It consists of two sets of symbols. The first denotes the pitch diameter tolerance. The second, which follows immediately, denotes the crest diameter tolerance. Each of these two sets of symbols is composed of a number denoting the grade tolerance, followed by a letter (capital for internal threads and lowercase for external threads) denoting the tolerance position. If the symbols for the pitch diameter tolerance are the same as those for the crest diameter tolerances, you may leave out one set (Fig. 10-22).

Two classes of fits are generally used in the ISO metric system. The first is a general-purpose class and is specified as 6H/6g. The second is for fits that require a closer tolerance and is specified as 6H/5g6g. More detailed information may be obtained from ISO and ANSI standards. However, unless a very specific tolerance class is necessary, many industries have adopted a simplified system for designating thread fits. If the general-purpose fit is acceptable, no tolerance class is specified. If a close fit is necessary, the letter C is placed after the pitch in the thread note. An example is M10 × 1.25C.

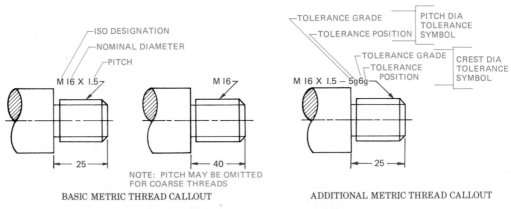

BASIC METRIC THREAD CALLOUT   ADDITIONAL METRIC THREAD CALLOUT

Fig. 10-22 *ISO metric thread specifications.*

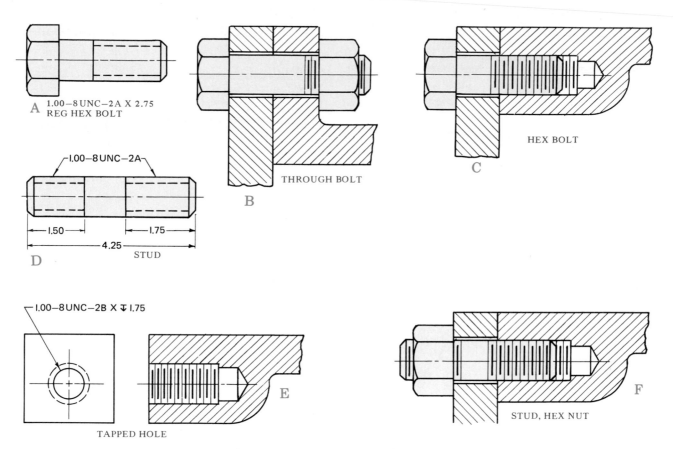

Fig. 10-23 *Bolts, studs, and threaded holes.*

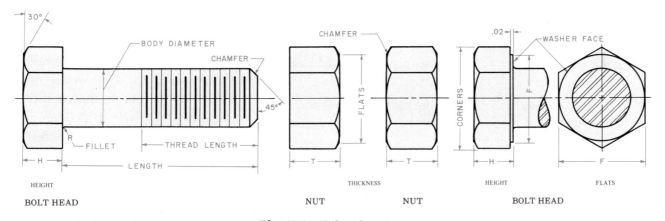

Fig. 10-24 *Bolt and nut terms.*

BOLTS AND NUTS

Various types of bolts and their uses are shown in Fig. 10-23. Figure 10-24 illustrates the terms used to describe standard bolts and nuts. In general, bolts and nuts are either regular or heavy (thick), and they are either square or hexagon. Regular bolts and nuts are used for the general run of work. Heavy (thick) bolts and nuts are somewhat larger. They are used where the bearing surface or the hole in the part being held must be larger. The types of bolts and nuts made in regular sizes include square bolts and nuts, hexagon bolts and nuts, and semifinished hexagon bolts and nuts. Types made in heavy (thick) sizes include hexagon bolts and nuts, semifinished hexagon bolts and nuts, finished hexagon bolts, and square nuts.

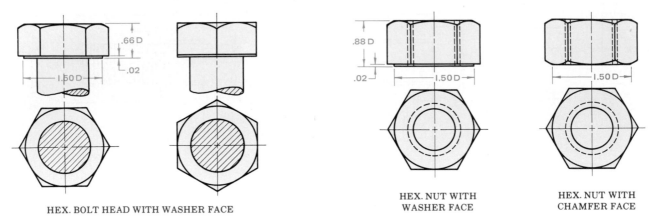

HEX. BOLT HEAD WITH WASHER FACE

HEX. NUT WITH
WASHER FACE

HEX. NUT WITH
CHAMFER FACE

Fig. 10-25 *Semifinished boltheads and nuts.*

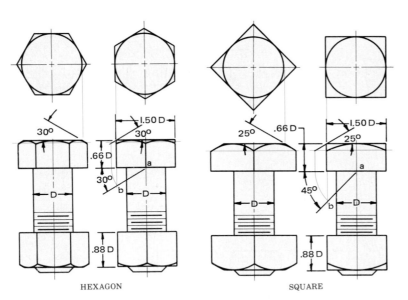

HEXAGON

SQUARE

Fig. 10-26 *Regular hexagon bolthead and nut.*

Fig. 10-27 *Regular square bolthead and nut.*

Regular bolts and nuts are not finished on any surface. Semifinished bolts and nuts are processed to have a flat bearing surface (Fig. 10-25). The bearing surface has either a washer face or a face with chamfered corners. Each face has a diameter equal to the distance across the flats (between opposite sides of the bolthead or nut). The thickness of the washer face is about .02 in. (0.4 mm). "Finished" bolts and nuts are so called only because of the quality of their manufacture and the closeness of their tolerance. They do not have completely machined surfaces.

Because boltheads and nuts come in standard sizes, they are generally not dimensioned on drawings. Instead, give the needed information

in a note (see Fig. 10-23A). The note specifies a 1.00-in. diameter, 8 threads per inch, Unified coarse-thread series, Class 2A fit, 2.75 in. long, regular hexhead bolt.

A stud or stud bolt (Fig. 10-23D) has threads on both ends. It is used where a bolt is not suitable and for parts that must be removed often. Dimension the length of thread from each end as shown. Dimension a tapped (threaded) hole as shown in (E). One way to use a stud is to screw it into a threaded hole permanently. The hole in another part is put over the stud and a nut is screwed onto the end, as shown in (F). In certain cases, a stud can be passed through two parts and a nut placed on each end.

Drawing Regular Boltheads and Nuts

When drawing regular boltheads and nuts, figure the dimensions from the proportions given in Figs. 10-26 and 10-27 or from Appendix A. Draw the chamfer angle at 30° for both the hexagon or the square forms. (The standard for the square form is actually 25°.) Find the radii for the chamfer arcs by trial. Note that you can find one half the distance across the corners, *ab,* by using the construction shown.

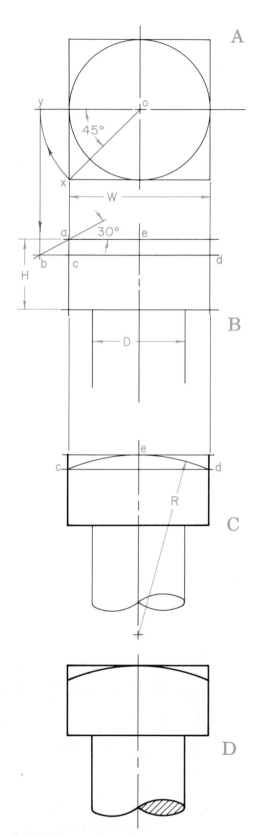

Fig. 10-28 *Drawing a regular square bolthead across the flats.*

Square Bolthead Across Flats

To draw a regular square bolthead across the flats, use the following proportions. If D is the major diameter of the bolt and W is the width of the bolthead across the flats, then $W = 1.5D$. The height of the head $H = .67D$.

Follow the method shown in Fig. 10-28 to draw the bolthead.

1. Draw centerlines and start the top view (A).

2. Start the top view by drawing a chamfer circle with a diameter equal to the distance across the flats.

3. Using a 45° triangle, draw a square about the circle.

4. Below the square, start the front view, as shown in (B) by drawing a horizontal line representing the bearing surface of the head. Lay off the height of the head H. Then draw the top line of the head.

5. To get the sides of the head, project lines down from the top view.

6. To draw the chamfer, begin by drawing line *ox*, (A). Revolve the line to make line *oy*, as shown.

7. Run a line from point *y* straight down to the front view (B).

8. From point *a* in the front view, draw a 30° chamfer line *ab* out to meet the line from point *y*. This forms the chamfer depth.

9. From point *b*, extend a line horizontally across to establish points *c* and *d*, as shown in (C).

10. Draw the chamfer arc through points *c*, *e*, and *d*, using radius R. Find the length of R by trial, or make it equal to W, the width across the flats.

11. Complete the view, as shown in (D).

Square Bolthead Across Corners

To draw a regular square bolthead across the corners, use the same proportions you would use to draw it across the flats ($W = 1.5D$, $H = .67D$). Use the following method to create the drawing, as shown in Fig. 10-29.

1. Draw centerlines.

2. Start the top view by drawing a chamfer circle with a diameter equal to W, the distance across the flats.

3. Using a 45° triangle, draw a square about the circle.

4. Start the side view by drawing a horizontal line to represent the bearing surface head H.

5. Project lines down from the diameter of the chamfer circle in the top view to meet the top line in the front view.

6. From this point, draw 30° chamfer lines to meet the lines at the sides of the head.

7. Find the two tangent points x shown in (A). Project these points to the top view to get points f and g, as shown in (B) and (C).

8. Draw line bcd.

9. Draw the chamfer arcs bfc and cgd using radius R. Find radius R by trial, or figure that $R = yo$ = half the distance across the corners.

10. Complete the view as shown in (D).

Use the same method to draw a nut, except that $T = .88D$ for regular nuts, and $T = D$ for thick nuts. Square bolts are not made in thick sizes.

Boltheads and nuts are usually drawn across corners on all views of design drawings, no matter what the projection. This is done to show the largest *clearance* (space needed for turning) that the bolt or nut must have. It is also done to keep hexagon heads and nuts from being confused with square heads and nuts.

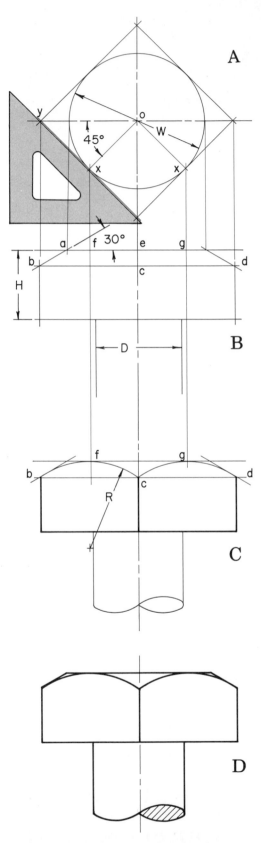

Fig. 10-29 *Drawing a regular square bolthead across the corners.*

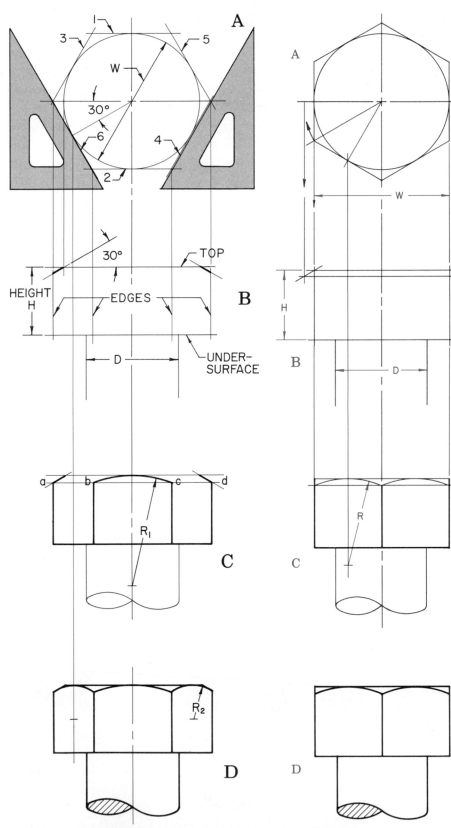

Fig. 10-30 *Drawing a regular hexagon bolthead across the corners.*

Fig. 10-31 *Drawing a regular hexagon bolthead across the flats.*

Hexagon Bolthead Across Corners

Use the same proportions as for the preceding drawings. Follow the steps shown in Fig. 10-30.

1. Start the top view by drawing centerlines and a chamfer circle with a diameter of W, the distance across the flats (A).

2. About this circle, draw a hexagon as shown by lines 1, 2, 3, 4, 5, and 6.

3. Begin the front view as shown in (B). Draw a horizontal line representing the bearing surface or undersurface of the head.

4. Lay off the height of the head and draw the top line.

5. Project the edges and the chamfer points from the corners of the top view. With these guides, draw in the chamfer line.

6. Draw line *abcd* to locate the chamfer intersections.

7. Draw arc *bc* using radius R_1. Find this radius by trial. Complete the view by drawing arcs ab and *cd* using radius R_2. Find this radius, too, by trial.

To draw a hexagon bolthead across flats, proceed as shown in Fig. 10-31. Draw hexagon nuts in the same way, but note the difference between the height of the head and the thickness of the nut.

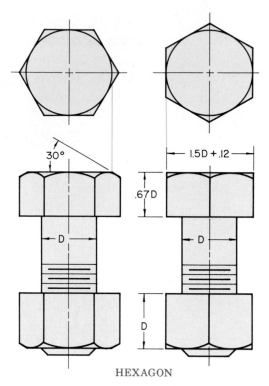

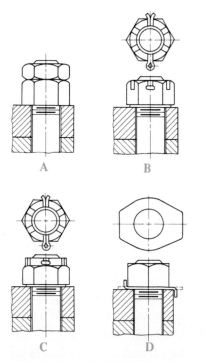

Fig. 10-32 *Thick hexagon bolthead and nut.*

Fig. 10-33 *Locking threaded fastenings.*

Drawing Thick Hexagon Boltheads and Nuts

When drawing thick hexagon boltheads and nuts, figure the dimensions from the proportions given in Fig. 10-32 or from Appendix A. See also ANSI B18.2 and B18.2.1.

A new standard covers hexagon structural bolts. These bolts are made of high-strength steel, and they are used for structural steel joints. The dimensions for drawing these high-strength steel bolts are listed in Appendix A.

OTHER THREADED FASTENERS

Several other kinds of fasteners are in common use for various purposes. Some of them work with nuts, bolts, and screws; others work alone. This section describes several threaded fasteners.

Lock Nuts and Lock Washers

Lock nuts and various devices such as lock washers are used to keep nuts, or bolts and screws, from working loose. Some forms of lock washers are shown in Fig. 10-33.

There are also many washer-type locking devices made to lock fasteners securely in place. Some of these are shown in Fig. 10-34. In addition, there are also self-locking fasteners that have threads coated with epoxy that is combined with a hardening agent before assembly. After assembly, the epoxy hardens and gives a strong, vibration-proof bond.

Fig. 10-34 *Various locking-type washers.*

Capscrews

A capscrew fastens two parts together by passing through a clearance hole in one and screwing into a tapped hole in the other (Fig. 10-35). In most cases, clearance holes are not shown on drawings. Capscrews have a naturally bright finish. This is in keeping with the machined parts with which they are used. Coarse, fine, or 8 threads may be used on capscrews. Socket-head capscrews may have Class 3A threads. The other head styles have Class 2A threads. Dimensions for drawing various sizes of American National Standard capscrews are given in Appendix A.

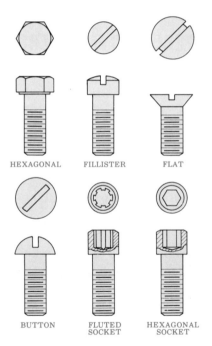

HEXAGONAL FILLISTER FLAT

BUTTON FLUTED SOCKET HEXAGONAL SOCKET

Fig. 10-35 *Cap screws.*

Machine Screws

Machine screws are used where the fastener must have a small diameter (Fig. 10-36). Sizes below .25 in. (6 mm) in diameter are specified by number. Machine screws may screw into a tapped hole or extend through a clearance hole and into a square nut. The finish on machine screws is bright. The ends are flat, as shown. Machine screws up to 2.00 in. (50 mm) long are threaded full length. Coarse or fine threads and Class 2 threads may be used on machine screws.

Setscrews

Setscrews are used to hold two parts together in a desired position. They do so by screwing through a threaded hole in one part and *bearing* (pushing) against the other (Fig. 10-37). There are two general types: square-head and headless. *NOTE:* Square-head setscrews can cause accidents when used on rotating parts. They may also violate safety codes. Headless setscrews can have either a slot or a socket. Any of the points shown can be used on any setscrew.

FLAT   OVAL   ROUND   FILLESTER   TRUSS   BINDING   PAN

Fig. 10-36 *Machine screws.*

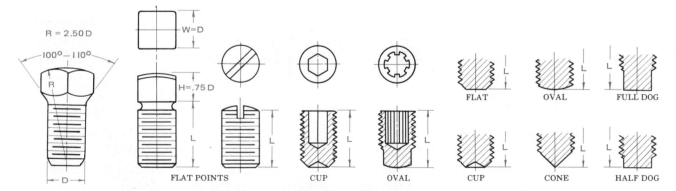

Fig. 10-37 *Setscrews.*

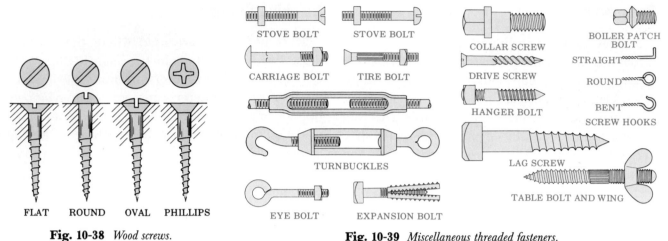

Fig. 10-38 *Wood screws.*

Fig. 10-39 *Miscellaneous threaded fasteners.*

Wood Screws

Wood screws are made of steel, brass, or aluminum. They are finished in various ways (Fig. 10-38). Steel screws can be bright (natural finish), blued, galvanized, or copper plated. Both steel and brass screws are sometimes nickel plated. Round-head screws are set with the head above the wood. Flat-head screws are set flush or countersunk. Draw wood screws as shown in Fig. 10-38. Specify them by number, length, style of head, and finish. For flat-head screws, dimension the overall length. Dimension round-head screws from under the head to the point. Dimension oval-head screws from the largest diameter of the countersink to the point. Sizes and dimensions are listed in Appendix A.

Miscellaneous Threaded Fasteners

Various other threaded fasteners are shown in Fig. 10-39. In most cases, the names denote the ways in which they are used.

Screw hooks and screw eyes are specified by diameter and overall length. A lag screw, or lag bolt, is used to fasten machinery to wood supports. It is also used in heavy wood constructions when a regular bolt cannot be used. It is similar to a regular bolt but has wood-screw threads. Specify a lag bolt by its diameter and the length from under the head to the point. The head of the lag bolt has the same proportions as a regular bolthead.

MATERIALS FOR THREADED FASTENERS

Threaded fasteners are usually made from steel, brass, bronze, aluminum, cast iron, wood, and nylon. Nylon screws and bolts are made in various bright colors, such as red, blue-green, yellow, and white.

DRAWING THREADS AND FASTENERS USING CAD

CAD software allows drafters to create repetitive tooth and thread forms quickly and accurately. The system allows you to create, group, and store various types of thread forms, fastener head forms, and other types and shapes of fasteners in a **symbol library.** The symbol library is actually just a drawing file–the same as any other drawing file. The symbols are the individual thread forms, fasteners, etc., that make up the library. For example, depending on your needs, you could place drawings of all the fasteners discussed in this chapter into a single symbol library named FASTENER.

The symbols in a symbol library are stored as *blocks.* A block is a permanent grouping of the lines, arcs, etc., that make up an individual symbol. In Auto-CAD, for example, enter the BLOCK command to create a new block. Give it a meaningful name (such as RIVET or SETSCREW) and use the cursor to tell the software which entities to include in the block.

From that point, the blocked symbol exists in the drawing file, even if it is not visible on the screen. To insert the RIVET block, enter the INSERT command and specify the block name, RIVET, as prompted. The CAD software allows you to place the block anywhere on the drawing, at any angle and at any scale.

To use the symbol library, insert the drawing file, and thus the symbols, into your current drawing. For example, suppose the symbol library is TOOLS.DWG. Begin by entering the INSERT command and specifying the name of the symbol library (TOOLS.DWG). *Important:* cancel the command when you get to the "Insertion point" prompt. The blocks will not appear on the screen, but they are now part of the drawing database, so you can insert the individual blocks into the current drawing as needed.

As an alternative, if you know you will be using the same symbols in most or all of your drawings, you can create a **prototype drawing** (template drawing) in which the symbols already exist. Do this by creating a new drawing. Set any parameters (units, snap, grid, and so on) that you know you will usually need. Then follow the procedure outlined above to insert the symbol library file into the drawing. Save the drawing, even though the screen appears blank. Name it something that will remind you that it is a prototype drawing–PROTO or PTYPE, for example. Then, when you begin a new drawing, specify your prototype drawing as the prototype for the new drawing. All of the settings you created in the prototype will be pre-established for you, but this is *not* the prototype drawing. It's a copy of the prototype, which you can use as the basis for your current drawing.

Many symbol libraries, including extensive libraries for fasteners, are commercially available. Before you take the time and trouble to create your own symbol library, you should check to see if similar libraries are available. However, it is important to understand how to create your own symbols and symbol libraries because CAD

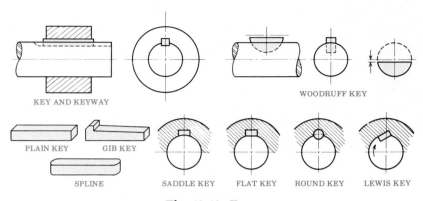

KEY AND KEYWAY • WOODRUFF KEY

PLAIN KEY • GIB KEY • SPLINE • SADDLE KEY • FLAT KEY • ROUND KEY • LEWIS KEY

Fig. 10-40 *Keys.*

standards, design standards, and even the library items themselves may change over time.

NONTHREADED FASTENERS

Some applications require fasteners in parts where threading cannot easily be accomplished. In other cases, there is simply no need to remove the fastener once it has been installed, so threads are unnecessary. Nonthreaded fasteners such as keys and rivets are used under these and other circumstances.

Keys

Keys are nonthreaded fasteners used to secure pulleys, gears, cranks, and similar parts to a shaft (Fig. 10-40). Keys are made in different forms for different uses. They range from the saddle key, for light duty, to special forms, such as two square keys, for heavy duty. The common sunk key can have a breadth of about one fourth the shaft diameter. Its thickness can vary from five eighths the breadth to the full breadth.

The Woodruff key is often used in machine-tool work. It is made in standard sizes. Specify it by number (see Appendix A). Special forms of pins have been developed to replace keys for some uses. These pins need only a drilled hole instead of the machining that is needed to make keyseats and keyways.

Rivets

Rivets are rods of metal with a preformed head on one end. They are used to fasten

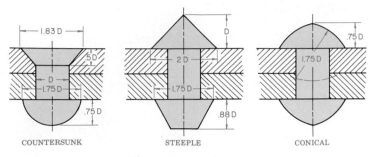

Fig. 10-41 *Large rivets.*

Fig. 10-42 *Small rivets.*

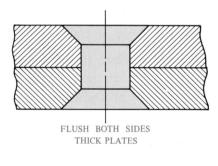

Fig. 10-43 *Explosive rivets.* (Courtesy of E.I. duPont de Nemours & Co.)

sheet-metal plates, structural steel shapes, boilers, tanks, and many other items permanently. The rivet is first heated red-hot. Then it is placed through the parts to be joined. It is held there in place while a head is formed on the projecting end. The rivet is then said to have been "driven."

Large rivets (Fig. 10-41) have nominal diameters ranging in size from .50 to 1.75 in. (12 mm to 45 mm). Small rivets (Fig. 10-42) range from .06 to .44 in. (2 or 3 mm to 11 mm) in diameter. See Appendix A for American National Standard rivet dimensions.

Some rivets are made especially for use where one side of the plates cannot be reached or where the space is too small to use a regular rivet. These are called *blind rivets*. One type is the duPont explosive rivet (Fig. 10-43). It has a small explosive charge in a cavity. After the rivet is inserted, the charge is exploded, forming a head. This rivet is thus excellent for blind riveting, since the head can be formed inside places that are closed or impossible to reach.

Sometimes plates that are riveted together need clear surfaces. This requires flush riveting (Fig. 10-44) on one or both sides. These rivets are used on airplanes, automobiles, spacecraft, etc.

Riveted joints (Fig. 10-45) are used for joining plates. They may have lap or butt joints. They may also have single or multiple riveting.

For some uses, as in tanks or steel buildings, high-strength structural bolts are used. Welding is also in wide use.

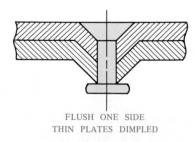

FLUSH BOTH SIDES
THICK PLATES

FLUSH ONE SIDE
THIN PLATES DIMPLED

Fig. 10-44 *Flush rivets.*

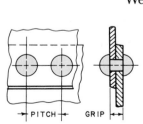

SINGLE-RIVETED LAP JOINT

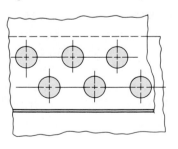

DOUBLE-RIVETED LAP JOINT
STAGGERED RIVETING

Fig. 10-45 *Riveted joints.*

CHAPTER 10

REVIEW

Learning Activities

1. Collect as many different types of fasteners as possible. Identify each by name. Determine the size and thread specifications of each if applicable.

2. Obtain technical drawings from industry. Locate threads and fasteners on the drawings, and study thread symbols and methods of specifying threads and fasteners. Convert some of the threads and specifications to ISO metric equivalents.

3. This chapter provides a brief history of the development of screws. At your school or public library, read more about the history of fasteners in general. Then suggest ways in which you think one or more specific fasteners could be further improved.

Questions

Write your answers on a separate sheet of paper.

1. A device for holding parts together is called a _____.

2. The true shape of a screw thread is based on a _____ curve.

3. The distance from a point on one thread form to a corresponding point on the next thread form is the _____.

4. The distance a threaded part advances with one complete revolution (turn) is called _____.

5. What type of thread is designed to transmit motion or power along the axis of the screw?

6. Name the three types of thread representation.

7. Double and triple threads may also be called _____ threads.

8. The three most common Unified thread series are coarse, _____, and _____.

9. What is the formula for the pitch of a thread?

10. A fastener with threads on both ends is called a _____.

11. What is the difference in designating metric fine- and coarse-thread series?

12. If a close fit is necessary for a metric screw thread, the letter _____ may be used in the note.

13. What is the difference between a Class 2A thread and a Class 2B thread?

14. Another name for the template drawing on which a new drawing is based in a CAD system is _____ drawing.

15. In CAD, what is a symbol library?

CHAPTER 10

PROBLEMS

The small CAD symbol next to each problem indicates that a problem is appropriate for assignment as a computer-aided drafting problem and suggests a level of difficulty (one, two, or three) for the problem.

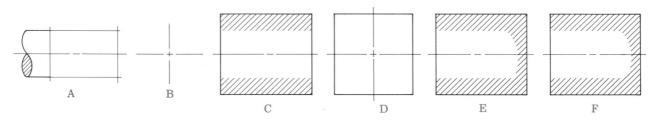

A   B   C   D   E   F

Fig. 10-46 *CAD2 Use dividers to take dimensions from the scale below. Draw the views as follows: (A) schematic representation of the 1.00-8UNC-2A threads; (B) end view of (A); (C) schematic representation of section through 1.00-8UNC-2B (internal) threads; (D) right-side view of (C); (E) schematic representation of section through Ø.88 × 1.50 deep, 1.00-8UNC-2B × 1.12 deep; (F) schematic representation of section through Ø.88 × 1.50 deep, 1.00-8UNC-2B × 1.50 deep.*

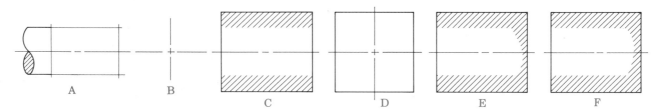

A   B   C   D   E   F

Fig. 10-47 *CAD2 Use dividers to take dimensions from the scale below. Draw the views as follows: (A) simplified representation of 1.00-8UNC-2A threads; (B) end view of (A); (C) simplified representation of section through 1.00-8UNC-2B (internal) threads; (D) right-side view of (C); (E) simplified representation of section through Ø.88 × 1.50 deep, 1.00-8UNC-2B × 1.50 deep.*

DECIMAL INCH

PROBLEMS

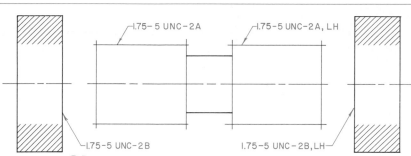

Fig. 10-48 *CAD2 Detailed representation of screw threads. Use dividers to take dimensions from the printed scale assigned at the bottom of the page. Draw the views as shown and complete them according to the specifications noted on each. Use detailed thread representation.*

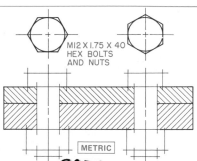

Fig. 10-49 *CAD2 Regular hexagon bolt and nut. Draw the views and complete the bolts and nuts in the sectional view. See Appendix A for bolt and nut detail sizes.*

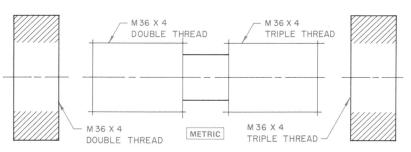

Fig. 10-50 *CAD2 Double and triple threads. Use dividers to take dimensions from the printed scale assigned at the bottom of the page. Draw the views as shown and complete them according to the specifications noted on each. Use detailed thread representation.*

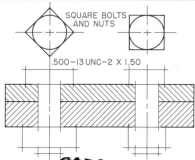

Fig. 10-51 *CAD2 Regular square bolt and nut. Draw the views and complete the bolts and nuts in the sectional view. See Appendix A for bolt and nut detail sizes.*

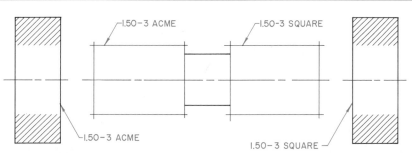

Fig. 10-52 *CAD2 Acme and square threads. Use dividers to take dimensions from the printed scale assigned at the bottom of the page. Draw the views as shown and complete them according to the specifications noted on each. Use detailed thread representation.*

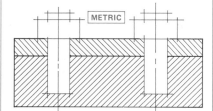

Fig. 10-53 *CAD2 Studs. Draw the view as shown and complete the Ø12 × 44 studs and hexagon nuts. Check Appendix A for specific nut sizes. Other dimensions may be taken from the printed scale assigned at the bottom of the page. Use schematic thread representation.*

PROBLEMS

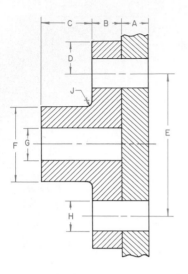

Fig.	Bolt Ø	A	B	C	D	E	F	G	H	J
U.S. customary (in.)										
10-54	.250	.25	.32	.62	.50	2.00	.75	.375	.281	.10
10-55	.375	.38	.44	.62	.62	2.25	.75	.375	.406	.12
10-56	.500	.50	.56	.88	.62	2.75	1.25	.625	.562	.12
10-57	.625	.62	.70	1.12	.75	3.25	1.75	.750	.688	.12
10-58	.750	.75	.80	1.25	.88	3.50	2.00	.875	.812	.20
10-59	1.000	1.00	1.12	1.50	1.12	4.00	2.50	1.125	1.125	.25
Metric (mm)										
10-60	6	6	8	16	12	50	20	9.6	6.3	3
10-61	10	10	12	20	20	60	20	9.6	10.3	4
10-62	16	16	18	24	24	80	40	16.5	16.5	4
10-63	24	24	26	32	26	100	60	29.0	25.0	6

Figs. 10-54 through 10-63 *CAD2 Take all dimensions from the table for the problem assigned, and draw the flange and head plate as shown. On the colored centerlines, draw American National Standard hex or square bolts and nuts or metric hex bolts and nuts as assigned. Place bolthead at the left and show bolthead across flats; nut across corners.* (Courtesy of Circle Design)

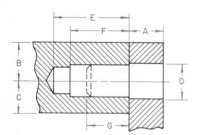

Fig.	Stud Ø	Nut	A	B	C	D	E	F	G
10-64	.750	Hex	.80	.88	.75	.812	1.75	1.38	1.00
10-65	.875	Sq	.94	1.25	.88	.938	2.00	1.56	1.12
10-66	1.000	Hex	1.12	1.12	1.00	1.125	2.25	1.75	1.50
10-67	1.125	Sq	1.25	1.50	1.12	1.250	2.75	2.12	1.50

Figs. 10-64 through 10-67 *CAD2 On the centerline shown, draw a stud with hexagon or square nut, across flats or corners, as directed by the instructor. Take dimensions from the table.* (Courtesy of Circle Design)

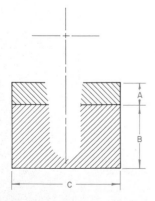

Fig.	Bolt Ø	Head Style	A	B	C
U.S. customary (in.)					
10-68	.500	Button (round)	.75	1.75	3.00
10-69	.500	Flat	.75	1.75	3.00
10-70	.500	Fillister	.75	1.75	3.00
Metric(mm)					
10-71	10	Flat	12	40	64
10-72	12	Fillister	12	40	64
10-73	14	Round	12	40	64

Figs. 10-68 through 10-73 *CAD2 Take dimensions from the table for the problem assigned, and draw the figure shown at the left. Refer to Appendix A for sizes and draw the assigned style of head and size of capscrew. Also, draw a top view of the screw head.* (Courtesy of Circle Design)

PROBLEMS

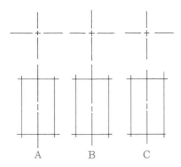

Fig. 10-74 *CAD2 Setscrews. Draw three setscrews: (A) Ø.75 × 1.25 long, square head, flat point; (B) Ø.75 × 1.25 long, slotted head, oval point; (C) Ø.75 × 1.25 long, socket head, cup point. Use schematic thread representation. Do not section.*

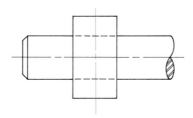

Fig. 10-76 *CAD2 Draw the Ø1.00 × 3.50 long shaft and Ø1.88 × 1.00 collar as shown. Draw the collar in full section. Add a No. 4 × 2.00 American National Standard taper pin on the colored centerline. Estimate sizes not given. Materials: shaft–steel; collar–cast iron.*

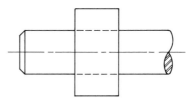

Fig. 10-78 *CAD2 Draw the Ø1.00 × 3.50 long shaft and Ø2.00 × 1.00 collar as shown. Draw the collar in half section and add an ANSI No. 608 Woodruff key at the top of the shaft. Estimate sizes not given. Materials: shaft–steel; collar–cast aluminum.*

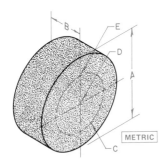

Fig. 10-75 *CAD2 Spacer. Draw two views of the spacer. Use schematic representation to show the threaded holes. A = Ø200, B = 36, C = Ø60, D = M24 (through), E = M10 (through). Add notes and all necessary dimensions.*

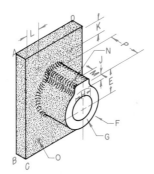

Fig. 10-77 *CAD2 Shaft support. Draw necessary views. At N, draw a Ø.31 setscrew (square head, flat point). At O (four places), draw Ø.38 coarse threads (simplified representation). All fillets and rounds = R.12. AB = 4.50, BC = .50, AD = 3.00, E = 1.50, F = R.88, G = Ø1.00, H = .75, J = .25, K = .62, L = .62, M = .50, P = 1.50.*

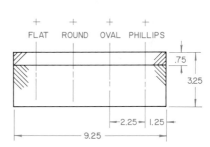

Fig. 10-79 *CAD2 Wood screws. Draw the view shown above. Add 2 1/2″ #12 wood screws on the colored centerlines. Show the four head types as specified and draw a top view of each on the center mark above the view.*

PROBLEMS

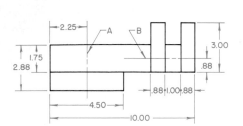

Fig. 10-80 *CAD2 Draw the view shown. On centerline A, draw Ø.50 , draw Ø.50 × 2.50 fillister-head capscrew (head at top). At B, draw a .38 × 4.00 flat-head capscrew. Show the view in section. Material: steel. Use simplified or schematic thread representation as assigned.*

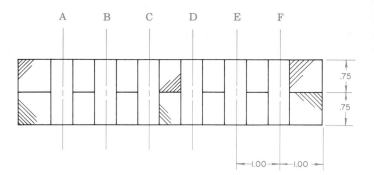

Fig. 10-81 *CAD2 Rivets. Draw the figure shown above. Overall sizes = 1.50 × 7.00. Refer to Chapter 10 and Appendix A and draw the heads (top and bottom) for Ø.50 rivets. Do not dimension. (A) Button head, (B) high button head, (C) cone head, (D) flat-top countersunk head, (E) round-top countersunk head, (F) pan head.*

═══ SCHOOL-TO-WORK ═══

Solving Real-World Problems

Fig. 10-82 *SCHOOL-TO-WORK problems are designed to challenge a student's ability to apply skills learned within this text. Be creative and have fun!*

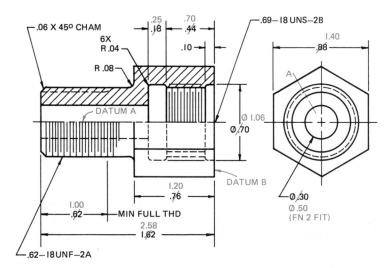

Fig. 10-82 *The thread adapter is designed to convert a .68-18UNS-2 spindle thread to a .62-18UNF-2 thread. However, a design change calls for the adapter to convert from Ø1.00-inch fine thread series with a Class 3 fit to Ø1.00-inch coarse threads with a Class 2 fit. A No. 12 socket head, cup point setscrew with fine threads is to be installed at centerline A. You decide the appropriate length. Also, datum A is to be perpendicular to datum B to within .003 in. at MMC.*

Make a working drawing of the thread adapter using the revised specifications listed above along with other changes marked on the drawing in color by the design engineer. Use simplified thread representation only. Scale: 1:1.

11

WORKING DRAWINGS

OBJECTIVES

Upon completion of Chapter 11, you will be able to:

○ Define *working drawing.*
○ Describe and produce detail drawings, assembly drawings, and assembly working drawings.
○ Develop and use tabulated (tabular) drawing for standard parts to be produced in a range of sizes.
○ Design and draw a title block for your drawings, incorporating standard items of information.
○ Develop a standard bill of materials or parts list.

VOCABULARY

In this chapter, you will learn the meanings of the following terms:

- assembly drawing
- assembly working drawing
- attributes
- bill of materials
- combination drawings
- detail drawing
- FAO
- reference assembly drawing
- tabulated drawing
- title block
- working drawing

A **working drawing** gives all the information needed to make a single part or a complete machine or structure. It tells precisely what the shape and size should be. It also tells exactly what kinds of material should be used, how the finishing should be done, and what degree of accuracy is needed. An illustration and a working drawing of a simple machine part are shown in Fig. 11-1. A working drawing can be a detail drawing, an assembly drawing, or an assembly working drawing.

WORKING DRAWINGS

Working drawings are usually multiview drawings with complete dimensions and notes added. They must provide all necessary information for the manufacture of parts and for assembly. Nothing must be left to guess.

A good working drawing follows the style and practices of the office or industry in which it is made. Most industries follow the standard recommended by the American National Standards Institute (ANSI). That way, plans can be easily read and understood from one industry to another.

No matter what special styles and practices a drafter uses in working drawings, he or she must follow certain basic rules. For example, the drafter must use proper line technique to make the contrast sharp and the drawing easy to read. Also, dimensions and notes must be clear and accurate. The drafter must use standard terms and abbreviations. When the drawing is finished, the drafter must check it carefully to avoid mistakes and to ensure accuracy. The drafter should check the drawing before submitting the drawing to the supervisor or checker for approval.

In the language of drawing, you describe an object by giving its shape and its size. All drawings follow this principle, whether they are for steam or gas engines, machines, buildings, airplanes, automobiles, missiles, or satellites.

A student's drawing may look rough and unfinished next to one made by an experienced drafter or engineer. The finished look of a drawing comes from a thorough knowledge of drafting and a great deal of practice. This chapter discusses the right order of going about creating a working drawing, as well as some of the procedures that drafters usually follow. You must become thoroughly familiar with these practices. Otherwise, your drawings will not have the style and good form that is demanded by today's industry.

Many drawings done for industry are drawn in pencil. However, pencil drawings must have good line technique. They must be neat and have few erasures. The lines must be dense and sharp.

There is also, however, a growing need for high-quality drawings in ink. This is because many drawings are stored using micro-reproduction, storage, and retrieval systems. Lines must therefore be thick and sharp, and lettering must be large and neat.

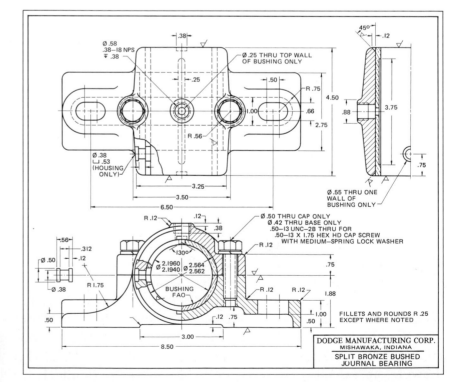

Fig. 11-1 *A working drawing of the split bronze-bushed journal bearing shown on the left.* (Courtesy of Dodge Manufacturing Corp.)

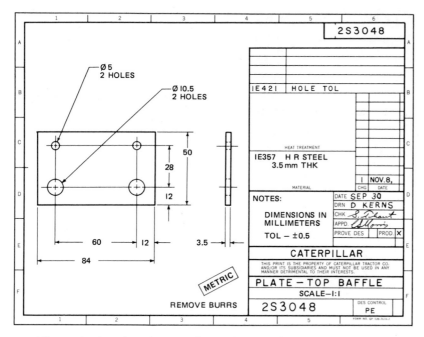

Fig. 11-2 *A working drawing of a simple part.* (Courtesy of Caterpillar, Inc.)

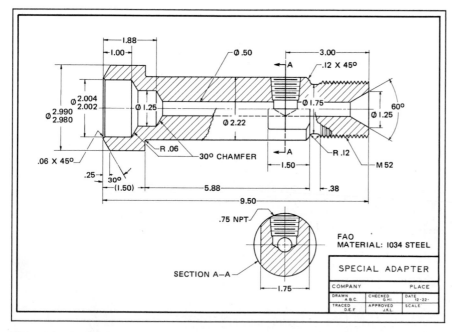

Fig. 11-3 *A single view and an extra section provide a complete description of this special adapter.*

Detail Drawings

A drawing of a single piece that gives all the information for making it is called a **detail drawing.** An example is the drawing of a simple part in Fig. 11-2. A detailed working drawing must be a full and exact description of the piece. It should show carefully selected views and include well-placed dimensions (Fig. 11-3).

When a large number of machines are to be produced, detail drawings of each part are often made on separate sheets. This is done especially when some of the parts may also be used on other machines. In some industries, however, several parts of a machine are usually detailed on the same sheet. Sometimes a separate detail drawing is made for each of several workers involved, such as the patternmaker, machinist, or welder. Such a drawing shows only the dimensions and information needed by the worker for whom it is made. Figure 11-4 shows an index-plunger operating handle. The illustration shows the piece as a forging (A) and after it has been machined (B).

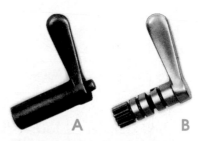

Fig. 11-4 *Index-plunger operating handle—forging (A) and finished part (B).* (Courtesy of The Hartford Special Machinery Co.)

Figure 11-5 shows a working drawing of the same piece. Notice how the drawing shows the parts to be removed after all the work is done. Notice also the detailed list of machine operations.

Figure 11-6 shows **combination drawings** (two-part detail drawings). One half gives the dimensions for the forging. The other half gives the machining dimensions and notes. A detail drawing can also contain calculated data.

Detail drawings are often made for standard parts that come in a range of sizes. Figure 11-7 is a **tabulated drawing,** or tabular drawing, of a bushing. In this kind of drawing, the different dimensions are identified by letters. A table placed

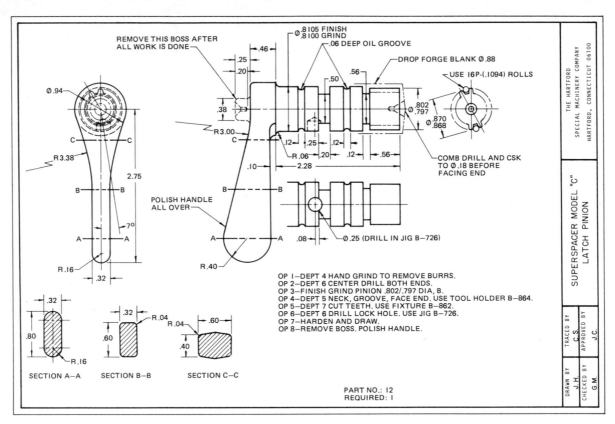

Fig. 11-5 *Working drawing of the part shown in Fig. 11-4.* (Courtesy of The Hartford Special Machinery Co.)

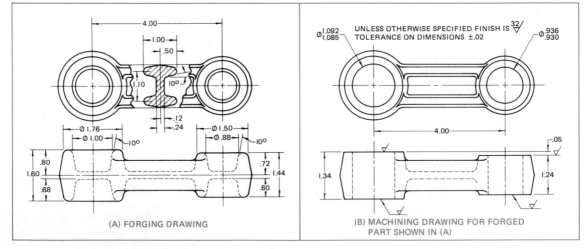

Fig. 11-6 *Two-part detail drawings showing separate information for forging and machining.*

on the drawing tells what each dimension is for different sizes of the part. Either all or some of the dimensions can be given this way. A similar kind of drawing is one in which the views are drawn with blank spaces for dimensions and notes (Fig. 11-8A). These are then filled in as needed with the required information (B). The views will not be to scale, except perhaps for one size.

Assembly Drawings

A drawing of a fully assembled construction is called an **assembly drawing.** Such drawings vary greatly in how completely they show detail and dimensioning. Their special value is that they show how the parts fit together, the look of the construction as a whole, the dimensions needed for installation, the space the construction needs, the foundation, the electrical or water connections, and so forth. When an assembly drawing gives complete information, it can be used as a working drawing. It is then called an **assembly working drawing.** This kind of drawing can be made only when there is little or no complex detail. An example is shown in Fig. 11-9. You can often show furniture and other wood construction in assembly working drawings by adding

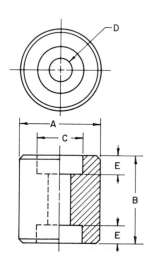

BUSHING					
PART NO.	A	B	C	D	E
CB 1	1.500	2.000	1.000	0.500	0.375
CB 2	1.625	2.125	1.125	0.625	0.437
CB 3	1.750	2.250	1.250	0.750	0.500
CB 4	1.875	2.375	1.375	0.875	0.562
CB 5	2.000	2.500	1.500	1.000	0.625

Fig. 11-7 *A tabulated (tabular) drawing.*

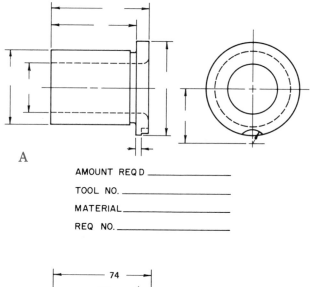

A

AMOUNT REQ D _____

TOOL NO. _____

MATERIAL_____

REQ NO. _____

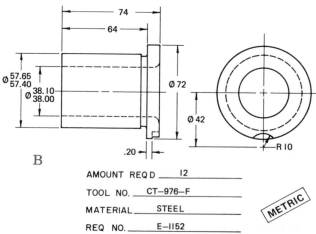

B

AMOUNT REQ D ____12____

TOOL NO. ___CT–976–F___

MATERIAL____STEEL____

REQ NO. ____E–1152____

METRIC

Fig. 11-8 *Detail drawing of a standard part with dimensions (A) blank and (B) filled in.*

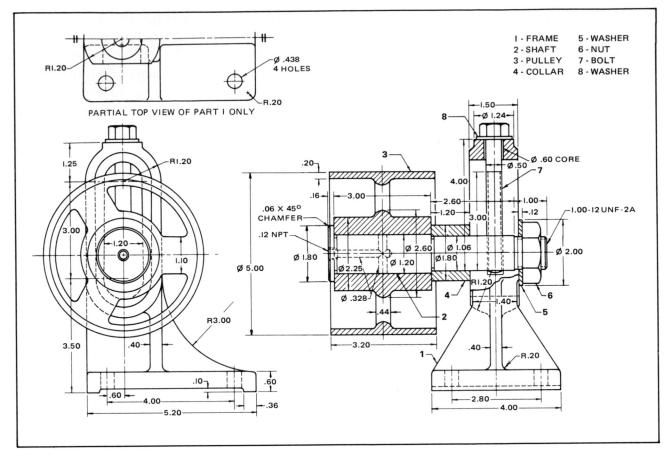

Fig. 11-9 *An assembly working drawing for a belt tightener.*

enlarged details or partial views as needed (Fig. 11-10). Note that fractional-inch dimensions are still commonly used in the furniture and woodworking industries. Note also that enlarged views are drawn in pictorial form, not as regular orthographic views. This method is peculiar to the cabinet-making trade and is not normally used in mechanical drawing.

Assembly drawings of machines are generally made to a small scale. Dimensions are chosen to tell overall distances, important center-to-center distances, and local dimensions. All, or almost all, hidden lines may be omitted. Also, if the drawing is made to a very small scale, unnecessary detail may be omitted. This has been done, for example, in the *outline assembly drawing* in Fig. 11-11. Either exterior or sectional views may be used. When the main aim is just to show the general look of the

construction, only one or two views need to be drawn. Because some assembled constructions are so large, you may need to draw different views on separate sheets. However, you must use the same scale on all sheets.

A **reference assembly drawing** is a special assembly drawing that identifies parts to be assembled (Fig. 11-12). Note the tabular list in the upper right-hand corner. Note also the dimensions shown.

Many other kinds of assembly drawings are made for special purposes. These include part assemblies for groups of parts, drawings for use in assembling or erecting a machine, drawings to give directions for upkeep and use, and so forth. A most important kind of assembly drawing is the *design layout*. It is from this kind of drawing that the detail drawings are made.

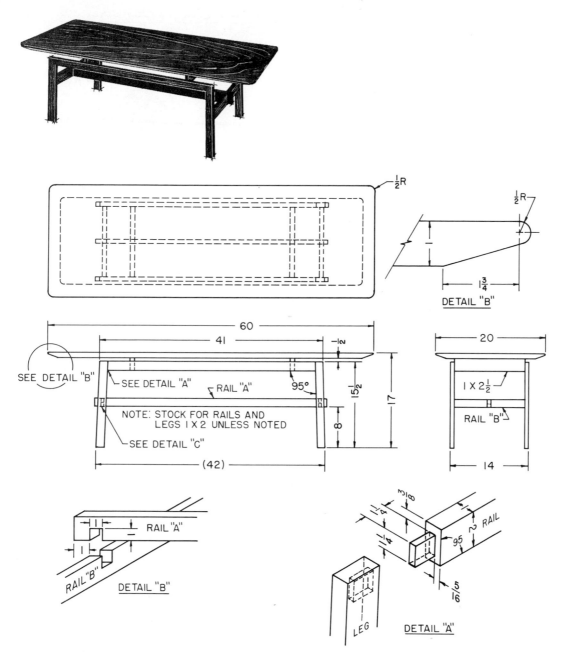

Fig. 11-10 *An assembly working drawing with enlarged details and partial views. NOTE: Fractional-inch sizes are often used on woodworking drawings.*

WORKING DRAWINGS IN CAD

CAD offers several advantages for creating a working drawing. For example, you can use a single solid model to provide all the views you need. (See Chapter 5, "Multiview Drawing.")

However, the biggest advantage of using a CAD system to create a working drawing is that you can use one drawing for several purposes. Instead of creating separate detail drawings for the welder and the machinist, for example, you can put the details each needs to see on a different layer of the drawing. To obtain a drawing tailored for the welder, turn off the layer or layers that do not pertain to the welding operation. (Turning off a layer removes it from the screen, but not from the drawing.) Then plot (print) the drawing. To create a drawing for the patternmaker, turn off the welder's layer, turn on the layer that gives the patternmaking details, and plot the drawing again.

SETTING UP A WORKING DRAWING

Your drawings will be easier to draw and use if you set them up properly. For example, the views and scales you choose greatly affect the readability of the drawing. Therefore, you must take the time to plan the drawing before you begin.

Choosing Views

To describe an object completely, you generally need two views. The views you choose should always be those that are easiest to read. Each view must add information to the description of the object. If it does not, it is not needed and should not be drawn.

In some cases, only one view is needed, as long as the shape and size are clearly understood. Also, you may choose to give information in a note rather than in a second view. For complex pieces, more than three views may be needed. As you plan the views you need, consider whether the additional views should be partial views, auxiliary views, or sectional views.

If you have any question about which views to draw, think of why you are making the drawing. A drawing is judged on how clearly and precisely it gives the information needed for its purpose.

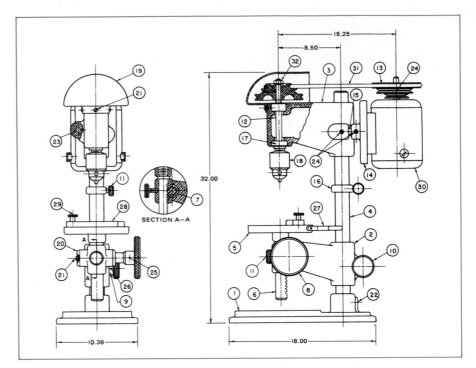

Fig. 11-11 *An outline assembly drawing.*

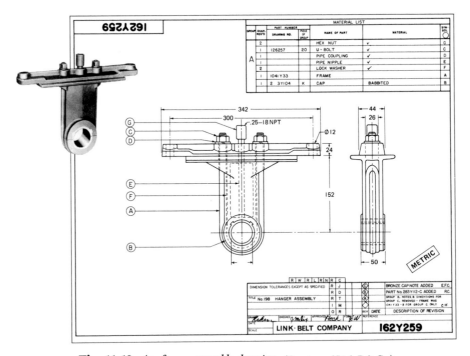

Fig. 11-12 *A reference assembly drawing.* (Courtesy of Link-Belt Co.)

Choosing a Scale

The scale for a detail drawing is chosen according to three factors:
- how large the drawing must be to show details clearly.
- how large it must be to include all dimensions without crowding.
- the size of the paper.

In most cases, the best choice is a full-size (1:1) scale. Other scales that are commonly used are half, quarter, and eighth. Avoid scales such as 1:6 (2.00″ = 1′-0), 1:3 (4.00″ = 1′-0), and 1:1.33 (9.00″ = 1′-0). If a part is very small, you can sometimes draw it to an enlarged scale, perhaps twice full size or more. See Chapter 3, "The Use and Care of Drafting Equipment," for more information about types of scales.

When you draw a number of details on one sheet, make them all to the same scale if possible. If you must use different scales, note the scale near each drawing. It is often useful to draw a detail, or part detail, to a larger scale on the main drawing. This saves both time and work in making separate detail drawings. For general assembly drawings, choose a scale that shows the details you want and one that looks good on the size of paper you are using. Sheet-metal pattern drawings are generally made full size for direct transfer to the sheet material in the shop.

For complete assemblies, a small scale is generally used. The scale is often fixed by the size of the paper the company has chosen for assemblies. For part assemblies, choose a scale to suit the purpose of the drawing. This might be to show how parts fit together, to identify the parts, to explain an operation, or to give other information.

Grouping and Placing Parts

Another part of planning a set of working drawings is to plan where each view or piece will be placed on the drawing sheet. When a number of details are used for a single machine, it is often a good idea to group them on a single sheet or set of sheets. A convenient method is to group the forging details together, the material details together, and so on. In general, show the parts in the position they will occupy in the assembled machine. That way, related parts will appear near each other. Long pieces such as shafts and bolts, however, may not work well using this method. They are often drawn with their long dimensions parallel to the long dimension of the sheet (Fig. 11-13).

TITLE BLOCKS

A **title block** is an area on a drawing that contains information about the drawing, the company, the drafter and so on. Every sketch or drawing must have some kind of title block. However, the form, completeness, and location of the title block can vary. On working drawings, it can be placed in a box in the lower right-hand corner (Fig. 11-14A), or it can be included in a record strip running across the bottom or end of the sheet (B) as far as needed.

The information included in a title block depends on company policy. Most title blocks contain some or all of the following information:
- the name of the construction, machine, or project.
- the name of the part or parts shown, or simple details.
- manufacturer, company, or firm name and address.
- the date (usually the date of completion).
- the scale or scales.
- heat treatment, working tolerances, and so forth.
- drawing number, shop order number, or customer's order number, according to the system used.
- drafting-room record: names or initials, with dates, of drafter, tracer, and checker, and approval of chief drafter, engineer, and so forth.
- a revision block or space for recording the changes made to a drawing, when needed, should be placed above or at the left of the title block.

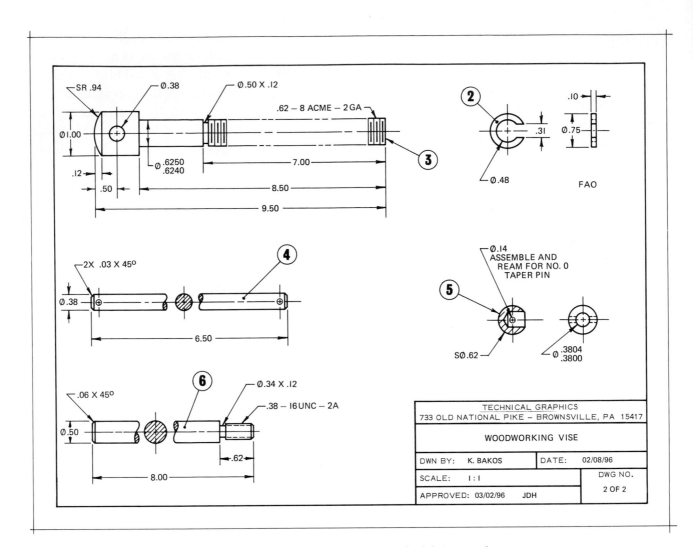

Fig. 11-13 *Title blocks. (A) A boxed title. (B) A strip title.*

A basic layout for a title block is shown in Fig. 11-15. In large drafting rooms, the title block is generally preprinted on the paper, cloth, or film, leaving spaces to be filled in. However, many firms use separate adhesive-backed title blocks. When CAD is being used, the title block may be included in the prototype drawings.

Fig. 11-14 *Basic layout for a title block. (ANSI Y 14.1)*

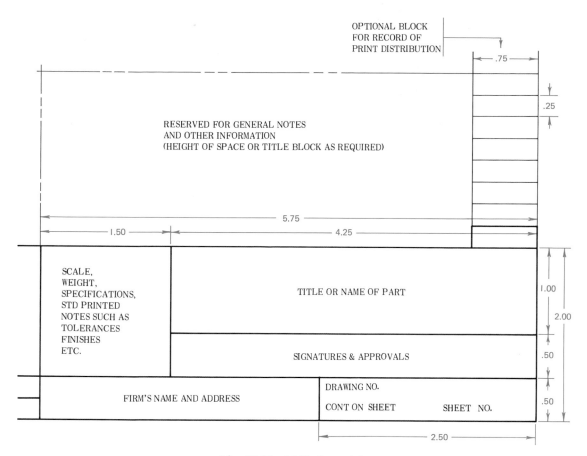

Fig. 11-15 *A bill of materials.*

BILL OF MATERIALS

Most working drawings include a list of parts, the materials of which they are made, identification numbers, and other important information. This information is especially important on assembly drawings of various kinds. It is also needed on detail drawings where a number of parts are shown on the same sheet.

The names of parts, material, number required, part numbers, and so forth are often given in notes near the views of each part. It is better, however, to place just the part numbers near the views, link them to the views with leaders, and then collect all the other information in tabulated lists. Such a list is called a **bill of materials.** Some companies also call it the *list of materials* or *parts list.* (Fig. 11-16A). It is usually placed above the title block or in the upper right corner of the sheet. It can also be written or typed on a separate sheet with a title such as "Bill of Materials for Drawing No. 00" to identify it. The recommended form is shown in Fig. 11-16B. The column widths and names may vary as needed.

CAD-GENERATED BILL OF MATERIALS

Creating a bill of materials can be a time-consuming, tedious task. However, many CAD programs allow you to generate the bill of materials for a part automatically. To do this, create blocks of all the structures you use in the drawing. (You should do this anyway if you are working on a drawing that requires the same symbol in many places, such as an electrical drawing or an architectural drawing.) Then you assign **attributes,** or information entities, to the blocks you create (Fig. 11-17).

BILL OF MATERIAL FOR IDLER PULLEY			
NAME	REQD	MATL	NOTES
IDLER PULLEY	I	C I	
IDLER PULLEY FRAME	I	C I	
IDLER PULLEY BUSHING	I	BRO	
IDLER PULLEY SHAFT	I	C R S	
Ø .62 HEX NUT	I		PURCHASED
WOODRUFF KEY # 405	I		PURCHASED
.12 OILER	I		PURCHASED

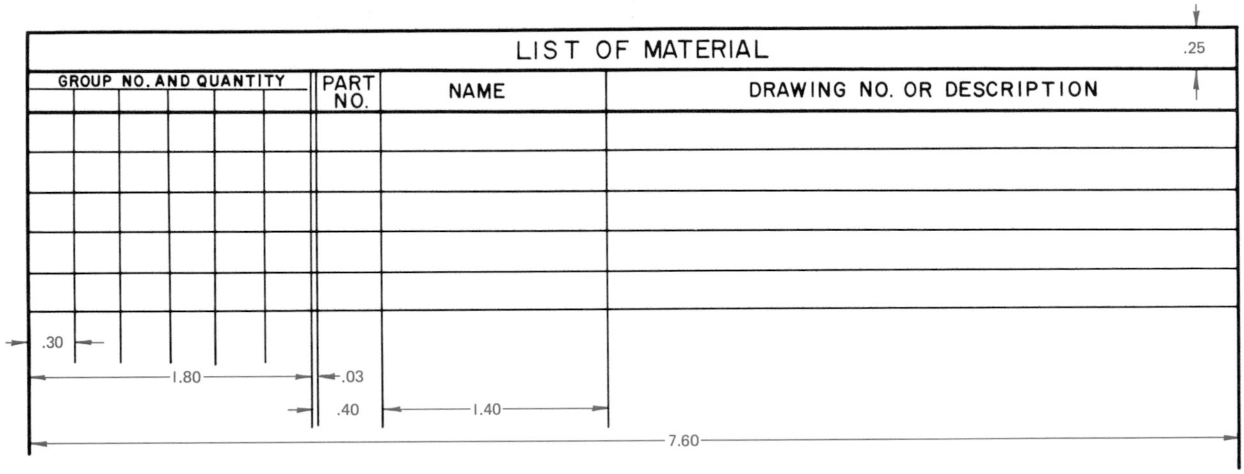

LIST OF MATERIAL

GROUP NO. AND QUANTITY					PART NO.	NAME	DRAWING NO. OR DESCRIPTION

.25 .30 1.80 .03 .40 1.40 7.60

Fig. 11-16 *Recommended form for a bill of materials.* (ANSI Y14.1)

Attributes can be anything you may need to track; examples include Model Number, Serial Number, Manufacturer, Date Purchased, Cost, and Current Value. After you have set up the attributes for a block, every time you insert that block into the drawing, the CAD program keeps track. When the drawing is finished, you can extract the attributes to create a bill of materials, a projected cost analysis, or other tabulated reports for the project.

Many standard part suppliers have ready-made pattern libraries, complete with attribute information. For example, cabinet makers need to include such information to supply prices on their line of products. In some cases, especially when only a few suppliers are used, it may be worth the effort to ask the supplier for electronic access to those libraries.

NOTES AND SPECIFICATIONS

Information that you cannot make clear in a drawing must be given in lettered notes and symbols. For example, special trade information is often given in this way. Notes are used for the following items:

- number (quantity) needed
- material
- kind of finish
- kind of fit
- method of machining
- kinds of screw threads
- kinds of bolts and nuts
- sizes of wire
- thickness of sheet metal

Other, similar information may also be included.

The materials in general use are wood, plastic, cast iron, wrought iron, steel, brass, aluminum, and various alloys. All parts to go together must be of the proper size so that they will fit. Pieces may be left rough, partly finished, or completely finished.

After a part is cast or forged, it must be machined on all surfaces that are to fit with other surfaces. Round surfaces are generally refined on a lathe. Flat surfaces are finished or smoothed on a planer, milling machine, or shaper. Holes are made with drill presses, boring mills, or lathes. Extra metal is allowed for surfaces that are to be finished. To specify such surfaces, place a √ on the lines that represent their edges. If the entire piece is to be finished, write a note such as FINISH ALL OVER, or **FAO.** No other mark is needed.

In the traditional drafting room, the kinds of machining, finish, or other treatment are specified in notes. Examples of these notes are: SPOT-FACE, GRIND, POLISH, KNURL, CORE, DRILL, REAM, COUNTERSINK, COUNTER-BORE, HARDEN, CASE HARDEN, BLUE, and TEMPER. Often other notes are added for special directions-to explain the assembly, for example, or the order of doing work.

In current dimensioning practice, the drafter usually does not specify the method or tool to be used. Only the finished size and shape are given. For additional information, review Chapter 6, "Dimensioning."

CHECKING A DRAWING

After a drawing is finished, it must be looked over very carefully before it is used. This is called *checking the drawing.* It is very important work. A drawing you have made should be checked by someone other than you. A person who has not worked on the drawing will be better able to spot errors.

To make a thorough check, drafters and checkers follow a set order of procedures. These checks should be made:

1. See that the views completely describe the shape of each piece.
2. See that there are no unnecessary views.

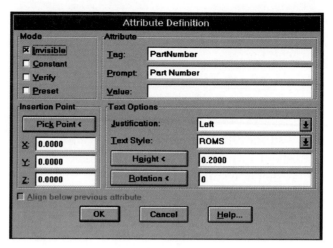

Fig. 11-17 *In AutoCAD, the Attribute Definition dialog box allows you to create block attributes. In this case, a part number attribute is being created. More than one attribute can be associated with each block.*

3. See that the scale used was large enough to show all detail clearly.
4. See that all views are to scale and that the right dimensions are given.
5. See that there will be no parts that will interfere with each other during assembly or operation and that necessary clearance space is provided around all parts that need it.
6. See that enough dimensions are given to define the sizes of all parts completely, and that no unnecessary or duplicate dimensions are given.
7. See that all necessary location or positioning dimensions are given with necessary precision.
8. See that necessary tolerances, limits and fits, and other precision information are given.
9. See that the kind of material and the quantity needed of each part are specified.
10. See that the kind of finish is specified, that all finished surfaces are marked, and that a finished surface is not called for where one is not needed.
11. See that standard parts and stock items, such as fasteners, handles, and catches, are used where suitable.
12. See that all necessary explanatory notes are given and that they are properly placed on the drawing.

Each drafter must inspect his or her own work for errors or omissions before having it further checked.

CHAPTER 11
REVIEW

Learning Activities

1. Visit industrial shops and study various methods of forming and machining materials such as metal, wood, and plastic.

2. Obtain different types of working drawings from industry. Study each to determine which are detail drawings, assembly drawings, or assembly working drawings.

3. Choose an object that has at least ten pieces that fit together. Imagine how an assembly drawing of the object might look. Then sketch the assembly drawing.

Questions

Write your answers on a separate sheet of paper.

1. A drawing that shows all the information necessary for manufacturing a part or complete product is a _____ drawing.

2. If an entire part is to be finished (machined), what general note should you place on the drawing?

3. Is Fig. 11-1 a detail drawing, an assembly drawing, or an assembly working drawing?

4. A drawing of a single part with all information needed for making it is called a _____ drawing.

5. What is another name for a parts list?

6. What is the outside diameter of Part No. CB 4 in Fig. 11-7?

7. What three factors help determine the scale to be used on a drawing?

8. What is another way of specifying the scale used for the drawing in Fig. 11-13?

9. What symbol is used to call for a finished surface?

10. What type of drawing might you create to represent a standard part that will be produced in a range of sizes?

11. Is Fig. 11-5 a detail drawing, an assembly drawing, or an assembly working drawing?

12. What factors should you take into consideration when setting up a working drawing? Explain.

13. To create a list of materials automatically using a CAD program, you define block _____ to hold information such as the part number and model number.

14. Explain how one CAD drawing can provide separate detail drawings of a part for several different purposes, such as welding, machining, and patternmaking.

CHAPTER 11

PROBLEMS

The small CAD symbol next to each problem indicates that a problem is appropriate for assignment as a computer-aided drafting problem and suggests a level of difficulty (one, two, or three) for the problem.

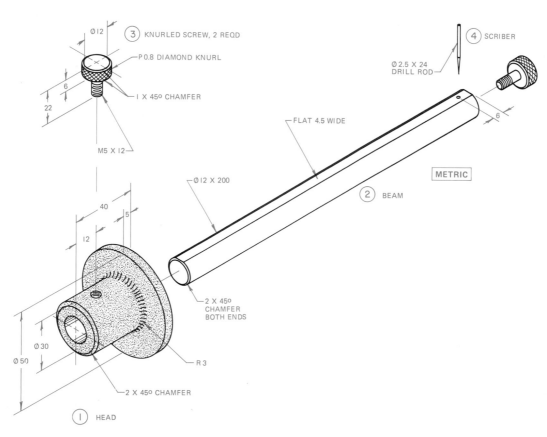

Fig. 11-18 *CAD3 Marking gage.*

Assignment 1: Make a working drawing of each part shown. Determine an appropriate scale, and dimension the drawing. Head: cast iron. Face of head is to be machined. Scriber is to be heat-treated after machining. Draw the knurled screw only once and specify "2 REQD."

Assignment 2: Make an assembly drawing of the marking gage. Dimension if required. Estimate all sizes and details not given.

PROBLEMS

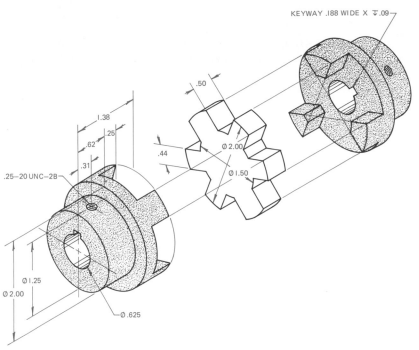

KEYWAY .188 WIDE X ⊽.09

.50

1.38

.25

.62

.25—20 UNC—2B

.31

.44

Ø 2.00

Ø 1.50

Ø 1.25

Ø 2.00

Ø .625

Fig. 11-19 *CAD3 Coupler.*

Assignment 1: Make a working drawing of each part shown. Determine an appropriate scale, and dimension the drawing. Ends: die-cast aluminum. Spacer: rubber.

Assignment 2: Make an assembly drawing of the coupler. Dimension if required. Estimate all sizes and details not given.

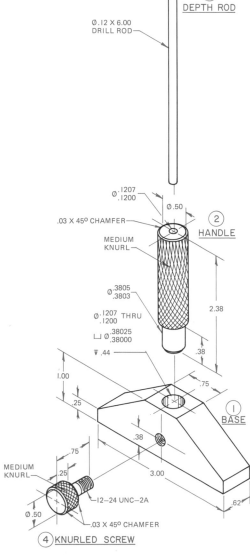

③ DEPTH ROD

Ø.12 X 6.00 DRILL ROD

Ø .1207/.1200

Ø.50

.03 X 45° CHAMFER

② HANDLE

MEDIUM KNURL

Ø .3805/.3803

Ø .1207/.1200 THRU

⌴ Ø .38025/.38000

⊽ .44

2.38

.38

1.00

.25

.75

① BASE

.38

MEDIUM KNURL

.75

.25

3.00

12—24 UNC—2A

Ø .50

.03 X 45° CHAMFER

.62

④ KNURLED SCREW

Fig. 11-20 *CAD3 Depth gage.*

Assignment 1: Make a working drawing of each part shown. Determine an appropriate scale, and dimension the drawing. All parts: cold-rolled steel.

Assignment 2: Make an assembly drawing of the depth gage. Dimension if required. Estimate all sizes and details not given.

PROBLEMS

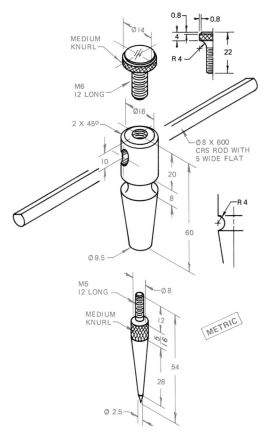

Fig. 11-21 *CAD3 Trammel.*

Assignment 1: Make a working drawing of each part shown. Determine an appropriate scale, and dimension the drawing. Specify "2 REQD" for the point, body, and knurled screw. The point is to be heat-treated after machining.

Assignment 2: Make an assembly drawing of the trammel. Estimate all sizes and details not given.

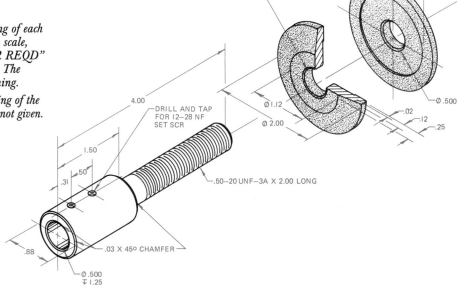

Fig. 11-22 *CAD3 Arbor.*

Assignment 1: Make a working drawing of each part shown. Determine an appropriate scale, and dimension the drawing. Flanges: die-cast aluminum. Shaft: cold-rolled steel.

Assignment 2: Make an assembly drawing of the arbor with a Ø6.00 × 1.00 grinding wheel between the flanges. Show sectional views where practical. Draw all fasteners. Estimate all sizes and details not given.

PROBLEMS

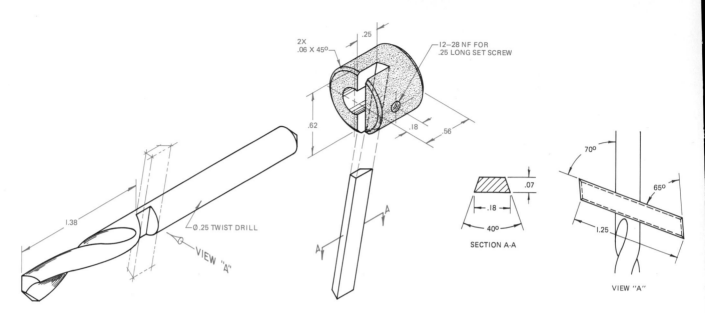

Fig. 11-23 *CAD3 Power expansion bit.*

    Assignment 1: Make a working drawing of each part shown. Determine an appropriate scale, and dimension the drawing. Cutter: tool steel. Body: cast iron. Use sectional views where necessary.

    Assignment 2: Make an assembly drawing of the power expansion bit. Dimension if required. Estimate all sizes and details not given.

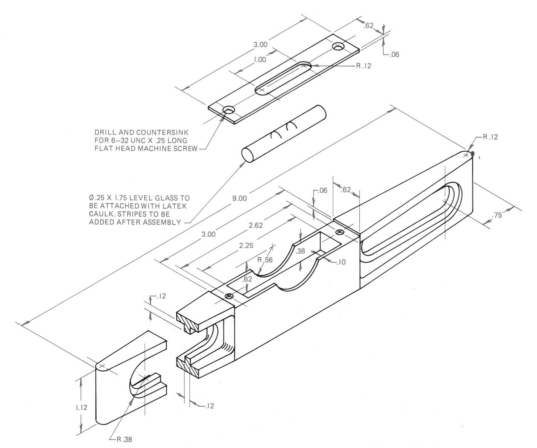

Fig. 11-24 *CAD3*
Level.

    Assignment 1: Make a working drawing of each part shown except the level glass. Dimension the drawing. Body: diecast aluminum. Top plate: cold-rolled steel. Use sectional views where necessary. Fillets = $\frac{1}{8}$ R.

    Assignment 2: Make an assembly drawing of the level. Dimension if required. Redesign the level to include vertical and 45° angle level glasses if desired. Estimate all sizes and details not given.

PROBLEMS

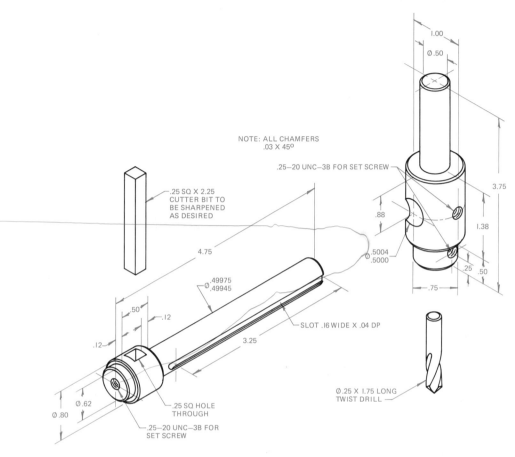

NOTE: ALL CHAMFERS
.03 X 45°

.25—20 UNC—3B FOR SET SCREW

1.00
Ø.50

.25 SQ X 2.25
CUTTER BIT TO
BE SHARPENED
AS DESIRED

3.75

.88

1.38

4.75

Ø .5004
.5000

.25
.50

.75

Ø .49975
.49945

.50
.12

.12

3.25

SLOT .16 WIDE X .04 DP

Ø .62

Ø .80

.25 SQ HOLE
THROUGH

.25—20 UNC—3B FOR
SET SCREW

Ø .25 X 1.75 LONG
TWIST DRILL

Fig. 11-25 *CAD3 Circle cutter.*
 Assignment 1: Make a working drawing of each part shown. Determine an appropriate scale, and dimension the drawing. Cutter: tool steel. Body and tool holder: cold-rolled steel.

 Assignment 2: Make an assembly drawing of the circle cutter. Show sectional views where necessary. Draw all fasteners. Estimate all sizes and details not given.

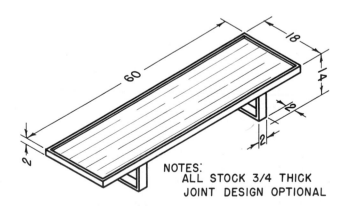

18

14

60

2

2

2

NOTES:
ALL STOCK 3/4 THICK
JOINT DESIGN OPTIONAL

Fig. 11-26 *CAD3 Coffee table. Make an assembly working drawing with all necessary details and partial views. Determine an appropriate scale. Dimension. Choose an appropriate wood and types of joints (see Fig. 11-10). Make a complete bill of materials. Estimate all sizes not given.*

PROBLEMS

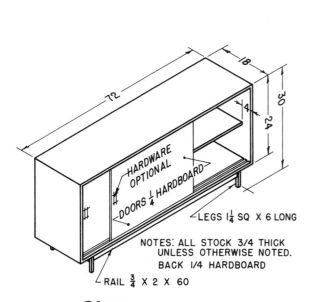

Fig. 11-27 CAD3 *Storage cabinet. Make an assembly working drawing with all necessary details and partial views. Determine an appropriate scale. Dimension. Choose an appropriate wood and types of joints (see Fig. 11-10). Make a complete bill of materials. Estimate all sizes not given.*

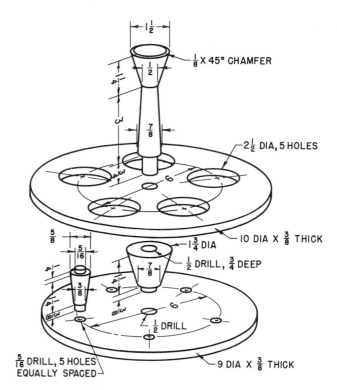

Fig. 11-28 CAD3 *Beverage server.*

Assignment 1: Make detail working drawings of each part shown. Determine an appropriate scale, and dimension the drawing. Material: hardwood.

Assignment 2: Make an assembly drawing of the beverage server. Dimension if required.

Assignment 3: Make a complete bill of materials for the beverage server.

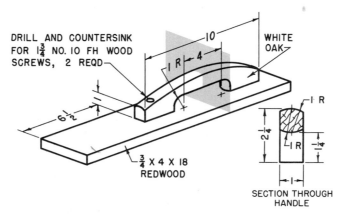

Fig. 11-29 CAD3 *Cement float. Make an assembly working drawing of the cement float. Determine an appropriate scale, and dimension the drawing. Materials: as noted. Use a 1.00″ grid for the curved outline of the handle. Add a bill of materials on the same drawing sheet.*

PROBLEMS

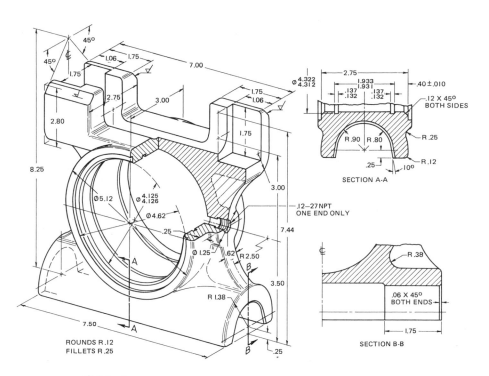

Fig. 11-30 *CAD3 Housing. Make a working drawing of the housing. Determine an appropriate scale. Use partial and sectional views where needed. Material: cast iron. Estimate all sizes not given.*

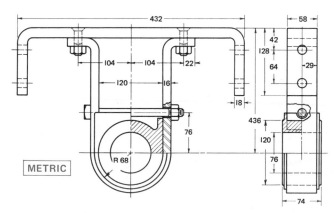

Fig. 11-31 *CAD3 Hung bearing. Make detail working drawings of each part of the hung bearing. Fully dimension each part. All bolts are 16 mm in diameter. Do not draw bolts and nuts. Estimate sizes not given.*

PROBLEMS

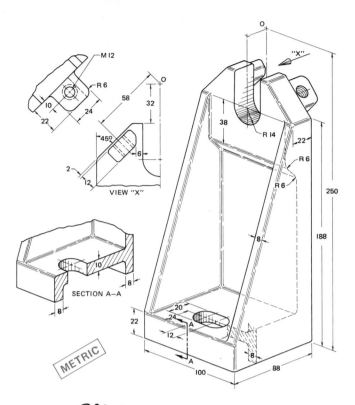

Fig. 11-32 CAD3 *End base. Make a working drawing of the end base. Determine an appropriate scale, and dimension the drawing. Use partial and sectional views where needed. Material: cast iron. Estimate all sizes not given.*

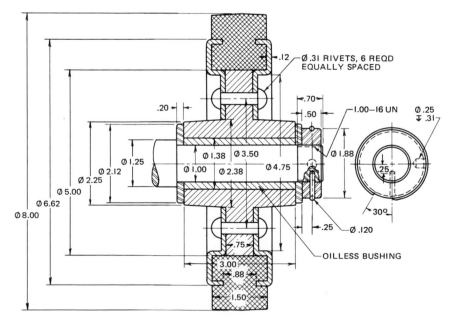

Fig. 11-33 CAD3 *Cushion wheel.*

Assignment 1: Make a front view and section of the cushion wheel, full size. This type of wheel is used on warehouse or platform trucks to reduce noise and vibration.

Assignment 2: Make a complete set of detail drawings, full size, with a bill of material, for the cushion wheel. Three sheets will be needed. Rivets are purchased and, therefore, need not be detailed but should be listed in the bill of materials.

PROBLEMS

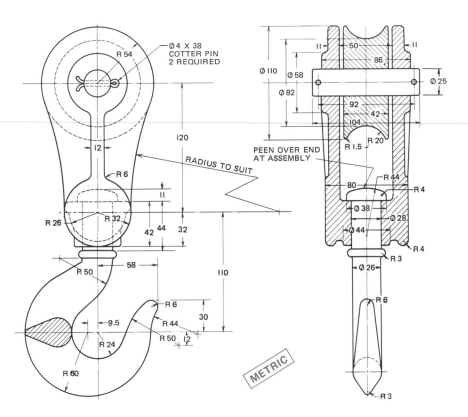

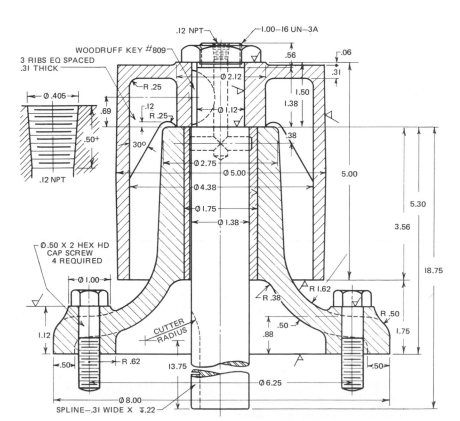

Fig. 11-34 *CAD3* *Crane hook. Make detail drawings of the crane hook parts.*

Fig. 11-35 *Flat-belly pulley. Make detail drawings of the base, pulley, bushing, and shaft, with a bill of materials for the complete pulley-and-stand unit. Scale: Full size. Use three sheets. Top view may be drawn as a half-plan view.*

PROBLEMS

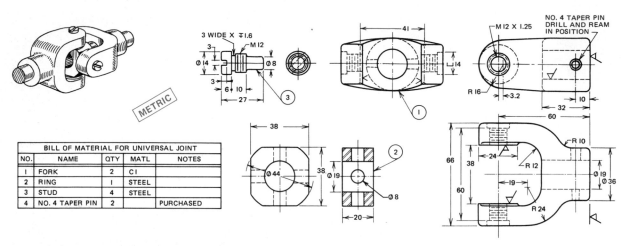

BILL OF MATERIAL FOR UNIVERSAL JOINT				
NO.	NAME	QTY	MATL	NOTES
1	FORK	2	CI	
2	RING	1	STEEL	
3	STUD	4	STEEL	
4	NO. 4 TAPER PIN	2		PURCHASED

Fig. 11-36 *Universal joint. Make a two-view assembly drawing of the universal joint in section.*

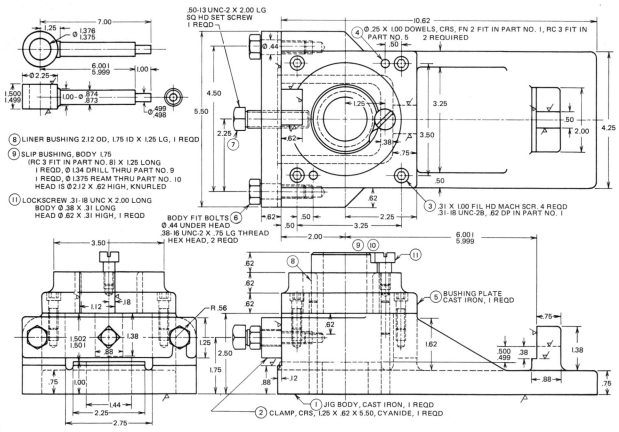

Fig. 11-37 *CAD3* *Production jig. A jig is a device used to hold a machine part (called the* work *or* production*) while it is being machined, or produced, so that all the parts will be alike within specified limits of accuracy. Notice the work, or production, in the upper left corner of the drawing. Notice also that the drawing shown does not follow present ANSI standards. Be sure to make necessary changes to update each drawing in the following assignments.*

Assignment 1: Make a detail working drawing of the jig body.

Assignment 2: Make a complete set of detail drawings for the jig, with a bill of materials. Use as many sheets as needed.

Assignment 3: Make a complete assembly drawing of the jig, three views. Give only such dimensions as are needed for putting the parts together and using the jig.

PROBLEMS

Fig. 11-38 CAD 3 *Refer to Fig. 11-1, journal bearing. Prepare a detail drawing of the journal base. Scale: Full size.*

Fig. 11-39 CAD 3 *Refer to Fig. 11-9, belt tightener. Prepare detail drawings of the frame, shaft, and pulley. Scale: Full size.*

Fig. 11-40 CAD 3 *Refer to Fig. 11-13, machine parts. Make detail drawings of the parts shown. Scale: Full size. Convert all dimensions to metric sizes (mm). Use simplified thread representation. Be sure to use current ANSI dimensioning standards.*

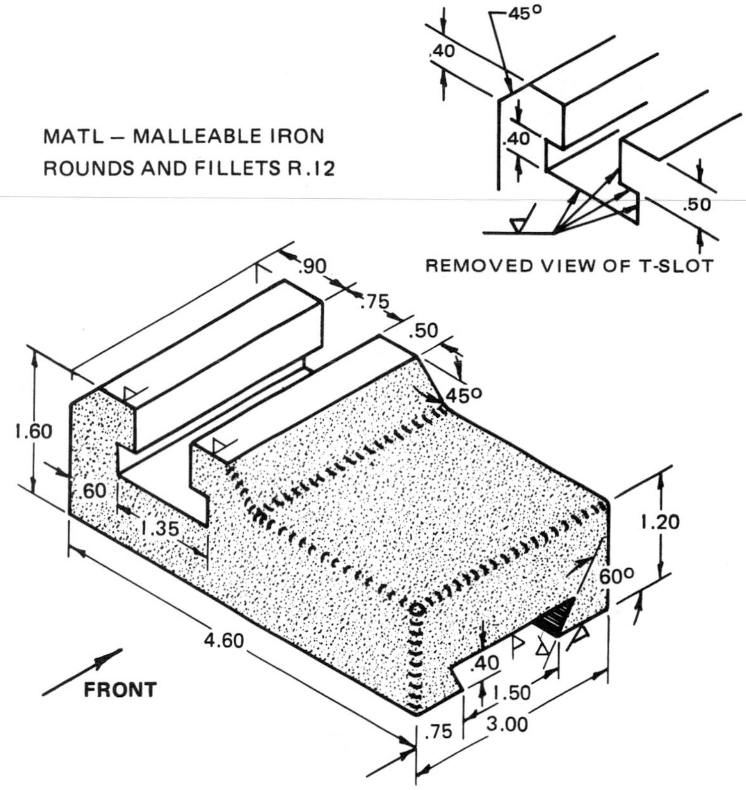

MATL – MALLEABLE IRON
ROUNDS AND FILLETS R.12
45°
.40
.40
.50
REMOVED VIEW OF T-SLOT
.90
.75
.50
45°
1.60
.60
1.35
1.20
60°
4.60
FRONT
.40
1.50
.75
3.00

Fig. 11-41 *Cross slide. Make a working drawing of the cross slide. Dimension. Scale: 1:1.*

Ø13
Ø24 SFACE
2 HOLES
25
38
METRIC
R19
15
15
50
18
25
10
32
10
2 RIBS
125
FRONT
ROUNDS AND FILLETS R2
MATL – MALLEABLE IRON
38
64
13

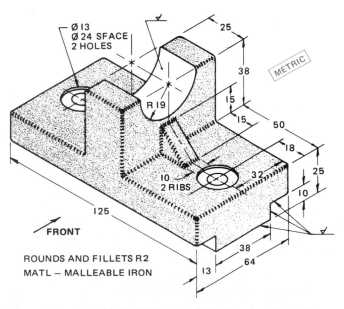

Fig. 11-42 *Guide bracket. Make a working drawing of the guide bracket. Dimension. Scale: 1:1.*

PROBLEMS

SCHOOL-TO-WORK

Solving Real-World Problems

Fig. 11-43 *SCHOOL-TO-WORK problems are designed to challenge a student's ability to apply skills learned within this text. Be creative and have fun!*

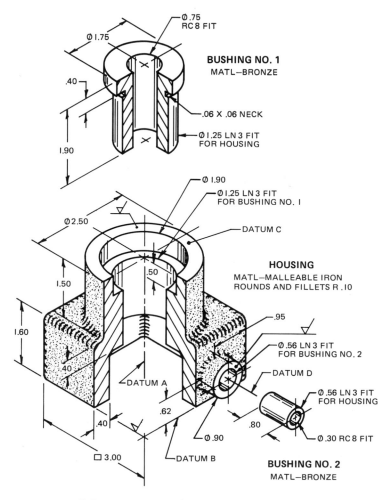

Fig. 11-43 **CAD 3** *Housing. On a B-size sheet, make a set of working drawings of the three parts of the housing. Scale: 1:1. Use sectional views as appropriate for clarity. Use the appendix to calculate limit dimensions. Include a bill of materials. Add geometric dimensioning and tolerancing symbols based on the following:*

1. *Datum A to be perpendicular to datum B and C within .002 at MMC.*
2. *Datum D to be perpendicular to datum A to within .003 at MMC and parallel to datums B and C within .002 at MMC.*
3. *Datum C to be parallel to datum B to within .001.*

PICTORIAL DRAWING–
TECHNICAL
ILLUSTRATION

OBJECTIVES

Upon completion of Chapter 12, you will be able to:

○ List various uses of pictorial drawings.
○ Describe/define the various types of pictorials.
○ Select and draw the most practical type of pictorial for a specific purpose.
○ Draw isometric circles (ellipses) using various methods.
○ Construct irregular curves in pictorial views.
○ Select and draw appropriate pictorial sections.
○ Make cavalier, normal, and cabinet oblique drawings.
○ Develop one-point and two-point perspective drawings.
○ Describe/define *technical illustrations*.
○ Render pictorial drawings using various techniques such as outline shading and surface shading.

VOCABULARY

In this chapter, you will learn the meanings of the following terms:

- airbrush
- axonometric projection
- cabinet oblique
- cavalier oblique
- dimetric projection
- frisket paper
- isometric axes

- isometric drawing
- isometric plane
- normal oblique
- perspective drawing
- photo retouching
- picture plane
- rendering

- scratchboard drawing
- technical illustration
- trimetric projection
- vanishing point
- wash rendering

Pictorial drawing is an essential part of the graphic language. It is very important in engineering, architecture, science, electronics, technical illustration, and many other professions. It appears in technical literature as well as in catalogs and assembly, service, and operating manuals. Architects use pictorial drawing to show what a finished building will look like. Advertising agencies use pictorial drawings to display new products.

Pictorial drawing is often used in exploded views on production and assembly drawings (Fig. 12-1). These views are made to illustrate parts lists (Fig. 12-2), to explain the operation of machines and equipment, and for many other commercial and technical purposes. In addition, most people use some form of pictorial sketches to help convey ideas that are hard to describe in words.

Nonisometric Lines

To draw nonisometric lines, first locate their two ends and then connect the points. Angles on isometric drawings also do not show in their true size. Therefore, you cannot measure them in degrees.

Figure 12-6 shows how to locate and draw nonisometric lines in an isometric drawing using the *box method*. The nonisometric lines are the slanted sides of the packing block shown in the multiview drawing in Fig. 12-6A. *NOTE: The colored lines are for instructional purposes only. You would not normally put them on your finished drawing.* To make an isometric drawing of the block, use the following procedure, referring as necessary to Fig. 12-6.

1. Block in the overall sizes of the packing block to make the isometric box figure as shown in (B).

2. Use dividers or a scale to transfer distances *AG* and *HB* from the multiview drawing to the isometric figure. Lay these distances off along line *AB* to locate points *G* and *H*. Then draw the lines connecting point *D* with point *G* and point *C* with point *H*. This is shown in (C).

3. Complete the layout by drawing *GJ* and *HI* and by connecting points *E* and *J* to form a third nonisometric line (D). Erase the construction lines to complete the drawing (E).

Angles

To draw the 40° angle shown in Fig. 12-7A, use the following procedure.

1. Make *AO* and *AB* any convenient length. Draw *AB* perpendicular to *AO* at any convenient place (A).

2. Transfer *AO* and *AB* to the isometric cube (B). Lay off *AO* along the base of the cube. Draw *AB* parallel to the vertical axis.

3. Connect points *O* and *B* to complete the isometric angle. If you check this angle with a protractor, you will find that it measures 40°.

Follow the same steps to construct the angle on the top of the isometric cube. This method can be used to lay out any angle on any isometric plane.

Figure 12-8 is a multiview drawing of an object with four oblique surfaces. An isometric view of this object can be made using either the box or the skeleton method, as shown in Fig. 12-8B and C.

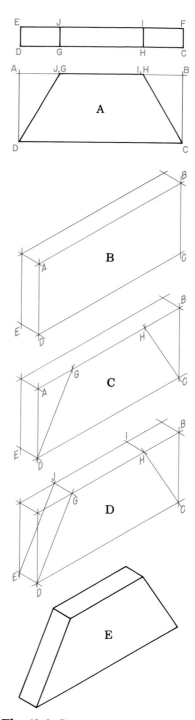

Fig. 12-6 *Drawing nonisometric lines.*

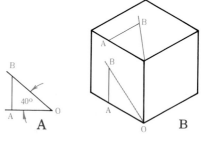

Fig. 12-7 *Constructing angles in isometric drawings.*

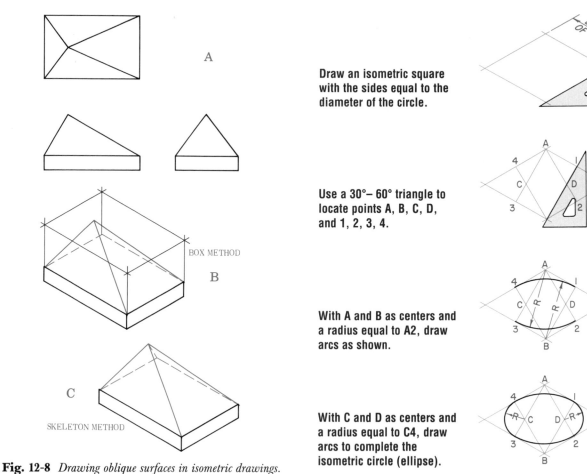

A

BOX METHOD

B

C

SKELETON METHOD

Fig. 12-8 *Drawing oblique surfaces in isometric drawings.*

Draw an isometric square with the sides equal to the diameter of the circle.

A

Use a 30°– 60° triangle to locate points A, B, C, D, and 1, 2, 3, 4.

B

With A and B as centers and a radius equal to A2, draw arcs as shown.

C

With C and D as centers and a radius equal to C4, draw arcs to complete the isometric circle (ellipse).

D

Fig. 12-9 *Steps in drawing an isometric circle.*

Isometric Circles

In isometric drawing, circles appear as ellipses. Since it takes a long time to plot a true ellipse, a four-centered approximation is generally drawn. Figure 12-9 describes how to create a four-centered approximation of an ellipse. Figure 12-10 shows isometric circles drawn on three surfaces of a cube.

Figure 12-11 shows how to make an isometric drawing of the cylinder shown as a multiview drawing in (A). First, draw an ellipse of the 3.00-in. circle following the procedure shown in Fig. 12-9. Next, drop centers *A, C,* and *D* a distance equal to the height of the cylinder (in this case, 4.00 in.) as shown in Fig. 12-11B. Draw lines *A′C′* and *A′D′*. Complete the drawing as shown in (C). Construct a line through *C′D′* to locate the points of tangency. Draw the arcs using the same radii as in the ellipse at the top. Finally, draw the vertical lines to complete the cylinder. Notice that the radii for the arcs at the bottom match those at the top.

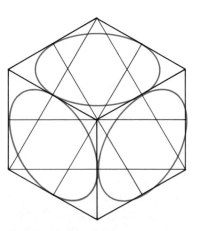

Fig. 12-10 *Isometric circles on a cube.*

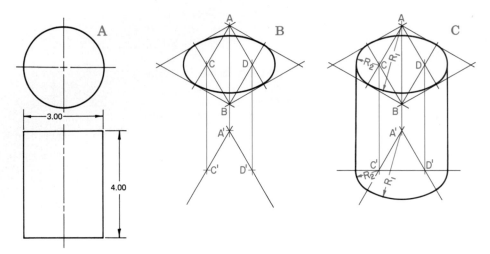

Fig. 12-11 *Steps in drawing an isometric cylinder.*

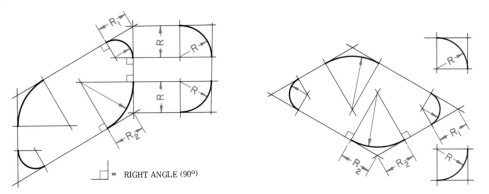

□ = RIGHT ANGLE (90°)

Fig. 12-12 *Drawing quarter rounds in isometric drawings.*

To draw quarter rounds, use the same method you use to draw quarters of a circle (Fig. 12-12). In each case, measure the radius along the tangent lines from the corner. Then draw the actual perpendiculars to locate the centers for the isometric arcs. Observe that the method for finding R_1 and R_2 is the same as that for finding the radii of an isometric circle. When an arc is more or less than a quarter circle, you can sometimes plot it by drawing all or part of a complete isometric circle and using as much of the circle as needed.

Figure 12-13 shows how to draw outside and inside corner arcs. Note the tangent points T and the centers 1 and 1' and 2 and 2'.

Irregular Curves

Irregular curves in isometric drawings cannot be drawn using the four-center method. To draw irregular curves, you must first plot points and then connect them using a French curve, as shown in Fig. 12-14.

Isometric Templates

Isometric templates are made in a variety of forms. They are convenient and can save time when you have to make many isometric drawings. Many of them have openings for drawing ellipses, as well as 60° and 90° guiding edges. Simple homemade guides (Fig. 12-15) are convenient for straightline work in isometric. Ellipse templates are very convenient for drawing true ellipses (Fig. 12-16). If you use these templates, your drawings will look better and you will not have to spend

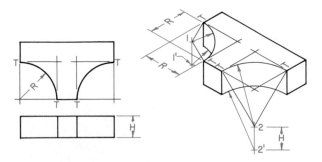

Fig. 12-13 *Constructing outside and inside arcs.*

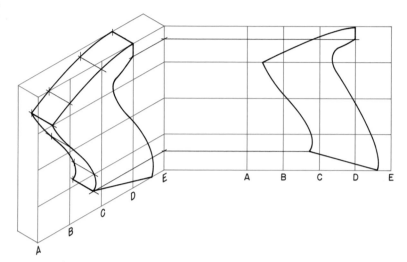

Fig. 12-14 *Constructing irregular curves in isometric drawings.*

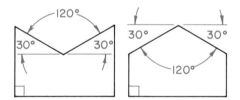

Fig. 12-15 *Simple isometric templates.*

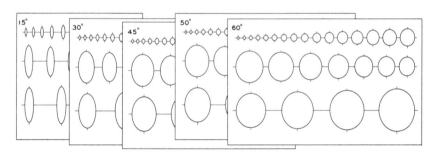

Fig. 12-16 *Ellipse templates.*

time plotting approximate ellipses. See Chapter 3 for information on how to use templates.

Creating an Isometric Drawing

Figure 12-17A shows a multiview drawing of a filler block. To make an isometric drawing of the block, begin by drawing the isometric axes in the first position as shown in (B). They represent three edges of the block. Draw them to form three equal angles. Draw axis line *OA* vertically. Draw axes *OB* and *OC* with the 30°-60° triangle. The point at which the lines meet represents the upper front corner *O* of the block, as shown in (C).

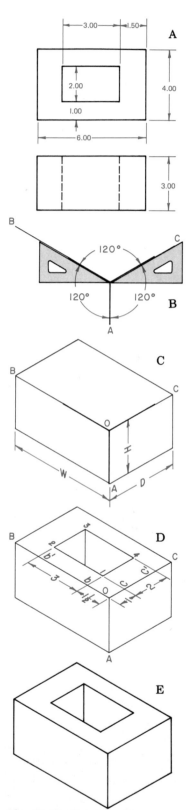

Fig. 12-17 *Steps in making an isometric drawing.*

Measure off the width *W,* the depth *D,* and the height *H* of the block on the three axes. Then draw lines parallel to the axes to make the isometric drawing of each block. To locate the rectangular hole shown in Fig. 12-17D, lay off 1.00 in. along *OC* to *c.* Then from *c* lay off 2.00 in. to *c'.* Through *c* and *c'* draw lines parallel to *OB.* In like manner, locate *b,* and *b'* on axis *OB* and draw lines parallel to *OC.* Draw a vertical line from corner 3. *NOTE: The dimensions, letters, and numerals are for instructional purposes; you would not normally put them on your drawing.* Darken all necessary lines to complete the drawing (Fig. 12-17E).

Pictorial drawings, in general, are made to show how something looks. Since hidden edges (lines) are not "part of the picture," they are normally left out. However, you might need to include them in special cases to show a certain feature.

The steps required to create an isometric drawing in the second position are similar to those for the first position. Fig. 12-18 shows the steps to draw a simple object in the second position.

Figure 12-19 shows how to make an isometric drawing of a guide. The guide is shown in a multiview drawing in (A). This drawing is more complex because it includes a circular hole and several rounded surfaces. Study the size, shape, and relationship of the views in (A) before you proceed. As you perform the following steps, refer to Fig. 12-19 as necessary.

1. Draw the axes *AB, AC,* and *AD* in the second position (B).

2. Lay off the length, width, and thickness measurements given in (A). That is, starting at point *A,* measure the length (3.00 in.) on *AB;* the width (2.00 in.) on *AC;* and the thickness (.62 in.) on *AD.*

3. Through the points found, draw isometric lines parallel to the axes. This "blocking in" produces an isometric view of the base.

4. Block in the upright part in the same way, using the measurements (2.00 in. and .75 in.) given in the top view in (A).

5. Find the center of the hole and draw centerlines as shown in (B).

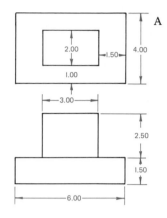

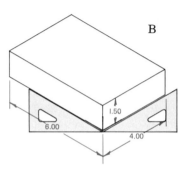

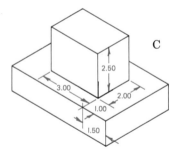

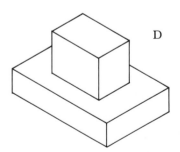

Fig. 12-18 *Isometric drawing with the axes in the second position. See also Fig. 12-4.*

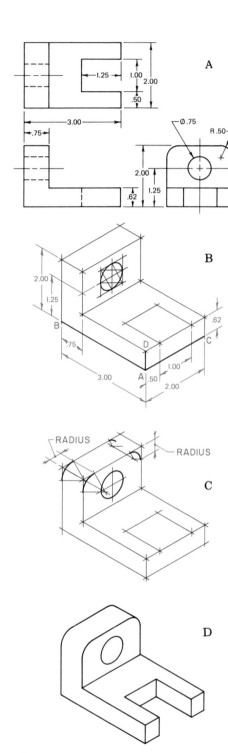

Fig. 12-19 *Steps in making an isometric drawing.*

6. Block in a .75-in. isometric square and draw the hole as an approximate ellipse.

7. To make the two quarter rounds, measure the .50-in. radius along the tangent lines from both upper corners (C). Draw real perpendiculars to find the centers of the quarter circles. See Fig. 12-14 for more information on drawing isometric quarter rounds.

8. Darken all necessary lines. Erase all construction lines to complete the isometric drawing, as shown in (D).

Reversed Axes

Sometimes you will want to draw an object as if it were being viewed from below. To do so in isometric drawing, reverse the position of the axes. (See Fig. 12-4). To draw an object using reversed axes, follow the example in Fig. 12-20. Consider how an object appears in a regular multiview drawing, as shown in (A). Then begin the isometric view by drawing the axes in reversed position (B). Complete the view with dimensions taken from the multiview drawing (C). Darken the lines to finish the drawing.

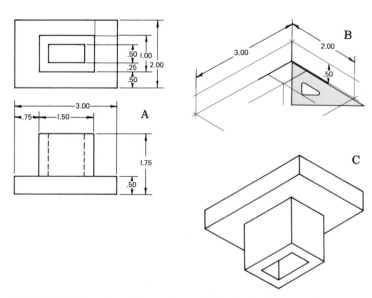

Fig. 12-20 *Steps in making an isometric drawing with reversed axes.*

Long Axis Horizontal

When long pieces are drawn in isometric, make the long axis horizontal. A drawing of this kind is illustrated in Fig. 12-21. In (A), a long object is shown in a multiview drawing. Fig. 12-21B shows the start of an isometric drawing, with the axes shown by thick black lines. In (C), the view is completed with dimensions taken from the drawing in (A). Remember, in isometric drawing, draw circles first as isometric squares; then complete them by the four-center method or by using an ellipse template.

Dimensioning Isometric Drawings

There are two general ways to place dimensions on isometric drawings. The older method is to place them in the isometric planes, or extensions of them, and to adjust the letters, numerals, and arrowheads to isometric shapes, as shown in Fig. 12-22A. The newer unidirectional system, shown in (B), is simpler. In this system, numerals and lettering are read from the bottom of the sheet. However, since isometric drawings are not usually used as working drawings, they are seldom dimensioned at all. Refer to Chapter

6, "Dimensioning," for more information about the aligned and unidirectional methods of dimensioning.

Isometric Sections

Isometric drawings are generally "outside" views. Sometimes, however, a sectional view is needed. To create a sectional view, take a section on an **isometric plane** (a plane parallel to one of the faces of the isometric cube). Figure 12-23 shows isometric full sections taken on a different plane for each of three objects. Note the construction lines showing the parts that have been cut away. Figure 12-24 illustrates isometric half sections. The construction lines in (A) and (C) are for the complete outside views of the original objects. Notice the outlines of the cut surfaces in both views. There are two ways to make the sectional views shown in (B) and (D). In the *cut method,* draw the complete outside view as well as the isometric cutting plane. Then erase the part of the view that the cutting plane has cut away. In the other method, first draw the section on the isometric cutting plane. Then work from it to complete the view.

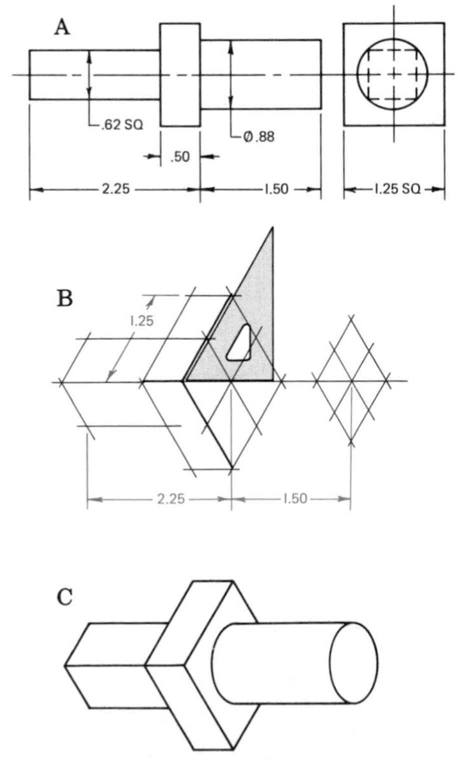

Fig. 12-21 *Steps in making an isometric drawing with the long axis horizontal.*

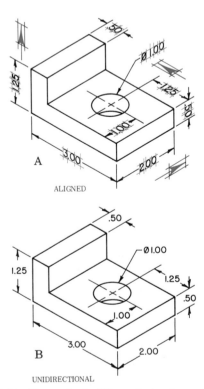

Fig. 12-22 *Two methods of dimensioning isometric views.*

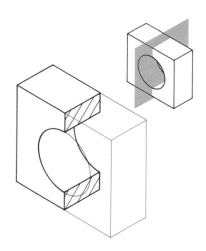
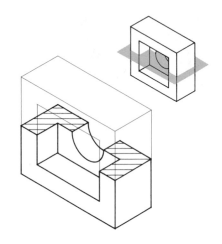
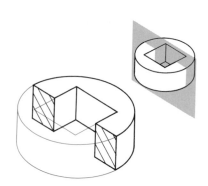

Fig. 12-23 *Examples of isometric full sections.*

Isometric Projection and Isometric Drawing

Isometric projection and isometric drawing are not the same thing, even though many people think they are. This section discusses the differences. While you probably will not have to make pictorial drawings using isometric projection, it is a good idea to understand the theory behind it.

One way to make an isometric projection is by *revolution* (drawing an object as if it were revolved or turned). Figure 12-25 shows a cube in the three normal views of a multiview drawing. In Fig. 12-25B, each of the three views has been revolved 45°. Notice that the front and side views now show as two equal rectangles. On the side view, draw a diagonal from point *O* to point *B*. This is called the *body diagonal*. The body diagonal is the longest straight line that can be drawn in a cube.

Revolve the cube upward until the body diagonal is horizontal, as shown in the side view (Fig. 12-25C). Notice that you

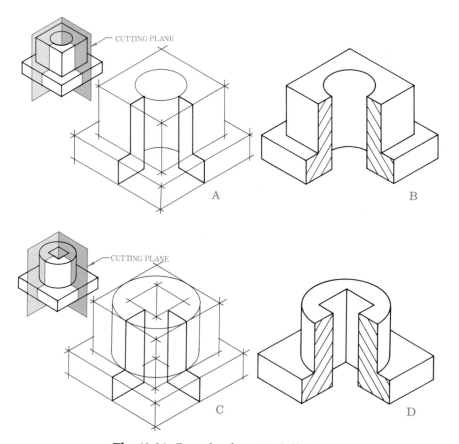

Fig. 12-24 *Examples of isometric half sections.*

must revolve the cube 35°16′ to achieve this. The front view now forms an isometric projection. Its lower edges form an angle of 30° to the horizontal.

Since the front view (isometric projection) of the revolved cube is made by projection, its lines are foreshortened. The actual difference is

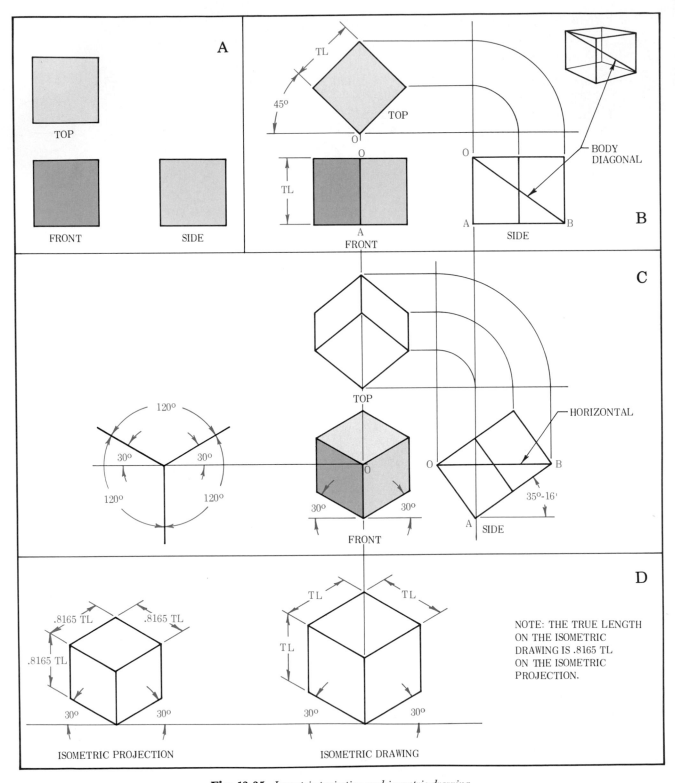

Fig. 12-25 *Isometric projection and isometric drawing.*

0.8165 to 1.00 in. In other words, 1.00 in. on the cube in Fig. 12-25A has been reduced to 0.8165 in. on the isometric projection in Fig. 12-25C.

Figure 12-25D shows an isometric drawing and an isometric projection of the same cube. In the isometric drawing, all edges are drawn their true length instead of the shortened length. The drawing shows the shape of the cube just as well as the projection. Its advantage is that it is easier to draw, because all its measurements can be made with a regular scale. It can also be drawn without projecting from other views and without using a special scale.

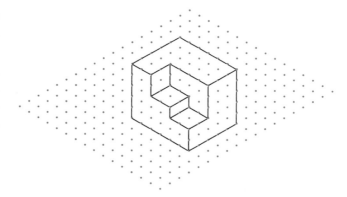

Fig. 12-26 *AutoCAD provides an isometric grid to help the drafter create isometric drawings quickly. In this drawing, the small dots represent the grid that displays on the screen. They do not appear on the printed drawing.*

CAD-GENERATED ISOMETRIC DRAWINGS

CAD systems provide drawing aids that help the drafter create two-dimensional isometric drawings quickly. The most important of these are the grid and snap features. Using AutoCAD, for example, you can use an isometric grid and snap to guide the isometric lines in a drawing. To create the isometric drawing in Fig. 12-26, for example, follow these steps.

1. Enter the SNAP command and choose the Style option.
2. Enter I to choose the isometric snap style.
3. Specify vertical and horizontal snap spacing of .50. Notice that the position of the crosshairs changes.
4. Turn on the grid by pressing the F7 key. Notice that the grid, too, reflects the isometric axes (Fig. 12-26).
5. Enter the ISOPLANE command to change the isometric plane reflected by the crosshairs. Notice that you can change the crosshairs to reflect the left, top, or right surface of the isometric object. *NOTE:* You can also toggle the isometric plane by pressing the CONTROL key and E simultaneously.
6. Enter the LINE command and draw the object. Notice how easy it is to draw the lines at the correct angles when snap is on.

The object in Fig. 12-26 contains no nonisometric lines. If it did, you could create them just as you would manually: find the endpoints of the line, and then draw the line from endpoint to endpoint. Be sure to use the Endpoint object snap to increase the accuracy of the drawing.

To dimension the drawing using the unidirectional method, change the snap style back to Standard. To do this, reenter the SNAP command, select the Standard option, and set the snap spacing. The snap and grid return to their standard formation.

Another way to form an isometric drawing is to create a solid model using the same methods described earlier in this book. Then use the CAD software's view function to define an isometric view of the model.

AXONOMETRIC PROJECTION

Isometric projection is one form of **axonometric projection** (projection that uses three axes at angles to show three sides of an object). Other forms are *dimetric* and *trimetric* projection. All three are made according to the same theory; the difference is the angle of projection (Fig. 12-27). In isometric projection, the axes form three equal angles of 120° on the plane of projection. Only one scale is needed for measurements along each axis.

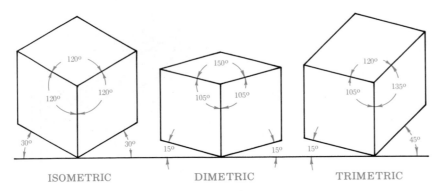

Fig. 12-27 *The three types of axonometric projection.*

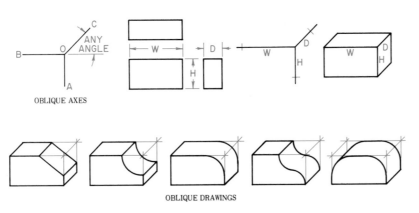

Fig. 12-28 *The oblique axes and oblique drawings.*

Isometric drawings are the easiest type of axonometric drawing to make. In **dimetric projection,** only two of the angles are equal, and two special foreshortened scales are needed to make measurements. In **trimetric projection,** all three angles are different, and three special foreshortened scales are needed.

Dimetric and trimetric drawings are complicated to draw. Therefore, drafters use them less often than other types of pictorial drawing.

OBLIQUE DRAWING

Oblique drawings are plotted in the same way as isometric drawings; that is, on three axes. However, in oblique, two axes are parallel to the **picture plane** (the plane on which the view is drawn), rather than just one, as in isometric.

These two axes always make right angles with each other (Fig. 12-28). As a result, oblique shows an object as if viewed face on. That is, one side of the object is seen squarely, with no distortion, because it is parallel to the picture plane.

The methods and rules of isometric drawing apply to oblique drawing. However, oblique also has some special rules. The first rule is:

Place the object so that the irregular outline or contour faces the front (Fig. 12-29A).

The second rule is:

Place the object so that the longest dimension is parallel to the picture plane (Fig. 12-29B).

Oblique Projection and Oblique Drawing

Oblique projection, like isometric projection, is a way of showing depth (Fig. 12-30). Depth is shown, as in isometric, by *projectors* (lines representing receding edges of the object). These lines are drawn at an angle other than 90° from the picture plane to make the receding planes visible in the front view. And, as in isometric, lines on these receding planes that are actually parallel to each other are drawn parallel. Figure 12-30 shows how an oblique projection is developed. You probably will never have to develop an oblique projection in this way, but as with isometric projection, it is a good idea to understand the theory behind it.

Usually, no distinction is made between oblique projection and oblique drawing. This is another difference from isometric. Because oblique drawing can show one face of an object without distortion, it has a distinct advantage over isometric. It is especially useful for showing objects with irregular outlines.

Figure 12-31 shows several positions for oblique axes. In all cases, two of the axes, *AO* and *OB*, are drawn at right angles. The oblique axis *OC* can be at any angle to the right, left, up, or down, as illustrated. The best way to draw an object is usually at the angle from which it would normally be viewed.

Types of Oblique Drawings

Oblique drawings are classified according to the length of the receding lines of an object along the oblique axis. Drawings in which the receding lines are drawn full length are known as **cavalier oblique.** Some drafters use three-quarter size. This is sometimes called **normal oblique,** or *general oblique.* If the receding lines are drawn one-half size, the drawing is **cabinet oblique.** Figure 12-32 shows a bookcase drawn in cavalier, normal, and cabinet drawings. Cabinet drawings are so named because they are often used in the furniture industry.

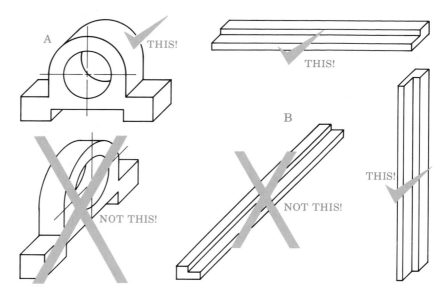

Fig. 12-29 *Two general rules for oblique drawings.*

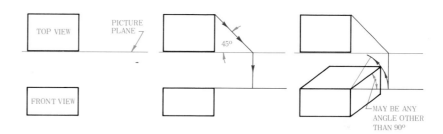

Fig. 12-30 *Oblique projection.*

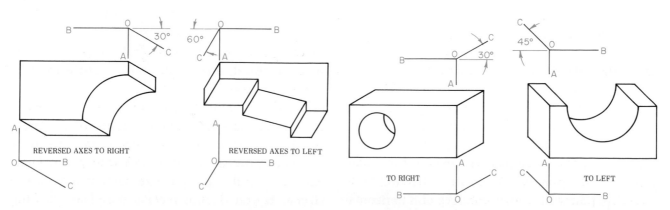

Fig. 12-31 *Positions for oblique axes.*

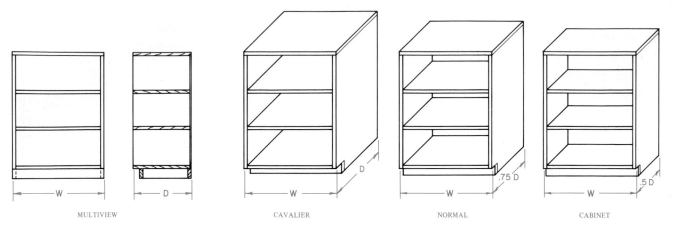

MULTIVIEW   CAVALIER   NORMAL   CABINET

Fig. 12-32 *Three types of oblique drawings.*

Oblique Constructions

As with isometric drawing, you should understand how to draw the geometry in an oblique drawing before you begin a complete drawing. The techniques used in oblique drawing are described below.

Angles and Inclined Surfaces

Angles that are parallel to the picture plane show in their true size. For all other angles, lay the angle off by locating both ends of the slanting line.

Figure 12-31 shows a plate with the corners cut off at angles. In Fig. 12-33B, C, and D, the plate is shown in oblique drawings. In (B), the angles are parallel to the picture plane. In (C), they are parallel to the profile plane. In each case, the angle is laid off by measurements parallel to one of the axes. These measurements are shown by the construction lines.

Oblique Circles

On the front face, circles and curves show in their true shape (Fig. 12-34). On other faces, they show as ellipses. Draw the ellipses using the four-center method or an ellipse template. Fig. 12-34A shows a circle as it would be drawn on a front plane, a side plane, and a top plane. Figure 12-34B and C show an oblique drawing with some arcs in a horizontal plane and in a profile plane, respectively.

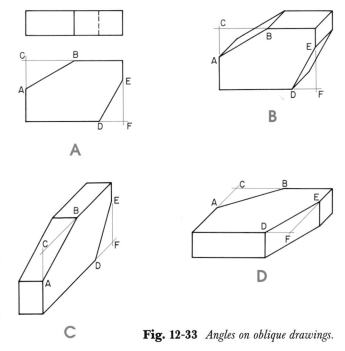

Fig. 12-33 *Angles on oblique drawings.*

When you draw oblique circles using the four-center method, the results will be satisfactory for some purposes, but not pleasing. Ellipse templates, when available, give much better results. If you use a template, first block in the oblique circle as an oblique square. This shows where to place the ellipse. Blocking in the circle first also helps you choose the proper size and shape of the ellipse. If you do not have a template, plot the ellipse as shown in Fig. 12-35.

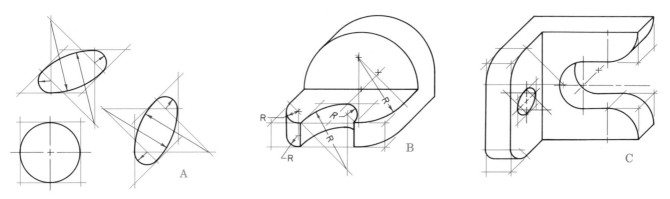

Fig. 12-34 *Circles parallel to the picture plane are true circles; on other planes, they appear as ellipses.*

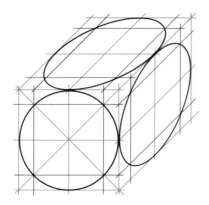

Fig. 12-35 *Plotting oblique circles.*

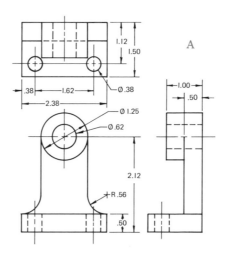

Multiview drawing of the object to be drawn in oblique.

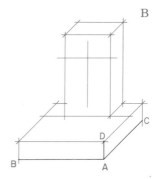

Draw the axes AB, AC, and AD for the base in second position and on them measure the length, width, and thickness of the base. Draw the base. On it, block in the upright, omitting the projecting boss as shown.

Creating an Oblique Drawing

Figure 12-36 shows the steps in making an oblique drawing. The procedure is much the same as that for creating an isometric drawing. Notice that this oblique drawing can show everything but the two small holes in true shape.

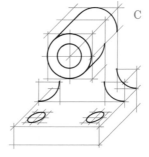

Block in the boss and find the centers of all circles and arcs. Draw the circles and arcs.

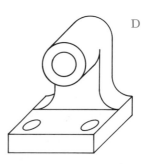

Darken all necessary lines, and erase construction lines to complete the drawing.

Fig. 12-36 *Steps in making an oblique drawing.*

Oblique Sections

Oblique drawings are generally "outside" views. Sometimes, however, you need to draw a sectional view. To do so, take a section on a plane parallel to one of the faces of an oblique cube. Figure 12-37 shows an oblique full section and an oblique half section. Note the construction lines indicating the parts that have been cut away.

CAD-GENERATED OBLIQUE DRAWINGS

To create a two-dimensional oblique drawing using a CAD system, it is generally a good idea to create the front face first. If the front face of an object has the same shape as the back face, you can then simply copy the front face to create the back of the object. Position the back at an angle to the front, as shown in Fig. 12-38. Draw lines connecting the front and back faces at the edges.

If the back face of the object is different from the front, you should still create the front face first. Then copy only those lines that are the same on the back face and work from there.

Creating a three-dimensional oblique CAD drawing, use the same procedure as for 3D isometric drawings. Create the solid model, and then define an oblique view using the CAD software's view function.

PERSPECTIVE DRAWING

A **perspective drawing** (Fig. 12-39) is a three-dimensional representation of an object as it looks to the eye from a particular point. Of all pictorial drawings, perspective drawings look the most like photographs. The distinctive feature of perspective drawing is that in perspective, lines on the receding planes that are actually parallel are not drawn parallel, as they are in isometric and oblique drawing. Instead, they are drawn as if they were *converging* (coming together.)

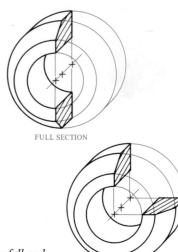

FULL SECTION

HALF SECTION

Fig. 12-37 *Oblique full and half sections.*

 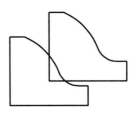

First create the front face of the object.

Next, copy the front face to create the back face. The position of the back face determines the angle of the third axis.

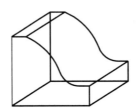 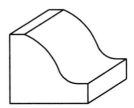

Using the LINE command, connect the front and back edges of the object.

Use the ERASE, TRIM, or BREAK command as necessary to remove hidden lines.

Fig. 12-38 *Steps for creating an oblique drawing using a CAD system.*

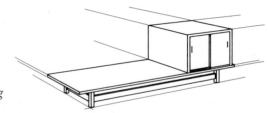

Fig. 12-39 *Perspective drawing of a music center.*

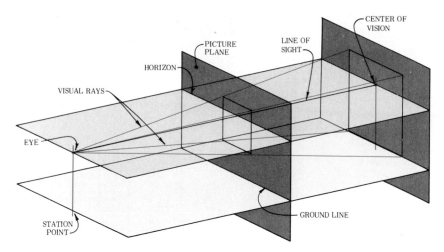

Fig. 12-40 *Some perspective terms.*

Where the ground plane on which the observer stands meets the picture plane, it forms the *ground line (GL)*. The *center of vision (CV)* is the point at which the line of sight pierces the picture plane. The *line of sight (LS)* is the visual ray from the eye perpendicular to the picture plane.

Figure 12-41 shows how, in perspective drawing, the projectors (receding axes) converge. The point at which they meet is called the **vanishing point.**

Figure 12-41 also shows how the observer's eye level affects the perspective view. This eye level can be anywhere on, above, or below the ground. If the object is seen from above, the view is an *aerial,* or *bird's-eye view.* If the object is seen from underneath, the view is a *ground,* or *worm's-eye view.* If the object is seen face on, so that the line of sight is directly on it rather than above or below, the view is a *normal view.* The view in Fig. 12-40 is a normal view.

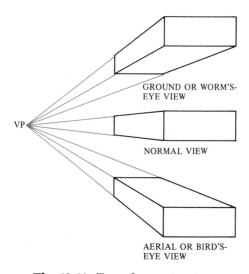

Fig. 12-41 *Types of perspective views.*

Definition of Terms

Figure 12-40 illustrates terms used in perspective drawing. A card appears on the plane at the right. The sight lines leading from points on this card and converging at the eye of the observer are called *visual rays.* The *picture plane (PP)* is the plane on which the card at the right is drawn. The *station point (SP)* is the point from which the observer is looking at the card. A horizontal plane passes through the observer's eye. Where it meets the picture plane, it forms the *horizon line (HL).*

Factors That Affect Appearance

Two factors affect how an object looks in perspective. The first is its distance from the viewer, and the second is its position (angle) in relation to the viewer.

The Effect of Distance

The size of an object seems to change as you move toward or away from it. The farther away from the object you go, the smaller it looks. The closer you get, the larger it seems to grow. Figure 12-42 shows a graphic explanation of this effect of distance. An object is placed against a scale at a normal reading distance from the viewer. In that position, it looks to be the size indicated by the scale. However, if it is moved back from the scale to a point twice as far away from the viewer, it looks only half as large. Notice that each time the distance is doubled, the object looks only half as large as before.

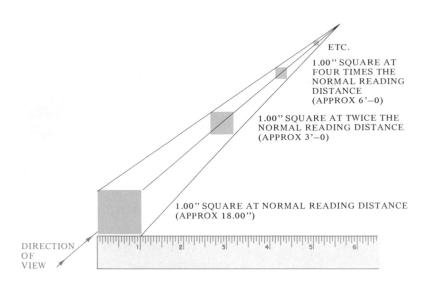

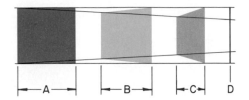

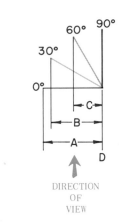

Fig. 12-42 *The size of an object appears half as large when the distance from the observer is doubled.*

Fig. 12-43 *The position of the object in relationship to the observer affects its appearance.*

The Effect of Position

The shape of an object also seems to change when it is viewed from different positions (angles). This is illustrated in Fig. 12-43. If you look at a square from directly in front, the top and bottom edges are parallel. But if the square is rotated so that you see it at an angle, these edges seem to converge. The square also appears to grow narrower. This foreshortening occurs because one side of the square is now farther away from you.

One-Point Perspective

One-point perspective (also called *parallel perspective*) is a perspective view in which there is one vanishing point (Fig. 12-44). Notice that if the lines of the building in Fig. 12-44 were extended, they would converge at a single point. Figure 12-45 shows an object in multiview and isometric drawings. Figure 12-46 shows how to draw the same object in a one-point bird's-eye view perspective. In Figs. 12-47 and 12-48, the object is drawn in one-point perspective in the other positions. Notice that in all three cases, one face of the object is placed on the picture plane (thus the name *parallel perspective*). Therefore, this face appears in true size and shape. True-scale measurements can be made on it.

Fig. 12-44 *The lines of the sidewalk, roof, and building's side appear to converge at a distance.* (Courtesy of Mishima)

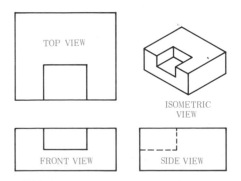

Fig. 12-45 *Multiview and isometric drawings of an object to be drawn in single-point perspective.*

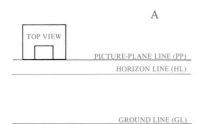

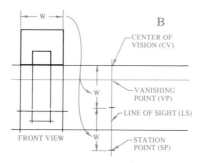

Decide on the scale to be used and draw the top view near the top of the drawing sheet. A more interesting view is obtained if the top view is drawn slightly to the right or to the left of center. Draw an edge (top) view of the picture plane (PP) through the front edge of the top view. Draw the horizontal line (HL). The location will depend upon whether you want the object to be viewed from above, on, or below eye level. Draw the ground line. Its location in relation to the horizon line will determine *how far* above or below eye level the object will be viewed.

Locate the station point (SP). (a) Draw a vertical line (line of sight) from the picture plane toward the bottom of the sheet. Draw the line slightly to the right or to the left of the top view. (b) Set your dividers at a distance equal to the width (W) of the top view. (c) Begin at the center of vision of the picture plane and step off 2 to 3 times the width (W) of the top view, along the line of sight, to locate the station point (SP). Project downward from the top view to establish the width of the front view on the ground line. Complete the front view.

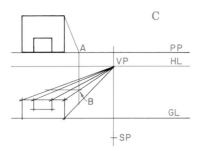

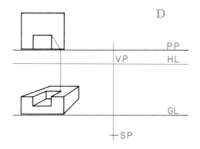

The vanishing point (VP) is the intersection of the line of sight (LS) and the horizon line (HL). Project lines from the points on the front view to the vanishing point. Establish depth dimensions in the following way: (a) Project a line from the back corner of the top view to the station point. (b) At point *A* on PP, drop a vertical line to the perspective view to establish the back edge. (c) Draw a horizontal line through point *B* to establish the back top edge.

Proceed as in the previous step to lay out the slot detail. Darken all necessary lines, and erase construction lines as desired to complete the drawing.

Fig. 12-46 *Procedure for making a single-point, or parallel, perspective drawing (bird's-eye view).*

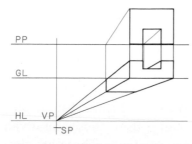

Fig. 12-47 *Single-point perspective, worm's-eye view.*

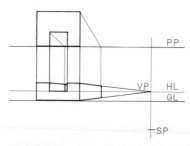

Fig. 12-48 *Single-point perspective, normal view.*

Two-Point Perspective

Two-point perspective drawings have two vanishing points. This type is also called *angular perspective,* since none of the faces are drawn parallel to the picture plane. Figure 12-49 shows a typical two-point perspective drawing.

Figure 12-50 shows an object in multiview and isometric drawings. Figure 12-51 shows how to draw this same object in two-point bird's-eye view perspective. Figures 12-52 and 12-53 show the object drawn in the other positions.

Fig. 12-49 *When a building is viewed at an angle, two sides can be seen. The top and ground lines on each side appear to converge toward points. This is the effect in two-point, or angular, perspective.* (Courtesy of Mishima)

Perspective Constructions

Perspective drawing involves techniques similar to those used for isometric and oblique. The techniques for inclined surfaces, circles, and arcs are described below.

Inclined Surfaces

Plot inclined surfaces in perspective by finding the ends of inclined lines and connecting them. This method of drawing is shown in Fig. 12-54.

Circles and Arcs

Figure 12-55 shows how to make a perspective view of an object with a cylindrical surface. First locate points on the front and top views. Then project them to the perspective view. Where the projection lines meet, a path is formed. Draw the perspective arc along the path using a French curve or an ellipse template.

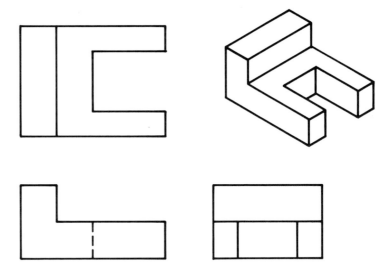

Fig. 12-50 *Multiview and isometric drawings of an object to be drawn in two-point perspective.*

Perspective Drawing Shortcuts

Perspective drawing can take a lot of time. This is because so much layout work is needed before you can start the actual perspective view. Also, a large drawing surface is often needed in order to locate distant points. However, you can offset these disadvantages by using various shortcuts. These are described on the following pages.

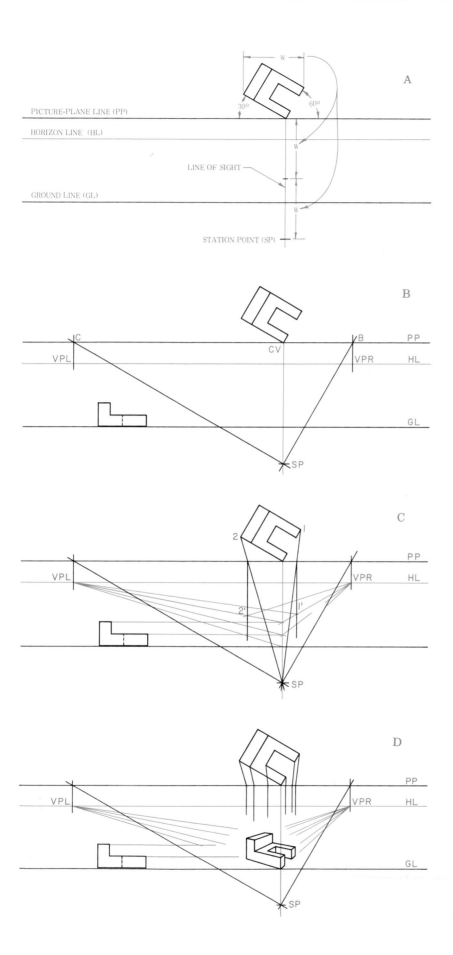

Draw an edge view of the picture plane (PP). Allow enough space at the top of the sheet for the top view. Draw the top view with one corner touching the PP. In this case, the front and side of the top view form angles of 30° and 60°, respectively. Other angles may be used, but 30° and 60° seem to give the best appearance on the finished perspective drawing. The side with the most detail is usually placed along the smaller angle for a better view. Draw the horizon line (HL) and the ground line (GL). (Follow the procedure given in Fig. 12-44.) Draw a vertical line (line of sight) from the center of vision (CV) toward the bottom of the sheet to locate the station point. (Follow the procedure given in Fig. 12-44.)

Draw line *SP-B* parallel to the end of the top view and line *SP-C* parallel to the front of the top view. (Use a 30°60° triangle.) Drop vertical lines from the picture plane (PP) to the horizon line (HL) to locate vanishing point left (VPL) and vanishing point right (VPR). Draw the front or side view of the object on the ground line as shown.

Begin to block in the perspective view by projecting vertical dimensions from the front view to the line of sight (also called *measuring line*) and then to the vanishing points. Finish blocking in the view as follows: (a) Project lines from points 1 and 2 on the top view to the station point. (b) Where these lines cross the picture plane (PP), drop vertical lines to the perspective view to establish the length and width dimensions. (c) Project point 1¹ to VPL and 2¹ to VPR.

Add detail by following the procedure described in the previous two main steps. Darken all necessary lines and erase construction lines as desired.

Fig. 12-51 *Procedure for making a two-point perspective drawing (bird's-eye view).*

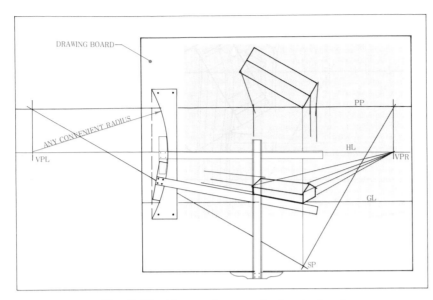

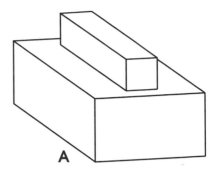

Fig. 12-58 *A homemade perspective drawing board.*

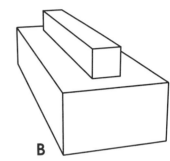

Fig. 12-59 *A simple 3D object created in AutoCAD. (A) An oblique view of the object created using the VPOINT (viewpoint) command. (B) A perspective view of the object created using the DVIEW (dynamic view) command.*

2. Place a drawing board within the layout, as shown in Fig. 12-56. For guides, use heavy cardboard or thin wood. Fasten the guide material in place. Then strike the arcs, using the vanishing points as centers and any convenient radius.

3. Cut the arcs with a sharp knife to form the guides. Draw the PP, HL, GL, LS, and SP on the board. You can now remove the board from the layout sheet.

4. Construct the T-square as shown. Notice that the top edge of the blade falls on the centerline of the heads. You must use thin material for the head as well as for the blade.

To use the perspective board:

1. Place a sheet of tracing paper in position between the guides (Fig. 12-58). Tracing paper will allow you to use the lines drawn on the board without redrawing them.

2. Draw the top view of the desired object in its proper position.

3. Proceed as in Fig. 12-51. Draw the lines that project toward VPR (vanishing point right) with the T-square head against the left guide. For those that project toward VPL (vanishing point left), use the right-hand guide. Draw the vertical lines with the T-square head against the bottom edge of the board.

PERSPECTIVE DRAWINGS IN CAD

In a good CAD system, perspective drawings do not require any special techniques, providing that you create the drawing in 3D. You create the drawing just as you would any other 3D drawing. After you have finished drawing the object (or objects), you can use special commands to create a perspective view of the object.

In AutoCAD, for example, you would enter the DVIEW command. DVIEW (Dynamic View) uses a camera and target analogy. You can manipulate the appearance of the drawing by changing the position of the camera (point of view) and target (object being viewed). You can also change the distance from which the viewer sees the object. Figure 12-59 shows a simple three-dimensional object before and after applying the DVIEW command.

TECHNICAL ILLUSTRATION

Generally, **technical illustration** is defined as a pictorial drawing that provides technical information by visual methods. Technical illustrations are usually created by turning a multiview drawing into a three-dimensional pictorial drawing. The drawing must show shapes and relative positions in a clear and accurate way. Shading may be used to bring out the shape. A technical illustration, however, is not necessarily a work of art. Therefore, the shading must serve a practical purpose, not an artistic one.

Technical illustrations range from simple sketches to rather detailed shaded drawings. They may be based on any of the pictorial methods: isometric, perspective, oblique, and so forth. The complete project, parts, or groups of parts may be shown. The views may be exterior, interior, sectional, cutaway, or phantom. The purpose in all cases is to provide a clear and easily understood description.

In addition to pictorials, technical illustrations include graphic charts, schematics, flowcharts, diagrams, and sometimes circuit layouts. Dimensions are not a part of technical illustrations because they are not working drawings.

Purposes of Technical Illustration

Technical illustration has an important place in all areas of engineering and science. Technical illustrations form a necessary part of the technical and service manuals for machine tools, automobiles, machines, and appliances. In technical illustration, pictorial drawings are used to describe parts and the methods for making them. They show how the parts fit together. They also show the steps that need to be followed to complete a product on the assembly line. They may even be used to set up an assembly line.

Drawings for use within a company's plant can sometimes be made by a drafter with artistic talent. However, the special skills needed for such drawings call for a professional technical illustrator. Technical illustrations have been used for many years in illustrated parts lists, operation and service manuals, and process manuals (Fig. 12-60).

The aircraft industry in particular has found production illustration especially valuable. In aircraft construction, pictorial drawings are used when the plan is first designed. These drawings are also used throughout the many phases of a plane's production and completion on the assembly line. When the plane is delivered to the customer, the service, repair, and operation manuals are also illustrated with pictorial drawings.

Choosing a Drawing Type

Most technical illustrations are pictorial line drawings. Therefore, you should have a complete understanding of the various types of pictorial line drawings and their uses. The first half of this chapter describes the various types of pictorial drawings and the procedure for drawing each. Usually, any type of pictorial drawing can be used as the basis for a technical illustration. However, some types are more suitable than others. This is especially true if the illustration is to be *rendered* (shaded).

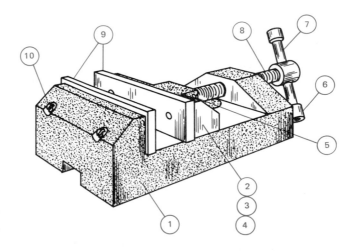

PART NO.	PART NAME	NO. REQD
1	BASE	1
2	MOVABLE JAW	1
3	MOVABLE JAW PLATE	1
4	MACHINE SCREW	1
5	LOCKING PIN	1
6	HANDLE STOP	2
7	HANDLE	1
8	CLAMP SCREW	1
9	JAW FACE	2
10	CAP SCREW	2

Fig. 12-60 *An illustrated parts list.*

Figure 12-61 shows a V-block drawn in several types of pictorial drawing. Notice the difference in the appearance of each. Isometric is the least natural in appearance. Perspective is the most natural. This might suggest, then, that all technical illustrations should be drawn in perspective. This is not necessarily true. While perspective is more natural than isometric in appearance, it takes more time to do, and it is also more difficult to draw. Thus it is often a more costly method to use.

The shape of the object also helps to determine the type of pictorial drawing to use. Figure 12-62 shows a pipe bracket drawn in isometric and oblique. This object can be drawn easily and quickly in oblique. Also, in many cases, the oblique looks more natural than isometric for objects of this shape.

If an illustration is to be used only in-plant, the illustrator usually makes the pictorial drawing in isometric or oblique. These are the quickest and least costly pictorials to make. If the illustration is to be used in a publication such as a journal, operator's manual, or technical publication, dimetric, trimetric, or perspective may be used.

Tools and Tips

Technical illustrators use the tools and regular drafting equipment described in Chapter 3. An H or 2H pencil, with the point kept well sharpened, is the most useful tool. A few other useful items include felt-tip pens, technical pens, masking tape, Scotch® tape, an X-Acto® knife, paper stomps, two or three artist's brushes, and an airbrush.

Craftint, Zip-a-tone, Chart-Pak, and similar press-on section linings, screen tints, and letters are also useful timesaving materials. You should know about the various methods of graphic reproduction.

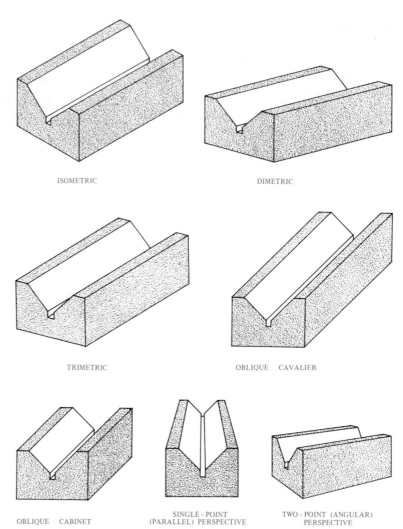

Fig. 12-61 *A V-block in various types of pictorial drawing.*

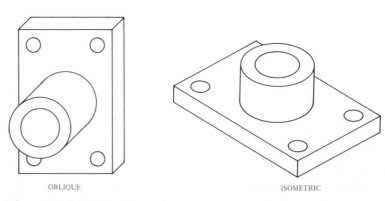

Fig. 12-62 *The shape of an object helps to determine the most suitable type of pictorial drawing to use.*

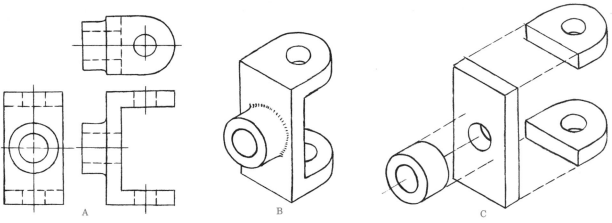

Fig. 12-63 *How a view is exploded.*

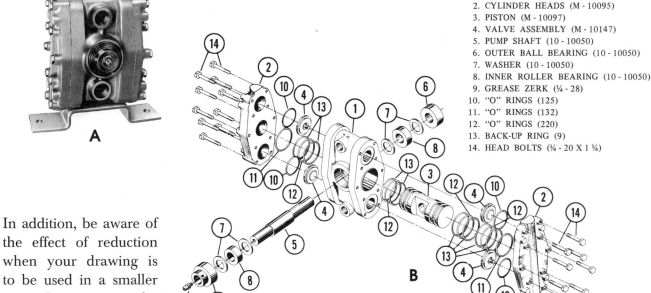

1. PUMP BODY (M - 10091)
2. CYLINDER HEADS (M - 10095)
3. PISTON (M - 10097)
4. VALVE ASSEMBLY (M - 10147)
5. PUMP SHAFT (10 - 10050)
6. OUTER BALL BEARING (10 - 10050)
7. WASHER (10 - 10050)
8. INNER ROLLER BEARING (10 - 10050)
9. GREASE ZERK (¼ - 28)
10. "O" RINGS (125)
11. "O" RINGS (132)
12. "O" RINGS (220)
13. BACK-UP RING (9)
14. HEAD BOLTS (¼ - 20 X 1 ¼)

Fig. 12-64 *(A) A high-pressure pump (B) An exploded view that shows and identifies the parts of a high-pressure piston pump.* (Courtesy of Industrial Division, Standard Precision, Inc.)

In addition, be aware of the effect of reduction when your drawing is to be used in a smaller size. Lines must be solid, sharp, and black. Erasures must be clean. The part of a drawing not being worked on should be kept covered with paper or sheet plastic.

Lettering is also an important element of technical illustration. In some cases, you can use templates. However, good freehand lettering is a required skill for a technical illustrator.

Exploded Views

Perhaps the easiest way to understand an exploded view is this: Take an object and separate it into its individual parts, as in Fig. 12-63.

Three views are shown in (A), and a pictorial view is shown in (B). In (C), an "explosion" has projected the elementary parts away from each other. This illustrates the principle of exploded views.

All exploded views are based upon the same principle: projecting the parts from the positions they occupy when put together. Simply, they are just pulled apart. The exterior of a high-pressure piston pump is shown in Fig. 12-64A. An exploded illustration of the pump is shown in Fig. 12-64B. Note that all parts are easily identifiable.

Identification Illustrations

Pictorial drawings are very useful for identifying parts. They save time when the parts are manufactured or assembled in place. They are also useful for illustrating operating instruction manuals and spare-parts catalogs.

Identification illustrations are usually presented in exploded views. If an object consists of only a few parts, identify them by their names and leaders. If the object contains several parts, number them as shown in Fig. 12-65. In this example, a tabulation shows names and quantities of the numbered parts.

Rendering

Surface shading, or **rendering,** of some kind may be used when shapes are difficult to read. For most industrial illustrations, accurate descriptions of shapes and positions are more important than fine artistic effects. You can often achieve satisfactory results without any shading. In general, you should limit surface shading when possible. Shade the least amount necessary to define the shapes that are being illustrated. Shading takes time and is expensive.

Different ways of rendering technical illustrations include the use of screen tints, pen and ink, wash, stipple, felt-tip pen and ink, smudge, and edge emphasis, among others. These means of rendering are used in the technical illustrations of aircraft companies, automobile manufacturers, and machine-tool makers. They can even be found in the directions that come with your TV set.

Outline Shading

Outline shading may be done mechanically or freehand. Sometimes a combination of both methods is used. The light is generally considered to come from in back of and above the left shoulder of the observer, across the diagonal of the object, as shown in Fig. 12-66A. This is a *convention,* or a standard method, used by drafters and renderers. In (B), the upper left and top edges are in the light, so they are drawn with thin lines. The lower right and bottom edges are in the shadow. They should be drawn with thick lines. In (C), the edges meeting in the center are made with thick lines to accent the shape. In (D), the edges meeting at the center are made with thin lines. Thick lines are used on the other edges to bring out the shape.

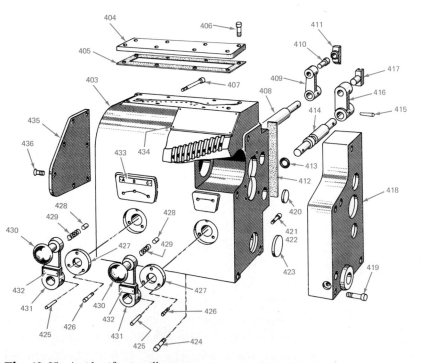

Fig. 12-65 *An identification illustration.* (Courtesy of The R. K. LeBlond Machine Tool Co.)

PART NO.	PART NAME	QTY.
403	QUICK CHANGE BOX	1
404	COVER, TOP	1
405	GASKET, COVER	1
406	SCREW, SOCKET HEAD CAP	8
407	SCREW	2
408	SHAFT, SHIFTER	1
409	LINK, SHIFTER	1
410	PIN	1
411	SHOE, SHIFTER	1
412	GASKET (MAKE IN PATTERN SHOP-BOX TO BED)	2
413	"O" RING	2
414	SHAFT, SHIFTER	1
415	PIN, TAPER	2
416	LINK, SHIFTER	1
417	SHOE, SHIFTER	1
418	COVER, SLIP GEAR	1
419	SCREW	4
420	PLUG	2
421	SCREW	3
422	SCREW	1
423	PLUG (NOT USED WITH SCREW REVERSE)	1
424	SCREW	3
425	PIN	2
426	SCREW	6
427	COLLAR	2
428	PLUNGER	2
429	SPRING	2
430	KNOB	2
431	LEVER	2
432	PLATE, FEED-THD.	1
433	PLATE, COMPOUND	1
434	PLATE, ENGLISH INDEX	1
435	COVER	1
436	SCREW	7

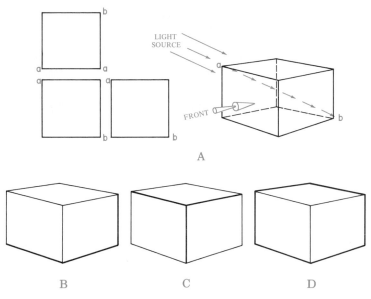

Fig. 12-66 *Light source and line-shaded cubes.*

Examples of the use of a small amount of line shading are shown in Figs. 12-67 and 12-68. In these cases, the shading is used to outline important parts of the drawings.

Surface Shading

With the light rays coming in the usual conventional direction, as in Fig. 12-69A, the top and front surfaces of a cube should be lighted. Therefore, the right-hand surface should be shaded. Fig. 12-69B, C, and D show various methods of shading. In (B), the front surface is unshaded and the right surface is lightly shaded. If the front surface has light shading, then the right side should have heavier shading, as in (C). Solid shading may sometimes be required to avoid confusion. If the front is shaded, then solid black may be used on the right-hand side, as shown in (D).

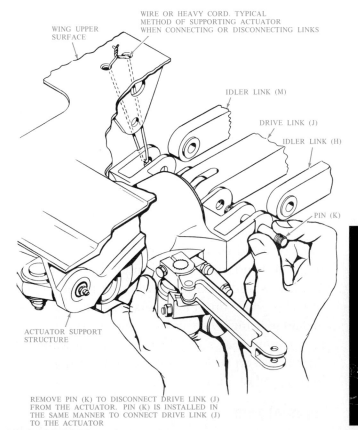

Fig. 12-67 *A maintenance manual illustration. Notice that only the necessary detail is shown and that just enough shading is used to emphasize and give form to the parts.* (Courtesy of Technical Illustrators Association)

Fig. 12-68 *Outline emphasis by a thick black or white line is an effective method of making a shape stand out.* (Courtesy of Rockford Clutch Division, Borg-Warner)

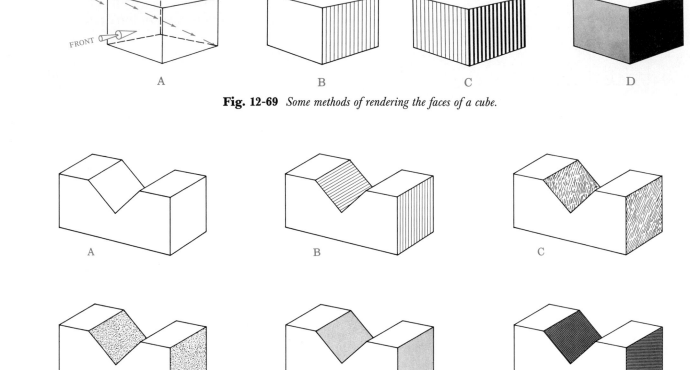

Fig. 12-69 *Some methods of rendering the faces of a cube.*

Fig. 12-70 *Examples of various kinds of rendering.*

Some Shaded Surfaces

Some methods of shading surfaces are shown in Fig. 12-70. An unshaded view is shown in (A) for comparison. Ruled-surface shading is shown in (B), freehand shading in (C), stippled shading in (D), and pressure-sensitive overlay shading at (E) and (F).

Stippling (D) consists of dots. Short, crooked lines can also be used to produce a shaded effect. It is a good method when it is well done, but it takes quite a bit of time and practice.

In some cases, you may not have the time to shade an illustration using these methods. When time is a factor, you may use pressure-sensitive shading overlays. These overlays are available in a great variety of patterns and can be applied easily.

Wash Rendering

Wash rendering (also called *wash drawing*) is a form of watercolor rendering that is done with watercolor and watercolor brushes. It is commonly used for rendering architectural drawings (Fig. 12-71). It is also used for advertising furniture and similar products in newspapers (Fig. 12-76A). This technique is highly specialized and is usually done by a commercial artist. However, some technical illustrators and drafters are, at times, required to do this kind of illustrating.

Scratchboard

Scratchboard drawing is a form of line rendering. Scratchboard is coated with india ink, and a sharp instrument is used to make the lines.

Fig. 12-71 *Wash rendering of an architectural drawing.* (Courtesy of Circle Design)

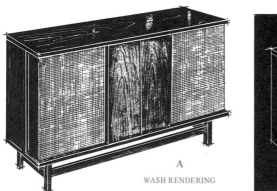

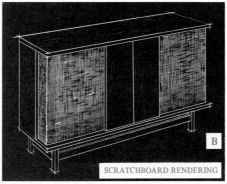

A
WASH RENDERING

B
SCRATCHBOARD RENDERING

Fig. 12-72 *Two methods for illustrating the same product: (A) wash rendering; (B) scratchboard.*

Fig. 12-73 *Airbrush rendering.*

To use this technique, draw the image on the inked surface of the scratchboard, and then scratch through the ink to expose the lines or surfaces (Fig. 12-72B).

Airbrush Rendering

Airbrush rendering produces illustrations that resemble photographs (Fig. 12-73). The **airbrush** (Fig. 12-74) is a miniature spray gun. It is used primarily to render illustrations and to retouch photographs. In airbrush rendering, compressed air is used to spray a solution (usually watercolor) that adds various shading effects on a drawing.

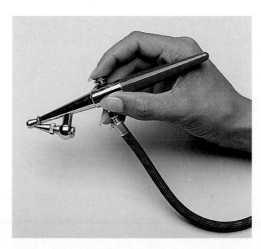

Fig. 12-74 *An airbrush.* (Courtesy of Ann Garvin)

Types of Airbrushes

Airbrushes are classified according to the size of their spray pattern. It is important to select your airbrush carefully. The size and style should match the work that you will be doing.

An *oscillating-needle airbrush* (one in which the needle vibrates) produces the smallest spray pattern. This type is capable of spraying very thin lines (hairlines) and small dots. However, it is the most expensive and is used only by professionals who do highly detailed rendering and retouch work.

A slightly larger airbrush, often called the *pencil type,* is a general-purpose illustrator's airbrush. It can be adjusted to spray thin lines. In addition, it can be opened up to spray large surfaces and backgrounds. This airbrush is most popular for student use.

The largest airbrush suitable for use in technical illustrating will spray a pattern large enough to do posters, displays, models, etc. This large brush is often called a *poster-type* airbrush.

Airbrushes are also classified by the type of spray control mechanism they use. An airbrush spray control has either a single action or a double action. Oscillating-needle and most pencil-type brushes have double-action mechanisms. Most poster-type brushes have single-action mechanisms. Single action simply means that when you press the finger lever, both color and air are expelled at the same time. Double action means that two motions are necessary. When you press the lever, air is released. Then when you pull the lever back, color is released. Much greater control is possible with a double-action airbrush. It is the kind usually used for rendering technical illustrations.

Air Supply

A constant supply of clean air is needed to produce a high-quality spray pattern. An air supply can be obtained from a carbonic gas (CO_2) unit or an air compressor.

An air transformer (regulator and filter) must be installed between the air supply and the airbrush.

The regulated pressure for most airbrush rendering is 32 pounds per square inch (lb./in.^{2}) [22 kilopascals (kPa)]. Less pressure may be used for special effects.

Supplies and Materials

A variety of supplies and materials is needed for airbrush work. Some are common art supplies. Others are special and can be obtained through an art or engineering supply house. The following list is in addition to standard drafting supplies and equipment:

- airbrush and hose
- rubber cement
- rubber-cement pickup
- razor knife (X-Acto® or similar)
- white illustration board (not pressed)
- frisket paper
- watercolor brushes
- medicine dropper
- designer's watercolors (black and white)
- palette
- photo retouching set

Procedure for Airbrushing

The following procedure is generally used for airbrushing. *NOTE:* This procedure creates standard effects used by most technical illustrators. To do special effects well, you must experiment with equipment and materials.

1. Prepare a line drawing of the desired object. Transfer it to the surface of the illustration board. Do not use standard typing carbon paper for transferring the image. Either buy a special transfer sheet or make one by blackening one side of a sheet of tracing vellum with a soft lead pencil.

2. Cover the image area with **frisket paper.** This material is available in two forms: prepared and unprepared. Prepared frisket paper has one adhesive side protected with wax paper. Unprepared frisket paper must be coated with thinned rubber cement. Prepared frisket paper is more convenient to use and is recommended for the beginner.

A

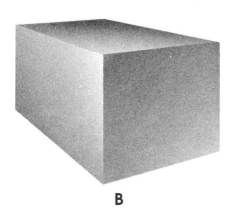

B

Fig. 12-75 *(A) Removing the frisket paper from the area to be airbrushed. (B) Finished rendering of a cube.* (A–Courtesy of Ann Garvin)

3. Cut and remove the frisket paper from the area to be airbrushed first (Fig. 12-75A). Use the rubber-cement pickup to remove any particles of rubber cement left on the surface. Cover all other areas of the illustration board not covered by the frisket paper.

4. Mix the black watercolor in the palette. Place water in the palette cup with a medicine dropper. Then squeeze about .12 inch of black watercolor onto the edge of the palette. Use a watercolor brush to mix the color into the water.

5. Transfer the mixed color from the palette to the color cup on the airbrush. You can use the watercolor brush to do this. Fill the cup about half full.

6. Render the exposed surface as desired.

7. Open a second portion of the frisket and cover the rendered surface. Continue in this way until all surfaces are rendered.

8. Remove the frisket. Figure 12-75B shows the finished rendering. Figure 12-76 shows examples of other objects that have been rendered. Highlights may be added using white watercolor.

Fig. 12-76 *Examples of airbrushed objects.*

Photo Retouching

Photo retouching is a process used to change details on a photograph. Details may be added, removed, or simply repaired. This process is often needed in preparing photographs for use in publications. It can also be used for changing the appearance of some detail.

Photo retouching is usually done on a glossy photograph. Photo retouching kits include the gray tones needed to work on black-and-white photographs. Use the same basic procedure outlined above for standard airbrush work. Take care not to damage the finish on the photo when cutting the frisket paper. You may also use a watercolor brush to touch up fine details. Figure 12-77 shows a photograph before and after it has been retouched.

DIGITAL TECHNICAL ILLUSTRATION

Many companies now do at least some of their technical illustration using a computer. This is not considered CAD; rather, it is a form of digital illustration.

As computers have become more powerful and their storage more plentiful, many illustration programs have appeared on the market for personal computers. Examples include CorelDRAW™ and Adobe Illustrator™. These programs offer tools that parallel manual artist's tools to allow technical illustrators to create their illustrations directly on the computer.

The biggest advantage of this is similar to the advantage of CAD: drawings can be saved, modified, printed, and reprinted as necessary. The illustrator does not have to begin again every time a design change occurs.

A

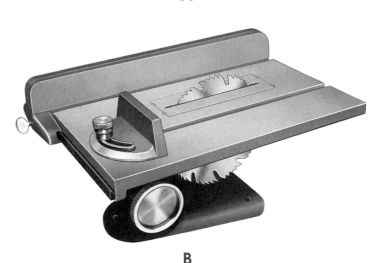

B

Fig. 12-77 *Before and after retouching.*

Digital cameras and scanners have introduced yet another capability to digital technical illustration. Photographs—both black-and-white and color—can now be scanned into computer programs and retouched professionally. An example of a program of this type is Adobe Photoshop™. These programs allow you to crop (cut away unwanted areas), change or enhance colors, correct for under- or overexposure, and hide flaws in the photograph.

CHAPTER 12

Learning Activities

1. Collect technical illustrations from newspapers, magazines, owner's manuals, and service manuals. Select the best examples and make a bulletin board display.

2. Collect sample pictorial line drawings and try to identify the type (axonometric, oblique, perspective) represented in each case. These may also be used for display.

Questions

Write your answers on a separate sheet of paper.

1. Describe at least two uses for pictorial drawings.
2. Name three types of axonometric projection.
3. What are the three most common types of pictorial drawing?
4. What do you call a line that is not parallel to any of the normal isometric axes?
5. Is an isometric projection larger or smaller than an isometric drawing?
6. In an isometric drawing, measurements can be made only along isometric lines. Explain why measurements cannot be made along nonisometric lines.
7. What are reversed axes in an isometric drawing? Describe a situation in which they might be useful.
8. In what way can the grid and snap features in a CAD program help a drafter create an isometric drawing?
9. Name the three most common types of oblique drawings.
10. In what way or ways is an oblique drawing different from an isometric drawing?
11. When creating an oblique drawing, how should you place the object?

12. What is the difference between a cabinet drawing and a cavalier drawing?
13. Which face of an object should you draw first when you use a CAD system to create an oblique drawing? Why?
14. Describe the general procedure for drawing a circle on an oblique drawing. Explain why the procedure for the front face differs from that for the other faces of the object.
15. Surface shading is also called _____.
16. Changing details on a photograph is called _____.
17. Which type of pictorial drawing is most natural in appearance?
18. Which type is least natural in appearance?
19. What is the *vanishing point* in a pictorial drawing?
20. What two factors affect how an object looks in perspective?
21. A small spray gun used for rendering is called an _____.
22. What is another name for two-point perspective? Why is it sometimes called this?

CHAPTER 12
PROBLEMS

The small CAD symbol next to each problem indicates that a problem is appropriate for assignment as a computer-aided drafting problem and suggests a level of difficulty (one, two, or three) for the problem.

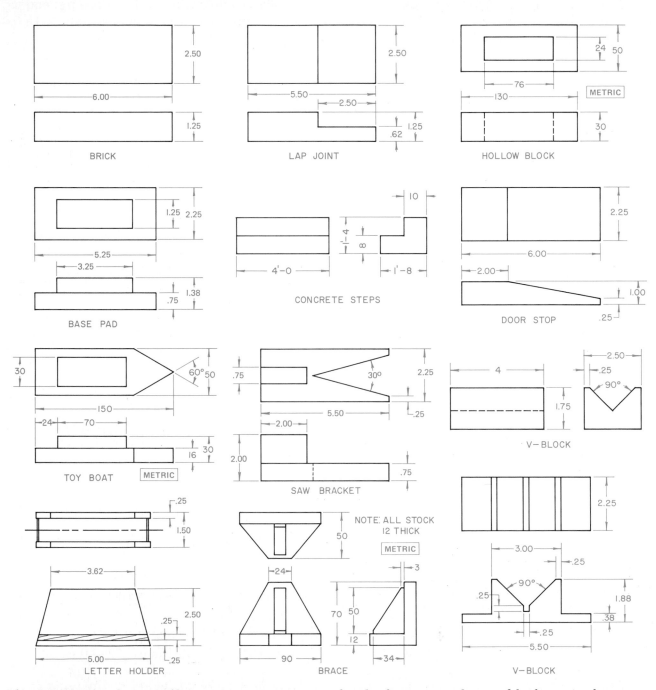

Fig. 12-78 *Isometric drawing problems. Determine an appropriate scale and make an isometric drawing of the object assigned.*
NOTE: These problems may also be used for oblique and perspective drawing practice.

PROBLEMS

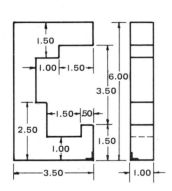

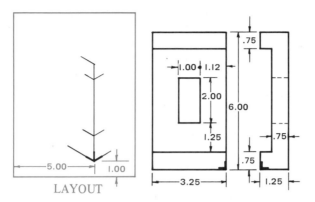

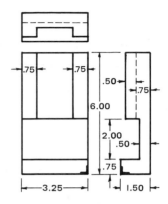

Fig. 12-79 *CAD2 Make an isometric drawing of the plate. Start at the corner indicated by thick lines.*

Fig. 12-80 *CAD2 Make an isometric drawing of the notched block. Start at the corner indicated by thick lines.*

Fig. 12-81 *CAD2 Make an isometric drawing of a babbitted stop. Start at the corner indicated by thick lines.*

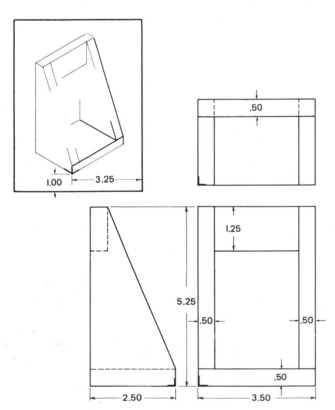

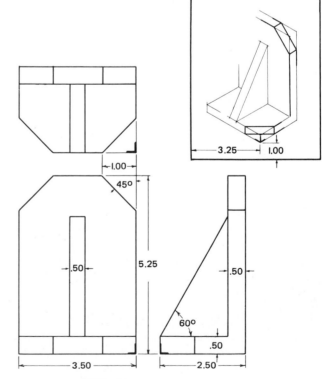

Fig. 12-82 *CAD2 Make an isometric drawing of the stirrup. Start the drawing at the lower left. Note the thick starting lines.*

Fig. 12-83 *CAD2 Make an isometric drawing of the brace. Start the drawing at the lower right. Note the thick starting lines.*

PROBLEMS

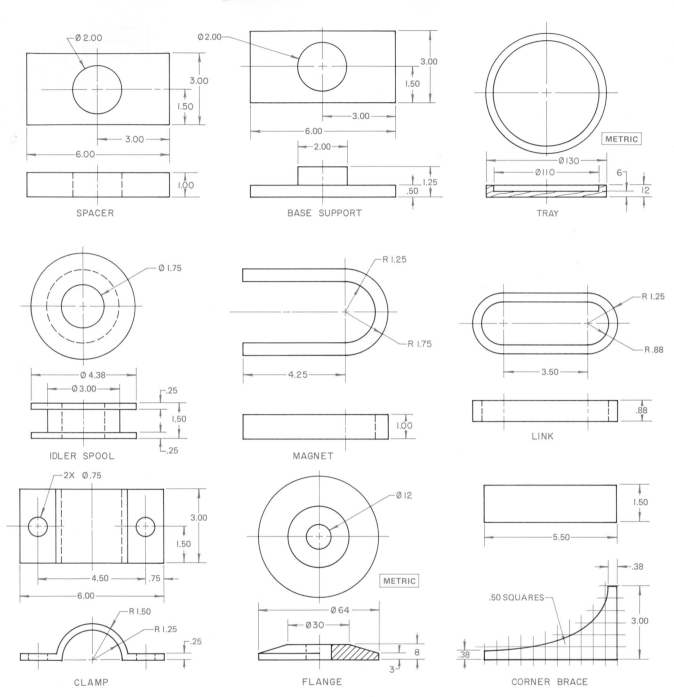

Fig. 12-84 *CAD2 Isometric drawing problems. Determine an appropriate scale for each drawing.*

Assignment 1: Make an isometric drawing of the object assigned.

Assignment 2: Make an isometric half or full section as assigned. Do not dimension unless instructed to do so. NOTE: These problems may also be used for oblique and perspective drawing practice.

PROBLEMS

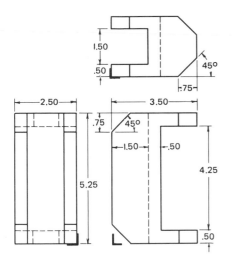

Fig. 12-85 *CAD3* *Make an isometric drawing of the cross slide. Use the layout of Fig. 12-82.*

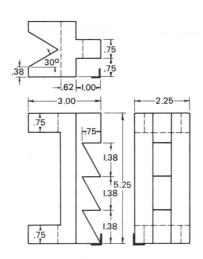

Fig. 12-86 *CAD3* *Make an isometric drawing of the ratchet. Use the layout of Fig. 12-82.*

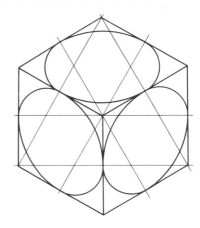

Fig. 12-87 *CAD3* *Make an isometric drawing of a 3″ cube with an isometric circle on each visible side.*

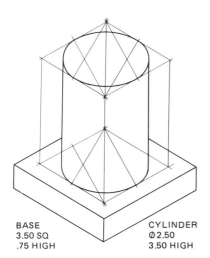

BASE
3.50 SQ
.75 HIGH

CYLINDER
Ø2.50
3.50 HIGH

Fig. 12-88 *CAD3* *Make an isometric drawing of a cylinder resting on a square plinth (square base).*

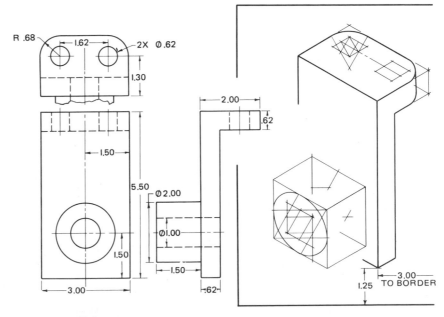

Fig. 12-89 *CAD3* *Make an isometric drawing of the hung bearing. Most of the construction is shown on the layout. Make the drawing as though all corners were square, and then construct the curves.*

PROBLEMS

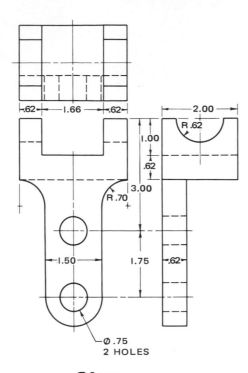

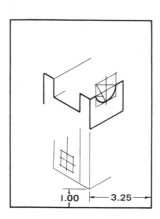

Fig. 12-90 *CAD3 Make an isometric drawing of the bracket. Some of the construction is shown on the layout. Make the drawing as though the corners were square, and then construct the curves.*

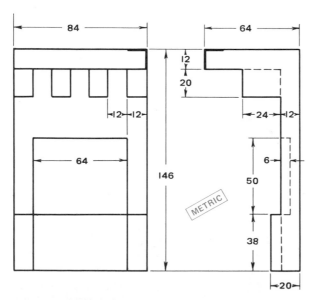

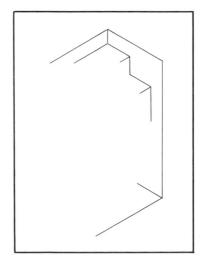

Fig. 12-91 *CAD3 Make an isometric drawing of the tablet. Use reversed axes. Refer to the layout to the right.*

PROBLEMS

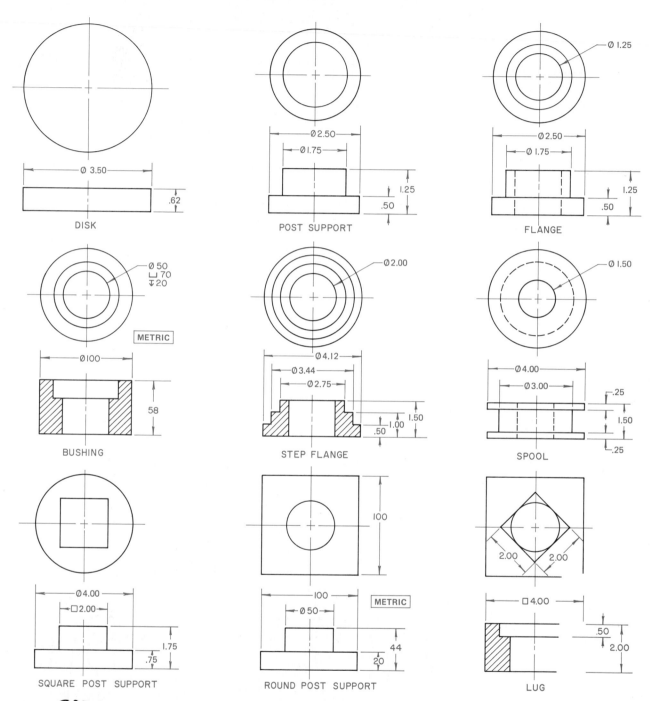

Fig. 12-92 CAD3 *Oblique drawing problems. Determine an appropriate scale for each drawing.*

Assignment 1: Make an oblique drawing of the object assigned.

Assignment 2: Make an oblique half or full section as assigned. Do not dimension unless instructed to do so. NOTE: These problems may also be used for isometric and perspective drawing.

PROBLEMS

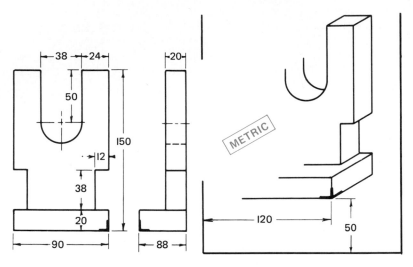

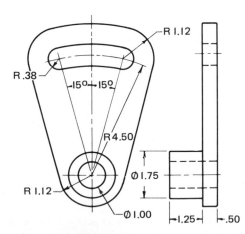

Fig. 12-93 *CAD3 Make an oblique drawing of the angle support.*

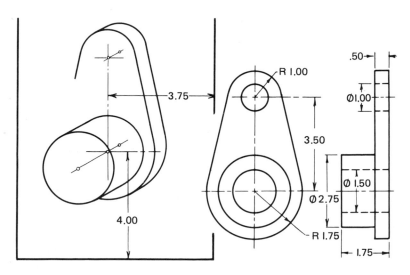

Fig. 12-94 *CAD3 Make an oblique drawing of the crank.*

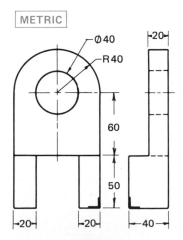

Fig. 12-95 *CAD3 Make an oblique drawing of the forked guide.*

Fig. 12-96 *CAD3 Make an oblique drawing of the slotted sector.*

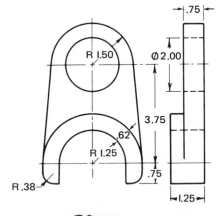

Fig. 12-97 *CAD3 Make an oblique drawing of the guide link.*

PROBLEMS

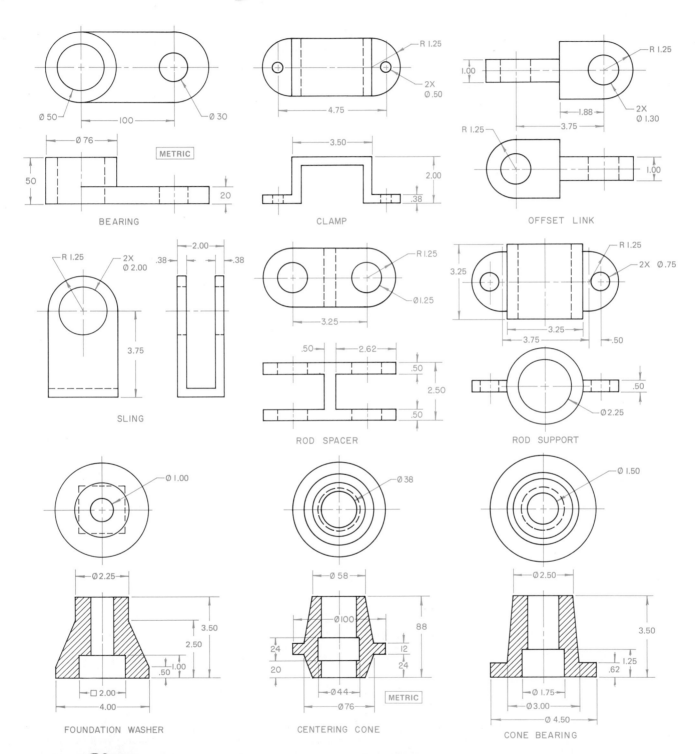

Fig. 12-98 *CAD3 Oblique drawing problems. Determine an appropriate scale for each drawing.*

Assignment 1: Make an oblique drawing of the object assigned.

Assignment 2: Make an oblique half or full section as assigned. Do not dimension unless instructed to do so. NOTE: These problems may also be used for isometric and perspective drawing practice.

PROBLEMS

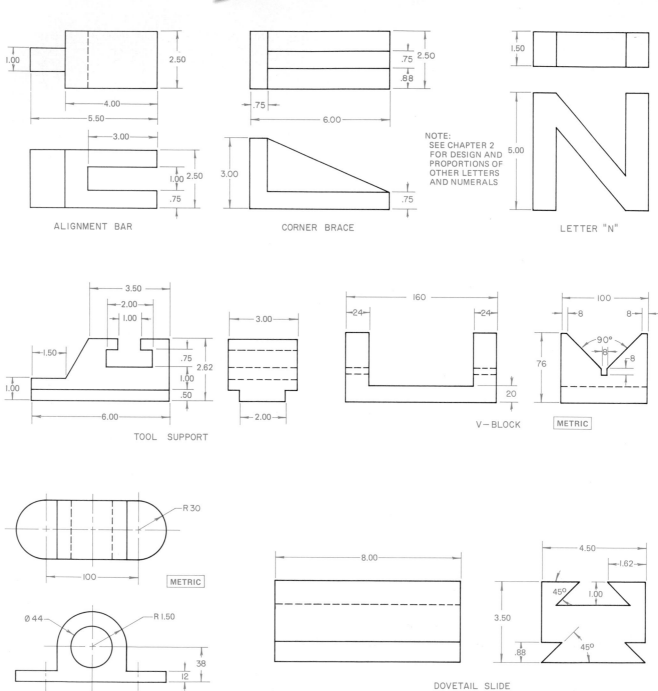

ALIGNMENT BAR

CORNER BRACE

NOTE:
SEE CHAPTER 2
FOR DESIGN AND
PROPORTIONS OF
OTHER LETTERS
AND NUMERALS

LETTER "N"

TOOL SUPPORT

V—BLOCK

METRIC

BEARING

METRIC

DOVETAIL SLIDE

Fig. 12-99 *CAD2 Perspective drawing problems. Make a one-point perspective or a two-point perspective drawing of the object assigned. Use light, thin lines for the construction lines and do not erase them. Brighten the finished perspective drawing lines. Use any suitable scale. The locations of the PP, HL, GL, etc., are optional.*

PROBLEMS

Fig. 12-100 *CAD3 Pictorial sketching and drawing.*

Assignment 1: Most technical illustrations are basically pictorial line drawings. In order to develop a good understanding of the relationship of the various types, make an isometric, oblique-cavalier, oblique-cabinet, single-point perspective, and two-point perspective sketch of the tool support in Fig. 12-99. Determine an appropriate scale.

Assignment 2: Make instrument drawings of the same tool support (Fig. 12-99) in isometric, oblique-cavalier, oblique-cabinet, single-point perspective, and two-point perspective. Determine an appropriate scale. Compare the sketches with the instrument drawings. Are they similar? Which type of pictorial drawing gives the most natural appearance?

Fig. 12-101 *CAD3 Exploded-view drawings. Make an isometric exploded-view drawing of the letter holder in Fig. 5-75. Determine an appropriate scale. Draw your own initials as an overlay .03" (1 mm) thick. Estimate the height and width.*

Fig. 12-102 *CAD3 Surface shading. Make a pictorial line drawing of the toy boat in Fig. 12-78. Redesign as desired, and determine an appropriate scale. Refer to Fig. 12-67 and render the boat using the technique shown in (C) or (D). Maintain sharp, clean lines for contrast.*

Fig. 12-103 *CAD3 Pictorial assembly drawing. Make a pictorial assembly drawing of the trammel in Fig. 11-21. Number the parts and add a parts list for identification. Do not show sectional views. Determine an appropriate scale, and redesign as desired. Add outline shading for accent if instructed to do so. Ink tracing is optional.*

Fig. 12-104 *CAD3 Identification illustration. Make an oblique-cavalier drawing of the mini-sawhorse in Fig. 5-78. Add part numbers and a parts list to make an identification illustration. Determine an appropriate scale. Render if required. Trace in ink if required.*

Fig. 12-105 *Airbrush rendering. Make two-point perspective line drawings of several basic geometric shapes (solids). Examples: cube, cylinder, sphere, cone. Determine an appropriate scale. Transfer each to a piece of hot-pressed illustration board and render in watercolor, using an airbrush.*

Fig. 12-106 *Scratchboard rendering. Make a two-point perspective drawing of the knife rack in Fig. 5-76. Determine an appropriate scale. Transfer the line drawing to a piece of scratchboard coated with india ink. Use any sharp instrument to scratch through the ink to expose the desired lines.*

Fig. 12-107 *CAD2 Exploded-view drawing. Make an isometric exploded-view drawing of the tic-tac-toe board in Fig. 5-81. Determine an appropriate scale. Use outline shading to add contrast.*

Fig. 12-108 *CAD2 Identification illustration. Make an isometric exploded-view drawing of the note box in Fig. 5-79. Determine an appropriate scale. Add part numbers and a parts list to make an identification illustration. Design your initials as an inlay attached to a circular disk. Redesign the notebox as desired. Render if required.*

Fig. 12-109 *Wash rendering. Make a one-point or two-point perspective drawing of any piece of wood furniture. Determine an appropriate scale. Transfer the line drawing to a piece of cold-pressed illustration board and render it, using watercolor and a watercolor brush. Use a touch of white or black to sharpen the edges. Keep wood grain and other fine detail lines sharp and clean.*

Fig. 12-110 *CAD3*

Assignment 1: Identification illustration. Determine an appropriate scale, and make an isometric assembly drawing of the trammel in Fig. 11-21. Show the full length of the beam and show two complete assemblies of the point, body, and knurled screw on the beam. Estimate any sizes not given. Redesign as desired. Add part numbers and a parts list to make an identification illustration.

Assignment 2: Render the trammel in pencil or transfer the line drawing to hot-pressed illustration board and render in watercolor with an airbrush. Add the part numbers and parts list on an overlay sheet.

Fig. 12-111 *Photo retouching. Obtain a glossy photograph of a simple machine or machine part. Study the individual features of the object and list all imperfections. Retouch the photograph using a hand watercolor brush or an airbrush, or both. If possible, have two copies of the original photograph so that a before and after comparison can be made.*

Fig. 12-112 *Airbrush rendering. Make a two-point perspective drawing of the hammer head in Fig. 5-82. Determine an appropriate scale. Transfer the line drawing to a piece of hot-pressed illustration board and render it in watercolor, using an airbrush. Add a touch of white or black to edges for contrast.*

Fig. 12-113 *Surface shading. Make a single-point perspective drawing of one of your own initials. Determine an appropriate scale. Render it using the technique shown in Fig. 12-67C or D. NOTE: This problem may also be used for practice in wash rendering or airbrush rendering. For wash rendering, transfer the line drawing to cold-pressed illustration board. For airbrush rendering, transfer the line drawing to hot-pressed illustration board.*

PROBLEMS

Fig. 12-114 *CAD2 Exploded-view drawing. Make an isometric exploded-view drawing of the garden bench in Fig. 5-80. Determine an appropriate scale. Use Ø.50 threaded rods, flat washers, and hex nuts to fasten the parts together. If assigned, add part numbers and a parts list to make an identification illustration. NOTE: This problem may also be used for practice in wash rendering.*

Fig. 12-115 *Surface shading. Make a two-point perspective drawing of the V-block in Fig. 12-78. Determine an appropriate scale. Refer to Fig. 12-68 and render the line drawing, using the technique assigned. Keep all edges crisp and sharp.*

Fig. 12-116 *Wash rendering. Make a two-point perspective drawing of a house. Scale: 1:50 or 1:100 ($\frac{1}{4}'' = 1'$-0 or $\frac{1}{8}'' = 1'$-0). Transfer the line drawing to a piece of cold-pressed illustration board and render it, using watercolor and a watercolor brush.*

Fig. 12-117 *Scratchboard rendering. Determine an appropriate scale, and make a two-point perspective drawing of any piece of wood furniture. Examples: coffee table, end table, desk, bench. Transfer the line drawing to a piece of scratchboard coated with india ink. Use any sharp instrument to scratch through the ink to expose the desired lines. See Fig. 12-76.*

Fig. 12-118 *Airbrush rendering. Make a single-point perspective drawing of the object of your choice in Figs. 5-75 through 5-82. Determine an appropriate scale. Transfer the drawing to a piece of hot-pressed illustration board and render it in watercolor, using an airbrush. NOTE: This problem may also be used for practice in outline or surface shading.*

Fig. 12-119 *Wash rendering. Scale: 1:50 or 1:100 ($\frac{1}{4}'' = 1'$-0 or $\frac{1}{8}'' = 1'$-0). Make a line drawing of the front elevation of a house. Transfer the line drawing to a piece of cold-pressed illustration board and render it, using watercolor and a watercolor brush. NOTE: This problem may also be used for practice in surface shading in pencil or charcoal.*

Fig. 12-120 *Surface shading. Make a pictorial line sketch or instrument drawing of the edge protector in Fig. 5-65. Determine an appropriate scale. Add surface shading as assigned by your instructor. Keep all edges sharp and crisp. NOTE: This problem may also be used for practice in airbrush rendering.*

Fig. 12-121 *Design and surface shading. Design and make a pictorial sketch of a "car of the future." Determine an appropriate scale. Render in pencil.*

Fig. 12-122 *Airbrush rendering. Make a pictorial assembly drawing of the coupler in Fig. 11-19. Determine an appropriate scale. Transfer the drawing to hot-pressed illustration board. Render in watercolor, using an airbrush. Add a touch of white or black to keep the edges crisp and sharp.*

SCHOOL-TO-WORK

Solving Real-World Problems

Fig. 12-123 *SCHOOL-TO-WORK problems are designed to challenge a student's ability to apply skills learned within this text. Be creative and have fun!*

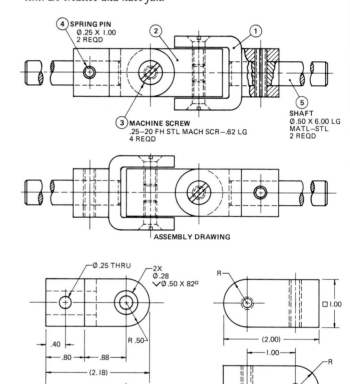

Fig. 12-123 *CAD3 Universal joint. An assembly drawing of a universal joint is shown in Fig. 12-123, along with detail drawings of the yoke and swivel. Standard stock parts such as screws and pins are shown only on the assembly drawing.*

This product is now ready to be manufactured and sold. However, a pictorial drawing is needed to show the customer the relationship of parts for assembly.

Complete the following on an A or B size drawing sheet:

1. *Make an exploded isometric assembly drawing showing all parts in their correct relative positions for assembly. Add part numbers.*

2. *Include a parts list with quantity, item name, material, description (material size, etc.) and part numbers.*

3. *Render using the shading technique of your choice.*

13

DESKTOP PUBLISHING WITH CAD

OBJECTIVES

Upon completion of Chapter 13, you will be able to:

○ Explain the role of desktop publishing in industry.
○ Describe how desktop publishing skills can enhance the value of a CAD drafter on the engineering team.
○ Define the basic terminology used in desktop publishing.
○ List the basic steps used in creating a desktop-published document.
○ Understand the benefits of using templates.

VOCABULARY

In this chapter, you will learn the meanings of the following terms:

- camera-ready copy
- desktop publishing
- hanging indent
- indent
- kerning
- leading
- macros
- pixels
- service bureaus
- tag
- template
- tracking
- type
- type family
- typeset

The role of drafters continues to change for successful companies who must compete with engineering teams from all over the world. Today, many drafters are using *desktop publishing* computer software to create documents that contain CAD drawings. These documents include proposals, reports, and manuals. Drafters who are in the greatest demand know how to use both CAD and desktop publishing software programs (Fig. 13-1).

Fig. 13-1 *Many of the documents companies produce to support their products include CAD drawings.*
(Courtesy of Keith Berry)

WHAT IS DESKTOP PUBLISHING?

Desktop publishing is the process of producing **typeset** pages, which include text, illustrations, and photographs arranged in an attractive format that is ready for printing. The publishing process involves three major activities: preparation, page makeup, and printing (Fig. 13-2). Preparation includes the writing, illustrating, and photography work that makes up the content of a document. Page makeup is the process of assembling this material onto a page. When page makeup occurs on a personal computer, it is called **desktop publishing.**

Printing in desktop publishing should not be confused with making a paper copy on a laser printer. Printing for desktop publishers involves high-speed machinery that produces many high-resolution pages of a document. The machinery then binds the pages into a book, booklet, or brochure.

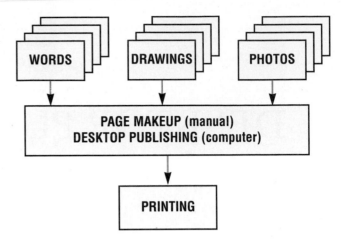

Fig. 13-2 *The publishing sequence.* (Courtesy of Circle Design)

THE ROLE OF DESKTOP PUBLISHING IN INDUSTRY

Before personal computers (PCs), each step of the publishing process was handled by a different person with expensive, specialized skills. Want ads in the publishing industry include positions for writers, editors, proofreaders, illustrators, typesetters, paste-up people, photographers, and press operators. It was nearly impossible for companies to produce their own publications without dedicating resources to a full publishing operation. Therefore, most of those companies that required documentation had to hire outside services or agencies to produce it.

Desktop publishing software automates some publishing activities so that one skilled person can perform the functions of several specialized jobs. This makes in-house publishing a reasonable alternative for many manufacturers who need to create documentation.

The need for documentation varies from company to company. Most companies need marketing materials such as brochures and product advertisements. Manufacturers must provide user's manuals to accompany their products (Fig. 13-3).

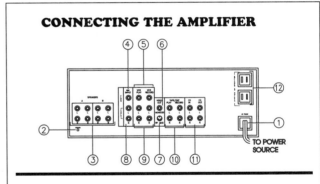

Fig. 13-3 *User's manuals explain to customers how to connect or assemble a company's product.* (Courtesy of Jody James)

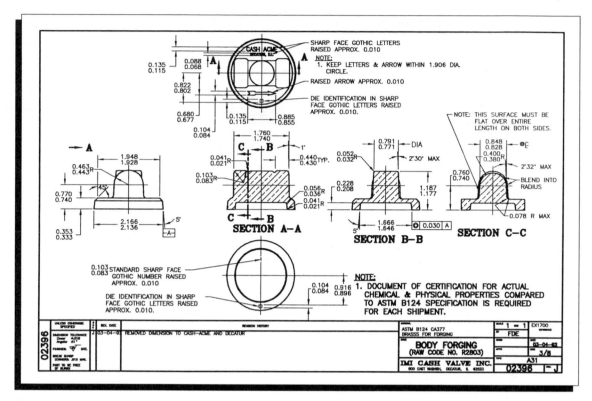

Fig. 13-4 *Using desktop publishing, these working drawings can be included in an electronic "paper trail" that documents the development and production of a product.*

In addition to these published materials, many companies maintain internal documentation for operation of machinery, safety procedures, annual reports, employee benefits, and so on. Engineering firms create and maintain a "paper trail" for products that begin with the initial design sketches and follow through production (Fig. 13-4).

Desktop publishing provides an excellent solution to these and most other documentation needs. Industry has found that desktop publishing has several advantages over traditional publishing methods:

- It is much less expensive than traditional publishing methods.
- It can be done "in-house," which decreases the number of people outside the company that see confidential information.
- It gives the company more control over the appearance of the material being printed.
- It allows the company to produce documentation much more quickly.
- It allows actual design sketches and CAD drawings to be incorporated into the documents, reducing the chance of error as well as duplication of effort.

Desktop publishing specialists create such documents on personal computers. Desktop publishing software allows a skilled operator to create documents quickly and at low cost. For companies who must compete in a world economy, this capability can make the difference between profit and loss.

For desktop publishing to be successful in industry, companies must commit to the proper resources. Simply buying a personal computer and desktop publishing software and then hiring someone to run it is not enough. The company must make sure that the person in charge of desktop publishing understands not only the documentation needs of the company, but also the typesetting standards and rules that must be followed to produce professional-looking documents. Without this understanding, the documentation may be a poor reflection on the company.

The remainder of this chapter explains how desktop publishing works. It also provides a quick overview of the knowledge and skills needed to create documents in an industrial setting.

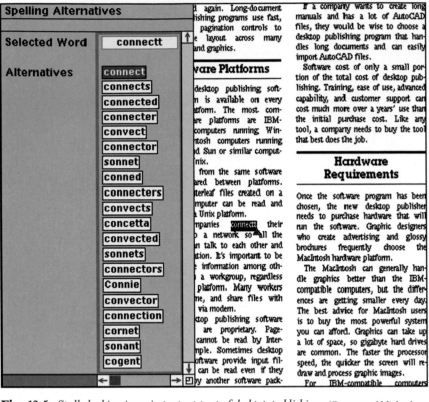

Fig. 13-5 *Spell checking is an important part of desktop publishing.* (Courtesy of Michael Eleder, Instructional Design Systems)

word processing programs check for spelling, report on the writing style, and can even determine the reading level of the text (Fig. 13-5). The main function is to help the writer manipulate words to create paragraphs that are easy to read and understand.

Large companies may employ technical writers to create the text for their documentation. Smaller companies often hire a single person who must write the text, edit it, and incorporate the appropriate illustrations using desktop publishing. In either case, the employee should be very familiar with the word processing software being used to take full advantage of its features.

CREATING FILES FOR DESKTOP PUBLISHING

Desktop publishing operators typically assemble pages from existing word processing, CAD, or graphics files. These files are created by writers and artists using word processing, computer drawing, and CAD programs. Some desktop publishing programs provide integrated word processing, drawing, and image editing capabilities, making other software unnecessary. At many large companies, teams of technical writers and illustrators create entire manuals using the word processing and illustrating capability of the advanced desktop publishing programs.

Word Processing Files

Words, sentences, and paragraphs are usually created by writers and modified by editors using word processing programs. WordPerfect®, Microsoft® Word, and IBM/Lotus Word Pro® are common word processing programs. The better

Using Macros

Most word processing software can be customized using **macros** (short scripts that accomplish tasks that will be needed over and over again). Although some macros can be complicated and difficult to create without programming knowledge, other macros take seconds to create and can save the employee—and the company—a great amount of time. Some word processors allow you to create a macro just by "recording" key strokes as you perform the task. By saving these key strokes as a macro, you can later perform the same task by pressing just a couple of keys on the keyboard.

Typesetting Rules

There are many special typesetting features that desktop publishing users need to learn about. In fact, entire courses are available that teach these specialized concepts. You should know about the typesetting rules that apply to text before you begin to create the word processing document.

in-line .... a hyphen dash
1998–99 .... an EN dash
for you—for me .. an EM dash
5 – 2 = 3 .... minus (math)

Fig. 13-6 *Examples of four different types of dashes.* (Courtesy of Circle Design)

12345 .... no space
12345 .... hairline space
12345 .... normal space
12345 .... EN space
　12345 .... EM space

Fig. 13-7 *Examples of different types of spaces used in desktop publishing.* (Courtesy of Circle Design)

HAIRLINE
0.5 POINT
0.75 POINT
1 POINT
1.5 POINTS
2 POINTS
3 POINTS
4 POINTS
6 POINTS

Fig. 13-8 *Common line thicknesses in desktop publishing.* (Courtesy of Circle Design)

in desktop publishing, just like there are in CAD. However, in desktop publishing, their thickness is measured in points (Fig. 13-8).

One mistake made by many people who are new to desktop publishing is the use of two spaces after a period. This was necessary on typewritten documents to aid readability. However, desktop publishing software controls the space after a period automatically. Typesetting requires only one space after the period. If you enter two spaces on the keyboard, the sentences in the document will be separated by too much white space.

Most word processing programs have the ability to create special typographic effects such as the dashes and spaces described above. If you do not use them in the original text file, you will have to go through the entire document in the desktop publishing software and change them.

CAD and Other Illustration Files

In engineering and design firms, many drawings exist within CAD *databases* (organized collections of CAD drawings): design sketches, working drawings, exploded assembly drawings, and so on (Fig. 13-9). These drawings are created by skilled CAD operators or technical illustrators using common CAD programs such as AutoCAD and CADKEY (Fig. 13-10).

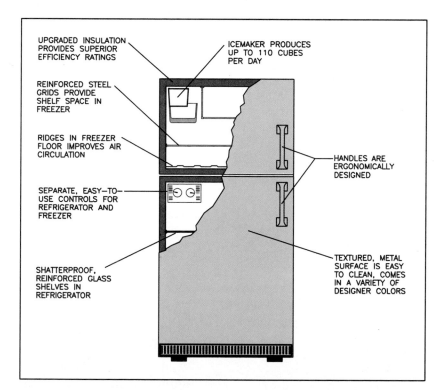

UPGRADED INSULATION PROVIDES SUPERIOR EFFICIENCY RATINGS

ICEMAKER PRODUCES UP TO 110 CUBES PER DAY

REINFORCED STEEL GRIDS PROVIDE SHELF SPACE IN FREEZER

RIDGES IN FREEZER FLOOR IMPROVES AIR CIRCULATION

SEPARATE, EASY-TO-USE CONTROLS FOR REFRIGERATOR AND FREEZER

SHATTERPROOF, REINFORCED GLASS SHELVES IN REFRIGERATOR

HANDLES ARE ERGONOMICALLY DESIGNED

TEXTURED, METAL SURFACE IS EASY TO CLEAN, COMES IN A VARIETY OF DESIGNER COLORS

Fig. 13-9 *This broken-out section of a refrigerator-freezer has been adapted for use as a marketing tool. Using the original CAD section avoids redrawing time and helps prevent errors.* (Courtesy of Jody James)

For example, typesetting has several different kinds of dashes (Fig. 13-6) and several types of spaces (Fig. 13-7). To produce professional documents, you should learn when to use each type of dash and space. There are also many types of lines

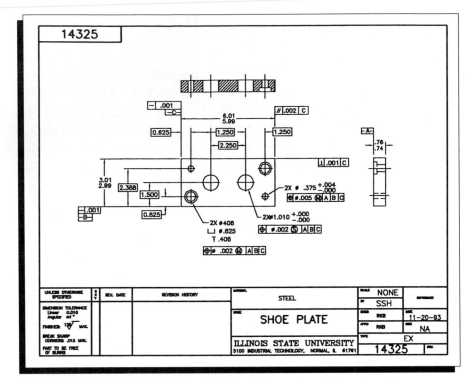

Fig. 13-10 *This CAD drawing, created in AutoCAD, can be used in various company documents with a desktop publishing system.*

In addition to CAD files, companies that produce documentation may need more artistic drawings of their products. In some cases, these can be created using the CAD software, but more often an illustration program is used. MacDraw, Corel-DRAW, and Adobe Illustrator are common examples of illustration packages for personal computers (Fig. 13-11). When a CAD drawing needs to be enhanced for marketing, the illustrator can import the CAD drawing into the illustration program and create the desired artistic effects (Fig. 13-12). To create other supporting art, such as simple line drawings or block diagrams, the illustrator usually begins a new drawing.

Most illustration programs store drawings in a *raster* format. That is, geometry is defined using the tiny dots called **pixels** (picture elements) on the computer screen. In CAD programs, geometry is defined using a set of mathematical instructions. This type of line art is said to be in a *vector* format. The desktop publishing operator must understand these differences and the processes needed to convert files of one type to the other without losing drawing information.

Fig. 13-11 *A typical drawing created using an illustration program.*

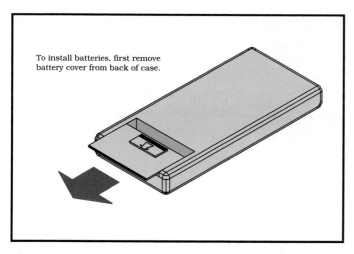

Fig. 13-12 *Illustration programs can be used to enhance CAD illustrations to help make a point, as shown in this excerpt from an owner's manual.* (Courtesy of Jody James)

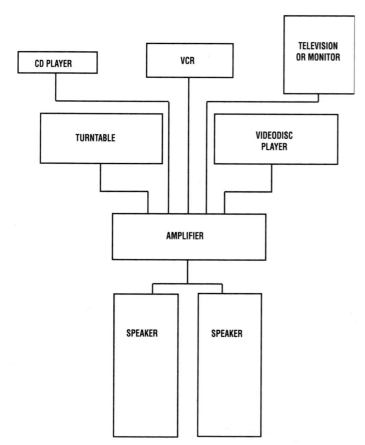

Fig. 13-13 *An example of a drawing in a technical manual that does not require the use of CAD art.* (Courtesy of Jody James)

Photographic Files

Some industrial documentation, such as promotional brochures and annual reports, may require photographs as well as CAD drawings (Fig. 13-13). Photographs are images that contain continuously varying shades of gray or color. To import a photograph into a desktop system, you can scan them into a computer using a high-resolution scanner. Scanning converts the continuous tones on a photograph into pixels (Fig. 13-14). The more dots per inch the scanner can produce, the more realistic the computer photo will appear. Photographs that have been scanned can then be manipulated using image editing software programs such as Adobe Photoshop or Micro-Grafx Designer.

File Formats

Many types of file formats are available to the desktop publisher. For example, every CAD program creates its own file format. In order for a desktop publishing program to read a file, the file must be converted into a format that the desktop publishing program recognizes. Identifying and converting file formats is an important part of the desktop publisher's job.

Fig. 13-14 *Scanned image of state-of-the-art word processing circa 1907.* (Courtesy of Michael Eleder, Instructional Design Systems)

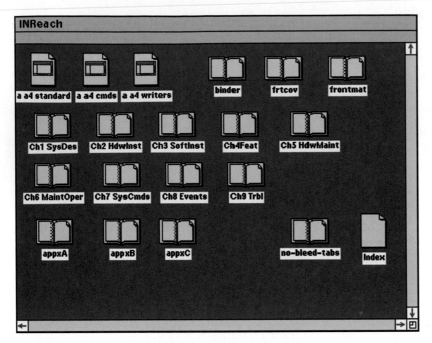

Fig. 13-15 *Advanced desktop publishing software controls formatting across many chapters in a book.* (Courtesy of Michael Eleder, Instructional Design Systems)

DESKTOP PUBLISHING SYSTEMS

A desktop publishing system consists of a combination of software and hardware. Both the software and the hardware requirements are somewhat different from those for CAD systems.

Software

A variety of desktop publishing software is available. Prices range from very inexpensive to very expensive, depending on the capability of the software. For industrial documentation, you must be sure that the software can perform the tasks needed to create technical documents.

Low-cost software such as Microsoft Publisher provides basic layout capabilities, such as those necessary to produce a church flyer or club newsletter. These programs can import text and pictures, but the file formats they recognize are often limited, and they have fewer user-definable properties. They are not as precise as some of the more expensive software packages, and they are generally not used in industry.

More expensive software, such as Adobe PageMaker®, Quark Xpress™, FrameMaker™, Interleaf™, and Corel Ventura®, offer much more extensive page makeup and layout tools (Fig. 13-15). They provide very precise control of the positioning of text and graphics. They include advanced color capabilities and can prepare a 4-color piece for offset printing. Many print shops create printing plates directly from these files, eliminating intermediate production steps.

When selecting a desktop publishing software program, it is important to consider what type of documents you need to create and what types of files you wish to import. For companies that want to create manuals that contain a lot of AutoCAD files, it would be wise to choose a desktop publishing program that can easily import AutoCAD files.

Like any tool, the desktop publishing program selected should be the one that meets the company's needs. Remember that the actual software costs only a small portion of the total cost of desktop publishing. Equipment, materials and supplies, and especially training can cost much more over a year's use than the initial purchase cost. Therefore, it's a good idea to look for a program that both meets the company's requirements and is relatively easy to learn and use.

Hardware

The hardware needed for a typical desktop publishing system includes the basic computer, a large-screen monitor, a high-resolution scanner, and a laser printer. Some systems also include high-end color printers to allow the operator to print color proofs of documents. In addition, if the company creates very large documents with many graphics files, a removable disk drive system (such as Syquest or Bernoulli) provides a method of transporting the finished document to the printer.

Hardware Platforms

Not every desktop publishing software program is available on every hardware platform. The most common hardware platforms are IBM-compatible computers running Windows, Macintosh computers running System 7, and Sun or similar computers running Unix.

Files created using the same software can often be shared among platforms. For example, FrameMaker and Interleaf files created on a Windows computer can be read and modified in FrameMaker or Interleaf on a Unix platform.

Most companies connect their computers to a network so that all the computers can communicate and share information. Many operators work at home and share files with other operators via modem. It's important to be able to share information among people in a work group, regardless of hardware platform. Remember, however, that although a company network can allow you to access files created on other platforms, in many cases you must still convert the files to a format usable in the company's desktop publishing system.

Hardware Requirements

The type of computer a company dedicates to desktop publishing should depend on the types of files being generated elsewhere throughout the company, the types of computers already in use, and the documentation needs of the company. Traditionally, IBM-compatible computers are easier to use when CAD files are to be included in the documentation. This is because most CAD files are created on either the IBM or the Unix platform. Although CAD files can be converted across platforms, it is much easier to remain within the original platform.

On the other hand, graphic designers who create advertising and glossy brochures frequently choose the Macintosh hardware platform. The Macintosh can generally handle graphics better than IBM-compatible computers, but the differences are rapidly getting smaller.

Whichever platform the company chooses, the best advice for buying desktop publishing hardware is to buy the most powerful system you can afford. Graphics can take up a lot of space, so the hard drive must be very large. The faster the processor speed, the quicker the screen will redraw and process graphic images. A high-end *graphics accelerator,* which increases the speed with which the computer processes graphic images, is also a good idea. In addition, many desktop publishing and illustration programs require a large amount of *RAM* (random access memory within the computer) to function properly. A large-screen monitor is almost a necessity for page layout because it can display a full page at one time. Most desktop systems have at least 20- or 21-inch monitors.

USING DESKTOP PUBLISHING SOFTWARE

Every desktop publishing software uses different command procedures. There are, however, many functional similarities among programs. Learning the basics of one program makes it easy to learn new programs.

Document Templates

Before placing words, illustrations, and photographs on the page, the desktop publishing operator usually applies a template. A **template** is a special blank page that controls the layout of the page.

Templates control typographic elements such as the number of columns, the size and style of the body type, the size and style of the headings, the size of the margins, the space between columns, and the spacing between letters, words, lines, and paragraphs. Templates make sure that similar documents for a company have a consistent look.

Companies that produce documentation often use templates that include their corporate identity. The corporate identity is the unique style of business documentation, including business cards, letterhead, envelopes, reports, brochures, advertisements, proposals, catalogs, and the company logo (Fig. 13-16). The corporate identity also includes the look of on-line documents such as CD-ROMs and Internet Web home pages.

Fig. 13-16 *A company logo, created using CAD or an illustration program, helps identify a company to potential clients.*

Most companies use a different template for each type of documentation they produce. When the desktop publisher creates an annual report, for example, he or she uses the annual report template.

Templates ensure consistency among documents because they enforce predetermined *formatting* rules on every file that is imported into the desktop publishing program. When desktop publishing programs first appeared on personal computers, unskilled operators tried to use too many features and cluttered the page with too many typographic variations. Today, skilled desktop publishers create templates to make sure page styles are attractive (Fig. 13-17). Templates make it easy to create attractive documents with minimal effort (Fig. 13-18).

Templates contain layout rules for the variable design elements of desktop publishing. These variable elements include type properties, paragraph properties, and page properties.

A

B

Fig. 13-17 *Without an understanding of design principles and typesetting techniques, a new desktop publishing operator may create designs like the example in (A). This is an example of poor design and page layout. The page would look much better as shown in (B), with fewer fonts and more color coordination.*

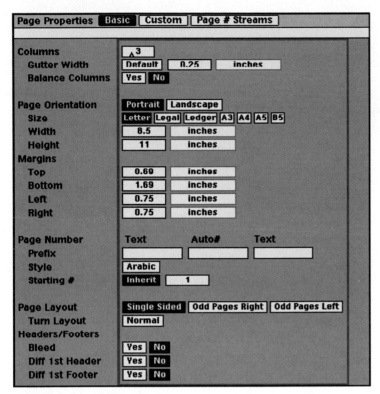

Fig. 13-18 *Typical page property controls that can be customized with desktop publishing software.* (Courtesy of Michael Eleder, Instructional Design Systems)

Defining Properties

Desktop publishing software controls the look of text, paragraphs, and pages with properties that can be defined by the user. Each type of property controls a different part of a document. You can apply properties separately for each document, or you can store them in template files so that they can be applied automatically to any document associated with the template.

Type Properties

Type refers to the letters and other characters as they appear on the page. Type is specified by family, font, and size. The **type family** describes the general appearance of the type. Some families contain special foreign letters, and some contain no letters at all (Fig. 13-19).

The **font** is the detailed appearance of the type within that family. Roman (normal), bold, and italic are the three most common fonts within a type family (Fig. 13-20).

Fig. 13-19 *Common type families.* (Courtesy of Michael Eleder, Instructional Design Systems)

The Helvetica Family of Type

Helvetica roman font
Helvetica bold font
Helvetica italic font
Helvetica bold/italic font
Helvetica black font
Helvetica light font
Helvetica narrow font

The Times Family of Type

Times roman font
Times bold font
Times italic font
Times bold/italic font

Fig. 13-20 *Two font families.* (Courtesy of Michael Eleder, Instructional Design Systems)

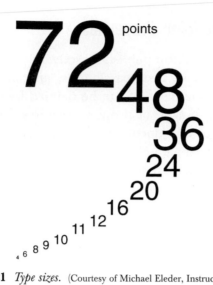

Fig. 13-21 *Type sizes.* (Courtesy of Michael Eleder, Instructional Design Systems)

This paragraph is set using 10-point Helvetica on 11-point leading.

leading

This paragraph is set using 10-point Helvetica on 15-point leading.

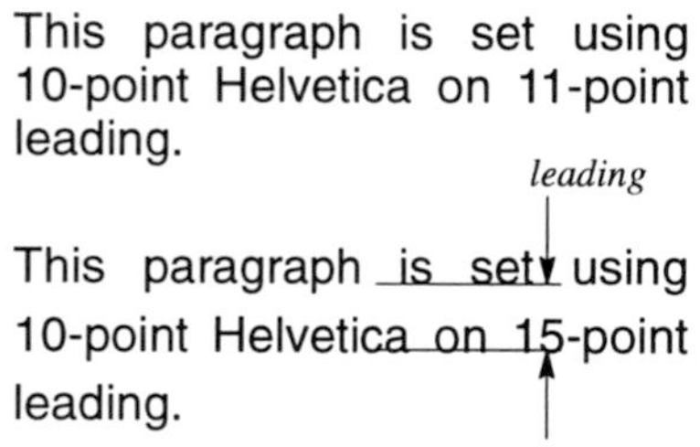

Fig. 13-22 *Two types of leading for a 10-point type size.*
(Courtesy of Michael Eleder, Instructional Design Systems)

Size is another characteristic of type. Type is sized in points. There are approximately 72 points per inch. Paragraphs often use 10- or 11-point type. Titles and headings are usually larger (Fig 13-21).

Desktop publishing programs allow users to control the spacing between letters and words. **Kerning** is the name for the horizontal space between letters, and **tracking** is the name for the horizontal spacing between words.

Another spacing control for type is the vertical spacing between lines called **leading.** Typewriters are usually limited to single- and double-spaced lines for the vertical spacing control between lines. Desktop publishing has much finer control of this spacing.

Leading is measured in points from the bottom of one line to the bottom of the next (Fig. 13-22). Type size and leading are typically specified as a fraction. For example, 11-point type with 12-point leading is specified as 11/12, and is called "eleven on twelve."

When the leading is the same size as the type, the type can be difficult to read. As leading increases, it becomes easier to read. Leading is often one to two points larger than the type size.

Paragraph Properties

Paragraphs have formatting properties that determine how each line begins and ends. Each type of paragraph can be given a **tag,** which controls all the formatting characteristics of the paragraphs to which the tag is assigned.

In the paragraph above, the lines end evenly on the right. That format is called *justified* text. The paragraph you are now reading has lines that are ragged (uneven) on the right. This format is called *flush left.*

This paragraph is ragged both left and right, and is *centered* on the column. That is, each line extends evenly to the right and left of the exact middle of the column. Centered text is harder to read and is usually reserved for titles or headings.

Notice that there is no extra vertical space between this paragraph and the one above it. That's because this paragraph has a top margin of zero.

However, this paragraph has extra space above it. This extra space is the paragraph's top margin. Top and bottom paragraph margins allow the desktop publisher to separate paragraphs from each other or allow the text to flow together. Note that some desktop publishing software refers to this as "space above" and "space below" the paragraph.

This paragraph is justified, but it has an additional left margin of .5 inch. Paragraphs can have additional margins on the left, the right, or both left and right. This is usually done to make text stand out or to distinguish it from the surrounding text. For example, long quotes are often typeset with extra left and right margins.

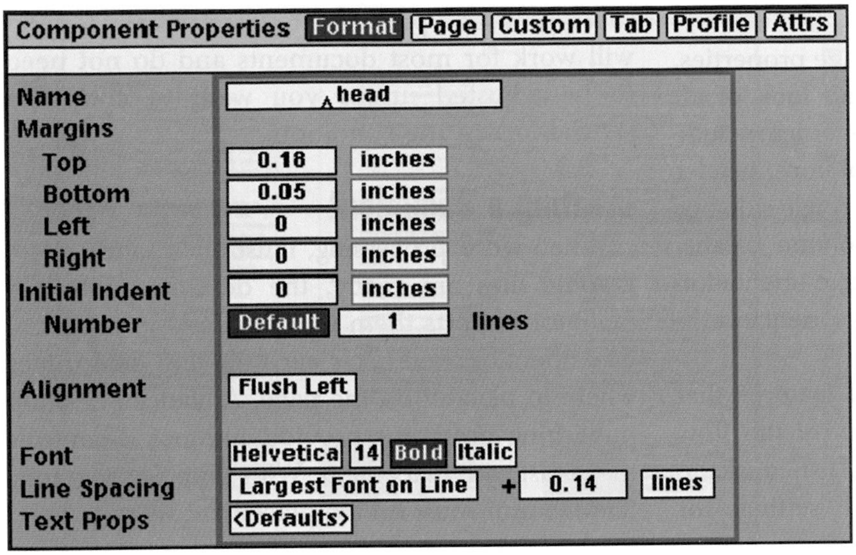

Component Properties [Format] [Page] [Custom] [Tab] [Profile] [Attrs]

Name	△head	
Margins		
Top	0.18	inches
Bottom	0.05	inches
Left	0	inches
Right	0	inches
Initial Indent	0	inches
Number	[Default] 1	lines
Alignment	[Flush Left]	
Font	[Helvetica] [14] [Bold] [Italic]	
Line Spacing	[Largest Font on Line] + [0.14]	lines
Text Props	[<Defaults>]	

Fig. 13-23 *Dialog boxes such as this one allow the operator to control the attributes of paragraph text.* (Courtesy of Michael Eleder, Instructional Design Systems)

Some paragraphs, like this one, have an extra margin on the first line only. This is called an **indent.** When the paragraph has no extra top and bottom margin, the paragraphs must have an indent so that the reader knows where each paragraph starts.

Some paragraphs use an indent that lasts for several lines to make the paragraph more interesting. When a large first letter is placed in the indented space, it is called a *drop cap.*

Another way to make a paragraph stand out is to let the first line hang out to the left, while the rest of the paragraph has an extra left margin. This is called a **hanging indent.** Hanging indents are often used for numbered lists. (See the Learning Activities at the end of this chapter for an example.) Using a hanging indent for numbered or bulleted lists creates a cleaner appearance in the document and makes the text easier to read.

To monitor editing and version control, paragraphs can be given special properties that indicate that changes have been made to the text.

(*Version control* is necessary to ensure that an older file is not saved over a newer file.) The large bar to the left is called a *change bar.* It indicates that changes or additions have been made to this portion of the paragraph since the last time it was published or printed. ~~A line through text indicates that this material will be removed in the next version of the document.~~ It is called a *strikethrough.*

Usually, tags that define paragraph properties are contained in the template and can be applied automatically by assigning the tags to individual paragraphs in the document. For example, the first level heading for a document might have a tag called *head,* and the body text might be called *para.* The tags for each paragraph define the font, leading, font size, and all the paragraph properties (Fig. 13-23).

When a word processing file is imported into a desktop publishing file and assigned a tag, the properties of the text are applied automatically. When the desktop publisher uses predefined tags in the company's templates, the company's documentation becomes more consistent. In addition, it is easier and faster to create documentation using predefined paragraph tags (Fig. 13-24).

Graphics Files

Importing graphics files is similar to importing text files. Use the following generic procedure.

1. Some programs require you to create a graphics box to "hold" the illustration before you import the graphics file. If you are using one of these programs, create a box about the size of the graphic in the location where you want to place the illustration. Make sure the box is "selected" before continuing.

2. Use the File Open command (or Get Picture or Import Graphic File) to select the first graphics file you wish to import.

3. If your program does not require a predefined graphics box, move the editing cursor to the location in the document where you want to place the first graphics file.

4. In some programs, you will need to paste the graphics image using a Paste command. Those that require graphics boxes, however, place the graphic directly into the selected graphics box.

5. Most desktop publishing programs allow for a variety of positioning options. Select the positioning method that you want and position the graphic on the page.

6. Crop (cut down) and size the graphic in its final position.

The Virtual Desktop Page

The *virtual page* that appears on the monitor screen is an electronic representation of what the paper page will look like. The margins, font, columns, headings, headers, and footers appear as a close approximation of the printed page (Fig. 13-26).

It is important to understand why the virtual page is an approximation of the printed page. Most monitor screens have a resolution of only 72 dots per inch (dpi). The printed page on a laser printer can produce between 300 and 1200 dpi. Typesetting image equipment can produce 2400 dpi, resulting in crisper, clearer images on the typeset page.

This means that a line that is $\frac{1}{5}$ of an inch long is exactly 480 dots long on a typeset page (2400 dpi $\times \frac{1}{5}$ inch = 480 dots). But the same line on the screen is represented by 14 dots because 72 dpi $\times \frac{1}{5}$ = 14.4 dots. There is no way to show a line 14.4 dots long on the screen, so the virtual page approximates the length as 14 dots long.

Most of the time this approximation is not a problem. However, some line endings and other alignment conditions may be slightly incorrect on the screen.

Fig. 13-26 *A newsletter as it appears on the virtual page.* (Courtesy of Michael Eleder, Instructional Design Systems)

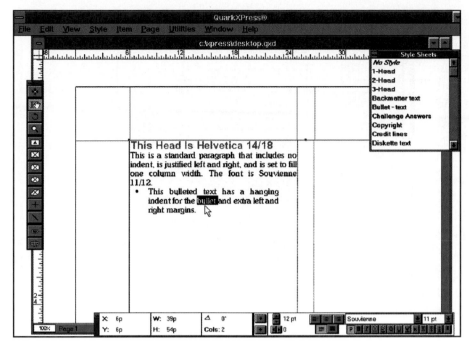

Fig. 13-27 *A sample virtual page on the QuarkXpress® desktop publishing software showing text, tags, and command menus.* (Courtesy of Jody James)

Some desktop publishing programs maintain a dynamic link between the imported text and the original word processing file. This means that when you change the word processing file, the text on the desktop page also changes. Other desktop publishing programs sever all links with the original program. The word processing file must be imported again if changes are made to it using the original word processing program. If your program is one of the latter, it is best to perform as much of the text editing as possible from within the desktop document.

Sophisticated desktop publishing programs provide powerful line drawing and photographic image editing capabilities. Drawings that require changes and scanned photographs that require adjustment can be edited directly from the desktop publishing program. Other desktop publishing programs require the use of separate line drawing and image editing programs to make changes.

The basic parts of the virtual page are shown in Fig. 13-27. The large white area represents the printed paper. The menus are usually located at the top of the screen. Tags are often positioned down the left side of the screen. (In many programs, the position of these elements is user-definable. Desktop publishers usually arrange them to suit their personal tastes.) A small triangle, I-beam shape, or arrow is used as the editing cursor. Changes to text occur at the cursor.

To edit text, you must first select it using the cursor. The highlighted text is usually shown in reverse video (white text against a black background).

Notice in Fig. 13-27 how the headings are larger and bold. The bullet tag is set up for a hanging indent with an extra left margin for the remainder of the text.

Editing Text and Graphics

Some text and graphics will need editing. Some words may be spelled incorrectly. Hyphenation must be monitored. Line endings must be adjusted to create a pleasing appearance on the page.

Checking Document Layout

In addition to editing the text and illustrations in a document, you must check the page layout. You should do this after all the editing has been done because changes to the text or graphics often changes the appearance of the entire page (Fig. 13-28). Follow these steps:

1. Adjust the margins, gutter (space between the columns), word spacing, letter spacing, line leading, and margins around illustrations.

2. Check that odd pages will print as right-hand pages, and even numbers will print as left-hand pages.

3. Make sure none of the text is running off the bottom of the page.

Proofing the Document

When the page is completely formatted, it must be proofed by the original author, by the editor, and by the manager in charge of the project. Sometimes the the text looks OK when it's not. *(Did you catch the error in the previous sentence?)*

In industry, nothing is ever published unless it has been proofed and signed off by a manager. Printing and distribution to thousands of people with an error is costly and makes the company look bad in the eyes of its customers. More importantly, if a mistake in a document causes customers to be unable to use a product without calling the manufacturer for help, the company may lose thousands of dollars due to product returns and decreased sales (Fig. 13-29). Therefore, it makes sense to have more than one person sign off on the documentation before it is printed.

Preparing for the Printer

The final step in the publishing process is to produce a paper or electronic product that the printer can use to print the document. In many cases, companies choose to create **camera-ready copy** for the printer. The printer photographs this paper master and creates printing plates from it. Therefore, the camera-ready copy must be produced on good paper at the highest resolution possible, usually on a laser printer at least 600 dpi.

Some printers make printing plates directly from the electronic file, eliminating a production step. This reduces production time,

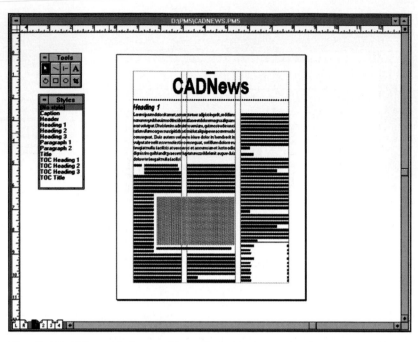

Fig. 13-28 *What might happen if you edited this page so that the text required two more lines in the third column—after you had already checked the page layout? Depending on the software settings, the chart at the bottom would probably jump to the next page, throwing off your page layouts for the next several pages.* (Courtesy of Michael Eleder, Instructional Design Systems)

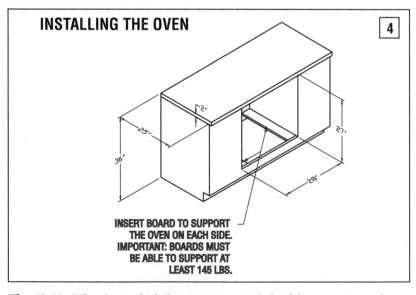

Fig. 13-29 *What do you think the consequences might be if the annotation at the bottom of this illustration erroneously said "BOARDS MUST BE ABLE TO SUPPORT AT LEAST 45 LBS"?* (Courtesy of Jody James)

reduces cost, and improves image quality. Even when you use this method, though, you should still supply a paper copy of the document, printed on a laser printer, for the printer to use as reference. In fact, some printers require it.

Many companies do the page layout in-house and then send the document files to a service bureau for processing. **Service bureaus** are companies that can take completed desktop publishing files and create film for printing. They have specialized equipment that can produce higher-quality output than a standard laser printer. If your company plans to use a service bureau, you should inquire to see which service bureaus in your area can process documents created by your desktop publishing program. Ask them what items they need to complete the job. Many service bureaus require a hard (paper) copy of the document and duplicate or triplicate diskettes. For large manuals and books, find out what type of removable disk drive cartridge their equipment has.

SECURITY AND THE MASTER FILE

Once you have created a desktop publishing master document file, you must keep accurate records about the locations of all files referenced by the master document. You should note any active links to existing word processing or graphics files. You will need this information because the next time the document requires a change, these files must be readily available.

Some companies *archive* finished documents in a special place, such as a safe deposit box or other fire-proof location. Archiving to a safe location is a good idea, especially for companies that would lose important information if a disaster occurred at the company site. Even those companies that don't archive their completed documents back up their files onto a storage medium such as a magnetic tape or recordable CD-ROM (Fig. 13-30).

JOB OPPORTUNITY AND SECURITY

In a competitive world market, only the strongest companies can survive. Employees who can add value and provide needed services are the ones who will be in demand and command the highest wages.

Fig. 13-30 *Removable hard disk drives allow you to store from 44 to 250 megabytes (MB) of data on a single cartridge for transfer to a service bureau.* (Courtesy of Keith Berry)

While few workers in the 21st century will enjoy complete job security, those with the best skills will have the least to fear from recession, international competition, and advancing technology. Desktop publishing is another tool that CAD operators can use to add value to a company and enhance their own career.

ADDITIONAL REFERENCES

One of the best references for operators who are new to desktop publishing is the manual that comes with the software. Become very familiar with these manuals; often they can guide you not only in the capabilities of the software, but also on standard usages and publishing protocols. In addition, desktop publishers use a number of references to improve their style and conform to established standards. The following are a few examples.

Parker, Roger, *Looking Good in Print,* Ventanna Press, 1990. Provides basic guidelines for creating almost any kind of desktop publishing piece.

Parker Roger, *The Makeover Book,* Ventanna Press, 1989. Shows many before and after examples of what looks good and what doesn't.

Felici, James, *The Desktop Style Guide,* Bantam Books, 1991. A brief handbook on the basics of desktop publishing.

Pocket Pal, International Paper, 1989. A primer on typesetting, printing, and paper produced for printing professionals by the International Paper Company, 6400 Popular Avenue, Memphis, TN 38197.

CHAPTER 13
REVIEW

Learning Activities

1. Visit a desktop publishing service bureau and examine the different types of documents they produce. Make a list of the different type of file formats that desktop publishers must use to create documents.

2. Ask permission to monitor the "junk mail" that comes to the school or department office. Group the "junk mail" into categories such as brochures, flyers, reports, etc. Check the "junk mail" at home. Make a list of the different types of documents that could have been created using desktop publishing.

3. Write to a company and ask for its annual report. Make up tag names for each type of heading and paragraph.

4. If you have access to a desktop publishing program and a CAD program, take an attractive three-dimensional CAD drawing and save it in various formats. Import the files into the desktop publishing program. Make a list of which formats work and which ones do not.

5. If you have access to desktop publishing, CAD, and word processing programs, write a letter to a friend or relative using the word processing program. Make a CAD drawing of your classroom. Import the word processing file and CAD file into a desktop publishing program and send the result to your friend or relative as a newsletter.

Questions

Write your answers on a separate sheet of paper.

1. Explain why some CAD operators now use desktop publishing software.

2. What are the three types of files typically imported into desktop publishing programs?

3. What is the difference between programs that use raster graphics and those that use vector graphics?

4. What is the advantage of using a template in a desktop publishing application?

5. What are the three types of properties used in desktop publishing?

6. What variables do type properties control? paragraph properties? page properties?

7. What is the purpose of defining tags in a template?

8. Why does the virtual page on a computer screen sometimes not mirror the printed page exactly?

9. Explain the importance of proofing a completed desktop publishing page.

CHAPTER 13
PROBLEMS

1. Import an exploded-view CAD drawing onto a desktop publishing page. Using the desktop publishing program's text editor, add a parts list showing item number, description, and quantity. Create a title **PARTS LIST** using bold type. Draw a box around the parts list.

2. Use a desktop publishing program to create a two-page newsletter for your school's CAD club or other organization.

3. Review several periodicals such as *Time* or *Newsweek*. Pick a page layout that is attractive to you. Create a template duplicating the layout as closely as you can using desktop publishing software. Then write a movie or book review using the new template.

4. Select your favorite CAD drawing. Import it so that it fills two-thirds of a desktop publishing page. In the remaining space, write a description of how you created the CAD drawing.

SCHOOL-TO-WORK

Solving Real-World Problems

Fig. 13-31 *SCHOOL-TO-WORK problems are designed to challenge a student's ability to apply skills learned within this text. Be creative and have fun!*

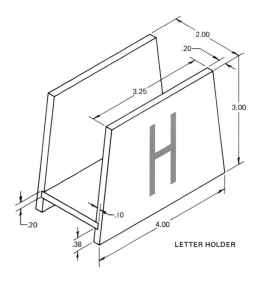

LETTER HOLDER

Fig. 13-31 *The company you work for has decided to manufacture and market a do-it-yourself kit for a letter holder. Your assignment is to develop a complete instruction sheet for the customer to follow in the assembly of the final product. Develop the instruction sheet on your CAD system to include the following:*

1. *A small-scale pictorial drawing of the finished product.*
2. *An exploded-view pictorial drawing illustrating the assembly.*
3. *A set of instructions on assembly.*
4. *A finishing schedule (if information is available).*
5. *An attractive sheet design with a .50-in. border, your company logo, and the company name and address. Use color if available on your system.*

and the dimensions. Not only can the various trade disciplines add to the drawing, but any combination of the layers can be shown at one time. The drafter can "hide" the rest of the layers without losing the information they contain. For example, the electrical designer doesn't necessarily need to see the placement of plumbing fixtures. By hiding the plumbing layer, the designer can reduce confusion—and the chance of making an error—while working on the electrical design.

The same procedure is used for the working drawings for a mechanical part. For example, layers may be created for the base periphery (outline) of the part, the rivet holes, and the dimensions (Fig. 14-9). When the drawing is used for production, each member of the production team hides or shows the layers as needed. The sheet-metal fabricator, who needs only the periphery and dimensions layers, can hide the hole layer to reduce confusion. The machinist drilling the holes into the sheet metal displays the periphery, hole, and dimension layers.

To further streamline communication among departments, most companies assign layering standards to be followed on all CAD drawings. For example, the dimensioning layer may always be a red layer called DIM. Product outlines may be assigned to a yellow layer named PROD, and electrical details may be assigned to a blue layer named ELEC (Fig. 14-10). By following conventions such as these, drafters know at a glance that the blue portion of any drawing is the electrical detail.

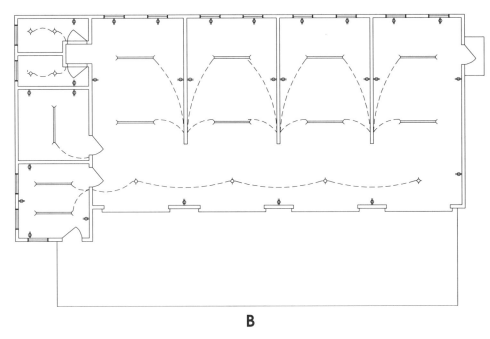

A

B

Fig. 14-9 *Drafters can display all the layers (levels) of a CAD drawing simultaneously (A) or select one or more layers to be "hidden" (B).*

BAKER ARCHITECTURAL, INC. DRAWING LAYER SPECIFICATIONS		
CONTENTS	**COLOR NO.**	**SAMPLE**
PLOT PLAN	10	
SITE WORK	32	
RESTRICTIONS/EASEMENTS	210	
FLOOR PLAN	138	
STRUCTURAL	18	
ROOFING	170	
DOORS/WINDOWS	140	
PLUMBING	20	
ELECTRICAL	190	
HVAC	14	
LANDSCAPING	96	

Fig. 14-10 *A layering standard such as this one can help reduce confusion and improve graphic communication within a company.*

Fig. 14-11 *Traditional vertical (A) and horizontal (B) storage cabinets for original drawings.* (Courtesy of Mayline Co.)

DRAWING CONTROL AND STORAGE

The great number of drawings done in a typical drafting room each day must be carefully controlled to avoid misplacing or causing irreparable damage to the drawings. Therefore, most companies develop a comprehensive system for cataloging and storing original drawings. For example, many large companies keep the original drawings for a contract together under a contract number or job number. This makes the drawings easily available for copying or revising (Fig. 14-11).

File Systems

Most companies employ one of two basic filing systems. In a **closed catalog system,** one person is responsible for keeping the drawings safe. Although this person may have a staff that files and copies the documents, the number of people allowed to handle original drawings is tightly controlled.

Other companies favor an **open file system.** In this system, the original drawings are always available to anyone in the department. The open file system is generally not preferred because easy access to the original drawings can lead to mishandling.

Drawing Storage

Storing drawings in their original form consumes considerable space. The drawings are often difficult to manage because of their size, and the overhead expenses (such as rent and temperature control) are often enormous. Therefore, most companies now store manual drawings on microfilm (Fig. 14-12). **Microfilm** is a storage medium on which a large drawing is reduced to a very small size and recorded on film. Microfilm storage greatly reduces the amount of space necessary to store drawings.

A number of photographic formats have been used in microfilming. However, 16 mm and 35 mm are the most popular film sizes. The film has high contrast, high resolution, and is fine-grained to produce high-quality drawings with good detail. When a drawing stored on microfilm needs to be copied or revised, a copy is made directly from the microfilm. Copies from microfilm can be scaled to fit any standard size paper.

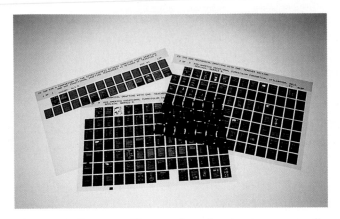

Fig. 14-12 *Since microfilm stores large drawings in a very small format, many drawings can be stored on a single microfilm data card.* (Courtesy of Keith M. Berry)

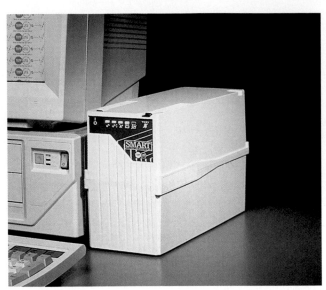

Fig. 14-13 *An uninterruptible power supply protects unsaved work in a CAD drawing when power fluctuates or fails.* (Courtesy of TrippLite)

CAD DATA CONTROL

CAD drawings require protection even while they are being created. A momentary power outage, or even a power fluctuation, can destroy any portion of a drawing that has not been "saved" to a storage medium such as a floppy disk or the computer's hard drive. Although saving a drawing is ultimately the responsibility of the individual drafter, companies usually have standard policies and procedures to help protect the files.

Almost every company that uses computers must implement standard file backup procedures. For example, a complete backup strategy for a CAD department might be:

• CAD operators are required to save their drawings at least once per hour to the CAD system's hard drive. If the drafter's work is very complex, the drawing should be saved once every half hour.

• At the end of each day, the CAD operator must back up (copy) all of the work done that day onto a floppy disk.

• At the end of every week, the department head must make a master copy of all the CAD files on the system. This master copy will be stored on a magnetic tape storage system and will be stored off-site in a fireproof location.

Some large companies that have networked CAD systems prefer to back up the CAD files at night. This can be done by a separate operator or group, or it can be programmed to occur automatically.

Many companies also install an *uninterruptible power supply (UPS)* to protect their computer systems from power outages and surges (Fig. 14-13). The UPS plugs into the power line between the power outlet and the computer. It protects the computer by detecting power failures and switching on a backup battery within microseconds. UPS units are capable of sustaining computer systems long enough for drafters to save their work and shut down the computers.

UPS units are available in several sizes. Some companies prefer to buy small, one-computer UPSs and supply each workstation individually. In other companies, one or more very large, high-capacity UPSs may be used to protect entire networks. Some of the larger UPS units can be programmed to shut down an entire network automatically if the power fails when no one is in the building.

DRAFTING MEDIA

The graphic communications prepared in the drafting room today are usually created on paper, illustration board, cloth, or film. These basic **media** (materials on which to draw; singular: *medium*) have improved greatly in recent years because industry constantly demands better quality in drafting materials. Note, however, that drafters now obtain better results from all types of media because better lead, ink, and erasers have become available.

Paper

The variety of opaque and translucent papers available today gives designers and drafters a broad spectrum from which to choose. Each type is capable of performing different tasks.

High-grade opaque paper (paper that cannot be seen through) is used for pictorial renderings that are later photographed for reproduction. Opaque paper can be lined for drawing graphs or diagrams or for plotting mathematical data.

Tracing papers are used for developing original detail drawings, so they must have a good surface quality. They must take pencil well and withstand repeated erasing. They must also withstand a lot of handling by the engineering team.

Translucent Tracing Papers

Natural tracing papers are made of rag. These papers are strong, but they are not really transparent enough to be used for some applications. Advances in papermaking have produced tracing paper that does not contain rag but is reasonably transparent and retains pencil and ink well. However, natural translucent paper without rag should be used only for sketching because it is not very durable.

Transparentized Tracing Papers

Transparentized paper, often called **vellum,** is paper that has been treated with resins to make it transparent. Vellum is ideal for quick reproduction because it contains rag and is very durable. Good erasability, long life, and high strength are other qualities of vellum. In fact, most types of vellum are considered ageproof. Nevertheless, vellum has one big disadvantage: it shows fold marks.

Cloth

Cloth (or linen, as it used to be called) is of higher quality than paper. It is an expensive medium made of cotton fiber sized with starch. In spite of its cost, it is often used for high-quality pencil and ink drawings because it is durable and erasable, and it reproduces well.

Film

The best medium developed for the engineering team in recent years is polyester film. It does not shrink, and it is transparent, ageproof, and waterproof. Pencil and ink produce a clear image on the surface of film. Its surface is stable even after repeated erasing. Film is classified by thickness and according to whether the surface is matte (not glossy) on one side or both.

Fade-Out Grid and Lined Stock

High grades of tracing paper and film are available with a grid pattern. This grid is very useful in drawing because it guides the drafter but does not show up on copies (Fig. 14-14). The grids generally have 10 squares, 8 squares, or 4 squares to the inch, although millimeter grids are also available. This grid paper, also called cross-section paper, is available in rolls, sheets, and pads. In addition to the standard square grids, special kinds of cross-section paper are made for orthographic and isometric drawings.

CAD MEDIA

CAD systems generally include printers or plotters that can produce a **hard copy** (final drawing on one of the media types described above). However, media for CAD drawings also includes removable magnetic and optical media such as floppy disks, tapes and CD-ROMs (Fig. 14-15).

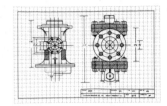

DRAWING PREPARED ON
PREPRINTED GRID

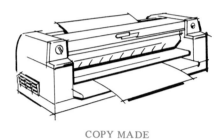

COPY MADE

PRINT DOES NOT SHOW
GRID LINES

Fig. 14-14 *Fade-out grid and film provide the drafter with guidelines that do not show on the final print.*

Fig. 14-15 *The medium used to store CAD drawings depends on the capabilities of the CAD system and the size of the files to be stored.* (Courtesy of Keith M. Berry)

Data stored on these media can be shipped inexpensively to other locations or sent electronically via a telecommunications network such as the Internet. If a hard copy is required, it can be printed or plotted directly from the tape or disk.

Currently, the most commonly used CAD-specific media are:

- **Magnetic tape** Tape drives in use today range from large reel-to-reel units to small, specialized units that accept **data cartridges.** The data cartridges, which resemble standard cassette tapes, can hold large volumes of data. – up to 4 gigabytes (4000 megabytes). Therefore, these cartridges can be used to store the drawings for entire projects, even when the projects are large and complex.

- **Floppy disk drives** Most CAD systems have floppy disk drives that accept floppy diskettes. These diskettes are removable, so

they are a good medium for distributing CAD drawings. Although diskettes are available in both high and low density formats, high-density diskettes are used almost exclusively. The major disadvantage of using floppy diskettes is that the amount of data they can hold is currently limited. A 5.25″ high-density diskette holds about 1.2 megabytes (MB) of data; a 3.5″ high-density diskette holds about 1.44 MB.

- **CD-ROM** Compact disc read-only memory (CD-ROM) is an excellent medium for distributing or storing larger volumes of data. Typical storage capacity of today's CD-ROM discs is a little more than 600 MB. However, new technology is increasing the amount of data that can be stored on a single disc.

- **Hard (fixed) disk drive** Hard disk drives can be considered a CAD medium, although they are generally used only for temporary storage of drawing files because they cannot be removed easily from the system. These drives are usually located inside the computer, although some systems also have external drives. The capacity of the hard disks commonly used in CAD systems ranges from 540 MB to more than 1.6 gigabytes (1600 MB).

Some companies are now moving toward a **paperless environment** to reduce or eliminate the use of hard-copy media. The goals for a paperless environment are to reduce the number of trees used to produce paper products, reduce the cost of supplies for the company and improve the environment.

Companies that have implemented a paperless environment complete their communications and presentations using only the computer. Computer-based messaging systems allow these companies to incorporate drawings, charts, graphs, and animations for better communication in-house to members of the engineering design team, as well as outside the company to the client or consumer. Computer-based messages also have another advantage over their paper counterparts: they can easily incorporate color for emphasis or special effects.

REVISIONS

When a drawing must be revised, the company must usually keep a copy of the original as well as the revision. Instead of recreating the entire drawing for the revision, drafters generally use an *intermediate*. An **intermediate,** also called a *secondary original,* is a copy of a drawing that is used in place of the original (Fig. 14-16). The designer and drafter can change the intermediate without changing the original. Using intermediates avoids tracing and redrawing.

The eight methods that follow are only some of the timesaving techniques used by professional drafters to revise intermediates.

- **Scissors Drafting** The unwanted portions of an intermediate are removed with scissors, a knife, or a razor blade. This is called *editing.* Then a new intermediate is made to incorporate the change. Other changes can then be made on the new secondary original.

- **Correction Fluids** Sometimes drafters remove unwanted data from an intermediate by applying correction fluid. New data can then be drawn in the cleaned area. Prints are then made from the corrected intermediates.

- **Erasable Intermediates** Erasable intermediates can be erased with ease. Thus, corrections can be made quickly without fluid. Redrawing can be done equally well with pencil or ink.

- **Block-Out or Masking** This system is best used when limited time is available for reworking a drawing. First, the area to be blocked out

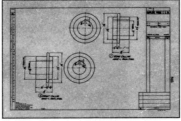

Say you have an existing drawing:

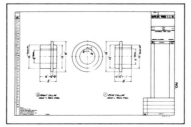

And you want to revise it like this.

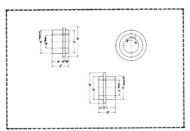

First you make a clear film reproduction of the original and cut out the elements.

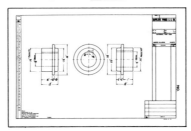

Then tape the elements in their new positions on a new form.

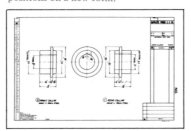

Photograph it on film with a matte finish (the tapes and film edges will disappear). Draw in whatever extra detail you want—and you have a new original drawing.

PASTEUP DRAFTING

Fig. 14-16 *Using an intermediate to rearrange the placement of drawings on a drawing sheet.* (Courtesy of Eastman Kodak Co.)

is covered with opaque tape. Next, a print is made in which the masked areas produce blank spots. Changes are then entered into these areas to produce the revised original.

- **Transparent Tape**   Data can be added to a drawing by placing the data on paper, cloth, or film, and then taping the data in place on an intermediate. Drafters often add dimensions, notes, and other data symbols in this way. Alternatively, the new data can be typewritten on transparent press-on material, which is then attached to the intermediate.

- **Composite Grouped Intermediates**   When several drawings are to be combined, a composite grouping can be cut from intermediates and taped in place. A new intermediate is then made of the composite.

- **Composite Overlays**   Translucent original drawings can be combined into a composite intermediate by placing the originals in the desired places. The composite is then photographed.

- **Photodrafting**   A picture not only saves a lot of words; it also limits the time needed for drafting. Architectural and engineering teams often use photo drawings because they are generally easier to understand (Fig. 14-17). In this intermediate drafting process, called **photodrafting,** photographs are overlaid with line work to create enhanced, annotated pictures of the product.

CAD REVISIONS

When a company uses CAD to create drawings, intermediates are not needed. Since the data is in digital format, the CAD operator merely retrieves a digital copy of the original design. *NOTE:* As a matter of good practice, the first thing the CAD operator should do is rename the file to protect the original drawing. The name change usually consists of adding a revision letter to the name so that members of the design team know which drawing is the latest version (Fig. 14-18).

Make a halftone print of the photograph.

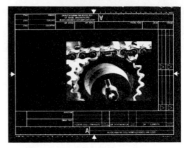

Tape the halftone print to a drawing form, and photograph it to produce a negative.

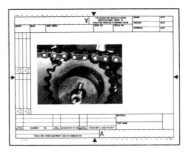

Make a positive reproduction on matte film.

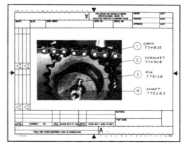

Now draw in your callouts— and the job is done.

PHOTO DRAFTING

Fig. 14-17 *Photodrafting can complement and clarify technical data.*

After specifying the new revision number, the CAD operator makes any required changes on the drawing. Once the changes have been made, the operator saves the new version of the drawing and then reprints or plots a new original showing the updated information. The original design remains unchanged.

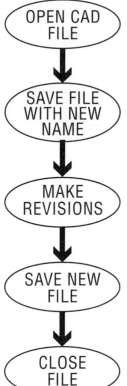

Fig. 14-18 *Careful drafters rename drawing files before making any changes. This helps keep the original safe and provides a logical record of revisions.* (Courtesy of Circle Design)

REPRODUCTION

Once a drawing is completed, many copies of it are often needed for various purposes. For example, the design engineering team needs copies they can refer to or use to develop or revise drawings. The factory needs copies to make and inspect parts. *Subcontractors* (other companies that supply parts) also need copies to make and inspect parts. The purchasing department needs copies to order standard parts not made in the shop. These include bolts, nuts, washers, gaskets, and so on. If the item is assembled in a place other than the shop, additional copies may be needed for assembly.

Fast, accurate, and economical methods of **reproduction** (making copies) have been available for many years. In choosing a method, the drafter must take the following factors into consideration.

- **Purpose of the copies**   What is the purpose of the copies? Will they be used for presentation, manufacturing, assembly, or publication?
- **Input of the originals**   What is the size, weight, color, and opacity of the medium used?
- **Output quality**   How readable must the copies be? Should they be transparent or opaque? Must they be workable?
- **Size of reproduction**   Are the copies to be enlarged, reduced, or the same size? Are they to be folded by machine?
- **Speed of reproduction**   How many copies of each original are needed? How quickly are they needed?
- **Color of reproduction**   Should the copy be blue-line, blueprint, sepia brown or some other color (or combination of colors)?
- **Cost of reproduction**   What are the estimated costs of materials, operations, and overhead?

The methods available today for copying engineering drawings include blueprinting, diazo, electrostatic copying, thermographic copying, photography, wax color transfer, and laser printing.

Blueprints

Blueprinting is an inexpensive process that makes a copy the same size as the original on a light-sensitive photographic paper. The copy is like a negative in that it has white lines on a blue background. Blueprint is the oldest copying process. The prints are washed in water after being exposed to light. Next, they are treated with a potash solution to make the blue darker, and then they are rewashed.

Blueprinting is a continuous machine process. The only handwork involved is trimming the finished copies.

Diazo

Diazo is a reproduction method in which prints have dark lines on a white background. For this reason, the diazo process is sometimes called *whiteprint*. Diazo prints are made from drawings created on a translucent medium and exposed to diazo paper that has a light-sensitive coating. Diazo printing is available in a few colors, including blue, black, and red.

The light-sensitive diazo coating is *developed* (treated chemically to bring out the image) to create the diazo prints. The three methods used to develop the coating include the dry process, the semi-moist process, and the pressure process.

Dry Process

The print paper for *dry process* developing has a dye coating. The original drawing is placed in contact with the print paper and the two are passed through a light source (Fig. 14-19A). The light cannot pass through the lines of the original drawing, but it can pass through all other areas of the paper. The light removes the dye coating on the print paper everywhere except where the light does not pass through the original drawing. The original drawing is then removed, and the print paper passes through ammonia gas to develop the remaining dye into a line image that is a faithful copy of the original.

Semi-Moist Process

The *semi-moist process* is similar to the dry process in that the original drawing and the diazo print paper are placed in contact and exposed to light. Then the original drawing is removed, and the print paper is exposed to a liquid developing chemical. As the paper is fed through fine-grooved rollers, the developing liquid brings out the image. Then the damp diazo paper passes over heating rods to dry the print (Fig. 14-19B).

Pressure Process

The *pressure process* was developed to make dry, odorless whiteprints. In this process, an amine chemical, or *activator,* develops the print. The operator adjusts the developing speed to accommodate the coating on the print material. The speed adjustment is necessary to make prints that have good contrast (Fig. 14-20).

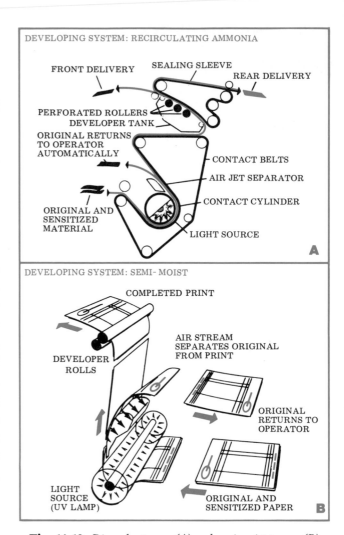

Fig. 14-19 *Diazo dry process (A) and semi-moist process (B).*

Fig. 14-20 *A pressure-process diazo developing system.* (Courtesy of Oce-Bruning)

The vellum and films for the pressure process are available in matte finish and are workable on either side. Copies made using the pressure process are of such high quality that they can be used in place of originals or as secondary originals for design changes.

Electrostatic Reproduction

Electrostatic reproduction (sometimes called *xerography*) uses electrostatic force (static electricity) to reproduce original drawings (Fig. 14-21). Electrostatic machines have a positively charged plate. To make a copy, the original drawing is placed between the plate and a light source. Where light strikes the charged surface, the charge *dissipates,* or goes away. Where the original drawing obscures (hides) the light source, the charge remains on the plate. A negatively charged powder

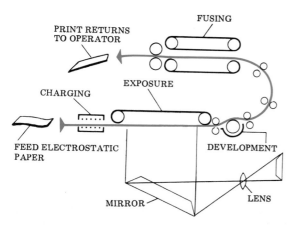

Fig. 14-21 *Electrostatic reproduction.*

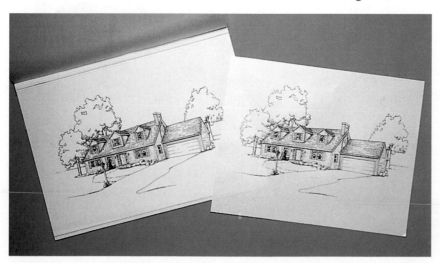

called **toner** adheres to the charged surfaces of the plate because it is attracted to the opposite (positive) charge on the plate. The machine then produces an electrical discharge in which the toner particles "jump" to the copy paper. The paper is then exposed to an intense heat source to fuse (bond permanently) the toner to the paper.

Photographic Reproduction

Film companies also provide ways to reproduce engineering drawings. Copy cameras make images of the drawings on light-sensitive material, such as film coated with silver halide. High-contrast films are used to give lines uniform widths and blackness (Fig. 14-22). The images are developed and fixed using procedures similar to those used to develop ordinary camera film.

Film makes better prints than the commonly used diazo process. The film has a matte finish, and changes can be made with ink or pencil. Photographic reproduction also allows drafters to reduce or enlarge images easily.

CAD REPRODUCTION

CAD drawings are naturally easier to reproduce than manually-created drawings because a new copy can be printed or plotted at any time from the drawing file. When CAD drawings are reproduced by creating a print from the original drawing file, each "copy" is actually an original.

The machines used to create hard copy from an electronic drawing file are necessarily different from those described in the preceding section. CAD materials must be printed using a machine (printer or plotter) that can communicate directly

Fig. 14-22 *The rendering on the right has been photographically reproduced from the one on the left. Notice that line clarity has not been lost on the photographic copy.* (Courtesy of Brent Phelps and Circle Design)

with the CAD system. The printers and plotters commonly used to create hard copies of CAD drawings include:

- **Electrostatic plotters** use the same electrostatic technology described in "Electrostatic Reproduction," with minor differences. The biggest difference is that the electrostatic plotter receives the drawing image directly from the CAD system.

- **Pen plotters** use ink pens in various colors to draw the CAD design on a variety of media. Different pens are used depending on the medium chosen. See "Pens for CAD Plotters" in Chapter 3 for additional information about pens for these plotters.

- **Thermal wax transfer printers** use wax to blend the colors used in the CAD drawing (Fig. 14-23). The wax is then transferred to paper and fused with heat. Wax transfers turn brittle with age, so they are not used for long-term storage.

- **Laser printers** produce hard copy on plain copy paper using a technology that is similar to electrostatic reproduction. Both black-and-white and color laser printers are available. Currently, however, color laser printers are not cost-effective. Also, most laser printers are currently limited to A- and B-size drawing paper.

- **Ink-jet printers** produce black-and-white or color images of good quality by spraying fine jets of ink onto paper. Because ink-jet printers are inexpensive, they are being used more and more in drafting rooms.

- **LED plotters** provide full-color images on paper, vellum, or film using LED technology (Fig. 14-24). They can plot drawings on B-size through E-size drawing sheets. LED plotters are relatively new, but their versatility, good quality, and value make them a good choice for many companies.

When a large number of copies of a CAD drawing are needed, none of the above machines are cost-effective. They use expensive materials and supplies, and most of them are too slow to produce large quantities quickly enough to be feasible.

Some companies use black-and-white or color photocopiers when they need a large number of copies. However, color photocopiers are expensive and require expensive supplies. Therefore, many companies prefer to send their original drawing files to a reproduction service company for reproduction. The service company reproduces the CAD drawings using the method and medium requested by the company that needs the copies.

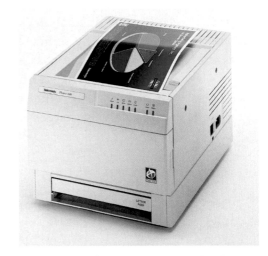

Fig. 14-23 *Wax transfer provides brilliant color, but it is expensive and tends to become brittle with age.* (Courtesy of Tektronix)

Fig. 14-24 *The LED plotter produces good-quality drawings of almost any size.* (Courtesy of JDL)

CHAPTER 14

REVIEW

Learning Activities

1. Obtain several technical reports from industry. Carefully review them to determine if they are concise and yet clearly written. Separate them according to the five kinds of technical reports described early in the chapter.

2. Obtain various kinds of sketches and drawings from industry. Separate them according to the four levels of graphic communication.

3. Study the title blocks and revision blocks on industrial drawings. Why are revision blocks located at the bottom of the sheet numbered from the bottom up?

4. Try making changes on drawings using one or more of the seven methods described in the chapter (scissors drafting, correction fluids, etc.). Use either industrial drawings or drawings you created for this activity.

Questions

Write your answers on a separate sheet of paper.

1. What are the five basic types of reports generally produced by engineering design teams during the design and manufacture of a product?

2. Explain how Gantt and PERT charts can help organize a project.

3. What is the purpose of a revision block?

4. Explain how pin-bar drafting helps control graphic communication within a drafting department.

5. Why are pin bars not necessary for CAD drawings?

6. Describe the difference between a closed catalog system and an open file system. Which system provides more control over original drawings?

7. Describe the advantages of using microfilm to store original drawings.

8. Name two things a company can do to protect information in drawings against temporary power outages.

9. Name at least four good qualities of vellum.

10. What is the one major disadvantage of using vellum?

11. Name at least five advantages of using polyester film for drafting.

12. In addition to traditional media, what other media can be used for CAD drawings?

13. Explain the purpose of an intermediate.

14. Describe at least three methods of revising intermediates.

15. What is the first thing a CAD drafter should do after opening a drawing file for revision? Why?

16. List several factors that affect the selection of a reproduction process.

17. What is the major difference between a copy made on a blueprint machine and one made on a diazo machine?

18. What is the major difference between the dry and semi-moist diazo processes?

19. Name the various methods of reproducing a CAD drawing.

15

UNDERSTANDING COMPUTER-AIDED DRAFTING

OBJECTIVES

Upon completion of Chapter 15, you will be able to:

○ Describe common CAD functions.
○ Identify entity selection techniques.
○ Identify geometric placement techniques.
○ Describe entity editing techniques.
○ Describe entity modification techniques.
○ List advantages of CAD vs. manual drafting.

VOCABULARY

In this chapter, you will learn the meanings of the following terms:

- **branching menu**
- **computer-aided manufacturing (CAM)**
- **crossing window**
- **dialog box**
- **entities**
- **entity attributes**
- **filtering**
- **icons**
- **intelligent attributes**
- **main menu**
- **masking**
- **objects**
- **point of origin**
- **pop-up window**
- **pull-down menus**
- **solid models**
- **syntax**
- **wireframe models**

CAD systems allow you to use a computer and CAD software to create the same shapes and drawings that you would create manually. The only difference is that you tell the CAD software how to draw the shapes. This chapter uses the CADKEY and AutoCAD software to demonstrate CAD tools you can use to produce a final design or drawing.

Think of what it would be like if you had to give instructions to another drafter to create a drawing. You would have to be very precise in your directions.

The same is true of a CAD system. The computer does exactly what you tell it to do—no more, and no less.

CADKEY displays commands at the top of the screen and requests more information at the bottom of the screen (Fig. 15-1A). AutoCAD displays commands at the *command line* in a *floating command window* at the bottom of the screen (Fig. 15-1B). Look in these two places to find the information required to create the results shown in the figures throughout this chapter.

ADVANTAGES OF USING CAD

Data that has been entered into a computer is called *digital data*. The engineering design team can transmit digital data over telephone lines, or wires to other computers. Therefore, the engineering design team can distribute CAD drawings, like all other information in digital form, anywhere in the world very quickly.

In most cases, a final drawing often requires less time to produce using CAD than drawing by hand once you understand the functions of the CAD software. Therefore, an experienced CAD operator is often more productive than a drafter that relies strictly on manual methods. Computer-aided drafting and design is now widely used because it reduces drawing production time.

LEARNING CAD

To learn a CAD system completely, you must become familiar with CAD functions and know the vocabulary that describes entities and their relationships. **Entities** (called **objects** in some CAD systems) are the lines, arcs, circles, and polygons used to define the shape of the part being drawn.

The trick to learning a CAD system is to concentrate not on which exact button to push, but on understanding the result you want to accomplish. Then you can find a command or function that will achieve that result.

When learning a CAD system, get into the habit of reading the CAD screen. The screen displays valuable information about the commands and functions in progress.

Just remember the saying—to design in CAD, open your eye LIDS:

- **L**ook for words that describe the result you want.
- **I**dentify any relationship to other entities.
- **D**ecide the method for placing the entities on the screen.
- **S**tructure the command.

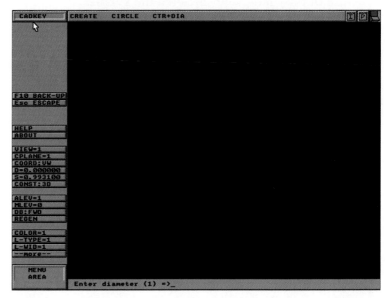

Fig. 15-1A *CADKEY for DOS displays the operator's commands at the top of the screen and requests for information at the bottom of the screen.*

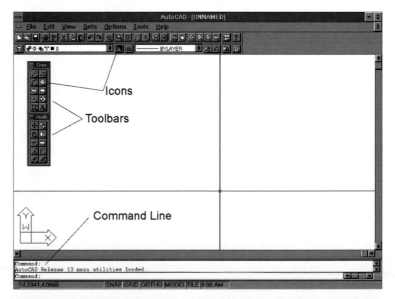

Fig. 15-1B *By default, AutoCAD displays all information at the command line at the bottom of the screen.*

COMMUNICATING ACTIONS

Drafters must give the computer system precise instructions to create a CAD drawing. Learning CAD software is like learning a foreign language. In the foreign language, you have to learn how to structure complete sentences. In CAD, you have to learn how to structure the CAD commands to get the result you expect on the computer screen. The structure of this "language" is called the **syntax.**

CAD systems use different methods to tell the system which function to perform. Some CAD systems use a noun-verb syntax in which you enter the noun first and then the verb. For example, you might enter "Line Insert" to place a line in the drawing. The noun (*Line*) occurs before the verb (*Insert*). Other CAD systems require the drafter to place the verb first. Using one of these systems, you might enter "Draw Line" to place a line in the drawing.

The three most common ways for a drafter to activate CAD functions are:
- entering the command at the command line.
- selecting the command from a menu.
- choosing an icon that represents the command.

Using the Command Line

To use the command line, the drafter uses the keyboard to enter the commands in a certain form. Most CAD systems allow you to abbreviate common commands. For example, in CADKEY you might enter INS LIN PAR for "insert line parallel." For many commands, you must add a numeric value to specify the length or size of the entity being created. In the previous example, you might enter INS LIN PAR5 to insert a parallel line five units long. (Units can be feet, inches, centimeters, meters, miles, or any other unit you establish in the drawing.)

Using Menus

CAD systems also have menus of functions from which you can choose. The menu displayed when you begin a design drawing is called the **main menu** or *default menu.* When you select a function from the main menu, additional choices may be displayed in a **branching menu.** Most CAD systems use a complete menu tree composed of branching menus. Reviewing this menu tree will help you become familiar with the CAD system quickly.

Some CAD systems display the main menu items across the top of the computer screen. These menus are called **pull-down menus.** When you select a function from a pull-down menu, additional choices may display below the main menu item.

If a menu appears within the drawing window, it is called a **pop-up window** or a **dialog box.** Pop-up windows usually require the operator to enter a value or make specific selections from choices listed in the menu or window.

Using Icons

Icons are small pictures that represent CAD functions. In some CAD programs, icons are grouped by function. In Fig. 15-1B, for example, the icons are located in the Draw and Modify toolbars on the AutoCAD screen.

When you select an icon, the function or command associated with that icon becomes active. For example, to draw a circle by specifying the radius, select the picture of a circle with a line from the circle to the center point.

Icon menus can also branch. Some CAD systems call this *paging.* After you select the main icon, a menu branches out or a window page of additional icons pops up.

CONTROLLING CAD SETTINGS

CAD software allows you to control the appearance of entities. You can set the color, linetype, line width, font (text) style, font height, character slant, and pen number. If you will be using the same settings for many different drawings, you can save the information in a start-up or *prototype* file.

An *attribute* is any characteristic that can be associated with an entity, such as color or function. The color, linetype, line width, and so on of an entity are known collectively as **entity attributes.** Entity attributes should be set to conform to the American National Standards Institute (ANSI)

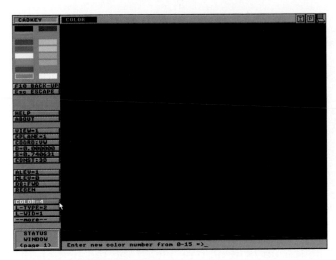

Fig. 15-2 *The drafter or designer selects a color from the selections displayed above the menu area.*

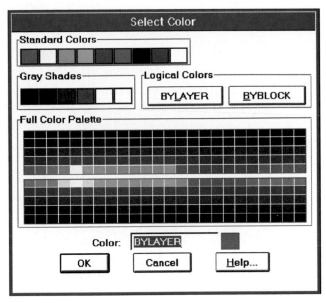

Fig. 15-3 *AutoCAD's Select Color dialog box allows you to select a color for an entity. However, many drafters use the BYLAYER option to set the color by layer rather than by entity.*

drafting standards. However, some large companies may allow minor deviations to the ANSI standards for special reasons, such as requirements of the product being developed or limitations of the CAD software in use.

Entities can also have other attributes, known as **intelligent attributes,** that describe the function of the entity. For example, an intelligent attribute might describe the function of an electronic component. CAD operators assign intelligent attributes to the entity during the design process. The engineering design team uses intelligent attributes to analyze the part or circuit being designed.

Different CAD programs handle entity attributes in various ways. Examples from two representative CAD programs, CADKEY and AutoCAD, are used in this discussion.

Color

CAD systems have different ways of letting the drafter or designer set the color of entities, text, and dimensions. For example, CADKEY has a menu setting called *color*. When you select the color function, a pop-up window appears so that you can pick the desired color for entities (Fig. 15-2). AutoCAD allows you to choose colors from a color chart in a pop-up window (Fig. 15-3). However, many experienced drafters choose to assign colors to entire layers rather than individual entities.

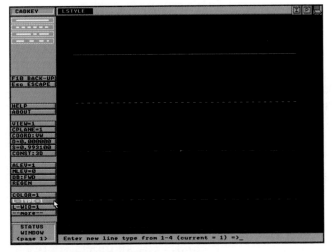

Fig. 15-4 *Selecting L-TYPE in CADKEY displays the solid, hidden, center, and phantom line options.*

This helps keep the drawing more consistent and makes editing easier.

Linetype

Selecting L-TYPE from the menu sets the linetype in CADKEY. CADKEY then displays four standard linetypes in the upper corner of the menu. You can select solid, hidden, center, or phantom linetypes (Fig. 15-4). Any entities created after the selection appears with the chosen linetype.

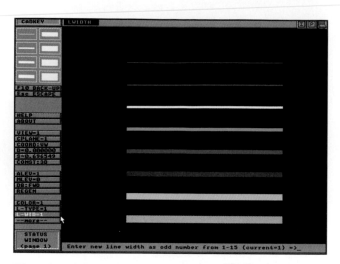

Fig. 15-5 *Selecting L-WID in CADKEY displays the line widths available. However, the drafter must enter a number at the bottom of the screen to set the number of pen strokes for the plotter.*

Other CAD systems offer a greater variety of linetypes to meet the needs of various industries. For example, AutoCAD has more than 40 linetypes that meet ANSI standards, ISO standards, or both.

Line Width

In CADKEY, you can set the line width by selecting L-WID from the menu. The display in the upper corner of the menu shows solid lines of different thickness. Line thicknesses vary in CADKEY from 1, 3, 5, 7, 9, 11, 13, and 15 pen strokes thick (Fig. 15-5). The numbers represent the amount of strokes the plotter pen makes to produce a line of the chosen thickness.

AutoCAD, on the other hand, applies line thickness by pen number. You cannot see the various line thicknesses on the screen, but by specifying pens of different widths, you can plot lines of various thicknesses.

Font

The font (text) entity attributes set or change the appearance of notes, labels, and dimensions in the final drawing. Font attributes include font style, character height, and character slant.

Most companies prefer to use plain, easy-to-read fonts on final production drawings. Other, fancier fonts are also available on CAD systems. However, these are used mostly for drawings created for use in desktop publishing systems. See Chapter 13 for more information about fonts and CAD files for desktop publishing.

Pen Definition

Pen definition assigns a pen number to entities, dimensions, notes, and labels. Most CAD systems assign the number 1 for the entity pen attribute in the startup or prototype file. If the red pen on the plotter is in pen holder number 5 and you want your dimensions to be red, you must assign an entity pen attribute of 5 to your dimensions. (Hint: To do this in one operation, place all the dimensions on a layer by themselves. Then apply the pen attribute to the layer rather than to individual entities.)

TYPES OF ENTITIES

CAD software includes commands that allow the drafter to create various types of entities. This section describes some of the basic entities common to most CAD programs.

Points

Points are markers for locations in space. They do not have a width, height, or depth, but they are considered entities because CAD systems visually display points on the screen. You can place point entities using the same methods used to place other entities on the screen. You can copy, move, or delete points just like any other type of entity.

In CADKEY, point entities display as a plus symbol (+). Other CAD systems, such as Auto-CAD, allow you to choose from several different methods of displaying points (Fig. 15-6).

Lines

In general, a *straight line* is defined as the shortest distance between two points. CAD systems allow you to define the start and end of a line using the Cartesian coordinate system. You can also use any

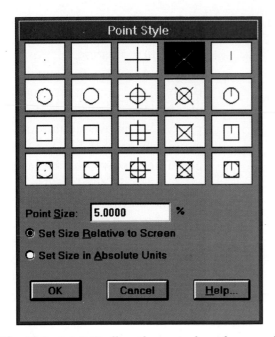

Fig. 15-6 *AutoCAD allows the user to choose from several different point styles.*

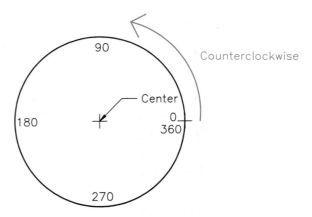

Fig. 15-7 *CAD systems draw circles counterclockwise from the three o'clock position.*

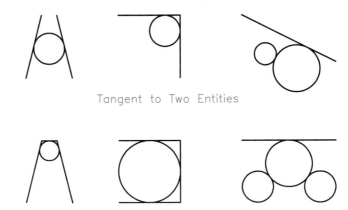

Fig. 15-8 *Circles can also be drawn tangent to two or three existing entities.*

of the positioning methods described later in this chapter to place the line at an exact location.

Circles and Arcs

As you may remember from Chapter 4, the total number of degrees in a full circle is 360°. Most CAD software creates circles in the counterclockwise direction from the three o'clock position (Fig. 15-7), although you can change the starting position if necessary. Most CAD systems allow you to create circles and arcs using several different methods. The most common methods are:

* specifying the location of the center of the circle and the diameter.
* specifying the location of the center of the circle and the radius.
* identifying two points on the circumference of the circle that describe the circle's diameter.
* identifying three points on the circumference of the circle.

* identifying two or three entities to which the circle will be tangent (Fig. 15-8).

In CADKEY, the three o'clock position is also the starting point for arcs (Fig. 15-9). Define the radius or diameter and the starting and ending angles using methods similar to those used for circles.

Other CAD systems may offer additional methods for creating arcs. Some common methods include:

* specifying the start and end points of the arc, as well as one other point on the arc.
* specifying the center of the arc and its start and end points.
* specifying the start point and the angle of the arc.

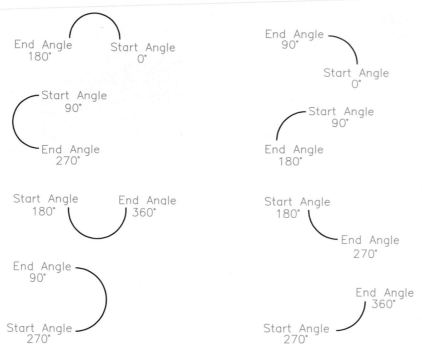

Fig. 15-9 *CAD systems draw arcs beginning at the three o'clock position. However, the drafter may specify different beginning and ending angles.*

Polygons

Closed figures that have three or more sides are called *polygons*. CAD systems treat all the sides of a polygon as a single entity. In other words, if you select one side of a polygon for deletion, the CAD system erases the entire polygon.

The drafter must specify how many sides the polygon will have. The diameter or radius entered by the drafter determines the size of the polygon. The drafter also specifies whether the CAD system will measure the diameter or radius from the corners or the flat sides of the polygon. (See Chapter 4 for more information about inscribed and circumscribed polygons.)

PLACING ENTITIES

CAD systems allow the drafter to place entities using several different methods. Each method is useful for specific situations. CAD operators need to know all of the methods to connect entities because mastering all of the positioning techniques increases designing production, improves data accuracy, and reduces manufacturing errors.

Using a Cursor

The first method of positioning entities is to use the cursor. Most drafters use the cursor when they are making a simple sketch of a part. This allows them to create simple shapes quickly, but the data generated may not be accurate enough for the final drawing.

CAD software can display an adjustable grid on the screen to help you place entities. Using the grid alone does not place the entity accurately, however. You must also use the snap function, which allows the cursor to attach (snap) to each grid point. For placing some entities, the grid and snap functions are accurate enough to allow you to place the entity correctly. However, do not get into a habit of relying entirely on the grid and snap functions.

Using Coordinates

Many manufacturers now use various types of **computer-aided manufacturing (CAM)** processes. In CAM processes, robots and other computerized machines cut, move, drill, and stock production parts. Industry calls these machines *computer numerical control (CNC)* machines. CNC machines cut the raw material into the shape defined by the CAD data.

There is little room for programmer error when a CAD file will be used with a CNC machine. Most CNC machines cannot span a gap or overlap (due to programmer error) of greater than .0004 in. in the entities. The machine simply stops at that point and the raw material becomes scrap metal. This is very costly to the company producing the part.

To prevent these expensive machining errors, the placement tolerance allowed for entities must be within .0004 in. (To give you an idea of the thickness, the average size of a human hair is .003 in. If you take a human hair and slice it lengthwise

10 times, the approximate thickness of each piece is .0003 in.)

Therefore, just using your eyes to connect entities is not accurate enough to prevent gaps or overlaps within the placement tolerance. It is very important that you master the positioning techniques (methods of placing entities) to ensure accurate final drawings.

To specify the location of a point, the endpoint of a line, etc., drafters use the Cartesian coordinate system described in Chapter 5. As you may remember, the top right quadrant of the Cartesian coordinate system is called the *World Coordinate System (WCS)* in AutoCAD. Other software refers to this as the *Absolute Coordinate System (ACS)*. They may include just the top right quadrant or the entire Cartesian coordinate system in their default displays (Fig. 15-10).

Entering the Cartesian coordinates varies from CAD system to CAD system. Some CAD systems require you to enter X=0 Y=0 Z=0 when prompted by the system. Other CAD systems need the coordinates only. In these systems, you use a comma to separate each value. For example, in AutoCAD you would enter 0,0,0. The value of X is first, then Y, and then Z.

When working in two dimensions on a CAD system that is capable of three dimensions, some CAD systems require only the values for X and Y. In two-dimensional drawings, the Z coordinate has a default value of zero. Therefore, the drafter only has to concentrate on the horizontal and vertical (X and Y) values.

CAD systems also allow you to specify other coordinate systems. Systems that use the same direction for X, Y, and Z coordinates in all three-dimensional views are called *View Coordinate Systems (VCS)*. Some CAD systems allow you to define one or more custom coordinate systems called *User Coordinate Systems (UCS)*.

Referencing Other Entities

All entities have a **point of origin,** or insertion point, which is defined when you create them. CAD systems allow you to place the point of origin by referencing an endpoint or center point

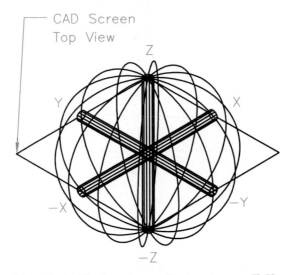

Fig. 15-10 *The Cartesian coordinate system uses X, Y, and Z coordinates for specific locations in three-dimensional space.*

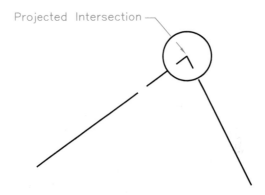

Fig. 15-11 *CAD systems can project the position in space at which two entities would intersect if they were extended.*

of another entity, as well as the real or projected intersection of two or more existing entities and other points of reference (Fig. 15-11). When you use these methods, you do not have to select the exact endpoint, center, or intersection on the screen. The CAD software calculates the position for you.

You can also position an entity by placing its point of origin at a point entity. (Note that a point entity is different from a Cartesian point location in 2D or 3D space.)

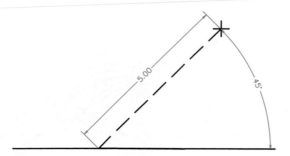

Fig. 15-12 *In this example, the Polar Coordinate System (PCS) is used to place the point entity 5″ and 45° from the right end of the existing line.*

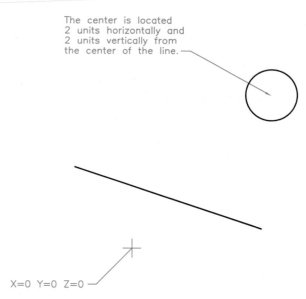

The center is located 2 units horizontally and 2 units vertically from the center of the line.

X=0 Y=0 Z=0

Fig. 15-13 *The center of this circle is positioned 2″ horizontally and 2″ vertically from the center of the line.*

In some cases, you will need to place entities at locations other than an endpoint, center, intersection, etc. CAD systems handle these situations in different ways. For example, CADKEY allows you to position an entity along a line or arc. To place a new entity using this method, you specify how far along the line the entity should be placed and from which end the measurement should start.

Using Polar Placement

Using polar placement is similar to positioning an entity along a line or arc. However, using polar coordinates places the entity at a given angle and distance relative to a drafter-identified location in space (Fig. 15-12). CADKEY calls this the *Polar Coordinate System (PCS)*. As an example, you can tell the CAD system to place a point 5 inches away at an angle of 45°, from a specific reference position.

CADKEY also allows you to temporarily set the origin (X=0 Y=0 Z=0) to a position located on an existing entity. This is called the *delta reference position*. You can think of *delta* as "from." From point A, go to point B. Then enter the X, Y, and Z coordinate values for the position of the new entity. Some CAD systems call this the *Relative Coordinate System (RCS)* (Fig. 15-13).

Some CAD systems, including CADKEY, use the notation dX, dY, and dZ to let you know that the relative coordinate system is in use. In other CAD systems, such as AutoCAD, an @ symbol in front of the coordinate values indicates relative coordinates.

AutoCAD has a similar function called From. The From function allows you to place a point of origin at a specific X and Y distance from an existing point on the drawing. However, the From function is used for new objects and will not function when you drag an existing object to a new location.

SELECTING ENTITIES

To edit existing entities, you must first select one or more entities to edit. In general, CAD systems allow you to use one of three basic selection methods.

Individual Selection

Cursor selection involves placing the cursor near the entities to be selected and picking them individually. This is the quickest method to use when you only need to select a few entities.

Window Selection

When you need to choose many entities, use the window selection method. In this method, you use the cursor to place a rectangular window around the entities to be selected. To create the window, you select two opposite corners of the rectangle (Fig. 15-14).

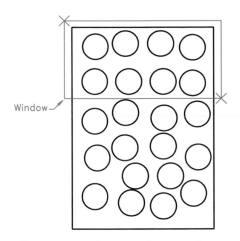

Fig. 15-14 *To place a window around entities, pick two diagonal positions to define a rectangular window.*

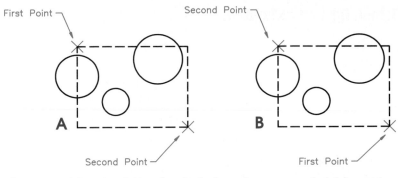

Fig. 15-15 *(A) In AutoCAD, when the drafter picks a point on the left first, only the small circle is selected because it is the only one that lies entirely within the window. (B) When the drafter picks a point on the right first, a crossing window is formed, and all three circles are selected.*

Most CAD systems offer several selection filters for use with the window selection method. For example, CADKEY provides the following options:

- **All in** selects all entities that are completely contained within the window.
- **Part in** selects all entities contained completely or partially within the window.
- **All out** selects all entities that lie completely outside the window boundary.
- **Part out** selects all entities that lie completely or partially outside the window boundary.

In AutoCAD, the type of window is determined by the order in which you select the corners of the rectangle. If you create the rectangle by entering a point on the left side first, followed by the diagonal corner on the right side, the window selects all entities that fall *completely* within its boundaries. If you begin the rectangle on the right side, a **crossing window** is formed. A crossing window selects all entities that fall completely or partially within its boundaries (Fig. 15-15).

Chain Selection

To use the chain selection method, choose entities that connect to each other, such as four lines that create the shape of a rectangle, and specify the beginning and ending entities of the chain.

The CAD system goes through the digital data and determines which entities connect. If any entities intersect any portion of the desired chain, most CAD systems stop the selection process and request more specific information.

Masking

You can also select entities by specifying the settings (attributes) assigned to the entity. CAD systems call this **masking** or **filtering.** Masking directs the CAD system to select only those entities that match the specified attribute settings. Commonly selected settings include entity type, linetype, line weight, color, pen number, and layer name. The CAD system filters through the digital data to choose all the entities that match the attribute settings. Masking can be used in combination with any of the three selection methods described above.

CHANGING ENTITIES

One advantage of using a CAD system instead of manual drafting procedures is that you can change the entities in a drawing easily. CAD systems call this *editing, transforming,* or *modifying* entities. Editing an entity changes its mathematical definition in the CAD database.

This section describes, in general terms, a few of the various CAD functions that allow you to change entities. For a more complete discussion, refer to the manual that came with your CAD software.

Trimming and Extending

To fine-tune a drawing, entities often need to be trimmed or extended. When trimming, some CAD systems require that you pick the side that remains after the trimming operation. Other systems require that you select the side to be trimmed away. To prevent unexpected results, you should be familiar with the selection method used by your CAD software.

Entities such as lines and arcs can be extended easily. To extend a line, for example, enter the appropriate command and choose a boundary to which it should be extended (Fig. 15-16).

Most CAD systems allow you to choose more than one entity to trim or extend at a time. In CADKEY, for example, if you choose two intersecting lines to be trimmed, you can choose to trim one or both of the lines in the same operation. If you choose to trim both lines, both are trimmed to the point of intersection.

Dividing

In CADKEY, dividing an entity removes the section of the entity found between two other entities. The drafter chooses the entity to be divided and the two entities that will form the boundary for the removed section.

CADKEY's Break option is similar to trim and divide except that the "broken-out" portion remains on the CAD screen. The entity is broken into two distinct mathematical definitions, so you can edit each new entity separately.

AutoCAD's BREAK command performs a function similar to the Divide and Break functions in CADKEY. This command allows you to break the entity into two pieces. During the function, you specify whether or not the CAD system should delete the broken-out section.

Stretching

Most CAD systems allow you to stretch part or all of an object on the screen. This function is especially useful for drawing revisions. If a part needs to be longer, for example, you simply stretch the part to the desired size. You can also shrink the part (Fig. 15-17).

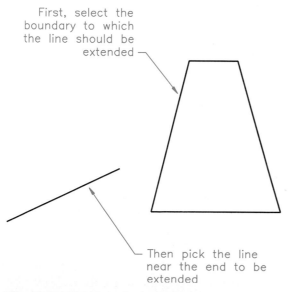

First, select the boundary to which the line should be extended —

Then pick the line near the end to be extended

Fig. 15-16 *To extend a line in AutoCAD, enter the EXTEND command, pick the boundary to which the line should be extended, and then pick the line near the end to be extended.*

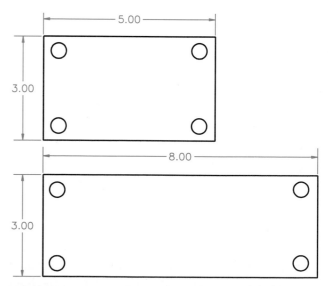

Fig. 15-17 *The objects at the bottom of the figure show the results of stretching a copy of the objects above. The associated dimensions are updated upon completion of the stretch function.*

Segmenting

CAD systems allow you to segment existing entities automatically into equal parts or into sections of a given length (Fig. 15-18). In AutoCAD, for example, the MEASURE command segments an entity into specified lengths. The DIVIDE command segments an entity into a specified number of equal parts.

Do not confuse segmentation with the divide function used to trim or break an entity. Segmenting creates multiple sections, unlike the divide function used to trim or break, which creates and/or removes just one section at a time.

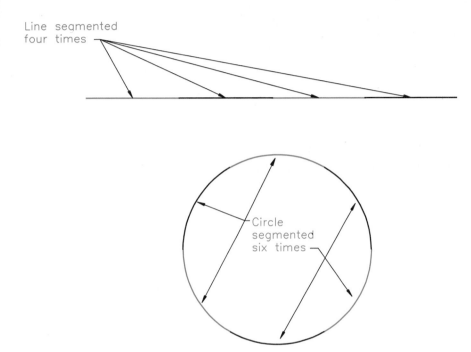

Fig. 15-18 *Entities can be segmented into equal sections. Any of the segments can be moved, copied, or deleted. Each entity can also be dimensioned if necessary.*

Copying and Moving

One CAD function that saves drafters a large amount of time is the Copy function. CAD systems allow you to make one or more copies of any number of existing entities. The original entity remains the same, and you can place the copies anywhere on the drawing using absolute coordinates or relative positioning techniques (Fig. 15-19).

You can also move entities around in the drawing. This operation is similar to copying, except that the original entity is moved to the new location you specify (Fig. 15-20).

Rotating

The Rotate function allows you to move or copy entities around an axis that you define. The selected entities revolve around the axis based on the angle you specify. Most CAD systems allow you to rotate one or more objects in two or three dimensions.

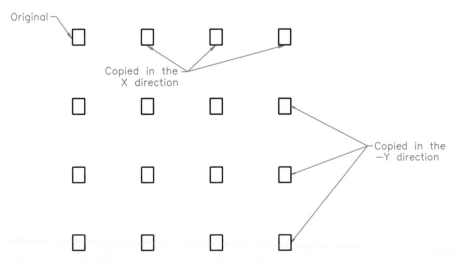

Fig. 15-19 *Rectangles were copied horizontally first, then vertically. Note that some CAD systems provide an Array function that allows the drafter to create a rectangular array such as this one in one operation.*

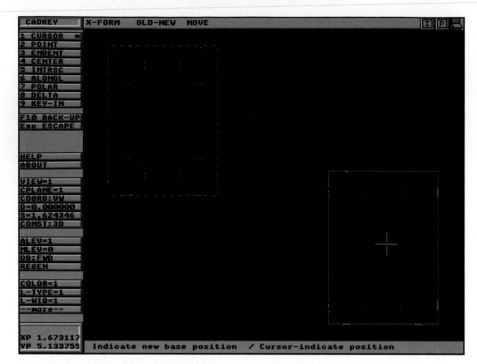

Fig. 15-20 *The objects in dashed lines show the original position of the entities.*

Scaling

As you may recall, *scaling* means to change the size of part or all of a drawing. CAD systems allow you to scale, or change the size of, one or more entities simultaneously. Note that scaling is different from stretching. When you scale an object, you generally scale the entire shape. Stretching applies to only part of a shape.

You can scale entities to make them larger or smaller by specifying a point of origin as a base point for the scaling. In most CAD systems, you can drag the cursor to make the entities larger and smaller, or you can enter a specific relative size, such as 1.50, to scale them.

Mirroring

Another useful CAD function is the Mirror function. This allows you to create a mirror image of one or more entities in a drawing file. The Mirror function has many uses. For example, you can create half of a symmetrical part and use Mirror to create the other half automatically. You can also use it to create one piece of a complex drawing, such as the left pedal of a bicycle. Then you can use the Mirror function to create a copy of the left pedal and place it in the correct position for the right pedal.

Deleting and Undeleting

Deleting an entity removes its mathematical definition from the CAD drawing's database. Some CAD systems call this *erasing*.

Many CAD systems also offer an Undelete function that undoes a prior deletion. These systems keep the digital data in memory until the CAD file is saved or until the CAD session ends. Therefore, if you change your mind about deleting an entity, you can usually undelete it as long as you have not closed the drawing file.

FINISHING A DRAWING

Most drawings require dimensions, notes, and labels. These are generally added to a drawing after all the shapes or parts in the drawing are complete.

Dimensions

Dimensions in CAD drawings are very similar to those you create manually. (See Chapter 6 for a detailed discussion of dimensioning.) By default, CAD dimensions are associated with the entities they describe. Some CAD systems call this *associative dimensioning*. If you stretch or scale a dimensioned entity, for example, the dimension updates automatically. You do not have to change or redraw the dimensions for the changed entities. Figure 15-17 shows an example of how associative dimensioning works.

Notes

Notes include any text on a drawing that gives information that applies to the entire drawing, rather than to one specific entity. For example, a note on a drawing might specify the type of material to be used or the radius of fillets and rounds on a cast part.

To add notes to a drawing, some CAD systems require you to enter the note at the command line and then insert it into the drawing. More advanced software allows you to see the note in the drawing as you are creating it. In other systems, you have the option of creating a note using separate word processing software. You can then save it as a text file and then read it into the drawing file.

Labels

Labels are similar to notes, but they include a leader line and arrow to point to the item that requires the label (Fig. 15-21). The drafter identifies where the leader line begins and ends.

SPECIAL FUNCTIONS

CAD systems have special functions that help you change the way entities are displayed on the computer screen and in the final print of the drawing. These functions make it easier to work with the drawing file. A few of the special functions common to most CAD systems are described in this section.

Display

The functions that change the way entities are displayed on the screen are the Zoom and Pan functions. The Zoom function allows you to "zoom" in on entities that are too small to be seen or edited when the full drawing is shown on the screen. It also allows you to zoom out to see the

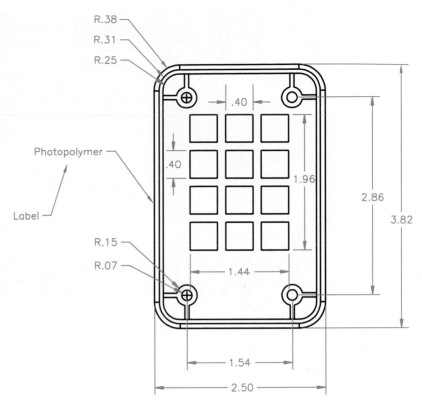

Fig. 15-21 *Labels are placed where needed to identify special features of the part.*

entire drawing at one time. Depending on the CAD system, Zoom may be divided into several separate functions.

At high magnifications (close zooms) the Pan function moves the drawing around so that the drafter can see different portions of the drawing. Pan does not change the size of the entities on the screen.

Views

Three-dimensional objects must often be viewed from more than one viewpoint. Views in a CAD drawing are the viewpoints from which the drawing is displayed. Most CAD systems can display all of the orthographic views of a 3D object (Fig. 15-22). In addition, advanced systems allow the user to define custom views. These views can make it easier for the drafter to work on inclined surfaces of the 3D part.

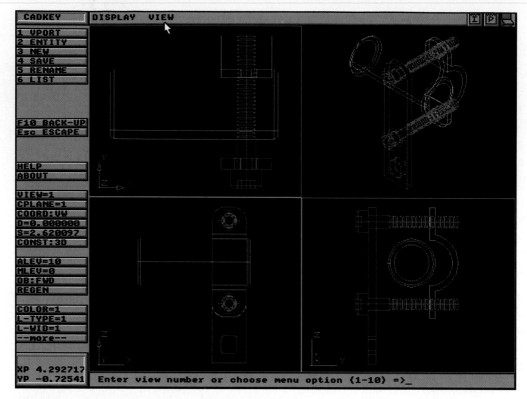

Fig. 15-22 *CAD systems allow the drafter to display the standard orthographic views of three-dimensional designs.* (Courtesy of Cadkey, Inc.)

Levels

Many CAD systems allow you to place entities on different levels, or layers, as described in Chapter 14. During the editing process, you can move entities from one layer to another, change the properties of layers, and use the layers as filters for selection procedures. You can also hide (or "freeze") one or more layers.

Blocks

When you need more than one of a group of entities, you can select them and save them as a pattern, or block. The block can be local (to be used only in the current drawing) or global (to be used in all drawings). You can also save several blocks in a single drawing file to create a symbol library. Commercial symbol libraries are also available for standard patterns, including fasteners, cabinets, windows, doors, and many others.

Modeling

When a company uses manual drafting methods to create working drawings for a new product, the product must be tested before mass production begins. To test for interference and other potential problems, they build a full-scale model of the product. They call the model a *mockup*. The engineering design team uses the mockup to find design interference problems.

CAD modeling allows the drafter to create a true, full-size, three-dimensional representation of the part being designed. The engineering design team uses the 3D model to analyze interference problems, mass properties, stress, and thermal analysis (Fig. 15-23). The 3D model can also be used for animation to show a potential customer how the part functions before it is produced. In addition, CAD models can be used with rapid prototyping equipment to produce a physical model of the part. (See Chapter 1 for more information

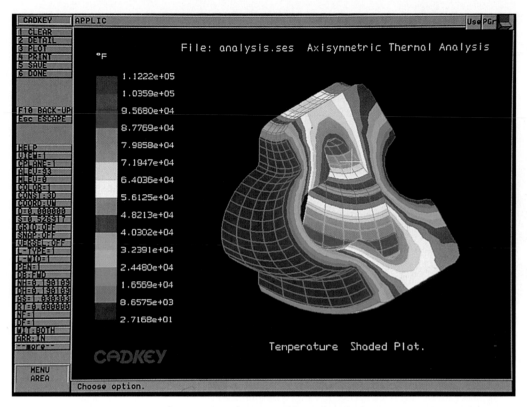

Fig. 15-23 *Axisymmetric thermal analysis.*

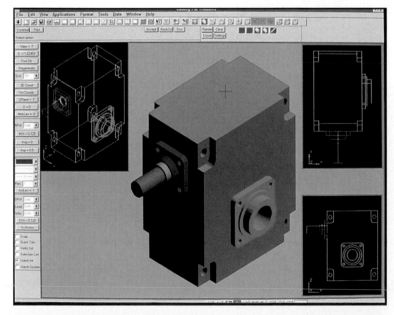

Fig. 15-24 *Hidden line removal hides the geometric entities that are not visible in the view being displayed.*

about rapid prototyping.) The modeling function, therefore, eliminates the need to build a mockup, which reduces the cost of production and aids in marketing the product.

Three-dimensional models can be wireframe models or solid models. **Wireframe models** exist in three dimensions but have no surfaces, so all of the edges are visible. Members of the engineering design team help people visualize the true shape of these models by:

- removing hidden lines so that lines that would normally not be seen in the real object are hidden from view in the drawing (Fig. 15-24).
- rendering the model to make it look as though it has a surface with color, and texture (Fig. 15-25).

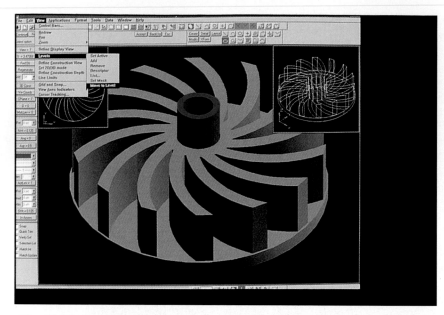

Fig. 15-25 *Rendering the impeller gives a clearer picture of the final production part.*

Solid models, rather than wireframe models, are used for most types of analysis and testing. The difference between a wireframe model and a solid model is that a solid model has mass properties, but a wireframe model does not (Fig. 15-26). Because a wireframe model has no mass properties, it cannot be used for true part analysis.

DRAWING LAYOUT

The drawing layout function allows you to position the orthographic views of a 3D model (Fig. 15-27). The CAD system arranges the orthographic views in an area that corresponds to the paper size you specify (Fig. 15-28). Some CAD systems call this *paper space*. This function saves time in producing the drawings needed for manufacturing because the orthographic views can be taken directly from the 3D model.

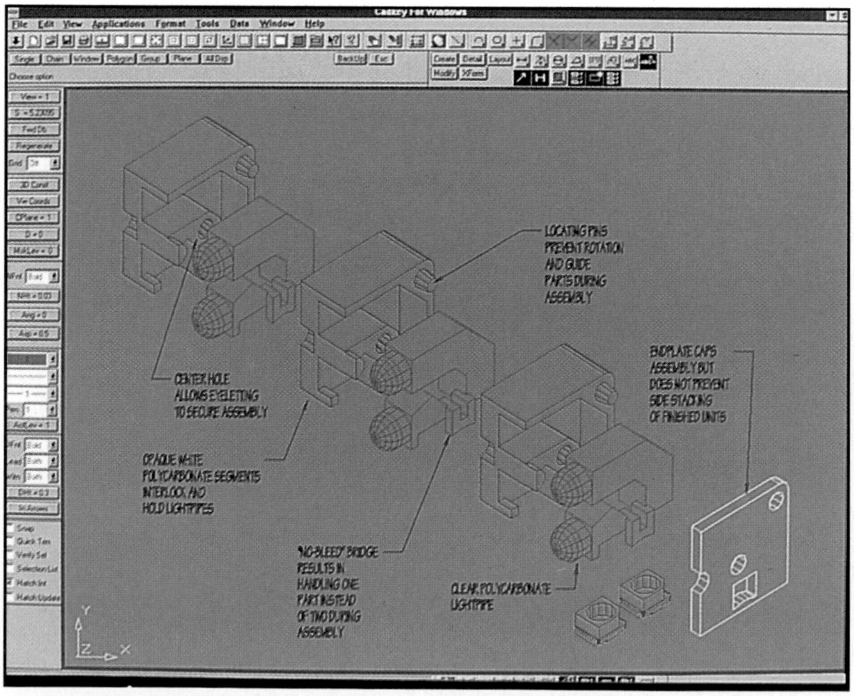

Fig. 15-26 *Until it has been rendered, a solid model looks much like a wireframe model.*

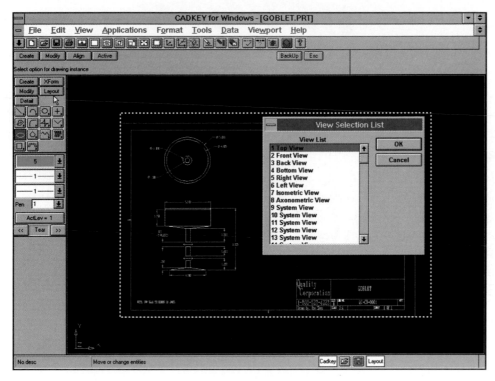

Fig. 15-27 *In CADKEY for Windows, the six standard orthographic views are available, in addition to multiple system- or user-generated views.* (Courtesy of Quality Corporation)

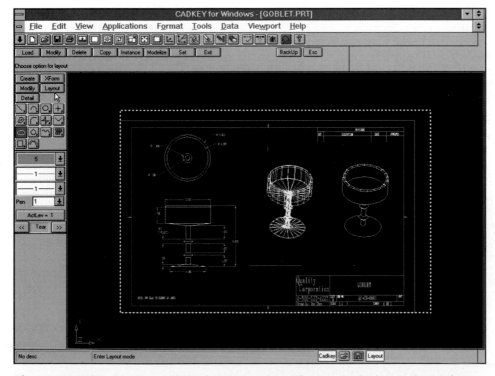

Fig. 15-28 *Goblet showing CADKEY for Windows layout function. In this layout, the top, front, and isometric views are shown, as well as an isometric with hidden lines removed.* (Courtesy of Quality Corporation)

CHAPTER 15

Learning Activities

1. Write the menu tree for one main menu function in your CAD software.
2. Work in teams of two for this activity. Simulate giving instructions to a CAD system by following this procedure: Team member #1 gives verbal instructions to team member #2, who follows only those verbal instructions (much like a computer follows the drafter's instructions). Team member #1 should describe the steps to draw a 5″ square with a 2.5″-diameter circle located at the center of the square. Then reverse the roles and give directions for other simple drawing tasks. What can you conclude about the necessity for clarity and accuracy in giving CAD instructions?
3. Prepare and give a demonstration of one CAD function to the class.

Questions

Write your answers on a separate sheet of paper.

1. List at least two advantages of using a CAD system to create drawings instead of creating them manually.
2. What does LIDS stand for?
3. What is syntax?
4. List the three common ways for a drafter to activate CAD functions.
5. What is the purpose of a branching menu?
6. List at least six entity attributes.
7. What is the difference between entity attributes and intelligent attributes?
8. List at least four entities that most CAD systems allow the drafter to create.
9. Describe at least two common placement techniques used for positioning entities.
10. List the selection techniques used to choose entities.
11. What is the point of origin of an entity?
12. What three types of text are used to detail drawings?
13. List the two main functions that are used to change the way entities are displayed on the screen.
14. What is a 3D wireframe?
15. What is the major difference between a wireframe model and a solid model? Why is this difference significant?
16. What is the purpose of rendering a model?
17. What two procedures can a drafter perform to make a wireframe drawing look more realistic?
18. What is the purpose of drawing layout?
19. Explain why it is important to place lines very accurately when you create a drawing for use with CNC equipment.
20. What is paper space?

DRAWING WITH AUTOCAD

OBJECTIVES

Upon completion of Chapter 16, you will be able to:

○ Create a drawing of a two-dimensional part using AutoCAD.
○ Use various methods of placement for AutoCAD entities.
○ Dimension objects in AutoCAD.
○ Prepare a standard title block.

VOCABULARY

In this chapter, you will learn the meanings of the following terms:

- **dock**
- **flyout**
- **icon**
- **model space**
- **paper space**
- **prototype drawings**
- **toolbars**

One of the most common CAD software programs in use in industry today is AutoCAD. This chapter shows you how to use AutoCAD Release 13 for Windows to create the part shown in Fig. 16-1. *NOTE:* To get the most benefit from this chapter, you should be at a CAD station as you work through the chapter.

NAVIGATING IN AUTOCAD

The AutoCAD screen is designed to provide all the information and commands you need to create a drawing. The standard (default) screen configuration is shown in Fig. 16-2.

In this configuration, pull-down menus appear at the top of the screen. These menus are similar to those in all Windows-compatible programs. When you click on the menu name, the pull-down menu appears (Fig. 16-3). Those menu items that have arrows on their right side produce submenus when you click on them. The items on these menus are meant mostly for file maintenance and utilities, rather than for drawing and editing tasks.

AutoCAD provides two ways to enter drawing and editing commands. The first is a complete set of **toolbars** (groups of related icons).

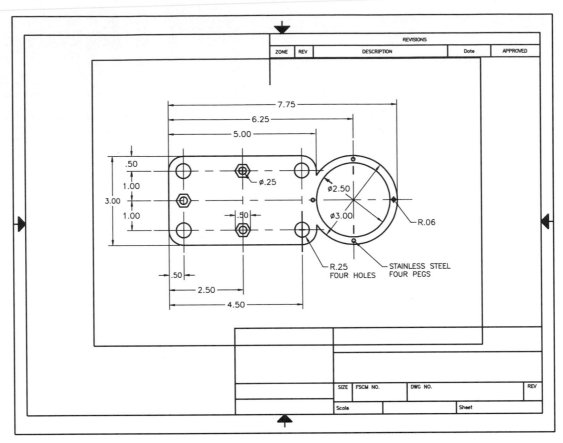

Fig. 16-1 *When you finish the procedures in this chapter, your drawing will look similar to this one.*

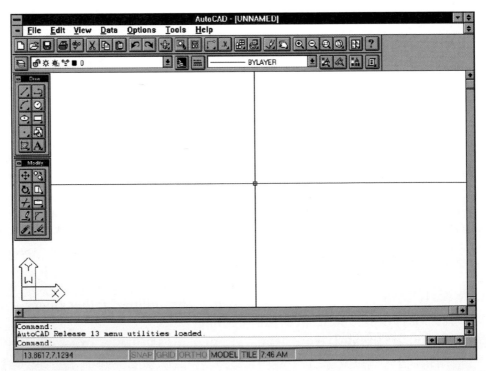

Fig. 16-2 *The components of the AutoCAD screen are shown here in their standard (default) configuration.*

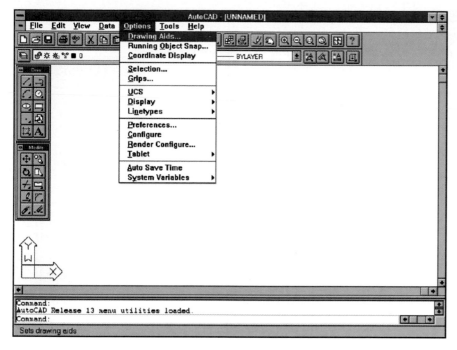

Fig. 16-3 *An example of a pull-down menu.*

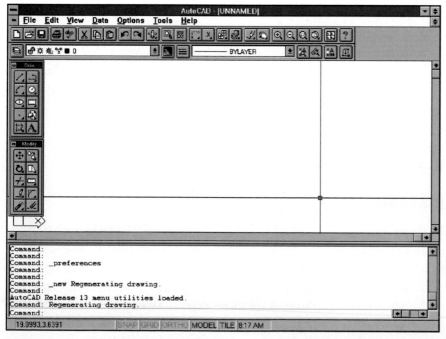

Fig. 16-4 *You can change the size of the floating command window to show as many lines as you need.*

(An **icon** is a small picture that, when you click on it, enters a command or command option.) By default, the Draw and Modify toolbars appear on the screen as floating toolbars. You can move floating toolbars around on the screen to place them conveniently for your current task. You can also **dock** the toolbars (attach them to the top or on either side of the drawing screen). In the standard configuration, the Standard and Object Properties toolbars are docked just below the pull-down menus at the top of the screen. Later in this chapter, you will use several of the icons and commands on these and other AutoCAD toolbars.

The other way to enter drawing and editing commands in AutoCAD is to type the command at the command line. The command line is located in a floating command window near the bottom of the screen. By default, this command window shows the current command and the previous two commands. However, you can change the size of the window to show more of the command history (a log of the commands you have entered) (Fig. 16-4).

Experienced drafters use a combination of entry methods to create drawings. As you work with AutoCAD, experiment with methods of entering commands to find a combination that feels comfortable for you.

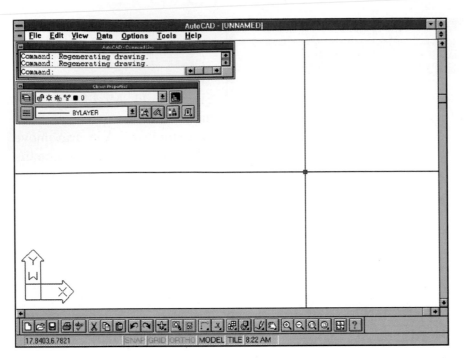

Fig. 16-5 *Some drafters prefer to change the appearance of AutoCAD to suit their needs. If your screen does not look like the one shown in Fig. 16-2, you should rearrange it, because the procedures in this chapter assume that you are using the standard configuration.*

Fig. 16-6 *When you point to an icon in AutoCAD, a tool tip shows the name of the icon.*

Because the screen in AutoCAD Release 13 for Windows is completely user-configurable, you may feel confused at first if you use a CAD station that has been customized by another user. If you become confused, take a moment to identify the basic parts: the pull-down menus, the toolbars (docked and floating), the floating command window, and the drawing area (Fig. 16-5). Also, to learn the function of any icon, position the mouse over the icon (do not click). After about a second, the name of the icon appears in a Windows-standard tool tip format (Fig. 16-6).

PREPARING A PROTOTYPE DRAWING

Most companies that use CAD have created **prototype drawings** to standardize the settings used for various types of drawings used by the company. A prototype drawing is a drawing file in which all or part of the drawing attributes are set, but in which a drawing has not been placed. After it has been saved, it can be used as a template on which new drawings are based.

This section describes a method to create a prototype drawing in AutoCAD. The prototype drawing will set up six layers typical of those used for mechanical parts. It will also set the linetype and color for each layer. *NOTE:* The procedures in this chapter assume that the AutoCAD screen is in its standard configuration, as shown in Fig. 16-2, and that the standard ACAD prototype drawing is the prototype for the new drawing.

Creating a New Drawing

To create a new drawing in AutoCAD, follow these steps:

1. From the Standard toolbar docked at the top of the screen, click on the New icon (Fig. 16-7).

The Create New Drawing dialog box appears, asking for the name of the new drawing and the name of the prototype drawing on which it should be based (Fig. 16-8).

2. Enter a name for the new drawing, such as aaaproto, where "aaa" represents your initials.

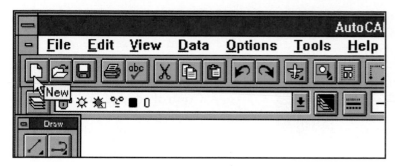

Fig. 16-7 *The location of the New icon.*

Fig. 16-8 *The Create New Drawing dialog box.*

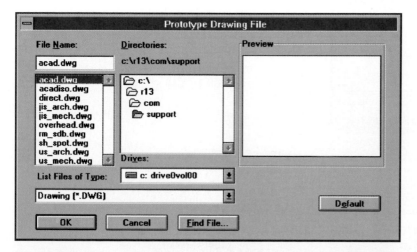

Fig. 16-9 *Find and double-click on the acad.dwg file in the Prototype Drawing File dialog box to set AutoCAD's standard prototype drawing as the prototype for your new drawing.*

3. If the prototype drawing is not acad.dwg, click on the Prototype... button and then double-click on acad.dwg in the resulting list box (Fig. 16-9). This sets AutoCAD's default prototype drawing as the prototype for the new drawing.

4. Click on the OK button (or press RETURN on the keyboard).

AutoCAD starts a blank drawing on the screen.

Creating the Layers

To create layers in AutoCAD, you can use the LAYERS command. However, the fastest way to create multiple layers is to use the Layers icon. Follow these steps:

1. Click on the Layers icon on the Object Properties toolbar (Fig. 16-10).

The Layer Control dialog box appears. The default layer (layer 0) appears in the list box near the top left of the dialog box. Notice that the cursor appears in a text entry box near the bottom of the dialog box.

2. Type the word OUTLINE.

3. Pick the New button.

The new layer name appears in the list box (Fig. 16-11).

4. Using steps 2 and 3, create five additional layers. Name them:

CENTER

DIMENSIONS

HIDDEN

HOLES

FASTENERS

The layers appear in the list box in alphabetical order. Notice that all the new layers have been assigned the continuous linetype and the color white. These are AutoCAD's defaults.

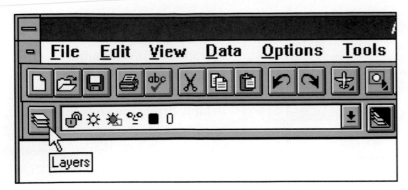

Fig. 16-10 *The location of the Layers icon.*

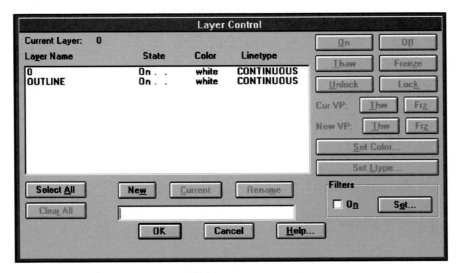

Fig. 16-11 *The new OUTLINE layer appears in the Layer Name list box.*

Setting Color and Linetype

You can set the color and linetype for each new layer before you close the Layer Control dialog box. Follow these steps:

1. Select (click on) the OUTLINE layer in the list box.

2. On the right side of the dialog box, click on the Set Color... button.

3. The Select Color dialog box appears (Fig. 16-12). Select the color red from the Standard Colors selection at the top of the dialog box, and pick the OK button.

4. Pick the OUTLINE layer again in the list box, or pick the Clear All button at the bottom of the dialog box to deselect the OUTLINE layer.

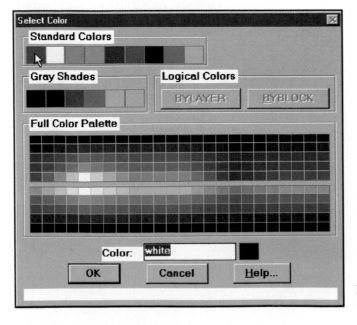

Fig. 16-12 *The Select Color dialog box allows you to set the color for each layer independently.*

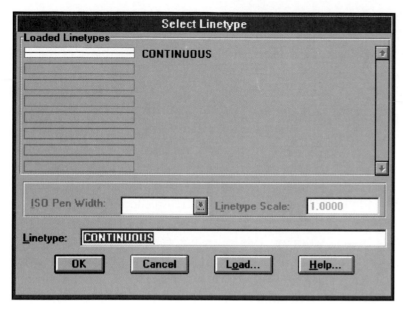

Fig. 16-13 *The Select Linetype dialog box.*

Fig. 16-14 *The Load or Reload Linetypes dialog box allows you to define linetypes for the current drawing file.*

5. Use the same procedure to set the other layers to the following colors:

CENTER	Yellow
HIDDEN	Cyan
HOLES	Green
DIMENSIONS	Magenta
FASTENERS	Blue

Now set the linetypes. The linetypes for OUTLINE, HOLES, DIMENSIONS, and FASTENERS will be continuous (the default). For the CENTER and HIDDEN layers, follow this procedure:

6. Select the CENTER layer in the list box.

7. Pick the Set Ltype... button (just below the Set Color... button).

The Select Linetype dialog box appears (Fig. 16-13).

8. Pick the Load... button at the bottom of the dialog box.

9. From the Load or Reload Linetypes dialog box, pick the Select All button (Fig. 16-14).

All the linetypes in the list box become selected.

10. Pick the OK button.

This loads all the linetypes and makes them available to the current drawing.

11. Select the CENTER linetype from the list box and pick the OK button. (Note that you may have to scroll through the list using the up and down arrows or the scroll bar at the right of the list box to find the CENTER linetype.)

12. Deselect the CENTER layer and select the HIDDEN layer in the list box.

13. Pick the Select Ltype... button again.

14. Select the HIDDEN linetype from the list box and pick the OK button.

15. Check the layers against Fig. 16-15 to make sure you created them and set their attributes correctly.

16. Pick OK in the Layer Control dialog box to finish the procedure.

17. Pick the Save icon to save the prototype drawing (Fig. 16-16).

PLANNING THE DRAWING

Before you begin a drawing, you should plan how you will proceed. In this case, you will be drawing the mechanical part shown in Fig. 16-1. Decide which parts of the object should be placed on which layer, and plan which parts to draw in which order.

You should also look for features that occur more than once on the object. As you discovered in Chapter 10, when an object has more than one of a feature, you can draw the feature once and create a block of it. You can then insert the block as many times as necessary. Examine the part or object you will be drawing to see if any of the features qualify for blocking.

Notice the nuts in the drawing. These nuts are examples of features that can be blocked. You will use this approach to save time in creating the drawing. Also notice the basic rectangular and circular shapes in the drawing. In the following procedure, you will first create the basic shapes; then you will refine them to make the finished drawing.

One additional note before you begin: As you work on this drawing, remember to save it frequently. Depending on how much you have accomplished, you should save your work every 5 to 15 minutes. To save your work in AutoCAD Release 13 for Windows, enter the SAVE command at the keyboard or pick the Save icon near the top left corner of the screen.

DRAWING THE PART

Now it's time to begin the actual drawing. Some drafters prefer to begin by building all the blocks they need before they draw the actual part. Others prefer to outline the part first. This procedure follows the latter approach.

Begin by creating the basic rectangle for the left side of the part. As you can see from Fig. 16-1, the rectangle is five inches wide and three inches high. Follow these steps:

1. Open a new drawing named PART1. Select your new prototype drawing as the prototype for the new drawing.

2. On the Object Properties toolbar (docked at the top of the screen), click on the Layer Control bar (Fig. 16-17A).

A drop-down list of layers defined in the drawing appears, as shown in Fig. 16-17B.

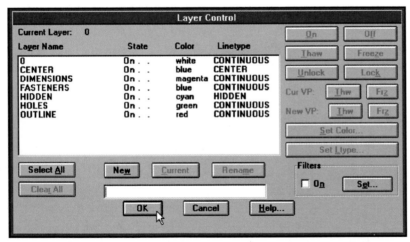

Fig. 16-15 *Check the layers on your drawing to make sure the layers, their colors, and their linetypes match those shown here.*

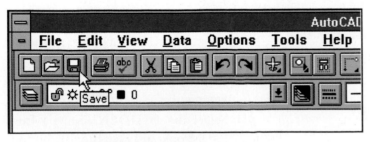

Fig. 16-16 *The location of the Save icon.*

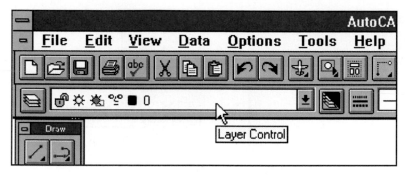

Fig. 16-17A *The location of the Layer Control bar.*

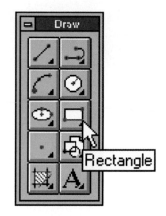

Fig. 16-18 *The location of the Rectangle icon.*

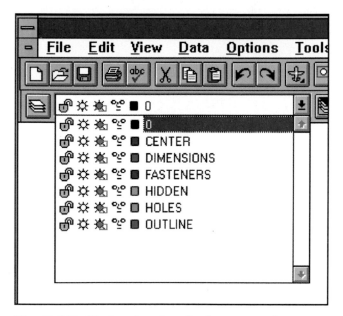

Fig. 16-17B *The drop-down layer list that appears when you click on the Layer Control bar.*

TABLE 16-1	AutoCAD Commands	
CREATE AN ARC		A
CREATE A CIRCLE		C
CREATE A LINE		L
ERASE AN ENTITY		E
COPY AN ENTITY		CP
MOVE AN ENTITY		M
REDRAW THE SCREEN		R
ZOOM TO A DIFFERENT MAGNIFICATION		Z

Table 16-1. *AutoCAD Commands.*

3. Click on the OUTLINE layer to change the current layer to OUTLINE. *NOTE:* On some computer systems, you may have to click in the drawing area after clicking on the OUTLINE layer to make the Layer Control drop-down list disappear.

4. From the Draw toolbar, pick the Rectangle icon (Fig. 16-18).

5. Enter the coordinates 0,0 to place the lower left corner of the rectangle at the origin. (Refer to Chapter 5 for more information about the Cartesian coordinate system.)

6. For the upper right corner, enter the coordinates 5,3.

The rectangle appears. Now you can use the ZOOM command to change the size of the drawing on the screen (without changing the actual dimensions of the object). This will make the drawing easier to work with.

7. Enter the ZOOM command by pressing the Z key and then RETURN on the keyboard.

8. At the resulting prompt, enter .5x and press RETURN to make the rectangle appear half as large on the screen.

As you can see, you can enter some commands in AutoCAD very quickly by entering just the first one or two letters of the command name. Other commands for which this is possible are listed in Table 16-1.

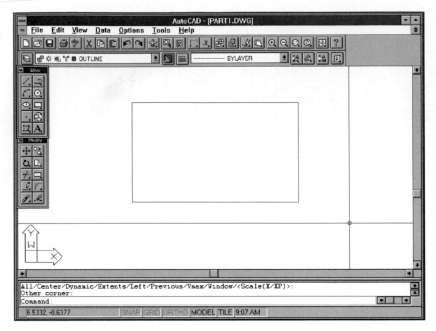

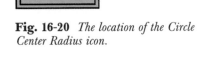

Fig. 16-20 *The location of the Circle Center Radius icon.*

Fig. 16-19 *The new drawing magnification. Note that the actual dimensions of the rectangle have not changed. It still measures 5 inches by 3 inches.*

At this screen resolution, you may find that the rectangle is not quite large enough to edit easily.

9. Press the space bar or the RETURN key to reenter the ZOOM command.

This demonstrates another shortcut for working in AutoCAD. The fastest way to reenter a command in AutoCAD is to press either the space bar or the RETURN key. The command is reentered automatically.

10. This time, instead of responding to the prompt at the keyboard, use the cursor to pick a point just above and to the left of the rectangle in the drawing area.

11. Move the cursor away from the point you picked. Notice that a "rubberband" box begins to form.

12. Pick a second point below and to the right of the rectangle.

The current drawing magnification changes to the rectangular viewing area you just specified (Fig. 16-19).

13. Pick the Save icon on the Standard toolbar to save your work.

Now you can draw the circles that represent the four holes at the corners of the rectangle. First, however, you must change the current layer and display an additional toolbar on the screen.

1. Pick the Layer Control bar on the Object Properties toolbar to see the drop-down list of layers.

2. Click on the HOLES layer to set the current layer to HOLES.

3. Click on the Tools pull-down menu and then click on the Toolbars item to see a list of available toolbars. Pick the Object Snap toolbar.

The Object Snap toolbar appears as a floating toolbar. (If its location is not convenient, click on the top bar of the toolbar, which contains its name, and hold the pick button down while you move the toolbar on the screen.)

Now draw the hole in the upper left corner of the rectangle. Follow these steps:

4. From the Draw toolbar, pick the Circle Center Radius icon (Fig. 16-20).

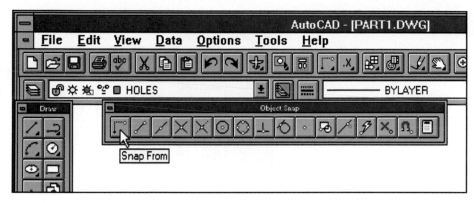

Fig. 16-21A *The location of the Snap From icon.*

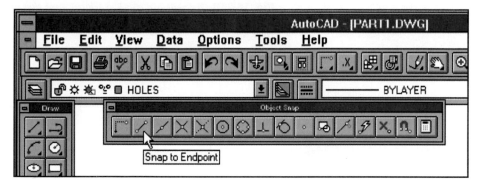

Fig. 16-21B *The location of the Snap to Endpoint icon.*

5. Instead of picking a point on the screen for the center of the circle, click on the Snap From icon in the Object Snap toolbar (Fig. 16-21A).

This enters the From object snap mode. (See Chapter 4 for more information about object snap modes.) The From object snap allows you to use a reference location to set the location of the current point, in this case the center of the hole.

6. Click on the Snap to Endpoint icon in the Object Snap toolbar (Fig. 16-21B).

Endpoint is an object snap mode that selects the endpoint of a line or arc.

7. In the drawing area, pick the upper left corner of the rectangle.

8. Place the center of the circle ½ inch below the corner by specifying @.5,-.5 and pressing RETURN.

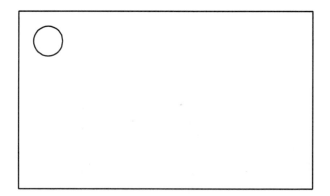

Fig. 16-22 *The first hole in its proper location on the part.*

9. When AutoCAD prompts for the radius, specify .25 and press RETURN (Fig. 16-22).

For the other three circles, you can use the MIRROR command.

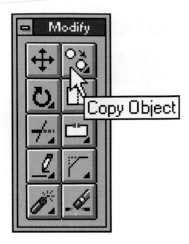

Fig. 16-23A *The location of the Copy Object icon.*

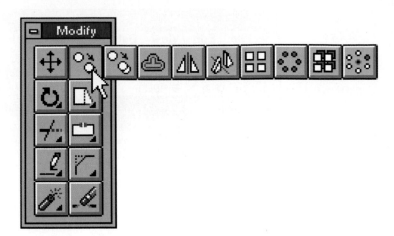

Fig. 16-23B *A flyout is a group of related icons that appears when you hold down the mouse button on icons that have small arrows in their lower right corner.*

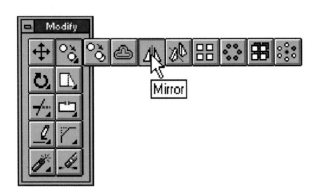

Fig. 16-23C *The location of the Mirror icon.*

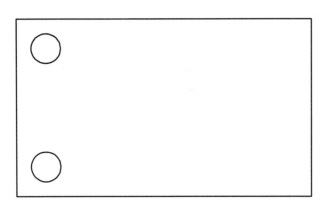

Fig. 16-24 *The result of the MIRROR operation: the second hole appears in its proper location on the part.*

10. On the Modify toolbar, pick and hold down the mouse button over the Copy Object icon (Fig. 16-23A).

A group of related icons appears (Fig. 16-23B). This is called a **flyout.** (All icons that have a small arrow in the lower right corner have flyouts.)

11. Still holding down the mouse button, move the cursor slowly over the icons in the flyout to see their names. Release the mouse button over the Mirror icon (Fig. 16-23C).

12. At the "Select objects" prompt, pick the circle you just completed and press RETURN (or click on the right mouse button to enter a RETURN).

13. For the first point of the mirror line, choose the Snap to Midpoint icon from the Object Snap toolbar.

14. At the "of" prompt, select the left vertical line of the rectangle.

15. When AutoCAD requests the second point, pick the Snap to Midpoint icon again, and then pick the right vertical line of the rectangle.

16. Enter N (for No) when AutoCAD asks if you want to delete the original object.

The circle representing the lower left hole appears. The object should look like the illustration in Fig. 16-24. Now use the MIRROR command again to create the other two holes in one operation.

17. Notice that the Mirror icon has replaced the Copy Object icon in the Modify toolbar (Fig. 16-25). Pick the Mirror icon to reenter the command.

18. At the "Select objects" prompt, select both of the holes you have already created, and then press RETURN.

19. For the first point of the mirror line, use the Snap to Midpoint icon and specify the upper horizontal line of the rectangle.

20. For the second point of the mirror line, use the Snap to Midpoint icon and specify the lower horizontal line of the rectangle.

The other two holes appear (Fig. 16-26).

21. If you have not done so recently, pick the Save icon to save your work.

Now create the nuts. Follow this procedure:

1. Click on the Layer Control bar in the Object Properties toolbar to see the drop-down list of layers.

2. Click on the FASTENERS layer and then click in the drawing area to change the current layer to FASTENERS.

3. From the Draw toolbar, pick the Circle Center Radius icon.

4. When AutoCAD prompts for the center of the circle, click on the Snap From icon in the Object Snap toolbar.

5. In response to the next prompt, click on the Snap to Midpoint icon in the Object Snap toolbar.

6. Pick the top horizontal line of the rectangle.

7. Specify @0,−.5 to place the center of the circle half an inch below the top center of the rectangle.

8. Specify a radius of .125.

The circle appears on the drawing. Then, to create the outside of the nut, use the POLYGON command as follows:

9. On the Draw toolbar, hold the mouse button down over the Rectangle icon to see the flyout icons.

10. Pick the Polygon icon from the flyout (Fig. 16-27).

11. For the number of sides, enter 6.

Fig. 16-25 *For icons that have flyouts, the last icon you used replaces the default icon in the toolbar.*

Fig. 16-26 *Your drawing should now contain four holes, spaced as shown here.*

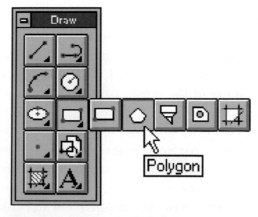

Fig. 16-27 *The location of the Polygon icon.*

12. To position the center of the polygon, pick the Snap to Center icon from the Object Snap toolbar and pick the circle that you created in steps 3 through 8 above. (*NOTE:* Do not pick inside the circle. You must pick a point on the circle itself.)

13. Enter the letter I for inscribed measurement.

14. For the radius, enter .25.

15. Save your work.

This completes the nut representation. At this point, you could use the COPY command to place the other two nuts on the drawing. However, for this drawing, you will create a block of the existing nut and then insert it into the other locations. (*NOTE:* The more copies of a block you need for a drawing, the more effective and efficient the blocking procedure becomes.) To create the block, follow these steps:

1. Enter the BLOCK command at the keyboard.

2. Name the block NUT.

3. At the "Insertion base point" prompt, pick the Snap to Center icon on the Object Snap toolbar and then pick the small circle that represents the inside diameter of the nut.

4. At the "Select Objects" prompt, use the cursor to pick a point above and to the left of the nut, and a second point below and to the right of the nut. Be sure the entire nut (represented by the circle and the hexagon) is contained in the resulting box or "window."

When you press RETURN, the nut is blocked and disappears from the screen. In this case, you have already placed the nut in its exact location. Rather than reinserting it and specifying the location again, you can bring it back to the screen using the OOPS command. This does not affect the creation of the block, which is still present in the drawing file.

5. At the keyboard, enter the OOPS command.

The nut reappears on the screen (Fig. 16-28). Now you can proceed to place the other two nuts on the drawing.

6. From the Draw toolbar, pick the Insert Block icon (Fig. 16-29) to make the Insert dialog box appear.

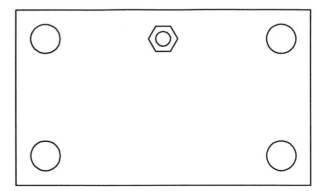

Fig. 16-28 *Your drawing should now have four holes and a nut, as shown in this illustration.*

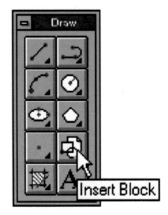

Fig. 16-29 *The location of the Insert Block icon.*

7. Pick the Block... button to see a list of blocks defined in the current drawing. (In this case, the list should contain one block named NUT.)

8. Pick NUT and then pick OK.

9. Pick OK again to close the Insert dialog box.

The block appears on the crosshairs.

10. Pick the Snap to Midpoint icon from the Object Snap toolbar and then select the left vertical line on the rectangle.

11. Press RETURN three times to accept the default X and Y scales of 1 and no rotation.

12. Pick the Move icon from the Modify toolbar, select the block, and press RETURN.

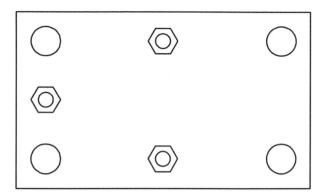

Fig. 16-30 *Your drawing should now look like the one shown in this illustration.*

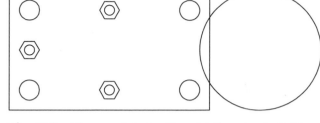

Fig. 16-31 *The outer circle should overlap the rectangle slightly, as shown here.*

13. For the base point, pick the Snap to Center icon from the Object Snap toolbar and then select the block.

14. For the second point of displacement, enter the relative coordinates @.5,0 to move the block into its proper position.

15. Pick the Insert Block icon again.

AutoCAD remembers the last block you inserted and presents it as the default for this insertion. Notice that the word NUT appears in the text box after the Block... button in the Insert dialog box.

16. Pick OK to accept the NUT block default.

17. For the insertion point, pick the Snap to Midpoint icon and then the lower horizontal line of the rectangle.

18. Press RETURN three times to accept the default X and Y scales of 1 and no rotation.

19. Pick the Move icon on the Modify toolbar, select the block you just inserted, and press RETURN.

20. For the base point, pick the Snap to Center icon and then pick the block.

21. Enter the relative coordinates @0,.5 to place the nut ½″ above the bottom of the rectangle.

22. Press the R key and then press RETURN to enter the REDRAW command.

This refreshes the drawing, eliminating the "blips" that remain on the screen from the previous drawing and editing commands. You can use the REDRAW command at any time during these procedures. Your drawing should now look similar to that in Fig. 16-30.

23. Save your work.

Next draw the two large circles that represent the right side of the part. Follow these steps:

1. Press the Z and RETURN keys to enter the ZOOM command.

2. Enter .5x to make the drawing appear smaller on the screen. This will make room on the screen for the right portion of the part.

3. Pick the Layer Control box in the Object Properties toolbar, and pick OUTLINE to make it the current layer.

4. Pick the Circle Center Radius icon from the Draw toolbar.

5. For the center point, pick the Snap From icon, the Snap to Midpoint icon, and then the right vertical line of the rectangle.

6. For the offset, specify the relative coordinates @1.25,0.

7. Enter a radius of 1.5.

The outer circle appears on the screen (Fig. 16-31). To create the inner circle, you could use the circle again, but drafters often find it more convenient to use the OFFSET command, as shown in the steps below.

1. Place the cursor over the Mirror icon on the Modify toolbar, and press and hold the mouse button to see the flyout icons. From the flyout, pick the Offset icon (Fig. 16-32).

2. Enter an offset distance of .25.

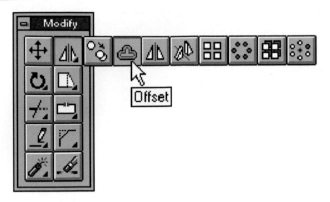

Fig. 16-32 *The location of the Offset icon.*

Fig. 16-34 *The location of the Trim icon.*

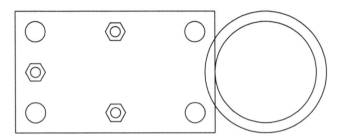

Fig. 16-33 *The result of the OFFSET operation.*

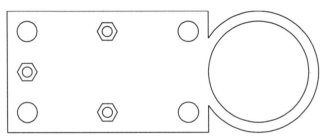

Fig. 16-35 *After the TRIM operation, your drawing should look like the one shown here.*

3. For the object to offset, select the large circle you just created.

4. At the "Side to offset?" prompt, pick a point anywhere inside the large circle.

5. Press RETURN to end the OFFSET command.

The second circle appears (Fig. 16-33). Notice that the right vertical line of the rectangle passes through the larger circle. Use the following steps to trim away the unwanted portions of the rectangle and circle.

1. Pick the Trim icon from the Modify toolbar (Fig. 16-34).

2. Select the large outside circle and the rectangle and press RETURN.

3. At the next prompt, pick on the part of the rectangle that passes through the circle to remove that part of the rectangle.

4. Pick on the part of the circle that extends into the rectangle to remove that part of the circle.

5. Press RETURN to end the TRIM command.

Your drawing should now look similar to that in Fig. 16-35. While you are editing the outline of the part, round the corners of the rectangular portion of the part by following these steps:

1. Pick the Explode icon on the Modify toolbar, select all the lines of the rectangle, and press RETURN.

This breaks the rectangle into individual line segments for editing.

2. Place the cursor over the Chamfer icon on the Modify toolbar and hold down the mouse button to see the flyout icons. Release the mouse button over the Fillet icon (Fig. 16-36).

3. Press the R key and RETURN to enter the Radius option.

4. Enter a new radius of .50.

5. Press the space bar or RETURN to reenter the FILLET command.

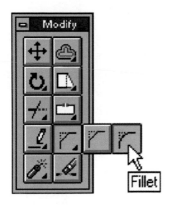

Fig. 16-36 *The location of the Fillet icon.*

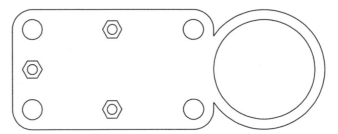

Fig. 16-37 *When you have completed all four fillets, your drawing should look like the one shown here.*

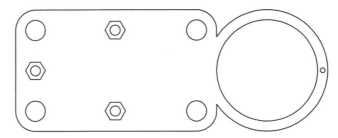

Fig. 16-38 *The first stainless steel peg in its proper location.*

6. Pick the upper horizontal line of the rectangle near the left side.

7. At the "Select second object" prompt, pick the left vertical line of the rectangle near the top (even though it may already appear to be "selected").

The top left corner of the rectangle becomes rounded.

8. Repeat the fillet operation for each of the other three corners.

When you finish, your drawing should look like the one in Fig. 16-37.

9. Save your work.

Now draw the small stainless steel pegs in the right side of the part next by following these steps:

1. Pick the Circle Center Radius icon from the Draw toolbar.

2. For the center point, pick the Snap From and Snap to Quadrant icons, and then pick the large outside circle at the 3 o'clock position.

3. For the offset, enter the relative coordinates @−.125,0.

4. Enter a radius of .06.

The first peg appears (Fig. 16-38). You could use several different methods to place the other three pegs. Because the pegs in this part are placed at equal intervals on the part, one of the most efficient methods is to use the ARRAY command. Follow these steps.

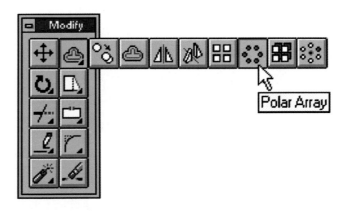

Fig. 16-39 *The location of the Polar Array icon.*

5. Place the cursor over the Offset icon and hold down the mouse button to see the flyout icons. Release the mouse button over the Polar Array icon (Fig. 16-39).

6. Select the small circle you just created and press RETURN.

7. For the center point of the array, pick the Snap to Center icon from the Object Snap toolbar and then pick either of the two large circles on the right side of the part.

8. For "Number of items," enter 4.

9. Accept the 360-degree default for "Angle to fill" by pressing RETURN.

10. Press the N key and RETURN to avoid rotating the circles as they are copied to their new positions.

The other three pegs appear (Fig. 16-40).

11. Save your work.

DIMENSIONING THE DRAWING

Now that the 2D drawing of the part itself is complete, it is time to dimension it and add any necessary notes and labels. First, however, you must change a few settings in the drawing.

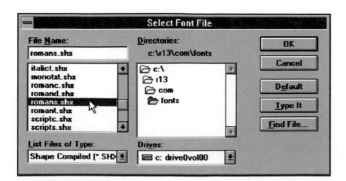

Fig. 16-40 *After the ARRAY operation, your drawing should look like the one shown here.*

In the procedure below, you will change the text style to one of the styles most often used by drafters, set the dimensioning units to two decimal places, and set the Grid and Snap features appropriately. Follow these steps:

1. Enter the STYLE command at the keyboard.

2. For the text style name, enter ROMANS.

3. In the resulting Select Font File dialog box, scroll through the file names until you see "romans.shx." Double-click on this file name to select it (Fig. 16-41).

4. Press RETURN repeatedly until the command line shows that ROMANS is now the current text style and the "Command" prompt reappears.

5. Enter the DDUNITS command at the keyboard.

6. In the lower left corner of the resulting Units Control dialog box, pick the drop-down box below "Precision:" to see a list of possible precisions (Fig. 16-42A).

7. Pick the 0.00 option.

8. In the lower right corner of the dialog box, pick the drop-down box below "Precision:" and pick the 0 option for angle precision (Fig. 16-42B).

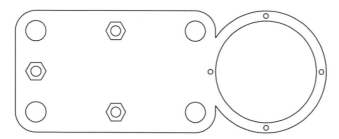

Fig. 16-41 *Scroll through the names in the list box until you see the romans.shx file.*

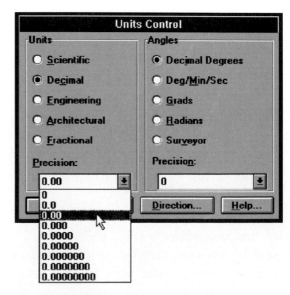

Fig. 16-42A *When you click on the Precision box on the Units side of the Units Control dialog box, a drop-down list of precisions appears.*

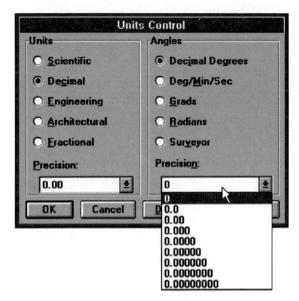

Fig. 16-42B *When you click on the Precision box on the Angles side of the Units Control dialog box, a similar list appears.*

Fig. 16-43 *The location of the Dimension Styles icon.*

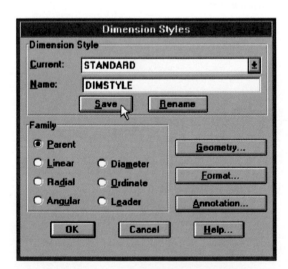

Fig. 16-44 *Be sure to pick the Save button after you name a new dimension style. Also remember to pick Save again after you make any changes to the style.*

9. Pick OK to set the new precisions and remove the dialog box.

10. On the Tools pull-down menu, place the cursor over the Toolbars item to see a list of available toolbars. Pick the Dimensioning toolbar.

11. Pick the Dimension Styles icon on the Dimensioning toolbar (Fig. 16-43).

12. In the text box after Name, change STANDARD to DIMSTYLE and pick the Save button (Fig. 16-44).

13. Pick the Annotation... button near the bottom right of the resulting Dimension Styles dialog box.

14. In the Primary Units section of the resulting Annotation dialog box, pick the Units... button.

15. In the Primary Units dialog box, set the precisions to match the precisions you set using DDUNITS (0.00 and 0).

16. Pick the check box next to Leading in the Zero Suppression area for both the units and the angles.

17. Pick OK to return to the Annotation dialog box.

18. Pick the drop-down box next to Style to see a list of text styles defined in this drawing. Pick the ROMANS style and then pick OK to return to the Dimension Styles dialog box.

19. Pick the Save button.

20. Pick the OK button.

21. Enter the GRID command at the keyboard and enter a value of .25.

22. Enter the SNAP command at the keyboard and enter a value of .25.

(For more information about the grid and snap functions, see Chapter 12.)

23. Save your work.

Now move the entire part to the middle of the drawing to make room for the dimensions.

1. Pick the Move icon from the Modify toolbar.

2. At the "Select objects" prompt, type ALL and press RETURN twice.

The entire part becomes highlighted.

3. For the base point, enter the absolute coordinates 0,0.

4. For the second point of displacement, enter the absolute coordinates 2,3.

5. Press the Z and RETURN keys to enter the ZOOM command.

6. Press the A key to show the entire drawing on the screen. The grid shows the currently defined drawing area.

7. If necessary, move some of the toolbars into more convenient locations on the screen.

8. Save your work.

Next, insert the centerlines into the drawing. Follow these steps:

1. Pick the Layer Control bar and make CENTER the current layer.

2. Double-click on the word ORTHO at the bottom of the screen to force straight vertical and horizontal lines.

3. Pick the Line icon from the Draw toolbar.

4. Using the snap and grid as guides, begin the line ½″ to the left of the part and extend it to ½″ to the right of the part, as shown in Fig. 16-45. Press RETURN to end the LINE command.

5. Place the remaining centerlines as shown in Fig. 16-46.

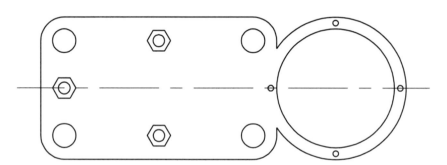

Fig. 16-45 *The location of the first centerline.*

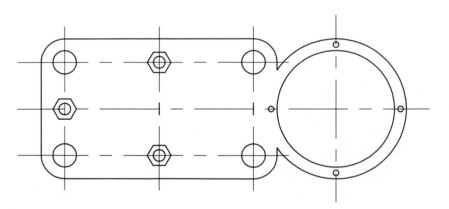

Fig. 16-46 *The centerlines for this part should be placed as shown here.*

Now you can begin to dimension the part. Follow these steps:

1. Click on the Layer Control bar in the Object Properties toolbar and select DIMENSIONS to be the current layer.

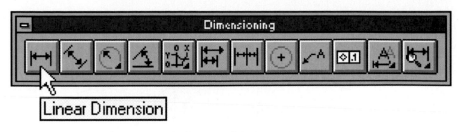

Fig. 16-47 *The location of the Linear Dimension icon.*

2. If Ortho is still on, double-click on the word ORTHO at the bottom of the screen to turn it off.

3. From the Dimensioning toolbar, pick the Linear Dimension icon to enter the DIMLINEAR command (Fig. 16-47).

4. Pick the Snap to Endpoint icon and then click on the left vertical line (near the top).

5. For the second extension line origin, pick the Snap to Endpoint icon and then click on the top portion of the right vertical line.

6. When the rubberband dimension appears on the cursor, move the cursor up to a point .75 inch above the top of the part and press the mouse button (Fig. 16-48A).

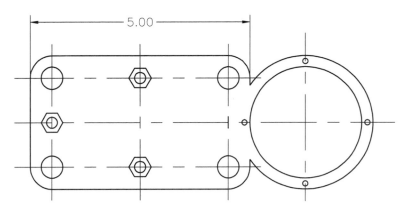

Fig. 16-48A *Place the first horizontal dimension as shown in this illustration.*

7. Reenter the DIMLINEAR command and use the Snap to Endpoint icon to snap to the left vertical line again.

8. For the right extension line, pick the Snap to Center icon, and then pick either of the large circles on the right of the part.

9. Move the cursor up to a point .25 inch above the previous dimension and press the mouse button.

10. Reenter the DIMLINEAR command, pick the Snap to Endpoint icon, and then pick the left vertical line again.

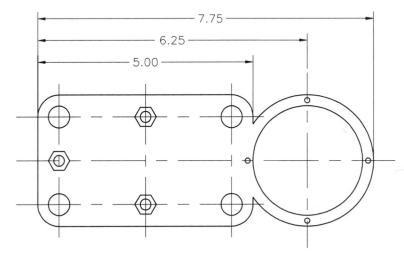

Fig. 16-48B *After you have finished step 11, your drawing should contain three horizontal dimensions, as shown here.*

11. For the right extension line, pick the Snap to Endpoint icon, and then pick the top of the vertical centerline that runs through the large circles.

12. Position the dimension .50 inch above the previous dimension (Fig. 16-48B).

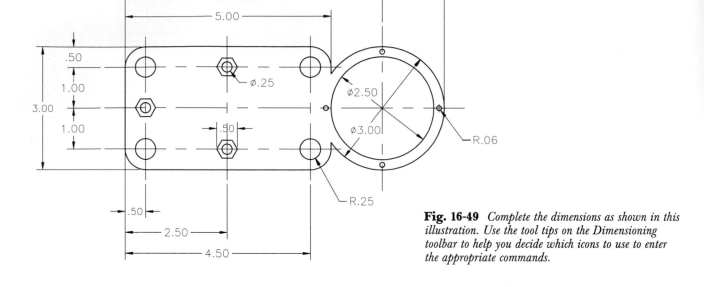

Fig. 16-49 *Complete the dimensions as shown in this illustration. Use the tool tips on the Dimensioning toolbar to help you decide which icons to use to enter the appropriate commands.*

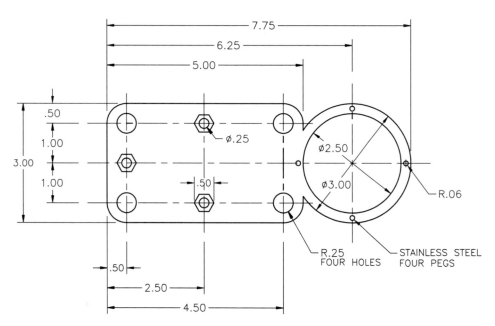

Fig. 16-50 *Use the Leader icon on the Dimensioning toolbar to create the note regarding the stainless steel pegs. Then use the DTEXT (Dynamic TEXT) command at the keyboard to insert the "FOUR HOLES" text.*

13. Continue dimensioning the part as shown in Fig. 16-49. Use appropriate commands from the Dimensioning toolbar, using the tool tips as necessary to find out which icon you should use for each dimension. Note that the DIMDIAMETER command is a flyout from the Radius Dimension icon.

14. Use the Leader icon in the Dimensioning toolbar to insert the "STAINLESS STEEL" label shown in Fig. 16-50.

15. Enter the DTEXT command at the keyboard.

16. Pick a point just below the R.25 dimension, enter a height of .18, and press RETURN to accept a rotation angle of 0. Then enter the text "FOUR HOLES."

17. Save your work.

Your finished drawing should look like the one in Fig. 16-50.

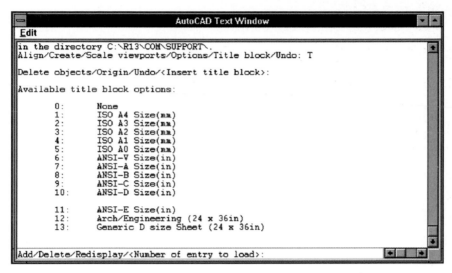

Fig. 16-51 *An AutoCAD Text window appears to list AutoCAD's standard title blocks.*

DRAWING LAYOUT

After completing the dimensions and labels, you can lay out the drawing. Many companies use their own title blocks or make changes to the title block provided by AutoCAD and save them. For this exercise, however, you will use one of Auto-CAD's standard border/title block combinations.

To lay out a drawing in AutoCAD, you must understand the difference between AutoCAD's model space and paper space. **Model space** is the mode in which you do most of your actual drawing. **Paper space** is the mode in which you lay out the drawing. In paper space, you can have as many views of the drawing as necessary. For three-dimensional drawings, for example, you might lay out a front, top, and right-side view. However, even for single-view drawings such as the one you are creating, paper space helps you position the drawing correctly on a border and title block. Follow these steps to lay out the drawing.

1. Notice the word MODEL near the bottom of the screen. This indicates that you are current-ly in model space. To change to paper space, double-click on the word MODEL.

The drawing seems to disappear from the screen, and the word PAPER replaces MODEL at the bot-tom of the screen. You have entered paper space.

2. Make layer 0 the current layer.

3. From the View pull-down menu, pick Floating Viewports and pick 1 Viewport from the list that appears.

4. Pick a point near the top left corner of the drawing area, and pick another point near the bottom right corner to define the size of the viewport. After the drawing reappears, click on the viewport border and use the grips to make the viewport the same size as the defined drawing area (as shown by the grid).

NOTE: Grips are "handles"–little blue boxes–that allow you to move or resize an object. When you click on a grip, it turns bright red to show that it has been selected. When you move the cursor and click again, the grip moves to the new location, resizing or moving the entity to which it is attached.

5. Change the current layer to OUTLINE.

6. From the View pull-down menu, pick Floating Viewports again. This time, select MV Setup from the resulting list.

7. Press the T key and RETURN to choose the Title Block option.

8. Press RETURN again.

A list of available title block options appears (Fig. 16-51).

9. Press the 7 key and RETURN to add the A-size title block to your drawing.

10. If AutoCAD asks if you want to create a separate drawing file for the title block, enter N for "No."

11. Press the S key and RETURN to scale the viewport so that the drawing fits within the title block.

12. At the "Select Objects" prompt, pick the viewport border and press RETURN.

The next two prompts allow you to scale the ratio of paper space units to model space units. When the number of paper space units and model space units are both 1, then paper space units are exactly the same as model space units.

13. Experiment with the ratios to find a size at which the drawing fits well within the border. For this drawing, you might try .50 for the first prompt and 1 for the second prompt.

14. When you are satisfied with the appearance of the drawing, press RETURN to end the MVSETUP command.

15. You may also wish to reposition the viewport within the title block. To do so, pick the Move icon and select the viewport border. The entire part drawing moves with the viewport.

16. Pick the Layer Control box and click on the yellow sun next to layer 0.

The sun changes to a snowflake. When you close the Layer Control list box, layer 0 will become *frozen* (no longer visible on the screen).

17. Save your work.

Finally, add the text for the title block to identify the drawing.

1. Use the ZOOM command (press Z and RETURN) to zoom in on the upper right corner of the drawing.

2. Pick the Explode icon on the Modify toolbar.

3. Pick any line on the border or title block and press RETURN.

This separates the border and title block into individual, editable components.

4. Pick the Properties icon on the Object Properties toolbar.

5. At the "Select Objects" prompt, pick the word DESCRIPTION on the screen and press RETURN.

6. In the resulting Modify Text dialog box, notice that the word DESCRIPTION is highlighted in the text box next to Text. Replace the word DESCRIPTION with a description of the current drawing and pick the OK button.

7. Press the space bar or RETURN to reenter the command. Then pick the word Date on the screen and press RETURN again.

8. In the text box, replace Date with the current date and pick OK.

9. Use the ZOOM command to zoom in on the lower right corner of the drawing.

10. Using the same procedure, change the text in the lower right corner of the drawing to reflect the current drawing (as necessary).

11. From the Standard toolbar, pick the Zoom All icon to see the entire drawing.

12. Save your work.

This completes the AutoCAD drawing. Review your work at this time, and check it against the drawing shown in Fig. 16-1. Make any changes necessary and resave the file.

CHAPTER 16

Learning Activities

1. Choose an AutoCAD command that is not used in this chapter and explain its use to the class.

2. Create and use an AutoCAD version of Pictionary®.

Questions

Write your answers on a separate sheet of paper.

1. In what part of the AutoCAD screen does the drawing appear?

2. What is the main purpose of the pull-down menus shown at the top of the screen?

3. When you enter a command using the keyboard, in what part of the AutoCAD screen does the command appear?

4. What is a tool tip? How can tool tips be helpful to you as you learn AutoCAD?

5. Briefly describe the process of defining layers in AutoCAD.

6. Which AutoCAD command changes the screen magnification without changing the actual dimensions of the objects on the screen?

7. What is the purpose of the From object snap?

8. Explain how using the grid and snap functions can help you create a drawing.

9. When you are trimming part of a circle from a part, which part of the circle should you select at the "Select object to trim" prompt?

10. What is the difference between the following two coordinates: 0,.5 and @0,.5?

CHAPTER 16
PROBLEMS

The small CAD symbol next to each problem indicates that a problem is appropriate for assignment as a computer-aided drafting problem and suggests a level of difficulty (one, two, or three) for the problem.

CAD problems may be generated from any chapter in this textbook. The following problems have been prepared specifically for CAD levels 1 and 2.

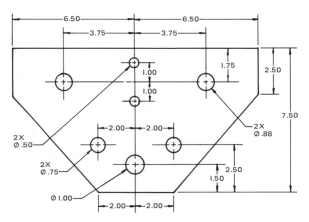

Fig. 16-52 *CAD1* *Prepare a one-view drawing of the trolley bracket, with dimensions. Use an appropriate scale.*

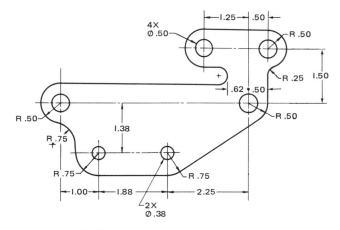

Fig. 16-53 *CAD1* *Prepare a one-view drawing of the holding bracket. Scale: Full. Material: Sheet brass. Thickness: No. 9 Brown and Sharp gage (0.1144″).*

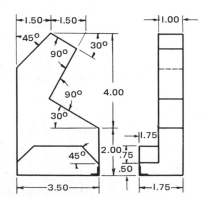

Fig. 16-54 *CAD1* *Make an isometric drawing of the notched slide. Prepare the three views necessary for the transfer of data.*

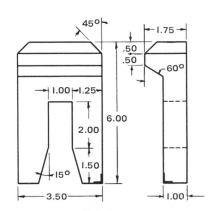

Fig. 16-55 *CAD1* *Make an isometric drawing of the slotted stop. Prepare the three views necessary for transfer to pictorial.*

PROBLEMS

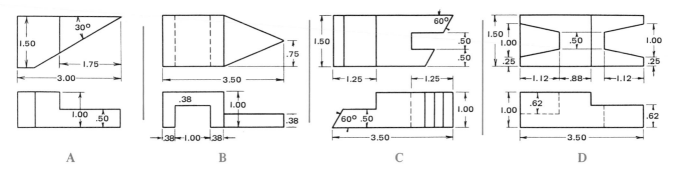

A   B   C   D

Fig. 16-56 **CAD1** *Draw three views with dimensions of the wedge (A), the V-lock (B), the lock stop at (C), or the middle stop at (D), as per your instructor's direction.*

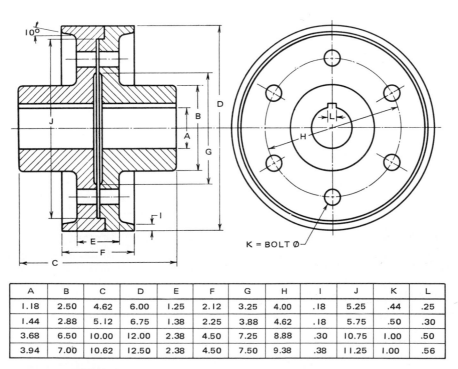

A	B	C	D	E	F	G	H	I	J	K	L
1.18	2.50	4.62	6.00	1.25	2.12	3.25	4.00	.18	5.25	.44	.25
1.44	2.88	5.12	6.75	1.38	2.25	3.88	4.62	.18	5.75	.50	.30
3.68	6.50	10.00	12.00	2.38	4.50	7.25	8.88	.30	10.75	1.00	.50
3.94	7.00	10.62	12.50	2.38	4.50	7.50	9.38	.38	11.25	1.00	.56

Fig. 16-57 **CAD2** *Make a completed working drawing of one of the sizes of the flange coupling. Choose a suitable scale. Select bolts from Bolt Tables in Appendix and fasten units.*

PROBLEMS

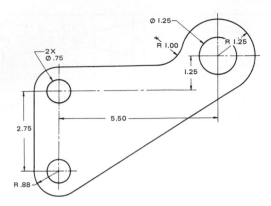

Fig. 16-58 *CAD2 Create the one-view drawing of the pivot guide. Scale: Full size. Dimension completely with decimal units.*

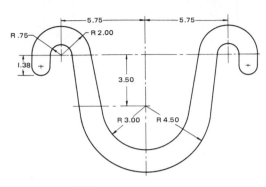

Fig. 16-59 *CAD2 Draw the single view of the hanging support. Select a suitable scale. Dimension completely.*

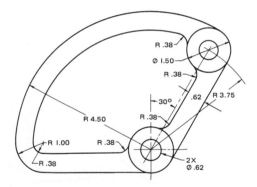

Fig. 16-60 *CAD2 Prepare a drawing of the sector. Dimension the casting and create a right-side view. Bottom hub 1.5"; top hub .75"; rib .75" centered.*

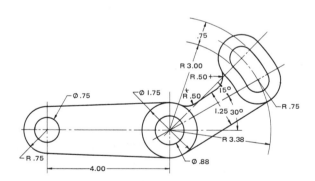

Fig. 16-61 *CAD3 Create two views of the adjustable arm. Dimension the detailed drawing of the casting and design (select) the thicknesses of the part as needed.*

PROBLEMS

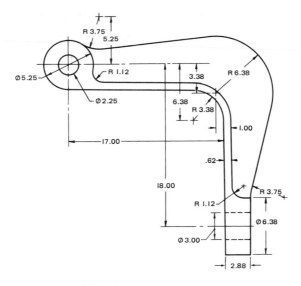

Fig. 16-62 *CAD3* *Create a single-view drawing of the end-loading hook. Select a suitable scale. Dimension the casting.*

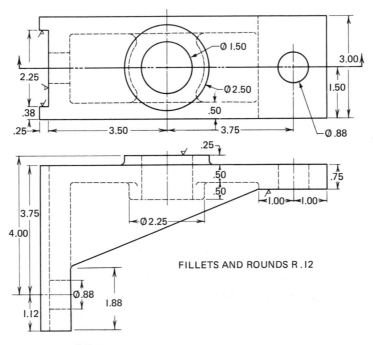

FILLETS AND ROUNDS R .12

Fig. 16-63 *CAD3* *Prepare a working drawing of the bearing bracket. Show three views, one in section. Update the finish marks on this casting.*

PROBLEMS

── SCHOOL-TO-WORK ──
Solving Real-World Problems

Fig. 16-64 through 16-65 *SCHOOL-TO-WORK problems are designed to challenge a student's ability to apply skills learned within this text. Be creative and have fun!*

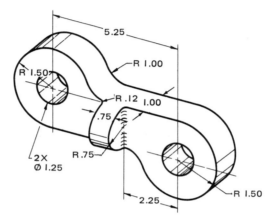

Fig. 16-64 **CAD 1** *Make a working drawing of the link stop.*

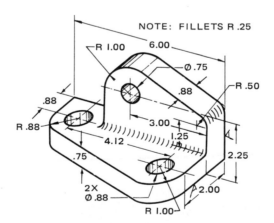

Fig. 16-65 **CAD 3** *Make a working drawing of the base anchor. Dimensions required. Update surface finish marks for this casting.*

17

WELDING DRAFTING

OBJECTIVES

Upon completion of Chapter 17, you will be able to:

○ List and describe various welding procedures.
○ Identify and use weld symbols in conjunction with other data to develop a complete welding symbol.
○ Determine the appropriate joint preparation for a specific weld application.
○ Prepare a comprehensive welding drawing for a fabricated part or product.

VOCABULARY

In this chapter, you will learn the meanings of the following terms:

- arc welding
- fillet welds
- flash welding
- fusion welding
- gas and shielded arc welding
- gas welding
- groove welding
- plug weld
- projection weld
- resistance welding
- seam weld
- slot weld
- spot weld
- thermit welding
- welding

Welding is a way of joining metal parts together. The art of welding is very old. In prehistoric times, it was used to make rings, bracelets, and other jewelry. Today it is very important in industry. Welded steel parts are lighter, stronger, and longer-lasting than parts made by forging or casting. Figure 17-1A shows a pulley housing made by casting. Compare it with the similar part made by welding (Fig. 17-1B).

Welding has become a major assembly method in industries that use steel, aluminum, and magnesium to build cars, trucks, airplanes, ships, or buildings (Fig. 17-2). In a single dump truck, there are hundreds of welds. The basic framework of the truck is assembled by welding.

Various welding processes and techniques have been developed for use in different situations and with different materials. However, most welding processes can be classified as either fusion welding or resistance welding. In **fusion welding,** the welder applies heat to create the weld. In **resistance welding,** the welder uses a combination of heat and pressure to make the weld.

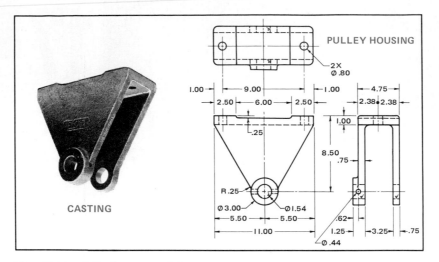

Fig. 17-1A *Pulley housing made by casting.* (Courtesy of Wellman Engineering Co. and Lincoln Electric)

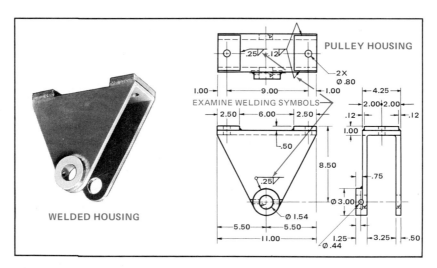

Fig. 17-1B *Pulley housing made by welding.* (Courtesy of Wellman Engineering Co. and Lincoln Electric)

TYPES OF JOINTS AND WELDS

Industrial processes require a variety of joints and welds because welding is used for so many different purposes. The various types of joints and welds can be combined in many different ways to achieve the desired strength and characteristics for each application. To specify a weld completely, the drafter or engineer must specify both the type of joint to be achieved and the type of weld to be used.

Joints

The five basic joints used in welding are butt joints, lap joints, corner joints, edge joints, and T-joints (Fig. 17-3). To choose the right joint, the drafter or engineer must be familiar with the specified materials and their conditions. He or she must also have sufficient practical welding experience to make a knowledgeable decision.

Fig. 17-2 *Welding is used in industry to join metal parts.* (Courtesy of Coachman Industries, Inc. Photo by Joe Hilliard)

Welds

The right weld for a specific job depends on the type of material specified, the tools to be used, and the cost of preparation. Some of the basic welds are described below.

- **Groove welds** are located in a groove or notch in the work material. The grooves are classified according to their shape (Fig. 17-4). Although Fig. 17-4 shows the welds applied to a butt joint, they can also be applied to any other joint. Note that the grooves may be single or double.
- **Fillet welds** are similar to groove welds, except that no groove is made in the material. Instead, the weld rests on top of the joint. An example of a fillet weld is shown in Fig. 17-5.
- **Plug welds** are welds that fit into a small hole in the work material (Fig. 17-5).
- **Slot welds** are similar to plug welds, except that the shape of the opening for the weld is slot-shaped. The welding symbol for a slot weld is the same as that for a plug weld.

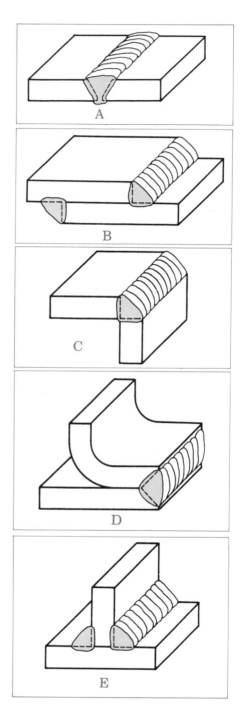

Fig. 17-3 *Five basic types of welded joints: (A) butt joint, (B) lap joint, (C) corner joint, (D) edge joint, and (E) T-joint.*

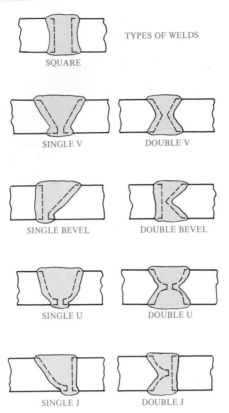

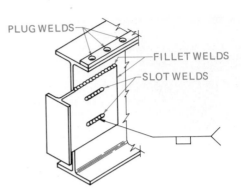

Fig. 17-5 *Plug and slot welds. The symbol shown applies to both kinds.*

Fig. 17-4 *Nine basic types of grooved welds as applied to a butt joint.*

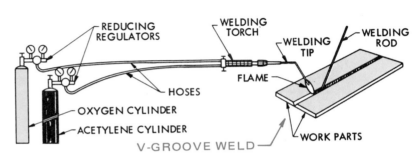

Fig. 17-6 *The gas-welding process.* (Courtesy of General Motors)

FUSION WELDING

The drafter who draws the parts to be welded works with a design engineer who knows what kinds of welding to use with different metals. Nevertheless, the drafter must be familiar with the various welding processes and their associated drafting symbols. Standard drafting symbols for welding have been established by the American Welding Society (Appendix Table A-26).

Fusion welding processes include gas, arc, thermit, and gas and shielded arc welding. Soldering and brazing, although called by their separate names, are also forms of fusion welding.

Fusion welding requires a welding filler material in the form of a wire or rod. The welder heats the material with a gas flame or a carbon arc. When it melts, it fills in a joint and combines with the metal being welded.

Gas Welding

Gas welding incorporates the use of a combination of gases to create the heat for welding (Fig. 17-6). This process was developed in 1885, when two gases—oxygen, from liquid air, and acetylene, from calcium carbide—were brought into use. Burning acetylene supported by oxygen gives temperatures of between 5000° and 6500°F (2760° and 3595°C).

Arc Welding

In 1881, De Meritens in France first performed a process known as *arc welding*. **Arc welding** is a form of fusion welding in which an electric arc forms between the *work* (part to be welded) and an electrode (Fig. 17-7). The source of the arc is generally a direct current (DC) generator or other

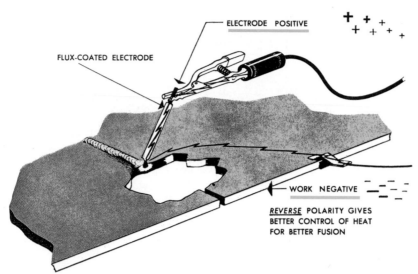

Fig. 17-7 *The arc-welding process.* (Courtesy of Republic Steel Corp.)

power source. The arc causes intense heat to develop at the tip of the electrode. This heat is used to melt a spot on the work and on a rod of filler material so that the two can be fused together.

Thermit Welding

Thermit welding is based on the natural chemical reaction of aluminum with oxygen. A mixture, or charge, made of finely divided aluminum and iron oxide is ignited by a small amount of special ignition powder. The charge burns rapidly, producing a very high temperature. This melts the metal, which then flows into molds and fuses mating parts.

Gas and Shielded Arc Welding

Aluminum, magnesium, low-alloy steels, carbon steels, stainless steel, copper, nickel, Monel, and titanium are some of the metals that can be welded using **gas and shielded arc welding.** As its name implies, this welding process combines arc welding and gas welding. Two forms of gas can be used in gas and shielded arc welding: *tungsten-inert gas (TIG)* and *metallic-inert gas (MIG).*

TIG Welding

In tungsten-inert gas welding, the electrode that provides the arc for welding is made of tungsten. Since it provides only the heat for fusion, some other material must be used with it for filler.

MIG Welding

In metallic-inert gas welding, the electrode contains a consumable metallic rod. It provides both the filler material and the arc for fusion.

Symbols for Fusion Welding

Drafters use special symbols to specify welds on a welding drawing. Figure 17-8 shows an example of the use of welding symbols in technical drawings. In (A), welding symbols are included on a machine drawing. In (B), they are included on a structural drawing.

Figure 17-9 describes the standard welding symbols approved by the American Welding Society. The notes in the illustration explain how to place symbols and data in relation to the reference line. By combining the symbols in Fig. 17-9, you can describe any welded joint, from the simplest to the most complex.

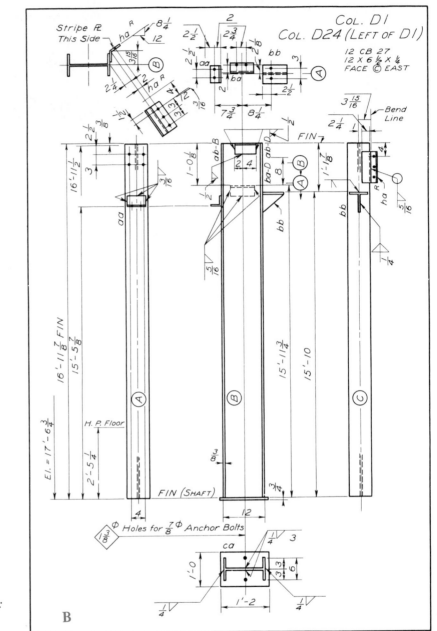

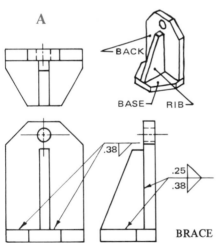

Fig. 17-8 *The application of welding symbols: (A) on a machine part, (B) on a structural drawing.*

Standard Symbols

The welding symbol tells what kind of weld to use at a joint and where to place it. Figure 17-10 shows five typical groove welds, along with the symbols used for each. The symbol can be drawn on either side of the joint as space permits. *NOTE:* When drawing a fillet, bevel, or J-grooved weld symbol, place the perpendicular leg of the symbol to the left (Fig. 17-11).

An arrow leads from the symbol's reference line to the joint. The side of the joint to which the arrow points is called the *arrow side*. The opposite side is called the *other side*. If the weld is to be on the arrow side of the joint, draw the type-of-weld part of the symbol below the reference line (Fig. 17-12A and D). If the weld is to be on the other side, draw the type-of-weld part of the symbol above the reference line (Fig. 17-12B and E).

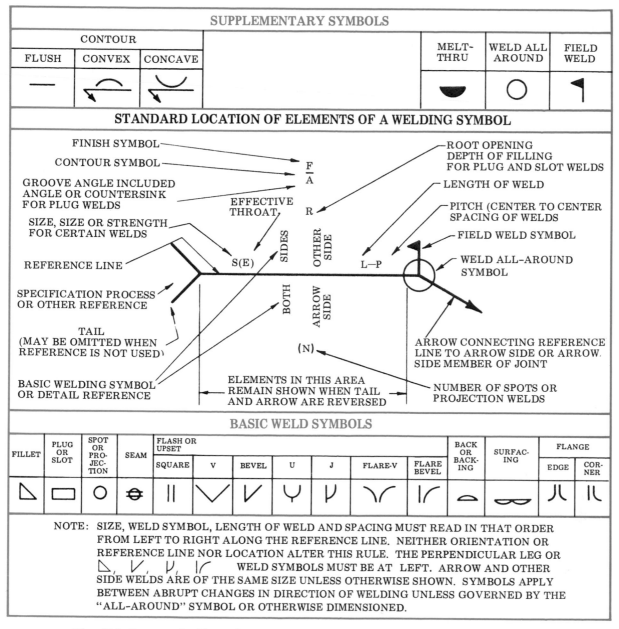

Fig. 17-9 *Location of welding information on a welding symbol.* (Courtesy of American Welding Society)

If the weld is to be on both sides of the joint, draw the type-of-weld part of the symbol both above and below the reference line (Fig. 17-12C).

When the weld is to be a J-groove weld, you must place the arrow from the welding symbol correctly to avoid confusion. For example, in Fig. 17-13A, it is not clear which piece is to be grooved. In (B), the arrow has been redrawn to show clearly that the vertical piece is to be grooved, as shown in (C). In (D), two welds are called for. The symbol below the reference line indicates a J-grooved weld on the arrow side. The arrow shows that it is the horizontal piece that is to be grooved. The symbol above the reference line indicates a fillet weld on the other side. The drawing in (E) shows how the completed welds would look.

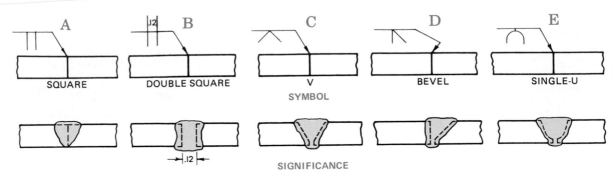

Fig. 17-10 *Symbols for five typical groove welds.*

Fig. 17-11 *The perpendicular leg on a weld symbol is always drawn to the left.*

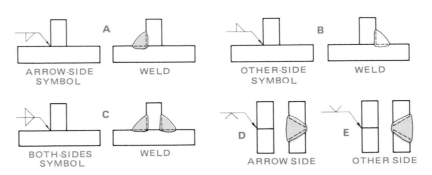

Fig. 17-12 *Weld symbols show the arrow side and the other side.*

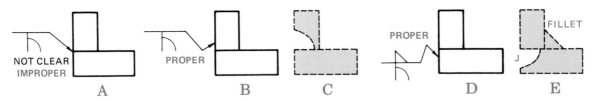

Fig. 17-13 *Welding symbols show the J-grooved weld.*

In Fig. 17-14A, the reference dimensions have been included with the welding symbols. Also, the typical specification A2 has been placed in the tail of the reference line. (See the next section for explanation of A2 and other supplementary symbols.) The drawing in (B) shows how this data is used.

Figure 17-14B shows the joint made according to the reference specifications in (A). This joint can be described as follows: a double filleted-welded, partially grooved, double-J T-joint with incomplete penetration. The J-groove is of standard proportion. The radius *R* is .50 in. (13 mm) and the included

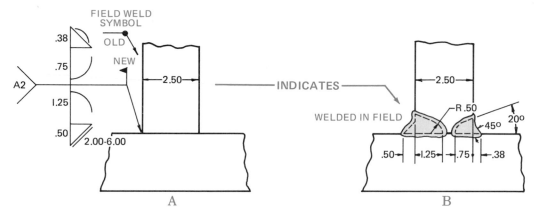

Fig. 17-14 *Interpretation of welding symbols and specifications.*

angle is 20°. The penetration is .75 in. (19 mm) for the other side and 1.25 in. (32 mm) deep for the arrow side. There is a .50-in (13-mm) fillet weld on the arrow side and a continuous .38-in. (10-mm) fillet weld on the other side. The fillet on the arrow side is 2.00 in. (50 mm) long. The pitch of 6.00 in. (150 mm) indicates that it is spaced 6.00 in. (150 mm) center-to-center. All fillet welds are standard at 45°.

Supplementary Symbols

In Fig. 17-14A, notice the solid black dot on the elbow of the reference line. This dot or a black flag is a supplementary symbol for a field weld. This means that the weld is to be made in the field or on the construction site rather than in the shop.

In the tail of the reference line in Fig. 17-14A is the typical specification A2. Its meaning is as follows: The work is to be metal-arc process, using a high-grade, covered, mild-steel electrode; the root is to be unchipped and the welds unpeened, but the joint is to be preheated.

In Fig. 17-14A, notice the flush contour symbol over the .50-in. (13-mm) fillet weld symbol. This indicates that the contour of this weld is to be flat-faced and unfinished. Over the .38-in. (10-mm) fillet weld on the same reference line, there is a convex contour symbol. This indicates that this weld is to be finished to a convex contour. Figure 17-9 shows the supplementary welding symbols to be used for finished welding techniques.

RESISTANCE WELDING

In 1857, James Prescott Joule developed *resistance welding,* a form of welding that combines heat and pressure to create a weld. In spite of its early development, however, resistance welding depends on electricity. Therefore, industry did not begin using this process until after the 1880s when electric power became available in large quantities.

Resistance welding is a good method for fusing thin metals. To join two metal pieces, an electric current is passed through the points to be welded. At those points, the resistance to the charge heats the metal to a plastic state. Pressure is then applied to complete the weld.

Spot Welds

When the current and pressure are confined to a small area between electrodes, the resulting weld is called a **spot weld.** Resistance spot welding is done on lapped parts, and the welds are relatively small. Figure 17-15A shows a reference symbol in the top view. The arrow points to the working centerline of the weld. In (A), the minimum diameter of each weld is specified at .30 in. (7.6 mm). In (B), the minimum shearing strength of each weld is specified at 800 lb. [3.558 kilonewtons (kN)]. In (C) and (D), the reference data indicates that the first weld is centered 1.00 in. (25 mm) from the left end and the welds are spaced 2.00 in. (50 mm) from centerline to centerline.

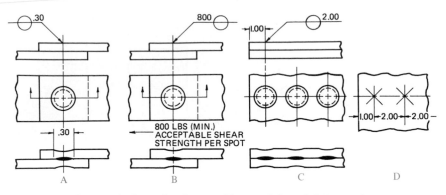

Fig. 17-15 *Examples of spot-welding symbols and their meanings.*

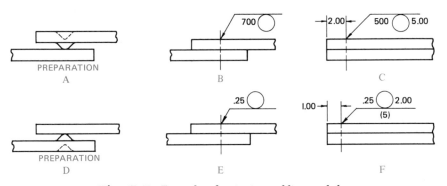

Fig. 17-16 *Examples of projection-welding symbols.*

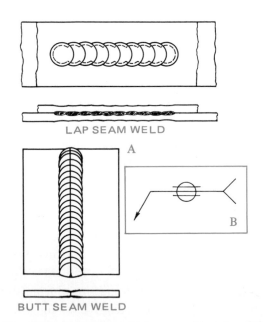

Fig. 17-17 *Butt-seam and lap-seam welds and symbols.*

Projection Welds

A **projection weld** is identified by strength or size. Figure 17-16A and D shows parts set up for such a weld. In each case, one part has a boss projection. In (B), the reference 700 lb. (3.10 kN) means that the acceptable shear strength per weld must be at least 700 lb. (3.10 kN). In (C), the reference data 500 lb. (2.22 kN) specifies the strength of the weld, the 2.00 means that the first weld is located 2.00 in. (50 mm) from the left side, and the 5.00 specifies a weld every 5.00 in. (125 mm) center to center. In (E), the number .25 in. (6.0 mm) specifies the diameter of the weld. In (F), the diameter of the weld is .25 in. (6.0 mm), there is a weld every 2.00 in. (50 mm) beginning 1.00 in. (25 mm) from the left side, and the (5) specifies that there is to be a total of five welds. Notice the arrow-side and other-side indications.

Seam Welds

A **seam weld** is a weld along the seam of two adjacent parts. Figure 17-17A shows butt-seam and lap-seam welds. The symbol for seam welding is shown in (B). The side view shows the two pieces positioned edge to edge for butt-seam welding and overlapping for lap-seam welding, which is done with a series of tangent spot welds.

Flash Welding

Flash welding is a special kind of resistance welding. It is done by placing the parts to be welded in very light contact or by leaving a very small air gap. The electric current then flashes, or arcs. This melts the ends of the parts, and the weld is made.

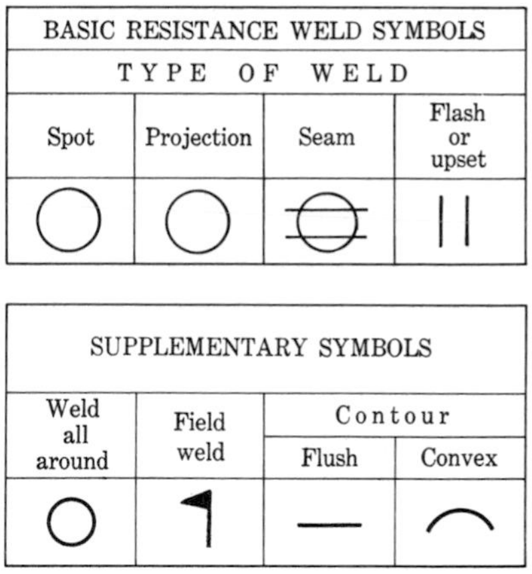

BASIC RESISTANCE WELD SYMBOLS			
TYPE OF WELD			
Spot	Projection	Seam	Flash or upset
◯	◯	⊖	‖

SUPPLEMENTARY SYMBOLS			
Weld all around	Field weld	Contour	
		Flush	Convex
◯	🚩	—	⌒

Fig. 17-18 *Basic resistance-weld symbols.*

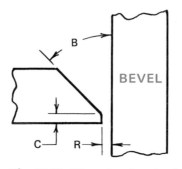

Fig. 17-20 *Dimensions for a bevel-grooved weld. B = 45° min., C = 0 to ⅛ in., R = ⅛ to ¼ in.*

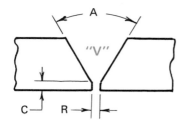

Fig. 17-19 *Dimensions for a V-joint weld. A = 60° min., C = 0 to ⅛ in., R = ⅛ to ¼ in. Stock: ½ to ¾ in.*

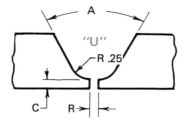

Fig. 17-21 *Dimensions for a U-grooved weld. A = 45° min., C = ¹⁄₁₆ to ³⁄₁₆ in., R = 0 to ⁹⁄₁₆ in.*

Symbols for Resistance Welding

Four basic symbols are used in resistance welding. They signify spot, projection, seam, and flash or upset welds (Fig. 17-18). The basic reference line and arrow used for arc- and gas-weld symbols are also used for resistance-weld symbols. However, in general, there is no arrow side or other side. The same supplementary symbols also apply, as Fig. 17-18 shows.

DIMENSIONING WELDS

Figure 17-19 shows the typical dimensions for a butt joint with a V-grooved weld. The dimensions for a T-joint with a bevel-grooved weld are shown in Fig. 17-20. Typical dimensions for a U-grooved weld on a butt joint are given in Fig. 17-21. Figure 17-22 shows the typical dimensions for a J-grooved weld on a T-joint.

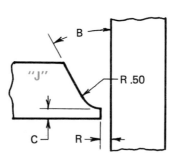

Fig. 17-22 *Dimensions for a J-grooved weld. B = 25° min., C = ¹⁄₁₆ to ³⁄₁₆ in., R = 0 to ⁹⁄₁₆ in.*

WELDING DRAFTING WITH CAD

In fields that require welding drafting, companies that use CAD software have a great advantage over those using manual methods. Special symbol libraries provide these companies with extensive, highly accurate welding symbols. The CAD operator uses overlay templates such as the one shown in Fig. 17-23 to provide quick access to the symbols. The detail development process takes much less time because the drafter does not have to draw each symbol. Instead, the drafter inserts the prepared symbols from the symbol library. Because of the time it saves, the CAD software available for steel detailing today has become an integral part of the mechanical drawing process.

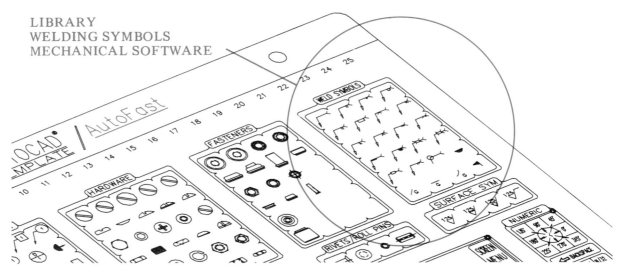

Fig. 17-23 *Weld symbol libraries are readily available for quick call-out details that meet ANSI standards. The drafter uses overlay templates on a digitizer to choose the weld symbols from the library.* (Courtesy of AutoFAST, DataStor)

CHAPTER 17

REVIEW

Learning Activities

1. Visit an industrial welding shop for a better understanding of the similarities and differences among the various welding processes. Carefully observe the finished welds produced by each. Also observe joint preparation and clamping devices used to hold parts in accurate positions for welding.

2. Obtain welding drawings from industry. Compare welding symbols with those shown throughout the chapter. Are the industries following ANSI standards? If not, use a translucent overlay on which to sketch corrections.

3. Invite an engineer from industry to visit your school to describe and demonstrate how a part can be converted from a casting to a weldment. Sketch the part as the demonstration takes place.

Questions

Write your answers on a separate sheet of paper.

1. Which kind of welding was developed first, resistance or fusion? Who was the developer?

2. Describe briefly the difference between gas welding and arc welding.

3. Name the gases that are normally used in gas welding.

4. What is the basic principle of arc welding?

5. What do MIG and TIG mean?

6. Name the five basic welding joints.

7. Name nine basic types of groove welds.

8. Draw a welding symbol and identify five features.

9. CAD drafters can access symbols in a symbol library easily by using a(n) _____ template.

10. Name at least one advantage of using a CAD system for steel detailing.

The small CAD symbol next to each problem indicates that a problem is appropriate for assignment as a computer-aided drafting problem and suggests a level of difficulty (one, two, or three) for the problem.

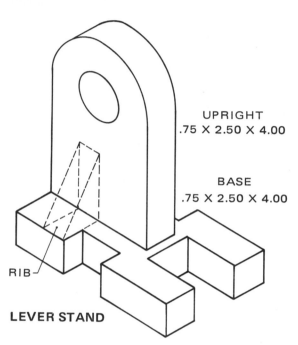

UPRIGHT
.75 X 2.50 X 4.00

BASE
.75 X 2.50 X 4.00

RIB

LEVER STAND

Fig. 17-24 *CAD2 Make a three-view drawing of the lever stand. Determine your own dimensions. Provide a support rib for the upright member. Provide dimensions and welding symbols.*

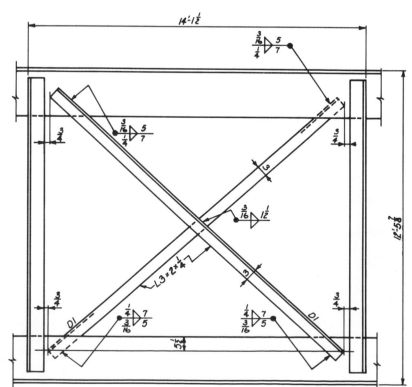

Fig. 17-25 *CAD2 Make a welding drawing of the double-angle cross bracing. Select a suitable scale. The horizontal members are 18 in.*

PROBLEMS

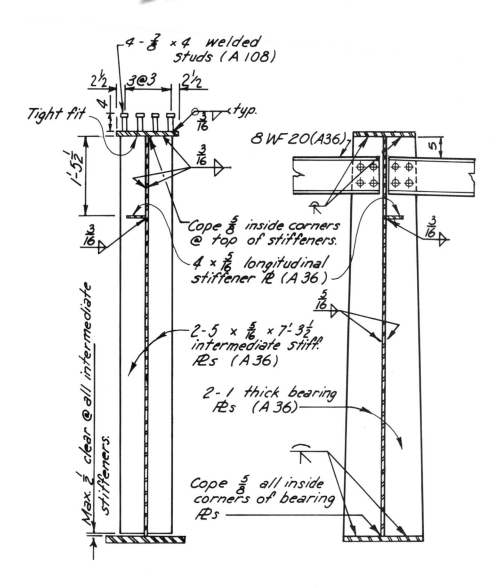

4 - ⅞ × 4 welded studs (A 108)

2½ 3@3 2½

4

Tight fit

○³⁄₁₆ typ.

1'-5½'

³⁄₁₆

8 WF 20(A36)

5

³⁄₁₆

³⁄₁₆

³⁄₁₆

Cope ⅝ inside corners @ top of stiffeners.

4 × ⁵⁄₁₆ longitudinal stiffener ℝ (A 36)

5⁄16

Max. ½ clear @ all intermediate stiffeners.

2-5 × ⁵⁄₁₆ × 7'-3½ intermediate stiff. ℝs (A36)

2-1 thick bearing ℝs (A 36)

Cope ⅝ all inside corners of bearing ℝs

TYPICAL GIRDER SECTIONS
Scale: ¾" = 1'- 0

Fig. 17-26 *CAD3* *Prepare a drawing of each girder section and place the symbols for welding in appropriate locations with dimensioning.*

PROBLEMS

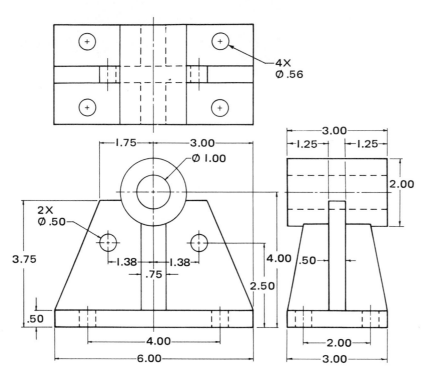

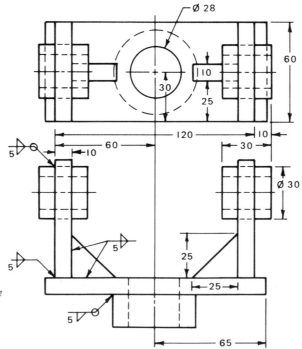

Fig. 17-27 *CAD3 Develop a three-view drawing of the bearing support. Apply appropriate welding symbols to assemble the five parts that form the bearing support.*

Fig. 17-28 *CAD3 Prepare three views of the double-bearing swivel support. Prepare a parts list. Dimension in decimals or in millimeters.*

METRIC

PROBLEMS

── SCHOOL-TO-WORK ──
Solving Real-World Problems

Fig. 17-29 through 17-31 *SCHOOL-TO-WORK problems are designed to challenge a student's ability to apply skills learned within this text. Be creative and have fun!*

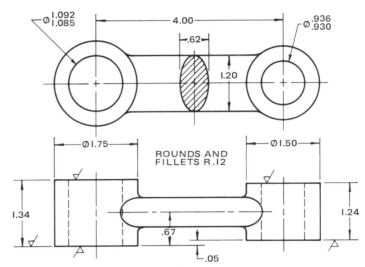

Fig. 17-29 *CAD2 Pin link. Convert the casting to a weldment. Scale 1:1 on an A-size sheet. Call for .19 in. welds and use .25-in.-thick flat stock to connect the two ends.*

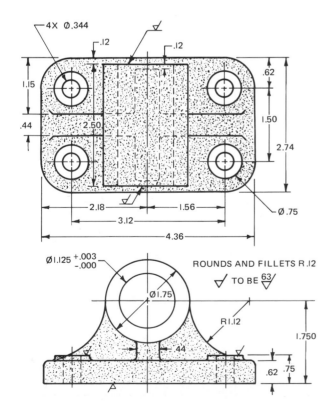

MATERIAL NO. 30 ASTM GREY IRON

Fig. 17-30 *CAD2 Shaft support. The shaft support shown in Fig. 17-30 has been cast in gray iron. This is a rather costly method for producing the part. The engineering division of your company has decided to convert the part from a casting to a weldment as a cost-saving measure.*

On an A-size sheet, make a welding drawing of the shaft support by converting the casting to a weldment. The thickness of the base plate can be reduced to .25 inch thick because steel is tougher and stronger than gray iron. Call for .19-in. welds.

PROBLEMS

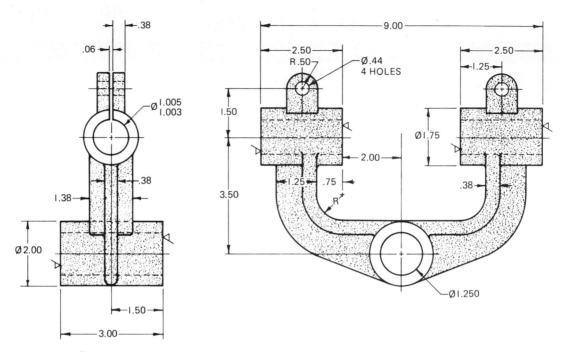

Fig. 17-31 *CAD3 Connecting link. Convert the casting shown in Fig. 17-31 to a weldment. Change round parts to square stock where practical to reduce joint preparation. Reduce the thickness of flat stock from .38 to .25 in. Number all parts and prepare a parts list. Welds are to be .19 in. Make your drawing on a B-size sheet. Scale: 1:1 or as assigned.*

18

SURFACE DEVELOPMENTS AND INTERSECTIONS

OBJECTIVES

Upon completion of Chapter 18, you will be able to:

○ Define surface development.
○ Prepare patterns using parallel-line development, radial-line development, and triangulation.
○ Prepare patterns for intersecting prisms.
○ Describe, select, and specify seams, laps, and hems.

VOCABULARY			
In this chapter, you will learn the meanings of the following terms:	• **crease lines**	• **parallel-line**	• **surface**
	• **developments**	**development**	**development**
	• **die stamping**	• **pattern**	• **transition pieces**
	• **elbow**	• **radial-line**	• **triangulation**
	• **element**	**development**	• **wiring**
	• **hemming**	• **stretchout line**	
	• **measuring lines**	• **stretchouts**	

A **surface development** is a full-size layout of an object made on a single flat plane. The book cover shown in Fig. 18-1 is an example of a surface development. In (A), the cover is laid out flat. In (B), it is wrapped around a book to make a protective covering. Notice that it fits neatly around all surfaces. It does so because each part has been carefully measured and laid out in relation to other parts. Surface developments are also called **stretchouts** or simply **developments.**

Making surface developments is an important part of industrial drafting. Many different industries use surface developments. Familiar items such as pipes, ducts for hot- or cold-air systems, parts of buildings, aircraft, automobiles, storage tanks, cabinets, boxes and cartons, frozen food packages, and countless other items are designed using surface development.

To make any such item, a surface development is first drawn as a **pattern.** This pattern is then cut from flat sheets of material that can be folded, rolled, or otherwise formed into the required shape. Materials used include paper, various cardboards, plastics and films, metals (such as steel, tin, copper, brass, and aluminum), wood, fiberboard, fabrics, and so on.

CREASE (FOLD) LINES

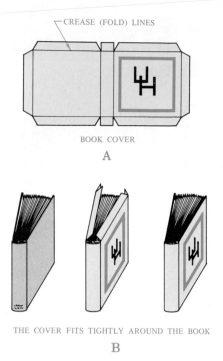

BOOK COVER

A

THE COVER FITS TIGHTLY AROUND THE BOOK

B

Fig. 18-1 *A book cover is an example of a surface development.*

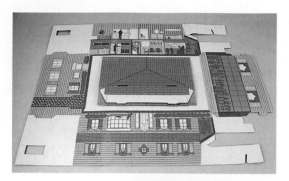

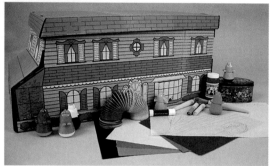

Fig. 18-2 *This container was made by cutting and folding a flat sheet.* (Courtesy of Ann Garvin)

The Packaging Industry

Packaging is a very large industry that uses surface developments. Creating packages takes both engineering and artistic skill because each package design must meet many requirements. For example, packages and containers for industrial goods are designed to be mass-produced at a reasonable cost. However, they must also perform their assigned jobs adequately. Packages for fragile objects, for instance, must be designed so that they protect their contents during shipment. Packages must also look attractive for sales appeal.

Another consideration is the durability of the package. Some packages are meant to be used just briefly and then thrown away. Others are made to last a long time.

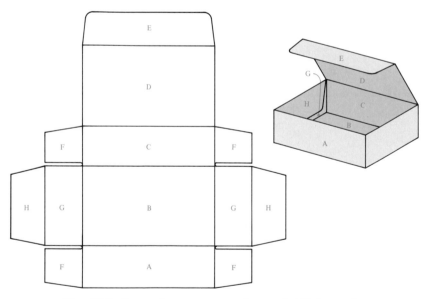

Fig. 18-3 *Pattern for a one-piece package with fold-down tabs.*

To meet these requirements, designers use many different materials in various thicknesses. Some are made of thin or medium-thick paper stock (Figs. 18-2 and 18-3). This material can be folded easily into the desired shape or form. Some are designed so that no glue is needed. Others may need glue on their tabs.

Packages made of cardboard, corrugated board, and many other materials require allowances for thickness. Examples are boxes made up of a separate container and cover (Fig. 18-4) and a slide-in box (Fig. 18-5).

Designing package pattern layouts, lettering, color, and artwork provides jobs for many people. Look at various cartons and packages. You will see many interesting and challenging problems in the development of surfaces.

Sheet-Metal Pattern Drafting

Many different metal objects are made from sheets of metal that are laid out, cut, and formed into the required shape and then fastened together. The objects can be shaped by bending, folding, or rolling. They can be fastened by riveting, seaming, soldering, or welding.

For each sheet-metal object, two drawings are usually made. One is a pictorial drawing of the finished product. The other is a development, or pattern, that shows the shape of the flat sheet that, when rolled or folded or fastened, will form the finished object (Fig. 18-6).

A great many thin metal objects without seams are formed by **die stamping** (pressing a flat sheet into shape under heavy pressure). Examples range from brass cartridge cases and household utensils to steel wheelbarrows and parts of automobiles and aircraft. Other kinds of thin metal objects are made by spinning.

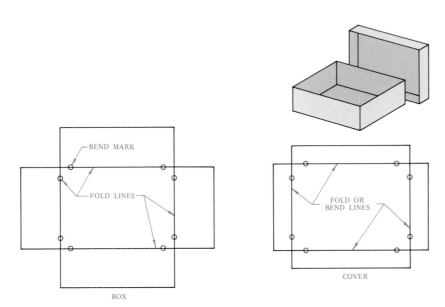

Fig. 18-4 *Pattern for a box and cover. Notice that the fold lines on the cover are positioned so that the cover will fit over the box after assembly.*

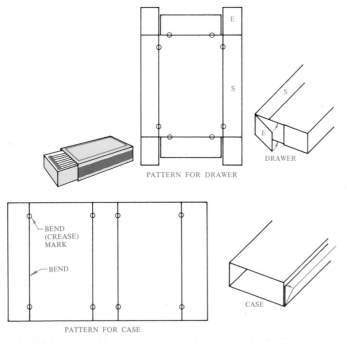

Fig. 18-5 *A two-part package with a slide-in box. The fold lines on the drawer are positioned so that the box will slide in correctly after assembly.*

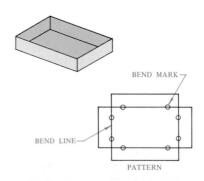

Fig. 18-6 *Pictorial drawing and stretchout of a sheet-metal object.*

Squaring Shears

Used for trimming and squaring sheet metal.

Box and Pan Brake

A small bench-mounted brake for straight bends up to full length of machine or for box and pan work up to 3 inch depth.

A — This shows how bottom edge of body and bottom of can are prepared by Burring Machine for Setting-Down.

B — The Pexto Setting-Down Machine closes the seam as shown here. It works both speedily and accurately.

Setting-Down Machine

The Setting-Down Machine prepares the seams in body of vessels for double seaming.

Folding Machine

The Folding Machine is used extensively for edging sheet metal or the forming of locks or angles.

AA — Showing edges turned by Pexto Burring Machine. Note right-angle burr on body of can and a still more pronounced burr on the bottom piece. The edge on bottom is turned smaller than on the body.

Burring Machine

A difficult operation to master, but practice will produce uniform flanges on sheet metal bodies. Prepares the burr for bottoms preparatory to setting down and double seaming.

AA — Seats for Wire — made on the Turning Machine.

Turning Machine

Used to prepare a seat in bodies to receive a wire. The operation is completed with use of Wiring Machine.

Wiring Machine

Works the metal completely and compactly around wire. Depending on shape of work, seats to receive wire are prepared on Folder, Brake or Turning Machine.

Forming Machines

Used for forming flat sheets into cylinders of various diameters, such as stove pipe, the bodies of vessels, cans, etc. Made in a variety of sizes and capacities.

Doubling-Seaming Machine

Offering in various styles and follows the setting-down operation.

Beading Machine

For ornamenting and stiffening sheet metal bodies.

Fig. 18-7 *Some of the machines used in sheet-metal work.* (Courtesy of Peck, Stow, and Wilcox Co.)

Such objects include some brass-, copper-, pewter-, and aluminumware. However, stamping and spinning stretch the metal out of its original shape.

Making sheet-metal objects can involve many operations. These include cutting, folding, wiring, forming, turning, beading, and so forth.

All are done with machines. Some machines and operations are shown in Fig. 18-7. The machines shown are for hand operations. For industrial mass production, complex automatic equipment is generally used to perform these operations.

Pattern Development

Sheet-metal patterns, like all other patterns, are developed using principles of surface geometry. There are two general classes of surfaces: plane (flat) and curved. The six faces of a cube are plane surfaces. The top and bottom of a cylinder are also plane surfaces. However, the side surface of the cylinder is curved (Fig. 18-8). There are also different kinds of curved surfaces. Those that can be rolled in contact with a plane surface, such as cylinders and cones, are called *single-curved surfaces*. Exact surface developments can be made for them.

Another kind of curved surface is the *double-curved surface*. It is found on spheres and spheroids. Exact surface developments cannot be made for objects with double-curved surfaces. However, drafters can make approximations.

Figure 18-9 shows how to cut a piece of paper so that it can be folded into a cube. The shape cut out is the pattern of the cube. Figure 18-10 shows the patterns for all five of the regular solids.

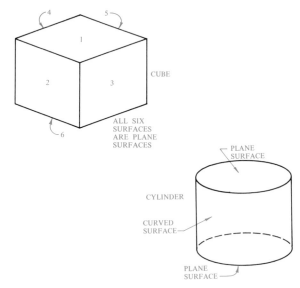

Fig. 18-8 *Plane and curved surfaces.*

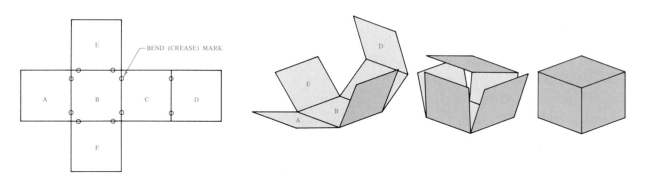

Fig. 18-9 *Pattern for a cube.*

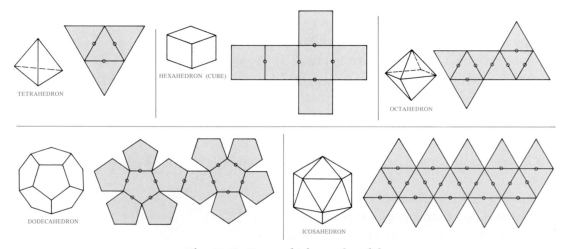

Fig. 18-10 *Patterns for five regular solids.*

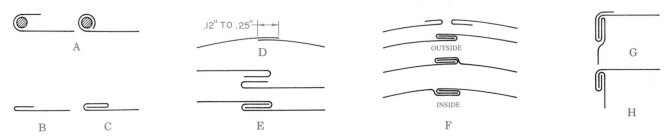

Fig. 18-11 *Methods of wiring, seaming, and hemming.*

To understand surface development better, lay these patterns out on stiff drawing paper. Then cut them out and fold them to make the solids. Secure the joints with tape.

Any solid that has plane surfaces can be made in the same way. However, each plane must be drawn in its proper relationship to the others in the development of the pattern.

Seams and Laps

Drawing developments is only part of sheet-metal pattern drafting. Drafters must also know about the processes of wiring, hemming, and seaming. In addition, they must know how much material should be added for each. **Wiring** involves reinforcing open ends of articles by enclosing a wire in the edge (Fig. 18-11A). To allow for wiring, a drafter must add a band of material to the pattern equal to 2.5 times the diameter of the wire. **Hemming** is another way of stiffening edges. Single- and double-hemmed edges are shown in Fig. 18-11B and C. Edges can also be fastened by soldering on lap seams (D), flat lock seams (E), or grooved seams (F). Other types of seams and laps are shown in (G) and (H). Each has its own general or specific use. How much material is allowed in each case depends on the thickness of the material, method of fastening, and application. In most cases, the corners of the lap are notched to make a neater joint.

PARALLEL-LINE DEVELOPMENT

Parallel-Line Development is a simple way of making a pattern by drawing the edges of an object as parallel lines. The patterns in Figs. 18-9 and 18-12 are made in this way.

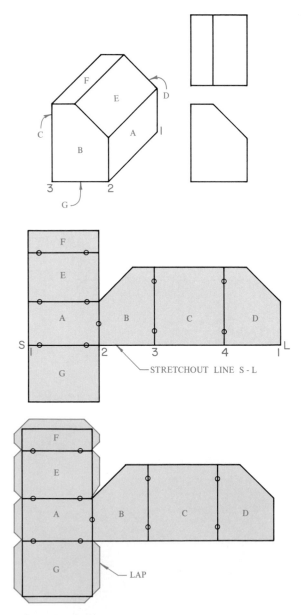

Fig. 18-12 *A pattern for a prism, showing stretchout line and lap.*

Prisms

Figure 18-13 is a pictorial view of a rectangular prism. In Fig. 18-14, a pattern for this prism is made by parallel-line development. To draw this pattern, proceed as follows:

1. Draw the front and top views full size. Label the points as shown (Fig. 18-14A).

2. Draw the **stretchout line** (SL). This line shows the full length of the pattern (when it is completely unfolded). Find the lengths of sides *1-2*, *2-3*, *3-4*, and *4-1* in the top view. Measure off these lengths on SL (Fig. 18-14B).

3. At points *1*, *2*, *3*, *4*, and *1* on the stretchout line, draw vertical **crease lines** (also called *fold* or *bend lines*). Make them equal in length to the height of the prism (Fig. 18-14C).

4. Project the top line of the pattern from the top of the front view. Make it parallel to SL. Darken all outlines until they are thick and black. You can use a small circle or X to identify a fold line (Fig. 18-14D).

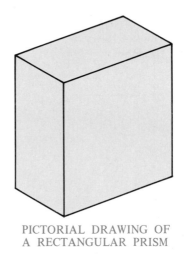

PICTORIAL DRAWING OF
A RECTANGULAR PRISM

Fig. 18-13 *Pictorial drawing of a rectangular prism.*

5. Add the top and bottom to the pattern by transferring distances *1-4* and *2-3* from the top view, as shown in Fig. 18-14E. Add laps or tabs as necessary for assembly of the prism. The size of the laps will vary depending on how they are to be fastened and the kind of material used.

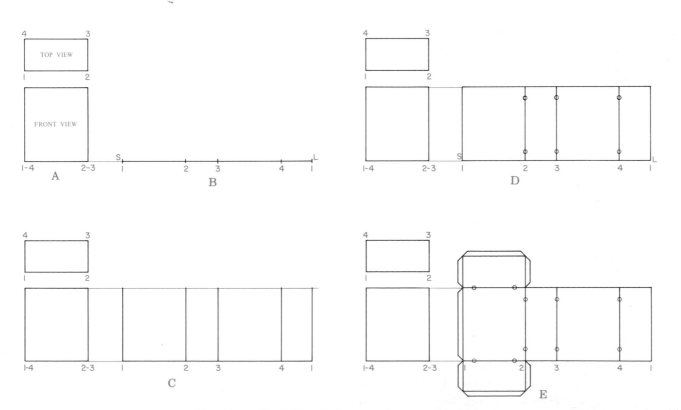

Fig. 18-14 *Parallel-line development of a rectangular prism.*

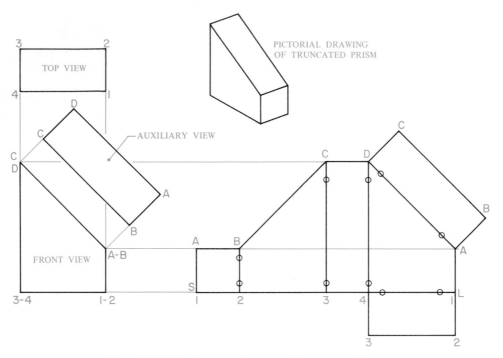

Fig. 18-15 *Development of a pattern for a truncated prism.*

A slight variation is the pattern for a truncated prism shown in Fig. 18-15. To draw it, first make front, top, and auxiliary views full size. Label points as shown. The next two steps are the same as steps B and C in Fig. 18-14. Then project horizontal lines from points *A-B* and *C-D* on the front view to locate points on the pattern. Connect the points to complete the top line of the pattern. Add the top and bottom as shown.

Cylinders

Figure 18-16 shows a surface development for a cylinder. It is made by rolling the cylinder out on a plane surface.

In the pattern for cylinders, the stretchout line is straight and equal in length to the circumference of the cylinder (Fig. 18-16B). If the base of the cylinder is perpendicular to the axis, its rim will roll out to form the straight line. If the base is not perpendicular to the axis, you will have to make a right section to get the stretchout line.

To develop a cylinder, imagine that the cylinder is actually a many-sided prism. Each side forms an edge called an **element.** Because there are so many elements, however, they seem to form a smooth

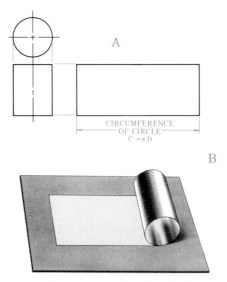

Fig. 18-16 *Developed surface of a right circular cylinder.*

curve on the surface of the cylinder. Imagining the cylinder in this way will help you find the length of the stretchout line. This length will equal the total of the distances between all the elements. Technically, of course, the elements are infinite in number. But, for your purposes, you need only mark off elements at convenient equal spaces around the circumference of the cylinder. (Figure 18-17 shows various methods

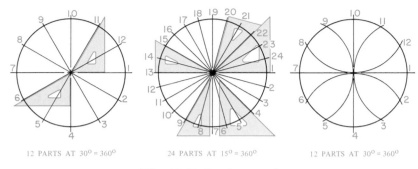

12 PARTS AT 30° = 360°   24 PARTS AT 15° = 360°   12 PARTS AT 30° = 360°

Fig. 18-17 *Dividing a circle.*

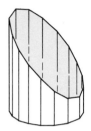

Fig. 18-18 *Pictorial drawing of a truncated right cylinder.*

of dividing a circle.) Then add up these spaces to make the stretchout line.

Figure 18-18 is a pictorial view of a truncated right cylinder, showing the imaginary elements. Figure 18-19 shows how to develop a pattern for this cylinder. To draw this pattern, proceed as follows:

1. Draw the front and top views full size. Divide the top view into a convenient number of equal parts (12 in this case). This locates a set of equally spaced points (the ends of the elements around the edge in the top view).

2. Begin the stretchout line. You will determine its actual length later, when you mark off the elements.

3. Using dividers, find the distance between any two consecutive elements in the top view. Then mark off this distance along SL as many times as there are parts in the top view. Label the points thus found as shown. Then, from each point, draw a vertical construction line upward. *NOTE:* In this and subsequent steps, the colored arrows on the figure show the direction in which the various lines are projected.

4. Project other lines downward from the elements on the top view to the front view. Label the points where they intersect the front view.

5. From these intersection points, project horizontal construction lines toward the development.

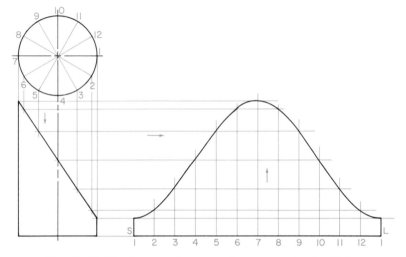

Fig. 18-19 *Development of a pattern for a truncated right cylinder.*

6. Locate the points where the horizontal construction lines intersect the vertical lines from SL. Connect these points in a smooth curve.

7. Darken outlines and add laps as necessary.

Since the surface of a cylinder is a smooth curve, your pattern will not be wholly accurate. This is because it was made by measuring distances on a straight line (chord) rather than on a curve. Figure 18-20 represents part of the top view of the cylinder discussed above. It shows that the distance from point to point is slightly less along the chord than along the arc (radial distance). The difference can be found by figuring the actual length of the arc using the following formula:

$$Circumference = \pi D$$

As long as you include enough elements to represent the cylinder adequately, however, the difference is so slight that the inaccuracy is not critical.

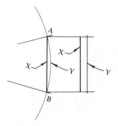

Fig. 18-20 *A straight line is the shortest distance between two points.*

A	B	C
Arc length: 1.0472	Arc length: .5236	Arc length: .2618
Chord length: 1.000	Chord length: .5176	Chord length: .2611
Difference: .0472	Difference: .0060	Difference: .0007

Fig. 18-21 *As the number of elements increases, accuracy increases also. However, using too many elements wastes time and increases the difficulty of the development. In (A), the number of elements is not sufficient to represent the cylinder adequately. In (B), the cylinder is represented adequately using 12 elements. In (C), the number of elements (24) is excessive.*

The number of elements you choose to use in cylinder development will vary depending on the size of the cylinder, the application, and other factors. However, be sure to avoid using too few or too many elements. If you use too few, the pattern approximation will not be accurate enough; if you use too many, the development becomes much more difficult, and it takes a much longer time to create (Fig. 18-21).

A slightly different method for developing a cylinder is shown in Fig. 18-22. In this case, a front and a half-bottom view are used. Attaching the two views saves time and increases accuracy. Notice that both methods produce the same results.

Elbows

An **elbow** is a joint in a pipe or duct—a place at which two pieces meet at an angle other than 180°. The simplest type of elbow is a square, or two-piece, elbow. Other, more complicated elbows include more pieces to provide a smoother curve.

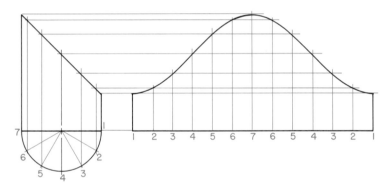

Fig. 18-22 *Development of a pattern for a truncated right cylinder, using a front and half-bottom view.*

Square Elbows

A square elbow consists of two cylinders cut off at 45°. Therefore, only one pattern is needed, as shown in Fig. 18-23. Allow a lap for the type of seam to be made, if required.

If a lap is not needed on the curved edges, both parts can be developed on one stretchout, as shown in Fig. 18-24. Notice that the seam on part A is on the short side, while on part B it is on the long side. In Fig. 18-23, the seam on both pieces is on the short side. In most cases, this is not critical.

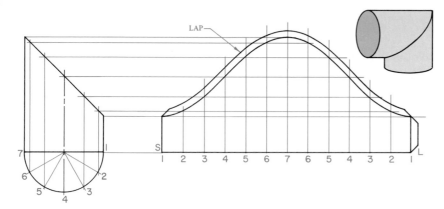

Fig. 18-23 *Pattern for a square (right-angle) elbow.*

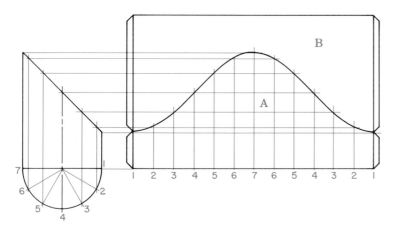

Fig. 18-24 *Both parts of the pattern may be made on one stretchout.*

Four-Piece Elbows

The pattern for a four-piece elbow is shown in Fig. 18-25. To draw it, proceed as follows:

1. Draw arcs with the desired inner and outer radii to produce an elbow to fit the desired pipe size, as shown in Fig. 18-25A. Divide the outer circle into six equal parts.

2. Draw a half-bottom view and divide it as shown in (B). Project the elements from this view to the front view.

3. Develop the pattern for part A just as you developed the pattern for the square elbow in Figs. 18-23 and 18-24. If you do not have to allow for laps on the curved edges, you can draw the patterns for all four parts as shown in (C). To find what lengths to mark off on the vertical lines, work from the front view. In the front view of part B, for example, continue the construction lines from the half-bottom view, but make them

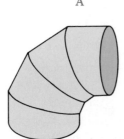

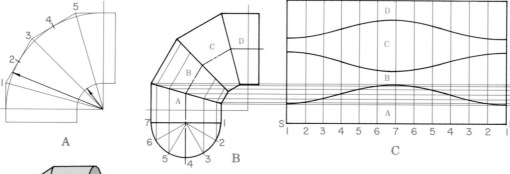

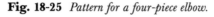

Fig. 18-25 *Pattern for a four-piece elbow.*

parallel to the surface lines. Then measure all the lines within part B. The curves for all the parts are the same. Therefore, you need plot only one of them. It can then be used as a template (pattern) for the others.

RADIAL-LINE DEVELOPMENT

In the patterns for prisms and cylinders, the stretchout line is straight, and the **measuring lines** (vertical construction lines) are perpendicular to it and parallel to each other. Hence the name "parallel-line development." On cones and pyramids, however, the edges are not parallel. Therefore the stretchout line is not a continuous straight line. Also, the measuring lines, instead of being parallel to each other, radiate from a single point. This type of development is called **radial-line development.**

Cones

Imagine the curved surface of a cone as being made up of an infinite number of triangles, each running the height of the cone. To understand the development of the surface, imagine rolling out each of these triangles, one after another, on a plane. The resulting pattern would look like a sector of a circle. Its radius would be equal to an element

of the cone; that is, a line from the cone's tip to the rim of its base. Its arc would be the length of the rim of the cone's base. The developed surface of a cone is shown in Fig. 18-26.

Right Circular Cones

A *right circular cone* is one in which the base is a true circle and the tip is directly over the center of the base (Fig. 18-27A). A *frustum* of a cone is made by cutting through the cone on a plane parallel to the base (Fig. 18-27C). The pattern for a cone is shown in Fig. 18-27B. To draw it, proceed as follows:

1. Draw front and half-bottom views to the desired size.

2. Divide the half-bottom view into several equal parts. Label the division points as shown.

Fig. 18-26 *Developed surface of a cone.*

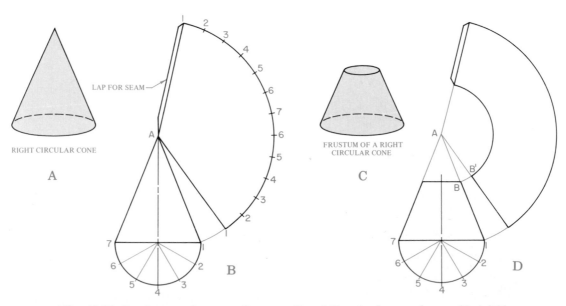

Fig. 18-27 *Development of a pattern for a cone (A and B) and a frustum of a cone (C and D).*

3. On the front view, measure the *slant height* of the cone; that is, the true distance from the *apex* (tip) to the rim of the base (line *A1*). Using this length as a radius, draw an arc of indefinite length as a measuring arc. Draw a line from apex *A* to the arc at any point a short distance from the front view.

4. With dividers, find the straight-line distance between any two division points on the half-bottom view. Then use this length to mark off spaces *1-2, 2-3, 3-4,* and so forth, along the arc. Label the points to be sure none have been missed. Complete the development by drawing line *A1* at the far end.

5. Add laps for the seam as required. How much to allow for the seam depends on the size of the development and the type of joint to be made.

Figure 18-27D shows the development for a frustum of a cone. To draw it, use the same method as in Fig. 18-27B, but draw a second arc *AB* from point *B* on the front view.

Truncated Circular Cone

A *truncated circular cone* is a circular cone that has been cut along a plane that is not parallel to the base (Fig. 18-28A). The pattern for such a cone is shown in Fig. 18-28B. To draw it, proceed as follows:

1. Draw the front, top, and bottom (or half-bottom) views.

2. Proceed as in Fig. 18-27 to develop the overall layout for the pattern.

3. Project points *1* through *6* from the bottom view to the front view and then to the apex. Label the points where they intersect the *miter line* (cut line) to avoid mistakes. These lines, representing elements of the cone, do not show in true length on the front view. Their true length shows only when they are projected to the points on the arc.

4. Project the elements of the cone from the apex to the points on the arc.

5. In the front view, find the points on the miter line that were located in step 3. Project horizontal lines from them to the edge of the front view. Continue these lines as arcs through the development. Mark the points where they

intersect the element lines. Join these points in a smooth curve. Complete the pattern by adding a lap.

Pyramids

Before you can begin to develop a pattern for a pyramid, you must find the true length of its edges. For example, in the pyramid shown in Fig. 18-29A, you need to find the true length of *OA*. Fig. 18-29B shows the top and front views of the pyramid. In neither view does the edge *OA* show in true length. However, if the pyramid were in the position shown in (C), the front view would show *OA* in true length. In (C), the pyramid has been revolved about a vertical axis until *OA* is parallel to the vertical plane. In (D), the line *OA* is shown before and after revolving.

The construction in Fig. 18-29D is a simple way to find the true length of the edge line *OA*. Revolve this view to make the horizontal line *OA'*. Project *A'* down to meet a base line projected from the original front view. Draw a line from this intersection point to a new front view of *O*. This line will show the true length of *OA*.

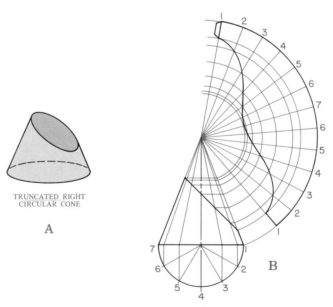

TRUNCATED RIGHT
CIRCULAR CONE

A

B

Fig. 18-28 *Development of a pattern for a truncated right circular cone.*

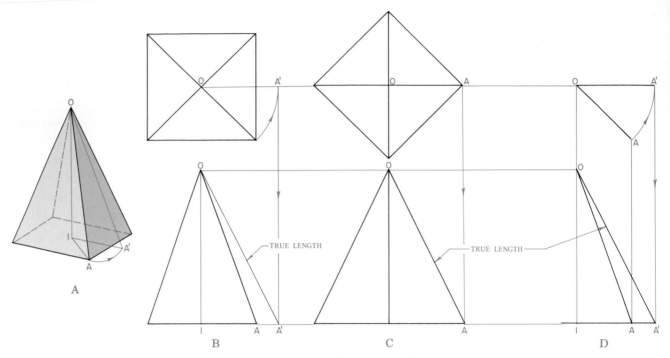

Fig. 18-29 *Finding the true length of a line.*

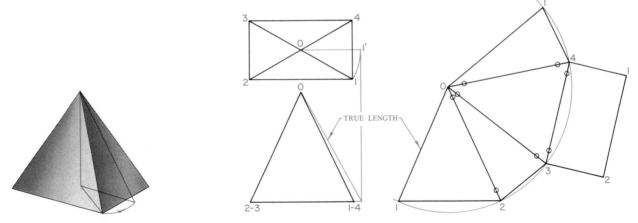

Fig. 18-30 *Development of a pattern for a right rectangular pyramid.*

Right Rectangular Pyramids

Figure 18-30 shows the pattern for a right rectangular pyramid. To draw it, proceed as follows:

1. Find the true length of one of the edges (*O1* in this case) by revolving it until it is parallel to the vertical plane (*O1'*).

2. With the true length as a radius, draw an arc of indefinite length to use as a measuring arc.

3. On the top view, measure the lengths of the four base lines. Mark these lengths off as the straight-line distances along the arc.

4. Connect the points and draw crease lines. Mark the crease lines.

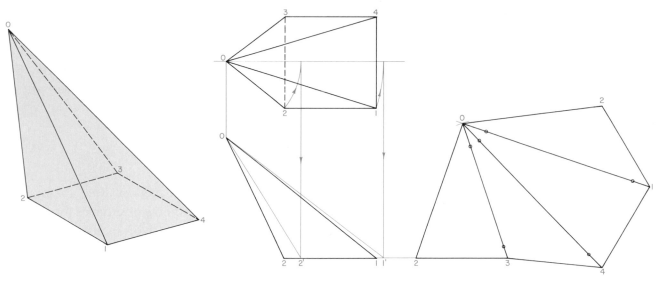

Fig. 18-31 *Developing a pattern for an oblique pyramid.*

Oblique Pyramids

Figure 18-31 shows a development of an oblique pyramid. To draw it, proceed as follows:

1. Find the true lengths of the lateral edges. Do this by revolving them parallel to the vertical plane, as is shown for edges *O2* and *O1*. These edges are both revolved in the top view, then projected to locate *2'* and *1'*. Lines *O2'* and *O1'* in the front view are the true lengths of edges *O2* and *O1*. Edge *O2* = edge *O3*. Edge *O1* = edge *O4*.

2. Start the development by laying off *2-3*. Since edge *O2* = edge *O3*, you can locate point *O* by plotting arcs centered on *2* and *3* and with radii the true length of *O2* (*O2'*). Point *O* is where the arcs intersect.

3. Construct triangles *O-3-4, O-4-1,* and *O-1-2* with the true lengths of the sides to complete the development of the pyramid as shown.

Truncated Oblique Pyramids

Figure 18-32 shows a pattern for a truncated oblique pyramid. The pyramid has an inclined surface *ABCD*. If it were not truncated, it would extend to the apex point *O*. To draw the pattern, proceed as follows:

1. Find the true lengths of *OA, OB, OC,* and *OD*. For this pyramid, *OA = OD* and *OB = OC*.

To locate these lengths, locate *B'* and *A'* in the front view. Lines *OB'* and *OA'* will give the true lengths.

2. Make a development of the pyramid as if it extended to point *O*.

3. Lay off the true lengths found on the corresponding edges of this development. Join them to make the pattern for the edge of the frustum.

The inclined surface is shown in the auxiliary view and could be attached to the rest of the pattern as indicated.

CADD DEVELOPMENTS

Although traditional drafting procedures are still serving the design staff, many major industries are converting the standards for development into computer data for use on CADD systems. The drafter must still understand how to develop various surfaces and intersections; in fact, surface development is exactly the same using a CADD system. However, the computer provides shortcuts for procedures that take time (and therefore cost money) when done manually.

Some companies use a CADD system to generate the basic two-dimensional pattern drawing from the flat pattern and then to convert it into a three-dimensional design. The CADD system can also control CNC (computer-numerical control) machines to make the physical part.

However, the reverse can also be achieved using a CADD system and specialty software. The designer can create a three-dimensional model and then reduce it to the two-dimensional form, ready for the fabricator, who shapes new products on the production line. Special software that works with the CADD software can automatically determine bend allowances and make calculations for bending and stretching for any specified material. The program also determines appropriate tools and punches that can be entered into the database for a graphic solution and designer approval.

Examine the simple solution developed with a CADD system in Fig. 18-33. In developing patterns, the design team needs to visualize in three dimensions. Can you see what the CADD-generated drawing of the fan mounting bracket in Fig. 18-34 would look like in a pattern layout?

TRIANGULATION

Triangulation is a method used for making approximate developments of surfaces that cannot be developed exactly. It involves dividing the surface into triangles, finding the true lengths of the sides, and then constructing the triangles in regular order on a plane. Because the triangles have one short side, on the plane they approximate the curved surface. Triangulation may also be used for single-curved surfaces.

Figure 18-35 shows how triangulation is used in developing an oblique cone. To draw this development, proceed as follows:

1. Draw elements on the top- and front-view surfaces to create a series of triangles. Number the elements *1, 2,* and so forth. For a better approximation of the curve, use more triangles than are shown here.

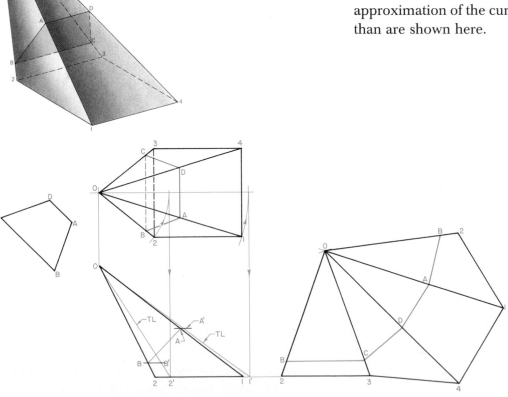

Fig. 18-32 *Developing a pattern for a truncated oblique pyramid.*

2. Find the true lengths of the elements by revolving them in the top view until each is horizontal. From the tip of each, project down to the front-view base line to get a new set of points *1, 2,* and so forth. Connect these with the front view of point *O* to make a true-length diagram, as shown in Fig. 18-35C.

3. To plot the development shown in Fig. 18-35D, construct the triangles in the order in which they occur. Take the distances *1-2, 2-3,* etc., from the top view. Take the distances *O1, O2,* etc., from the true-length diagram.

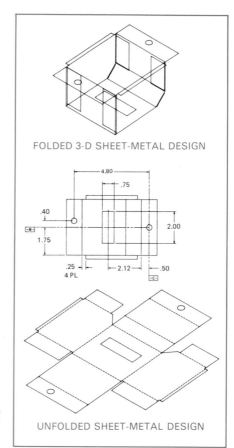

FOLDED 3-D SHEET-METAL DESIGN

UNFOLDED SHEET-METAL DESIGN

Fig. 18-33 *CAD systems can support design and drafting of tooling for sheet-metal development.* (Courtesy of Computervision)

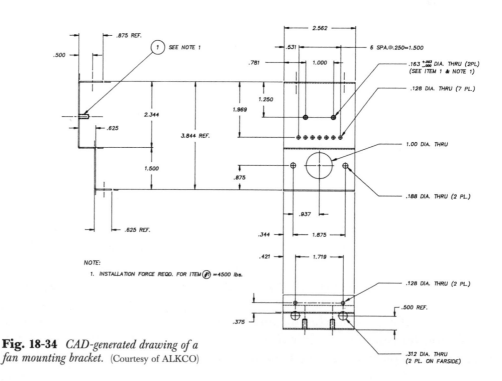

NOTE:
1. INSTALLATION FORCE REQD. FOR ITEM ⓕ =4500 lbs.

Fig. 18-34 *CAD-generated drawing of a fan mounting bracket.* (Courtesy of ALKCO)

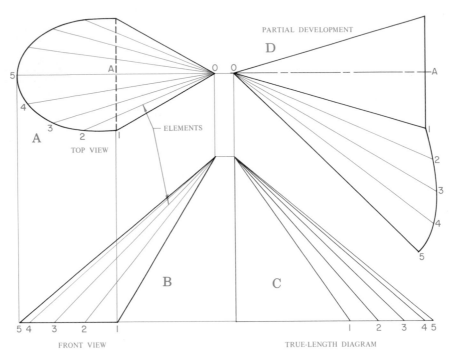

Fig. 18-35 *Triangulation used in developing an oblique cone.*

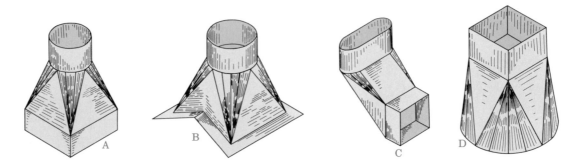

Fig. 18-36 *Examples of transition pieces.*

TRANSITION PIECES

Transition pieces are used to connect pipes or openings of different shapes, sizes, or positions. Transition pieces have a surface that is a combination of different forms, including planes, curves, or both. A few transition pieces are shown in Fig. 18-36.

Figure 18-37 shows the making of a square-to-round transition piece. According to the manufacturer, the piece can be made easily in two parts.

The illustration shows how one of the parts is positioned for making the last of eight partial bends (creases) on the second conical corner. For this conical blending, one end of the plate is moved out the proper distance for each bend. At the other end, meanwhile, the point for the square corner remains fixed.

Transition pieces have a surface that is a combination of different forms. These forms are developed by triangulation. Figure 18-38 shows a transition piece connecting two square ducts, one of which is at 45° with the other. This piece is made up of eight triangles—four of one size and four of another. To draw the development, proceed as follows:

1. On the top and front views, number the various points as shown. Lines *1-2, 2-3, 3-4,* and *4-1* show in their true size in the top view. Lines *AB, BC, CD,* and *DA* show in their true size in both views.

2. Find the true length of one of the slanted lines, in this case *A4*. Do this by revolving it in the top view until it is parallel to the vertical plane. Then project down to the front view, where the true length will show in line *4A'*. This will also be the true length of *D4, D3,* and so forth.

Fig. 18-37 *Forming a square-to-round transition piece.*

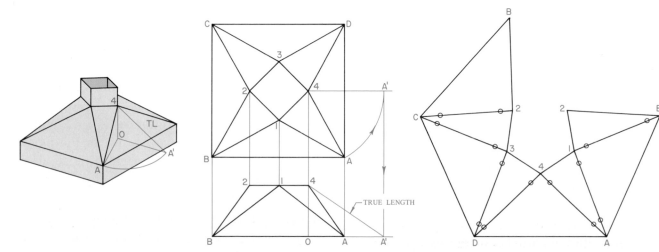

Fig. 18-38 *Development of a square-to-square transition piece.*

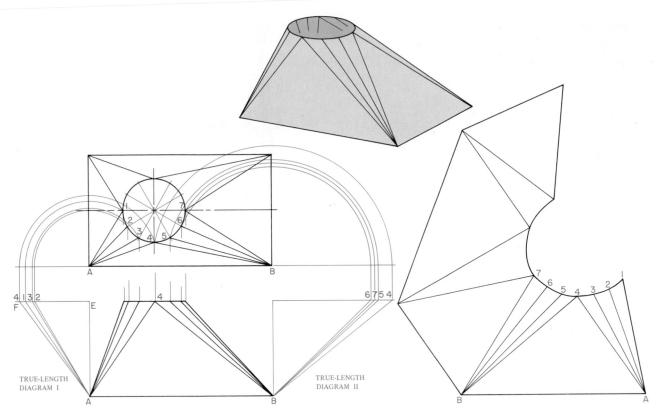

Fig. 18-39 *Development of a rectangular-to-round transition piece.*

3. Start the development by drawing line *DA*.

4. Using *A* and *D* as centers and a radius equal to the true length of *A4* or any of the other slanted lines, draw intersecting arcs to locate point *4* on the development of the piece.

5. Draw another arc using point 4 as a center and a radius equal to line *4-3*. Where this arc intersects the arc drawn in step 4 with *D* as a center, point *3* is located.

6. Proceed to lay off the remaining triangles until the transition piece is completed.

Figure 18-39 shows a transition piece made to connect a round pipe with a rectangular one. This piece contains four large triangles. Between them are four conical parts with apexes at the corners of the rectangular opening and bases, each one quarter of the round opening. To draw the development, proceed as follows:

1. Start with the partial cone whose apex is at *A*. Divide its base, *1-4*, into a number of equal parts, such as *1-2*, *2-3*, and *3-4*. Then draw lines *A2* and *A3* to give triangles approximating the cone.

2. Find the true length of each of these lines. In practice, this is done by constructing a separate diagram (diagram I). The construction is based on the fact that the length of each line is the hypotenuse of a triangle whose altitude is the altitude of the cone and whose base is the length of the top view of the line.

3. Begin diagram I by drawing the vertical line *AE* (the altitude of the cone) on the front view. On base *EF*, lay off distances *A1*, *A2*, and so forth, taken from the top view. In the figure, this is done by swinging each distance about point *A* in the top view, then dropping perpendiculars to base *EF*.

4. Connect the points thus found with point *A* in diagram I to find the desired true lengths. Diagram II, constructed in the same way,

gives the true lengths of triangle lines *B4, B5,* and so forth, on the cone whose apex is at *B.*

5. Start the development with the seam line *A1.* Draw line *A1* equal to the true length of *A1* (taken from true-length diagram I).

6. With *1* as a center and a radius equal to distance *1-2* in the top view, draw an arc. Then draw another arc using *A* as a center and a radius equal to the true length of *A2.* Where the two arcs intersect will be point *2* on the development.

7. With *2* as the center and radius *2-3,* draw an arc. Then draw another with center *A* and a radius equal to the true length of *A3.* Where these two arcs intersect will be point *3.*

8. Proceed similarly to find point *4.* Then draw a smooth curve through points *1, 2, 3,* and *4.*

9. Attach triangle *A4B* in its true size. Find point *B* by drawing one arc from *A* with radius *AB* taken from the top view and another arc from *4* with a radius the true length of *B4.* Where these arcs intersect will be point *B.*

10. Continue until the development is completed.

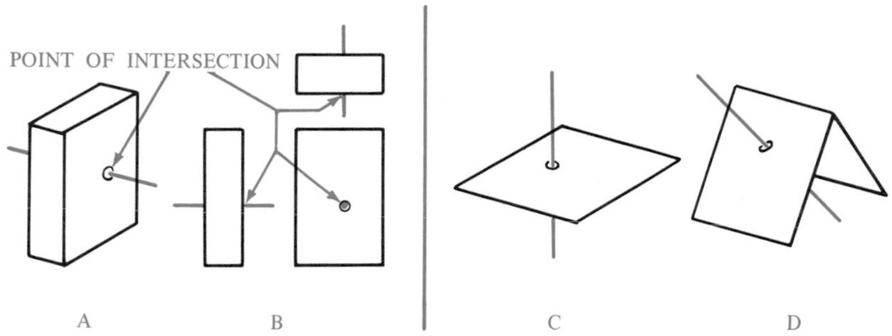

INTERSECTIONS

As you may recall from Chapter 8, a line intersects a plane at the *piercing point,* also called the *point of intersection* (Fig. 18-40). When two plane surfaces meet, the line where one passes through the other is called the *line of intersection* (Fig. 18-41). When a plane surface meets a curved surface, or where two curved surfaces meet, the line of intersection may be either a straight line or a curved line, depending on the surfaces and their relative positions.

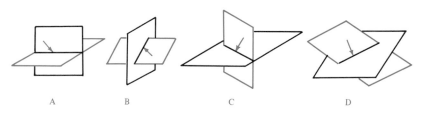

POINT OF INTERSECTION

A   B   C   D

Fig. 18-40 *The intersection of a line and a plane is a point.*

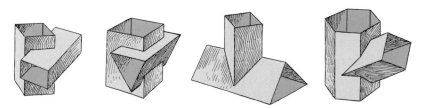

A   B   C   D

Fig. 18-41 *The intersection of two planes is a line. The arrow points to the line of intersection.*

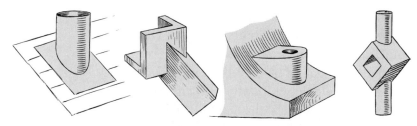

Fig. 18-42 *Examples of intersections between prisms.*

Fig. 18-43 *Other examples of intersections.*

Package designers, sheet-metal workers, and machine designers all must be able to find a point at which a line pierces a surface and the line where two surfaces intersect. Figures 18-42 and 18-43 show some of the ways in which different surfaces intersect.

Intersecting Prisms

To draw the line of intersection of two prisms, first start the regular top and front views. Fig. 18-44 shows an example in which a square prism passes through a hexagonal prism. Through the front edge of the square prism, pass a plane parallel to the vertical plane. The top view of this plane appears as line *AA*. The cut this plane makes in one of the faces of the hexagonal prism shows in the front view as cutting line *aa*. This line meets the front edge of the square prism at point *1*. Point *1* is a point on both prisms and, therefore, a point in the desired line of intersection.

Next, through the top edge of the square prism, pass plane *BB*, also parallel to the vertical plane. This plane makes one cutting line through the lateral edges of the hexagonal prism. The intersection of the two lines determines point *2* on the front view. Points *1* and *2* are on both prisms. Therefore, a line joining them lies on both prisms and thus is a part of the line of intersection. Plane *BB* also determines point *3*.

The planes in the above examples are called *cutting planes.* You can use them to solve most problems in intersections. For intersecting prisms, pass a plane through each edge on both prisms that touch the line of intersection. Each such plane makes cutting lines on both prisms. The intersection of the cutting lines identifies a point on the required line of intersection. In Fig. 18-45, four cutting planes are required. No planes are needed in front of *AA* or behind *DD*, because there they would cut only one of the prisms. Planes *AA* and *DD* are thus called *limiting planes.*

Intersecting Cylinders

Figure 18-46 shows how to draw the line of intersection of two cylinders. Since there are no edges on the cylinders, you have to assume positions for the cutting planes. Draw plane *AA* to contain the front line (element) of the vertical cylinder. This plane will also cut a line (element) on the horizontal cylinder. The intersection of these two lines in the front view identifies a point on the required curve. Similarly, the other planes cut

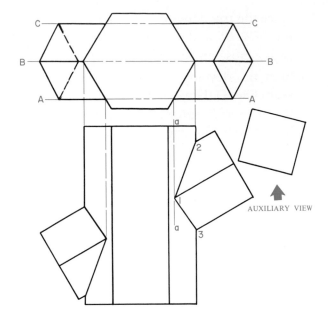

Fig. 18-44 *Intersecting prisms.*

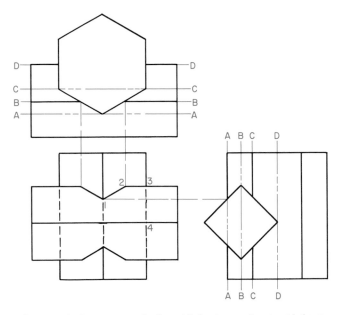

Fig. 18-45 *Intersection of a four-sided prism and a six-sided prism.*

other lines on both cylinders that intersect at points common to both cylinders.

The figure also shows the development of the vertical cylinder. Figure 18-47 shows how to draw the line of intersection of two cylinders joined at an angle. Here the cutting planes are located by an auxiliary view. To make the development of the inclined cylinder, take the length of the stretchout line from the circumference of the auxiliary view.

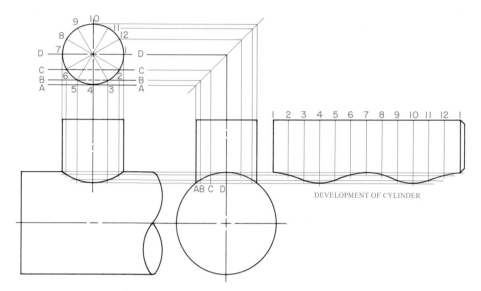

DEVELOPMENT OF CYLINDER

Fig. 18-46 *Intersection of cylinders at a right angle.*

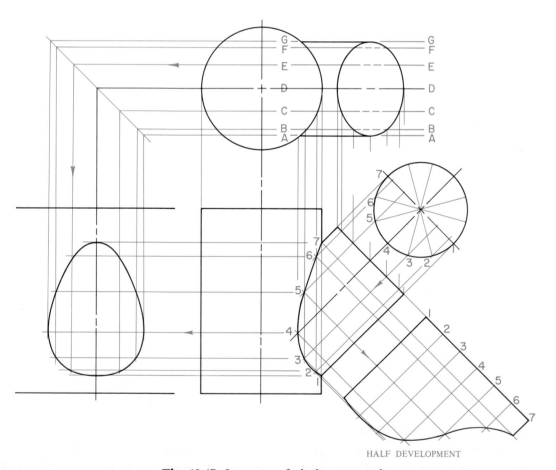

HALF DEVELOPMENT

Fig. 18-47 *Intersection of cylinders at an angle.*

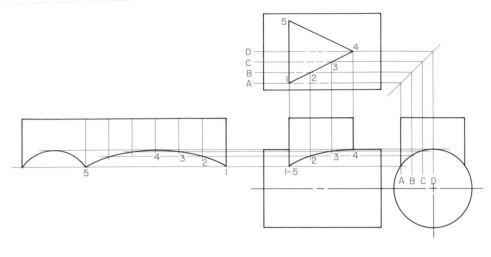

Fig. 18-48 *Intersection of a prism and a cylinder.*

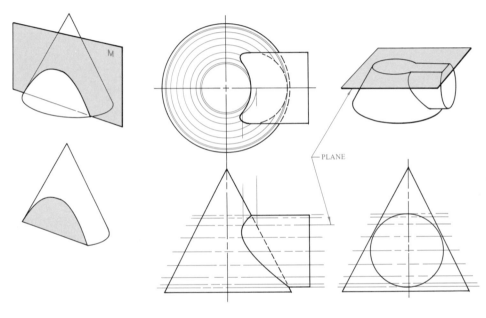

Fig. 18-49 *Intersection of a cylinder and a cone.*

Choose a cutting plane that divides this circumference into equal parts so that the measuring lines are equally spaced along the stretchout line. Project the lengths of the measuring lines from the front view. Join their ends with a smooth curve using a French curve.

Intersection of Cylinders and Prisms

You can also find the intersection line of a cylinder and a prism by using cutting planes. In Fig. 18-48, a triangular prism intersects a cylinder. Planes *A, B, C,* and *D* cut lines on both the prism and the cylinder. These lines cross in the front view at points that determine the curve of intersection. To make the development of the triangular prism, take the length of the stretchout line from the top view. Take the lengths of the measuring lines from the front view. Note that one plane of the triangular prism (line *1-5* in the top view) is perpendicular to the axis of the cylinder. Make the curve of the intersection line on that face using the radius of the cylinder. Create the other curve with a French curve.

Intersection of Cylinders and Cones

To find the intersection line of a cylinder and a cone, use horizontal cutting planes, as shown in Fig. 18-49. Each plane cuts a circle on the cone and two straight lines on the cylinder. Points of intersection occur where the straight lines of the cylinder cross the circles of the cone in the top view. Project these points onto the front view to get the intersection line. Figure 18-50 shows this construction for a single plane. Use as many planes as are needed to make a smooth curve.

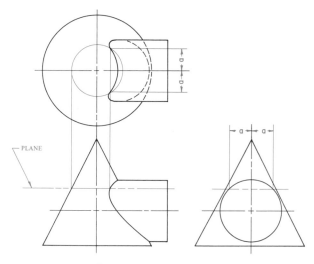

Fig. 18-50 *A cutting plane.*

Intersection of Planes and Curved Surfaces

Figure 18-51 shows the intersection of a plane *MM* and the curved surface of a cone. To find the line of intersection, use horizontal cutting planes *A, B, C,* and *D*. Each plane cuts a circle from plane *MM*. Thus, you can locate points common to *MM* and the cone, as in the top view. Project these points onto the front view to get the curve of intersection.

Figure 18-52 shows the intersection at the end of a connecting rod. To find the curve of intersection, use cutting planes perpendicular to the rod's axis. These planes cut circles as shown in the end view. The circles, in turn, cut the "flat" at points that can be projected back as points on the curve.

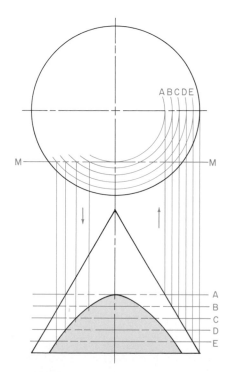

Fig. 18-51 *Intersection of a plane and a curved surface.*

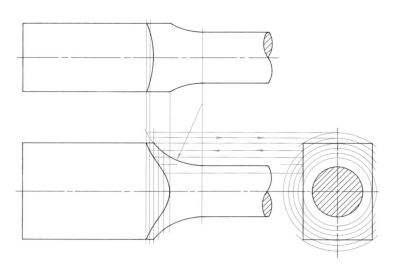

Fig. 18-52 *Intersection of a plane and a turned surface.*

CHAPTER 18
REVIEW

Learning Activities

1. Collect a variety of packages (cartons). Take them apart and lay them out flat to see how the original pattern was developed.

2. Select several odd-shaped objects and design packages to suit. Trace the patterns onto Kraftboard, cut them out, and assemble them. Be sure to provide tabs for fastening corners.

3. Develop patterns for a tetrahedron, hexahedron, octahedron, dodecahedron, and icosahedron. Add tabs for fastening corners. Transfer the patterns to Kraftboard, cut them out, and assemble them. Use your own judgment for sizes. See Fig. 18-10 for the design of these items.

4. Obtain sample surface development drawings from industry. Are they drawn full size or to a reduced scale? Why? Can you visualize the final shape of the assembled parts without seeing pictorials of them? Do the drawings follow the drafting standards shown in the chapter?

5. Visit an industrial heating, ventilating, and air-conditioning (HVAC) shop. Observe how parts are drawn, transferred to sheet material, and fabricated. Are drawings fully dimensioned?

Questions

Write your answers on a separate sheet of paper.

1. Name the three basic types of surface developments.

2. Name the two general classes of surfaces.

3. A sphere has a _____ curved surface.

4. A development that goes from round to square is called a _____ piece.

5. Lines on a stretchout that show where to make a fold are called _____ lines.

6. Using a series of triangles to develop a pattern is called _____.

7. Cone shapes are produced by using _____-line development.

8. Another name for development is _____.

9. Pressing a flat sheet into shape under heavy pressure is called _____.

10. Another name for pattern drafting is _____.

11. The stretchout line may also be called a _____ line.

12. A template may also be called a _____.

13. A cone in which the base is a true circle and the tip (point) is directly over the center of the base is a _____ cone.

14. What is another name for the tip or point of a cone?

15. The point at which a line pierces a plane (surface) is called the _____ or the _____.

CHAPTER 18
PROBLEMS

The small CAD symbol next to each problem indicates that a problem is appropriate for assignment as a computer-aided drafting problem and suggests a level of difficulty (one, two, or three) for the problem.

Fig. 18-53 *CAD1 Developments. Scale: Full size. Problems A through L are planned to fit on an 11.00″ × 17.00″ or 12.00″ × 18.00″ drawing sheet. Draw the front and top views of the problem assigned. Develop the stretchout (pattern) as shown in the example at the right. For problems A through F, add the top in the position it would be drawn for fabrication. Include dimensions and numbers if instructed to do so. Patterns may be cut out and assembled.*

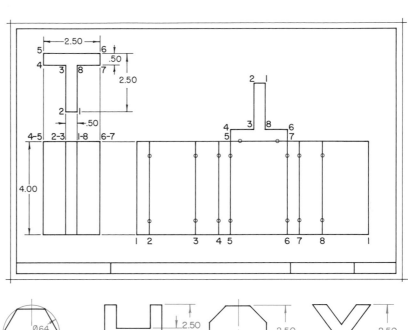

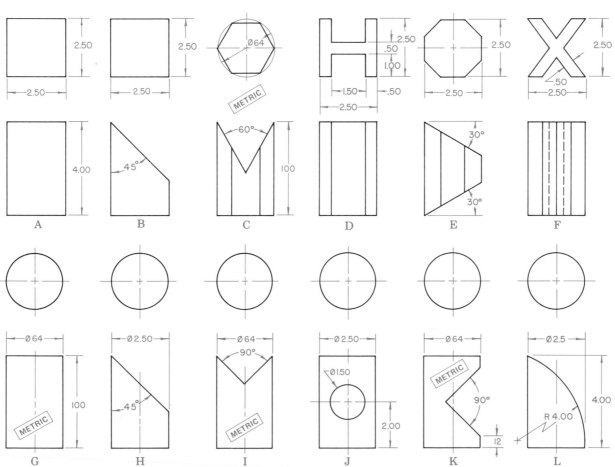

PROBLEMS

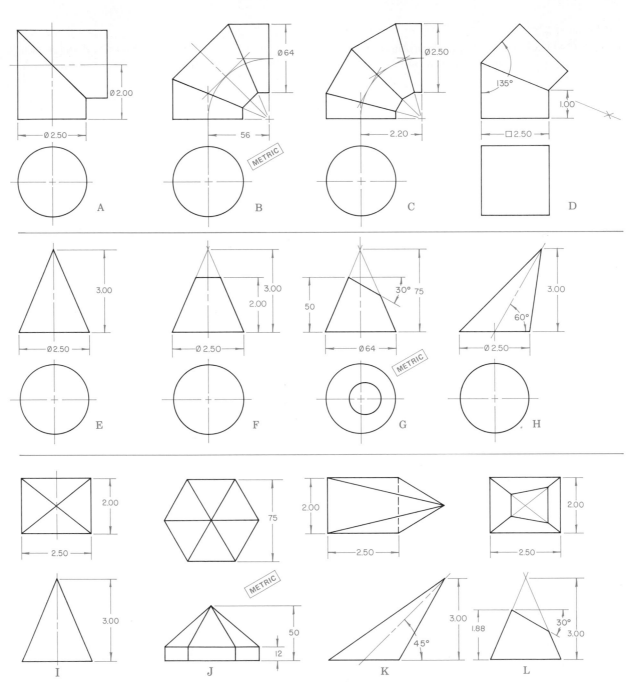

Fig. 18-54 *CAD* *Make two views of the problem assigned and develop the pattern. Scale: Full size.*

PROBLEMS

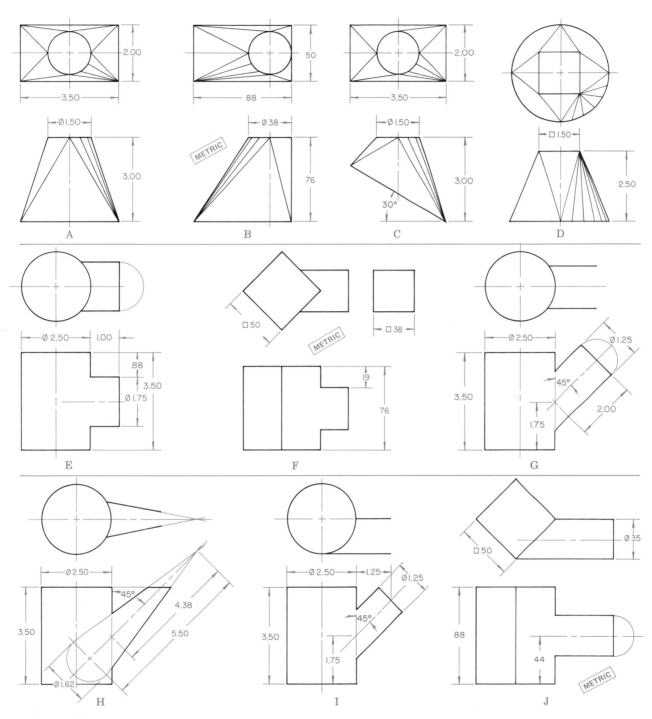

Fig. 18-55 **CAD3** *Make two views of the problem assigned. In problems A through D, complete the top views and develop the pattern. In problems E through J, complete views where necessary by developing the line of intersection and completing the top views in G, H, and I. Develop patterns for both parts in problems E through J.*

PROBLEMS

Make pattern drawings for Figs. 18-56 and 18-57. No other views are necessary. Make complete working drawings of Figs. 18-59 through 18-60, including all necessary views and patterns. Take dimensions from the printed scales at the bottom of the page for Figs. 18-58 through 18-60.

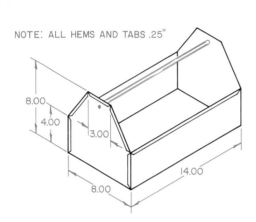

Fig. 18-56 *CAD1 Tool tray.*

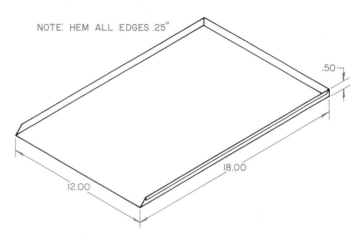

Fig. 18-57 *CAD1 Cookie sheet.*

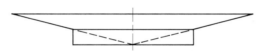

Fig. 18-58 *CAD1 Brass candy tray.*

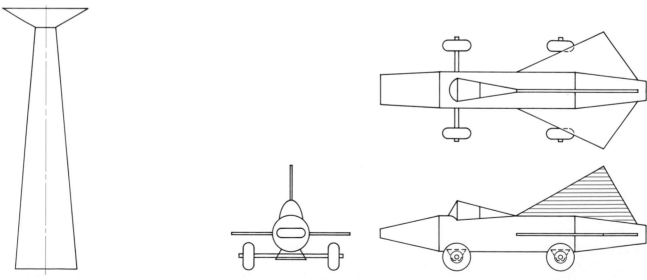

Fig. 18-59 *CAD2*
Candlestick.

Fig. 18-60 *CAD2 Model racer.*

PROBLEMS

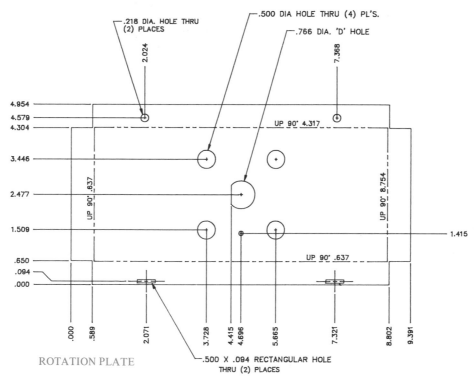

ROTATION PLATE

Fig. 18-61 *Make an orthographic multiview drawing of the CAD pattern drawing of the rotation plate.* (Courtesy of ALKCO Mfg.)

Fig. 18-62 *Develop a pattern drawing for the three-view CAD drawing illustrated in Figure 18-34.* (Courtesy of ALKCO Mfg.)

Fig. 18-63 *Create a CAD drawing of Fig. 18-33. Dimension the drawing in a manner similar to that used in Fig. 18-62.*

SCHOOL-TO-WORK

Solving Real-World Problems

Fig. 18-64 *SCHOOL-TO-WORK problems are designed to challenge a student's ability to apply skills learned within this text. Be creative and have fun!*

Fig. 18-64 *Package design. Design a package (carton) to be used in shipping the model racer shown in Fig. 18-60. The overall dimensions of the racer are 4.50 × 5.00 × 11.00 inches. The carton should be designed as a one-piece development, easily assembled. Try to design it in a way that will require no adhesive for assembly.*

If available, use desktop publishing or CAD to design the outside surface of the carton. Use various colors if possible. Trace the pattern onto Kraftboard, cut it out, and assemble it. Does it look as professional as a carton you would expect to see on a shelf at a hobby shop?

19

CAMS AND GEARS

OBJECTIVES

Upon completion of Chapter 19, you will be able to:

○ Explain the purpose of cams and gears.
○ Develop a displacement diagram.
○ Develop a profile of a cam.
○ Describe cam motion.
○ Define common gear terminology.
○ Draw gear teeth using the simplified method.

VOCABULARY

In this chapter, you will learn the meanings of the following terms:

- addendum
- bevel gear
- cam
- circular pitch
- cycloidal curve
- dedendum
- diametral pitch
- displacement diagram

- follower
- gear
- harmonic motion
- involute curve
- pinion
- pitch diameter
- pressure angle
- root diameter

- spur gear
- uniform motion
- uniformly accelerated and decelerated motion
- worm gear

Cams and gears are machine parts that do specific jobs. They can transmit motion, change the direction of motion, or change the speed of motion in a machine (Fig. 19-1). The ability to draw cams and gears and to understand their function is a first step in becoming a machine designer. This chapter introduces the skills that are basic to drawing and understanding these important machine parts.

CAMS

A **cam** is a machine part that usually has an irregular curved outline or a curved groove. When the cam rotates, it transmits a specific, continuous motion to another machine part that is called the **follower.** The cam and the follower together make up the cam mechanism. The cam drives the follower. The design of the cam determines the motion and path of the follower.

Fig. 19-1 *Cams and gears are necessary parts of machines.* (Courtesy of USI-Clearing, a U.S. Industries Co.)

Kinds of Cams

All cams can be thought of as simple inclines that produce or transmit a predetermined motion. Cams of various sizes and shapes transmit motion to the follower in specific ways. For example, a plate cam like that in Fig. 19-3 has a follower that moves up and down as the cam turns on the shaft. A cylindrical cam such as the one in Fig. 19-4 has a follower that moves back and forth parallel to the axis of the shaft. The grooved cam in Fig. 19-5 has a follower that follows the groove, moving in an irregular pattern as the cam turns on a shaft.

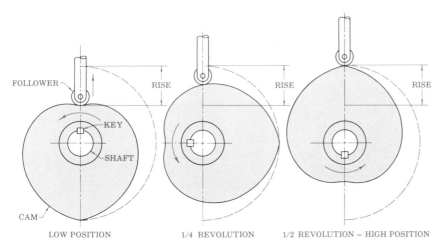

Fig. 19-2 *Cam action and terms.*

Fig. 19-3 *A plate cam.* (Courtesy of Camco)

This motion is needed in all automatic machinery. Therefore, cams are important to the automatic control and accurate timing found in many kinds of machinery. The unlimited variety of shapes makes the cam useful to the designer.

The illustrations in Fig. 19-2 are pictorial descriptions of how the cam works. The stroke, or *rise,* takes place within one half of a full revolution, or 180°. The movement is repeated every 360°, or once during each full revolution.

Fig. 19-4 *A cylindrical cam.* (Courtesy of Camco)

Fig. 19-5 *A grooved cam.*
(Courtesy of Camco)

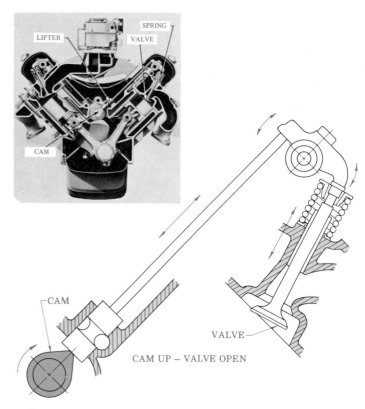

Fig. 19-6 *Automobile valve cam.* (Courtesy of Oldsmobile Div., General Motors Corp.)

Several other types of cams have been developed for specialized uses. For example, Fig. 19-6 shows a cam for operating the valve of an automobile engine. This cam has a flat follower that rests against the face of the plate cam. Figure 19-7 shows several other types of cams:

- a slider cam (A) moves the follower up and down as the cam moves back and forth.
- an offset plate cam (B) has a point follower that is off center.
- a cam with a pivoted roller-follower (C) transmits motion at a 90° angle.
- a cylindrical-edge cam may employ a swinging follower (D).

Cam Followers

Figure 19-8 shows three types of followers for plate cams. The *roller follower* reduces friction to a minimum, so it transmits force at high speeds. The *point follower* and the *flat-surface follower* are made with a hardened surface to reduce wear from friction. These followers are generally used with cams that rotate slowly.

Plate cams require a spring-loaded follower to ensure that contact is made throughout a full revolution. The *rise* (lifting) of the follower by the cam is made through direct contact

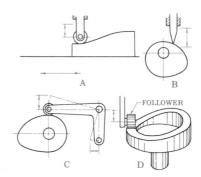

Fig. 19-7 *Kinds of cams: (A) slider; (B) offset plate; (C) pivoted roller; (D) cylindrical roller.*

between the cam and follower. However, the cam and follower are normally not in contact during a *drop* (fall) or *dwell* (rest) unless contact is brought about by an outside pressure. Outside pressure is exerted by a spring pushing the follower against the cam to ensure direct contact.

Displacement Diagrams

The shape of the cam determines the direction of motion in the follower as shown in 19-9A. A **displacement diagram** such as the one in Fig. 19-9B shows the motion the cam will produce through one revolution. The length of the displacement diagram represents one revolution of 360°. The height of the diagram represents the total displacement stroke of the follower from its lowest position (in this case, 1.87 in.).

Cam Motion

Cams can be designed so that the followers have three different types of motion. Displacement diagrams are used to plot the different kinds of motion.

Uniform Motion

Cams that follow **uniform motion** (steady motion) are suitable for high-speed operations. Technically, uniform motion is "straight-line" movement in which time and distance are directly proportional. In other words, equal distances on the rise (8 units) are made for equal distances on the travel (equal intervals of time, 8 units). However, to avoid a sudden jar at the beginning and end of the motion, designers use arcs to change it slightly. In Fig. 19-10A, the thin colored line represents uniform motion that has been modified in this way.

The cam in Fig. 19-9A displays uniform motion. The length, or time, of a revolution is divided into convenient parts. The parts are proportional to the number of degrees (based on one full 360° revolution) for each action. Each part or division is called a *time period*. These proportional parts are identified as *A, 1', 2', B', C, 4', 5', D', E, 6', 7'*, and back to *A*, as shown. Each proportional part or time period is a 30° angular division of the *base circle* shown in Fig. 19-9A.

In Fig. 19-9A, point *O* is the center of the cam shaft, and point *A* is the lowest position of the center of the roller follower. The center of the roller follower must be raised 1.875 in. with uniform motion during the first 120° of a revolution of the shaft. It must then dwell for 30°, drop 1.250 in. for 90°, dwell for another 30°, and drop .625 in. during the remaining 90°. The shaft is assumed to revolve uniformly (with constant speed).

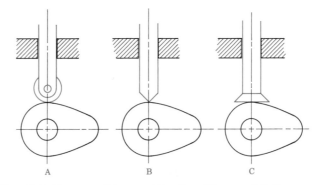

Fig. 19-8 *Plate cam followers: (A) roller; (B) point; (C) flat surface.* (Courtesy of Camco)

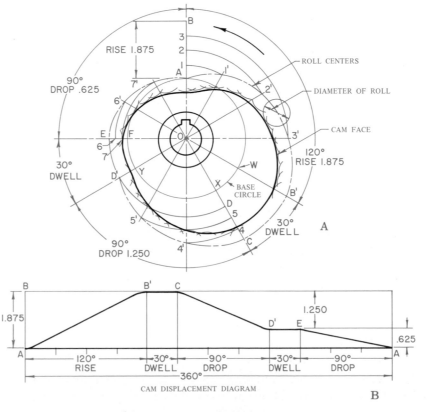

Fig. 19-9 *Cam displacement diagram.*

The displacement diagram in Fig. 19-9B illustrates the solution to the requirements listed above. Points *A, B', C, D', E*, and back to *A* relate to the travel pattern that will take place in one complete revolution.

Harmonic Motion

Other cams follow **harmonic motion,** which is a smoother-acting motion that follows a harmonic curve. Like uniform-motion cams, harmonic-motion cams are suitable for use with high speeds. They are also used when a smooth start-and-stop motion is needed.

The method for plotting a harmonic curve is shown in Fig. 19-10B. To draw harmonic motion, first draw a semicircle with the rise as the diameter. Divide the semicircle into eight equal parts, using radial lines. Then divide the travel into the same number of equal parts. Project the eight points horizontally from the semicircle until they intersect the corresponding vertical projections, as shown. Finally, draw a smooth curve through all eight points with an irregular curve.

Uniformly Accelerated and Decelerated Motion

Cams with **uniformly accelerated and decelerated motion** (steadily increasing and decreasing motion) do not operate at a constant speed. The motion produced by these cams is very smooth. The motion of these cams follows a parabola, or parabolic curve. A *parabola* is formed when a cone is sliced vertically at any place other than through the center. (See Fig. 18-51).

To design a cam with uniformly accelerated and decelerated motion, first divide the rise into parts proportional to 1, 3, 5, . . . , 5, 3, 1, as shown in Fig. 19-10C. Note that the parts are not of equal size.

Then divide the travel into the same number of parts, but divide it into equal parts. Project the points horizontally from the rise until they intersect the corresponding vertical projections. Finish by drawing a smooth curve through all the points of intersection, as shown.

Drawing Cam Profiles

The profile of the cam in Fig. 19-9A has five important features.

1. **Rise.** Divide the rise *AB,* or 1.875 in., into a number of equal parts. Four parts are used in the illustration, but eight parts would make the layout more accurate. The rise occurs from A to W (120°). Divide it into the same number of equal parts as the rise (four at 30°) with radial lines from *O.* Using center *O,* draw arcs with radii *O1, O2, O3,* and *B'* until they locate *1', 2', 3',* and *B'* on the four radial lines. Use an irregular curve to draw a smooth line through these four points.

2. **Dwell.** First draw an arc *B'C* (30°) using radius *O'B'.* This allows the follower to be at rest, because it will stay the same distance from center *O.*

3. **Drop.** At *C,* lay off *CD* (1.250 in.) on a radial line from *O.* Divide it into any number of equal parts (three are shown). Next, divide the arc *XY* (90°) on the base circle into the same number of equal parts (three). Draw three radial lines from center *O* every 30°. Then draw arcs with center *O* and radii *O4, O5,* and *OD* to locate points *4', 5',* and *D'* on the three radial lines. Using an irregular curve, draw a smooth line through points *4', 5',* and *D'.*

4. **Dwell.** Draw an arc *D'E'* (30°) with radius *OD.* This will provide the constant distance from center *O* to let the roller follower be at rest.

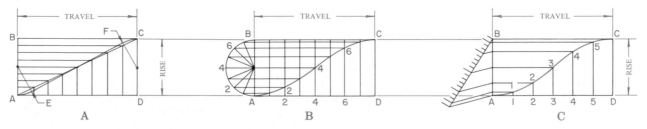

Fig. 19-10 *Kinds of cam motion: (A) uniform; (B) harmonic; (C) uniformly accelerated and decelerated.*

5. **Drop.** In the last 90° of a full revolution, the roller will return to point *A*. It will move through a distance *EF*, or .625 in. First divide *EF* into any number of equal parts (three are shown). Then divide arc *FA* into the same number of equal parts (three). Next, draw radial lines every 30°. Then draw arcs with radii *O6* and *O7* to locate points *6'* and *7'*. Using the irregular curve as a guide, draw a smooth curve through points *E, 6', 7',* and *A*. This finishes the roll centers.

Using the line-of-roll centers as a centerline, draw successive arcs (one after the other) with the radius of the roller, as shown. Then use an irregular curve to draw the cam profile. The profile should be a smooth curve tangent to the arcs you drew representing the roller.

Another drawing for a face, or plate, cam is shown in Fig. 19-11. Note that the amount of movement, or rise, is given by showing the radii for the dwells (4.50 in. and 7.00 in.). Harmonic motion is used, and there seem to be two rolls working on this cam. In Fig. 19-12, a drawing for a *barrel cam* (cylindrical cam) is shown with a displacement diagram. The diagram shows two dwells and two kinds of motion. Note that the distance traveled from center to center is 1.50 in. and is called *throw*.

CAD-GENERATED CAMS

The procedure for creating a cam profile using a CAD system is basically the same as that for creating it manually. The difference is the ease with

which various tasks can be completed. Because the CAD system can take care of basic drawing tasks such as creating the various diameters, the drafter can concentrate on the more complicated details.

Another time-saving option for CAD drafters is to use third-party parametric software that works with the CAD software. *Parametric software* is software that allows you to enter *parameters* (specific dimensions and other details), and then draws the required part according to the specifications you provide.

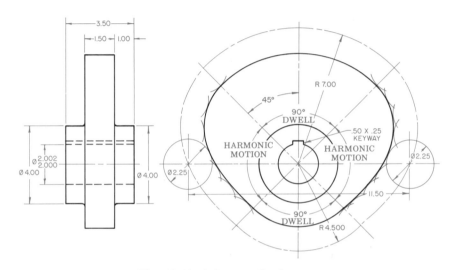

Fig. 19-11 *A drawing of a plate cam.*

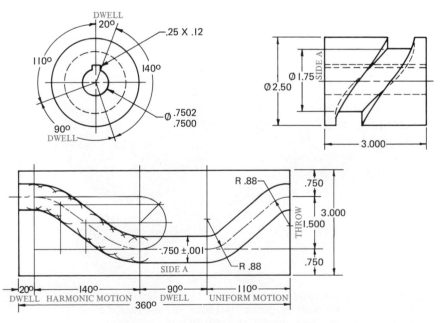

Fig. 19-12 *A drawing of a cylindrical cam.*

Parametric cam-generating software, for example, might allow the drafter to enter details such as the type of cam, requirements for periods or rise, dwell, and drop, and so on. Then the software would produce a cam based on the drafter's specifications and present it for the drafter's approval. The drafter can override any of the computer's decisions at any time.

GEARS

A **gear** is a device that transmits motion using a series of teeth. A few of the many kinds of gears are illustrated in Fig. 19-13.

Spur Gears

One of the most practical and dependable machine parts for transmitting rotary motion from one shaft to another is the **spur gear.** The operation of simple spur gears can be explained in this way: If the rims of two wheels are in contact, as in Fig. 19-14, both will revolve if only one is turned. If the small friction wheel is two thirds the diameter of the larger wheel, it will make 1.50 revolutions for every 1.00 revolution of the larger

wheel. This assumes that no slipping occurs. However, when the load on the driven wheel gets larger and the wheel is hard to turn, slipping begins to occur. Therefore, friction wheels cannot be counted on for a smooth transfer of rotary motion. When teeth are added to the wheels in Fig. 19-14, they become spur gears (Fig. 19-15). Teeth added to the wheels provide the same kind of motion as rolling friction wheels. Now, however, there is no slipping.

Figure 19-16 illustrates the parts of a spur gear. Refer to this illustration as you work with the gear formulas presented in this chapter.

Gear Teeth

The basic forms used for gear teeth are *involute* and *cycloidal* curves. These curves are explained in the following paragraphs.

The gears in Fig. 19-15 are good examples of involute gears. (Note that the small spur gear is called the **pinion.**) The teeth on these gears have a special shape that lets them mesh smoothly. This shape is an **involute curve.** Figure 19-17 explains an involute curve that is used in drawing gear teeth. As shown in the illustration, you can think of an involute of a circle as a curve made by taut string as it unwinds from around the circumference of a cylinder.

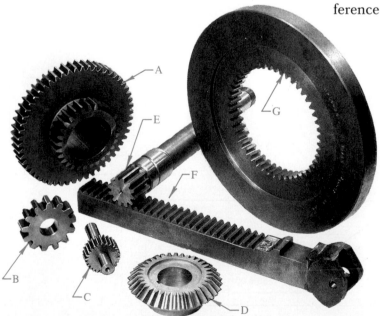

Fig. 19-13 *Several gears that are used as typical machine elements: (A, B, and C) spur gears; (D) bevel gear; (E) pinion (spur gear); (F) rack (E and F together are called a* rack and pinion*); (G) internal gear.*

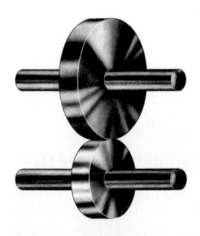

Fig. 19-14 *Friction wheels are a simple means of transmitting rotary motion from one shaft to another.*

Fig. 19-15 *The spur gear. Teeth added to friction wheels provide a more efficient means of transmitting rotary motion.*

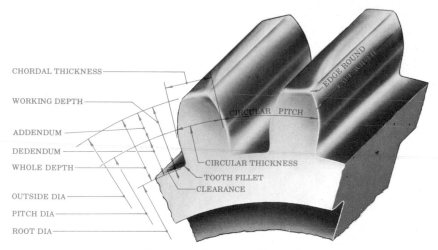

CHORDAL THICKNESS
WORKING DEPTH
ADDENDUM
DEDENDUM
WHOLE DEPTH
OUTSIDE DIA
PITCH DIA
ROOT DIA

EDGE ROUND
FACE WIDTH
CIRCULAR PITCH
CIRCULAR THICKNESS
TOOTH FILLET
CLEARANCE

Fig. 19-16 *Gear terms illustrated.*

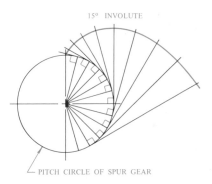

15° INVOLUTE

PITCH CIRCLE OF SPUR GEAR

Fig. 19-17 *Involute of a circle.*

A **cycloidal curve** can be thought of as the path of a curve formed by a point on a rolling circle (Fig. 19-18). The information given in this chapter is for the $14\frac{1}{2}°$ involute system. It can be used for a 20° involute system as well, since the only practical difference is the number of degrees of the **pressure angle.** The $14\frac{1}{2}°$ or 20° refers to the pressure angle (Fig. 19-19). The pressure angle and the distance between the centers of mating spur gears determine the diameters of the base circles (Fig. 19-19). The point of tangency of the gears is their **pitch diameter.** These are equal to the diameters of the rolling friction wheels that are replaced by the gears. The involute is drawn from the base circle, which is smaller than the pitch circle.

In Fig. 19-20, R_A is the radius of the pitch circle of the gear with the center at A. R_B is the radius of the pitch circle of the pinion with the center at B. The distance between gear centers is $R_A + R_B$. The line of pressure $T_A T_B$ is drawn through O (which is the point of tangency of the pitch circles). It makes the pressure angle φ (Greek letter phi) with the perpendicular to the

line of centers. This angle is $14\frac{1}{2}°$. Note that lines AT_A and BT_B are drawn from centers A and B perpendicular to the pressure-angle line $T_A T_B$. A point X on a cord (line of pressure line $T_A T_B$.) will describe the points that form the involute curve as the cord winds and unwinds. This represents the outlines of gear teeth outside the base circles. The profile of the gear tooth inside the base circle is a radial line. Figure 19-17 shows the cord unwinding off the surface on the base circle. To simplify the drawing of the involute curve, note that the radius in Fig. 19-21 is $\frac{1}{8}$ the pitch diameter.

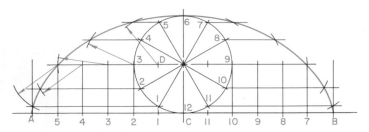

Fig. 19-18 *A cycloidal curve.*

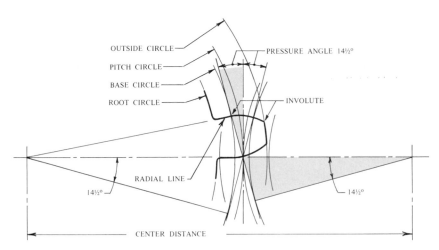

Fig. 19-19 *The pressure angle. Note that the center distance indicates the distance between shafts of mating gears.*

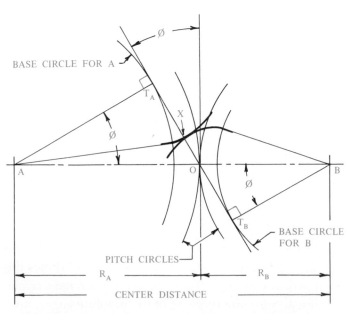

Fig. 19-20 *Gear tooth interaction; the rolling nature of surface contact.*

Applying Spur-Gear Formulas

Table 19-1 defines terms commonly used in gear formulas, and Table 19-2 presents the actual formulas. Refer to these two tables as you apply the gear formulas.

Drafters apply these formulas to the written descriptions and information they are given. The following information is typical of the instructions a drafter might receive.

A pair of involute gears is to be drawn according to the specifications that follow:

- *The pressure angle will be 14 ½°*
- *The distance between parallel shaft centers will be 12.00 in.*
- *The driving shaft will turn at 800 revolutions per minute (r/min) clockwise.*
- *The driven shaft will turn at 400 r/min.*
- *The diametral pitch equals 4 (number of teeth per inch of pitch diameter).*

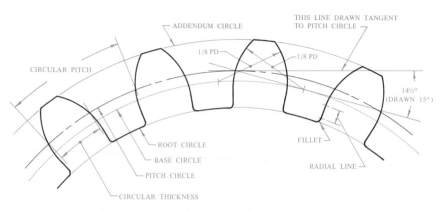

Fig. 19-21 *Simplified method of drawing a gear tooth.*

TABLE 19-1 SPUR-GEAR TERMS AND SYMBOLS		
Term	**Symbol**	**Definition**
	N	number of teeth
	N_G	number of teeth of gear
	N_P	number of teeth of pinion
pitch diameter	D	diameter of pitch circle
	D_G	pitch diameter of gear
	D_p	pitch diameter of pinion
diametral pitch	P	number of teeth per inch of pitch diameter
addendum	a	radial distance the tooth extends above the pitch circle
dedendum	b	radial distance the tooth extends below the pitch circle
outside diameter	D_o	overall gear size: pitch diameter plus twice the addendum
root diameter	D_R	pitch diameter minus twice the dedendum
whole depth	h_t	radial distance from the root diameter to the outside diameter; equal to addendum plus dedendum
circular pitch	p	distance from a point on one tooth to the same point on the next tooth measured along the pitch circle; the distance of one tooth and one space
circular thickness	t_c	thickness of a tooth measured along the pitch circle
clearance	c	the distance between the top of a tooth and the bottom of the mating space when gear teeth are meshing; the difference between the addendum and the dedendum
working depth	h_k	the distance a tooth projects into the mating space; twice the radial distance of the addendum
pressure angle	o	the direction of pressure between teeth at the point of contact
base-circle diameter	D_b	the circle from which the involute profile is developed

Table 19-1 *Definitions of terms and symbols used to describe spur gears.*

TABLE 19-2 SPUR GEAR FORMULAS	
number of teeth	$N = DP = \dfrac{\pi D}{p} = DO \times P - 2$
pitch diameter	$D = \dfrac{N}{P} = D_0 - 2a$
diametral pitch	$P = \dfrac{N}{D} = \dfrac{\pi}{p}$
addendum	$a = \dfrac{1}{P} = \dfrac{p}{\pi}$
dedendum	$b = \dfrac{1.157}{P} = \dfrac{1.157\,p}{\pi}$
outside diameter	$D_O = \dfrac{N+2}{P} = D + 2a = \dfrac{(N+2)p}{\pi}$
root diameter	$D_R = D - 2b = D_0 - 2(a+b)$
whole depth	$h_t = a + b = \dfrac{2.157}{P} = \dfrac{2.157\,p}{\pi}$
circular pitch	$\dfrac{\pi D}{N} = \dfrac{\pi}{P}$
circular thickness	$t = \dfrac{p}{2}$
chordal thickness	$t_C = D\sin\left(\dfrac{90^0}{N}\right)$
clearance	$c = \dfrac{.157}{P} = \dfrac{.157\,p}{\pi}$
working depth	$h_k = 2 \times a = 2 \times \dfrac{1}{P}$

Table 19-2 *Formulas used to calculate information for drawing spur gears.*

To find the rest of the information needed to draw the gears, follow these steps:

1. Find the pitch radius of the pinion and the spur gear. These calculations are based on the ratio of the velocity of the two cylinders. One cylinder drives the other. Thus, the ratio is obtained by dividing the velocity (in r/min) of the driver by the velocity (r/min) of the driven member:

pitch radius of pinion:

$$R_p = \frac{400}{400+800} \times 12.00 \text{ in.} = 4.00\text{-in. radius}$$

pitch radius of spur gear:

$$R_x = \frac{800}{400+800} \times 12.00 \text{ in.} = 8.00\text{-in. radius}$$

The velocity ratio therefore equals $\frac{1}{2}$(4.00 in.:8.00 in.).

2. Find the number of teeth on the pinion:

Number of teeth on the pinion:

$$N_p = DP = 4 \times 8 = 32$$

3. Find the number of teeth on the spur gear:

Number of teeth on spur gear:

$$N_s = DP = 4 \times 16 = 64$$

4. Find the addendum:

Addendum:

$$a = \frac{1}{P} = \frac{1}{4}\text{in.} = .25 \text{ in.}$$

5. Find the dedendum:

Dedendum:

$$b = \frac{1.157}{p} = \frac{1.157}{4} = .289 \text{ in.}$$

Involute gears are interchangeable as long as they have the same diametral pitch, pressure angle, addendum, and dedendum.

Involute Rack and Pinion

A rack and pinion is shown in Fig. 19-22. A *rack* is a gear with a straight pitch line instead of a circular pitch line. The tooth profiles become straight lines. These lines are perpendicular to the line of action.

Worm and Wheel

Figure 19-23 shows how the worm and wheel mesh at right angles. The **worm gear** is similar to a screw. It may have single or multiple threads. This system is used to transmit motion between two perpendicular, nonintersecting shafts. The *wheel* is similar to a spur gear in design except that the teeth must be curved to engage the worm gear

Fig. 19-22 *A rack and pinion.* (Courtesy of Brad Foote Gear Works, Inc.)

Fig. 19-23 *Application of a worm and wheel.* (Courtesy of Industrial Drives Division, Eaton Corp.)

Bevel Gears

When two gear shafts intersect, **bevel gears** are used to transfer motion. Figure 19-24 shows four sets of rolling friction cones. Think of bevel gears as replacing the friction cones, just as the spur gear replaced the circular friction wheels. Figure 19-25 illustrates mating bevel gears. The smaller gear is called the *pinion*. When the gears are the same size and the shafts are at right angles, bevel gears are referred to as *miter gears*.

Some basic information about bevel gears is given in Fig. 19-26. Look for similarities and differences in the design of spur and bevel gears.

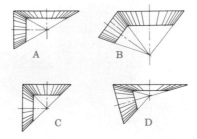

Fig. 19-24 *Rolling cones that represent bevel gearing.*

Fig. 19-25 *Mating bevel gears.* (Courtesy of Arrow Gear Co.)

TABLE 19-3	SYMBOLS FOR BEVEL GEARS
Symbol	**Definition**
α	addendum angle
δ	dedendum angle
Γ	pitch angle
Γ_R	root angle
Γ_O	face angle
a	addendum
b	dedendum
a_n	angular dedendum
A	cone distance
F	face
D	pitch diameter
D_O	outside diameter
N	number of teeth
P	diametral pitch
R	pitch radius

Table 19-3 *Symbols used to describe bevel gears.*

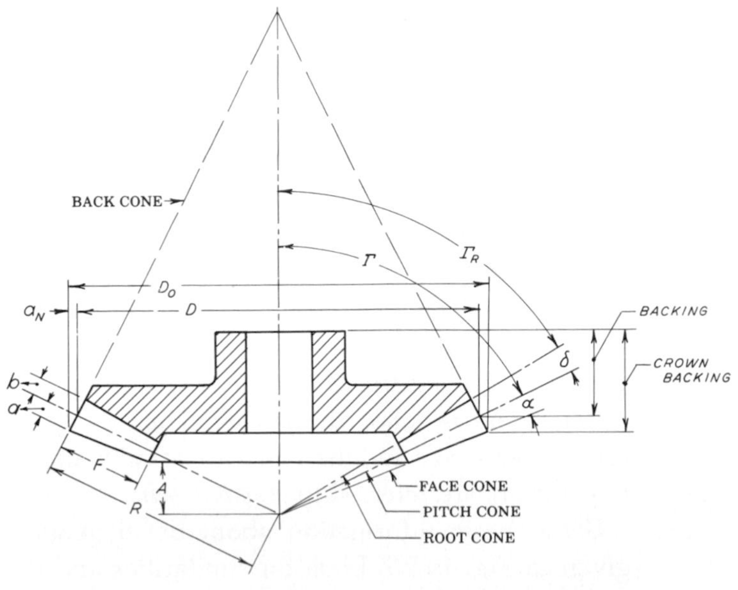

The circular pitch and the diametral pitch are based on the pitch diameter just as they are in spur gears. The pitch diameter is the diameter of the pitch cone in the bevel-gear design. The symbols for important features, such as the angles, are listed in Table 19-3. (The three Greek letters used to represent bevel angles are α (alpha), δ (delta), and Γ (gamma).)

Drawing Gears

It is not necessary to show the teeth on typical gear drawings. A drawing for a cut spur gear is shown in Fig. 19-27. The gear blank should be drawn with dimensions for making the pattern and for the machining operations. The spur-gear drawing should include information for cutting the teeth.

Fig. 19-26 *Bevel-gear terms. Note the cone shape of the gear.*

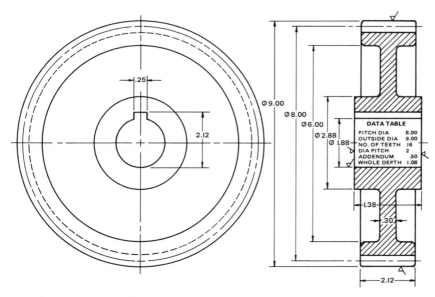

DATA TABLE	
PITCH DIA	8.00
OUTSIDE DIA	9.00
NO. OF TEETH	16
DIA PITCH	2
ADDENDUM	.50
WHOLE DEPTH	1.08

Fig. 19-27 *Simplified profile and cross section of a spur-gear working drawing.*

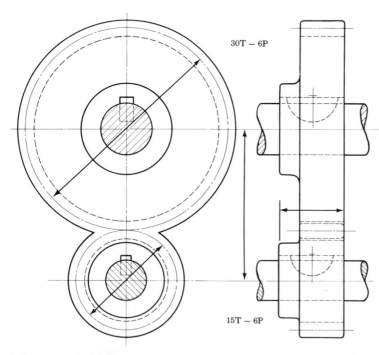

Fig. 19-28 *Simplified drawing of mating spur gears. Note the tangent pitch circles, the number of teeth, and the pitch.*

It should also include information for the tolerances required and a notation of the material to be used. On assembly drawings, a simplified gear may be used, as shown in Fig. 19-28. Even though the drawing is simplified, however, it should include all necessary notes for making the gear.

On working drawings of bevel gears, the needed dimensions for machining the blank must be given, as well as all the gear information. An example of a working drawing for a cut bevel gear is shown in Fig. 19-29.

The American National Standards Institute has established standards for satisfactorily detailing gear drawings. ANSI Y14.7, Section Seven, and B6.1 and B6.5 are standard references that can be used for further study of the subject of gear detailing.

CAD

CAD-GENERATED GEARS

Most CAD programs contain many commands that aid the drafter in designing and detailing gears. For example, commands for editing, copying, arraying, and grouping provide quick, easy ways to draw repeating tooth forms. In addition, symbol libraries that contain the basic forms allow the drafter to insert ready-made gears or gear symbols into a drawing.

A CAD-detailed gear drawing is shown in Fig. 19-30. However, this drawing does not meet ANSI standards. Study the drawing and decide what you could change to correct the drawing. Also think about how you would create the corrected drawing using a CAD system. What elements could be done more easily using CAD software and symbol libraries?

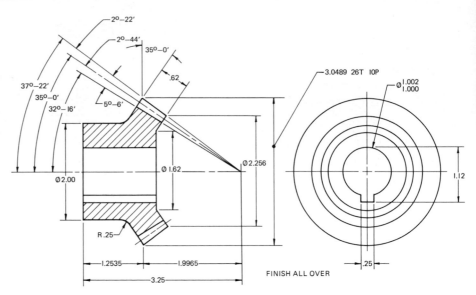

Fig. 19-29 *Simplified profile and detailed section of a bevel gear.*

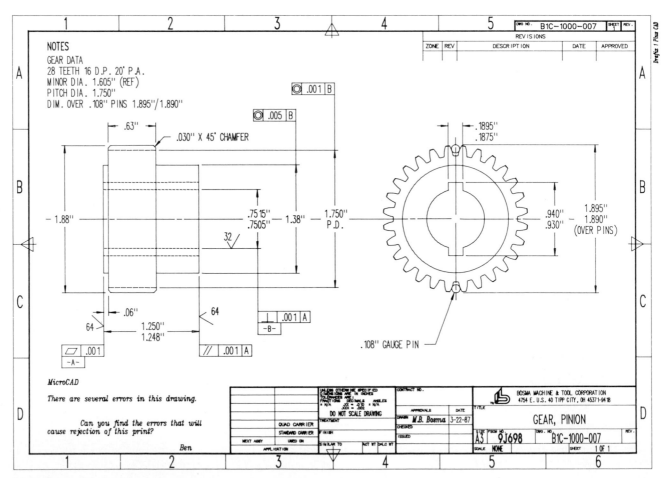

Fig. 19-30 *This CAD drawing does not meet basic industry standards. Check the drawing for proper techniques. Check the title block, the tolerances, and the parts list. Also note basic drafting techniques.* (Courtesy of BOSMA Machine Tool Corp. Drawn with Drafix One-Plus Software.)

CHAPTER 19

REVIEW

Learning Activities

1. Collect different types of gears and cams. Identify and classify each as described in the chapter. Try to visualize and describe the function of each. (Obtain large and small spur gears similar to those shown in Fig. 19-15.)

2. Make a set of friction wheels with diameters that closely approximate the pitch diameters of the spur gears.

3. Obtain drawings from industry that contain cam and gear applications. Do they follow the drafting standards shown in this chapter? Can you visualize the applications?

4. Visit an industry to observe the design and manufacture of cams and gears. Why are gear teeth generally not drawn in their true size and shape?

Questions

Write your answers on a separate sheet of paper.

1. Other than time, what is important in the design of a cam?

2. What is the purpose of a displacement diagram?

3. List three types of cams and three types of followers.

4. List three uses of a cam and the kinds of cam used.

5. Why is harmonic motion used in a cam?

6. How is the curve on a spur-gear tooth developed?

7. What technical phrase describes the ratio of the number of teeth to the pitch diameter?

8. Rolling cones are used to describe the meshing of what kind of gears?

9. What two bits of information are needed to find circular pitch?

10. List two applications of bevel gears.

11. Name the four circles used by the drafter in drawing gears.

12. List the changes necessary to improve the overall techniques in the detailed CAD drawing in Fig. 19-30.

13. What is a parametric software program?

14. Describe the advantages of using parametric software in combination with CAD software when working with cams and gears.

CHAPTER 19

PROBLEMS

The small CAD symbol next to each problem indicates that a problem is appropriate for assignment as a computer-aided drafting problem and suggests a level of difficulty (one, two, or three) for the problem.

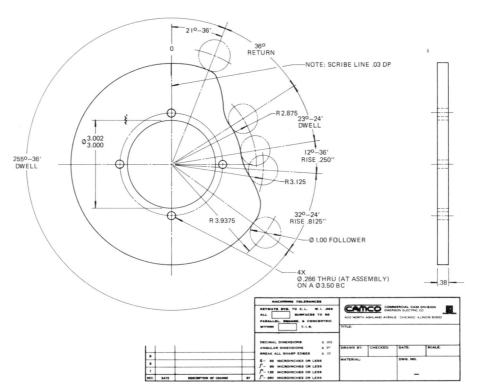

Fig. 19-31 CAD3 *Make a profile of the radial plate cam. Prepare a displacement diagram similar to the one in Fig. 19-9 to illustrate the travel patterns from the base circle.*

Fig. 19-32 CAD3 *Prepare a drawing of the displacement diagram and the cam profile in Fig. 19-9. Shaft: Ø0.62″; roller: Ø0.56″; base circle: Ø2.50″; hub: Ø1.12″. Diagram line = 2.50 × 3.14.*

Fig. 19-33 CAD3 *Prepare the three diagrams in Fig. 19-10 to illustrate the three types of cam motion. Travel: 2.50″; rise: 1.25″.*

Fig. 19-34 CAD3 *Draw the two views of the plate cam in Fig. 19-11. Prepare a displacement diagram to illustrate the rises.*

PROBLEMS

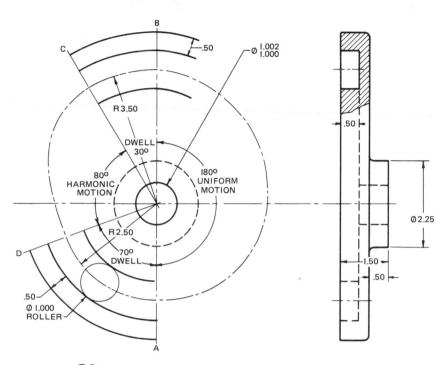

Fig. 19-35 _CAD3_ _Draw a box (grooved) cam for the conditions given in the illustration. Draw the path of the roll centers and lay off the angles. From A to B, rise from 2.50″ to 3.50″, with modified uniform motion. From B to C, dwell, radius 3.50″. From C to D, drop from 3.50″ to 2.50″, with harmonic motion. From D to A, dwell, radius 2.50″. Draw the groove, using a roller with a 1.00″ diameter in enough positions to fix outlines for the groove. Complete the cam drawing. Keyway .50″× .12″. Leave construction lines._

Fig. 19-36 _Prepare a two-view drawing of the spur gear in Fig. 19-27. Make a simplified profile and section as shown. Using the given data, draw in two gear teeth after calculating circular thickness._

PROBLEMS

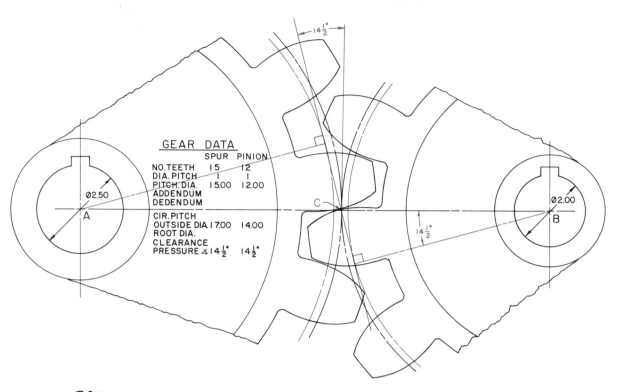

Fig. 19-37 *CAD3 Complete the gear data, using formulas from this chapter. Make an enlarged drawing of a mating spur and pinion as shown. Select a suitable scale. Use an involute to draw the gear tooth profile or ⅛ PD, as directed by the instructor.*

Fig. 19-38 *Prepare a working drawing of the mating gears in Fig. 19-23. Calculate the data necessary to draw the gear teeth and prepare a data table.*

PROBLEMS

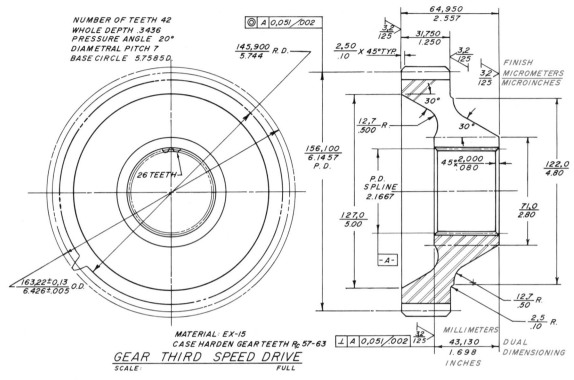

NUMBER OF TEETH 42
WHOLE DEPTH .3436
PRESSURE ANGLE 20°
DIAMETRAL PITCH 7
BASE CIRCLE 5.7585 D.

MATERIAL: EX-15
CASE HARDEN GEAR TEETH Rc 57-63

GEAR THIRD SPEED DRIVE
SCALE: FULL

Fig. 19-39 *CAD3 Prepare a spur-gear detail drawing, using the data given to determine necessary dimensions. Note that the gear is designed in metric dimensions (using a comma in place of a decimal point) and is dual-dimensioned with decimal inches.* (Courtesy of International Harvester Co.)

Fig. 19-40 *Prepare the two views of the bevel gear in Fig. 19-29. Review the simplified profile and insert the proper pitch circles as centerlines in the front view. Dimension the drawing.*

Fig. 19-41 *Prepare a corrected drawing of Figure 19-30.*
(Gear, Pinion–Monorail Tractor Drive. Courtesy of BOSMA Machine & Tool Corp.)

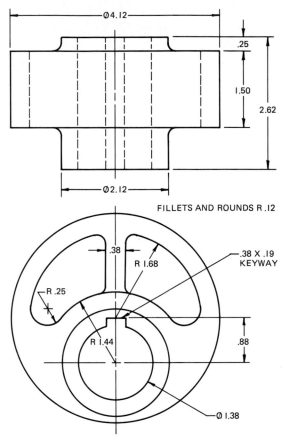

Fig. 19-42 **CAD3** *Prepare a detailed drawing of the feeder cam.* (Courtesy of VersaCAD Corp.)

SCHOOL-TO-WORK

Solving Real-World Problems

Fig. 19-43 *SCHOOL-TO-WORK problems are designed to challenge a student's ability to apply skills learned within this text. Be creative and have fun!*

Fig. 19-43 *Complex cam displacement diagram. The design engineers in your company need a displacement diagram for a complex cam that will be used to change the direction of motion in a mechanism from rotating to reciprocating motion. They have given you the following information to use in developing the displacement diagram:*

During the first 90°, the cam uniformly rises through one third of its total rise. It then dwells for 30°. Beginning at 120°, the cam moves through the rest of its rise using harmonic motion. After another 30° dwell, the cam falls using uniform acceleration-deceleration. The total rise is 3.60 inches.

20

ARCHITECTURAL AND STRUCTURAL DRAFTING

OBJECTIVES

Upon completion of Chapter 20, you will be able to:

○ Describe career opportunities in architectural and structural drafting.
○ Prepare a complete set of house plans.
○ Name the parts of a structure.
○ Identify structural steel members that form the framework of buildings, bridges, etc.
○ Prepare details of structural steel components.
○ Describe various materials used in structural design.

VOCABULARY

In this chapter, you will learn the meanings of the following terms:

- **A/E/C**
- **balloon framing**
- **cross section**
- **dead load**
- **detail section**
- **elevations**
- **finite element analysis (FEA)**
- **laminating**
- **live load**
- **nominal size**
- **plank and beam framing**
- **prestressed concrete**
- **reinforced concrete**
- **schedules**
- **site plan**
- **specifications**
- **structure**
- **truss**
- **western framing**

The work that architects do greatly affects the way we live. Our everyday lives take place within an *environment* (setting) designed by architects and city planners. Architects design buildings and the spaces around them to form rural communities, suburbs, or cities. They also design other **structures** (anything that has been constructed) such as bridges and dams that make life easier for people.

Architects have been planning, designing, and supervising the construction of buildings around the world for centuries. Today they continue to provide functional buildings that encourage and heighten all human activity.

The architect's job is to give people a better and richer life—to design a better tomorrow.

Architecture is not just buildings. It is people—how they live, work, play, and worship. Good architectural environments can make any human activity better. Thus, architecture can be thought of as any type of physical environment. It is the sprawling suburb, the crowded inner city, the mighty industrial complex, the bright lights of Broadway, and main street U.S.A.

You can easily examine architects' work (Fig. 20-1). Look all around your community. Evaluate some of the new and old styles of housing.

549

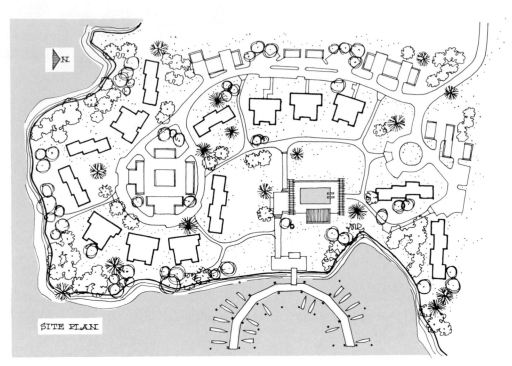

SITE PLAN

Fig. 20-1A *What are four major features planned in this new community?*

Fig. 20-1B *Condominiums such as these add geometric design elements as well as a large number of residences to a community.* (Courtesy of Ann Garvin)

Look closely at the apartments, condominiums, churches, banks, stores, and offices in your neighborhood. Are they useful and attractive? Did the architect do a good job of planning? Were durable, attractive materials selected?

CAREER OPPORTUNITIES

Many different types of jobs are related to the building industry. The rapid growth of small towns and big cities demands the creation of new environments. The team includes:
- architects
- city planners
- landscape architects
- interior designers
- specification writers
- drafting technicians
- illustrators
- construction supervisors
- structural engineers
- mechanical engineers
- office personnel

All of these people work together on large architectural projects. The entire design team is faced with an exciting and difficult job. Some employees work with the clients. Others offer design services. Someone has to estimate costs.

The team must also consider new materials and ways of building. One major concern today is energy conservation. The design team is constantly challenged to plan structures that minimize energy use or to use alternative forms of energy. Another challenge is planning and constructing buildings that are compatible with or even beneficial to the environment. As laws are passed requiring stricter measures, the design team must explore new ways to comply.

Professional organizations can provide information about career opportunities. One example is the American Institute of Architects, located at 1735 New York Avenue N.W., Washington, D.C. 20006.

THE ARCHITECT'S OFFICE

Like other types of businesses, architectural firms can be partnerships, sole proprietorships, or corporations. In a typical architectural partnership, there are generally two to four main partners. The partners employ six to twelve people. However, some partnerships may have a hundred or more employees. Sole proprietorships have just one owner. Corporations are owned by people who buy stock in the company. Whatever the form of business organization, however, an architectural group works as a team. Each member of the architectural team has a well-defined position.

A large firm may offer all the basic architectural services, including architectural design, structural design, mechanical engineering, civil engineering, landscape design, interior design, and urban planning. These services may be used in the design and building of one structure or of an entire city. Some firms provide only one service. They may make subcontracts with other specialized firms in working on a project. The National Council of Architecture Licensing Boards in Washington, D.C., promotes professional registration of architects within each state.

Architectural offices usually have an overall project director to coordinate activity on several different projects. Individual team leaders might help find clients or work on a particular project.

COMMUNITY ARCHITECTURE

You can see the effects of the architect's planning by looking carefully at any town or city. Many residential areas have a distinctive architectural style, such as a traditional or contemporary (modern) style. Others combine a number of styles. Some styles may be built in various materials to create different effects. A town may have wooden, or frame, cottages and brick bungalows. The larger residential buildings usually have rental apartments. Some may be *condominiums*. These are buildings in which the units are individually owned, not rented (Fig. 20-1B).

Different architectural styles can be identified by geometric forms, size, roof shape, and materials. Several kinds of houses are shown in this chapter. California contemporary, ranch, French contemporary, Colonial, French Chateau, Georgian, and English Tudor are popular styles of residential buildings.

Neighborhoods

Neighborhoods are residential areas that contain housing as well as other buildings and spaces people need to live. Residential buildings are arranged in geometric patterns. When the pattern is rectangular, the basic unit is called a *block*. Churches and synagogues may be found on the corners of blocks or in the center of a community. Parks or recreation centers are conveniently placed throughout each neighborhood. Figure 20-2 shows a contemporary recreation center for a neighborhood that has a lot of space. The more important, busy streets often form the boundaries of a neighborhood. These streets are frequently lined with small stores, small offices, and apartment buildings.

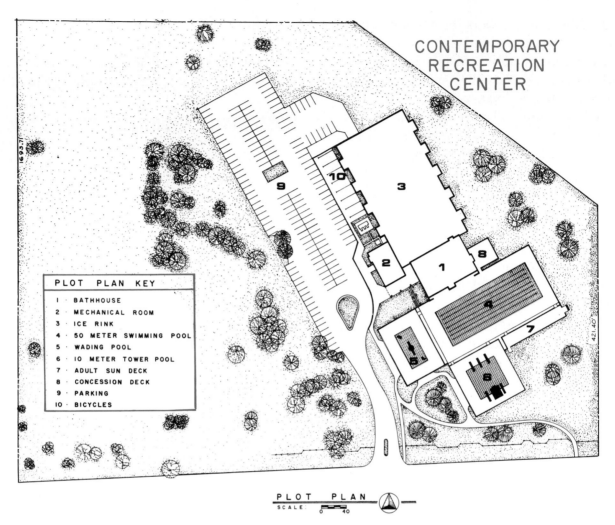

CONTEMPORARY RECREATION CENTER

PLOT PLAN KEY

1 · BATHHOUSE
2 · MECHANICAL ROOM
3 · ICE RINK
4 · 50 METER SWIMMING POOL
5 · WADING POOL
6 · 10 METER TOWER POOL
7 · ADULT SUN DECK
8 · CONCESSION DECK
9 · PARKING
10 · BICYCLES

PLOT PLAN
SCALE   0   40

Fig. 20-2 *The plot plan and major features of a recreation center in a proposal to a park board.* (Courtesy of J. E. Barclay, Jr., & Associates)

Urban Development

Neighborhoods combine to make up towns and cities. One of the main forces affecting the human environment today is the movement of people from *rural* (farm) areas to *urban* (city) areas.

In the United States, many people have also moved from "inner" cities and farms to *suburbs*. These are communities on the edges of cities. The growing cities and suburbs need *municipal* (public) buildings to serve their residents. Municipal buildings often include the city hall and facilities for services such as police, fire protection, water, and sanitation. Health centers such as medical offices and hospitals are also important. Cultural facilities such as libraries, museums, and theaters are significant needs. Architectural services are needed to plan and develop these communities.

Traffic Patterns

Whether a community's streets form rectangular blocks or wandering lanes with "dead ends," they must be connected to major thoroughfares. The major streets are usually lined with large stores, shopping centers, and office buildings. Apartment buildings, banks, and hospitals are also generally located on or near key roads. Government bodies known as building and zoning commissions decide what kinds of buildings will be in different areas. Traffic patterns should be designed to let industrial employees get to and from work easily. The success of a community may rely on good planning for vehicles and pedestrians.

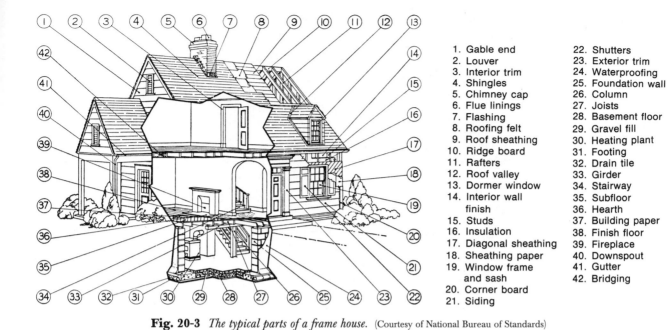

1. Gable end
2. Louver
3. Interior trim
4. Shingles
5. Chimney cap
6. Flue linings
7. Flashing
8. Roofing felt
9. Roof sheathing
10. Ridge board
11. Rafters
12. Roof valley
13. Dormer window
14. Interior wall finish
15. Studs
16. Insulation
17. Diagonal sheathing
18. Sheathing paper
19. Window frame and sash
20. Corner board
21. Siding
22. Shutters
23. Exterior trim
24. Waterproofing
25. Foundation wall
26. Column
27. Joists
28. Basement floor
29. Gravel fill
30. Heating plant
31. Footing
32. Drain tile
33. Girder
34. Stairway
35. Subfloor
36. Hearth
37. Building paper
38. Finish floor
39. Fireplace
40. Downspout
41. Gutter
42. Bridging

Fig. 20-3 *The typical parts of a frame house.* (Courtesy of National Bureau of Standards)

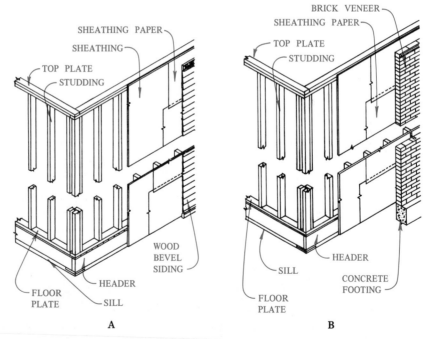

Fig. 20-4 *A frame wall (A) with siding and (B) with brick veneer.*

RESIDENTIAL CONSTRUCTION

The main parts of a house are shown in Fig. 20-3. Every house does not have all of these parts. Some parts may be made of different materials. The typical wood framing of an exterior wall is shown in Fig. 20-4A. The framing begins on the foundation wall with the sill, header, and floor plate. The stud wall is then erected. Next, sheathing, plywood, or insulation board is put on. Many kinds of facing can be used. Horizontal or vertical siding is typical. In addition, shakes or shingles are often used. Also common today is brick veneer, shown in Fig. 20-4B.

Framing

The framework of a residential building must be strong and rigid to ensure low maintenance costs over many years. Even a prefabricated home and a custom-designed home have some features in common.

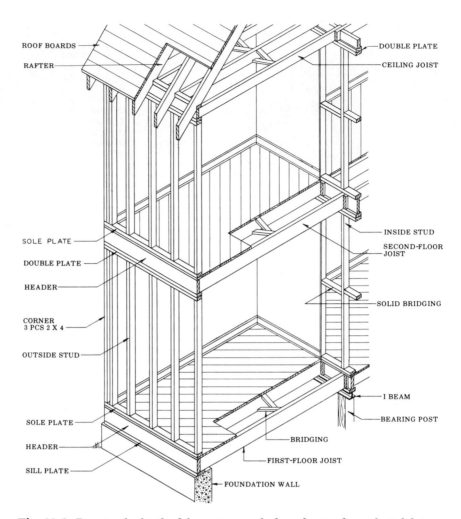

ROOF BOARDS

RAFTER

DOUBLE PLATE

CEILING JOIST

SOLE PLATE

DOUBLE PLATE

HEADER

CORNER
3 PCS 2 X 4

OUTSIDE STUD

INSIDE STUD

SECOND-FLOOR
JOIST

SOLID BRIDGING

SOLE PLATE

HEADER

SILL PLATE

I BEAM

BEARING POST

BRIDGING

FIRST-FLOOR JOIST

FOUNDATION WALL

Fig. 20-5 *Examine the details of the western, or platform, framing for residential design.*

Western Framing

In **western framing,** or *platform framing*, each floor is framed separately (Fig. 20-5). The first floor is a platform built on top of the foundation wall. Studs are one story high. They are used to develop and support the framework for the second story and the load-bearing interior walls.

Balloon Framing

In **balloon framing,** the studs are two stories high, as shown in Fig. 20-6. A false girt inserted into the stud wall carries the second-floor joists. A box sill is used. Diagonal bracing brings rigidity to the corners. While this system is not commonly used today, architects must know about it to be able to remodel older homes.

Plank and Beam Framing

Plank and beam framing (Fig. 20-7) uses heavier posts and beams than the other systems. These members carry a deck of continuous planking. This kind of framing allows ceilings to be higher and more open, with fewer supporting members. It also generally costs less to build.

Sill Construction

Figure 20-8 shows different types of sill and wall construction. In (A), the frame wall is set up on a box-sill construction. Note the metal termite shield on top of the foundation wall. In (B), a brick veneer starts below the floor line on a stepped foundation wall. In (C), the slab construction is reinforced with a wire mesh.

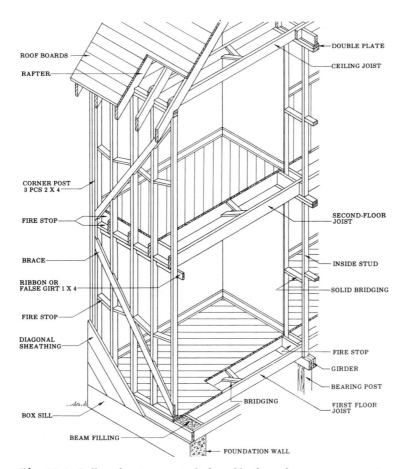

ROOF BOARDS

RAFTER

CORNER POST
3 PCS 2 X 4

FIRE STOP

BRACE

RIBBON OR
FALSE GIRT 1 X 4

FIRE STOP

DIAGONAL
SHEATHING

BOX SILL

BEAM FILLING

DOUBLE PLATE

CEILING JOIST

SECOND-FLOOR
JOIST

INSIDE STUD

SOLID BRIDGING

FIRE STOP

GIRDER

BEARING POST

FIRST FLOOR
JOIST

BRIDGING

FOUNDATION WALL

Fig. 20-6 *Balloon framing is typical of an older form of two-story construction.*

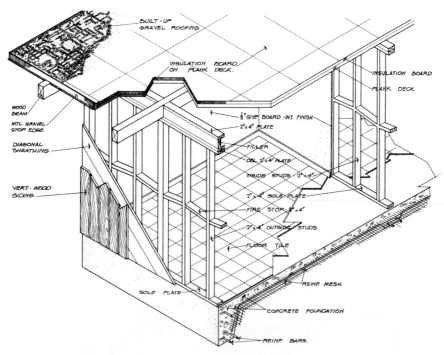

BUILT-UP
GRAVEL ROOFING

INSULATION BOARD
ON PLANK DECK.

INSULATION BOARD

PLANK DECK

WOOD
BEAM

MTL. GRAVEL
STOP EDGE.

DIAGONAL
SHEATHING

VERT. WOOD
SIDING.

½" GYP BOARD -INT FINISH.

2"x 4" PLATE

FILLER.

DBL. 2"x4" PLATE

INSIDE STUDS 2"x 4"

2"x 4" SOLE PLATE

FIRE STOP-2"x 4"

2"x 4" OUTSIDE STUDS

FLOOR TILE

REINF MESH.

SOLE PLATE

CONCRETE FOUNDATION

REINF. BARS.

Fig. 20-7 *A plank and beam framing with a low-profile roof.*

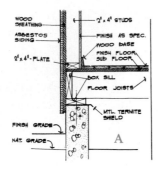

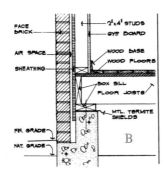

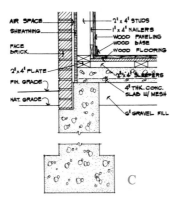

Fig. 20-8 *Various types of sill construction. (A) Box-sill with a frame wall; (B) stepped footing with a brick-veneer wall; (C) stepped footing with a brick-veneer wall and poured-concrete slab floor.*

Corner Studs and Sheathing

Some typical corner bracing is shown in Fig. 20-9. Diagonal *sheathing* (covering) was formerly the most common kind of bracing. Today, however, to save labor, builders use horizontal sheathing and plywood along with modular insulation board. Plywood is used not only on exterior walls but also for *interior decking* (subflooring) and roof sheathing. Anchor bolts secure the superstructure to the substructure. These are normally ½- to ¾-in. (12- to 19-mm) diameter bolts. They are placed about 8 ft. (2440 mm) apart and extend 18 in. (457 mm) into the concrete.

Roof Designs

Some basic roof types are shown and named in Fig. 20-10. The shape of the roof is often the key to a building's architectural style. The vertical measurement of a roof is called the *rise*. The horizontal dimension is called the *run*. Together, they determine the *roof pitch* (slope).

The most common roof shapes are the gable, hip, flat, and shed shapes. The mansard, gambrel, butterfly, combination, clerestory, and A-frame shapes are used more with specific design styles. (See Fig. 20-27, later in this chapter, for an example of a home with a French mansard roof.) This kind of roof is shaped to fit and hug a building. It is textured in natural wood shakes to give a feeling of warmth.

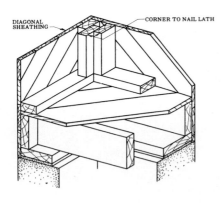

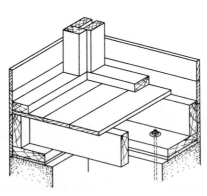

Fig. 20-9 *Arrangement of corner studs in forming a frame wall.*

Stairway Framing and Detail

Three types of stairs are the straight run, the platform, and the circular. Stairs are made up of *risers* (the vertical part of each step) and *treads* (the horizontal part of each step). These parts are illustrated in Fig. 20-11. The rise of a flight of stairs is the height measured from the top of one floor to the top of the next floor. The run is the horizontal distance from the face of the first riser to the face of the last riser in one stairwell. It also equals the sum of the width of the treads.

Risers are generally $6\frac{1}{2}$ to $7\frac{1}{2}$ in. (165 to 190 mm) high. The width of treads is such that the sum of one riser and one tread is about 17 to 18 in. (430 to 480 mm). A 7-in. (180-mm) riser and an 11-in. (280-mm) tread is considered a general standard. You can easily use a scale to divide the floor-to-floor height into the number of risers. A good stairway feels comfortable and safe to use. A simple rule for building a safe stairway is to keep the angle of incline between 28° and 35°.

On working drawings, stairways are usually not drawn in their entirety. Instead, break lines are used, and the drawing shows what is on the level beneath the stairs.

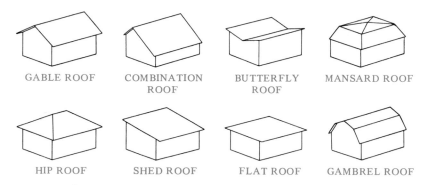

GABLE ROOF   COMBINATION ROOF   BUTTERFLY ROOF   MANSARD ROOF

HIP ROOF   SHED ROOF   FLAT ROOF   GAMBREL ROOF

Fig. 20-10 *Some typical residential roof types that define style.*

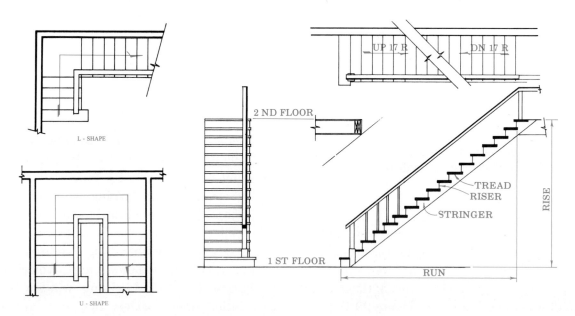

Fig. 20-11 *Stair details with terms and layouts.*

Fig. 20-12 *Typical front-door patterns.*

Doors

Doors are usually 6'-8 or 7'-0 (2000 or 2100 mm) high. The width may vary from 2'-0 to 3'-0 (600 to 900 mm), but it is usually 2'-8 or 3'-0 (800 to 900 mm). The thickness varies from $1\frac{3}{8}$ to $1\frac{3}{4}$ in. (35 to 45 mm) for interior doors, and from $1\frac{3}{4}$ to $2\frac{1}{2}$ in. (45 to 65 mm) for exterior doors. The head, jamb, and sill details may vary depending on whether a swinging, sliding, or folding door is used. The door must fit its frame closely. Yet it must also open and close easily. Figure 20-12 shows various patterns for doors.

Doors to the outside are usually larger than others to allow for heavier use and for bringing in furniture. These doors are usually 3'-0 (900 mm) wide. Bedroom doors are usually 2'-6 (762 mm). Bathroom doors run 2'-4 (710 mm). Bifold and folding accordion doors have special features.

Windows

The style of a house determines what style of window is used and how windows are placed. Double-hung and casement windows are practical in most kinds of houses. However, horizontal sliding windows are very popular. Figure 20-13 shows the various types available to the architect.

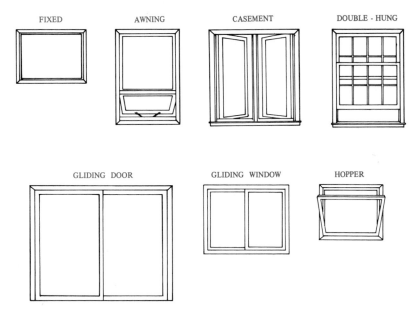

Fig. 20-13 *Some typical windows and sliding doors. CAD libraries can store window details for selection from a software database.*

Casement windows are popular for French and English designs. They are hinged at the sides to swing open (in or out). Hopper windows are hinged at the bottom. Awning windows are hinged at the top. Both hopper and awning windows are also called *projected windows*. Sliding windows move sideways. Double-hung windows are commonly used for Colonial and American-style structures. Each double-hung window contains two independent sashes that can move in a vertical track. A counterbalance holds them at any position desired. Newer windows can have a press-in, spring-loaded track that is very convenient. Fixed windows and *jalousies* (windows made of adjustable glass slats) are special kinds of windows. Figure 20-14 shows sectional details and some technical terms for a window. Normally, windows are placed in walls so that their tops line up with the top of the door.

WORKING DRAWINGS

Working drawings include plans, elevations, sections, schedules, schematics, and details. Along with the specifications for materials and finish, they are the guides used in the construction and erection of a building.

Plans

A *plan* is a section that cuts horizontally through walls and shows room arrangements (Fig. 20-15). For example, the three main areas of a residential plan are the living, sleeping, and service areas. In general, a plan is made for each floor (story) of a building, as well as for the foundation and/or basement.

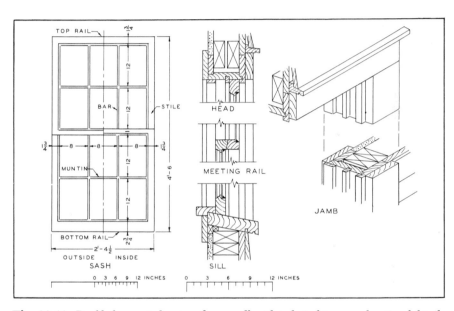

Fig. 20-14 *Double-hung window in a frame wall with technical terms and sectional details.*

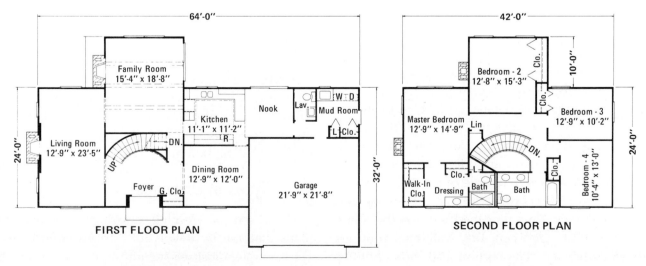

Fig. 20-15 *The floor plan is the first of four basic types of drawings.* (Courtesy of Inland Steel Co.)

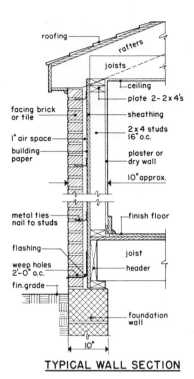

Fig. 20-16 *An elevation, the second of the four basic drawings.* (Courtesy of Inland Steel Co.)

TYPICAL WALL SECTION

Fig. 20-17 *A cross section detail, the third of the basic drawings.*

Elevations

Elevations are vertical projections of buildings that help define their structural form and architectural style. The roof shape, sides, and side openings are shown in the architectural style chosen and in relationship to the plan. With such drawings, the need for changes to provide balance or symmetry can be easily studied (Fig. 20-16).

To draw an elevation, start with the grade line. Then locate the centerlines that indicate the finished working dimensions. Plot the doors, windows, and other openings from the floor plan.

Sections

Architects create *sections* to show internal features that are not visible on the plan view or elevations. The two basic kinds of sections are detail sections and cross sections.

Detail Sections

In a **detail section,** a vertical plane cuts through walls to show construction details (Fig. 20-17). The material details are labeled as shown. Detail sections are generally drawn to a much larger scale than that of the elevation. It shows the wall in a much clearer and more detailed view.

Figure 20-18 is an example of a detail section. It contains two typical roof-framing sections showing cornice details. Note that the members that finish off the joints between the wall and roof are carefully labeled. The section also shows built-up flat roofing for plank and beam construction.

Cross Sections

The second kind of section is a full section, or **cross section.** A cross section cuts across the entire structure to aid in interpreting the relationships of the important spaces. In full sections, it is easier to see the proportions between spaces and how these relate to construction and use.

Site Plan

A **site plan** shows the lot and locates the house on it. It should give complete and accurate dimensions. It should also show all driveways, sidewalks, utility easements, and other pertinent information required by the building inspector.

Other Types of Drawings

Many sets of working drawings also include plans that specify details for various contractors. For example, most sets include an electrical wiring plan. Other plans that might be included are

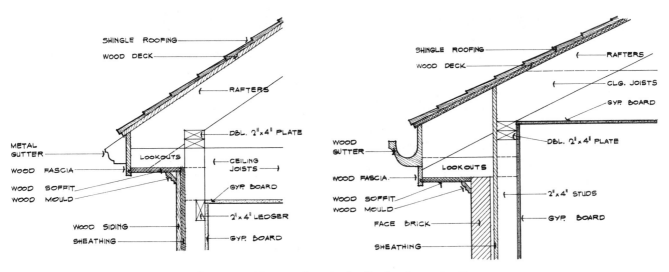

Fig. 20-18 *Two typical cornice details, showing construction.*

Fig. 20-19 *A perspective, the fourth type of basic drawing.* (Courtesy of Inland Steel Co.)

plumbing plans and heating, ventilation, and air-conditioning (HVAC) plans. However, when the specifications for these specialties are fairly simple and straightforward, these details are often included on the basic plan.

Schedules and Specifications

Schedules are lists that define and describe details shown by symbols on the actual drawings. A list of all the types of doors and their sizes is an example of a schedule. The schedules are usually fitted on an unused area of one of the drawing sheets. However, if there are many schedules or if they are particularly complex, they may be placed on a separate drawing sheet.

Architectural **specifications** are more detailed written instructions that should accompany any set of plans. They are essential for turning these plans into a building. They note the general conditions of the site. They also specify the materials to the client's and contractor's mutual agreement. Guidelines for specifications have been drawn up by the American Institute of Architects (AIA). These guidelines are available through local and state chapters of the AIA.

Perspective Drawings

Although they are not formally part of the working drawings, perspective drawings are often created for sales and marketing purposes, as well as for client approval. It is easiest to see architectural form in a pictorial presentation drawing. One-, two-, or three-point perspective drawings show a building in a realistic way. Architectural perspective studies do not have to be mechanically accurate. Therefore, freehand techniques, along with good proportions, can be used in making these drawings (Fig. 20-19).

EXAMPLES OF WORKING DRAWINGS

Figure 20-20 is a preliminary design for a ranch house. Figures 20-21 through 20-26 are the working drawings developed from this design. They form a complete set of plans for the house.

The basement plan in Fig. 20-21 serves as a guide for constructing the foundation. Therefore, it must be completely dimensioned. It should be checked with the first-floor plan; in fact, you can develop the basement plan directly from the first-floor plan. Note the foundations for the porch and garage. Window placement depends on structural needs.

The first-floor plan in Fig. 20-22 is a horizontal section taken above the floor. It shows all walls, doors, windows, and other structural features. It also shows fixed features such as the cabinets, stairways, heating and plumbing fixtures, lighting outlets in the walls, and ceilings. Frame walls are drawn to what is shown on the scale as 6 in. thick. Windows and doors are located properly, then represented by conventional symbols. Their sizes are listed on schedules that accompany the plans.

Figure 20-23 shows front and side elevations. The elevations show the exterior look of a house, floor and ceiling heights, openings for windows and doors, roof pitch, and selected materials. A complete set of plans includes at least four elevations, one for each side of the building.

Fig. 20-24 details a typical wall section from substructure to superstructure. Door and window details are also included. They are developed as the building progresses. That way, the millwork and custom framework will assemble readily.

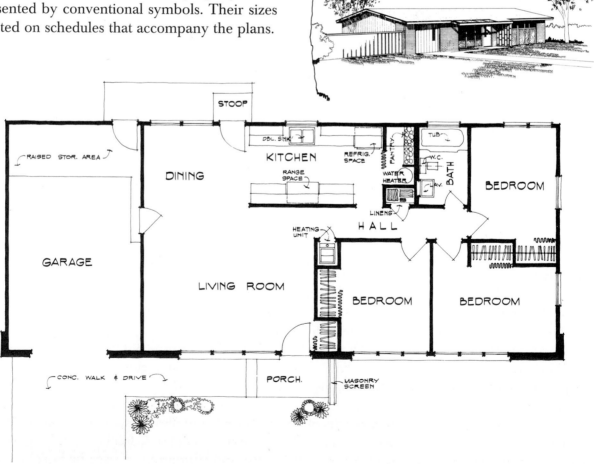

Fig. 20-20 *A preliminary study (undimensioned) of a one-story ranch developed for client approval.*

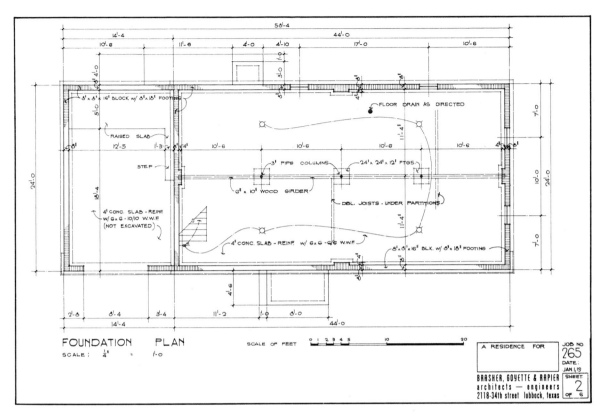

Fig. 20-21 *A foundation plan with structural notes and dimensions ready for contractor approval.*

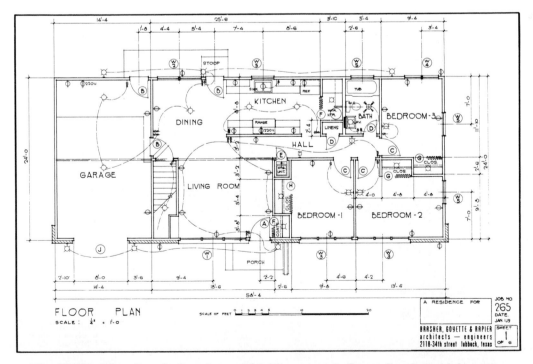

Fig. 20-22 *A floor plan of a frame house with brick veneer on the front wall and around the corners.*

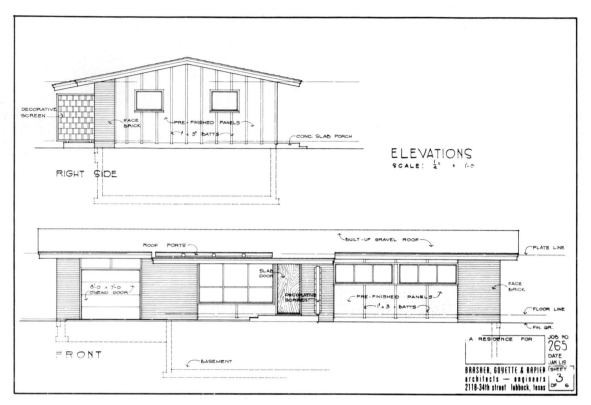

Fig. 20-23 *Elevations showing foundation walls and roof pitch of 2.5:12 (gravel roof).*

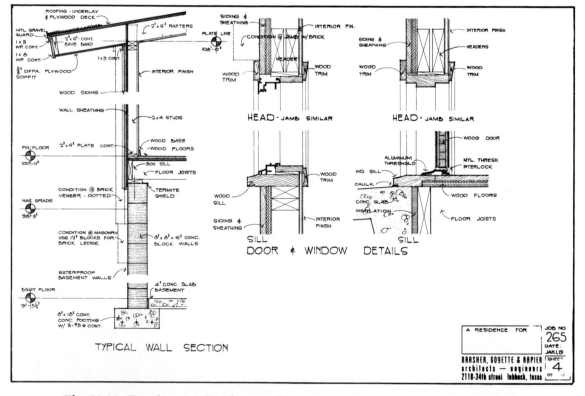

Fig. 20-24 *Typical sections. Foundation walls could be poured concrete, as shown in Fig. 20-8B.*

A plan of the electrical wiring is shown in Figure 20-25. It shows the circuits for 110-V and 220-V service. It also shows the electric outlets and switches located for the major appliances. A key is included for the various symbols. Additional data is listed in the formal specifications.

Figure 20-26 is the site plan. The site plan in the figure is for an ordinary urban lot. Note that the site plan also contains the roof plan. Its center ridge represents a gable roof. What scale would be best for drawing this site plan? Figure 20-26 also includes schedules describing the five types of windows and nine types of doors used in the house. Another, perhaps even more important, schedule lists the finishes required for the floors, walls, and ceilings.

Another example of a set of working drawings is shown in Figs. 20-28 through 20-36. These drawings were developed for the home shown in Figure 20-27. The site plan (Fig. 20-28) is contoured to show the gentle roll of the land. The family-room (lower level), main-level (Fig. 20-29), and upper-level (Fig. 20-30) plans are arranged gracefully around a central core of stairs and hallways. The front elevation (Fig. 20-31) shows the rolling contour of the site. Note the line weights on the cedar shakes that make up the roof. These add depth to the view. The rear elevation (Fig. 20-32) shows contoured stairs between the main and lower levels. The right and left elevations are shown in Figs. 20-33 and 20-34. The longitudinal section (Fig. 20-35) explains the arrangement of levels and rooms. The sections in Fig. 20-36 are identified by symbols on the main floor plan (Fig. 20-29).

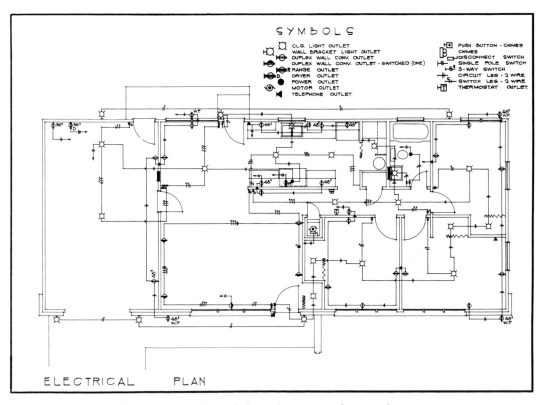

Fig. 20-25 *An electrical plan indicates circuit layout and services.*

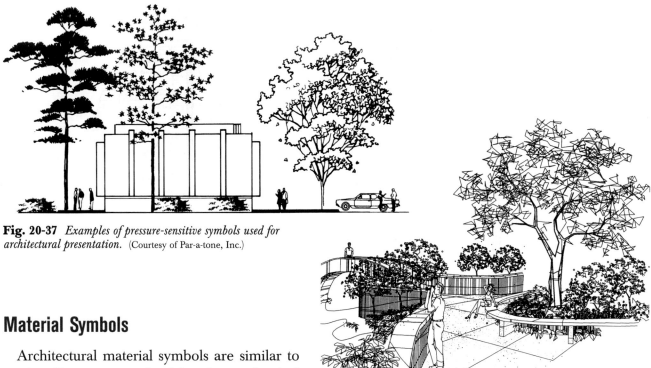

Fig. 20-37 *Examples of pressure-sensitive symbols used for architectural presentation.* (Courtesy of Par-a-tone, Inc.)

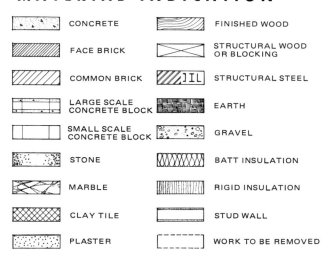

Material Symbols

Architectural material symbols are similar to section lines or crosshatching in mechanical drafting. However, instead of merely indicating sections, they also indicate the materials used in building. Examples of material symbols are shown in Fig. 20-38. These symbols are used in plan, section, and elevation drawings. See Fig. 20-39 for examples of how these symbols are used in drawings.

Symbols for Door and Window Openings

Some door symbols are shown in Fig. 20-40. The plan shows the outside door ajar in a single line. An arc shows the direction in which the doors swing. Interior doors are shown in a variety of ways.

Typical windows are shown in Fig. 20-39. Symbols help the reader interpret window movement. They also show what type of wall material frames the window.

Templates

Templates have improved the quality of drawings and increased efficiency in the architect's office by providing standard symbols that can be drawn more quickly. Templates are usually made of lightweight plastic with openings that form

MATERIAL INDICATION

CONCRETE		FINISHED WOOD	
FACE BRICK		STRUCTURAL WOOD OR BLOCKING	
COMMON BRICK		STRUCTURAL STEEL	
LARGE SCALE CONCRETE BLOCK		EARTH	
SMALL SCALE CONCRETE BLOCK		GRAVEL	
STONE		BATT INSULATION	
MARBLE		RIGID INSULATION	
CLAY TILE		STUD WALL	
PLASTER		WORK TO BE REMOVED	

Fig. 20-38 *Typical material symbols used in architectural detail. CAD software makes these symbols easy to retrieve and use.*

structural shapes or fixtures. You follow the outline of the opening with your pencil or pen to make the graphic image. Templates are available in ⅛″ and ¼″ scales.

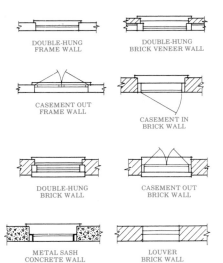

Fig. 20-39 *Window symbols for various types of window openings (see Fig. 20-13).*

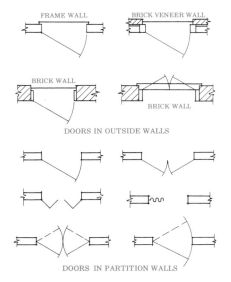

Fig. 20-40 *Door openings with swing symbols put in place using architect's template.*

Fig. 20-41 *CAD software can provide the designer with architectural symbols on a template fixed to a system digitizer.* (Courtesy of Autodesk, Inc.)

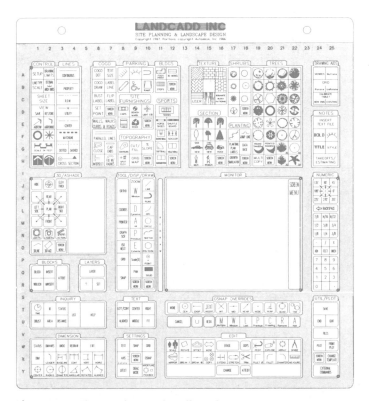

Fig. 20-42 *This template overlay allows the CAD operator to insert standard 2D and 3D landscaping symbols into a drawing by picking the symbols on the template.* (Courtesy of LANDCADD)

CADD TEMPLATES

Most CADD systems include a digitizing tablet that allows the drafter to digitize, or create an electronic form of, a manually created drawing. In addition to this capability, however, digitizers can be used with template overlays to insert standard symbols into drawing files. The general architectural template in Fig. 20-41 and the landscaping template shown in Fig. 20-42 are only two of many examples of templates that speed up the CAD operator's work.

ARCHITECTURAL DRAFTING TECHNIQUES

A good line technique is a very important skill for architectural designing and detailing. In some cases, architectural drafting practices differ slightly from mechanical drafting practices. For example, Figure 20-43 shows the proper technique for drawing lines on an architectural drawing. Notice that the lines intersect and stop after forming a flared corner.

Using line weights correctly is also part of good line technique. The visual weights of lines express spatial relationships. Good lines are needed for good drawing reproduction. They are needed to make the design appealing. They are needed so that the third dimension of depth can be imagined on a two-dimensional medium, as shown in Fig. 20-44. In well-prepared details, line tone and texture complement each other.

Lettering

Both traditional and contemporary styles of lettering are commonly used by architects. Traditional lettering is based on the Old Roman alphabet, as shown in Fig. 20-45. These letters have *serifs* (small lines extending from the main strokes). Elaborate and important titles are often designed with Old Roman letters. However, plain, single-stroke Gothic lettering is better for architectural working drawings (Fig. 20-46). The condensed and extended styles shown are often used by experienced professionals. They use the right style for inserting notes or balancing the drawing.

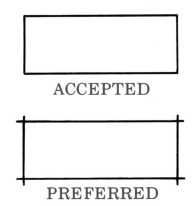

ACCEPTED

PREFERRED

Fig. 20-43 *Architectural line technique.*

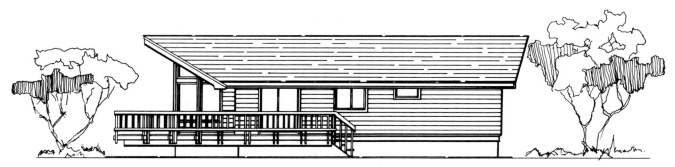

Fig. 20-44 *Line techniques that feature the major proportions and add three dimensions.* (Courtesy of Justus Company, Inc.)

Fig. 20-45 *Old Roman lettering applied in an architectural style.*

OLDTOWN PUBLIC LIBRARY
NINTH AVENUE AND STANTON STREET

DRAWN BY	DATE	YURIKO OHASHI	REVISED BY	DATE
TRACED BY	DATE	ARCHITECT	JOB NO	
CHECKED BY	DATE	LEA BLDG. – OLDTOWN, TEX	SHEET **2** OF 8	

Preparing Title Blocks

Like other drafting companies, architectural firms usually create a quickly identifiable image with a logo, as shown in Fig. 20-47. Some firms use forms with preprinted borders, title, and revision blocks. The blocks are ready to be filled in with appropriate pencil or ink lettering. The typical title blocks shown in Fig. 20-48 were created by students preparing for professional roles.

Fig. 20-46 *(A) Single-stroke freehand architectural lettering: condensed (narrow), extended (bold expanded), and general Gothic Stroke. (B) Some CAD software is capable of providing the drafter with architecturally styled lettering.* (Courtesy of E-Z Word LANDCADD, Inc.)

Fig. 20-47 *Title block and bold logo for Douglass Storm and Associates, Architectural Designers. Formal title blocks may be stored in a CAD database.*

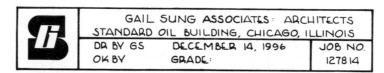

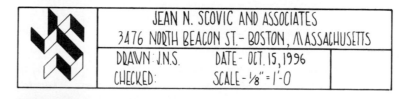

Fig. 20-48 *Title blocks created by students to convey a designer's image.*

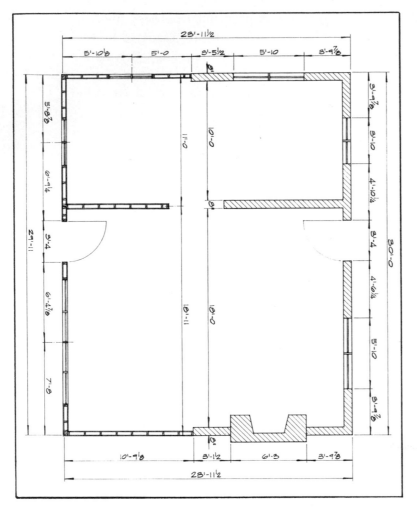

Fig. 20-49 *Techniques in architectural dimensioning for a frame wall and a brick wall.*

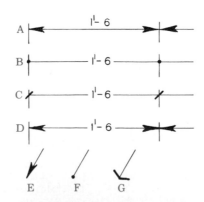

Fig. 20-50 *Arrowheads, heavy dots, and heavy 45° slashes are used in architectural dimensioning.*

Dimensioning Techniques

With a few exceptions, the general dimensioning techniques discussed in Chapter 6 are used to dimension architectural drawings. However, the dimension line is unbroken. In addition, the dimensions appear above the dimension line and are given in feet and inches, as shown in Fig. 20-49.

The dimension lines are placed about $\frac{3}{8}$ in. (10 mm) apart and end with the usual arrowhead or with one of several other symbols, as shown in Fig. 20-50. Note that dimension lines may cross when sizes of interior rooms are given. The plan has only width and depth dimensions, as shown. Any needed heights are shown as floor-to-ceiling dimensions on one of the elevations or on a suitable wall section (Fig. 20-51).

Construction and Finish Dimensions

The dimensions used on architectural drawings are either finish or construction dimensions. The elevations generally have the expected finished dimensions. The floor plan and sectional details have either construction or finish dimensions, as shown in Fig. 20-51.

Dimensioning Exterior Walls

Structural wall openings on the plan drawings for frame and brick-veneer structures are dimensioned in similar ways, but again, there are some differences. Fig. 20-52 shows standard dimensioning practice for residential structures with various types of exterior walls. The crosshatched masonry wall in (A) is dimensioned to the outside face of the wall. This is done with solid brick, concrete, and concrete-block walls. The frame wall in (B) is dimensioned to the face of the stud. The brick-veneer wall in (C) with crosshatched brick and a sole plate is dimensioned to the face of the stud. The remaining material is assumed to be exterior wall facing.

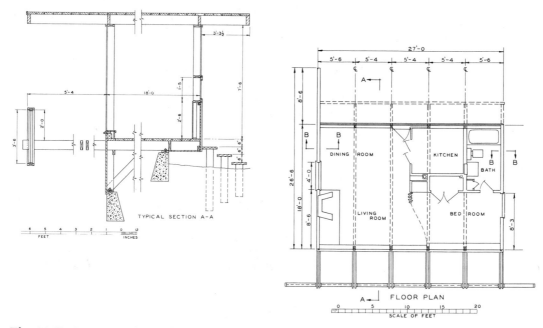

Fig. 20-51 *Dimensions that apply to section and plan drawings.* (Adapted from drawings of Henry Hill, architect, AIA and John W. Kruse, associate, AIA, San Francisco, CA.)

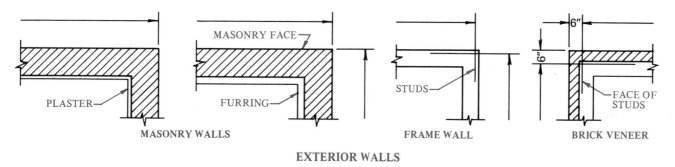

EXTERIOR WALLS

Fig. 20-52 *Wall symbols and four dimensioning techniques.*

Modular Coordination

To help simplify and standardize building design, the building industry and the American National Standards Institute established a standard: Modular Coordination, A62.1. This standard was sponsored by the American Institute of Architects. According to this standard, the basic customary-inch module is 4 in. for all U.S. building materials and projects. The International Standards Organization has established 100 mm (almost 4 in.) as the module for countries using the metric system. Modular components are typically designed on a 4-in. or 100-mm centerline. All buildings currently designed for customary modular specifications are planned with multiples of 4 in. Modular planning reduces building costs and time because materials do not have to be cut to many different sizes and there is less waste.

Architectural Scales

Residential designs and details are usually developed with the aid of reduction scales. The architect's scale is discussed in Chapter 3. The $\frac{1}{4}''$ = 1'-0 (1:50 in metric) scale is best suited for plans for houses and small buildings. The usual scale for larger buildings is $\frac{1}{8}''$ = 1'-0 (1:100 in metric). Plot plans may be drawn at $\frac{1}{10}''$ = 1'-0 or $\frac{1}{32}''$ = 1'-0, but it is better to use an engineer's scale at 1" = 20', 1" = 30', or 1" = 40' (1:200, 1:300, or 1:400 in metric). Enlarged details are developed at 1" = 1' or $1\frac{1}{2}''$ = 1'. Sectional details are defined at $\frac{1}{2}''$ = 1' or $\frac{3}{4}''$ = 1'. Some details may require half-size or full-size drawings. Metric scale for enlarged details may be 1:5, 1:10, or 1:20.

ARCHITECTURAL CADD

Architectural CADD has grown to include much more than just an electronic means of creating a set of working drawings. Using **A/E/C** (architectural, engineering, and construction) drawings, architects can now integrate various functions with CADD systems for complete documentation (Fig. 20-53A). Advanced systems for the integrated architectural office generally include:

- 3D architectural modeling
- architectural production drawings
- space planning/facility management drawings
- engineering production drawings

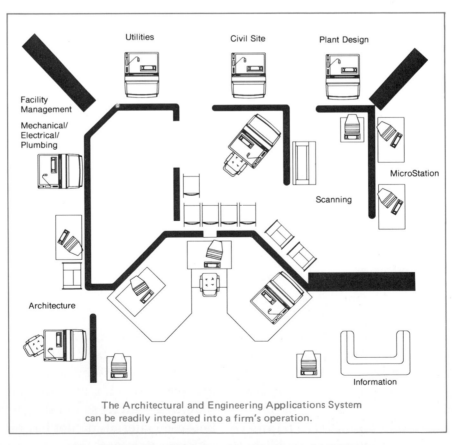

The Architectural and Engineering Applications System can be readily integrated into a firm's operation.

Fig. 20-53A *An architect's studio with integrated A/E/C system.*

In addition to A/E/C applications, software companies have developed a variety of architectural applications to meet the demands of architectural companies. These programs form a distinct subset of the drafting software available for general use. While many architects use general-purpose CAD programs such as AutoCAD, MicroStation, and CADKEY, most of them supplement the basic program with add-on software designed especially for architectural design and drafting. Some architectural firms use nothing but the specialty software, foregoing the large, expensive, general-purpose packages.

The software available for architectural use is as varied as the needs of the individual firms (Fig. 20-53B). Some architects work entirely in two dimensions. They create 2D working drawings that look much like those created by traditional methods. Inexpensive 2D programs such as AutoCAD LT provide all the functionality these companies may need.

However, more and more companies are starting to provide advanced three-dimensional visualization of their product designs. For these companies, 2D software is no longer sufficient. They are more likely to use CADD software designed specifically for 3D architectural design, such as DataCAD or ArchiCAD (Fig. 20-54).

Fig. 20-53B *CAD stations provide the architect with a new design environment.* (Courtesy of Brent Phelps & Dr. Benedict Wong, University of North Texas)

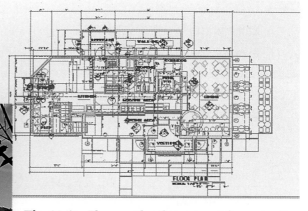

Fig. 20-54 *The same plan that was used for construction of this Boston restaurant was used to generate all of the interior and exterior elevations. The architect used ArchiCAD to create 3D views as he drew the 2D plan.* (Courtesy of PDE Associates, Inc.)

ARCHITECTURAL MATERIALS

The new environments designed by architects are created from materials. Detailed drawings and specifications must show how these materials are to be fabricated and constructed. The *Sweets Architectural Catalog* for materials manufacturers is one of the most important tools of the design team. Many materials are available in standard units for the building trades. Others are custom-designed. A few standards are discussed here.

Lumber

Lumber may be specified by its **nominal size.** The nominal dimensions are the dimensions before the wood has been surfaced, or prepared for use. These dimensions differ from the actual dimensions of the surfaced wood.

For example, most of the lumber and boards for residential construction are *dressed* (finished or surfaced) on four sides. The dressed sizes are noted in Table 20-1.

Masonry

Figure 20-55 gives the sizes of brick building materials. The common brick has modular dimensions. Brick, block, stone, and stucco may be used for exterior and interior walls and floors. Most of these materials can also serve as structural load-bearing walls for supporting floors and roofs. They generally can be *bonded* (overlapped) to increase their structural strength, as shown in Fig. 20-56. Bonding also forms decorative patterns. Well-designed masonry needs little upkeep. Its colors and patterns are important to the overall architectural design.

TABLE 20-1

	LUMBER				
Nominal size	2 x 4	2 x 6	2 x 8	2 x 10	2 x 12
Dressed size	$1\frac{1}{2}''$ x $3\frac{1}{2}''$	$1\frac{1}{2}''$ x $5\frac{1}{2}''$	$1\frac{1}{2}''$ x $7\frac{1}{2}''$	$1\frac{1}{2}''$ x $9\frac{1}{2}''$	$1\frac{1}{2}''$ x $11\frac{1}{2}''$
Nominal size	4 x 6	4 x 8	4 x 10	6 x 6	6 x 8
Dressed size	$3\frac{9}{16}''$ x $5\frac{1}{2}''$	$3\frac{9}{16}''$ x $7\frac{1}{2}''$	$3\frac{9}{16}''$ x $9\frac{1}{2}''$	$5\frac{1}{2}''$ x $5\frac{1}{2}''$	$5\frac{1}{2}''$ x $7\frac{1}{2}''$
Nominal size	6 x 10	8 x 8	8 x 10		
Dressed size	$5\frac{1}{2}''$ x $9\frac{1}{2}''$	$7\frac{1}{2}''$ x $7\frac{1}{2}''$	$7\frac{1}{2}''$ x $9\frac{1}{2}''$		
	BOARDS				
Nominal size	1 x 4	1 x 6	1 x 8	1 x 10	1 x 12
Actual size, common boards	$\frac{3}{4}''$ x $3\frac{9}{16}''$	$\frac{3}{4}''$ x $5\frac{9}{16}''$	$\frac{3}{4}''$ x $7\frac{1}{2}''$	$\frac{3}{4}''$ x $9\frac{1}{2}''$	$\frac{3}{4}''$ x $11\frac{1}{2}''$
Actual size, shiplap	$\frac{3}{4}''$ x $3''$	$\frac{3}{4}''$ x $4\frac{15}{16}''$	$\frac{3}{4}''$ x $6\frac{7}{8}''$	$\frac{3}{4}''$ x $8\frac{7}{8}''$	$\frac{3}{4}''$ x $10\frac{7}{8}''$
Actual size, tongue-and-groove	$\frac{3}{4}''$ x $3\frac{1}{4}''$	$\frac{3}{4}''$ x $5\frac{3}{16}''$	$\frac{3}{4}''$ x $7\frac{1}{8}''$	$\frac{3}{4}''$ x $9\frac{1}{8}''$	$\frac{3}{4}''$ x $11\frac{1}{8}''$

Table 20-1 *Standard sizes of lumber and boards (customary U.S. measure).*

Concrete

Concrete is another type of structural material. As footing, foundation walls, and poured-in forms, it supports floor loads. Both interior and exterior walls can be formed from concrete. Concrete can also be precast to give certain finishes. Color, texture, and pattern all work together to create a desired "look" for a concrete wall.

EVALUATING ARCHITECTURE

Those who study and evaluate architectural structures look for three things:

- The structure has to satisfy a social purpose. That is, it must be suitable for the human activity for which it was designed. Begin by studying how people move through the structure to see how its spaces meet their needs (Fig. 20-57A).
- The structure must be well engineered. Structural members have to be well constructed. Materials must be well selected (Fig. 20-58).
- The structure must have *aesthetic value* (beauty). Successful architecture has to have a pleasing form based on appealing design qualities (Fig. 5-57B).

Trained architects can easily evaluate architectural form. Designing a human environment means merging function, structure, and beauty.

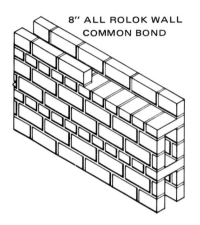

Fig. 20-56 *Common brick bond for building a wall.* (Courtesy of Structural Clay Products Institute)

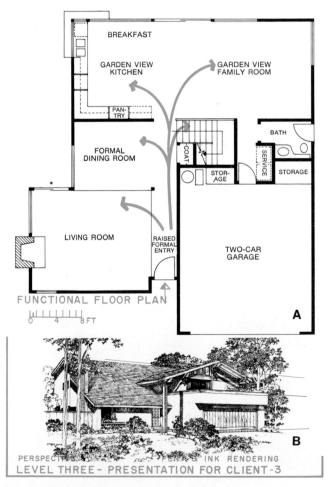

Fig. 20-57 *A functional traffic pattern matches a well-defined contemporary California residence.* (Courtesy of Larwin Co.)

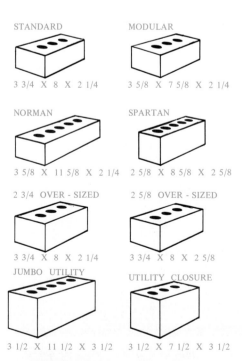

STANDARD
3 3/4 X 8 X 2 1/4

MODULAR
3 5/8 X 7 5/8 X 2 1/4

NORMAN
3 5/8 X 11 5/8 X 2 1/4

SPARTAN
2 5/8 X 8 5/8 X 2 5/8

2 3/4 OVER - SIZED
3 3/4 X 8 X 2 1/4

2 5/8 OVER - SIZED
3 3/4 X 8 X 2 5/8

JUMBO UTILITY
3 1/2 X 11 1/2 X 3 1/2

UTILITY CLOSURE
3 1/2 X 7 1/2 X 3 1/2

Fig. 20-55 *Types and sizes of brick.*

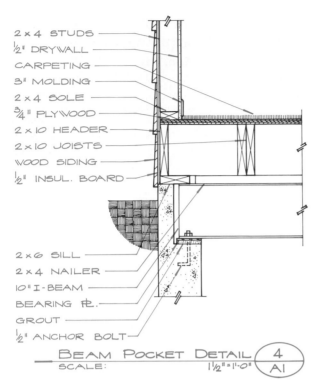

2 × 4 STUDS
½" DRYWALL
CARPETING
3" MOLDING
2 × 4 SOLE
¾" PLYWOOD
2 × 10 HEADER
2 × 10 JOISTS
WOOD SIDING
½" INSUL. BOARD

2 × 6 SILL
2 × 4 NAILER
10" I-BEAM
BEARING PL.
GROUT
½" ANCHOR BOLT

BEAM POCKET DETAIL 4 / AI
SCALE: 1½" = 1'-0"

Fig. 20-58 *A structural detail that defines good engineering.*
(Courtesy of Douglas Storm)

VIRTUAL EVALUATION

The advanced architectural software available today allows specialists in architectural visualization to provide *virtual* images and effects that help designers and clients "see" what a finished structure will look like—before construction on the structure has even started. For example, a graphic artist or designer can create a *photorealistic image* of the building based on a 3D CAD model. The artist renders the model using software such as 3D Studio, Alias PowerAnimator, or a similar special effects package (Fig. 20-59). The result is an image in which texture, light and shadow, and balanced proportions make the image look almost like a photograph.

Fig. 20-59 *This model of a toy store was rendered in Alias PowerAnimator by Eric Hanson of Gensler & Associates. Notice the special lighting effects and textures that make the model look realistic.* (Courtesy of Gensler and Associates, Architects)

A more radical departure from traditional architectural drafting is the virtual *walkthrough*. A walkthrough is a "tour" through a 3D model. Until recently, walkthroughs required extensive computer power and highly functional software. However, some software packages, such as the relatively low-cost DataCAD software, now make it much easier to "walk through" the computer model. By viewing the model from the inside – with windows, doors, and other structural features in place, the designer or client can tell if the windows in the room look unbalanced, or if they need to be moved up or down.

The simplest walkthroughs are very structured. The viewer has to look where the "camera" points, without the option of looking right or left. More advanced virtual reality systems allow the viewer to look to the right or left (or up, down, or behind) and actually "see" in that direction (Fig. 20-60).

Those architectural firms that offer interior design as well as structural design can benefit even more from virtual reality systems. On a color-calibrated system, for example, a designer can show a client a virtual room and change the details until the client is satisfied. Suppose a designer is creating a showroom for a client who sells fine art, for example. The client can choose carpets and rugs of various textures and colors, paint or wallpaper for the walls, and even the type of lighting for the showroom. The designer loads these textures, colors, and lights into the CADD design model, and the client "walks through" the showroom to determine the effects. Although this is still an expensive technique, using virtual reality to choose colors and lights give the designer and the client much more control over the final effect – no more unwelcome surprises because "that carpet looks simply awful with this wallpaper!"

Fig. 20-60 *This interactive model of Frank Lloyd Wright's Unity Temple in Oak Park, Illinois was rendered with Lightscape Visualization Technologies, San Jose, California. (©1992) You can "walk through" the temple to see what it looks like.* (Courtesy of Lightscape Technologies, Inc.)

Fig. 20-61 *The John Deere & Company administrative center, designed by Eero Saarinen and Associates, required many large sheets of structural details.* (Courtesy of John Deere & Co.)

THE STRUCTURAL DRAFTER

The structural drafter is usually a member of an engineering team. This team often works with other teams under the direction of a project manager or a job superintendent. As an example of this teamwork, structural and architectural designers often combine their efforts. The architectural designer designs the form of a building based on the function it will have (Fig. 20-61). Then the structural designer designs the frame of the building to fit this form (Fig. 20-62). The work of the structural drafter is very important in engineering. The construction of buildings, bridges, and other structures depends on the detailed instructions in structural drawings.

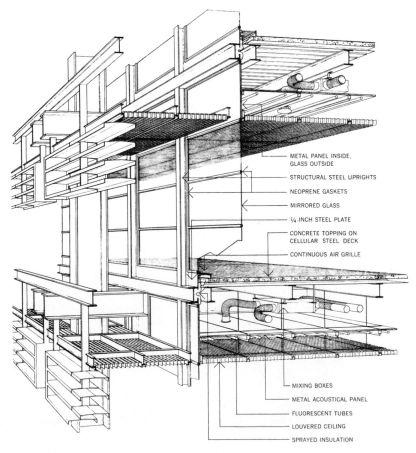

METAL PANEL INSIDE, GLASS OUTSIDE

STRUCTURAL STEEL UPRIGHTS

NEOPRENE GASKETS

MIRRORED GLASS

¼-INCH STEEL PLATE

CONCRETE TOPPING ON CELLULAR STEEL DECK

CONTINUOUS AIR GRILLE

MIXING BOXES

METAL ACOUSTICAL PANEL

FLUORESCENT TUBES

LOUVERED CEILING

SPRAYED INSULATION

Fig. 20-62 *A pictorial of structural detail.* (Courtesy of John Deere & Co.)

A structural drafter usually works at one of the following five jobs:

- Drafting details in an architect's or engineer's office.
- Preparing construction details for a contractor (making the shop drawings for a construction company).
- Drafting structural details for a manufacturer of structural materials.
- Working for the engineering department of a manufacturing plant that maintains its own engineering operations.
- Preparing drawings for government or other agencies that regulate the construction and design of public buildings, bridges, dams, and other structures.

Career opportunities for structural drafters generally depend on how much practical experience they gain as junior members of engineering teams. Structural drafters can be promoted to jobs such as structural detail checker, estimator, chief drafter, construction supervisor, or building inspector (Fig. 20-63).

Further career opportunities can be gained through continuing education. Formal training at a junior college or university is very helpful. To succeed as a structural drafter, you must have the ability to design and detail structural components and have an aptitude for mechanics.

STRUCTURAL MATERIALS

Designers and detail drafters must be familiar with a great many structural materials. In addition, all members of the engineering design team must learn about new construction materials and systems as they are developed. All structural materials have special ways of being assembled (fastened together). These ways must be considered whenever accurate drafting details are prepared (Fig. 20-64).

The basic structural materials used today are steel, wood, concrete, structural clay products, and stone masonry. Different materials have different characteristics. It is the designer's job to choose the right combination of materials to bear the stresses imposed by a building.

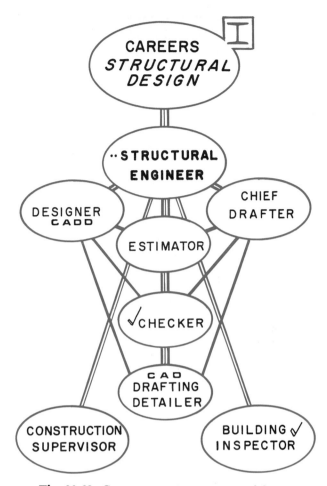

Fig. 20-63 *Career opportunities in structural design.*

Structural Steel

Steel shapes have framed the skyscrapers of our cities for nearly three quarters of a century. The American Institute of Steel Construction maintains regional offices from coast to coast to help provide guidelines for designing and building steel structures.

Steel makes a good construction material because of the shapes into which it can be formed at the mill. Steel shapes are produced in rolling mills. They are then shipped to fabrication shops where they are cut to specific lengths and where connections are prepared.

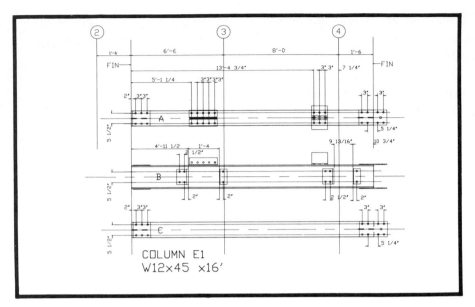

Fig. 20-64 *A detailed beam has accurate dimensions.* (Courtesy of Manual Details)

The American Institute of Steel Construction (AISC), 101 Park Avenue, New York, NY 10017, publishes the *Manual of Steel Construction (AISC Manual)*. The seventh edition of the *AISC Manual,* or any handbook published by a major steel company, lists all the major shapes available and the great variety of their sizes and weights. The *AISC Manual* contains tables for designing and detailing the various shapes in any combination. Figure 20-65 shows cross sections of various plain steel shapes. These shapes are grouped by the AISC as shown in Table 20-2.

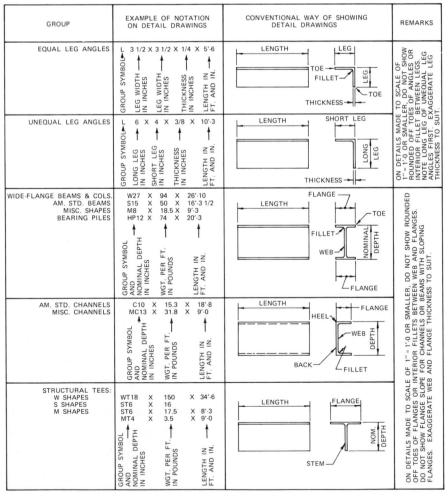

Fig. 20-65 *Steel shapes in section.* (Courtesy of American Institute of Steel Construction)

Examples of Steel Structures

Many different kinds of structures are framed in steel. Two examples–a building and a bridge–are examined in this section.

• **Steel-Framed Building**– A completed A-frame building is shown in Fig. 20-66. Each of the 13 steel A-frame structural members is approximately 220 ft. (67,000 mm) across at the base, 135 ft. (41 150 mm) across at the top, and 15 ft (4570 mm) across the vertical bents. It is built of tubes, wide-flanged sections, and chords made of 18 × 26 in. (455 × 660 mm) tubes. Each A-frame was assembled on the site and erected in five pieces. All the steel members are connected with high-strength bolts (HSB).

The A-frames were designed with the aid of a computer to hold up under many different combinations of *loads* (weights or pressures borne by a structure). These include wind loads, temperature changes, the **dead load** (weight of the building), and the **live load** weights added temporarily, such as furniture. If the building is designed correctly, these loads are carried through the structure to the ground. For the building in Fig. 20-66, which is located in Florida, the designer had to allow for wind loads from hurricane winds. Allowance must always be made for loads created by geographical location, the prevailing weather, the function of the building, and the size and shape of the structure.

TABLE 20-2	STRUCTURAL STEEL SHAPES	
Group Name	**Symbol**	**Comments**
American Standard beams	S	Used to be called I-beams.
American Standard channels	C	
Miscellaneous channels	MC	Include special-purpose channels that are not standard.
Wide flange shapes	W	Used both as beams and as columns.
Miscellaneous shapes	M	Lightweight shapes that look in cross section like W shapes.
Structural tees	ST, WT MT	Made by splitting S, W, and M shapes, usually along the middepth of their webs.
Angles	L	Consist of two legs of equal or unequal widths, at right angles to each other.
Plates and flat bars	PL, Bar	Rectangular in cross section.

Table 20-2 *The AISC groups plain steel shapes into these nine basic categories.*

Fig. 20-66 *The structural shapes in this A-frame resort hotel are dependent on structural steel detail.* (Courtesy of United States Steel Corp.)

- **Steel-Framed Bridge**—Figure 20-67 shows the schematic for the steel framework of a bridge over a deep gorge of the Rio Grande near Taos, New Mexico (Fig. 20-67). The bridge has a rigid structural framing made of high-strength steel fastened with high-strength bolts and welds. More than 1900 customary tons, or 1725 metric tons (t), of structural steel were used in its construction. The center span is 600 ft. (183 m) long, and the two side spans are each 300 ft. (91.5 m) long. The distance from the canyon floor to the bridge floor is 600 ft. (183 m). A detail drawing of a typical welded-steel member of this bridge appears in Fig. 20-68.

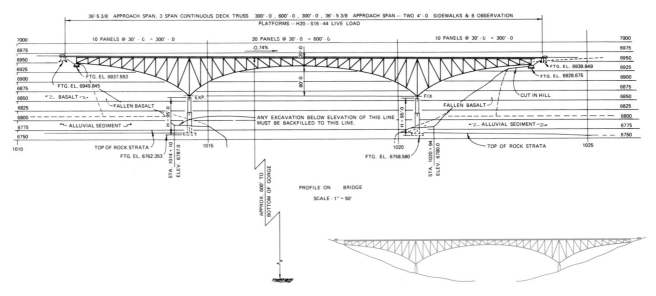

Fig. 20-67 *A schematic design of the bridge framework for a bridge over the Rio Grande in New Mexico.* (Courtesy of United States Steel Corp.)

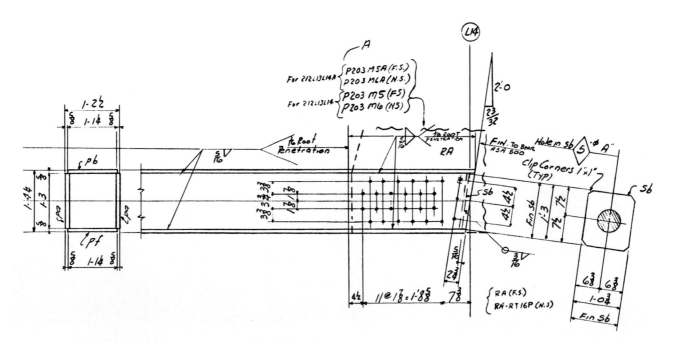

Fig. 20-68 *A welded beam for the bridge shown in Fig. 20-67.* (Courtesy of United States Steel Corp.)

Steel Systems

Using steel, engineers have developed a number of new structural systems. The Unistrut Space Frame shown in Fig. 20-69 consists of four or five modular units in a geometric pattern. The basic unit is made of four or five parts that are bolted together. The system is used mainly in canopies and roofs. Note how the geometric pattern of the exposed steel becomes a part of the overall design. This roof is in a mall in Columbia, Maryland.

Structural Domes

Figure 20-70 shows a dome over a theater in Reno, Nevada. The geometry used in this structure is *geodesic*. It is different from other dome geometry in that its strength is in all directions (omnidirectional). The three-dimensional triangulated framing of the dome (Fig. 20-71) makes it exceptionally strong. It also uses less material than other dome frames. Geodesic geometry was invented by Dr. R. Buckminster Fuller.

Fig. 20-69 *Geometrically arranged structural members help shape buildings.* (Courtesy of Unistrut Corp.)

Fig. 20-70 *A flexible geometric steel dome for a theater.* (Courtesy of Temcor)

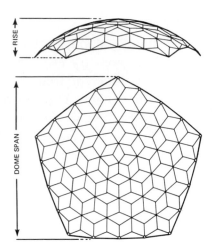

Fig. 20-71 *The dome in Fig. 20-70 has geometric characteristics.* (Courtesy of Temcor)

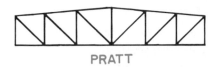

PRATT

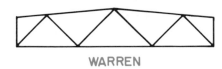

WARREN

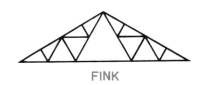

FINK

BOWSTRING

Fig. 20-72 *Roof- and bridge-truss diagrams.*

Trusses

A **truss** is a configuration of structural elements that adds strength to a structure. Roofs and bridges are examples of two structures that often depend on trusses. Figure 20-72 shows diagrams and names for some roof trusses and bridge trusses. The ones shown are only a few of those available. Each type can also be modified to carry specific loads.

Fastening Structural Steel

Structural steel requires high-strength fasteners. Examples of fasteners often used for structural members include rivets, high-strength steel bolts, and welds.

As you may recall from Chapter 10, *rivets* are permanent fasteners that have a preformed head on one end. After being driven through the materials to be fastened, a head is formed on the other end. The standard symbols for rivets are shown in Fig. 20-73. A typical buttonhead rivet is shown in (B). If the riveting is to be done in the shop, the drafter uses shop rivet symbols. These are open circles the diameter of the rivet head. If the riveting is to be done in the field (on the construction site), the drafter uses field rivet symbols. These are blacked-in circles the diameter of the rivet hole. Lines on which rivets are spaced are called *gage lines*. The distance between rivet centers on the gage lines is called the *pitch*.

High-strength steel bolts are rated by the American Society for Testing and Materials (ASTM). The bolt can be applied in the field or in the shop. The hole into which it fits is normally $\frac{1}{16}$ in. (2 mm) larger than the bolt. Fig. 20-74 shows two kinds of bolted connections: frame and seated. The bolt transmits the force of the beam load to the column. The stress this creates in the bolt is *shear* (cutting). The stress created in the column is *compressive* (pushing together).

Structural steel members can also be welded together. This is usually accomplished using the metal-arc process. The fillet weld is the most common on structural connectors. See Chapter 17 for a review of the standard welding processes and symbols.

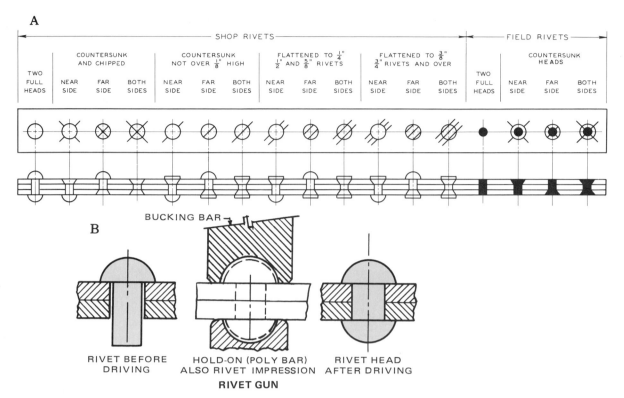

Fig. 20-73 *Rivet symbols and buttonhead rivet*

Drafting Structural Details

The plain shapes shown in Table 20-2 and Fig. 20-65 are basic to structural detailing. You must be familiar with them in order to make adequate drawings.

Structural details are prepared at a scale of 1″ = 1′-0 for beams up to 21 in. (533 mm) in depth. For beams of greater depth, a ¾″ = 1′-0 scale is preferred. The overall length of structural members can be shortened as long as details are shown adequately. Also, very small dimensions, such as a clearance, can be exaggerated to clarify views.

Designers place on their design drawings all the information needed to prepare shop drawings. Figure 20-75 shows a small part of a designed floor plan. There are enough notes and dimensions on the plan to prepare a shop drawing of the wide-flanged beam (W).

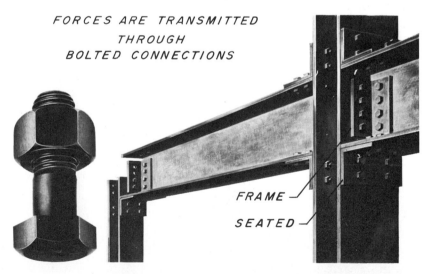

Fig. 20-74 *Structural bolt (left) and bolted connections.*

The 20-ft. dimension is presumed to be the structural bay, or distance from *A* to *B*. The structural members are at right angles to one another unless noted. The height given on the line diagram of a beam is significant. Height elevations are assumed to be level at the figure given. The figure shows two elevations, 98′-6 and 98′-9.

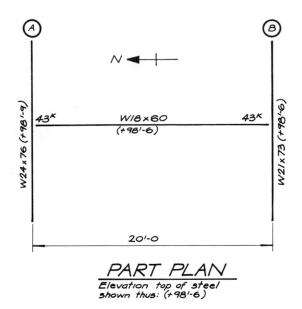

PART PLAN

Elevation top of steel
shown thus: (+98'-6)

Fig. 20-75 *A small part of a plan, arranging steel members between beams A and B.* (Adapted from AISC Handbook, with permission of the publisher.)

2 BEAMS - B1

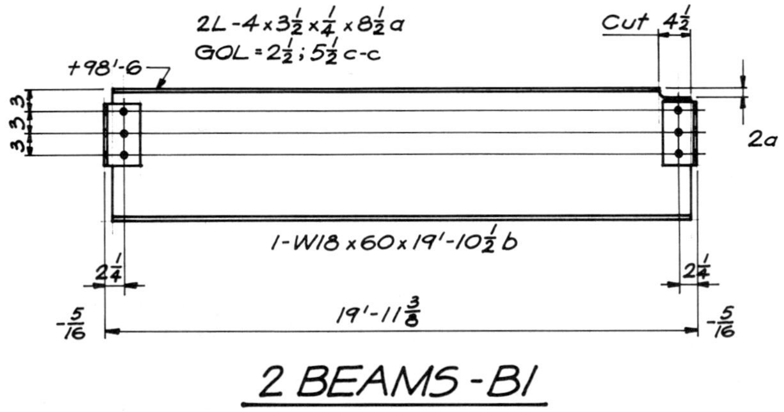

Fig. 20-76 *Typical beam detail.* (Adapted from AISC Handbook, with permission of the publisher.)

Detail drawings such as the one in Fig. 20-76 are essential for making structural pieces. The figure is a detail of a beam. This kind of drawing seldom describes the connections of mating parts (Fig. 20-77). Instead, it just shows those features (for example, connection angles) that will be involved when the piece is used in building.

In preparing structural details, the drafter refers to the handbook and the dimensions for detailing. Figure 20-78 shows both frame and seated connections.

Dimensioning Structural Steel Members

In structural drawings, dimensions are given primarily to working points. For beams, give dimensions to the centerline. For angles and (normally) to channels, give dimensions to the backs. Give vertical dimensions on beams and channels to the tops or bottoms. Generally, do not dimension the edges of flanges and the toes of angles. Make the dimension lines continuous and unbroken. Place the dimensions above the dimension lines. When dimensions are in feet and inches, use the foot symbol but not the inch symbol.

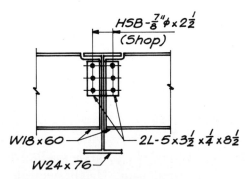

Fig. 20-77 *Typical connection or framing of mating beams.* (Adapted from AISC Handbook, with permission of the publisher.)

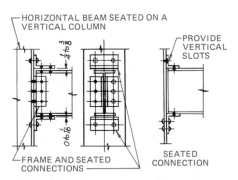

Fig. 20-78 *Frame and seated connections.* (Adapted from AISC Handbook, with permission of the publisher.)

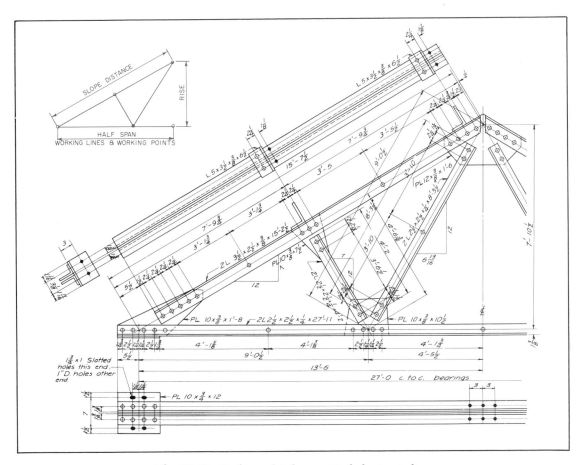

Fig. 20-79 *Roof-truss detail symmetrical about centerline.*

Figure 20-79 is a structural drawing of a small steel roof truss. This symmetrical piece is detailed about the left of a centerline. Study the drawing closely. On the drawing, each member is completely dimensioned or described. In addition, the dimensions shown adequately relate the fixed location of each structural member.

Wood for Construction

Wood is commonly used for the frames of homes and other small structures. Details for wood construction have been developed by the National Forest Products Association. They are now used as a standard method of construction.

Structural timber can be manufactured in many forms by **laminating** (cutting wood into thin slabs and gluing them together). Builders can buy these "factory-grown" timbers in any size or shape (Fig. 20-80). Some of the forms available include

Fig. 20-80 *Laminated wood forms take many shapes.* (Courtesy of American Institute of Timber Construction)

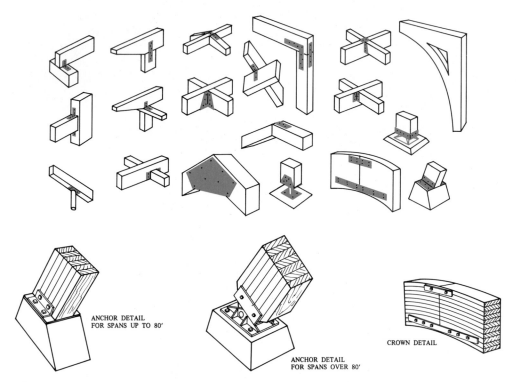

ANCHOR DETAIL
FOR SPANS UP TO 80'

ANCHOR DETAIL
FOR SPANS OVER 80'

CROWN DETAIL

Fig. 20-81 *Construction details for timber construction.* (Courtesy of American Institute of Timber Construction)

tudor arches, radial arches, parabolic arches, A-frames, and tapered beams. The American Institute of Timber Construction (AITC) has set up guidelines for makers of structural glued laminated timber.

Figure 20-81 shows some of the construction details that must be used with structural timber. These detail drawings show how timbers are joined together and how structural members are anchored to foundations.

Concrete Systems

Many of our buildings, bridges, and dams have only been made possible by concrete. The American Concrete Institute has prepared a manual of standard practice for concrete structures.

Concrete has only limited strength unless it is specially prepared. It is made from a mixture of gravel, sand, water, and portland cement. Various grades are produced, depending on the proportions of these ingredients. The concrete can also be reinforced or prestressed.

Reinforced concrete has steel bars embedded in it. These bars are arranged to bear the structural loads that the concrete could not support by itself. When concrete and steel are combined, they can be used in the monolithic form shown in Fig. 20-82. Fig. 20-83 shows a typical reinforced concrete detail.

Fig. 20-82 *Monolithic concrete forms shape new structures.* (Courtesy of Ceco Steel Products Corp.)

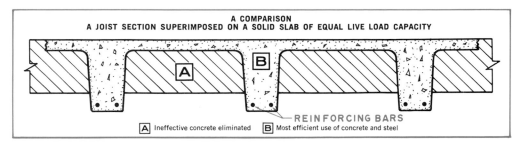

Fig. 20-83 *Reinforced concrete detail.* (Courtesy of Ceco Steel Products Corp.)

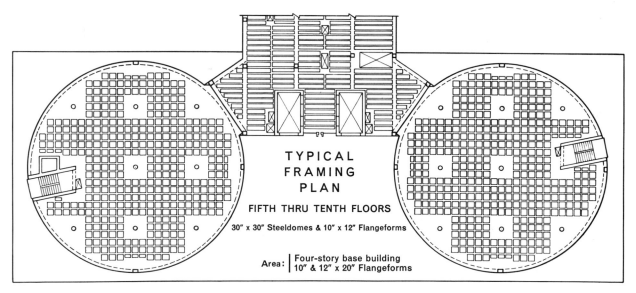

Fig. 20-84 *Concrete placement plan.* (Courtesy of Ceco Steel Products Corp.)

In **prestressed concrete,** the reinforcing bars are stretched before the concrete is poured over them. The prestressed form will then accept a predefined load. This combination of concrete and stretched steel is stronger than either plain concrete or reinforced concrete.

Concrete forms designed by a structural engineer are drawn for the manufacturer's use only. Construction drawings of these forms are prepared by the manufacturer. These drawings are made to show the contractor the location, placement, and connections. See Fig. 20-84 for typical details.

Structural Clay Systems

The solid brick wall is the oldest type of masonry construction known. Bricks are made from different types of clay in many shapes, forms, and colors. The common brick size is $2\frac{1}{4} \times 3\frac{3}{4} \times 8$ in. ($57 \times 95 \times 203$ mm). Brick walls are made to support floors and roofs. For the vertical walls to be able to carry the horizontal floors and roofs, the bricks must have high compressive strength.

Bricks in construction are arranged in overlapping and interlocking patterns and fastened together with connecting mortar joints. This produces a structural assembly that acts as a single structural unit. Figure 20-85 shows some of the common bonds and structural patterns used with bricks. Note the terms applied to the brick.

The strength of a structural clay system generally depends on the strength of the mortar joints. When the design limits this strength, the designer can call for *reinforced masonry*. This is masonry with steel rods or wire embedded in the mortar. Brick or concrete masonry can also be used to enclose a structural steel framework. This provides enough fireproofing to meet standard building codes.

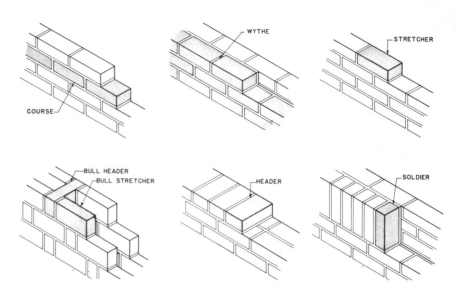

Fig. 20-85 *Brick bonding forms structural walls.* (Courtesy of Structural Clay Products)

STRUCTURAL ANALYSIS

Intended use and the environment in which the structure will stand are among the many factors that can affect the weight or force that a structure must withstand. Because the effects of failure can be disastrous, a structure's load-bearing capacity and structural strength are critical in structural design.

Many factors, such as the types of material and the method of fastening used, affect the strength of the structure. Obviously, it is impossible to test the structure directly before it is built. For straightforward analysis of simple structural elements, the structural designers use calculators and tables in engineering texts. However, in complicated structures that may contain several interacting forces, materials, and members, analysis quickly becomes more difficult. In these situations, structural designers use software such as **finite element analysis (FEA)** programs to test computer models of structural members. FEA programs such as AutoDesign allow the user to assign material properties to various parts of a CAD model. Then the

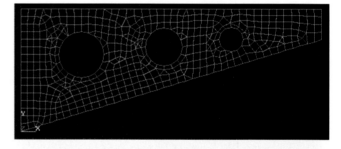

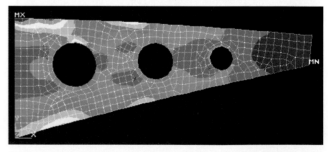

Fig. 20-86 *The results of a typical computerized finite element analysis, showing stress contours and possible weak spots in a design.* (Courtesy of Roopinder Tara)

user applies various "loads" to the model, and the software automatically creates stress contours that show any weak spots in the structural design (Fig. 20-86).

CHAPTER 20

REVIEW

Learning Activities

1. Obtain sets of drawings for the design of a residence and of a commercial building. Study them carefully. Do the drawings follow the standards shown in this chapter? Do they fully describe all aspects of the design for construction purposes?

2. Invite an architect to visit your class to describe the process of designing a structure and preparing the working drawings. How many people are involved in the process? How many drafters? How many professional engineers and architects?

3. Visit a construction site. Study the drawings associated with the structure and try to relate various aspects of the drawing to the construction details.

4. Construct scale models of various construction details, such as a section through a wall or a roof framing detail.

Questions

Write your answers on a separate sheet of paper.

1. Name the two styles of lettering used in architecture.

2. List four house styles.

3. What are the three main ways to judge a building's architecture?

4. What are the four basic types of drawings that an architect makes?

5. What major services does an architect render?

6. Define architecture in your own words.

7. Draw four material symbols used by the architect.

8. Illustrate a line technique important to architectural style.

9. What are pressure-sensitive symbols?

10. List six types of windows used in houses.

11. What are the preferred scales for plan and elevation drawings of small buildings?

12. What scale is preferred for enlarging details?

13. List three types of wall constructions common in residential design.

14. What is modular coordination?

15. Name five jobs for structural drafters.

16. Name five basic structural steel shapes.

17. What is the major advantage of a geodesic dome?

18. How do you dimension beams, angles, and channels?

19. How do you draw dimension lines on structural drawings?

20. What is the difference between reinforced concrete and prestressed concrete?

21. What symbols are used in structural dimensions, and when do you use them?

22. On what does the strength of structural clay systems depend?

23. What is the difference between a frame and a seated steel-beam connection?

24. What are the metric dimensions of a common brick?

CHAPTER 20

PROBLEMS

The small CAD symbol next to each problem indicates that a problem is appropriate for assignment as a computer-aided drafting problem and suggests a level of difficulty (one, two, or three) for the problem.

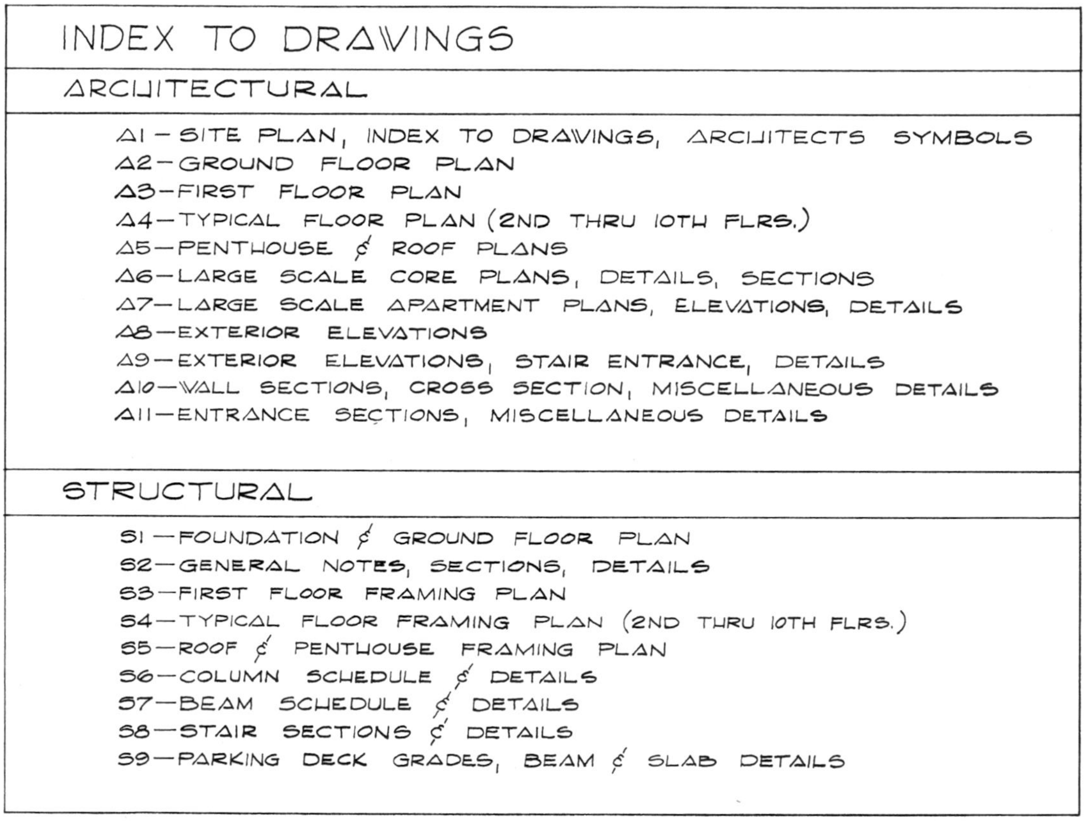

INDEX TO DRAWINGS

ARCHITECTURAL

A1 – SITE PLAN, INDEX TO DRAWINGS, ARCHITECTS SYMBOLS
A2 – GROUND FLOOR PLAN
A3 – FIRST FLOOR PLAN
A4 – TYPICAL FLOOR PLAN (2ND THRU 10TH FLRS.)
A5 – PENTHOUSE & ROOF PLANS
A6 – LARGE SCALE CORE PLANS, DETAILS, SECTIONS
A7 – LARGE SCALE APARTMENT PLANS, ELEVATIONS, DETAILS
A8 – EXTERIOR ELEVATIONS
A9 – EXTERIOR ELEVATIONS, STAIR ENTRANCE, DETAILS
A10 – WALL SECTIONS, CROSS SECTION, MISCELLANEOUS DETAILS
A11 – ENTRANCE SECTIONS, MISCELLANEOUS DETAILS

STRUCTURAL

S1 – FOUNDATION & GROUND FLOOR PLAN
S2 – GENERAL NOTES, SECTIONS, DETAILS
S3 – FIRST FLOOR FRAMING PLAN
S4 – TYPICAL FLOOR FRAMING PLAN (2ND THRU 10TH FLRS.)
S5 – ROOF & PENTHOUSE FRAMING PLAN
S6 – COLUMN SCHEDULE & DETAILS
S7 – BEAM SCHEDULE & DETAILS
S8 – STAIR SECTIONS & DETAILS
S9 – PARKING DECK GRADES, BEAM & SLAB DETAILS

Fig. 20-87 *CAD2 Using the information shown here, practice lettering the architectural style shown in the index to drawings in Fig. 20-28. The lettering sizes should be $\frac{1}{4}''$, $\frac{3}{16}''$, and $\frac{1}{8}''$. Use guide lines and show the box around the lettering.*

PROBLEMS

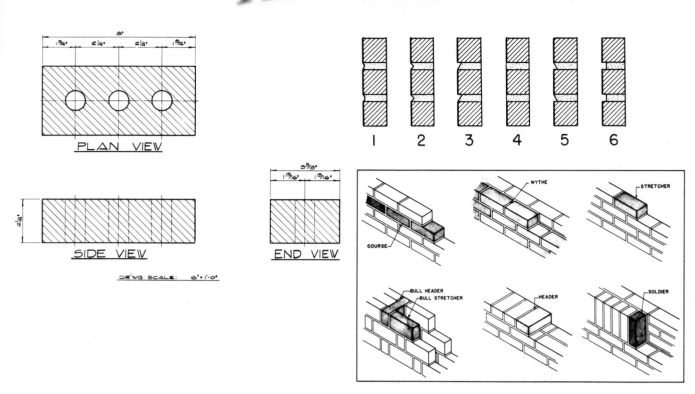

Fig. 20-88 *CAD2 Make a three-view drawing of the brick at the scale shown. Dimension and label. Label the stretcher, header, and face. Make a pictorial in isometric of the brick boards. The nominal dimensions are $2^1/_4'' \times 4'' \times 8''$ with $^1/_2''$ mortar joints for easy drafting. Scale: $1^1/_2'' = 1'-0$. Draw the six views of the bricks in section. Label the types of joints: (1) concave; (2) V-joint; (3) weathered; (4) flush; (5) struck; and (6) raked.*

PROBLEMS

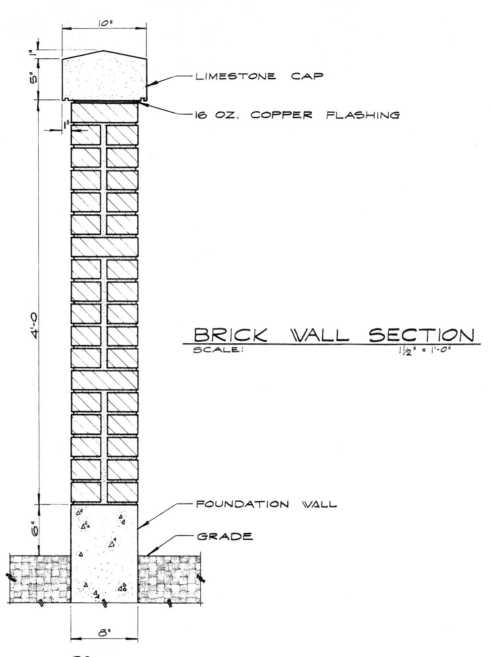

LIMESTONE CAP

16 OZ. COPPER FLASHING

BRICK WALL SECTION

SCALE: 1½" = 1'-0"

FOUNDATION WALL

GRADE

Fig. 20-89 *CAD2 Prepare an isometric drawing of a brick wall section 4'-0 long. You may show the brick bonding, or you may draw it as horizontal lines in pictorial (30°). Scale: 1½" = 1'-0.* (Courtesy of Doug Storm)

PROBLEMS

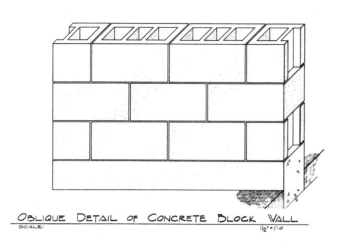

OBLIQUE DETAIL OF CONCRETE BLOCK WALL
SCALE: 1½" = 1'-0

Fig. 20-90 *CAD2 Prepare an oblique or isometric detail of a concrete block wall as assigned by the instructor. Investigate the sizes of block. Scale: 1½" = 1'-0.*

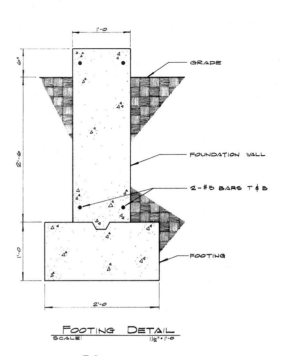

- GRADE
- FOUNDATION WALL
- 2-#5 BARS T&B
- FOOTING

FOOTING DETAIL
SCALE: 1½" = 1'-0

Fig. 20-91 *CAD2 Draw a footing detail at the scale shown and label the parts.*

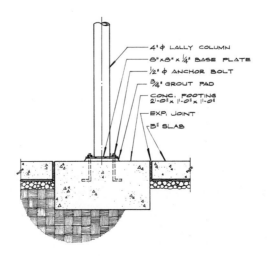

- 4" ∅ LALLY COLUMN
- 8"x8"x¼" BASE PLATE
- ½" ∅ ANCHOR BOLT
- ¾" GROUT PAD
- CONC. FOOTING 2'-0"x 1'-0"x 1'-0"
- EXP. JOINT
- 5" SLAB

TYPICAL COLUMN FOOTING
SCALE: 1" = 1'-0

Fig. 20-92 *CAD2 Prepare a section detail of the column footing, using the given sizes of the structural items.* (Courtesy of Doug Storm)

PROBLEMS

Fig. 20-93 through 20-98 *These are all part of the same structure. When preparing the details (Figs. 20-95 through 20-98), refer to the framing plan (Fig. 20-94).*

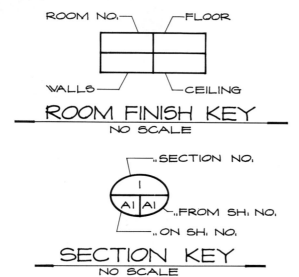

Fig. 20-93 *Prepare a sectional key diagram with an ellipse template and a room finish symbol. Note the section notes on framing diagram.* (Courtesy of Doug Storm)

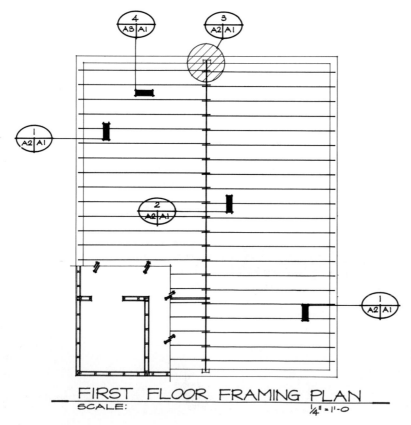

Fig. 20-94 *CAD2 Prepare a framing diagram as shown. Outside building dimensions are 24′-0 × 30′-0. Joists are 16″ on center. Include your sectional diagrams as ellipses or rectangles.* (Courtesy of Doug Storm)

PROBLEMS

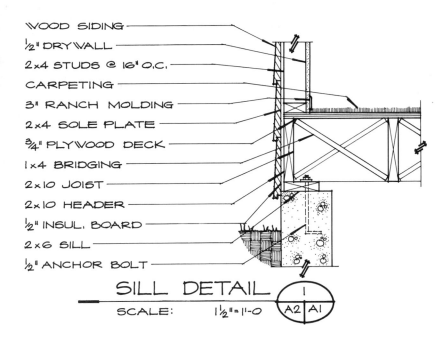

WOOD SIDING
½" DRYWALL
2×4 STUDS @ 16" O.C.
CARPETING
3" RANCH MOLDING
2×4 SOLE PLATE
¾" PLYWOOD DECK
1×4 BRIDGING
2×10 JOIST
2×10 HEADER
½" INSUL. BOARD
2×6 SILL
½" ANCHOR BOLT

SILL DETAIL
SCALE: 1½"=1'-0   ① A2 | A1

Fig. 20-95 *Using the nominal dimensions of the materials listed, construct the sill detail in section. The lapped, 1″ siding has an 8″ exposure. Scale as shown.* (Courtesy of Doug Storm)

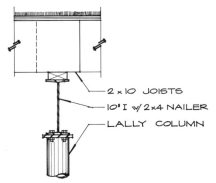

2×10 JOISTS
10" I w/ 2×4 NAILER
LALLY COLUMN

COLUMN DETAIL
SCALE: 1½"=1'-0   ② A2 | A1

Fig. 20-96 **CAD2** *Prepare a partial girder-column detail. The column is 4″ with a 6″ plate welded to the top. The 10″ I-beam is bolted to the plate. The 2 × 4 rests on the 4 ½″ flange of the beam.* (Courtesy of Doug Storm)

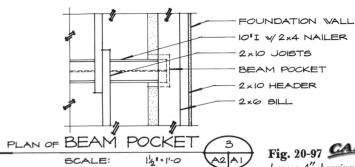

FOUNDATION WALL
10" I w/ 2×4 NAILER
2×10 JOISTS
BEAM POCKET
2×10 HEADER
2×6 SILL

PLAN OF BEAM POCKET
SCALE: 1½"=1'-0   ③ A2 | A1

Fig. 20-97 **CAD2** *Prepare a plan of the 6″ beam pocket. The beam is to have a 4″ bearing on the concrete wall.* (Courtesy of Doug Storm)

PROBLEMS

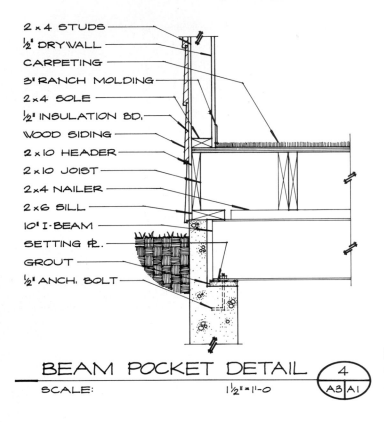

2 x 4 STUDS
½" DRYWALL
CARPETING
3" RANCH MOLDING
2 x 4 SOLE
½" INSULATION BD.
WOOD SIDING
2 x 10 HEADER
2 x 10 JOIST
2 x 4 NAILER
2 x 6 SILL
10" I-BEAM
SETTING PL.
GROUT
½" ANCH. BOLT

BEAM POCKET DETAIL   4

SCALE: 1½" = 1'-0   A3 | A1

Fig. 20-98 *CAD2* *Prepare a sectional beam-pocket detail showing the 10″ I-beam on the stepped foundation wall. Allow the beam flange to appear as a nominal ½″ thickness.* (Courtesy of Doug Storm)

Fig. 20-99 and 20-100 *CAD2* *Prepare a pictorial of a typical roof detail. Studs, ceiling joists, and rafters are 16″ on center. Joist and rafters: 2″ × 6″. Studs and top plates: 2″ × 4″. Ridge beam: 2″ × 8″. Scale: 1½″ = 1′-0. Draw a pictorial of the built-up girder (3) 2″ × 10″.*

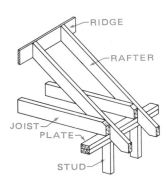

Fig. 20-99.

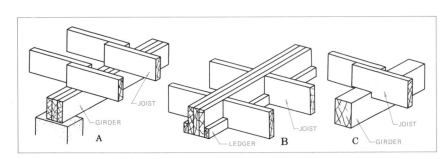

Fig. 20-100.

PROBLEMS

Fig. 20-101 through 20-106 *These are a partial set of plans for a two-story residence. Prepare drawings as assigned by the instructor.*

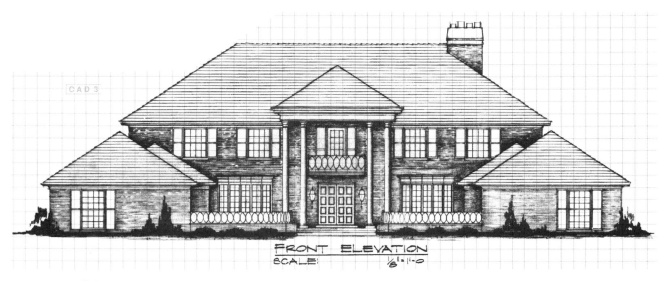

Fig. 20-101 *CAD3* *Draw the front elevation of the two-story residence. Use the modular grid to establish the sizes. (Each square on the grid is 2'-0 sq.) Draw at a scale of* $^1/_8'' = 1'$-0*. Illustrate the brick, masonry walls, and asphalt roofing with horizontal lines. Add windows, shutters, and decorative appointments to suit your own design.* (Courtesy of A. W. Wendell & Sons, architect-contractor)

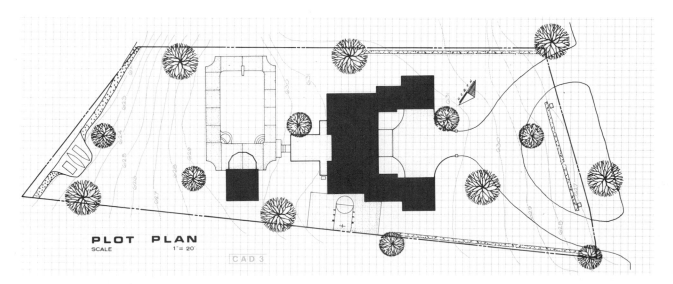

Fig. 20-102 *CAD3* *Examine the gridded plot plan, and develop the boundary lines at a scale of* $1'' = 20'$*, on a C-size sheet. (Each square on the grid is 2'-0 sq.) Dimension each boundary line. Complete the plan of the house on the site. Landscape with trees, and apply contour lines. All radii on the driveway must be 18'. The pool, cabana, and basketball court are optional.* (Courtesy of A. W. Wendell & Sons, architect-contractor)

PROBLEMS

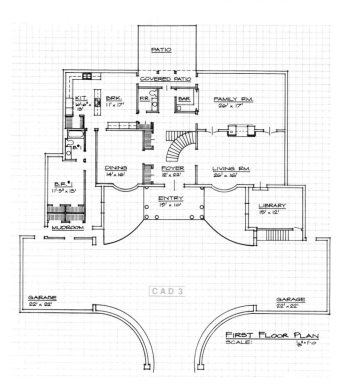

FIRST FLOOR PLAN
SCALE: ⅛"=1'-0

Fig. 20-103 *CAD3 Draw a first-floor plan at a scale of ⅛" = 1'-0. By examining the grid, locate windows on the plan from elevations. (Each square on the grid is 2'-0 sq.) Label the rooms and the appropriate sizes for this preliminary architectural study for your client.* (Courtesy of A. W. Wendell & Sons, architect-contractor)

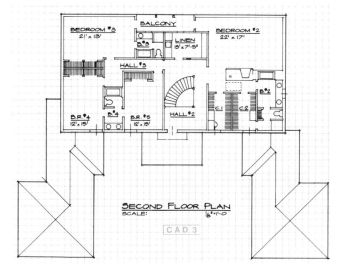

SECOND FLOOR PLAN
SCALE: ⅛"=1'-0

CAD 3

Fig. 20-104 *CAD3 Prepare a second-floor plan to include the roof plan over the single-story area. Scale: ⅛" = 1'-0. Label the rooms with names and dimensions. Note the hip roof design and intersecting inclined planes forming the roof. (Each square on the grid is 2'-0 sq.)*

PROBLEMS

RIGHT ELEVATION
SCALE: 1/8"=1'-0

Fig. 20-105 *CAD3 Draw a right-side elevation by examining the proportions on the modular grid. (Each square on the grid is 2'-0 sq.) Note the centerlines that locate the finished floor and ceiling lines. Find the common roof pitch and label it on the elevation.* (Courtesy of A. W. Wendell & Sons, architect-contractor)

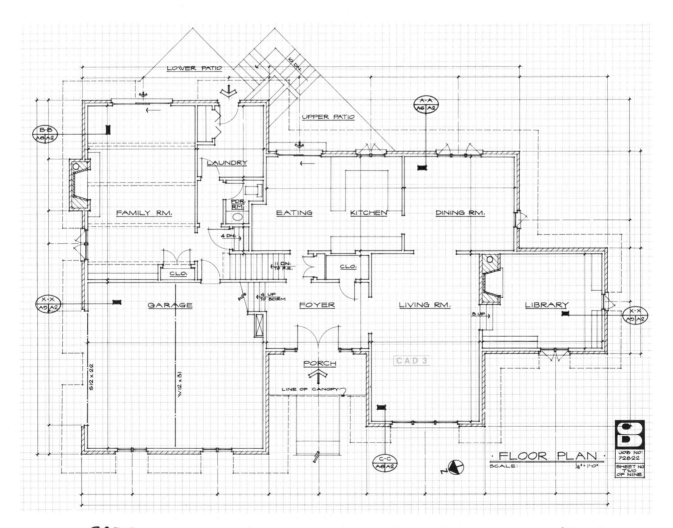

FLOOR PLAN
SCALE: 1/4"=1'-0"

Fig. 20-106 *CAD3 Note that this floor plan is Fig. 20-29 in the text. Prepare the first-floor plan at a scale of 1/2" = 1'-0. Add the dimensions to the plan. Examine the modular grid for room sizes and compare them to Fig. 20-29. (Each square on the grid is 2'-0 sq.)*
(Courtesy of A. W. Wendell & Sons, architect-contractor)

PROBLEMS

Fig. 20-107 through 20-110 *Examine the front, west, and south elevations of the chapel in relation to the floor plan. Draw the plan and lay out the narthex, nave, and chancel at a scale of ⅛″ = 1′-0. Use dividers and the scale on the floor plan to determine sizes. Render the plan with appointments and add site landscaping. Draw elevations with an 18′-high roof line using shingles. Remember that this is a preliminary study and dimensions are not necessarily accurate at this stage of presentation. Render the elevations.* (Courtesy of Doug Storm, architect)

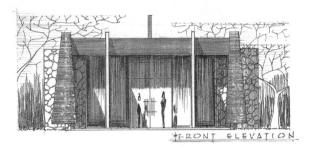

Fig. 20-107 *CAD3*

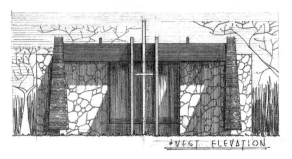

Fig. 20-108 *CAD3*

Fig. 20-109 *CAD3*

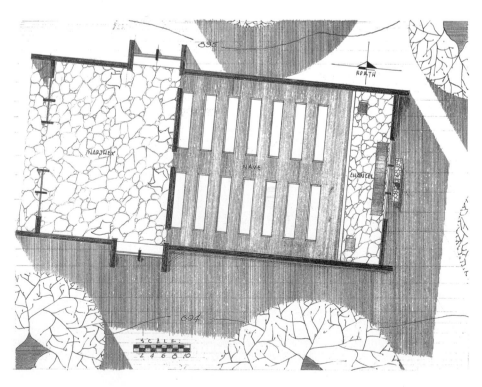

Fig. 20-110 *CAD3*

PROBLEMS

Fig. 20-111 *Develop an isometric drawing of the western, or flat form, framing in Fig. 20-5. Rafters: 2″ × 6″. Joists: 2″ × 8″. Studs: 2″ × 4″.*

Fig. 20-112 *Develop a detailed floor plan of the ranch-style house in Fig. 20-22, with complete dimensions. Scale: ¼″ = 1′-0. Optional: Complete a set of plans, elevations, and sections for the house. NOTE: The cross-corner technique adds style to the drawings.*

Fig. 20-113 *Develop a complete set of plans for the contemporary multilevel house in Fig. 20-27. Include the site plan, floor plans, elevations, and sections.*

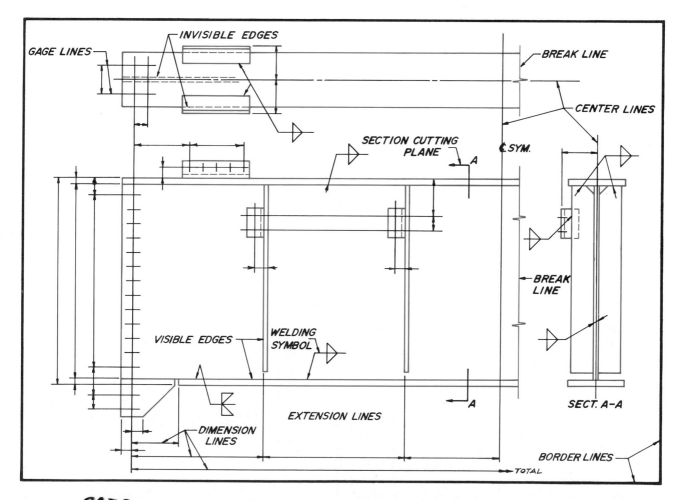

Fig. 20-114 *CAD3 Prepare a detail for a 22″-high girder with a 6″ flange. Develop the girder showing the conventional lines (shown in color) with heavy line weights. Identify angles, welds, stiffeners, and field bolts. Fill in missing dimensions with the instructor's guidance. Scale: 1½″ = 1′-0.*

PROBLEMS

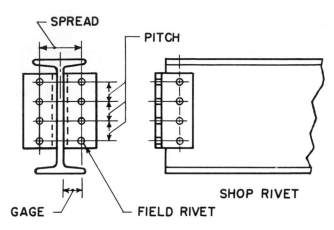

SPREAD

PITCH

GAGE

FIELD RIVET

SHOP RIVET

Fig. 20-115 *CAD3 Draw the detail of the frame connection on an S-beam that is 8″ × 18.4 lbs. The flange is 4″ wide. The web is .270″. The 3″ × 3″ connection angle is 6″ long. Pitch = 1½″. Gage = 1¾″. Rivet = ½″. Prepare the detail at half scale.*

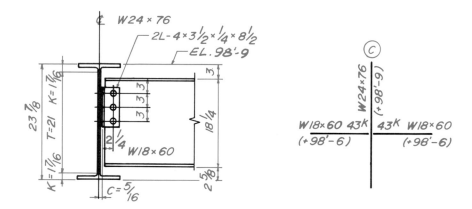

Fig. 20-116 *CAD3 Prepare a partial detail of the frame connection of an 18″-wide flanged beam with a 24″-wide flanged beam. Scale: 3″ = 1′-0. Prepare a partial part-plan diagram about column center C, with noted members, and elevations.*

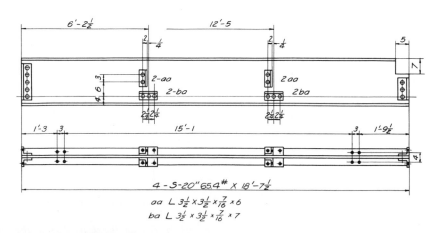

aa L 3½ × 3½ × 7⁄16 × 6
ba L 3½ × 3½ × 7⁄16 × 7

Fig. 20-117 *CAD3 Prepare a detail drawing of the standard S-beam, 20″ × 65.4 lbs. The flange is 6¼″ wide with web thickness of ½″. Scale: ¾″ = 1′-0.*

PROBLEMS

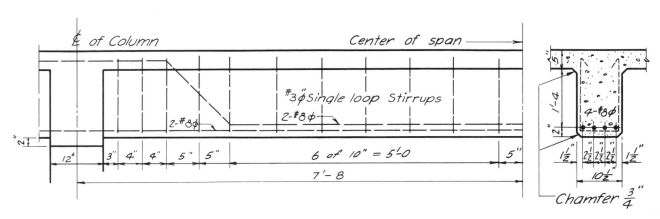

Fig. 20-118 *CAD3 Prepare a drawing of the reinforced concrete detail shown. Scale: 1½″ = 1′ 0.*

Fig. 20-119 *Prepare a truss detail of Fig. 20-79 at a scale of 1″ = 1′-0. (1:12). NOTE: Prepare the left half of the symmetrical beam as shown about centerline.*

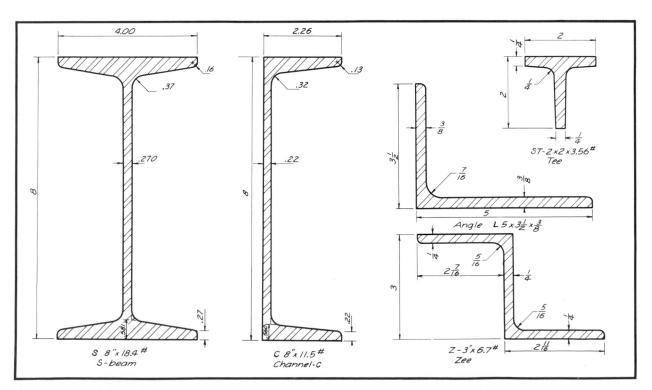

Fig. 20-120 *CAD3 Detail the standard steel shapes shown at an appropriate scale, and show sectioning.*

PROBLEMS

— SCHOOL-TO-WORK —

Solving Real-World Problems

Fig. 20-121 through 20-122 *SCHOOL-TO-WORK problems are designed to challenge a student's ability to apply skills learned within this text. Be creative and have fun!*

Fig. 20-121 Storage building. The company for which you work as an architectural drafter needs a set of plans for an outbuilding in which landscaping and gardening equipment, tools, and supplies can be stored and serviced. It is to accommodate two lawn tractors, a lawn cart, three trimming mowers, dry lawn chemicals and spreader, and hand tools (rakes, shovels, etc.) You are to design the storage building as follows:

Size: Not to exceed 450 square feet.

Base: Concrete slab with floor drains.

Exterior: Rustic design, wood siding, asphalt shingle roof, double exterior doors (minimum 6'-0 opening), adequate windows for natural light.

Interior: Open space (no partitions), wall and ceiling finish of your choice, fluorescent lighting, 115V electrical outlets on each wall, workbench with tool cabinet for servicing lawn and garden equipment.

Your assignment on this project is to develop the following:

1. Integral slab and foundation (footing) plan
2. Floor plan
3. Elevations
4. Roof plan
5. Typical wall sections
6. Electrical plan
7. Window and door schedule
8. Materials list

Fig. 20-122 Design of residence. You are now working for an architectural design firm that does custom home design work. A client has contracted with your firm to have a residence designed that will be placed on a lot 150'-0 wide × 300'-0 deep. It is a corner lot and is relatively level from side to side but slopes approximately 3'-0 uniformly from front to back. All utilities are underground and run parallel with the front property line.

The client has given the following general specifications:

Rooms: Kitchen, dining room, living room, family room, laundry, two-car garage, storage for athletic equipment, three bedrooms, two baths, full basement with exercise room.

Exterior: Two-story brick, cedar shingle roof.

Interior: Plaster walls, hardwood floors in all but baths, laundry, kitchen, and basement. Ceramic tile in baths and kitchen. Vinyl floor covering in laundry, and concrete floor in basement.

Size: Not to exceed 3200 square feet.

You have been assigned this project. The finished set of plans will include the following:

1. Site plan
2. Foundation plan
3. Floor plans
4. Roof plan
5. Elevations
6. Typical wall sections
7. Electrical plan
8. Plumbing plans
9. HVAC plans
10. Window and door schedules
11. Room finish schedule
12. Details list

21

MAP DRAFTING

OBJECTIVES

Upon completion of Chapter 21, you will be able to:

○ Describe the work and career opportunities of a cartographer.
○ Construct maps using standard symbols and techniques.
○ Identify various types of maps such as plats, structural maps, geological sections, block diagrams, and contour maps.
○ Define key terms used in cartography.

VOCABULARY

In this chapter, you will learn the meanings of the following terms:

• block diagram	• fault	• photogrammetry
• cartographers	• formations	• plat
• cartography	• geology	• sedimentary
• contour distance	• igneous	• stratum (*pl.* strata)
• contours	• metamorphic	• strike
• dip	• operations map	• topographic maps

M*apmakers* are people who have been trained to gather the necessary information and prepare maps. Mapmakers are also called **cartographers.** Skilled map drafters prepare maps in detail for civil engineers, scientists, geographers, and geologists.

Mapmaking, or **cartography,** is a pictorial method of representing facts about the surface of the earth or other bodies in the solar system. Satellites take high-altitude photographs that can help mapmakers work efficiently and accurately (Fig. 21-1). Mapmakers then use a method called **photogrammetry** to make three-dimensional measurements from the resulting photographs. Computers bring speed and efficiency to nearly every stage in the process, enabling cartographers to create specialized maps from existing measurements rather than through extensive and costly fieldwork.

Fig. 21-1 *Satellites can take high-resolution photographs from which scientists make map surveys of the earth and other planets in our solar system.* (Courtesy of NASA)

Many of the terms used in map drafting are defined briefly in this chapter; more are defined in the glossary of this textbook. For more complete descriptions of these terms, refer to books on surveying and geology.

CAREER OPPORTUNITIES

Many jobs are available for those who are skilled in map preparation. The field of civil engineering is always expanding. Railroads, highways, harbor facilities, airports, and space stations are just a few areas in this broad industry in which map planning is being used.

The drafter may prepare maps and charts under the direction of the design engineer or cartographer. There may be opportunities for advancement in job areas such as photogrammetry, surveying with laser beams, or research and development projects with the geographer. Additional information about mapmaking is available from the Association of American Geographers (AAG), 1146 16th Street N.W., Washington, D.C. 20036.

MAP SCALES

Some maps that indicate ownership of property, such as city plats, must be very accurate. They may be drawn to a large scale in order to note all the information of physical property. Other maps, such as those showing the geography of states or countries, may need to show boundary lines, streams, lakes, or coastlines over large areas. The scale on these maps must be quite small. In fact, some maps of very large areas may use a scale of several miles to the inch.

Satellite photos divide the United States into quadrangle maps no larger than 71 square miles, drawn at a scale of 1:24 000. At that level, 1 in. on the map equals 2000 feet on the ground. This allows the map to show buildings, roads, and other physical features.

The civil engineer's scale is used for map drawings. Distances are given in decimals of a foot, mile, meter, or kilometer. (See the maps in geography and history books for examples.)

Without knowing the scale of a map, it cannot be read accurately. Therefore, the scale of a map must be shown as part of the basic map information. The scale of a map is generally noted as 1 in. = 500 ft., or 1 part = 6000 parts, noted as 1:6000. The scale of 1 in. = 1 mile can also be shown as 1:63 360.

SCALING CAD MAPS

With a CAD system, designers can create any size map at full scale. Because a "unit" in a CAD drawing can be anything the drafter wants it to be—a foot, a meter, an inch, a mile, or any other unit—even the largest maps can be drawn at full size.

Maps of large areas that are drawn at full scale using a CAD system cannot, of course, be printed at full scale. Scaling in most CAD systems takes place just before printing or plotting. The drafter has several scaling options (Fig. 21-2). The most acceptable is to use one of the industry standard scales, such as 1 in. = 1 mile. However, for some noncritical applications, the drafter can calculate the largest scale that will fit on the available sheet size and print to that scale. Many CAD programs also have a "plot to fit" option that automatically calculates the largest scale for the chosen sheet size.

TYPES OF MAPS

Maps that are made for different reasons have distinct features and notations. Examples of several kinds of maps are illustrated and explained in this section.

Plats

A map used to show the boundaries of a piece of land and to identify it is called a **plat.** The amount and kind of information presented on a plat depend on the purpose of the map.

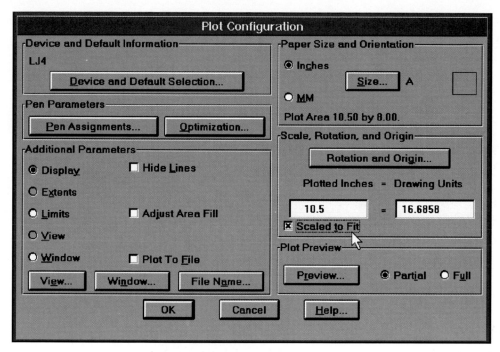

Fig. 21-2 *In AutoCAD, the drafter can plot to a specific scale or, as shown here, choose to let AutoCAD calculate the scale at which the drawing will fit on the paper.*

The plat of a plane survey that was made to accompany the legal description of a property is shown in Fig. 21-3. Accuracy of information on a plat is all-important; it must agree with the legal description exactly.

Another type of plat is a *city plat*. City plats are made for the following reasons:

- to keep a record of street improvements
- to show the location of utilities
- to record sizes and location of property for tax purposes

Part of a city plat is shown in Fig. 21-4. Notice the numbering of the lots and the location of streets, sidewalks, and other details found in a city.

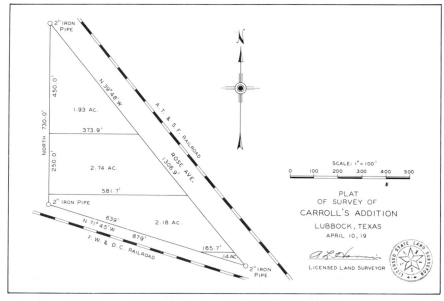

Fig. 21-3 *Plat of a survey. Note the parts that make up the plat: the acreage of each part, the iron pipe locating the corners of the graphic scale, the signature of the surveyor, and the official seal. (One acre = 43,450 feet.)*

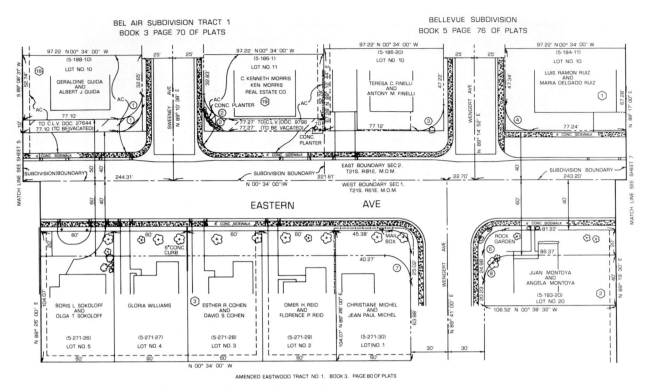

Fig. 21-4 *Part of a city map drawn with the aid of a computer.* (Courtesy of City of Las Vegas and CalComp)

Operations Maps

Operations maps are maps that show the relationship between the land's physical features and the operation that is to be performed. An example of an operations map is shown in Fig. 21-5. Well-executed operations maps help engineering, management, or government groups in the presentation. A presentation well done aids in the selling of a program.

Contour Maps

Since maps are one-view drawings, vertical distances and differences in ground level do not show. They can, however, be shown by lines of constant level called **contours,** as shown in Fig. 21-6. The contour lines represent the height of the ground above sea level. Contour lines that are close together indicate a steeper slope than lines that are far apart. This can be seen by projecting the intersections of the horizontal level lines with the profile section, as shown in Fig. 21-6.

Note that the contour map and the profile correspond to the plan and section of an ordinary drawing. Note the horizontal cutting plane *AA* on the contour map. This line shows exactly where the profile is taken. You can see how the profile would change if the cutting plane were moved toward the ocean or to some other new position.

The **contour distance** (vertical distance between contour lines) can be adjusted according to the cartographer's needs and the scale of the map. A 10-ft. distance may be quite satisfactory for some purposes, while a 5-ft. distance may be used on maps that require a larger scale. For close detail work, such as an irrigation project, the contour distances may be reduced to 2, 1, or even .5, ft. On small-scale maps with a high degree of relief, distances may be increased from 20 to 200 ft. or more. As an aid to reading the map, every fifth contour is usually emphasized by drawing a much heavier line (refer to Fig. 21-7).

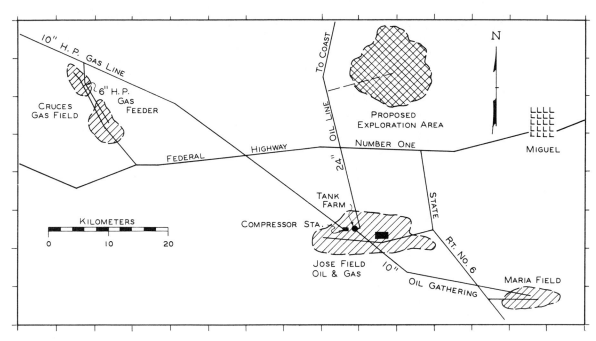

Fig. 21-5 *An oil-field operations map. Note that the scale is in kilometers (1 km = .621 statute mile).*

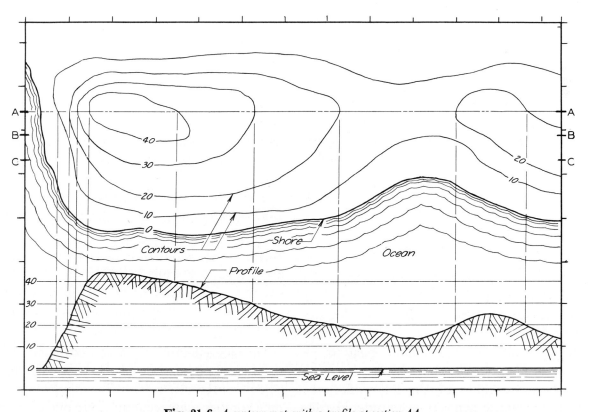

Fig. 21-6 *A contour map with a profile at section AA.*

Another example of a contour map is shown in Fig. 21-7. This map uses a contour distance of 20 ft. The elevation in feet is marked in a break (gap) in each contour line. Notice that the drainage is shown in intermittent streams.

Gathering Information for Contour Maps

Before a contour map can be drawn, elevations must be obtained in the field (map location) for several key points that control the drawing of the contours. The following methods are used:

- A grid system in which all intersection elevations are obtained, along with important elevations on grid lines
- Points located by transit and stadia rod, with the corresponding elevations figured by plane table
- Aerial photographic surveys (Fig. 21-8)

Experience in surveying is needed in all of these mapping methods.

The finished map in Fig. 21-8 was produced by photogrammetric methods. The equipment for making photogrammetric maps is shown in Fig. 21-9. The actual drawing of the map is usually done by scribing lines in the coating on a piece of film. An example of scribing on film is shown in Fig. 21-10.

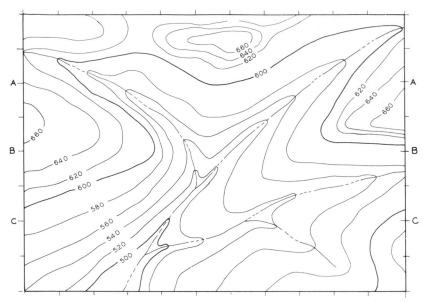

Fig. 21-7 *A contour map with intermittent streams indicated by the lines with three dashes.*

Fig. 21-8 *Portions of an aerial photo and a topographic map of Concepción, Chile, compiled by photogrammetric methods. Actual maps are on a scale of 1:2000 with 1-meter contours. Photos from which the maps were made are at a scale of 1:100,000.* (Courtesy of Aerial Service Corp., Philadelphia, a division of Litton Industries)

Techniques for Drawing Contour Maps

In the past, drafters used contour pens to ink contour lines. Contour pens have blades that swivel so that contour lines can be followed easily. However, a technical pen is now considered satisfactory; it is replacing the contour pen for most applications.

The curved lines on a contour map must be drawn with great accuracy because each line represents a precise elevation. To complete the curves on a contour map accurately, use one of the many types of flexible curves that can be bent to match the curves to be drawn on the map.

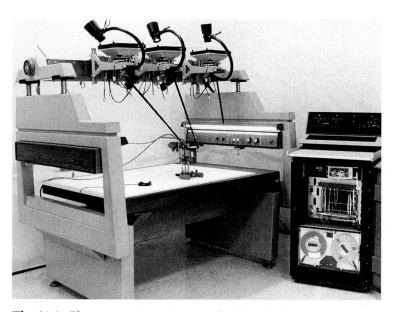

Fig. 21-9 *Photogrammetric equipment with computerized storage.* (Courtesy of Keuffel & Esser Co.)

Fig. 21-10 *Scribing by hand or with a computerized plotter with a jewel point on coated film is a very accurate process.* (Courtesy of Keuffel & Esser Co.)

Topographic Maps

Topographic maps present complete pictorial descriptions of the areas shown. These maps show information such as boundaries, natural features, structures, vegetation, and relief (elevations and depressions).

Symbols are used for many of the features shown on topographic maps. Some of these are shown in Fig. 21-11. A more complete list is given in Appendix E. Maps using topographic symbols can be obtained at a low cost from the Director, U.S. Geological Survey, Department of the Interior, or from the U.S. Coast and Geodetic Survey, Department of Commerce, Washington, D.C. Some symbols from the U.S. Coast and Geodetic Survey are shown in Fig. 21-12. Naval charts (maps) can be ordered from the Hydrographic Office, Bureau of Navigation, Department of the Navy. Aeronautical maps (Fig. 21-13) use special symbols that need to be understood in order to be read.

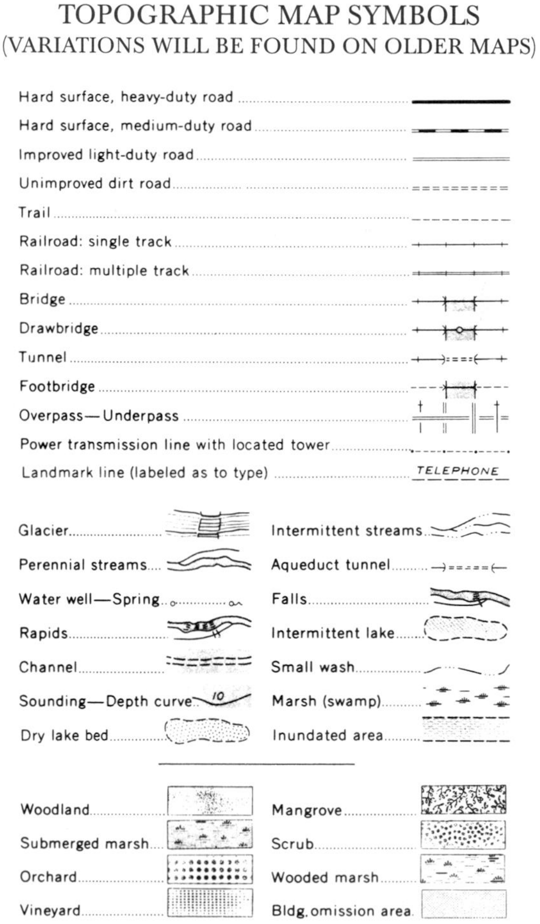

TOPOGRAPHIC MAP SYMBOLS
(VARIATIONS WILL BE FOUND ON OLDER MAPS)

Hard surface, heavy-duty road

Hard surface, medium-duty road

Improved light-duty road

Unimproved dirt road

Trail

Railroad: single track

Railroad: multiple track

Bridge

Drawbridge

Tunnel

Footbridge

Overpass—Underpass

Power transmission line with located tower

Landmark line (labeled as to type) — TELEPHONE

Glacier | Intermittent streams
Perennial streams | Aqueduct tunnel
Water well—Spring | Falls
Rapids | Intermittent lake
Channel | Small wash
Sounding—Depth curve | Marsh (swamp)
Dry lake bed | Inundated area

Woodland | Mangrove
Submerged marsh | Scrub
Orchard | Wooded marsh
Vineyard | Bldg. omission area

Fig. 21-11 *Some conventional symbols used on maps. Note the intermittent streams as used in Fig. 21-13.* (Excerpted from the U.S. Coast and Geodetic Survey)

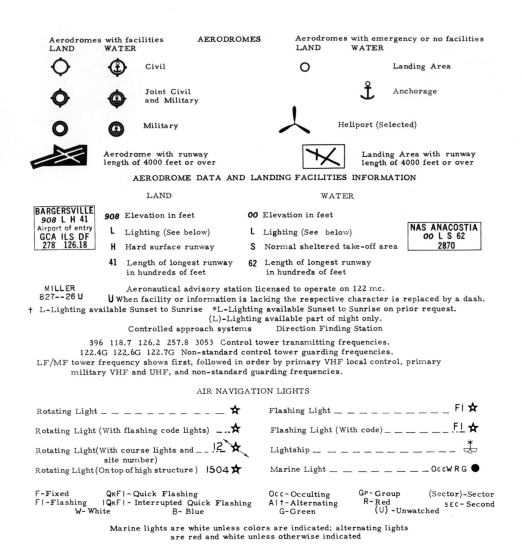

Aerodromes with facilities   **AERODROMES**   Aerodromes with emergency or no facilities
LAND   WATER   LAND   WATER

◯   ⚓   Civil   ◦   Landing Area

◉   ⊕   Joint Civil   ⚓   Anchorage
                  and Military

◎   ⊛   Military   ⋏   Heliport (Selected)

(aerodrome with runway symbol) Aerodrome with runway   (landing area square with X) Landing Area with runway
                  length of 4000 feet or over   length of 4000 feet or over

AERODROME DATA AND LANDING FACILITIES INFORMATION

LAND   WATER

| BARGERSVILLE |
| 908 L H 41 |
| Airport of entry |
| GCA ILS DF |
| 278 126.18 |

908 Elevation in feet   **00** Elevation in feet

L Lighting (See below)   **L** Lighting (See below) | NAS ANACOSTIA |
                                                                      | 00 L S 62 |
H Hard surface runway   **S** Normal sheltered take-off area | 2870 |

41 Length of longest runway   **62** Length of longest runway
     in hundreds of feet   in hundred's of feet

MILLER   Aeronautical advisory station licensed to operate on 122 mc.
827--26 U   **U** When facility or information is lacking the respective character is replaced by a dash.
† **L**-Lighting available Sunset to Sunrise   ***L**-Lighting available Sunset to Sunrise on prior request.
                  (L)-Lighting available part of night only.
            Controlled approach systems   Direction Finding Station

   396 118.7 126.2 257.8 3053 Control tower transmitting frequencies.
   122.4G 122.6G 122.7G Non-standard control tower guarding frequencies.
LF/MF tower frequency shows first, followed in order by primary VHF local control, primary
     military VHF and UHF, and non-standard guarding frequencies.

AIR NAVIGATION LIGHTS

Rotating Light _ _ _ _ _ _ _ _ _ ☆   Flashing Light _ _ _ _ _ _ _ _ _ Fl ☆

Rotating Light (With flashing code lights) _ .. ☆   Flashing Light (With code) _ _ _ _ _ _ Fl ☆

Rotating Light (With course lights and _ _ 12 ☆   Lightship _ _ _ _ _ _ _ _ _ _ _ ⚓
          site number)

Rotating Light (On top of high structure) 1504 ☆   Marine Light _ _ _ _ _ _ _ Occ W R G ●

F-Fixed   QkFl- Quick Flashing   Occ-Occulting   Gp- Group   (Sector)-Sector
Fl -Flashing   IQkFl - Interrupted Quick Flashing   Alt-Alternating   R-Red   sec-Second
     W- White   B- Blue   G-Green   (U) -Unwatched

          Marine lights are white unless colors are indicated; alternating lights
                    are red and white unless otherwise indicated

Fig. 21-12 *Some aeronautical symbols.* (Courtesy of U.S. Coast and Geodetic Survey)

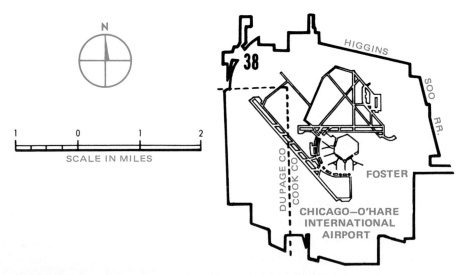

Fig. 21-13 *A single-line diagram of an airport. What symbols from Fig. 21-19 or Appendix
E could apply to this map? How wide is the airport from east to west?*

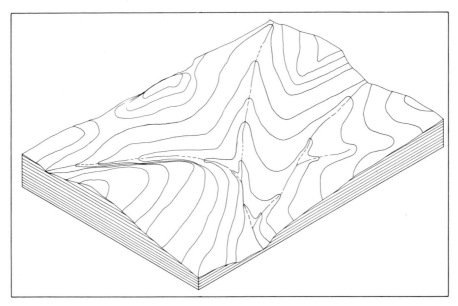

Fig. 21-14 *A block diagram shows a block of earth in an isometric view. This model was made from Fig. 21-7.*

Block Diagrams

Discussion thus far has concentrated on mapping in the horizontal plane and the vertical plane by profiles or sections. To help people see three-dimensionally, drafters can create a **block diagram** (a three-dimensional projection using the isometric view). The block diagram in Fig. 21-14 was developed from the contour map in Fig. 21-7. Keep in mind that each contour represents a level plane, similar to a card in a deck of cards. As in all isometric drawings, true lengths are measured on the isometric axes.

GEOLOGICAL MAPPING

To understand geological mapping, you must first understand a little about geology. **Geology** is the science that deals with the makeup and structure of the earth's surface and interior depths. The crust of the earth is made up of three groups of rock: igneous, sedimentary, and metamorphic.

Igneous rocks, for purposes of this discussion, are the basic crystalline materials that make up the earth's crustal ring. This rock was once molten (melted). It has cooled, but it has not been eroded, nor has its makeup changed.

Sedimentary rocks, as a rule, are deposited in water in layers of different thicknesses similar to the layers of an onion. If you cut an onion perpendicular to its axis, you will notice a series of concentric rings. In a slice of the earth's crust made in a sedimentary area, a similar pattern can be seen. A series of layers that can be identified by texture, color, and material is visible (refer to pictures of the Grand Canyon).

Metamorphic rocks are generally considered to be sedimentary rocks that have been deeply buried, heated to high temperatures, and subjected to great pressure. The combined heat and pressure recompose the rocks so that they can no longer be identified as sedimentary.

Nature, being ever-changing, folds, tips, and slices sedimentary layers in a number of ways. Sometimes sedimentary layers include large areas of crystalline (igneous) rock. Often, the whole mass may be tipped, perhaps for miles. In some places, this tipping raises the mass of rock high above sea level. In other places, the mass drops thousands of feet. The geologist making investigations has the problem of representing what has happened or what a particular area looks like.

Figure 21-15 is part of a geological surface map. The colored lines represent the line of surface exposure of the contact between two **formations** (like the line of two contacting layers of a cut onion). The geologist locates this in the field and notes the observation point with the "T" symbol. The **strike** (direction of this contact) is shown by the top of the T. The **dip** (slope of the contact) is shown by the figures at the T, such as 23° on the left side of Fig. 21-15, below section line *XX*. Since the stem of the T, in this case, is pointing to the east, or to the right, the dip is 23° to the east. In other words, this formation slopes 23° below the horizontal and to the east. Another T symbol on the left side shows a 30° dip to the west. The heavy, broken line near the left edge represents a **fault** (the line along which the layer broke).

GEOLOGICAL SECTIONS

Geological sections help in the interpretation of geological surface maps. The example in Fig. 21-16 shows what the geologist believes the area below the surface is like. This is a section along line *XX* of Fig. 21-15. The dips that the geologist noted are used in developing the curvature of the folds. By means of a typical section of the region, the geologist can determine the various normal thicknesses of each formation or **stratum** (distinct layer). These values are used

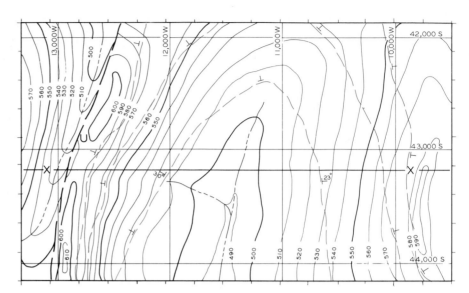

Fig. 21-15 *Part of a geological surface map. Where is the highest elevation on line XX?*

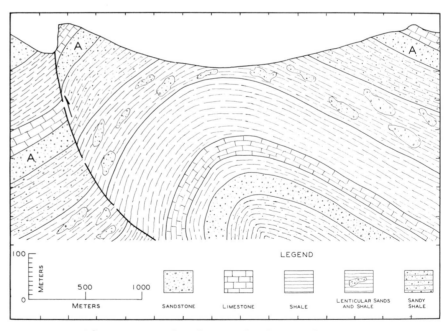

Fig. 21-16 *A geological section along line XX of Fig. 21-9.*

in making this section. The fault, as indicated, shows the area at the right to be upthrown. The displacement is easily seen by comparing the position of formation *A* on either side of the fault.

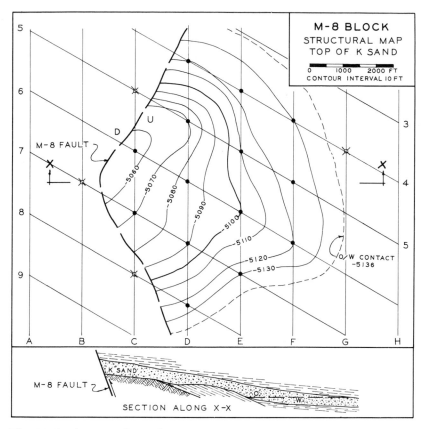

Fig. 21-17 *A structural map showing strata details below the surface of an oil field.*

SUBSURFACE MAPPING

Subsurface mapping is a means of showing details of strata lying below the surface of the earth. It can show the top or bottom of a given formation or possibly an assumed horizon. Information for constructing such a map is obtained from many sources. These sources may include core holes, electrically recorded logs, seismograph surveys, and so forth.

An example of information that was obtained from electrically recorded logs taken in a series of oil wells is shown in Fig. 21-17. The wells are located on a grid pattern. Producing wells are indicated by a solid black circle. Dry holes are indicated by an open circle with outward-extending rays. The top of a producing sand that is cut by a fault on the west is shown. Notice that the contours are numbered with negative values, or depths that are below sea level. The greater the value, the deeper

the point below sea level. Section *XX* shows the thickness of the sand and the level of the *oil-water contact*.

The readability of geological maps and sections can be greatly improved by the use of colors. In Fig. 21-15, colors may be applied to each of the formations that show between the (color) formation-contact lines. This can also be done in Fig. 21-16 by applying the same colors to the corresponding formations. The use of colors helps bring out the three-dimensional relationship of the surface and the shape of the structure. Color also helps people understand the geology of the area. The drafter creates a paper print of the tracing, adds color, and then rubs the print carefully to give smooth, even color texture.

The creation of geological maps and drawings is an important part of the extractive minerals industry, particularly the petroleum industry. With the aid of maps, it is possible to keep proper records and information so that activity in this economic field can be continued. Standards for records are different from company to company. However, general standards are well covered in technical literature such as publications of the AIME, Petroleum Branch, the AAG, U.S. Geological Survey, U.S. Bureau of Mines, and others.

CAD CAD-AUTOMATED CARTOGRAPHY

Mapping requires a tremendous amount of information, complicated calculations, and precise drafting skills. Computer-aided drafting with specialized software can meet the challenge with clear, accurate, and up-to-date maps. When the information that has to be illustrated graphically seems to be constantly changing, a computerized data system can relate quickly to change.

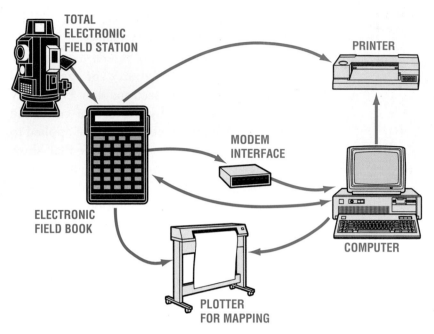

TOTAL ELECTRONIC FIELD STATION

PRINTER

MODEM INTERFACE

ELECTRONIC FIELD BOOK

COMPUTER

PLOTTER FOR MAPPING

Fig. 21-18 *A system flowchart starts with a total electronic station on the site and transfers data to the computer mapping center for processing.* (Courtesy of Jeff Stoecker)

The system for mapmaking supports the complete process from collecting initial data through map production and revisions (Fig. 21-18).

CAD systems accept both graphic and nongraphic information sources (Figs. 21-19 and 21-20). Survey and field data from electronic transfer devices, existing maps, aerial photographs, and polar or orthogonal data can be entered into the computer database. The transfer can include using the drawing tablet, keyboard, a variety of digitizers, and photo data conversion devices.

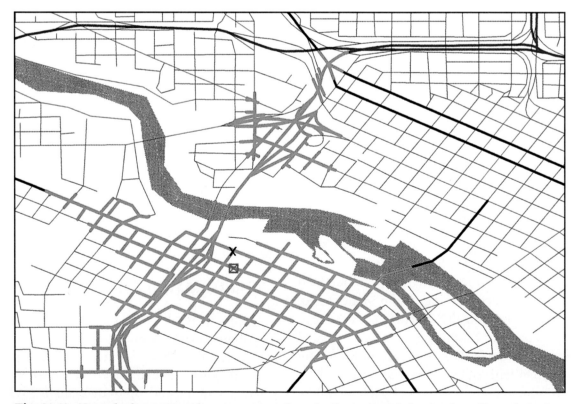

Fig. 21-19 *New technology can provide more accurate information for commercial planning than older methods of analysis. This street map was created using a computer program called ARC/View 3.0 Network Analyst. The progarm took factors such as distances, speed limits, and turn restrictions into account to arrive at a final map of all the streets within a three–minute drive time of a proposed retail site.* (Courtesy of Geographic Data Technology/ESRI)

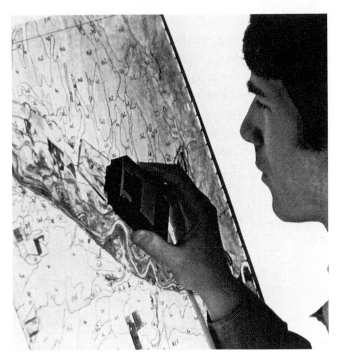

Fig. 21-20 *This man is using a digitizing puck to enter information into a computer system.* (Courtesy of ComputerVision)

For mapping applications, CAD systems allow drafters to use various line fonts (divided highway lines, single lanes, railroads, boundary lines, etc.), create new symbol libraries, and establish mapping conventions (such as using exact latitude and longitude coordinates). In addition, the software can use a variety of colors to distinguish geographic areas.

Using CAD software, cartographers can create, edit, and see more maps more quickly than traditional methods have allowed in the mapping studio. For example, the computer-generated natural contour block diagram in Fig. 21-21A can be redesigned and redrawn to reflect a cut and fill design for land development, as shown in Fig. 21-21B. (A *cut* is earth to be removed to obtain a desired level or slope in preparation for construction; *fill* is earth to be supplied and put in place for the same reason.)

Maps developed using a CAD system support industry standards for reproduction. For example, maps can be produced using output devices such as color plotters and printers (electrostatic and laser). Some software also produces registered separations for conventional printing. All of the data can be edited and rearranged easily on the computer screen prior to reproduction.

FROM THIS . . .

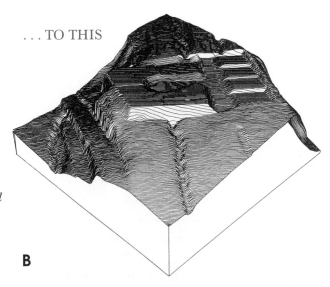

. . . TO THIS

A

Fig. 21-21 *The lines of this computer-generated block diagram (A) and its revision (B) would take many hours and a high level of skill if drawn manually.*

B

Land developers and landscape architects use CAD software to create details on their survey maps or plot plans (Fig. 21-22). The designer can construct drawings using preexisting blocks from symbol libraries to create and edit the plan with relative ease. Detailed symbols are available for arbors, curbs and parking stops, decks and bridges, drainage, erosion control, site design for the handicapped, lighting, paving, playground equipment, planting, site furniture, shelters, signage, steps and ramps, site utilities, and contoured walls, among many others (Fig. 21-23).

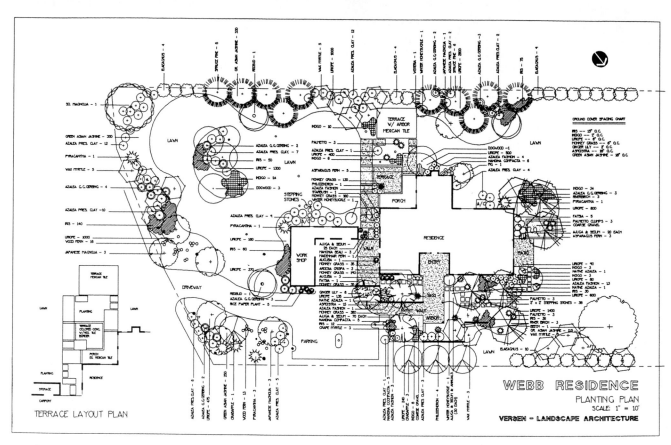

Fig. 21-22A *A site plan may be developed with the use of software programs designed for land developers.* (Courtesy of LANDCADD)

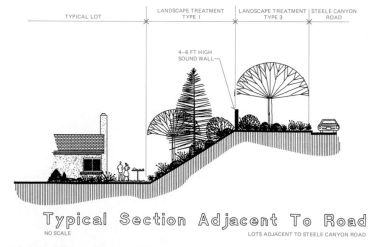

Fig. 21-22B *CAD symbol libraries also include symbols for elevation views.*

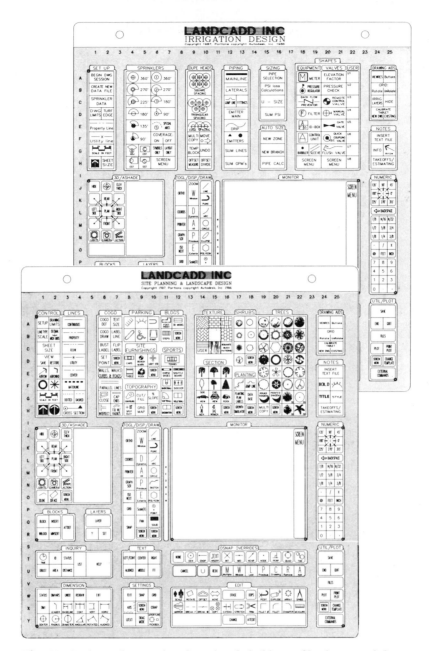

Fig. 21-23 *A site plan may be enhanced with the library of landscape symbols available on CAD systems.* (Courtesy of LANDCADD)

CHAPTER 21

REVIEW

Learning Activities

1. Invite your geography or science teacher to visit your class to "show and tell" about various kinds of maps and to describe career opportunities for specialists in cartography.

2. Invite a civil engineer to visit your class to explain about the field of civil engineering and the kinds of maps used by workers in this field.

3. Designate one person in your class to contact officials in the office of the U.S. Geological Survey by calling 1-800-USA-MAPS. Ask for information about topographic maps produced by the USGS.

4. Collect different types of maps and develop a bulletin board display by grouping them according to the various categories described in the chapter.

Questions

Write your answers on a separate sheet of paper.

1. If a map covers a large area, the scale is _____.
2. If a map covers a small area, the scale is _____.
3. Define *plat*.
4. Define *contour line*.
5. Name some reasons for making and using a contour map.
6. What is a topographic map?
7. What is a geological map?
8. Which field of engineering is most generally associated with maps and mapmaking?
9. Maps that show the relationship between the land's physical features and the purpose for which the land is to be used are _____ maps.
10. The vertical distance between contour lines is called _____.
11. Name the three groups of rocks that make up the crust of the earth.
12. A pictorial drawing (generally isometric) of a block of the earth showing profiles and contours is a _____.
13. Another name for slope is _____.
14. Earth to be supplied and put in place to change the contour of a piece of land is called _____.
15. Earth to be removed to change the contour of a piece of land is called _____.
16. A layer of rock, earth, etc., that is clearly different from the matter next to it is called _____.
17. A bend in a layer or layers of rock brought about by forces acting upon the rock after it was formed is a _____.
18. At what scale are maps created on a CAD system?
19. What is the benefit of using a CAD system to work with maps for which the data changes frequently?

CHAPTER 21

PROBLEMS

The small CAD symbol next to each problem indicates that a problem is appropriate for assignment as a computer-aided drafting problem and suggests a level of difficulty (one, two, or three) for the problem.

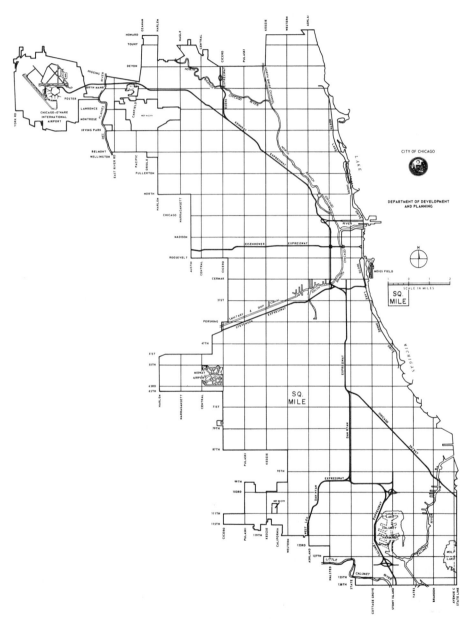

Fig. 21-24 *CAD3 Plot the map of Chicago on a C-size sheet. Scale: ³/₄″ = 1 mile, or other suitable scale. Calculate the number of square miles, acres, or square kilometers that make up this city. (1 square mile = 2,589,988 square meters)*

Fig. 21-25 *(A) Make a drawing of a plat survey as shown in Fig. 21-3. Use a working space of 11″ × 15″. Scale: 1″ = 100′. Start the lowest point 8.25″ from the left border and 1.50″ up from the bottom border. (B) Lay out a residential tract of land and divide it into lots. Plan most lots with 75′ frontage and 150′ or more in depth. Rose Avenue is a main street.*

Fig. 21-26 *Prepare a drawing of a contour as shown in Fig. 21-6. Use a working space of 8.50″ × 11.00″. Draw 2 times the size shown in the figure. Use dividers or a grid to enlarge the figure (vertical profile scale: 1″ = 20′; horizontal scale: 1″ = 500′). Draw the profile on line CC.*

Fig. 21-27 *Make a city map, using the data provided in Fig. 21-4, showing streets, sidewalks, and lots with dimensions.*

Fig. 21-28 *Make an operations map as shown in Fig. 21-5. Draw a grid sheet over the map. Short marks along the border are to assist in drawing the grid. Redraw on an 8.00″ × 15.00″ working space. How long is the 10.00″ gas line? How long is the 24.00″ oil line?*

Fig. 21-29 *Make a contour map as shown in Fig. 21-7. Use the grid on the border to enlarge the contours. Scale: grid = 1″ squares. Working space: 8.00″ × 11.00″.*

Fig. 21-30 *Prepare a contour map of Fig. 21-15 and a vertical profile through XX. The grids are 1.00″ apart. Scale: 1″ = 100′.*

PROBLEMS

SCHOOL-TO-WORK

Solving Real-World Problems

Fig. 21-31 *SCHOOL-TO-WORK problems are designed to challenge a student's ability to apply skills learned within this text. Be creative and have fun!*

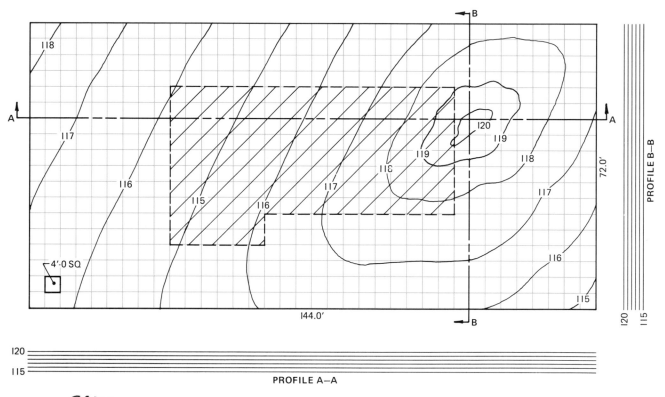

PROFILE A—A

Fig. 21-31 *CAD2 Contours and profiles. A client has purchased the building lot shown in Fig. 21-31 on which she plans to construct a house (see crosshatched area). Contours need to be changed slightly to provide for a more ideal building lot. As a drafter, your job is to do the following:*

1. *On a B-size sheet, use a grid (.25 in. squares) to draw the plot plan as shown.*
2. *Draw profiles as indicated by cutting-plane lines AA and BB.*
3. *Using colored lines or dashed lines, revise the contour to result in a gentle slope in all directions away from the outline of the house.*
4. *Draw revised profiles of AA and BB using colored lines or dashed lines.*
5. *Add landscaping symbols for shrubs, trees, etc., if assigned.*

22

GRAPHS, CHARTS, AND DIAGRAMS

OBJECTIVES

Upon completion of Chapter 22, you will be able to:

○ Describe the use and importance of graphic charts and diagrams.
○ Prepare various kinds of charts from established data.
○ Choose the appropriate charting format to communicate most clearly the data being presented.
○ Use color and creative symbology to enhance the presentation of charts.

VOCABULARY

In this chapter, you will learn the meanings of the following terms:

- bar chart
- comparison chart
- compound bar chart
- conversion chart
- curve
- flowchart
- line graphs
- nomograms
- omission chart
- pictograph
- pie chart
- progressive chart
- step chart
- strata graph

Graphic charts and diagrams are an important part of technical drawing. They are important to scientists, engineers, mathematicians, and nearly everyone else in everyday life. A great variety of graphic charts can be made to show many different types of information visually. The general characteristics and uses of a few types are discussed in this chapter.

USES OF CHARTS AND GRAPHS

Scientists use charts, graphs, and diagrams to record and study the results of research. Engineers use them to record information about materials and conditions. Mathematicians use them to record facts and numerical information. Doctors use charts to record body temperature, heart action, and other body functions. Other people use and read charts and diagrams to learn about the weather, the stock market, finances, and many other things. Because the information in Fig. 22-1A is in chart form, you can readily see that driver reaction distance and automobile breaking distance increase as speed increases. Figure 22-1B contains the same information, but it takes more time to read, study, and understand it. You can see and understand the relationship of speed and distance more easily in Fig. 22-1A because of its graphic presentation.

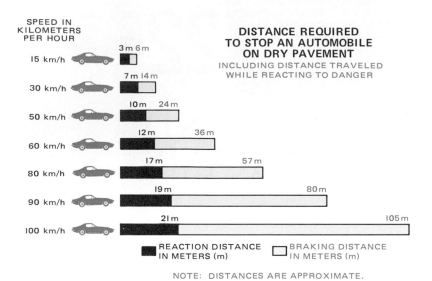

Fig. 22-1A *Chart showing stopping distances at different speeds for automobiles.*

KILOMETERS PER HOUR	DISTANCE IN METERS		
	REACTION DISTANCE	BRAKING DISTANCE	TOTAL DISTANCE
15	3	3	6
30	7	7	14
50	10	14	24
60	12	24	36
80	17	40	57
90	19	61	80
100	21	84	105

STOPPING DISTANCES AT DIFFERENT SPEEDS FOR AUTOMOBILES

Fig. 22-1B *The same information takes longer to read in this form.*

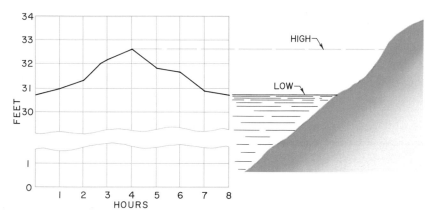

Fig. 22-2 *A graphic chart showing the rise and fall of water in a tidal basin.*

Charts, graphs, and diagrams are pictures of numerical information that show the relationship of one thing to another. For example, they can be used to show trends in areas such as economics. Using economic charts, you can tell at a glance whether the cost of living and wages are rising or falling over a period of time. This type of information can be determined quite easily because of the pictorial nature of a graph or chart.

Charts can also show ratios. Figure 22-1A is an example of a chart that shows the ratio of speed to distance. Other charts show percentages of a whole. Charts can also be used to explain information that is not numerical. For instance, a *flowchart* shows sequential information. That means

that the chart shows which operation comes first, second, third, and so on (see Fig. 22-35).

Graphic charts may also be used to solve various kinds of mathematical problems. You should be able to answer the following questions easily by studying the chart in Fig. 22-2.

- When during each tide cycle does the highest water level occur?
- Approximately what is the level of the water at low tide?
- Approximately what is the level of the water at high tide?
- For approximately how long during each tide cycle is the water higher than 31 feet?

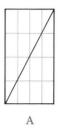
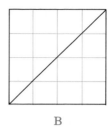
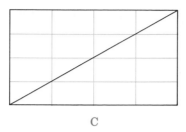

A   B   C

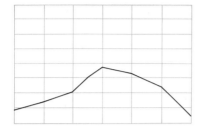

Fig. 22-3 *A false impression may result if vertical and horizontal scales are not properly selected.*

Fig. 22-4 *The vertical scale used here is too small. It gives the effect of very little change in values. The movement is slow.*

You probably had no trouble answering the questions because charts of this type are easy to understand.

SETTING UP A GRAPH OR CHART

Although there is no one "right" way to present a certain set of information in a chart, following some general guidelines can help you present your information in a clear, logical format. This section addresses some factors you need to consider when creating a graph or chart.

Choosing a Type of Chart

Some types of charts are better suited to certain types of information than others. Therefore, it is important to choose a type that presents your information clearly and concisely. For example, suppose you have interviewed the students in your school to find out the percentage of students who use various types of transportation. A pie chart is more suitable than a line graph for displaying this information. As you read about the various types of charts and graphs presented in this chapter, see if you can understand why.

Selecting the Paper

After you select the type of graph or chart, you should select a paper that is suited to the type of graph or chart you will draw. Printed grid or graph paper is available in many forms. As you

may recall from Chapter 2, graph paper may be purchased with lines ruled for drafting use at 4, 5, 8, 10, 16, or 20 to the inch, as well as in many other forms. Metric sizes are also available. Graph paper that has certain lines printed thicker than others is also available. This type of graph paper makes it easy to plot points and to read the finished chart. The lines on grid paper may form squares or rectangles, as shown in Fig. 22-3.

Setting Horizontal and Vertical Scales

The selection of proper vertical and horizontal scales is the next important task. The vertical and horizontal scales must give a true pictorial *impression* by the angle of the slope of the curve. In Fig. 22-3, notice that different impressions are given by the three charts. Chart A presents a very abrupt change. Chart B presents a normal change. Chart C shows a very slow or gradual change. You should choose the scale that gives the most accurate pictorial impression of the information you are trying to portray.

Figures 22-4, 22-5, and 22-6 show another example of the importance of selecting proper scales. The information in Fig. 22-4 would be much easier to read if the vertical scale were increased. In Fig. 22-5, the horizontal scale should be increased. The graph is easiest to read when the scales are drawn in the proportions shown in Fig. 22-6.

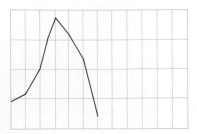

Fig. 22-5 *This curve is plotted from the same data used for Fig. 22-4. The horizontal scale is too small. The movement is fast.*

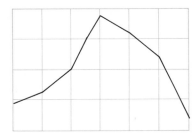

Fig. 22-6 *Scales should be selected with care so that the appearance of the curve plotted from the data will aid the understanding of the information.*

Labeling the Graph or Chart

Several different types of charts and graphs are illustrated and explained in this chapter. Each type of graph is adapted to show a certain type of information clearly. However, without the proper labels, graphs tell little about the intended subject. Every chart or graph needs certain labels to make its purpose clear.

First, you should label every chart or graph with an appropriate title. The title should be fairly short, but descriptive enough that the reader can tell at a glance what kind of information the chart contains. Do not use long, complicated titles–they require more time to read and are often more confusing than enlightening. The title should be well lettered and placed near the chart.

Also, you should provide a key to tell what each element of a chart represents. This can be done in many ways, depending on the type of chart. For example, in the transportation chart in Fig. 22-7 (right), small pictures identify each part of the graph. In Fig. 22-8, a formal key is shown in the top left corner of the graph.

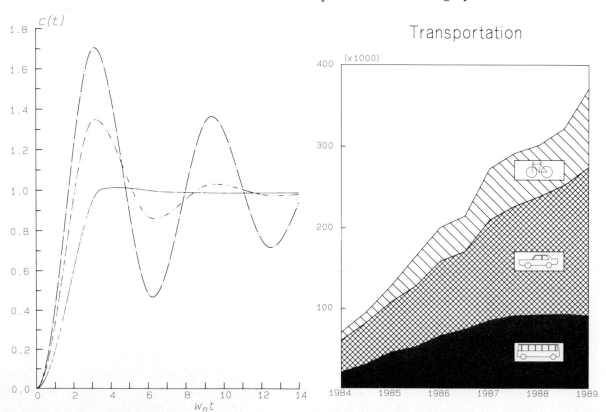

Fig. 22-7 *Computer-generated charts.* (Courtesy of Hewlett-Packard)

Using Color

Black-and-white charts are often used by scientists and mathematicians for recording information. Black-and-white charts can also be found in newspapers. The use of color, however, has become quite common, especially in the preparation of charts, graphs, and diagrams for magazines, books, pamphlets, and various other publications. Color is also used a great deal in making charts for display purposes.

The use of color adds a great deal to the appearance and emphasis of the chart and makes it easier to understand (Fig. 22-9). Color may be added in a variety of ways. Colored pencils, felt-tipped pens, watercolors, or other similar materials are easy to use. Most of these can be found in the drafting room, art room, or at home. Commercially prepared pressure-sensitive materials are available at art and engineering supply stores.

TYPES OF CHARTS AND GRAPHS

Various types of charts, graphs, and diagrams have been developed to present information clearly. This section reviews a few of the most common types.

Line Graphs

Line graphs (also called *line charts*) are most often used to show *trends* or how something changes over time. For example, changes in the weather, ups and downs in scales, or trends in population growth can be plotted and shown graphically on a line graph. A line graph can have one or several curves. *NOTE:* The **curve** on a chart is not necessarily a curved line. As shown in Fig. 22-10, a curve on a chart may be a straight line, a curved line, a broken line, or a stepped line. It may also be a straight line or curved line adjusted to plotted points.

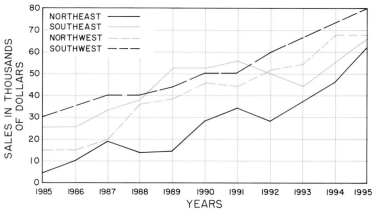

Fig. 22-8 *A multiline chart.*

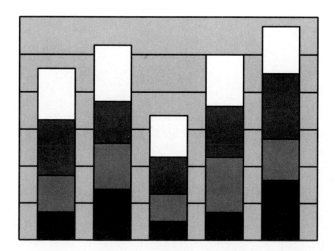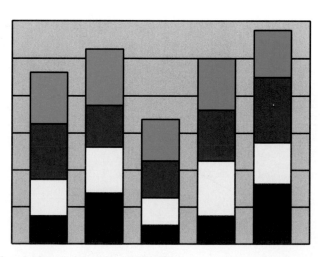

Fig. 22-9 *A multiple-bar chart in black and white and the same one in color.*

Special Types of Line Graphs

Special line graphs are often created for specific purposes. For example, Fig. 22-11 is a **conversion chart** that shows the conversion from Fahrenheit to Celsius degrees. Conversion charts are a form of one-curve line graphs that are convenient for changing from one value to another. Line graphs can also have more than one curve. The line graph in Figure 22-8 contains four curves that compare sales in various parts of the country.

Another type of line graph is a shaded-surface, or **strata graph.** Figure 22-12 shows a typical strata graph. This type of graph can compares two different items, such as use of different kinds of materials, over a period of time.

Figure 22-13 is an example of an **omission chart.** An omission chart is used when the vertical scale needs to be much larger than the horizontal scale to show the information adequately. If Fig. 22-13, for example, there are no values below 35, so a portion of the chart is "broken out."

A **step chart** is a special type of line graph which shows data that remains constant during regular or irregular intervals. For example, a step chart might show time periods during which a price remained constant or was raised or lowered. Figure 22-14 is an example of a step chart.

Drawing a Line Graph

Follow steps 1 through 9 to create a line graph.

1. Prepare and list the information to be presented (Fig. 22-15).

2. Select plain paper or ready-ruled graph paper.

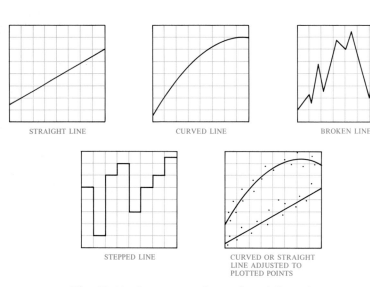

STRAIGHT LINE   CURVED LINE   BROKEN LINE

STEPPED LINE   CURVED OR STRAIGHT
                      LINE ADJUSTED TO
                      PLOTTED POINTS

Fig. 22-10 *Curves on graphs may have different forms.*

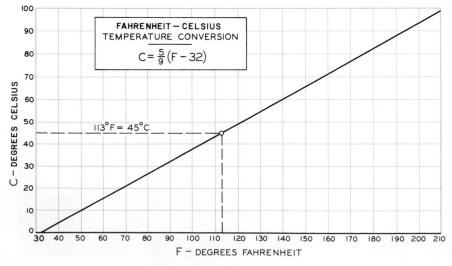

Fig. 22-11 *A conversion chart.*

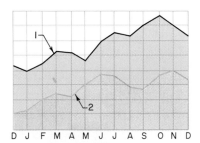

Fig. 22-12 *A shaded-surface, or strata, chart uses shaded areas for contrast.*

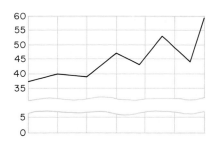

Fig. 22-13 *An omission chart may be used for some purposes, as shown, in order to use a larger vertical scale.*

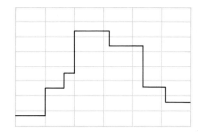

Fig. 22-14 *A step chart shows data that remains constant during regular or irregular intervals.*

SCORING INFORMATION EASTERN HIGH SCHOOL	
GAME NUMBER	POINTS SCORED
1	38
2	20
3	50
4	40
5	40
6	30
7	10
8	40
9	55
10	45

Fig. 22-15 *Information to be presented in a line chart.*

3. Select a suitable size and proportion for your chart so that the overall design will be effective.

4. Select an appropriate scale.

5. If you are not using graph paper, lay off and draw a thin horizontal line and a thin vertical line. These become the X axis and Y axis, respectively (Fig. 22-16A). The intersection of the X and Y axes is zero.

6. Lay off the scale divisions on the X axis and the Y axis (Fig. 22-16B).

7. Mark the scale values on the X and Y axes (Fig. 22-16B).

8. Plot the points accurately by using the information you have listed (Fig. 22-15). Use small circles, triangles, or squares rather than crosses or dots for plotting points (Fig. 22-16C).

9. Connect the points to complete the line graph (Fig. 22-16D).

If you draw more than one curve on a line graph, use different types of lines or different colors for each curve (Fig. 22-17). Use a full, continuous line of the brightest color for the most important curve. In general, the scales and other identifying notes, or captions, are placed below the X axis and to the left of the Y axis. For large charts, you may sometimes want to show the Y-axis scale at both the right and the left sides. You may also want to show the X-axis scale at the top and bottom. This will make the charts more convenient to read.

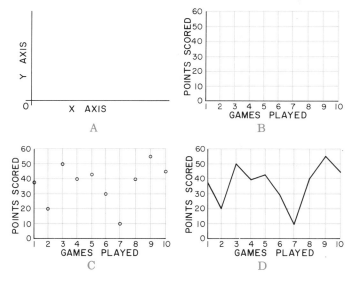

Fig. 22-16 *Steps in drawing a line chart.*

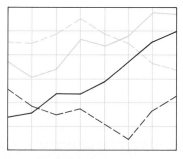

Fig. 22-17 *Use different types of lines and different colors to distinguish curves on a multiline chart.*

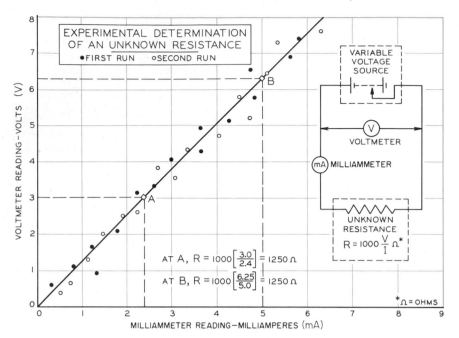

Fig. 22-18 *An engineering test chart.*

Engineering Charts

Experimental information may be plotted from tests and used to obtain an unknown value. For example, the graph in Fig. 22-18 plots the results of tests of an unknown resistance. After correcting the results for variable conditions, the points seem to show a straight-line curve. (Ω is the Greek letter *omega*.) Values can be taken from two points and inserted in the formula. In this way, a value for the unknown resistance can be obtained and checked. Notice that tests were made on two occasions, and the results for both tests were plotted on the chart. A straight-line curve has been drawn along the center of the path made by the dots.

Nomograms are engineering charts that show the solutions to problems containing three or more variables (kinds of information). Figure 22-19 is an example of a nomogram. A straight line from values on the outside scales cross the inside scale. The solution to the equation can be read at this point of intersection. *Nomography* is a special kind of chart construction that requires more than simple mathematics.

Bar Charts

A **bar chart** is probably the most familiar kind of graphic chart. Bar charts are easily read and understood.

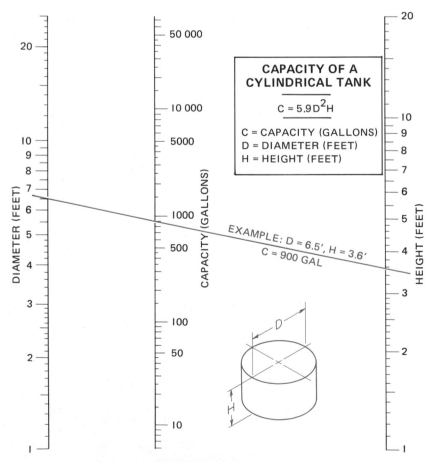

Fig. 22-19 *A nomogram.*

Types of Bar Charts

Like line graphs, bar charts can take many forms. For example, a one-column bar chart consists of a single rectangle representing 100 percent (Fig. 22-20F). The chart pictured in Fig. 22-20F represents the total number of games won, lost, and tied.

Other bar charts use one bar per item. A simple multiple-column bar chart is shown in Fig. 22-21. A multiple-column bar chart with horizontal bars is shown in Fig. 22-22. It gives speed ranges for Caterpillar tractors. Note that the bars do not start at the same line because they show different speed ranges. This kind of chart is called a **progressive chart.**

A **compound bar chart** is shown in Fig. 22-23, where the total length of the bars is made up of two parts. The black portion is the distance traveled at a given speed before application of the brakes. The blue portion is the distance traveled after applying the brakes. The two portions are added together graphically to give the total distance involved. Figure 22-24 shows another example of a compound-bar chart.

A **comparison bar chart** is one in which each item has two or more bars. The bars in Fig. 22-25, for example, might show the amount of a product manufactured (A) and the amount sold (B) by various companies during one year or by one company during several years.

The bar chart shown in Fig. 22-26 has negative as well as positive values. These values are represented by bars set below the X axis. The Y axis must have values less than 0 and be drawn past the X axis. The same information is given in the line graph in Fig. 22-27. The vertical scale is the same as in Fig. 22-26. The dashed line shows the net gain. It can be found by subtracting the negative value from the positive value for each month and plotting the resulting difference.

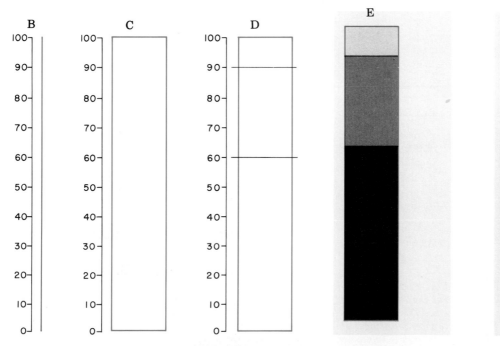

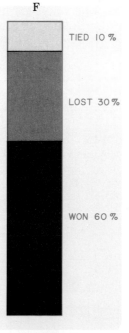

Fig. 22-20 *Steps in drawing a one-column bar chart.*

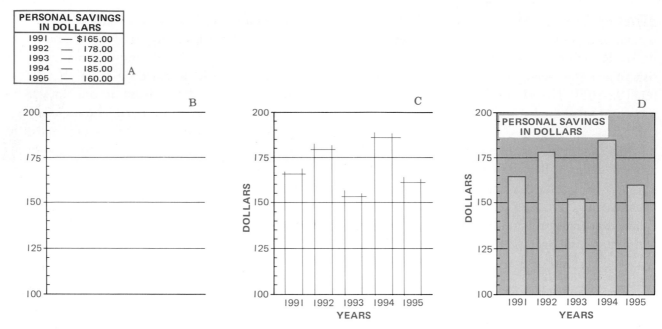

PERSONAL SAVINGS IN DOLLARS	
1991	— $165.00
1992	— 178.00
1993	— 152.00
1994	— 185.00
1995	— 160.00

Fig. 22-21 *Steps in drawing a multiple-column bar graph.*

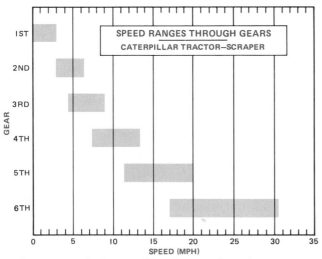

Fig. 22-22 *This horizontal bar chart is a form of progressive chart.* (Courtesy of Caterpillar Tractor Co.)

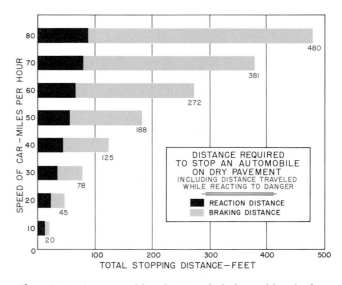

Fig. 22-23 *A compound bar chart in which the total length of each bar is the sum of two or more parts.*

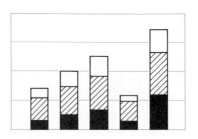

Fig. 22-24 *A multiple-bar chart with divided bars. The bars are divided to show the amount of each of three substances that make up the total.*

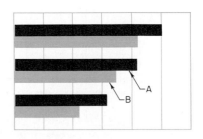

Fig. 22-25 *A comparison bar chart.*

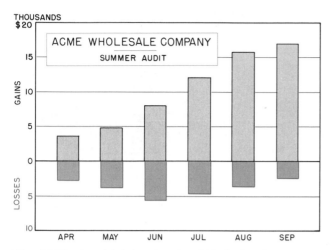

Fig. 22-26 *A multiple-bar chart in which the bars have positive and negative values.*

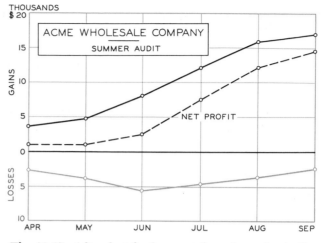

Fig. 22-27 *A line chart for the same information as that in Fig. 22-26.*

Drawing a One-Column Bar Chart

Follow steps 1 through 6 to make a one-column bar chart.

1. Prepare and list the information to be presented (Fig. 22-20A).

2. Lay off the long side equal to 100 units. Figure 22-20B shows how this is done.

3. Lay off a suitable width and complete the rectangle (Fig. 22-20C).

4. Lay off the percentage of the parts and draw lines parallel to the base (Fig. 22-20D).

5. Crosshatch, shade, or color the various parts, as shown in Fig. 22-20E.

6. Letter all necessary information in or near the parts so that it can be read easily (Fig. 22-20F).

Drawing a Multiple-Column Bar Chart

Follow steps 1 through 5 to make a multiple-column bar chart.

1. Prepare and list the information to be presented (Fig. 22-21A).

2. Select a suitable scale and lay off the X and Y axes.

3. Lay off the scale divisions (Fig. 22-21B).

4. Block in the bars using the information gathered in step 1. Allow enough space between the bars for all necessary lettering. Make the bars any convenient width so that the overall appearance is pleasing (Fig. 22-21C).

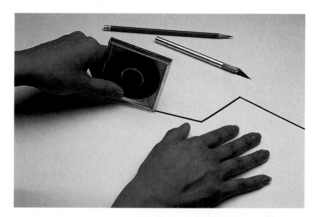

Fig. 22-28 *Applying adhesive tape to a graphic chart.* (Courtesy of Ann Garvin)

5. Complete the bar chart by adding shading or color to the bars, lettering, and any other lines and information. This adds to the appearance of the chart and the ease with which it can be read and understood. A three-dimensional appearance may also be added, as shown in Fig. 23-21D.

Tape Drafting

Adhesive tape provides a quick and simple method of preparing most types of bar charts. Adhesive tape comes in many colors, designs, and widths. It is applied from a roll dispenser (Fig. 22-28) and pressed onto the chart in the desired position. Tapes of different widths provide a quick and simple way of making bar charts.

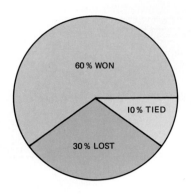

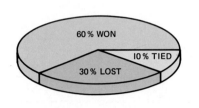

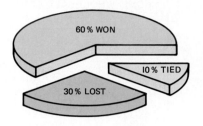

FOOTBALL SEASON RECORD
RELATIONSHIP OF GAMES WON, LOST, AND TIED

Fig. 22-29 *A 100% circular chart, or pie chart.*

Pie Charts

In a **pie chart,** a circle represents 100%. Various sectors of the "pie" represent percentages of the whole (Fig. 22-29). Every 3.6° of the circle = 1%. For example, 30% equals 30 × 3.6°, or 108°. If a circle is to be divided into a 24-hour day, each hour represents 15° on the circle (Fig. 22-30).

Pie charts may be drawn flat or in pictorial, as shown in Fig. 22-29. Follow steps 1 through 4 to create a pie chart.

1. Prepare and list the information to be presented (Fig. 22-31A).
2. Draw a circle of the desired size. Lay off and draw the radial lines representing the amount or percentage for each part on the circumference of the circle (Fig. 22-31B).
3. Crosshatch, shade, or color the various parts, as shown in Fig. 22-31C.
4. Complete the pie chart by adding all necessary information (Fig. 22-31D).

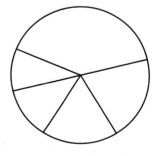

Fig. 22-30 *A pie chart representing 24 hours in a day may be used to show time relationships.*

DISTRIBUTION OF CLASS TREASURY		
ITEM	COST	%
DANCE	$ 72.00	40
PARTY	36.00	20
PICNIC	32.40	18
CLASS PLAY	21.60	12
PHOTOGRAPHS	18.00	10
TOTAL	$ 180.00	100%

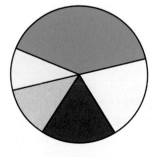

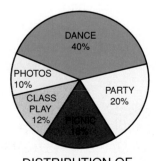

DISTRIBUTION OF
CLASS TREASURY

Fig. 22-31 *Steps in drawing a pie chart.*

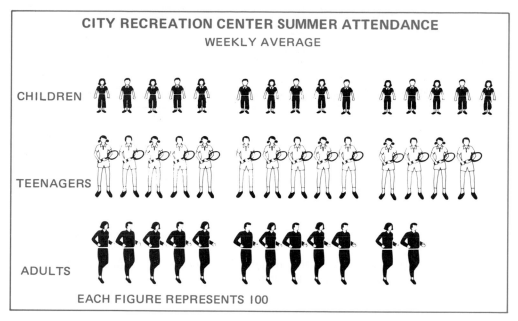

Fig. 22-32 *A pictograph.*

Pictographs

Pictorial graphic charts, or **pictographs,** are similar to bar charts except that pictures or symbols are used instead of bars. Figure 22-32 illustrates a pictograph. The chart pictured is a multiple-bar chart. In this chart, each figure represents 100 people. (A symbol or picture may represent any number that the maker of the chart assigns to it, as long as the chart is labeled to identify the value of each symbol.)

Adhesive symbols may be used on pictographs. They are available in many forms (Fig. 22-33). These symbols make pictorial charts easy to construct.

Organization Charts and Flowcharts

There are many kinds of organization charts. The most common kind, however, is the **flowchart** (Figure 22-34). A flowchart shows the path, or flow, of items through a procedure or other sequence of events. Figure 22-34, for example, traces the routing of drawings from the top engineer to the shop, as well as the organization of the drafting department.

Fig. 22-33 *Many styles of adhesive symbols are available for use on graphic charts.* (Courtesy of Chart-Pak, Inc.)

A flowchart may show the path or series of operations that it takes to manufacture a product or a material. An example of this is the flowchart of steelmaking in Fig. 22-35.

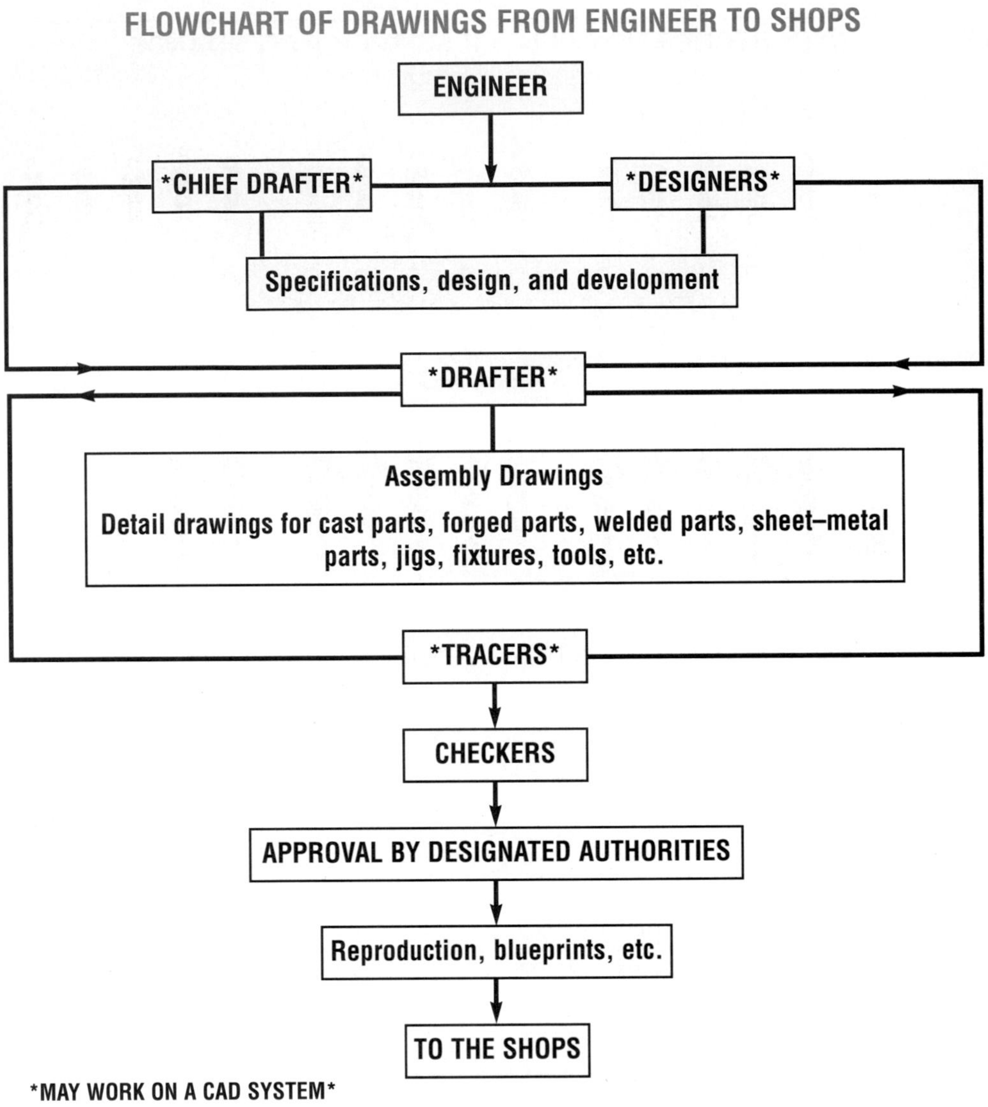

FLOWCHART OF DRAWINGS FROM ENGINEER TO SHOPS
ENGINEER
CHIEF DRAFTER
DESIGNERS
Specifications, design, and development
DRAFTER
Assembly Drawings
Detail drawings for cast parts, forged parts, welded parts, sheet–metal parts, jigs, fixtures, tools, etc.
TRACERS
CHECKERS
APPROVAL BY DESIGNATED AUTHORITIES
Reproduction, blueprints, etc.
TO THE SHOPS
MAY WORK ON A CAD SYSTEM

Fig. 22-34 *An organization chart and flowchart combined.*

FLOWCHART OF STEELMAKING
BENEFICIATION
PELLETS
IRON ORE
SINTER
COAL
COKE OVEN
LIMESTONE QUARRIES
CRUSHING & SCREENING
OXYGEN PLANT
BLAST FURNACE
SLAG
IRON
SCRAP
BESSEMER CONVERTER
BASIC OXYGEN FURNACE
OPEN HEARTH FURNACE
ELECTRIC FURNACE
STEEL
INGOTS
SOAKING PIT
BLOOMING OR SLABBING MILL
BLOOMS
BILLETS
SLABS
STRUCTURAL SHAPES
BARS
COLD DRAWN BARS
WIRE RODS
WIRE
HOT ROLLED SHEET & STRIP
COLD ROLLED SHEET STRIP & BLACK PLATE
TIN PLATE
PLATES
SKELP
PIPE

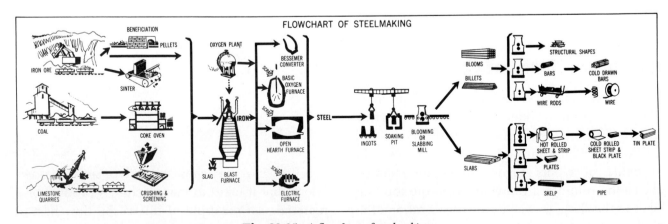

Fig. 22-35 *A flowchart of steelmaking.*

CREATING CHARTS AND DIAGRAMS WITH CAD

Because many desktop publishing packages accept input from CAD programs, CAD is often a good choice for creating charts, graphs, and diagrams to be included in materials published by the company. Charts, diagrams, and other CAD drawings can enhance desktop-produced pamphlets, brochures, and other presentation materials (Fig. 22-36). The desktop packages that accept CAD drawings merge high-quality text and color graphic output.

To be used in a desktop publishing program, CAD files have to be *translated* into a format the desktop software can read. The most common of these is the DXF (data exchange format) standard. However, many software programs can also import files in HPGL (Hewlett-Packard Graphic Language) format.

Fig. 22-36 *Computers can aid in the design and development of high-quality graphic charts and diagrams.* (Courtesy of Arnold & Brown)

CHAPTER 22

REVIEW

Learning Activities

1. Collect various types of graphs, charts, and diagrams from newspapers, magazines, etc. Separate them by categories and develop a bulletin board display. Label the different categories.

2. Gather information on any issue in your school or community. Decide how best to display this information in chart form, and produce a chart using the information gathered.

Questions

Write your answers on a separate sheet of paper.

1. What kind of chart is used to show the solutions to problems that contain three or more kinds of information?

2. Is the curve on a chart always a curved line? Explain.

3. Refer to the chart in Fig. 22-8. What was the dollar volume of sales in the Northeast region in 1987?

4. Refer to the chart in Fig. 22-23. How far does a car travel during reaction time at a speed of 60 mph?

5. In Fig. 22-23, what is the total stopping distance at a speed of 60 mph?

6. Refer to Fig. 22-19. What is the capacity in gallons of a cylindrical tank 10 ft. in diameter and 17 ft. high?

7. In Fig. 22-19, what is the height of a cylindrical tank that holds 100 gal. and is 2 ft. in diameter?

8. In Fig. 22-19, a tank holds 3000 gal. and is 10 ft. high. What is its diameter?

9. What is another name for a pictorial graphic chart?

10. A bar chart in which the total length of the bars consists of two or more parts is called a _____ chart.

11. What kind of chart is made to show solutions to problems containing three or more variables?

12. Refer to Fig. 22-11. Convert 65°C to Fahrenheit.

13. What kind of chart shows sequential information?

14. A false _____ may result if vertical and horizontal scales are not properly selected for a chart.

15. Paper with grid lines is called _____ paper.

16. Line graphs are generally used to show changes or _____.

17. A chart that shows various ranges of data is called a _____ chart.

18. On a pie chart, the full circle represents _____% of the data.

19. On a pie chart, a segment of 90 degrees represents _____% of the data.

20. On a pie chart, a section representing 75% of the data would require a segment of _____ degrees.

CHAPTER 22

PROBLEMS

The small CAD symbol next to each problem indicates that a problem is appropriate for assignment as a computer-aided drafting problem and suggests a level of difficulty (one, two, or three) for the problem.

NOTE: All charts and diagrams in the following problems can be developed on the CAD system. The level of difficulty is **CAD2** .

Fig. 22-37 *Draw a pie chart to show the population distribution in the following four regions of the United States:*

> *Northeast 24.2%*
>
> *North Central 27.8%*
>
> *South 31.2%*
>
> *West 16.8%*

Use color or various types of crosshatching for contrast.

Fig. 22-38 *Draw a pie chart showing that 67.4% of the population of the United States lives within metropolitan areas, while only 32.6% lives outside metropolitan areas. Use different colors or crosshatch areas for contrast.*

Fig. 22-39 *Make a one-column bar chart to show how your allowance was spent last month.*

Fig. 22-40 *Make a multiple-bar chart showing a comparison of how you spent your allowance over a period of 4 months.*

Fig. 22-41 *Assignment 1: The average cost per pound for beef varied over a period of 10 years as follows:*

Year	Cost per Pound
1986	*$2.30*
1987	*2.20*
1988	*2.30*
1989	*2.38*
1990	*2.25*
1991	*2.35*
1992	*2.42*
1993	*2.38*
1994	*2.65*
1995	*3.30*

Make a line graph showing this relationship.

Assignment 2: *The average cost of pork varied somewhat differently. Add a second line to your chart showing a comparison of the cost of beef to the cost of pork over the same period of time. Use colored pencils or different types of lines for contrast.*

Year	Cost per Pound
1986	*$2.30*
1987	*2.40*
1988	*2.60*
1989	*2.55*
1990	*2.48*
1991	*2.50*
1992	*2.70*
1993	*2.80*
1994	*2.90*
1995	*3.10*

Assignment 3: *Add another line to the chart showing the average cost of poultry for the same 10 years.*

Year	Cost per Pound
1986	*$1.44*
1987	*1.52*
1988	*1.60*
1989	*1.40*
1990	*1.46*
1991	*1.70*
1992	*1.80*
1993	*1.66*
1994	*1.76*
1995	*1.90*

PROBLEMS

Fig. 22-42 *Draw a flowchart showing how to mass-produce a project of your choice in the school shop.*

Fig. 22-43 *Make an organizational chart showing the administrative structure of your school.*

Fig. 22-44 Assignment 1: *Draw a pie chart or a one-column bar chart showing a breakdown of the average person's income if 24.5% goes to federal taxes, 5.8% goes to state taxes, and 1.3% goes to local taxes.*

Assignment 2: *Compute the dollar value of each category above for a gross income of $40,000. Mark each dollar value on your chart. Be sure the figures total $40,000.*

Fig. 22-45 Assignment 1: *Make a pictorial chart (pictograph) showing male and female population in your grade in school.*

Assignment 2: *Make a bar chart showing male and female population in your drawing class.*

Fig. 22-46 *Draw a pictorial chart showing the enrollment of technical drawing classes in your school. Your instructor can supply the information.*

Fig. 22-47 Assignment 1: *Make a line graph showing the hourly change in outside temperature for a 12-hour period during any day.*

Assignment 2: *Record similar information for several days and make a multiline chart to show a comparison.*

Fig. 22-48 *Draw a bar chart to show home consumption of electricity for 1 year as follows:*

Month	Kilowatt-Hours Used
January	900
February	885
March	800
April	783
May	722
June	600
July	494
August	478
September	525
October	650
November	735
December	820

Fig. 22-49 *Plot the batting averages of the players on your favorite baseball team.*

Fig. 22-50 *The number of cars per mile of road in the United States is growing. Draw a pictorial chart from the data below to show the growth and anticipated increase.*

Year	Cars per Mile of Road	Cars per Kilometer of Road
1945	9	2
1955	15	5
1965	20	8
1975	26	12
1985	38	17
1995	45	23

Fig. 22-51 *Draw a vertical bar chart showing the following student attendance for a given week of school. The total school enrollment is 925.*

Day	Attendance
Monday	625
Tuesday	715
Wednesday	800
Thursday	775
Friday	695

PROBLEMS

Fig. 22-52 Assignment 1: *Compute your daily calorie intake for 1 week. Make a line chart representing this information.*

Assignment 2: *Using the same information, prepare a horizontal bar chart.*

Fig. 22-53 *Make a multiline chart representing individual game scores of the top five players on the school basketball team for any given season.*

Fig. 22-54 Assignment 1: *From the stock-market listings in the newspaper, select any stock and record its daily status for 10 days. Plot the information on a line chart.*

Assignment 2: *Select several stocks and make a multiline chart showing a comparison of growth and decline.*

Fig. 22-55 *Make a pictorial chart or a pie chart showing a breakdown of the source of each dollar received by the federal government. Use the information given below.*

Individual income tax	*$0.28*
Employment tax	*0.24*
Corporate income tax	*0.16*
Borrowing	*0.15*
Excise tax	*0.12*
Other (miscellaneous)	*0.05*

Fig. 22-56 *Make a pictograph or a pie chart showing a breakdown of the expenditure of each dollar by the federal government. Use the information given below.*

National defense	*$0.31*
Income security	*0.27*
Interest	*0.08*
Health	*0.07*
Commerce, transportation, housing	*0.06*
Veterans	*0.05*
Education	*0.04*
Agriculture	*0.03*
Other (miscellaneous)	*0.09*

Fig. 22-57 *The following information includes five common foods and the number of calories and grams of carbohydrates in a 4-ounce serving of each. Make a bar chart illustrating these facts.*

Food	Calories	Carbohydrates
Chocolate ice cream	*150*	*14*
Peas	*75*	*14*
Pizza	*260*	*29*
Milk	*85*	*6*
Strawberries	*30*	*6*

Fig. 22-58 Assignment 1: *The data in the list below represents a percentage breakdown for a family budget. Make a pie chart illustrating this information.*

Food	*23.1%*
Housing	*24.0%*
Transportation	*8.8%*
Clothing	*10.9%*
Medical care	*5.6%*
Income tax	*12.5%*
Social Security	*3.8%*
Miscellaneous	*11.3%*

Assignment 2: *Compute the dollar value of each category for a gross income of $35,000. Mark each on your chart. Be sure the figures total $35,000.*

PROBLEMS

Fig. 22-59 *Accidents involving children occur in various places. Draw a horizontal bar chart or a pie chart using the places and percentages given below.*

At home	25%
Between home and school	8%
On school grounds	15%
In school buildings	21%
In other places	31%

——— SCHOOL-TO-WORK ———

Solving Real-World Problems

Fig. 22-60 *SCHOOL-TO-WORK problems are designed to challenge a student's ability to apply skills learned within this text. Be creative and have fun!*

Fig. 22-60 *Graphic charts. Here is your chance to be creative. The president of the company for which you work as a drafter is preparing his annual report to stockholders. Your assignment is to develop a bar chart and a line chart showing the company's gross sales, production costs, and net profits for the year based on the following information. Use color to enhance the appearance of your work.*

    Optional: Make overhead projection transparencies of your charts if equipment and materials are available.

MONTH	Gross Sales	Production Costs	Net Profit
JAN	$250,000	$200,000	$50,000
FEB	400,000	300,000	100,000
MAR	350,000	300,000	350,000
APR	500,000	375,000	125,000
MAY	700,000	500,000	200,000
JUN	500,000	400,000	100,000
JUL	650,000	525,000	125,000
AUG	750,000	600,000	150,000
SEP	800,000	650,000	150,000
OCT	900,000	700,000	200,000
NOV	950,000	765,000	185,000
DEC	700,000	600,000	100,000
Totals	**$7,450,000**	**$5,915,000**	**$1,535,000**

ELECTRICAL AND ELECTRONICS DRAFTING

OBJECTIVES

Upon completion of Chapter 23, you will be able to:

○ Describe career opportunities in electrical and electronics drafting.
○ Define basic electrical and electronics terminology.
○ Use standard ANSI symbols in the development of electrical and electronics diagrams.
○ Differentiate between block diagrams and schematic diagrams.

VOCABULARY

In this chapter, you will learn the meanings of the following terms:

- alternating current (AC)
- ammeter
- atom
- block diagram
- conductors
- current
- cycle
- direct current (DC)
- frequency
- insulators
- interconnection diagram
- parallel circuits
- resistance
- schematic diagram
- semiconductors
- series circuits
- single-line diagram
- voltage
- voltmeter
- wiring diagram

Progress in making electrical power started with Thomas Alva Edison's electricity generator in New York City in 1882. Since then, electricity has become one of humanity's practical servants. It has completely changed the communication, manufacturing, and utility industries. Electricity and electronics are a powerful team for space shuttles, computers, communications systems, automated machinery, and everyday appliances.

Imagine miniature circuits (microelectronics) on a chip of silicon that is smaller than a fingernail. The newest communication system in the United States has a digital (number) switching system that uses a single silicon chip to replace 150,000 transistors. This communication control system can make several million decisions in one second using a single chip. Of course, a few years ago, the transistor took the place of the vacuum tube.

Because new discoveries and applications for electronics occur daily, an electrical or electronic drafter on an engineering design team must constantly develop new skills, especially in the CAD area. The drafter must show good judgment, skill, and originality when working from design sketches or written instructions.

Scientists, along with electrical engineers, designers, technicians, and drafters, seem to be building an electrical and electronic world. The industrial robot in Fig. 23-1 is an example of electromechanical progress. The space shuttle is another example. It uses fuel cells for electric power. The shuttle is designed to use the power during ascent, descent, and in-orbit operations with carefully calculated energy requirements (Fig. 23-2).

CAREER OPPORTUNITIES

The electronics industry is one of the largest manufacturing industries in the country. This industry offers special opportunities to young women and men who can make the freehand, formal, and CAD drawings it needs. Electronic drafting uses the same basic rules as all other types of drafting. However, electronic drafting also requires the preparation of schematic diagrams, block diagrams, and technical illustrations.

Industry provides training programs in electrical and electronics drafting. However, it helps if you have learned drafting and electronics in high school before entering such programs. A few courses at technical colleges or junior colleges in electronic drafting and in electromechanical systems (moving things with electric power) will help if you are looking for a technician rating on the design team. To learn more about careers, write to the following groups:

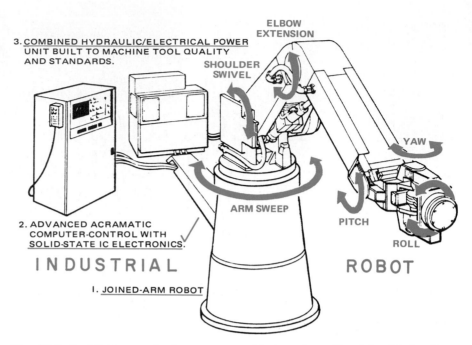

Fig. 23-1 *By definition, a robot is a reprogrammable electronic machine designed to handle materials or tools for a variety of tasks.* (Courtesy of Cincinnati Milacron)

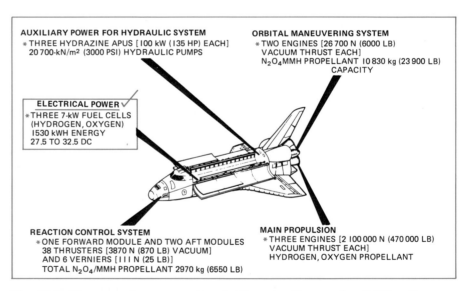

Fig. 23-2 *The power subsystem on a space shuttle can provide as much as 7,000 watts (average) to 12,000 watts (peak) for major energy-consuming payloads.* (Courtesy of NASA)

Institute of Electrical and Electronic Engineers (IEEE)

345 East 47 Street

New York, NY 10017

Electronic Industries Association (EIA)

2001 I Street N.W.

Washington, D.C. 20036

CAD CAREER OPPORTUNITIES

One of the most important uses of CAD is in high-tech electronics. Today, CAD technicians and electrical engineers work together on advanced systems dedicated to design, layout, and testing computer chips and circuit boards. CAD designers can design and describe any kind of electronic circuitry. They use high-speed computer programs capable of manipulating CAD models to form and test (using electronic impulses) the completed circuit.

Opportunities for technicians who understand the fundamentals of electronic circuitry increase as they develop CAD skills. They may join design teams in high-tech offices and serve as accredited technicians.

Fig. 23-3 *A room especially designed for an electronic environment. What would you put into this space?* (Courtesy of Motorola)

ELECTRICAL AND ELECTRONIC DRAFTING

Before you proceed with this chapter, you should understand the difference between electricity and electronics. *Electricity* has to do with the flow of electrons moving through wires or other metal conductors. Common examples are house wiring systems, generators, and transformers. Electricity refers to an energy source.

Electronics has to do with the flow of electrons moving through metal conductors and conductors other than metals. Some of these other conductors are gases, vacuums, and materials called **semiconductors.** The most common semiconductors are transistors and diodes made of germanium or silicon. Electronics refers to devices that make use of electricity.

Students who have had a basic course in electricity will find it easier to understand and make electrical or electronic drawings. This chapter introduces electrical and electronic symbols, wiring diagrams, and circuit diagrams and relates them to electrical and electronic drafting. For students who have not had a basic course in electricity, the paragraphs that follow give a brief explanation of the subject.

ELECTRONIC ENVIRONMENT

People have always tried to control their environment (surroundings). Today, electronic devices control the air we breathe at home and at school. Electronic devices have been made that control the cooling, heating, lighting, and sound systems for different ways of living.

The electronic room shown in Fig. 23-3 is inside a 25-ft. (7620 mm) fiberglass dome. The room has a recessed living area in the center. This arrangement makes it easier to relax. The room also has an electronic sight and sound system. *Mood lighting* (lighting that changes brightness or color) responds to the remote-controlled color television, the television recorder, and the sound system.

The pictorial sketch (Fig. 23-4) shows the major components (parts) of the electronic environment. The center table at (1) is the control center for the components of the room. The rectangle at (2) is a color television set. The rectangle at (3) is a videotape recorder that tapes television shows. The rectangles numbered (4) are speaker towers. They also contain the mood lighting that changes with the sound. The center table rises mechanically to show the controls for the sound system, color television, and VCR. The overhead lighting is *kinetic* (moving). It changes with the beat of the music. You can see three of the four sound towers that hold speakers and special lighting. Each tower has three speakers, one each for high-, low-, and middle-range sound.

The electric circuits for this room must be planned as a wiring diagram. The electrical design engineer makes this diagram. The electronic components for the room must be designed by the electronic design team.

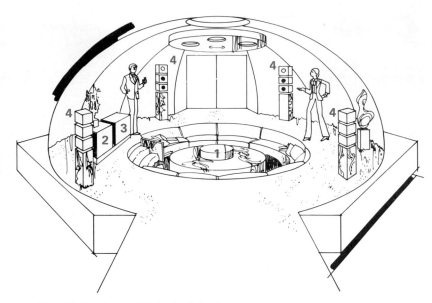

Fig. 23-4   *A pictorial sketch of the electronic equipment.*   (Courtesy of Motorola)

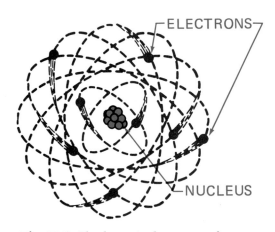

Fig. 23-5   *The electron in the structure of an atom.*

ELECTRICITY

The source of electrical energy is the tiny **atom.** All atoms are made up of many kind of particles. One of these particles is the electron. The electron is the most important particle in the study of electricity and electronics.

The electrons in an atom rotate around the nucleus, or center, of the atom (Fig. 23-5). All electrons in all atoms are the same. Each electron has what is called a *negative charge* of electricity.

Different kinds of atoms have different numbers of electrons and other particles. When atoms have the same number of electrons and the same number of *protons* (a particle in the nucleus), they are the same *element.* Copper, gold, and lead are examples of elements.

When atoms join together, they form *molecules.* When different kinds of atoms join, they form *compounds.* Water, acids, and salts are common compounds.

Voltage and Current

Sometimes electrons can be made to leave their "parent" atoms, the atoms of which they were originally part. This happens, for example, when a piece of wire is connected across the terminals (electrical connections) of a battery. The battery produces an electrical pressure called **voltage.**

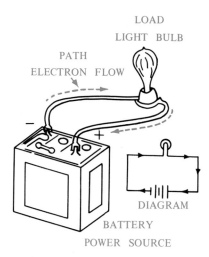

Fig. 23-6 *A simple electric circuit.*

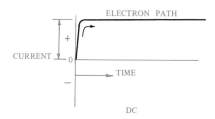

Fig. 23-7 *Direct current attains magnitude and keeps it as long as the circuit is complete.*

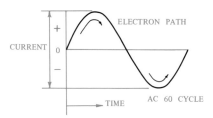

Fig. 23-8 *Alternating current builds up from zero to a maximum in a positive direction, falls to zero, then builds to a maximum in a negative direction and falls back to zero.*

The symbol for voltage is V. The voltage causes a steady stream of electrons (called a **current**) to flow through the wire. When you connect a light bulb (load) to the wire, electrons move through the lamp filament (thin wire in the bulb) from the battery (power source) (Fig. 23-6). The energy of the moving electrons is changed into heat energy as the filament becomes white-hot. The glow of the filament produces the light.

The electron pathway is formed by the battery, the wire, and the lamp filament. This is a simple form of *electric circuit.* In other circuits, electrical energy is changed into other kinds of energy. Some of these kinds of energy are magnetism, sound, and light.

A **direct current (DC)** is a flow of electrons through a circuit in one direction only (Fig. 23-7). An **alternating current (AC)** is a flow of electrons in one direction during a fixed time period, and then in the opposite direction during a similar time period (Fig. 23-8). Current is measured in amperes, and the symbol for current is I.

One complete AC alternation is called a **cycle.** The number of times this cycle is repeated in one second is called the **frequency** of the alternating current, such as 60 cycle. We measure frequency in *hertz (Hz).*

Resistance

Electrons can move through some materials more easily than through other materials. Electric current flows more easily through a copper wire than through a steel wire of the same size. We say that the steel offers more **resistance** than the copper. Materials with small resistance to the flow of electrons are called **conductors.** Silver is the best conductor known. However, it costs too much for general use. Copper and aluminum are also good conductors. They are the most widely used.

Materials through which electrons will not flow easily are called **insulators.** The insulators used most often are glass, porcelain, plastics, and rubber compounds. We measure resistance in *ohms,* and the symbol for resistance is R.

Basic Electrical Units

As we have seen, volts are used to measure pressure, ohms to measure resistance, and amperes to measure current. In addition, we use *watts* (W) to measure power. You can often tell the value of a unit used to show an electrical quantity by looking at the prefix (beginning) of the word. Take, for example, the unit *kilovolt.*

Kilo means "thousand." Thus, 1 kilovolt (kV) equals 1000 V. Similarly, 1 kilowatt (kW) equals 1000 W and 1 kilohm (kΩ) equals 1000 Ω. Another prefix is *milli,* which means "thousandth." Thus 1 milliampere (mA) equals .001 A. For other unit prefixes, see Appendix B.

Basic Formulas

The amounts of voltage, current, and resistance in a circuit are related. This relationship is called *Ohm's Law.* Ohm's Law may be expressed as shown in the 12 formulas in Fig. 23-9. In these formulas, V = volts (pressure), I = amperes (current), and R = ohms (resistance). These letters are used in mathematical formulas to mean "unknown amounts." They are not International System (SI) symbols. The SI unit symbols are V for electrical pressure, A for current, and R for resistance. These SI unit symbols should only be used with known amounts.

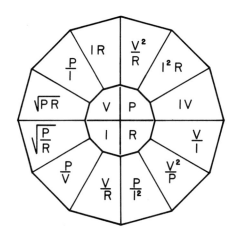

Fig. 23-9 *Power expressed as formulas in Ohm's Law.*

Basic Electric Circuits

The basic types of electric circuits include series circuits, parallel circuits, series and parallel circuits together. These terms are explained in the following paragraphs.

Series Circuits

Series circuits are those in which the current flows from the source (battery, generator, and so forth), through one resistance (lamp, motor, and so forth) after another. This is shown in Figs. 23-10 through 23-12.

In Fig. 23-10, a bell (A) gets its power from a battery (C) when the circuit is closed (when the electron pathway is complete). The circuit is normally open. The pushbutton (B) closes the circuit to make the bell ring.

In Fig. 23-11, a buzzer A gets its power by the current from a transformer C. What is item B? What does it do in this circuit?

In Fig. 23-12, four lamps (C, D, E, and F) get power from a generator (A) when the fused switch B is closed. All the lights must be on. If any one is not, the circuit will be open. This is why some strings of Christmas tree lights go out completely when just one lamp burns out. They are connected in series.

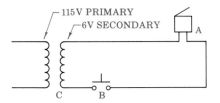

Fig. 23-11 *A series circuit for a buzzer.*

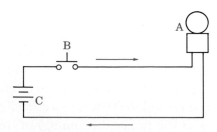

Fig. 23-10 *A series circuit diagram.*

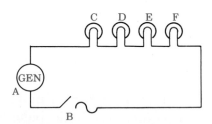

Fig. 23-12 *A series circuit for four lamps.*

Parallel Circuits

Parallel circuits let the current flow through more than one path simultaneously. This is shown in Figs. 23-13 and 23-14. There are three separate branches (paths *C, D,* and *E*) with lamps on Fig. 23-13. Each lamp is separate from the others. If one lamp burns out, the others still work. With a parallel string of lights on a Christmas tree, the remaining lights still burn if some are missing, loose, or burned out.

A siren is shown in Fig. 23-14. It can be turned on by any of the pushbuttons (*A, B, C,* or *D*). These buttons are all connected in parallel. A good use of this would be an alarm system to warn of an attempted holdup at a store. The pushbuttons, connected in parallel, would be under the counters and in the cashier's office.

Notice that the symbol for the siren is the same as for a loudspeaker. So that they will not be confused, the note SIREN has been added.

Combination Circuits

Combining series and parallel circuits permits many different arrangements. In Fig. 23-15, lamps *C* and *D* are in series. Lamps *E* and *F* are in parallel. Both lamps *C* and *D* must be on if switch *A* is closed, since they are in series. When switches *A* and *B* are closed, all the lamps *C, D, E,* and *F* are lighted. Lamps *E* and *F* will work separately. If one fails, the other stays lighted because they are in parallel. However, because lamps *C* and *D* are in series, when one fails, the other does not light.

MEASURING ELECTRICITY

Many kinds of electrical instruments can be used to measure electricity. Two main ones are the ammeter and the voltmeter. The **ammeter** measures electric current in amperes. To measure the amount of current flowing through a resistance (light, motor, and so on), connect the ammeter directly in series with the resistance you want to measure. This is shown in Fig. 23-16.

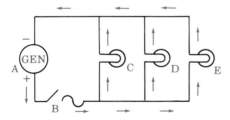

Fig. 23-13 *A parallel circuit for three lamps.*

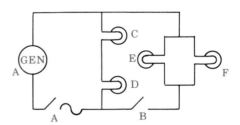

Fig. 23-15 *A combination series and parallel circuit.*

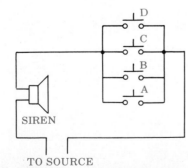

Fig. 23-14 *A parallel circuit diagram with four pushbuttons.*

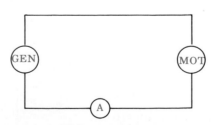

Fig. 23-16 *Ammeter connection in series in a circuit.*

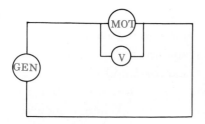

Fig. 23-17 *Voltmeter connection in parallel in a circuit.*

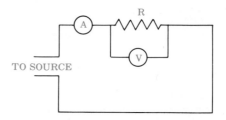

TO SOURCE

Fig. 23-18 *Ammeter and voltmeter connections.*

The **voltmeter** measures the electromotive force (pressure) in volts. Connect the voltmeter in parallel with the part of a circuit where you want to measure the voltage (Fig. 23-17).

Figure 23-18 shows both an ammeter and a voltmeter connected in a circuit. The ammeter measures the current flowing through the resistance *R*. The voltmeter measures voltage flowing across the resistance. The amperes and the voltage are then measured.

Other instruments are also useful for measuring electricity. An *ohmmeter* measures resistance, for example. A *multimeter* is an instrument that combines the functions of an ammeter, a voltmeter, and an ohmmeter (Fig. 23-19).

Fig. 23-19 *A digital multimeter can measure resistance, voltage, or current.* (Courtesy of Keith M. Berry and Radio Shack– a division of Tandy Corp.)

GRAPHIC SYMBOLS

Graphic symbols on electrical and electronic diagrams show the components in a circuit and how they are connected. Symbols can be drawn quickly and easily with templates (Fig. 23-20) or with symbol libraries available for CAD software.

The graphic symbols in Fig. 23-21 and the following sections about using graphic symbols are adapted from the American National Standard Graphic Symbols for Electrical and Electronics Diagrams (ANSI Y32.2) by permission of the Institute of Electrical and Electronic Engineers, Inc. (IEEE).

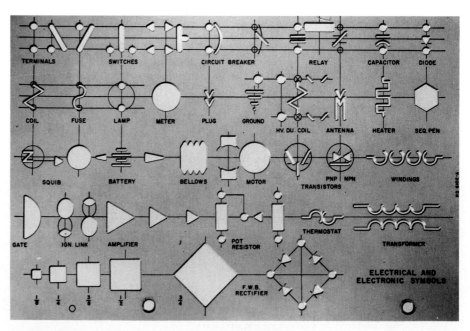

Fig. 23-20 *Template for electrical and electronic circuits.* (Courtesy of RapiDesigns, Inc.)

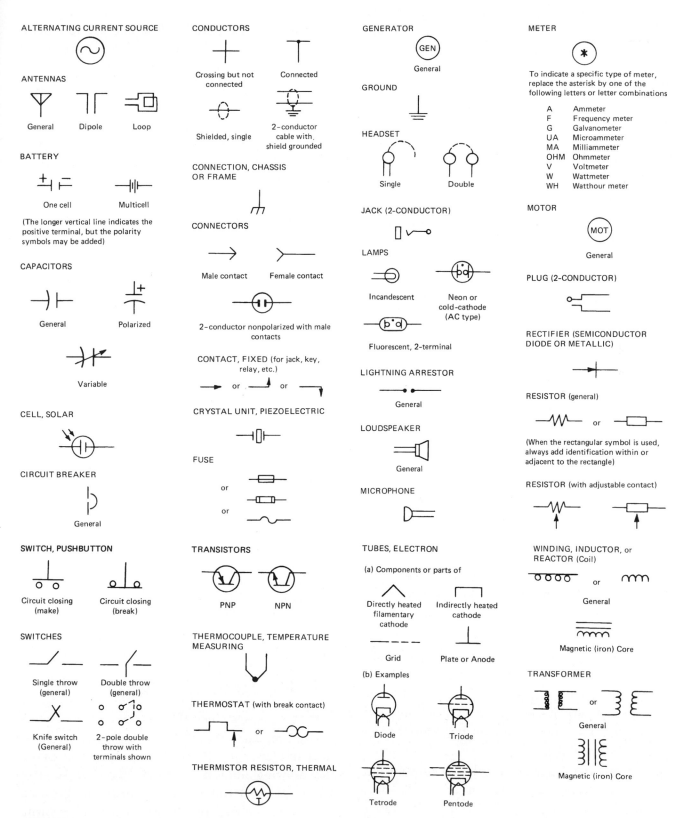

Fig. 23-21 *Standard symbols listed in American National Standard Graphic Symbols for Electrical and Electronic Diagrams, ANSI Y32.2.* (With permission of the Institute of Electrical and Electronic Engineers, Inc.)

Graphic symbols for electrical engineering are a shorthand way to show through drawings how a circuit works or how the parts of a circuit are connected. A graphic symbol shows what a part in the circuit does.

Drafters use graphic symbols on single-line (one-line) diagrams, on schematic diagrams, or on connection or wiring diagrams. You can relate graphic symbols with parts lists, descriptions, or instructions by marking the symbol.

Drawing Electrical and Electronic Symbols

When using electrical or electronic symbols on a drawing, keep the following rules in mind:

1. A symbol is made up of all its various parts.

2. The direction a symbol is facing on a drawing does not change its meaning. This is true even if the symbol is drawn backwards.

3. The width of a line does not affect the meaning of the symbol. Sometimes, however, a wider line can be used to show that something is important.

4. Each symbol shown in the standard is the right size related to all the other symbols. That is, if one of these symbols is drawn twice as big as shown in the standard, any other symbol should also be drawn twice as big.

5. A symbol can be drawn any size needed. Symbols are not drawn to scale. However, their size must fit in with the rest of the drawing.

6. The arrowhead of a symbol can be drawn closed ——▶ or open ——▷ .

7. The standard symbol for a terminal (○) can be added to any one of the graphic symbols used where connecting lines are attached. These are not part of the graphic symbol unless the terminal symbol is part of the symbol shown in the standard.

8. To make a diagram simpler, a symbol for a device may be drawn in parts. If this is done, the relationship of the parts must be shown.

9. Most of the time, an angle of a line connected to a graphic symbol does not matter. Generally, lines are drawn horizontally and vertically.

10. Sometimes it may be desirable to draw paths and equipment that will be added to the circuit later, or that are connected to the circuit but are not part of it. This is done by drawing lines made up of short dashes (- - - -).

11. If details of type, impedance, rating, and so on, are needed, they may be drawn next to a symbol. The abbreviations used should be from the *American National Standard Abbreviations for Use on Drawings* (Y1.1). Letters that are joined together and used as parts of graphic symbols are not abbreviations.

12. Use the ground symbol ⊥ only when the circuit ground is at a potential level equivalent to that of earth potential (Fig. 23-22). Use the symbol ⊅ when you do not get an earth potential from connecting the ground wire to the structure that houses or supports the circuit parts. This is known as a *chassis ground*.

Circuit Components

Figure 23-23 shows and names some of the electrical and electronic components most often used. The symbol for a component should look like that component. You should know what each component does and how it works.

CONVENTIONS FOR ELECTRICAL DRAWINGS

Drafters follow standard conventions when drawing electrical or electronic diagrams. These standards make it easier for people to read and understand the diagrams.

Line Conventions and Lettering

If there is a chance that the size of a drawing may be changed, remember to choose a line thickness and letter size that will let people understand the drawing after it is made larger or smaller. For most electrical diagrams meant to be used for manufacturing, or for use in a smaller form, draw symbols about 1.5 times the size of those shown in American National Standard Y32.2. A guide for drawing lines on electrical diagrams is shown in Fig. 23-24.

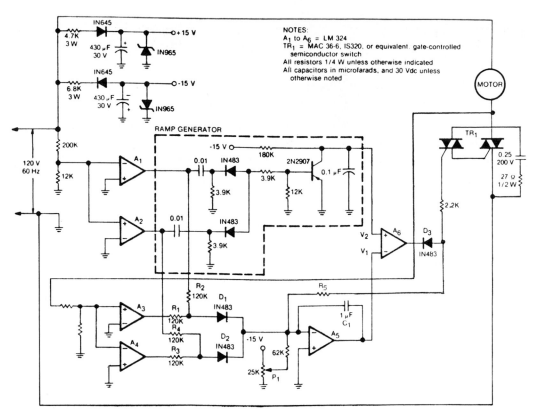

Fig. 23-22 *A diagram of a power-factor controller.*

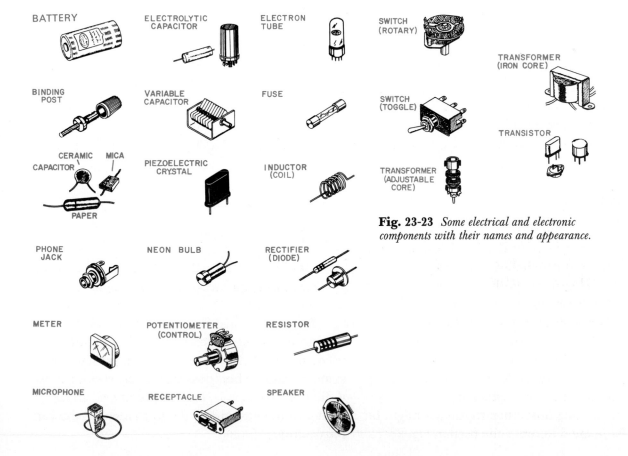

Fig. 23-23 *Some electrical and electronic components with their names and appearance.*

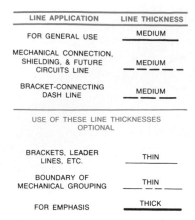

Fig. 23-24 *Line conventions for electrical diagrams.*

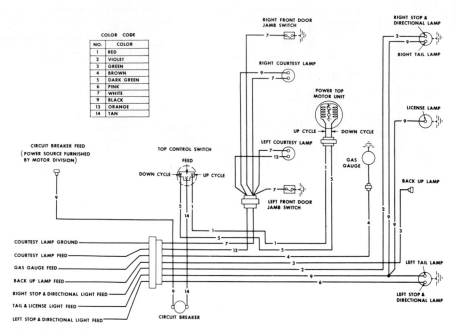

Fig. 23-25 *Wire color scheme used on an automobile wiring circuit.* (Courtesy of Chevrolet Motor Division, General Motors Corp.)

Line thickness and lettering used with electrical diagrams should conform with American National Standard Y14.2 (latest issue) and local needs so that microfilm of the diagrams can be made. Draw lines of medium thickness for general use on electrical diagrams. Use thin lines for brackets, leader lines, and so on. When something special needs to be set off, such as main or transmission paths, use a line thick enough to stand out from the general-purpose lines.

Color Codes

Color codes are an easy way to show information when drawing circuit diagrams. Color codes are also used on the actual wiring of the circuit. In electrical and electronic work, drafters use a color code to show certain characteristics of components, to identify wire leads, and to show where wires are connected. A color code may be included on the diagram (Fig. 23-25).

When using a color code, you should look up the Electronic Industries Association (EIA) standards. Also look up any other codes that might be needed. Note the different code used in Fig. 23-25.

THE EIA COLOR CODE STANDARD		
COLOR	**ABBREVIATION**	**NUMBER**
Black	BLK	0
Brown	BRN	1
Red	RED	2
Orange	ORN	3
Yellow	YEL	4
Green	GRN	5
Blue	BLU	6
Violet	VIO	7
Gray	GRA	8
White	WHT	9

Fig. 23-26 *The EIA color code standard.*

It is not the same code as the EIA standard code used in Fig. 23-26. You should always use EIA standard code when possible; when you must use a different code, it is very important to include a table like the one in Fig. 23-25 to identify the various colors.

Layout of Electrical Diagrams

Lay out electrical diagrams so that the main parts are easily seen. The parts of the diagram should have space between them. This is so that there will be an even balance between blank spaces and lines. Allow enough blank area around symbols so that notes or reference information will not be crowded. Avoid larger spaces, however. Only allow large spaces if circuits will be added there later.

TYPES OF ELECTRICAL DIAGRAMS

There are many kinds of electrical diagrams. Each type of diagram suits its purpose. The definitions that follow are from the *American National Standard Drafting Manual, Electrical Diagrams* (ANSI Y14-15) with the permission of the publisher, the American Society of Mechanical Engineers, 345 East 47 Street, New York, NY, 10017.

Single-Line Diagrams

A **single-line diagram** shows the course of an electric circuit and the parts of the circuit using single lines and graphic symbols (Fig. 23-27). It tells in a basic way how a circuit works, leaving out much of the detailed information usually shown on schematic or connection diagrams. Single-line diagrams make it possible to draw complex circuits in a simple way. For example, in a single-line diagram of a communication or power system, a single line may stand for a multiconductor communication or power circuit. Most of the time, single-line diagrams can be drawn with the same methods used to draw schematic diagrams.

Schematic Diagrams

A **schematic diagram,** or elementary diagram, shows the ways a circuit is connected and what the circuit does. The schematic diagram does not have to show the size or shape of the parts of the circuit. It does not have to show where the parts of the circuit actually are. This section provides some guidelines for creating schematic drawings.

Layout

Use a layout that follows the circuit, signal, or transmission path from input to output, from power source to load, or in the order that the equipment works. Do not use long, interconnecting lines. In general, lay out schematic diagrams so that they can be read from left to right (input on left and output on right), as shown in Fig. 23-28. Complex diagrams should generally be laid out to read from upper left to lower right. They may be laid out in two or more layers. Each layer should be read from left to right. Where possible, draw endpoints for outside connections at the outer edges of the circuit layout. Functional groups are often outlined with dashed lines to make the schematic easier to read (Fig. 23-29).

Connecting Lines

It is better to draw connecting lines horizontally or vertically. Use as few bends and crossovers as possible. Do not connect four or more lines at one point if they can just as easily be drawn another way.

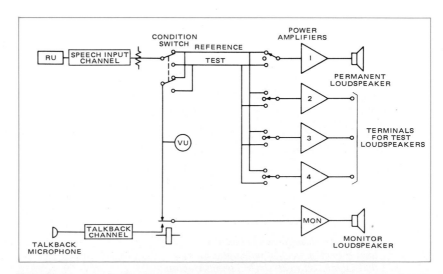

Fig. 23-27 *A single-line diagram that illustrates electronics and communication circuits.*

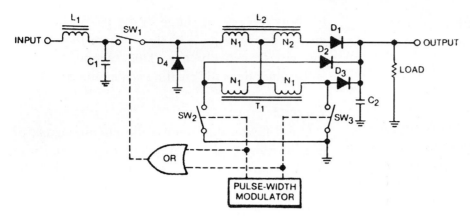

Fig. 23-28 *A voltage regulator or "Buck/Booster."* (Courtesy of NASA)

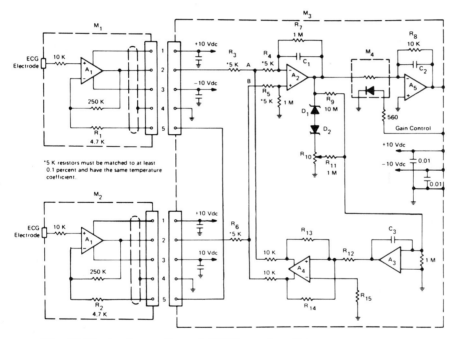

Fig. 23-29 *An electrocardiograph (EKG) signal conditioner.* (Courtesy of NASA)

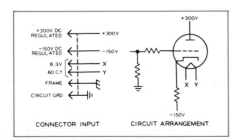

Fig. 23-30 *Identification of interrupted lines. At left, a group of interrupted lines on the diagram. At right, single lines interrupted on the diagram.*

When you draw connecting lines parallel to each other (side by side), the spacing between lines should be no less than .06 in. (2 mm) at the final drawing size. Group parallel lines according to what they do. It is best to draw them in groups of three. Allow double spacing between groups of lines.

Interrupted Single Lines

When a single line is interrupted, show where the line is going in the same place you identify it. This is shown in Fig. 23-30 for the power and filament circuit paths. The following section on interrupted group lines tells how to identify grouped and bracketed lines. Do the same for single interrupted lines.

Interrupted Group Lines

When interrupted lines are grouped and bracketed, identify the lines as shown in Fig. 23-31. You can show at the brackets where lines are meant to go or where they are meant to be connected. Do this by using notes outside the brackets, as shown in Fig. 23-31, or by using a dashed line between brackets, as shown in Fig. 23-32. When using a dashed line to connect brackets, draw it so that it will not be mistaken for part of one of the bracketed lines. Begin the dashed line in one bracket and end it in no more than two brackets.

The following tells in detail about schematic diagrams used with electronic and communication equipment. Use this material with the general standards of schematic diagrams already discussed.

Block Diagrams

A **block diagram** is usually made up of squares or rectangles, or "blocks," joined by single lines (Fig. 23-33). The blocks show how the components or stages are related when the circuit is working. Note the arrowheads at the terminal ends of the lines. These arrowheads show which way the signal path travels from input to output, reading the diagram from left to right. The identification of the stage is lettered within the block or just outside it. The blocks are often drawn along with symbols and a schematic diagram (see Figs. 23-41 and 23-42). The size of a block diagram is generally determined by the amount of information lettered on the components.

Engineers often draw or sketch block diagrams as a first step in designing new equipment. Because blocks are easy to sketch, the engineer can try many different layouts before deciding which to use. The overlay method of sketching discussed in Chapter 2 can be used with great success in drawing or sketching block diagrams.

Block diagrams are also used in catalogs, descriptive folders, and advertisements for electrical equipment. In technical service literature, block diagrams aid in the repair of equipment.

Connection (Wiring) Diagrams

A *connection diagram,* or **wiring diagram,** shows how the components of a circuit are connected. It may cover connections inside or outside the components. The connection diagram usually shows how a component looks and where it is placed.

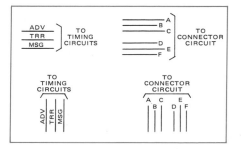

Fig. 23-31 *Typical arrangement of line identifications and circuit destinations.* (Courtesy of NASA)

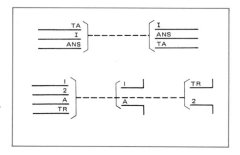

Fig. 23-32 *Typical interrupted lines interconnected by dashed lines. The dashed line shows the interrupted paths that are to be connected. Individual line identifications show matching connections.*

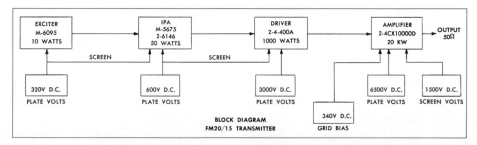

Fig. 23-33 *Block diagram of a 20,000-W broadcast transmitter.*

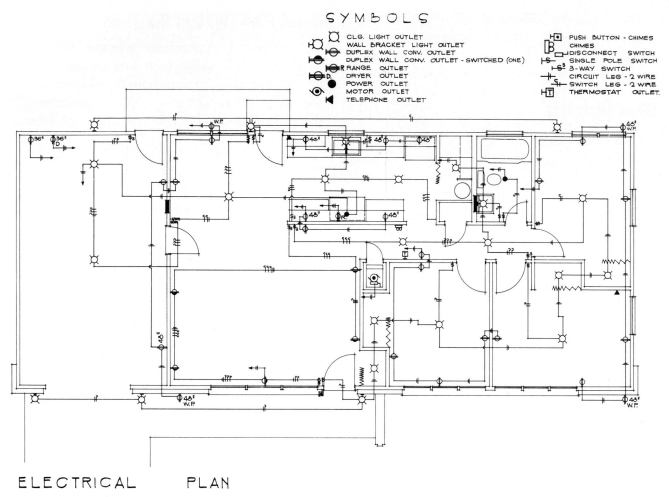

ELECTRICAL   PLAN

Fig. 23-34 *Electrical plan for a ranch house. (See Chapter 20.)*

An **interconnection diagram** is a type of connection or wiring diagram that shows only connections outside a component. An interconnection diagram shows connections between components. The connections inside the component are usually left out.

Figure 23-35 is an interconnection diagram that shows the electrical connections between the different components of an electrical or electronics system. Generally, the connections inside each component are not shown. The name of each component is given. Each component is shown on the diagram by a rectangle.

Printed Circuit Drawings

Printed circuit drawings are used in making printed circuit boards for use in electronic equipment. Each drawing is an exact layout of the pattern of the circuit needed. The drawing is made actual size or larger. If drawn larger, it can be made smaller by photography. The lines (conductors) on the pattern should be at least .03 in. (1 mm) wide. They should be spaced at least .03 in. (1 mm) apart.

The circuit layout pattern is transferred to a copper-clad insulating base. This is done by

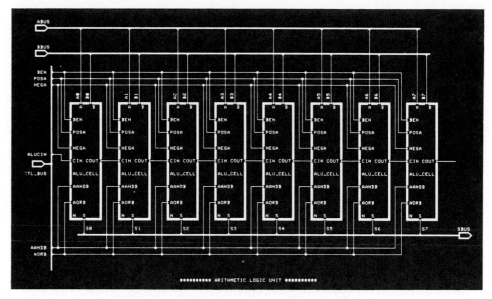

Fig. 23-35 *This schematic is an example of the detail contained with a block. Schematic pages consist of logic gates, interconnect and circuit parameters.* (Courtesy of Daisy Systems)

photography or in some other way. Etching is one way to remove the copper from all areas of the insulating base except for the circuits needed.

The components may be shown on the printed circuit board using symbols or other markings. This information is transferred to the printed circuit diagram from a component identification overlay (Fig. 23-36).

Electrical Layouts for Buildings

Figure 23-35 shows the usual way in which an architect locates electric outlets and switches in a building. This plan only shows where the lights, base plugs, and switches are to be placed. A list of the symbols used in architectural electrical drawings is shown in Fig. 23-37.

Note that you cannot build a good electrical system using this diagram. A complete and detailed set of electrical drawings is needed to create the actual system. These drawings must be made by someone who knows the engineering needs of the system.

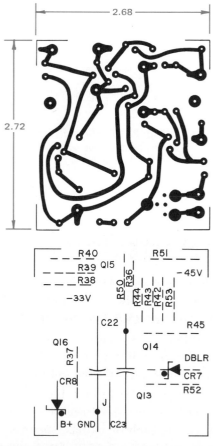

Fig. 23-36 *A printed circuit layout with a component identification overlay.*

Fig. 23-37 *Electrical wiring symbols for architectural design and floor plan layout. (See ranch house, Chapter 20.)*

THE ROLE OF CAD SYSTEMS

CAD systems are very useful for electrical and electronic drafting. As in other areas of drafting, symbol libraries are available for electrical and electronic symbols. In addition, CAD software can help the drafter organize and simplify electrical and electronic diagrams. For example, the drawing in Fig. 23-38 appears to be a high-level block diagram. However, using various capabilities of the CAD software, each block has been associated with detailed circuit information. In this way, the drafter can show just the high-level (overall) information, or as much of the detail as necessary. In addition, the interblock connections can be maintained automatically by using logic software such as LOGICIAN.

Engineers working on circuit design need high-performance software and hardware to support their test designs and circuitry. Figure 23-39 shows an example of a dedicated CADD system architecture for use in engineering design.

CAD software can also be interfaced with software that verifies a circuit's integrity. This software tests the circuit's ability to function using highly accurate simulated impulses timed with precision waveforms (Fig. 23-40).

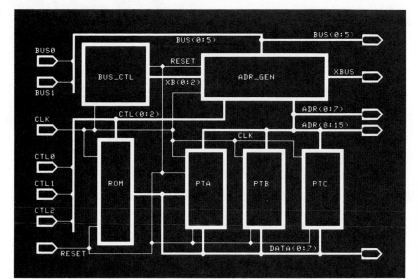

Features GRAPHICS EDITORS

- **Comprehensive set of software tools for the production of block and schematic diagrams**
- **Automatic electrical network creation**
- **Full range of circuit design elements**
- **Powerful, robust command language**
- **Multiple use of standard, predefined components**
- **Automatic naming of all design elements**
- **Complete library of graphic primitives to build complex shapes**
- **Immediate response to user commands**

This schematic represents a high-level block diagram that defines circuit functions; each block contains detailed circuit information. All interblock connections are maintained automatically by the LOGICIAN.

Fig. 23-38 *This CAD-generated schematic represents a high-level block diagram and defines circuit information. All interlock connections are maintained automatically.* (Courtesy of Daisy Systems)

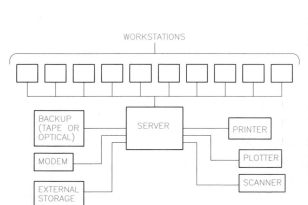

Fig. 23-39 *CADD system architecture for high-tech performance in circuit design.* (Courtesy of Daisy Systems)

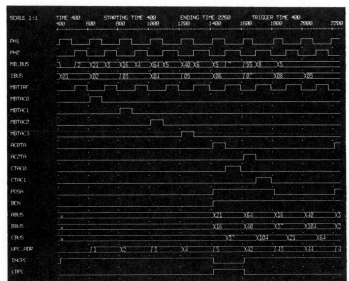

Fig. 23-40 *The waveform executed by an 8-bit CPU with more than 1500 simulation elements for 1700 100-nanosecond clocks, which corresponds to 170,000 simulated time units. Run time is 12 seconds.*

CHAPTER 23

REVIEW

Learning Activities

1. Obtain various kinds of electrical and electronic diagrams (drawings). Attempt to trace circuits and interpret symbology. Develop a bulletin board display of selected drawings.

2. Invite an electrical/electronics engineer to visit your class to describe career options in the fields of electricity and electronics. Be prepared to ask questions about your specific interests related to the field.

3. Visit an industry that designs and manufactures electrical or electronic products. Observe the process from design to the finished product.

4. Select one student in the class to write to the Institute of Electrical and Electronics Engineers (IEEE), 345 East 47 Street, New York, NY 10017. Ask for information about careers.

5. Select one student in the class to write to Electronics Industries Association (EIA), 2001 I Street N.W., Washington, D.C. 20036. Ask for information about careers.

Questions

Write your answers on a separate sheet of paper.

1. How many transistors have typically been replaced with a single silicon chip?

2. An electromechanical system involves _____.

3. The source of electrical energy is the _____.

4. Do electrons have a negative or positive charge of electricity?

5. When atoms join together, they form _____.

6. Terminals are electrical _____.

7. A battery, wire, and lamp filament in combination form a simple electric _____.

8. The flow of electrons through a circuit in one direction only is _____ current.

9. The flow of electrons in one direction for a fixed time period and then in the opposite direction for a similar time period is _____ current.

10. Frequency is measured in _____.

11. The ease with which electrons can move through materials describes _____.

12. Materials with low resistance are called _____.

13. Materials through which electrons will not flow easily are called _____.

14. Resistance is measured in _____.

15. In Ohm's Law, the letter _____ stands for volts; the letter I stands for _____; and the letter R stands for _____.

16. The two primary kinds of electrical circuits are series and _____.

CHAPTER 23

PROBLEMS

The small CAD symbol next to each problem indicates that a problem is appropriate for assignment as a computer-aided drafting problem and suggests a level of difficulty (one, two, or three) for the problem.

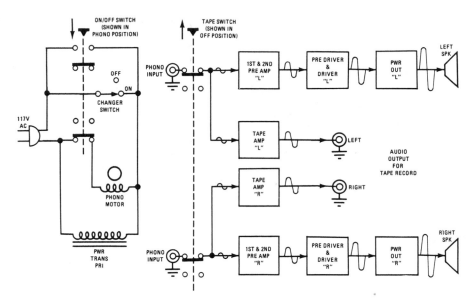

Fig. 23-41 *Prepare a complete single-flow diagram for an AM-FM stereo unit as shown. Estimate sizes and draw twice the size. Use a template if one is available or use a CAD system and appropriate symbol libraries.*

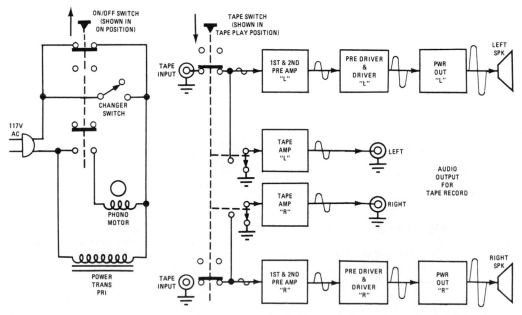

Fig. 23-42 *Draw the tape player signal-flow diagram. Note that when the tape button is depressed, all other functions (AM, FM, and phono) are disabled. Also, the changer switch cannot turn on the phono motor. Since there is no outlet on the left and right audio-output jacks, no direct tape recording can take place. The tape preamp is grounded.*

PROBLEMS

Fig. 23-43 *Prepare a CAD drawing of the voltage regulator, "Buck/Booster," in Fig. 23-28.*

Fig. 23-44 *Prepare a CAD drawing of the EKG signal conditioner in Fig. 23-29.*

Fig. 23-45 *Prepare a sketch of the floor plan in Fig. 23-34 on grid paper. Sketch in electrical symbols.*

SCHOOL-TO-WORK

Solving Real-World Problems

Fig. 23-46 *SCHOOL-TO-WORK problems are designed to challenge a student's ability to apply skills learned within this text. Be creative and have fun!*

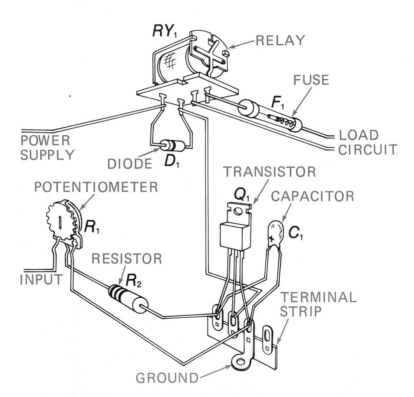

Fig. 23-46 **CAD2** *Pictorial to schematic diagram. Fig. 23-46 is a pictorial drawing showing various components in an electronic circuit. The design engineer has determined that this combination of components will serve a particular function in the electronic portion of a proposed product. However, for production purposes, a schematic diagram is required. Your assignment as drafter is to convert the pictorial diagram to a schematic diagram. Refer to Fig. 23-21 for schematic symbols.*

24

AEROSPACE DRAFTING

OBJECTIVES

Upon completion of Chapter 24, you will be able to:

○ Identify drafting and design careers within the aerospace industry.
○ Identify the various types of drawings that are used in aerospace design and drafting.
○ Describe the basic parts of an airplane.

VOCABULARY

In this chapter, you will learn the meanings of the following terms:

- aeronautics
- aerospace industry
- airfoils
- alloys
- bulkheads
- composites
- computer-integrated manufacturing (CIM)
- fixed-wing aircraft
- fuselage
- hydraulics
- lift
- spars

Aeronautics is the science of designing, building, and operating all types of flying vehicles. The **aerospace industry** includes the design and manufacture of commercial airplanes, leisure aircraft, corporate jets, helicopters, and various spacecraft, as well as many similar products. The aerospace industry also makes spacecraft that orbit the Earth or fly millions of miles to the planets Jupiter and Saturn (Fig. 24-1). In addition, the aerospace industry includes the equipment needed to navigate and communicate aboard an aircraft. Examples of these items include computers, radios, global positioning systems (GPS), and speed control equipment.

Many different kinds of drawings are needed for aircraft manufacture. Drawings range from casting and forging drawings to sheet-metal layouts, electrical schematics, and lofting layouts.

CAREER OPPORTUNITIES

New ways of making airplanes, missiles, and spacecraft are always being found. This means that challenging opportunities exist for young people interested in aerospace design and drafting. The basic aerospace team is made up of hundreds of scientists, engineers, designers, drafters, and technicians, as well as thousands of skilled workers (Fig. 24-2).

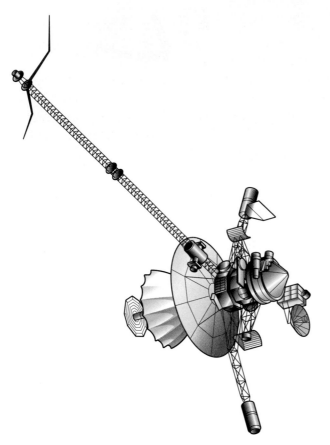

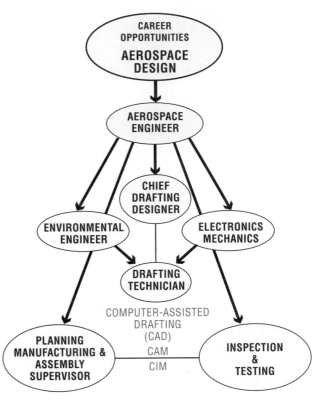

Fig. 24-2 *The aerospace industry offers many drafting opportunities.* (Courtesy of Circle Design)

Fig. 24-1 *Galileo is a space probe sent by the United States to explore parts of the solar system that are too far away for people to visit. After a six-year journey, Galileo reached Jupiter on December 7, 1995. Its two-year mission is to orbit Jupiter and send scientific data about the planet back to Earth.* (Courtesy of NASA/Jeff Stoecker)

Drafters who work in the aerospace industry usually begin as technicians or engineering technologists. As a member of a design team, drafters may make drawings of mechanical or electrical systems. These are drawn from the designer's sketch.

Many specialists are needed in the aerospace industry, from the person at the drawing board or CAD station to the person testing models in the wind tunnel. As a drafter gains experience in a certain area, he or she may move up to a job with more responsibility. For example, with some schooling and on-the-job training, a drafter may become a designer. Designers have more responsibilities than entry-level drafters.

If you are interested in aerospace drafting, you should study science and mathematics in high school and junior college. You should also take courses in technical drawing or engineering graphics. To learn more about careers in the aerospace industry, write to: American Institute of Aeronautics and Astronautics, 1290 Avenue of the Americas, New York, NY 10019.

AEROSPACE DRAFTING AND CADD

Computers now help enormously not only in the design phase of aircraft production, but also in the actual manufacturing. **Computer-integrated manufacturing (CIM)** is commonly used to manufacture aircraft parts and accessories directly from CAD drawings.

Fig. 24-3 *The Boeing 777 was the first commercial jet designed entirely by computers using CAD and solid modeling techniques. Design of the 777 took 238 teams nearly five years using CAD technology. The finished aircraft is 209'-1" (63.7 meters) long and has a wingspan of 199'-11" (60.9 meters).* (Courtesy of Boeing/Ron Shea)

Many of the larger aircraft companies also use specialized CAD software for applications such as testing and analysis. One example is Boeing's 777 Division, which produces the 777-200 commercial aircraft (Fig. 24-3). This division uses CATIA (computer-aided three-dimensional interactive application) and ELFINI (Finite Element Analysis System), both developed by Dassault Systèmes of France and licensed in the United States through IBM. Boeing's designers also use EPIC (Electronic Preassembly Integration) on CATIA and other digital preassembly applications developed by Boeing. Using these applications, Boeing produced its first 777 airplane just .023 inch (about the thickness of a playing card) away from perfect alignment. For comparison, most airplanes line up to within .50 inch (Fig. 24-4).

777 General Arrangement

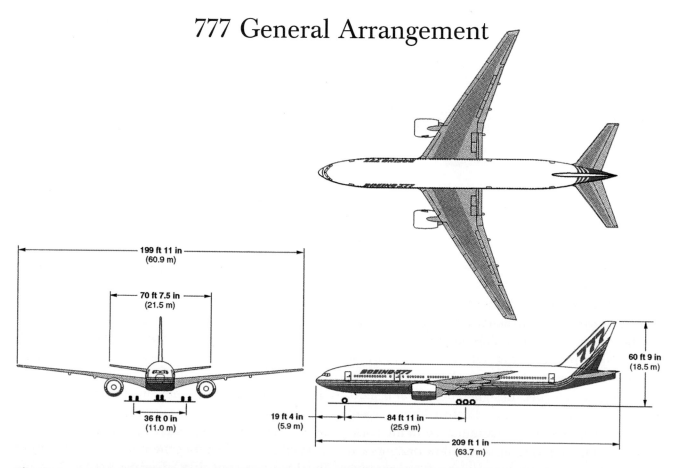

Fig. 24-4 *A three-view drawing of the Boeing 777 commercial jet. Computer technology enabled Boeing to produce the first 777 to within .023" of perfect alignment—more than 20 times better than the usual error of .50".* (Courtesy of Boeing/Ron Shea)

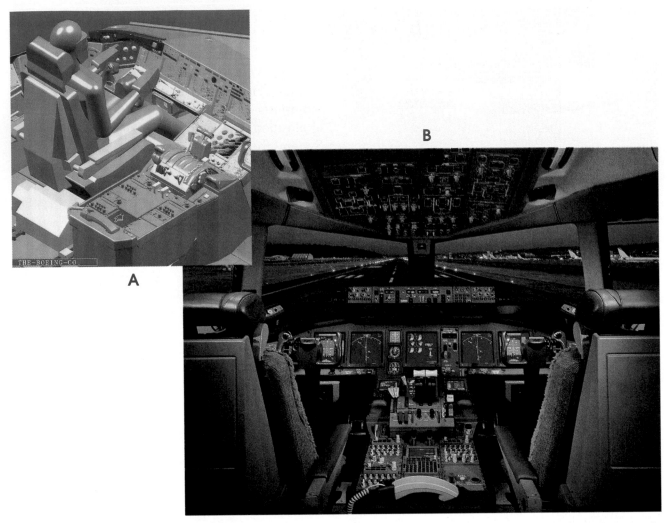

Fig. 24-5(A) *This computer image shows how parts of the 777 were preassembled as solid models on a computer. As the work progressed, the solid model took on the form of the finished cockpit.* **(B)** *The final cockpit after design changes.* (Courtesy of Boeing/Ron Shea)

The Boeing 777 was the first commercial jet to be designed entirely using CAD and three-dimensional solids technology. Throughout the design process, the airplane was "preassembled" on the computer, eliminating the need for a costly, full-scale mockup (Fig. 24-5). Preassembly of the three-dimensional solid models generated on the computer allowed designers to position parts correctly, assuring a proper fit. In addition, computer models of humans were placed in the cockpit and in maintenance areas to allow designers to check the ergonomic placement of instruments (Fig. 24-6).

AIRCRAFT COMPONENTS

Most aircraft contain an enormous number of parts. For example, the Boeing 777-200 has 132,500 unique parts, not counting fasteners. Each of these parts has to be designed and manufactured separately. Therefore, in addition to drafting skills, drafters and designers who work in the aerospace industry need a fundamental understanding of the various components of aircraft.

This section describes the major components of fixed-wing aircraft. **Fixed-wing aircraft** are all of those that have unmoving wings as a part of

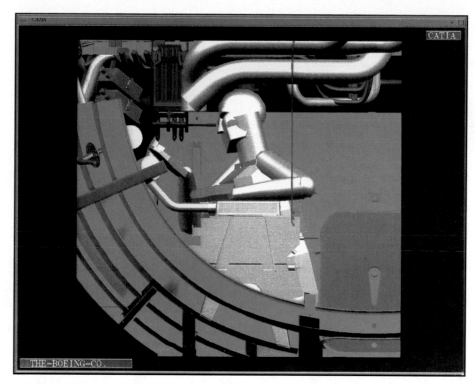

Fig. 24-6 *Boeing engineers used CATIA to insert human models into electronically pre-assembled sections of the airplane to ensure that airline mechanics would have proper access and functional working areas. Here, a CATIA human model is shown working on part of the flight controls underneath the airplane's nose section.* (Courtesy of Boeing/Ron Shea)

their basic structure. Commercial jets, business jets, and small planes are examples of fixed-wing aircraft.

Figure 24-7 shows a basic three-view drawing of the McDonnell Douglas F-4C Phantom jet. Figure 24-8 is an exploded view that shows the parts that make up the jet. Refer to these illustrations as you read about the basic parts of an aircraft in the following paragraphs.

Fuselage

The **fuselage** (central body) consists of the pilot, passenger, and cargo compartments. The fuselage structure is made up of a set of shaped **bulkheads** (walls) and rings. The fuselage also has *longitudinal* (lengthwise) members. Together, these form a strong framework. The outer skin of the fuselage for most commercial and private aircraft is made of sheet metal fastened to the framework with rivets or machine screws.

Wings and Airfoils

Aircraft wings are composed of ribs, which define their shape. The ribs are connected to **spars** (beams) that run toward the fuselage and away from it. The wing skins are attached to the ribs and spars with metal fasteners.

The shape of an aircraft's wings is critical to its ability to fly efficiently (Fig. 24-9). As the wing passes through the air at an angle, it pushes the air downward. This causes an equal and opposite upward push called **lift.** An aircraft can fly because of the angle of the wing as it moves through air. Look at the shape and position of the wing as you try to determine how well an aircraft can fly.

Airfoils are movable parts that help control the aircraft. They include ailerons, rudders, stabilizers, flaps, and tabs. (You can find these parts in the parts list in Fig. 24-8.) A *chord line* is a straight line between the leading edge and the trailing edge of an airfoil.

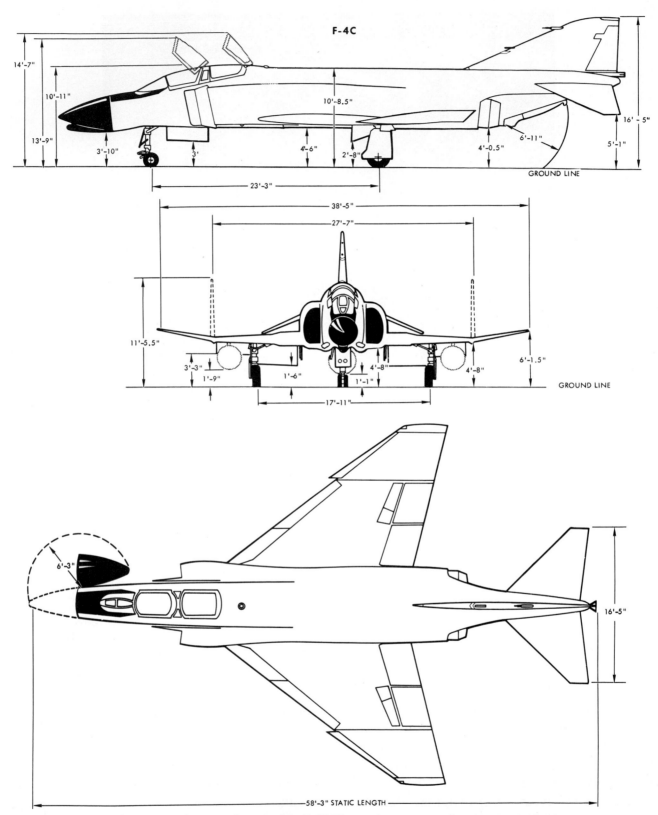

Fig. 24-7 *A three-view drawing of the F-4C Phantom jet.* (Courtesy of McDonnell Douglas Corp.)

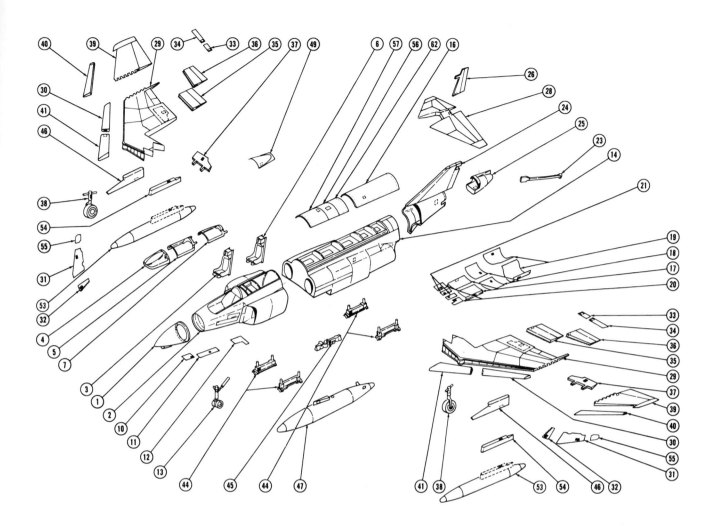

1. Radome
2. Forward fuselage
3. Pilot seat
4. Windshield
5. Forward canopy
6. Radar operation seat
7. Aft canopy
10. Nose landing gear door, forward
11. Nose landing gear door, aft
12. Hydraulic compartment access door
13. Nose landing gear shock strut
14. Center fuselage
16. Fuel tank door
17. Engine access door
18. Engine access door
19. Engine access door
20. Engine access door
21. Auxiliary engine air door
23. Arresting hook
24. Aft fuselage
25. Tail cone
26. Rudder
28. Stabilator
29. Center section wing
30. Leading edge flap
31. Main landing gear strut door
32. Main landing gear inboard door
33. Inboard spoiler
34. Outboard spoiler
35. Flap
36. Aileron
37. Speed brake
38. Main landing gear shock strut
39. Outer wing
40. Leading edge flap, outboard
41. Leading edge flap, inboard
44. Missile rack
45. Bomb rack
46. Missile pylon
47. External centerline fuel tank
49. Data link access door
53. External wing fuel tank
54. External wing fuel tank pylon
55. Landing gear door, outboard
56. Boom IFR receptacle access door
57. Fuel cell access door
62. Fuel cell access door

Fig. 24-8 *An exploded view helps to explain the parts of the McDonnell Douglas F-4C Phantom jet.* (Courtesy of McDonnell Douglas Corp.)

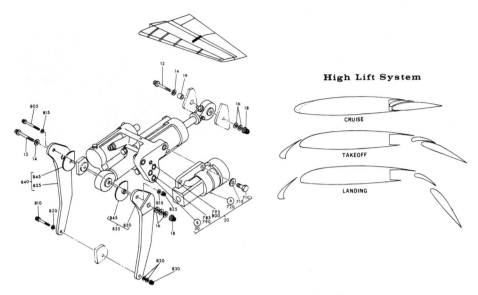

Fig. 24-9 *The wing design is critical to how efficiently an airplane flies.* (Courtesy of McDonnell Douglas Corp.)

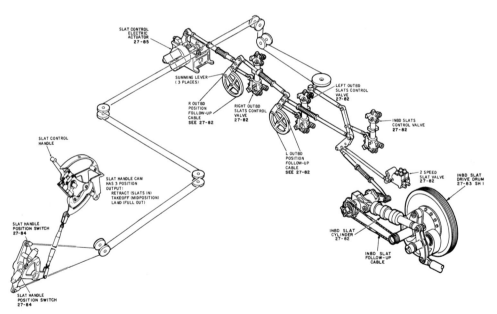

Fig. 24-10 *An assembly drawing of a landing gear that is controlled by a hydraulic system.* (Courtesy of McDonnell Douglas Corp.)

Landing Gear

The landing gear of most aircraft is raised and lowered by **hydraulics** (fluid under pressure) or by electricity. Hydraulic shock absorbers ease the shock of landing (Fig. 24-10).

Engines

The engines are one of the heaviest parts of an aircraft, so they must be positioned carefully to avoid a weight imbalance (Fig 24-11). On planes that have two engines, the engines are often

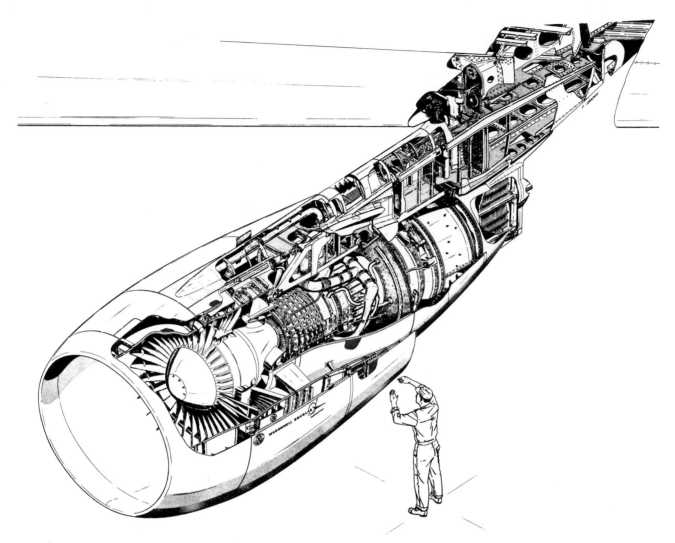

Fig. 24-11 *The engine assembly, or power plant, is one of the heaviest parts of an airplane.* (Courtesy of McDonnell Douglas Corp.)

placed under the wings. The engines on the Boeing 777, for example, are located under the wings. On the McDonnell Douglas DC-10 and the Lockheed L1011, which have three engines, two are located under the wings and the third is located at the back of the fuselage (Fig. 24-12). On aircraft such as the military transport C130, which has four engines, two engines are placed on each wing (Fig. 24-13).

Other Systems

Many other systems are also found on aircraft. These include air conditioning, compartment pressurization (to keep normal air pressure at high altitudes), and hydraulic, electronic, and plumbing systems, as well as specialized navigation systems. These and other systems are generally designed and manufactured by other companies under contract to the large aircraft companies such as Boeing, Lockheed Martin, and McDonnell Douglas.

AIRCRAFT MATERIALS

Today's aircraft have parts made of many different materials. Aluminum, magnesium, and lightweight steel are usually strong enough for the inside parts. However, friction caused by the movement of high-speed aircraft through the atmosphere can cause the exterior surfaces to become very hot.

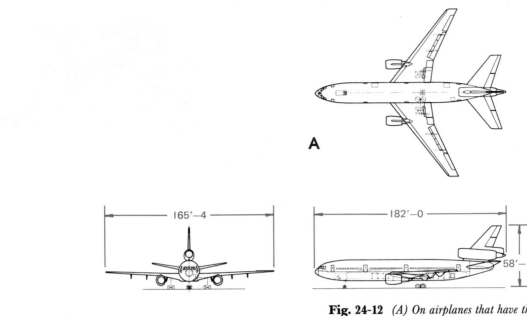

165'–4 182'–0 58'–1

Fig. 24-12 *(A) On airplanes that have three engines, one engine is located under each wing. The third is positioned at the back of the fuselage, as shown in this three-view drawing of a DC-10. (B) The Lockheed L-1011 is another example of a three-engine commercial jet.* (A–Courtesy of McDonnell Douglas Corp.; B–Courtesy of Corel Gallery)

This is especially true of spacecraft, although supersonic aircraft such as the *Concorde* and certain military jets are also affected (Fig. 24-14). Special materials that can withstand high temperatures are used for the exteriors of these aircraft. Many of these materials are **alloys** (mixtures) of more than one metal. Others are **composites** made of carbon fiber, specially developed plastics, and titanium, among other components.

AIRCRAFT DRAFTING PRACTICES

Aircraft manufacturers maintain engineering manuals for their workers. These manuals usually include all the various drawings needed to manufacture and maintain each aircraft the company sells, as well as specifications and directions for maintaining the aircraft properly. It is very important that the drawings in these manuals be precise and error-free.

Fig. 24-13 *The C130 military transport aircraft has two engines on each wing.* (Courtesy of Corel Gallery)

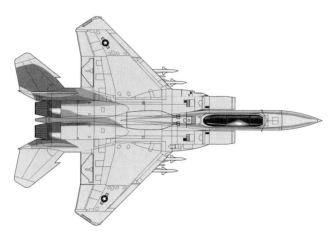

Fig. 24-14 *Military aircraft such as this F15 fighter have a special surface that can withstand the high temperatures generated by friction at speeds above Mach 1 (the speed of sound).* (Courtesy of Corel Gallery)

The Society of Automotive Engineers (SAE) publishes a book of aerospace and automotive drawing standards. Drafters in the aerospace industry must be familiar with these standards in order to produce acceptable drawings.

Aircraft are very complex products that require many types of drawings. Generally, two- or three-view drawings show the overall dimensions of the craft (Fig. 24-7). Exploded pictorials show how the aircraft or individual parts of the aircraft should be assembled. Figures. 24-8 and 24-9 show examples of typical exploded drawings.

Many detail drawings are also needed. Even the smallest subassemblies are critical to the safe operation of an aircraft and must be detailed correctly (Fig. 24-15). Three-dimensional drawings such as that of a landing gear in Fig. 24-10 are used mainly for identification. From this drawing, a mechanic can identify the individual components that make up the landing gear assembly. Cutaway sections such as the drawing of the engine assembly in Fig. 24-11 are also used for general identification purposes.

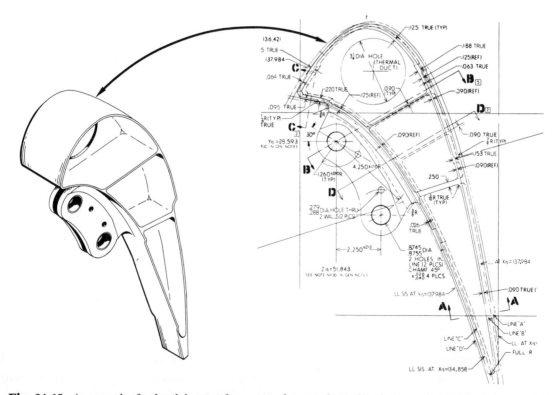

Fig. 24-15 *An example of a detail drawing for an aircraft part to be machined.* (Courtesy of McDonnell Douglas Corp.)

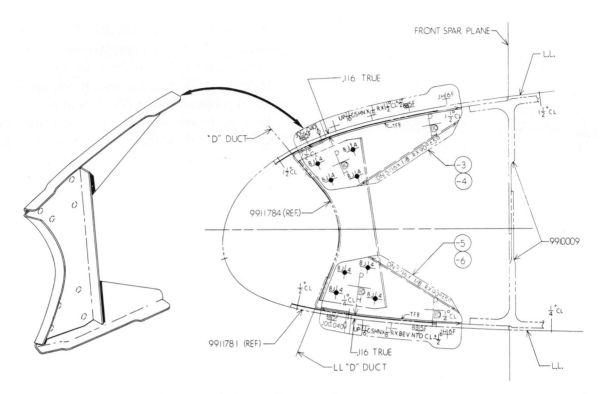

Fig. 24-16 *An example of a flat-pattern development for an aircraft support structure.* (Courtesy of McDonnell Douglas Corp.)

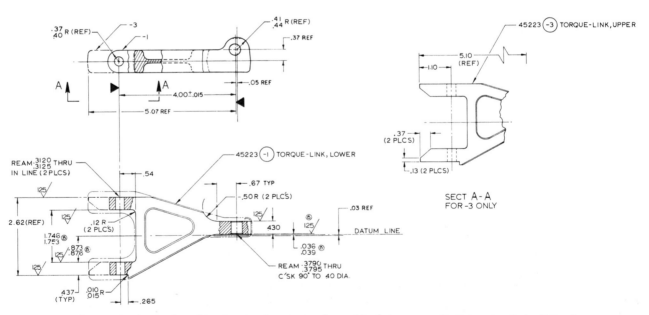

Fig. 24-17 *A typical working drawing for a part to be machined.* (Courtesy of North American Rockwell Corp.)

The types of drawings required for parts and subassemblies depend on the method that will be used to make them. For example, some parts require flat-pattern developments. Figure 24-16 shows a typical flat-pattern development for a support structure. Machining and forging drawings are required for other parts. Figure 24-17 shows a working drawing for a torque link arm to be machined. Figure 24-18 is a forging blank drawing for another torque link arm.

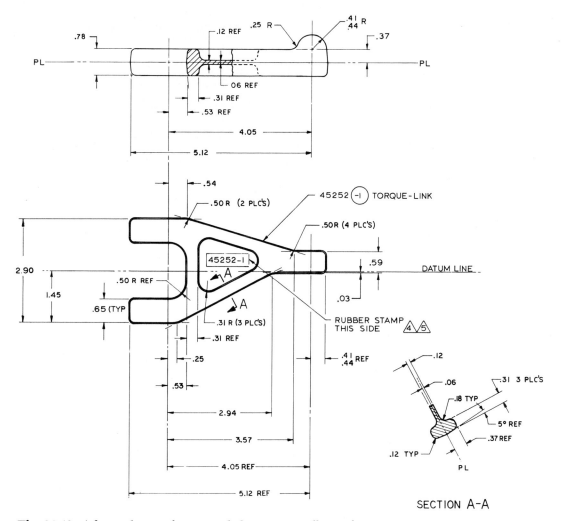

Fig. 24-18 *A forging drawing for a torque link arm on a small aircraft.* (Courtesy of North American Rockwell Corp.)

Drawings for Large Aircraft

Large commercial aircraft need drawings in addition to those used for smaller "light" aircraft. *Inboard profiles* show the relative location and use of various areas on the aircraft. An inboard profile of the DC-10 is shown in Figure 24-19. Seating arrangement and design are also important for commercial jets. Figure 24-20 is a seating diagram that shows possible arrangements for Boeing's 777.

Drawings for Smaller Aircraft

Smaller aircraft include products such as business jets, private (leisure) planes, and helicopters (Fig. 24-21). The drawings for these smaller aircraft are similar to those for large aircraft. Three-view drawings give the overall dimensions and features of the aircraft, as shown in Figs. 24-22 and 24-23. Detail drawings are needed for all the parts and subassemblies. The working drawings in Figs. 24-17 and 24-18 are for torque link arms used on smaller aircraft.

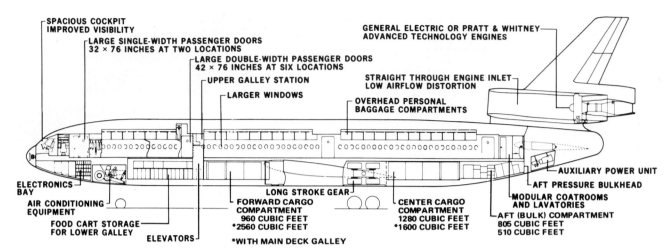

Fig. 24-19 *The inboard profile of a DC-10 reveals the intended use of space within the main body, or fuselage.* (Courtesy of McDonnell Douglas Corp.)

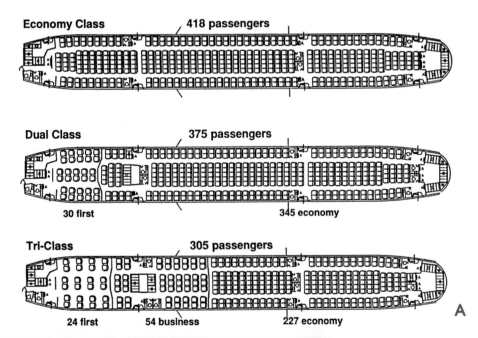

A

B

Fig. 24-20 *(A) Many commercial airlines use flexible seating arrangements to make the aircraft as versatile as possible. This illustration shows three possible seating arrangements for the Boeing 777. (B) The photo shows the actual seating that results from the Tri-Class model.* (Courtesy of Boeing/Ron Shea)

The drawings for small aircraft are just as important as those for larger commercial aircraft. The main difference is that smaller aircraft are less complex and may require fewer drawings. Seating arrangements and inboard profiles are generally not needed for small aircraft.

Fig. 24-21 *The Cessna 172 is an example of a small private plane used mostly for leisure activities.* (Photo courtesy of Cessna and Corel Gallery)

Drawings for Spacecraft

Spacecraft include all vehicles that are meant to fly outside the Earth's atmosphere. Examples include the space shuttle and space probes such as *Galileo, Voyager* and *Voyager II,* as well as missiles and satellites (Fig. 24-24). Complete sets of working drawings, including assembly drawings, sections, and details, are required to manufacture spacecraft. Figure 24-25 shows an exploded drawing of the space shuttle. Based on this illustration and the information in this chapter, what other drawings do you think are needed to manufacture a space shuttle?

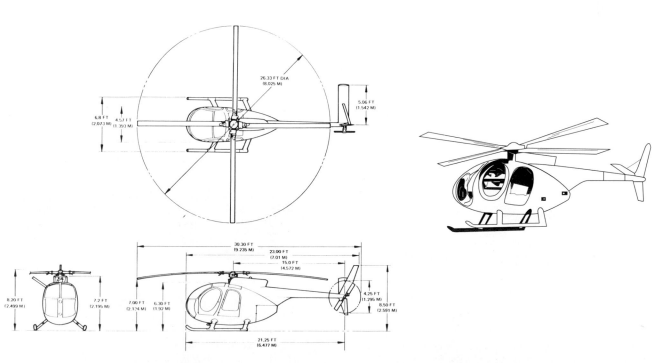

Fig. 24-22 *A typical civilian helicopter.* (Courtesy of Hughes Tool Co. and Corel Gallery)

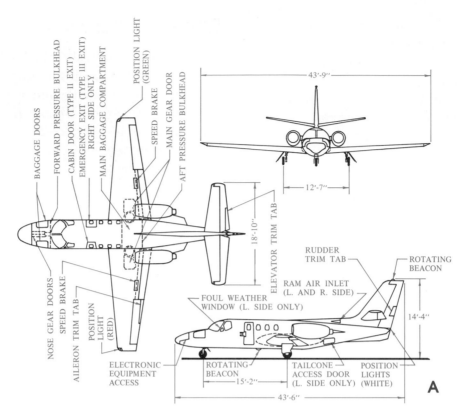

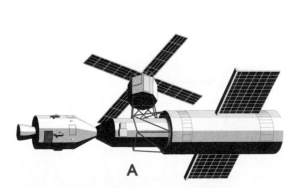

Fig. 24-23 *(A) A three-view drawing of a business jet. (B) Another example of a twin-engine business or corporate jet.* (A–Courtesy of Cessna Aircraft Co.; B–Courtesy of Corel Gallery)

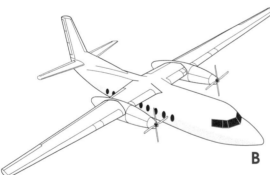

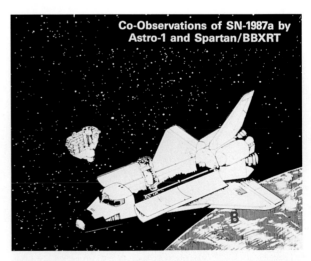

Fig. 24-24 A *Skylab, the first United States space station, orbited the earth from May 1973 to February 1974.* (Courtesy of Corel Gallery)

Fig. 24-24 B *(B) Astro-1 and Spartan spacecraft observe SN-1987A, a recently discovered supernova, using a broadband x-ray telescope.* (Courtesy of NASA)

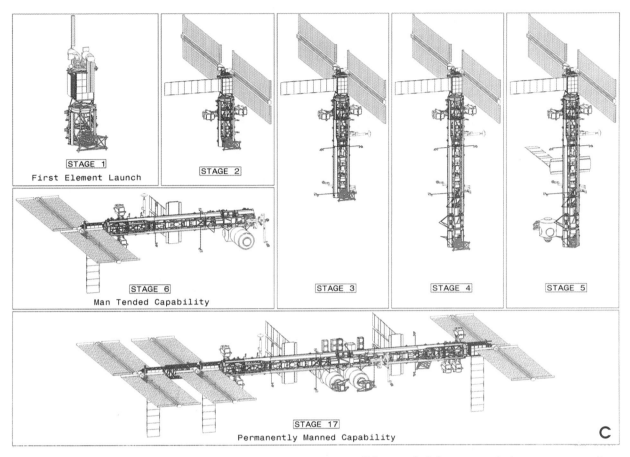

STAGE 1
First Element Launch

STAGE 2

STAGE 6
Man Tended Capability

STAGE 3

STAGE 4

STAGE 5

STAGE 17
Permanently Manned Capability

C

Fig. 24-24 C *By the year 2000 the American space station* Freedom *will have reached the stage at which it can permanently support human life.* (Courtesy of Grumman/NASA)

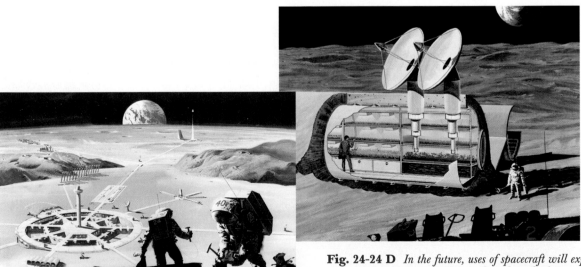

Fig. 24-24 D *In the future, uses of spacecraft will expand to include permanent settlements and working facilities outside the Earth's atmosphere. In these NASA concept drawings, people live and work in a township on the moon.* (Courtesy of Lockheed/NASA)

D

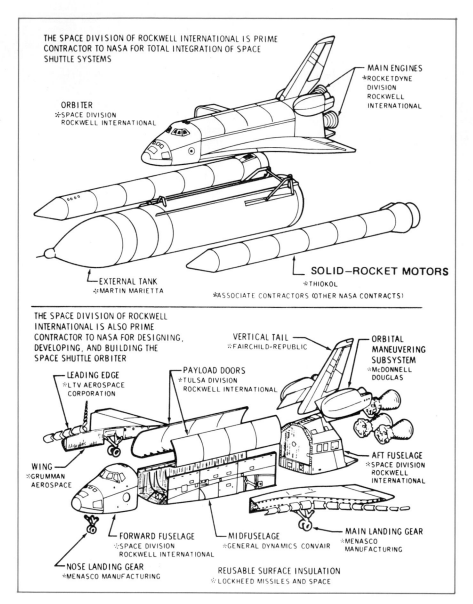

THE SPACE DIVISION OF ROCKWELL INTERNATIONAL IS PRIME CONTRACTOR TO NASA FOR TOTAL INTEGRATION OF SPACE SHUTTLE SYSTEMS

MAIN ENGINES
✻ROCKETDYNE DIVISION ROCKWELL INTERNATIONAL

ORBITER
✧ SPACE DIVISION ROCKWELL INTERNATIONAL

EXTERNAL TANK
✧ MARTIN MARIETTA

SOLID—ROCKET MOTORS
✻THIOKOL

✻ASSOCIATE CONTRACTORS (OTHER NASA CONTRACTS)

THE SPACE DIVISION OF ROCKWELL INTERNATIONAL IS ALSO PRIME CONTRACTOR TO NASA FOR DESIGNING, DEVELOPING, AND BUILDING THE SPACE SHUTTLE ORBITER

VERTICAL TAIL
✧ FAIRCHILD-REPUBLIC

ORBITAL MANEUVERING SUBSYSTEM
✧ McDONNELL DOUGLAS

LEADING EDGE
✧ LTV AEROSPACE CORPORATION

PAYLOAD DOORS
✧ TULSA DIVISION ROCKWELL INTERNATIONAL

WING
✧ GRUMMAN AEROSPACE

AFT FUSELAGE
✧ SPACE DIVISION ROCKWELL INTERNATIONAL

FORWARD FUSELAGE
✧ SPACE DIVISION ROCKWELL INTERNATIONAL

MIDFUSELAGE
✧ GENERAL DYNAMICS CONVAIR

MAIN LANDING GEAR
✧ MENASCO MANUFACTURING

NOSE LANDING GEAR
✧ MENASCO MANUFACTURING

REUSABLE SURFACE INSULATION
✧ LOCKHEED MISSILES AND SPACE

Fig. 24-25 *Component drawings for the space shuttle.* (Courtesy of NASA)

CHAPTER 24

Questions

Write your answers on a separate sheet of paper.

1. Define the word *aeronautics*.
2. List at least four types of aircraft discussed in this chapter.
3. Describe the basic career opportunities in aerospace drafting.
4. What is CIM?
5. List at least one advantage of using CAD and 3D solid models in designing and manufacturing an aircraft.
6. What is a fixed-wing aircraft?
7. What is the fuselage of an aircraft?
8. Explain the importance of the shape of the wing on a fixed-wing aircraft.
9. What are airfoils?
10. Why are some high-speed aircraft made of alloys or composites instead of traditional metals?
11. What types of drawings are needed to manufacture and maintain a commercial jet?

CHAPTER 24
PROBLEMS

The small CAD symbol next to each problem indicates that a problem is appropriate for assignment as a computer-aided drafting problem and suggests a level of difficulty (one, two, or three) for the problem.

SCHOOL-TO-WORK

Solving Real-World Problems

Fig. 24-26 *SCHOOL-TO-WORK problems are designed to challenge a student's ability to apply skills learned within this text. Be creative and have fun!*

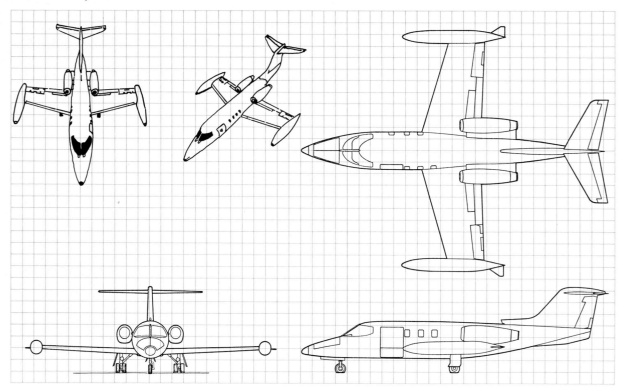

Fig. 24-26 **CAD3** *Using grid paper, sketch the views of the executive jet. Prepare one pictorial view of the Lear jet.*

PROBLEMS

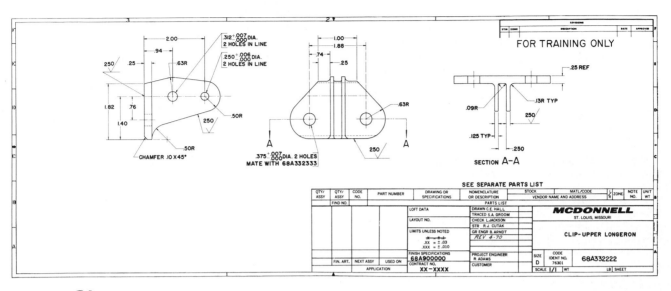

Fig. 24-27 **CAD3** *Prepare the three drawings of the clip. Incorporate the following change order: Change the 2.00 dimension in zone F3 to 2.625.*

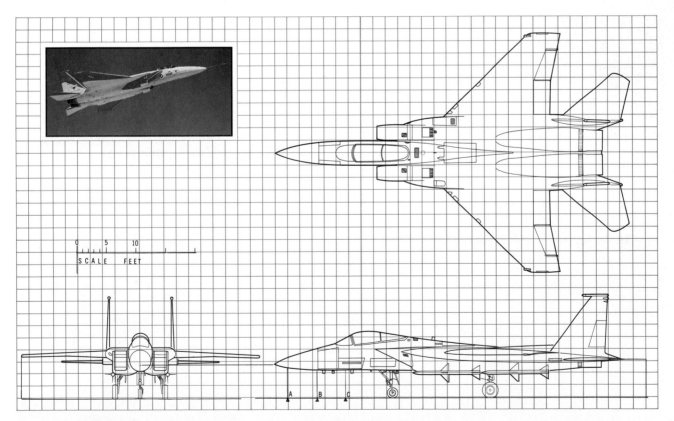

Fig. 24-28 **CAD3** *Draw one view of the F-15 fighter aircraft on grid paper. Using dividers, locate the contours accurately.*

PROBLEMS

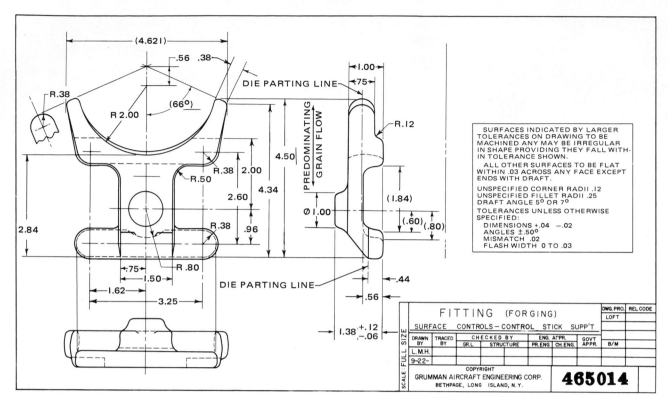

Within the drawing:

(4.621)

.56 .38

DIE PARTING LINE

R.38

R 2.00

(66°)

.75

1.00

PREDOMINATING GRAIN FLOW

R.12

4.50

R.38 2.00

R.50

2.60 4.34

Ø 1.00

(1.84)

(.60) (.80)

2.84

R.38 .96

.75 R .80

1.50

DIE PARTING LINE

.44

1.62

.56

3.25

1.38 +.12 −.06

SURFACES INDICATED BY LARGER TOLERANCES ON DRAWING TO BE MACHINED ANY MAY BE IRREGULAR IN SHAPE PROVIDING THEY FALL WITH-IN TOLERANCE SHOWN.
    ALL OTHER SURFACES TO BE FLAT WITHIN .03 ACROSS ANY FACE EXCEPT ENDS WITH DRAFT.
UNSPECIFIED CORNER RADII .12
UNSPECIFIED FILLET RADII .25
DRAFT ANGLE 5° OR 7°
TOLERANCES UNLESS OTHERWISE SPECIFIED:
    DIMENSIONS +.04 −.02
    ANGLES ±.50°
    MISMATCH .02
    FLASH WIDTH 0 TO .03

FITTING (FORGING)

SURFACE CONTROLS − CONTROL STICK SUPP'T

DWG. PRO. REL CODE
LOFT

DRAWN BY | TRACED BY | CHECKED BY | | ENG. APPR. | | GOVT APPR. | B/M
| | GR.L | STRUCTURE | PR.ENG | CH.ENG | |
L.M.H.
9-22-

COPYRIGHT
GRUMMAN AIRCRAFT ENGINEERING CORP.
BETHPAGE, LONG ISLAND, N.Y.

465014

SCALE: FULL SIZE

Fig. 24-29 *Make a working drawing of the control-stick support.*

Fig. 24-30 *Prepare a working drawing of the torque link in Fig. 24-17.*

Fig. 24-31 *Prepare a working drawing of the forged torque link in Fig. 24-18 and dimension completely.*

PROBLEMS

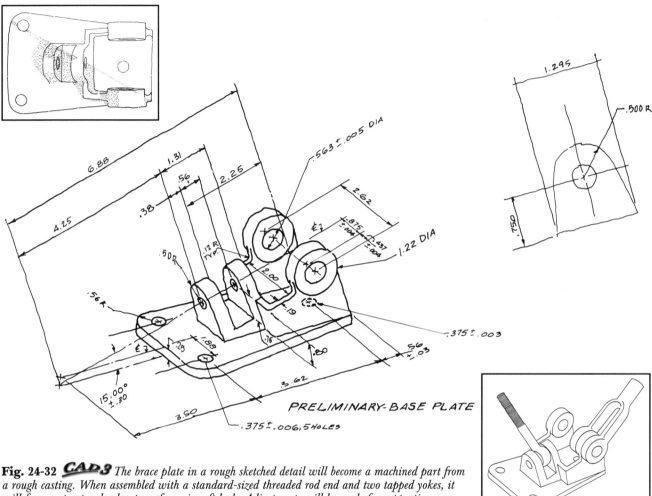

PRELIMINARY-BASE PLATE

Fig. 24-32 **CAD3** *The brace plate in a rough sketched detail will become a machined part from a rough casting. When assembled with a standard-sized threaded rod end and two tapped yokes, it will form a structural subsystem of an aircraft body. Adjustments will be made for supporting loading from this bracketing point. Prepare a detailed working drawing with necessary orthographic projections, auxiliary views and sections that best define the brace plate. Scale: Full. Media size: B.*
(Courtesy of Computervision)

REFERENCE TABLES

AMERICAN NATIONAL STANDARDS

A few standards that are useful for reference are listed below. All ANSI standards are subject to revisions, so be sure to consult the latest issues. A catalog with prices is published by the American National Standards Institute, Inc., 1430 Broadway, New York, NY 10018.

Abbreviations	Y1.1
Acme Screw Threads	B1.5
Graphical Electrical Symbols for Architectural Plans	Y32.9
Graphical Symbols for Electrical and Electronics Diagrams	Y32.2
Graphical Symbols for Heating, Ventilating and Air Conditioning	Z32.2.4
Graphical Symbols for Pipe Fittings, Valves, and Piping	Z32.2.3
Graphical Symbols for Plumbing	Y32.4
Graphical Symbols for Welding	Y32.3

Hexagon Head Cap Screws, Slotted Head Cap Screws, Square Head Set Screws, Slotted Headless Set Screws	B18.6.2
Keys and Keyseats	B17.1
Large Rivets	B18.4
Machine Tapers	B5.10
Pipe Threads	B2.1
Preferred Limits and Fits for Cylindrical Parts	B4.1
Round Head Bolts	B18.5
Slotted and Recessed Head Tapping Screws and Metallic Drive Screws	B18.6.4
Slotted and Recessed Head Wood Screws	B18.6.1
Small Solid Rivets	B18.1
Socket Cap, Shoulder, and Set Screws	B18.3
Square and Hex Bolts and Screws, Including Hex Cap Screws and Lag Screws	B18.2.1
Square and Hex Nuts	B18.2.2
Unified Screw Threads	B1.1
Woodruff Keys and Keyseats	B17.2

AMERICAN NATIONAL DRAFTING STANDARDS MANUAL

Section 1. Size and Format	Y14.1
Section 2. Line Conventions, Sectioning and Lettering (ISO R 128)	Y14.2
Section 3. Multi/Sectional View Drawings	Y14.3
Section 4. Pictorial Drawing	Y14.4
Section 5. Dimensioning and Tolerancing for Engineering Drawings (ISO R 129; R 406)	Y14.5
Section 6. Screw Threads	Y14.6
Section 7. Gears, Splines, and Serrations	Y14.7
Section 9. Forging	Y14.9
Section 10. Metal Stamping	Y14.10
Section 11. Plastics	Y14.11
Section 14. Mechanical Assemblies	Y14.14
Section 15. Electrical and Electronics Diagrams	Y14.15
Section 17. Fluid Power Diagrams	Y14.17

TABLE A-1. FRACTIONAL-INCH, DECIMAL-INCH, AND MILLIMETER EQUIVALENT CHART

Fractions					Decimal Inches			Millimeters		
4ths	8ths	16ths	32nds	64ths	To 2 places	To 3 places	To 4 places	To 1 place	To 2 places	To 3 places
				1/64	.02	.016	.0156	0.4	.40	.397
			1/32	...	.03	.031	.0312	0.8	.80	.794
				3/64	.05	.047	.0469	1.2	1.20	1.191
		1/16	...	...	.06	.062	.0625	1.6	1.59	1.588
				5/64	.08	.078	.0781	2.0	1.98	1.984
			3/32	...	.09	.094	.0938	2.4	2.38	2.381
				7/64	.11	.109	.1094	2.8	2.78	2.778
	1/8	...	...	...	.12	.125	.1250	3.2	3.18	3.175
				9/64	.14	.141	.1406	3.6	3.57	3.572
			5/32	...	.16	.156	.1562	4.0	3.97	3.969
				11/64	.17	.172	.1719	4.4	4.37	4.366
		3/16	...	...	.19	.188	.1875	4.8	4.76	4.762
				13/64	.20	.203	.2031	5.2	5.16	5.159
			7/32	...	.22	.219	.2188	5.6	5.56	5.556
				15/64	.23	.234	.2344	6.0	5.95	5.953
1/4	...	...	...	...	.25	.250	.2500	6.4	6.35	6.350
				17/64	.27	.266	.2656	6.8	6.75	6.747
			9/32	...	.28	.281	.2812	7.1	7.14	7.144
				19/64	.30	.297	.2969	7.5	7.54	7.541
		5/16	...	...	.31	.312	.3125	7.9	7.94	7.938
				21/64	.33	.328	.3281	8.3	8.33	8.334
			11/32	...	.34	.344	.3438	8.7	8.73	8.731
				23/64	.36	.359	.3594	9.1	9.13	9.128
	3/8	...	...	...	.38	.375	.3750	9.5	9.52	9.525
				25/64	.39	.391	.3906	9.9	9.92	9.922
			13/32	...	.41	.406	.4062	10.3	10.32	10.319
				27/64	.42	.422	.4219	10.7	10.72	10.716
		7/16	...	...	.44	.438	.4375	11.1	11.11	11.112
				29/64	.45	.453	.4531	11.5	11.51	11.509
			15/32	...	.47	.469	.4688	11.9	11.91	11.906
				31/64	.48	.484	.4844	12.3	12.30	12.303
				...	.50	.500	.5000	12.7	12.70	12.700
				33/64	.52	.516	.5156	13.1	13.10	13.097
			17/32	...	.53	.531	.5312	13.5	13.49	13.494
				35/64	.55	.547	.5469	13.9	13.89	13.891
		9/16	...	...	.56	.562	.5625	14.3	14.29	14.288
				37/64	.58	.578	.5781	14.7	14.68	14.684
			19/32	...	.59	.594	.5938	15.1	15.08	15.081
				39/64	.61	.609	.6094	15.5	15.48	15.478
	5/8	...	...	...	.62	.625	.6250	15.9	15.88	15.875
				41/64	.64	.641	.6406	16.3	16.27	16.272
			21/32	...	.66	.656	.6562	16.7	16.67	16.669

TABLE A-1. *(CONTINUED)*

Fractions					Decimal Inches			Millimeters		
4ths	8ths	16ths	32nds	64ths	To 2 places	To 3 places	To 4 places	To 1 place	To 2 places	To 3 places
				43/64	.67	.672	.6719	17.1	17.07	17.066
		11/16	. . .	. . .	.69	.688	.6875	17.5	17.46	17.462
				45/64	.70	.703	.7031	17.9	17.86	17.859
			23/32	. . .	.72	.719	.7188	18.3	18.26	18.256
				47/64	.73	.734	.7344	18.6	18.65	18.653
3/4	. . .	. . .	. . .	. . .	.75	.750	.7500	19.1	19.05	19.050
				49/64	.77	.766	.7656	19.4	19.45	19.447
			25/32	. . .	.78	.781	.7812	19.8	19.84	19.844
				51/64	.80	.797	.7969	20.2	20.24	20.241
		13/16	. . .	. . .	.81	.812	.8125	20.6	20.64	20.638
				53/64	.83	.828	.8281	21.0	21.03	21.034
			27/32	. . .	.84	.844	.8438	21.4	21.43	21.431
				55/64	.86	.859	.8594	21.8	21.83	21.828
	7/8	. . .	. . .	. . .	.88	.875	.8750	22.2	22.22	22.225
				57/64	.89	.891	.8906	22.6	22.62	22.622
			29/32	. . .	.91	.906	.9062	23.0	23.02	23.019
				59/64	.92	.922	.9219	23.4	23.42	23.416
		15/16	. . .	. . .	.94	.938	.9375	23.8	23.81	23.812
				61/64	.95	.953	.9531	24.2	24.21	24.209
			31/32	. . .	.97	.969	.9688	24.6	24.61	24.606
				63/64	.98	.984	.9844	25.0	25.00	25.003
					1.00	1.000	1.0000	25.4	25.40	25.400

TABLE A-2. THREADS PER INCH AND TAP DRILL SIZES

| Size Inches | | Graded Pitch Series | | | | | | Constant Pitch Series | | | | | |
| Number or Fraction | Deci-mal | Coarse UNC | | Fine UNF | | Extra Fine UNEF | | 8 UN | | 12 UN | | 16 UN | |
		Threads per Inch	Tap Drill Dia.	Threads per Inch	Tap Drill Dia.	Threads per Inch	Tap Drill Dia.	Threads per Inch	Tap Drill Dia.	Threads per Inch	Tap Drill Dia.	Threads per Inch	Tap Drill Dia.
0	.060	—	—	80	3/64	—	—	—	—	—	—	—	—
2	.086	56	No. 50	64	No. 49	—	—	—	—	—	—	—	—
4	.112	40	No. 43	48	No. 42	—	—	—	—	—	—	—	—
5	.125	40	No. 38	44	No. 37	—	—						
6	.138	32	No. 36	40	No. 33	—	—						
8	.164	32	No. 29	36	No. 29	—	—						
10	.190	24	No. 25	32	No. 21	—	—						
¼	.250	20	7	28	3	32	.219	—	—	—	—	—	—
5/16	.312	18	F	24	1	32	.281	—	—	—	—	—	—
3/8	.375	16	.312	24	Q	32	.344	—	—	—	—	UNC	—
7/16	.438	14	U	20	.391	28	Y	—	—	—	—	16	V
½	.500	13	.422	20	.453	28	.469	—	—	—	—	16	.438
9/16	.562	12	.484	18	.516	24	.516	—	—	UNC	—	16	.500
5/8	.625	11	.531	18	.578	24	.578	—	—	12	.547	16	.562
¾	.750	10	.656	16	.688	20	.703	—	—	12	.672	UNF	—
7/8	.875	9	.766	14	.812	20	.828	—	—	12	.797	16	.812
1	1.000	8	.875	12	.922	20	.953	UNC	—	UNF	—	16	.938
1⅛	1.125	7	.984	12	1.047	18	1.078	8	1.000	UNF	—	16	1.062
1¼	1.250	7	1.109	12	1.172	18	1.188	8	1.125	UNF	—	16	1.188
1⅜	1.375	6	1.219	12	1.297	18	1.312	8	1.250	UNF	—	16	1.312
1½	1.500	6	1.344	12	1.422	18	1.438	8	1.375	UNF	—	16	1.438
1⅝	1.625	—	—	—	—	18	—	8	1.500	12	1.547	16	1.562
1¾	1.750	5	1.562	—	—	—	—	8	1.625	12	1.672	16	1.688
1⅞	1.875	—	—	—	—	—	—	8	1.750	12	1.797	16	1.812
2	2.000	4.5	1.781	—	—	—	—	8	1.875	12	1.922	16	1.938
2¼	2.250	4.5	2.031	—	—	—	—	8	2.125	12	2.172	16	2.188
2½	2.500	4	2.250	—	—	—	—	8	2.375	12	2.422	16	2.438
2¾	2.750	4	2.500	—	—	—	—	8	2.625	12	2.672	16	2.688
3	3.000	4	2.750	—	—	—	—	8	2.875	12	2.922	16	2.938
3¼	3.250	4	3.000	—	—	—	—	8	3.125	12	3.172	16	3.188
3½	3.500	4	3.250	—	—	—	—	8	3.375	12	3.422	16	3.438
3¾	3.750	4	3.500	—	—	—	—	8	3.625	12	3.668	16	3.688
4	4.000	4	3.750	—	—	—	—	8	3.875	12	3.922	16	3.938

Note: The tap diameter sizes shown are nominal. The class and length of thread will govern the limits on the tapped hole size.

TABLE A-3. ISO METRIC SCREW THREADS

Nominal Size Dia (mm) Preferred	Coarse Thread Pitch	Coarse Tap Drill Size	Fine Thread Pitch	Fine Tap Drill Size	4 Thread Pitch	4 Tap Drill Size	3 Thread Pitch	3 Tap Drill Size	2 Thread Pitch	2 Tap Drill Size	1.5 Thread Pitch	1.5 Tap Drill Size	1.25 Thread Pitch	1.25 Tap Drill Size	1 Thread Pitch	1 Tap Drill Size	0.75 Thread Pitch	0.75 Tap Drill Size	0.5 Thread Pitch	0.5 Tap Drill Size	0.35 Thread Pitch	0.35 Tap Drill Size
1.6	0.35	1.25																				
1.8	0.35	1.45																				
2	0.4	1.6																				
2.2	0.45	1.75																				
2.5	0.45	2.05																			0.35	2.15
3	0.5	2.5																			0.35	2.65
3.5	0.6	2.9																			0.35	3.15
4	0.7	3.3																	0.5	3.5		
4.5	0.75	3.7																	0.5	4.0		
5	0.8	4.2																	0.5	4.5		
6	1	5.0															0.75	5.2				
8	1.25	6.7	1	7.0											1	7.0	0.75	7.2				
10	1.5	8.5	1.25	8.7									1.25	8.7	1	9.0	0.75	9.2				
12	1.75	10.2	1.25	10.8							1.5	10.5	1.25	10.7	1	11						
14	2	12	1.5	12.5							1.5	12.5	1.25	12.7	1	13						
16	2	14	1.5	14.5							1.5	14.5			1	15						
18	2.5	15.5	1.5	16.5					2	16	1.5	16.5			1	17						
20	2.5	17.5	1.5	18.5					2	18	1.5	18.5			1	19						
22	2.5	19.5	1.5	20.5					2	20	1.5	20.5			1	21						
24	3	21	2	22					2	22	1.5	22.5			1	23						
27	3	24	2	25					2	25	1.5	25.5			1	26						
30	3.5	26.5	2	28					2	28	1.5	28.5			1	29						
33	3.5	29.5	2	31					2	31	1.5	31.5										
36	4	32	3	33					2	34	1.5	34.5										
39	4	35	3	36					2	37	1.5	37.5										
42	4.5	37.5	3	39	4	38	3	39	2	40	1.5	40.5										
45	4.5	39	3	42	4	41	3	42	2	43	1.5	43.5										
46	5	43	3	45	4	44	3	45	2	46	1.5	46.5										

TABLE A-4. ACME AND STUB ACME THREADS*

ANSI preferred diameter-pitch combinations

Nominal (Major) Dia.	Threads/in.	Nominal (Major) Dia.	Threads/in.	Nominal (Major) Dia.	Threads/in.	Nominal (Major) Dia.	Threads/in.
¼ (.250)	16	¾ (.750)	6	1½ (1.500)	4	3 (3.000)	2
5⁄16 (.312)	14	⅞ (.875)	6	1¾ (1.750)	4	3½ (3.500)	2
⅜ (.375)	12	1 (1.000)	5	2 (2.000)	4	4 (4.000)	2
7⁄16 (.438)	12	1⅛ (1.125)	5	2¼ (2.250)	3	4½ (4.500)	2
½ (.500)	10	1¼ (1.250)	5	2½ (2.500)	3	5 (5.000)	2
⅝ (.625)	8	1⅜ (1.375)	4	2¾ (2.750)	3		

*ANSI B1.5 and B1.8. Diameters in inches.

TABLE A-5. SIZES OF NUMBERED AND LETTERED DRILLS

No.	Size	No.	Size	No.	Size	No.	Size
80	0.0135	53	0.0595	26	0.1470	A	0.2340
79	0.0145	52	0.0635	25	0.1495	B	0.2380
78	0.0160	51	0.0670	24	0.1520	C	0.2420
77	0.0180	50	0.0700	23	0.1540	D	0.2460
76	0.0200	49	0.0730	22	0.1570	E	0.2500
75	0.0210	48	0.0760	21	0.1590	F	0.2570
74	0.0225	47	0.0785	20	0.1610	G	0.2610
73	0.0240	46	0.0810	19	0.1660	H	0.2660
72	0.0250	45	0.0820	18	0.1695	I	0.2720
71	0.0260	44	0.0860	17	0.1730	J	0.2770
70	0.0280	43	0.0890	16	0.1770	K	0.2810
69	0.0292	42	0.0935	15	0.1800	L	0.2900
68	0.0310	41	0.0960	14	0.1820	M	0.2950
67	0.0320	40	0.0980	13	0.1850	N	0.3020
66	0.0330	39	0.0995	12	0.1890	O	0.3160
65	0.0350	38	0.1015	11	0.1910	P	0.3230
64	0.0360	37	0.1040	10	0.1935	Q	0.3320
63	0.0370	36	0.1065	9	0.1960	R	0.3390
62	0.0380	35	0.1100	8	0.1990	S	0.3480
61	0.0390	34	0.1110	7	0.2010	T	0.3580
60	0.0400	33	0.1130	6	0.2040	U	0.3680
59	0.0410	32	0.1160	5	0.2055	V	0.3770
58	0.0420	31	0.1200	4	0.2090	W	0.3860
57	0.0430	30	0.1285	3	0.2130	X	0.3970
56	0.0465	29	0.1360	2	0.2210	Y	0.4040
55	0.0520	28	0.1405	1	0.2280	Z	0.4130
54	0.0550	27	0.1440				

TABLE A-6. METRIC TWIST DRILL SIZES

METRIC DRILL SIZES		Reference Decimal Equivalent (Inches)	METRIC DRILL SIZES		Reference Decimal Equivalent (Inches)	METRIC DRILL SIZES		Reference Decimal Equivalent (Inches)
Preferred	Available		Preferred	Available		Preferred	Available	
—	0.40	.0157	—	2.7	.1063	14	—	.5512
—	0.42	.0165	2.8	—	.1102	—	14.5	.5709
—	0.45	.0177	—	2.9	.1142	15	—	.5906
—	0.48	.0189	3.0	—	.1181	—	15.5	.6102
0.50	—	.0197	—	3.1	.1220	16	—	.6299
—	0.52	.0205	3.2	—	.1260	—	16.5	.6496
0.55	—	.0217	—	3.3	.1299	17	—	.6693
—	0.58	.0228	3.4	—	.1339	—	17.5	.6890
0.60	—	.0236	—	3.5	.1378	18	—	.7087
—	0.62	.0244	3.6	—	.1417	—	18.5	.7283
0.65	—	.0256	—	3.7	.1457	19	—	.7480
—	0.68	.0268	3.8	—	.1496	—	19.5	.7677
0.70	—	.0276	—	3.9	.1535	20	—	.7874
—	0.72	.0283	4.0	—	.1575	—	20.5	.8071
0.75	—	.0295	—	4.1	.1614	21	—	.8268
—	0.78	.0307	4.2	—	.1654	—	21.5	.8465
0.80	—	.0315	—	4.4	.1732	22	—	.8661
—	0.82	.0323	4.5	—	.1772	—	23	.9055
0.85	—	.0335	—	4.6	.1811	24	—	.9449
—	0.88	.0346	4.8	—	.1890	25	—	.9843
0.90	—	.0354	5.0	—	.1969	26	—	1.0236
—	0.92	.0362	—	5.2	.2047	—	27	1.0630
0.95	—	.0374	5.3	—	.2087	28	—	1.1024
—	0.98	.0386	—	5.4	.2126	—	29	1.1417
1.00	—	.0394	5.6	—	.2205	30	—	1.1811
—	1.03	.0406	—	5.8	.2283	—	31	1.2205
1.05	—	.0413	6.0	—	.2362	32	—	1.2598
—	1.08	.0425	—	6.2	.2441	—	33	1.2992
1.10	—	.0433	6.3	—	.2480	34	—	1.3386
—	1.15	.0453	—	6.5	.2559	—	35	1.3780
1.20	—	.0472	6.7	—	.2638	36	—	1.4173
1.25	—	.0492	—	6.8	.2677	—	37	1.4567
1.3	—	.0512	—	6.9	.2717	38	—	1.4361
—	1.35	.0531	7.1	—	.2795	—	39	1.5354
1.4	—	.0551	—	7.3	.2874	40	—	1.5748
—	1.45	.0571	7.5	—	.2953	—	41	1.6142
1.5	—	.0591	—	7.8	.3071	42	—	1.6535
—	1.55	.0610	8.0	—	.3150	—	43.5	1.7126
1.6	—	.0630	—	8.2	.3228	45	—	1.7717
—	1.65	.0650	8.5	—	.3346	—	46.5	1.8307
1.7	—	.0669	—	8.8	.3465	48	—	1.8898
—	1.75	.0689	9.0	—	.3543	50	—	1.9685
1.8	—	.0709	—	9.2	.3622	—	51.5	2.0276
—	1.85	.0728	9.5	—	.3740	53	—	2.0866
1.9	—	.0748	—	9.8	.3858	—	54	2.1260
—	1.95	.0768	10	—	.3937	56	—	2.2047
2.0	—	.0787	—	10.3	.4055	—	58	2.2835
—	2.05	.0807	10.5	—	.4134	60	—	2.3622
2.1	—	.0827	—	10.8	.4252			
—	2.15	.0846	11	—	.4331			
2.2	—	.0866	—	11.5	.4528			
—	2.3	.0906	12	—	.4724			
2.4	—	.0945	12.5	—	.4921			
2.5	—	.0984	13	—	.5118			
2.6	—	.1024	—	13.5	.5315			

TABLE A-7. HEXAGON-HEAD BOLTS AND CAP SCREWS

U.S. CUSTOMARY (INCHES)			METRIC (MILLIMETERS)		
Nominal Bolt Size	Width Across Flats F	Thickness T	Nominal Bolt Size and Thread Pitch	Width Across Flats F	Thickness T
.250	.438	.172	M5 x 0.8	8	3.9
.312	.500	.219	M6 x 1	10	4.7
.375	.562	.250	M8 x 1.25	13	5.7
.438	.625	.297			
.500	.750	.344	M10 x 1.5	15	6.8
.625	.938	.422	M12 x 1.75	18	8
.750	1.125	.500	M14 x 2	21	9.3
.875	1.312	.578	M16 x 2	24	10.5
1.000	1.500	.672	M20 x 2.5	30	13.1
1.125	1.688	.750	M24 x 3	36	15.6
1.250	1.875	.844	M30 x 3.5	46	19.5
1.375	2.062	.906	M36 x 4	55	23.4
1.500	2.250	1.000			

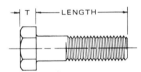

TABLE A-8. HEXAGON-HEAD NUTS

U.S. CUSTOMARY (INCHES)				METRIC (MILLIMETERS)			
Nominal Nut Size	Distance Across Flats F	Thickness Max.		Nominal Nut Size and Thread Pitch	Distance Across Flats F	Thickness Max.	
		Style 1 (Regular)–H	Style 2 (Thick)–H_1			Style 1 (Regular)–H	Style 2 (Thick)–H_1
.250	.438	.218	.281	M4 x 0.7	7	—	3.2
.312	.500	.266	.328	M5 x 0.8	8	4.5	5.3
.375	.562	.328	.406	M6 x 1	10	5.6	6.5
.438	.625	.375	.453	M8 x 1.25	13	6.6	7.8
.500	.750	.438	.562	M10 x 1.5	15	9	10.7
.562	.875	.484	.609	M12 x 1.75	18	10.7	12.8
.625	.938	.547	.719	M14 x 2	21	12.5	14.9
.750	1.125	.641	.812	M16 x 2	24	14.5	17.4
.875	1.312	.750	.906	M20 x 2.5	30	18.4	21.2
1.000	1.500	.859	1.000	M24 x 3	36	22	25.4
1.125	1.688	.969	1.156	M30 x 3.5	46	26.7	31
1.250	1.875	1.062	1.250	M36 x 4	55	32	37.6
1.375	2.062	1.172	1.375				
1.500	2.250	1.281	1.500				

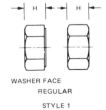
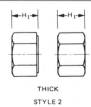

WASHER FACE
REGULAR
STYLE 1

THICK
STYLE 2

TABLE A-9. COMMON MACHINE AND CAP SCREWS

U.S. CUSTOMARY (INCHES)											METRIC (MILLIMETERS)										
Nominal Size	Hexagon Head		Socket Head		Flat Head		Fillister Head		Round or Oval Head		Nominal Size	Hexagon Head		Socket Head		Flat Head		Fillister Head		Round or Oval Head	
	A	H	A	H	A	H	A	H	A	H		A	H	A	H	A	H	A	H	A	H
.250	.44	.17	.38	.25	.50	.14	.38	.17	.44	.19	M3	5.5	2	5.5	3	5.6	1.6	6	2.4	5.6	1.9
.312	.50	.22	.47	.31	.62	.18	.44	.20	.56	.25	4	7	2.8	7	4	7.5	2.2	8	3.1	7.5	2.5
.375	.56	.25	.56	.38	.75	.21	.56	.25	.62	.27	5	8.5	3.5	9	5	9.2	2.5	10	3.8	9.2	3.1
.438	.62	.30	.66	.44	.81	.21	.62	.30	.75	.33	6	10	4	10	6	11	3	12	4.6	11	3.8
.500	.75	.34	.75	.50	.88	.21	.75	.33	.81	.35	8	13	5.5	13	8	14.5	4	16	6	14.5	5
.625	.94	.42	.94	.62	1.12	.28	.88	.42	1.00	.44	10	17	7	16	10	18	5	20	7.5	18	6.2
.750	1.12	.50	1.12	.75	1.38	.35	1.00	.50	1.25	.55	12	19	8	18	12						
											14	22	9	22	14						
											16	24	10	24	16						

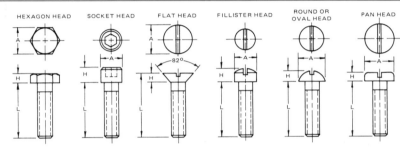

TABLE A-10. SETSCREWS

U.S. CUSTOMARY (INCHES)		METRIC (MILLIMETERS)	
Nominal Size	Key Size	Nominal Size	Key Size
.125	.06	M1.4	0.7
.138	.06	2	0.9
.164	.08	3	1.5
.190	.09	4	2
.250	.12	5	2
.312	.16	6	3
.375	.19	8	4
.500	.25	10	5
.625	.31	12	6
.750	.38	16	8

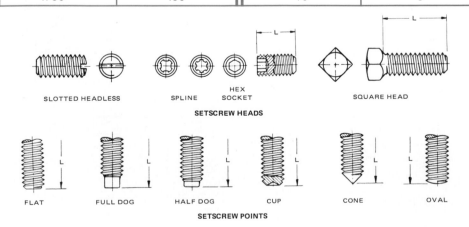

TABLE A-11. COMMON WASHER SIZES

U.S. CUSTOMARY (INCHES)						METRIC (MILLIMETERS)						
Bolt Size	Flat Washers Type A-N			Lockwashers Regular			Bolt Size	Flat Washers			Lockwashers		
	ID	OD	Thick	ID	OD	Thick		ID	OD	Thick	ID	OD	Thick
#6	.156	.375	.049	.141	.250	.031	2	2.2	5.5	0.5	2.1	3.3	0.5
#8	.188	.438	.049	.168	.293	.040	3	3.2	7	0.5	3.1	5.7	0.8
#10	.219	.500	.049	.194	.334	.047	4	4.3	9	0.8	4.1	7.1	0.9
#12	.250	.562	.065	.221	.377	.056	5	5.3	11	1	5.1	8.7	1.2
.250	.281	.625	.065	.255	.489	.062	6	6.4	12	1.5	6.1	11.1	1.6
.312	.344	.688	.065	.318	.586	.078	7	7.4	14	1.5	7.1	12.1	1.6
.375	.406	.812	.065	.382	.683	.094	8	8.4	17	2	8.2	14.2	2
.438	.469	.922	.065	.446	.779	.109	10	10.5	21	2.5	10.2	17.2	2.2
.500	.531	1.062	.095	.509	.873	.125	12	13	24	2.5	12.3	20.2	2.5
.562	.594	1.156	.095	.572	.971	.141	14	15	28	2.5	14.2	23.2	3
.625	.656	1.312	.095	.636	1.079	.156	16	17	30	3	16.2	26.2	3.5
.750	.812	1.469	.134	.766	1.271	.188	18	19	34	3	18.2	28.2	3.5
.875	.938	1.750	.134	.890	1.464	.219	20	21	36	3	20.2	32.2	4
1.000	1.062	2.000	.134	1.017	1.661	.250	22	23	39	4	22.5	34.5	4
1.125	1.250	2.250	.134	1.144	1.853	.281	24	25	44	4	24.5	38.5	5
1.250	1.375	2.500	.165	1.271	2.045	.312	27	28	50	4	27.5	41.5	5
1.375	1.500	2.750	.165	1.398	2.239	.344	30	31	56	4	30.5	46.5	6
1.500	1.625	3.000	.165	1.525	2.430	.375							

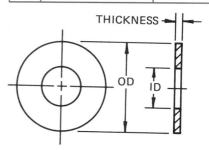

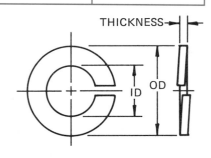

TABLE A-12. COTTER PINS

U.S. CUSTOMARY (INCHES)

Nominal Bolt or Thread-Size Range	Nominal Cotter-Pin Size (A)	Cotter-Pin Hole	Head Size (Min.) (B)	Min. End. Clearance*
.125	.031	.047	.062	.06
.188	.047	.062	.094	.08
.250	.062	.078	.125	.11
.312	.078	.094	.156	.11
.375	.094	.109	.187	.14
.438	.109	.125	.218	.14
.500	.125	.141	.250	.18
.562	.141	.156	.281	.25
.625	.156	.172	.312	.40
1.000–1.125	.188	.203	.375	.40
1.250–1.375	.219	.234	.438	.46
1.500–1.625	.250	.266	.500	.46

METRIC (MILLIMETERS)

Nominal Bolt or Thread-Size Range	Nominal Cotter-Pin Size (A)	Cotter-Pin Hole	Head Size (Min.) (B)	Min. End. Clearance*
–2.5	0.6	0.8	1.2	1.5
2.5–3.5	0.8	1.0	1.6	2.0
3.5–4.5	1.0	1.2	2.0	2.0
4.5–5.5	1.2	1.4	2.4	2.5
5.5–7.0	1.6	1.8	3.2	2.5
7.0–9.0	2.0	2.2	4.0	3.0
9.0–11.0	2.5	2.8	5.0	3.5
11.0–14.0	3.2	3.6	6.4	5.0
14.0–20.0	4.0	4.5	8.0	6.0
20.0–27.0	5.0	5.6	10.0	7.0
27.0–39.0	6.3	6.7	13.6	10.0
39.0–56.0	8.0	8.5	16.0	15.0
56.0–80.0	10.0	10.5	20.0	20.0

*End of bolt to center of hole

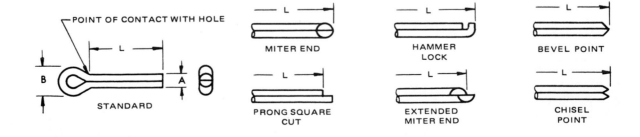

TABLE A-13. AMERICAN NATIONAL STANDARD TAPER PINS

TAPER PINS   *Maximum length for which standard reamers are available. Taper .25 in. (1:48) per ft.

Size No.	0000000	000000	00000	0000	000	00	0
Size (large end)	0.0625	0.0780	0.0940	0.1090	0.1250	0.1410	0.1560
Maximum Length*	0.625	0.750	1.000	1.000	1.000	1.250	1.250
Size No.	1	2	3	4	5	6	7
Size (large end)	0.1720	0.1930	0.2190	0.2500	0.2890	0.3410	0.4090
Maximum Length*	1.250	1.500	1.750	2.000	2.250	3.000	3.750
Size No.	8	9	10	11	12	13	14
Size (large end)	0.4920	0.5910	0.7060	0.8600	1.032	1.241	1.523
Maximum Length*	4.500	5.250	6.000	(Special Sizes. Special Lengths.)			

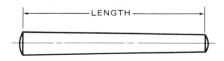

LENGTH

TABLE A-14. SQUARE AND FLAT STOCK KEYS

U.S. CUSTOMARY (INCHES)					METRIC (MILLIMETERS)				
Diameter of Shaft		Square Key		Flat Key		Diameter of Shaft		Square Key		Flat Key	
		Nominal Size		Nominal Size				Nominal Size		Nominal Size	
From	To	W	H	W	H	Over	Up To	W	H	W	H
.500	.562	.125	.125	.125	.094	6	8	2	2		
.625	.875	.188	.188	.188	.125	8	10	3	3		
.938	1.250	.250	.250	.250	.188	10	12	4	4		
1.312	1.375	.312	.312	.312	.250	12	17	5	5		
1.438	1.750	.375	.375	.375	.250	17	22	6	6		
1.812	2.250	.500	.500	.500	.375	22	30	7	7	8	7
2.375	2.750	.625	.625			30	38	8	8	10	8
2.875	3.250	.750	.750			38	44	9	9	12	8
3.375	3.750	.875	.875			44	50	10	10	14	9
3.875	4.500	1.000	1.000			50	58	12	12	16	10

SQUARE

FLAT

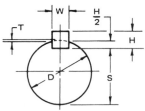

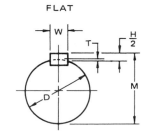

C = ALLOWANCE FOR PARALLEL KEYS = .005 in. OR 0.12 mm

$$S = D - \frac{H}{2} - T = \frac{D - H + \sqrt{D^2 - W^2}}{2} \qquad T = \frac{D - \sqrt{D^2 - W^2}}{2}$$

$$M = D - T + \frac{H}{2} + C = \frac{D + H + \sqrt{D^2 - W^2}}{2} + C$$

W = NOMINAL KEY WIDTH (INCHES OR MILLIMETERS)

TABLE A-15. WOODRUFF KEYS

U.S. CUSTOMARY (INCHES)					Key No.	METRIC (MILLIMETERS)				
Nominal Size	Key			Keyseat		Nominal Size	Key			Keyseat
A x B	E	C	D	H		A x B	E	C	D	H
.062 x .500	.047	.203	.194	.172	204	1.6 x 12.7	1.5	5.1	4.8	4.2
.094 x .500	.047	.203	.194	.156	304	2.4 x 12.7	1.3	5.1	4.8	3.8
.094 x .625	.062	.250	.240	.203	305	2.4 x 15.9	1.5	6.4	6.1	5.1
.125 x .500	.049	.203	.194	.141	404	3.2 x 12.7	1.3	5.1	4.8	3.6
.125 x .625	.062	.250	.240	.188	405	3.2 x 15.9	1.5	6.4	6.1	4.6
.125 x .750	.062	.313	.303	.251	406	3.2 x 19.1	1.5	7.9	7.6	6.4
.156 x .625	.062	.250	.240	.172	505	4.0 x 15.9	1.5	6.4	6.1	4.3
.156 x .750	.062	.313	.303	.235	506	4.0 x 19.1	1.5	7.9	7.6	5.8
.156 x .875	.062	.375	.365	.297	507	4.0 x 22.2	1.5	9.7	9.1	7.4
.188 x .750	.062	.313	.303	.219	606	4.8 x 19.1	1.5	7.9	7.6	5.3
.188 x .875	.062	.375	.365	.281	607	4.8 x 22.2	1.5	9.7	9.1	7.1
.188 x 1.000	.062	.438	.428	.344	608	4.8 x 25.4	1.5	11.2	10.9	8.6
.188 x 1.125	.078	.484	.475	.390	609	4.8 x 28.6	2.0	12.2	11.9	9.9
.250 x .875	.062	.375	.365	.250	807	6.4 x 22.2	1.5	9.7	9.1	6.4
.250 x 1.000	.062	.438	.428	.313	808	6.4 x 25.4	1.5	11.2	10.9	7.9

NOTE: Metric key sizes were not available at the time of publication. Sizes shown are inch-designed key sizes soft-converted to millimeters. Conversion was necessary to allow the student to compare keys with slot sizes given in millimeters.

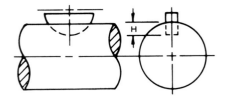

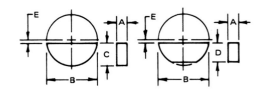

TABLE A-16. WIRE AND SHEET-METAL GAGES

Dimensions in Decimal Parts of an Inch

No. of Wire Gage	American, or Brown & Sharpe	Birmingham, or Stubs Wire	Washburn & Moen or American Steel & Wire Co.	W. & M. Steel Music Wire	New American S. & W. Co. Music Wire Gage	Imperial Wire Gage	U.S. Standard Gage for Sheet and Plate Iron and Steel
00000000	....	....	....	0.0083			
0000000	....	....	....	0.0087			
000000	....	....	....	0.0095	0.004	0.464	0.46875
00000	....	....	....	0.010	0.005	0.432	0.4375
0000	0.460	0.454	0.3938	0.011	0.006	0.400	0.40625
000	0.40964	0.425	0.3625	0.012	0.007	0.372	0.375
00	0.3648	0.380	0.3310	0.0133	0.008	0.348	0.34375
0	0.32486	0.340	0.3065	0.0144	0.009	0.324	0.3125
1	0.2893	0.300	0.2830	0.0156	0.010	0.300	0.28125
2	0.25763	0.284	0.2625	0.0166	0.011	0.276	0.265625
3	0.22942	0.259	0.2437	0.0178	0.012	0.252	0.250
4	0.20431	0.238	0.2253	0.0188	0.013	0.232	0.234375
5	0.18194	0.220	0.2070	0.0202	0.014	0.212	0.21875
6	0.16202	0.203	0.1920	0.0215	0.016	0.192	0.203125
7	0.14428	0.180	0.1770	0.023	0.018	0.176	0.1875
8	0.12849	0.165	0.1620	0.0243	0.020	0.160	0.171875
9	0.11443	0.148	0.1483	0.0256	0.022	0.144	0.15625
10	0.10189	0.134	0.1350	0.027	0.024	0.128	0.140625
11	0.090742	0.120	0.1205	0.0284	0.026	0.116	0.125
12	0.080808	0.109	0.1055	0.0296	0.029	0.104	0.109375
13	0.071961	0.095	0.0915	0.0314	0.031	0.092	0.09375
14	0.064084	0.083	0.0800	0.0326	0.033	0.080	0.078125
15	0.057068	0.072	0.0720	0.0345	0.035	0.072	0.0703125
16	0.05082	0.065	0.0625	0.036	0.037	0.064	0.0625
17	0.045257	0.058	0.0540	0.0377	0.039	0.056	0.05625
18	0.040303	0.049	0.0475	0.0395	0.041	0.048	0.050
19	0.03589	0.042	0.0410	0.0414	0.043	0.040	0.04375
20	0.031961	0.035	0.0348	0.0434	0.045	0.036	0.0375
21	0.028462	0.032	0.03175	0.046	0.047	0.032	0.034375
22	0.025347	0.028	0.0286	0.0483	0.049	0.028	0.03125
23	0.022571	0.025	0.0258	0.051	0.051	0.024	0.028125
24	0.0201	0.022	0.0230	0.055	0.055	0.022	0.025
25	0.0179	0.020	0.0204	0.0586	0.059	0.020	0.021875
26	0.01594	0.018	0.0181	0.0626	0.063	0.018	0.01875
27	0.014195	0.016	0.0173	0.0658	0.067	0.0164	0.0171875
28	0.012641	0.014	0.0162	0.072	0.071	0.0149	0.015625
29	0.011257	0.013	0.0150	0.076	0.075	0.0136	0.0140625
30	0.010025	0.012	0.0140	0.080	0.080	0.0124	0.0125
31	0.008928	0.010	0.0132	....	0.085	0.0116	0.0109375
32	0.00795	0.009	0.0128	....	0.090	0.0108	0.01015625

TABLE A-17. AMERICAN NATIONAL STANDARD WELDED AND SEAMLESS STEEL PIPE

Nominal Pipe Size	Nominal wall thickness				Tap Drill
	Outside Dia.	Standard Wall	Extra Strong Wall	Double Extra Strong Wall	
⅛	0.405	0.068	0.095	.....	.34
¼	0.540	0.088	0.119	.....	.44
⅜	0.675	0.091	0.126	.....	.58
½	0.840	0.109	0.147	0.294	.72
¾	1.050	0.113	0.154	0.308	.92
1	1.315	0.133	0.179	0.358	1.16
1¼	1.660	0.140	0.191	0.382	1.50
1½	1.900	0.145	0.200	0.400	1.74
2	2.375	0.154	0.218	0.436	2.22
2½	2.875	0.203	0.276	0.552	2.62
3	3.500	0.216	0.300	0.600	3.25
3½	4.000	0.226	0.318	.....	3.75
4	4.500	0.237	0.337	0.674	4.25
5	5.563	0.258	0.375	0.750	5.31
6	6.625	0.280	0.432	0.864	6.31
8	8.625	0.322	0.500	0.875	7.31

NOTE: To find the inside diameter, subtract twice the wall thickness from the outside diameter. Schedule numbers have been set up for wall thicknesses for pipe and the standard should be consulted for complete information. Standard wall thicknesses are for Schedule 40 up to and including nominal size 10. Extra strong walls are Schedule 80 up to and including size 8, and Schedule 60 for size 10.

AMERICAN NATIONAL STANDARD LIMITS AND FITS

The following tables are designed for use in the basic hole system of limits and fits described in Chapter 6, Dimensioning. Information for these tables is adapted from ANSI B4.1, "Preferred Limits and Fits for Cylindrical Parts." For larger sizes and additional information, refer to the standard.

There are five distinct classes of fits:

RC   Running or sliding clearance fits

LC   Locational clearance fits

LT   Transition clearance or interference fits

LN   Locational interference fits

FN   Force or shrink fits

These five classes of fits are placed in three general categories as follows.

RUNNING AND SLIDING FITS (TABLE A-18)

These fits provide a similar running performance, with suitable lubrication allowance, throughout the range of sizes. The clearances for the first two classes, used chiefly as sliding fits, increase more slowly than for the other classes, so that accurate location is maintained even at the expense of free relative motion.

RC 1: *Close sliding fits* accurately locate parts that must assemble without perceptible play.

RC 2: *Sliding fits* are for accurate location, but with greater maximum clearance than RC 1. Parts move and turn easily but do not run freely, and in the larger sizes may seize with small temperature changes.

RC 3: *Precision running fits* are the closest fits expected to run freely. They are for precision work at slow speeds and light journal pressures. They are not suitable under appreciable temperature differences.

RC 4: *Close running fits* are chiefly for running fits on accurate machinery with moderate surface speeds and journal pressures, where accurate location and minimum play are desired.

RC 5 and RC 6: *Medium running fits* are for higher running speeds, heavy journal pressures, or both.

RC 7: *Free running fits* are for use where accuracy is not essential, where large temperature variations are likely, or under both these conditions.

RC 8 and RC 9: *Loose running fits* are for materials such as cold-rolled shafting and tubing, made to commercial tolerances.

LOCATIONAL FITS (TABLES A-19 THROUGH A-21)

These fits determine only the location of mating parts and may provide rigid or accurate location, as in interference fits, or some freedom of location, as in clearance fits. They fall into the following three groups.

LC: *Locational clearance fits* are for normally stationary parts that can be freely assembled or disassembled. They run from snug fits for parts requiring accuracy of location, through the medium clearance fits for parts such as spigots, to the looser fastener fits where freedom of assembly is of prime importance.

LT: *Transitional locational fits* fall between clearance and interference fits for application where accuracy of location is important, but a small amount of clearance or interference is permissible.

LN: *Locational interference fits* are used where accuracy of location is of prime importance, and for parts needing rigidity and alignment with no special requirements for bore pressure. Such fits are not for parts that transmit frictional loads from one part to another by virtue of the tightness of fit; these conditions are met by force fits.

TABLE A-18. AMERICAN NATIONAL STANDARD RUNNING AND SLIDING FITS[a]

Basic hole system. Limits are in thousandths of an inch.

Nominal Size Range, Inches Over — To	Class RC 1 Close Sliding Fit — Limits of Clearance	RC1 Hole	RC1 Shaft	Class RC 2 Sliding Fit — Limits of Clearance	RC2 Hole	RC2 Shaft	Class RC 3 Precision Running Fit — Limits of Clearance	RC3 Hole	RC3 Shaft	Class RC 4 Close Running Fit — Limits of Clearance	RC4 Hole	RC4 Shaft
0–0.12	0.1	+0.2	−0.1	0.1	+0.25	−0.1	0.3	+0.4	−0.3	0.3	+0.6	−0.3
	0.45	−0	−0.25	0.55	−0	−0.3	0.95	−0	−0.55	1.3	−0	−0.7
0.12–0.24	0.15	+0.2	−0.15	0.15	+0.3	−0.15	0.4	+0.5	−0.4	0.4	+0.7	−0.4
	0.5	−0	−0.3	0.65	−0	−0.35	1.12	−0	−0.7	1.6	−0	−0.9
0.24–0.40	0.2	+0.25	−0.2	0.2	+0.4	−0.2	0.5	+0.6	−0.5	0.5	+0.9	−0.5
	0.6	−0	−0.35	0.85	−0	−0.45	1.5	−0	−0.9	2.0	−0	−1.1
0.40–0.71	0.25	+0.3	−0.25	0.25	+0.4	−0.25	0.6	+0.7	−0.6	0.6	+1.0	−0.6
	0.75	−0	−0.45	0.95	−0	−0.55	1.7	−0	−1.0	2.3	−0	−1.3
0.71–1.19	0.3	+0.4	−0.3	0.3	+0.5	−0.3	0.8	+0.8	−0.8	0.8	+1.2	−0.8
	0.95	−0	−0.55	1.2	−0	−0.7	2.1	−0	−1.3	2.8	−0	−1.6
1.19–1.97	0.4	+0.4	−0.4	0.4	+0.6	−0.4	1.0	+1.0	−1.0	1.0	+1.6	−1.0
	1.1	−0	−0.7	1.4	−0	−0.8	2.6	−0	−1.6	3.6	−0	−2.0
1.97–3.15	0.4	+0.5	−0.4	0.4	+0.7	−0.4	1.2	+1.2	−1.2	1.2	+1.8	−1.2
	1.2	−0	−0.7	1.6	−0	−0.9	3.1	−0	−1.9	4.2	−0	−2.4

Nominal Size Range, Inches Over — To	Class RC 5 (Medium Running Fits) — Limits of Clearance	RC5 Hole	RC5 Shaft	Class RC 6 — Limits of Clearance	RC6 Hole	RC6 Shaft	Class RC 7 — Limits of Clearance	RC7 Hole	RC7 Shaft	Class RC 8 (Loose Running Fits) — Limits of Clearance	RC8 Hole	RC8 Shaft	Class RC 9 — Limits of Clearance	RC9 Hole	RC9 Shaft
0–0.12	0.6	+0.6	−0.6	0.6	+1.0	−0.6	1.0	+1.0	−1.0	2.5	+1.6	−2.5	4.0	+2.5	−4.0
	1.6	−0	−1.0	2.2	−0	−1.2	2.6	−0	−1.6	5.1	−0	−3.5	8.1	−0	−5.6
0.12–0.24	0.8	+0.7	−0.8	0.8	+1.2	−0.8	1.2	+1.2	−1.2	2.8	+1.8	−2.8	4.5	+3.0	−4.5
	2.0	−0	−1.3	2.7	−0	−1.5	3.1	−0	−1.9	5.8	−0	−4.0	9.0	−0	−6.0
0.24–0.40	1.0	+0.9	−1.0	1.0	+1.4	−1.0	1.6	+1.4	−1.6	3.0	+2.2	−3.0	5.0	+3.5	−5.0
	2.5	−0	−1.6	3.3	−0	−1.9	3.9	−0	−2.5	6.6	−0	−4.4	10.7	−0	−7.2
0.40–0.71	1.2	+1.0	−1.2	1.2	+1.6	−1.2	2.0	+1.6	−2.0	3.5	+2.8	−3.5	6.0	+4.0	−6.0
	2.9	−0	−1.9	3.8	−0	−2.2	4.6	−0	−3.0	7.9	−0	−5.1	12.8	−0	−8.8
0.71–1.19	1.6	+1.2	−1.6	1.6	+2.0	−1.6	2.5	+2.0	−2.5	4.5	+3.5	−4.5	7.0	+5.0	−7.0
	3.6	−0	−2.4	4.8	−0	−2.8	5.7	−0	−3.7	10.0	−0	−6.5	15.5	−0	−10.5
1.19–1.97	2.0	+1.6	−2.0	2.0	+2.5	−2.0	3.0	+2.5	−3.0	5.0	+4.0	−5.0	8.0	+6.0	−8.0
	4.6	−0	−3.0	6.1	−0	−3.6	7.1	−0	−4.6	11.5	−0	−7.5	18.0	−0	−12.0
1.97–3.15	2.5	+1.8	−2.5	2.5	+3.0	−2.5	4.0	+3.0	−4.0	6.0	+4.5	−6.0	9.0	+7.0	−9.0
	5.5	−0	−3.7	7.3	−0	−4.3	8.8	−0	−5.8	13.5	−0	−9.0	20.5	−0	−13.5

[a]ANSI B4.1.

TABLE A-19. AMERICAN NATIONAL STANDARD CLEARANCE LOCATIONAL FITS[a]

Basic hole system. Limits are in thousandths of an inch.

Nominal Size Range, Inches Over To	Class LC 1			Class LC 2			Class LC 3			Class LC 4			Class LC 5		
	Limits of Clearance	Standard Limits		Limits of Clearance	Standard Limits		Limits of Clearance	Standard Limits		Limits of Clearance	Standard Limits		Limits of Clearance	Standard Limits	
		Hole	Shaft		Hole	Shaft		Hole	Shaft		Hole	Shaft		Hole	Shaft
0–0.12	0	+0.25	+0	0	+0.4	+0	0	+0.6	+0	0	+1.6	+0	0.1	+0.4	−0.1
	0.45	−0	−0.2	0.65	−0	−0.25	1	−0	−0.4	2.6	−0	−1.0	0.75	−0	−0.35
0.12–0.24	0	+0.3	+0	0	+0.5	+0	0	+0.7	+0	0	+1.8	+0	0.15	+0.5	−0.15
	0.5	−0	−0.2	0.8	−0	−0.3	1.2	−0	−0.5	3.0	−0	−1.2	0.95	−0	−0.45
0.24–0.40	0	+0.4	+0	0	+0.6	+0	0	+0.9	+0	0	+2.2	+0	0.2	+0.6	−0.2
	0.65	−0	−0.25	1.0	−0	−0.4	1.5	−0	−0.6	3.6	−0	−1.4	1.2	−0	−0.6
0.40–0.71	0	+0.4	+0	0	+0.7	+0	0	+1.0	+0	0	+2.8	+0	0.25	+0.7	−0.25
	0.7	−0	−0.3	1.1	−0	−0.4	1.7	−0	−0.7	4.4	−0	−1.6	1.35	−0	−0.65
0.71–1.19	0	+0.5	+0	0	+0.8	+0	0	+1.2	+0	0	+3.5	+0	0.3	+0.8	−0.3
	0.9	−0	−0.4	1.3	−0	−0.5	2	−0	−0.8	5.5	−0	−2.0	1.6	−0	−0.8
1.19–1.97	0	+0.6	+0	0	+1.0	+0	0	+1.6	+0	0	+4.0	+0	0.4	+1.0	−0.4
	1.0	−0	−0.4	1.6	−0	−0.6	2.6	−0	−1	6.5	−0	−2.5	2.0	−0	−1.0
1.97–3.15	0	+0.7	+0	0	+1.2	+0	0	+1.8	+0	0	+4.5	+0	0.4	+1.2	−0.4
	1.2	−0	−0.5	1.9	−0	−0.7	3	−0	−1.2	7.5	−0	−3	2.3	−0	−1.1

Nominal Size Range, Inches Over To	Class LC 6			Class LC 7			Class LC 8			Class LC 9			Class LC 10			Class LC 11		
	Limits of Clearance	Standard Limits		Limits of Clearance	Standard Limits		Limits of Clearance	Standard Limits		Limits of Clearance	Standard Limits		Limits of Clearance	Standard Limits		Limits of Clearance	Standard Limits	
		Hole	Shaft		Hole	Shaft		Hole	Shaft		Hole	Shaft		Hole	Shaft		Hole	Shaft
0–0.12	0.3	+1.0	−0.3	0.6	+1.6	−0.6	1.0	+1.6	1.0	2.5	+2.5	− 2.5	4	+ 4	− 4	5	+ 6	− 5
	1.9	−0	−0.9	3.2	−0	−1.6	3.6	−0	−2.0	6.6	−0	− 4.1	12	− 0	− 8	17	− 0	−11
0.12–0.24	0.4	+1.2	−0.4	0.8	+1.8	−0.8	1.2	+1.8	−1.2	2.8	+3.0	− 2.8	4.5	+ 5	− 4.5	6	+ 7	− 6
	2.3	−0	−1.1	3.8	−0	−2.0	4.2	−0	−2.4	7.6	−0	− 4.6	14.5	− 0	− 9.5	20	− 0	−13
0.24–0.40	0.5	+1.5	−0.5	1.0	+2.2	−1.0	1.6	+2.2	−1.6	3.0	+3.5	− 3.0	5	+ 6	− 5	7	+ 9	− 7
	2.8	−0	−1.4	4.6	−0	−2.4	5.2	−0	−3.0	8.7	−0	− 5.2	17	− 0	−11	25	− 0	−16
0.40–0.71	0.6	+1.6	−0.6	1.2	+2.8	−1.2	2.0	+2.8	−2.0	3.5	+4.0	− 3.5	6	+ 7	− 6	8	+10	− 8
	3.2	−0	−1.6	5.6	−0	−2.8	6.4	−0	−3.6	10.3	−0	− 6.3	20	− 0	−13	28	− 0	−18
0.71–1.19	0.8	+2.0	−0.8	1.6	+3.5	−1.6	2.5	+3.5	−2.5	4.5	+5.0	− 4.5	7	+ 8	− 7	10	+12	−10
	4.0	−0	−2.0	7.1	−0	−3.6	8.0	−0	−4.5	13.0	−0	− 8.0	23	− 0	−15	34	− 0	−22
1.19–1.97	1.0	+2.5	−1.0	2.0	+4.0	−2.0	3.0	+4.0	−3.0	5	+6	− 5	8	+10	− 8	12	+16	−12
	5.1	−0	−2.6	8.5	−0	−4.5	9.5	−0	−5.5	15	−0	− 9	28	− 0	−18	44	− 0	−28
1.97–3.15	1.2	+3.0	−1.2	2.5	+4.5	−2.5	4.0	+4.5	−4.0	6	+7	− 6	10	+12	−10	14	+18	−14
	6.0	−0	−3.0	10.0	−0	−5.5	11.5	−0	−7.0	17.5	−0	−10.5	34	− 0	−22	50	− 0	−32

[a]From ANSI B4.1.

TABLE A-20. AMERICAN NATIONAL STANDARD TRANSITION LOCATIONAL FITS[a]

Basic hole system. Limits are in thousandths of an inch.

Nominal Size Range, Inches Over To	Class LT 1 Fit	Standard Limits Hole	Shaft	Class LT 2 Fit	Standard Limits Hole	Shaft	Class LT 3 Fit	Standard Limits Hole	Shaft	Class LT 4 Fit	Standard Limits Hole	Shaft	Class LT 5 Fit	Standard Limits Hole	Shaft	Class LT 6 Fit	Standard Limits Hole	Shaft
0–0.12	−0.10	+0.4	+0.10	−0.2	+0.6	+0.2	...	...	...	...	...	...	−0.5	+0.4	+0.5	−0.65	+0.4	+0.65
	+0.50	−0	−0.10	+0.8	−0	−0.2	...	...	...	...	...	...	+0.15	−0	+0.25	+0.15	−0	+0.25
0.12–0.24	−0.15	+0.5	+0.15	−0.25	+0.7	+0.25	...	...	...	...	...	...	−0.6	+0.5	+0.6	−0.8	+0.5	+0.8
	+0.65	−0	−0.15	+0.95	−0	−0.25	...	...	...	...	...	...	+0.2	−0	+0.3	+0.2	−0	+0.3
0.24–0.40	−0.2	+0.6	+0.2	−0.3	+0.9	+0.3	−0.5	+0.6	+0.5	−0.7	+0.9	+0.7	−0.8	+0.6	+0.8	−1.0	+0.6	+1.0
	+0.8	−0	−0.2	+1.2	−0	−0.3	+0.5	−0	+0.1	+0.8	−0	+0.1	+0.2	−0	+0.4	+0.2	−0	+0.4
0.40–0.71	−0.2	+0.7	+0.2	−0.35	+1.0	+0.35	−0.5	+0.7	+0.5	−0.8	+1.0	+0.8	−0.9	+0.7	+0.9	−1.2	+0.7	+1.2
	+0.9	−0	−0.2	+1.35	−0	−0.35	+0.6	−0	+0.1	+0.9	−0	+0.1	+0.2	−0	+0.5	+0.2	−0	+0.5
0.71–1.19	−0.25	+0.8	+0.25	−0.4	+1.2	+0.4	−0.6	+0.8	+0.6	−0.9	+1.2	+0.9	−1.1	+0.8	+1.1	−1.4	+0.8	+1.4
	+1.05	−0	−0.25	+1.6	−0	−0.4	+0.7	−0	+0.1	+1.1	−0	+0.1	+0.2	−0	+0.6	+0.2	−0	+0.6
1.19–1.97	−0.3	+1.0	+0.3	−0.5	+1.6	+0.5	−0.7	+1.0	+0.7	−1.1	+1.6	+1.1	−1.3	+1.0	+1.3	−1.7	+1.0	+1.7
	+1.3	−0	−0.3	+2.1	−0	−0.5	+0.9	−0	+0.1	+1.5	−0	+0.1	+0.3	−0	+0.7	+0.3	−0	+0.7
1.97–3.15	−0.3	+1.2	+0.3	−0.6	+1.8	+0.6	−0.8	+1.2	+0.8	−1.3	+1.8	+1.3	−1.5	+1.2	+1.5	−2.0	+1.2	+2.0
	+1.5	−0	−0.3	+2.4	−0	−0.6	+1.1	−0	+0.1	+1.7	−0	+0.1	+0.4	−0	+0.8	+0.4	−0	+0.8

[a]ANSI B4.1.

TABLE A-21. AMERICAN NATIONAL STANDARD INTERFERENCE LOCATIONAL FITS[a]

Basic hole system. Limits are in thousandths of an inch.

Nominal Size Range, Inches Over To	Class LN 1 Limits of Interference	Standard Limits Hole	Shaft	Class LN 2 Limits of Interference	Standard Limits Hole	Shaft	Class LN 3 Limits of Interference	Standard Limits Hole	Shaft
0–0.12	0	+0.25	+0.45	0	+0.4	+0.65	0.1	+0.4	+0.75
	0.45	−0	+0.25	0.65	-0	+0.4	0.75	−0	+0.5
0.12–0.24	0	+0.3	+0.5	0	+0.5	+0.8	0.1	+0.5	+0.9
	0.5	−0	+0.3	0.8	-0	+0.5	0.9	0	+0.6
0.24–0.40	0	+0.4	+0.65	0	+0.6	+1.0	0.2	+0.6	+1.2
	0.65	−0	+0.4	1.0	-0	+0.6	1.2	−0	+0.8
0.40–0.71	0	+0.4	+0.8	0	+0.7	+1.1	0.3	+0.7	+1.4
	0.8	−0	+0.4	1.1	-0	+0.7	1.4	−0	+1.0
0.71–1.19	0	+0.5	+1.0	0	+0.8	+1.3	0.4	+0.8	+1.7
	1.0	−0	+0.5	1.3	-0	+0.8	1.7	−0	+1.2
1.19–1.97	0	+0.6	+1.1	0	+1.0	+1.6	0.4	+1.0	+2.0
	1.1	−0	+0.6	1.6	-0	+1.0	2.0	−0	+1.4
1.97–3.15	0.1	+0.7	+1.3	0.2	+1.2	+2.1	0.4	+1.2	+2.3
	1.3	−0	+0.7	2.1	-0	+1.4	2.3	−0	+1.6

[a]ANSI B4.1.

FORCE FITS (TABLE A-22)

A force fit is a special type of interference fit, normally characterized by maintenance of constant bore pressures throughout the range of sizes. Thus the interference varies almost directly with diameter. To maintain the resulting pressures within reasonable limits, the difference between its minimum and maximum value is small.

FN 1: *Light drive fits* require light assembly pressures and produce more or less permanent assemblies. These are suitable for thin sections, long fits, or cast-iron external members.

FN 2: *Medium drive fits* are for ordinary steel parts or for shrink fits on light sections. They are about the tightest fits that can be used with high-grade cast-iron external members.

FN 3: *Heavy drive fits* are suitable for heavier steel parts or for shrink fits in medium sections.

FN 4 and FN 5: *Force fits* are for parts that can be highly stressed, or for shrink fits where heavy pressing forces are impractical.

In each of the five tables on limits and fits, the range of nominal sizes (left column) is given in inches. All other sizes and values are given in thousandths of an inch to save space on the tables. To convert these values to inches, simply move the decimal point three places to the left. For example, 0.2 in the tables becomes .0002 in.

The basic diameter is given in the first column, "Nominal Size Range, Inches." For example, a basic diameter of 2.5000 in. falls between 1.97 and 3.15 on the tables. Read across the top row in Table A-18 to the next major heading, "Class RC 1, Close Sliding Fit." If an RC 1 fit is desired, the limits would be determined as follows: Read across from 1.97-3.15 to the next column, which shows that the minimum clearance, or allowance (tightest permissible condition), is .0004 in. and that the maximum clearance (loosest permissible condition) is .0012 in.

Under the "Standard Limits" column are the applicable limits for the "Hole" and the "Shaft." Using the example above, add .0005 to 2.5000 in. to obtain the upper limit or size of the hole. Since the lower limit for the hole is zero, the lower limit and the basic size are the same, or 2.5000 in. Therefore, on the drawing the hole would be dimensioned as

$$2.5005$$
$$2.0000$$

or

$$2.5000 \, ^{+.0005}_{0}$$

Limit dimensions applied to the shaft would be done in the same way, as follows:

Basic shaft size	2.5000
Upper limit	−.0004
	2.4996

Basic shaft size	2.5000
Lower limit	−.0007
	2.4993

Therefore, on the drawing the shaft would be dimensioned as

$$2.4996$$
$$2.4993$$

or

$$2.4996 \, ^{0}_{-.0003}$$

The allowance (minimum clearance) is 2.5000 (minimum hole size) minus 2.4996 (maximum shaft size), or .0004 in. The maximum clearance is 2.5005 (maximum hole size) minus 2.4993 (minimum shaft size), or .0012 in.

TABLE A-22. AMERICAN NATIONAL STANDARD FORCE AND SHRINK FITS[a]

Basic hole system. Limits are in thousandths of an inch.

Nominal Size Range, Inches Over To	Class FN 1 Light Drive Fit			Class FN 2 Medium Drive Fit			Class FN 3 Heavy Drive Fit			Class FN 4 Force Fit			Class FN 5 Force Fit		
	Limits of Interference	Standard Limits		Limits of Interference	Standard Limits		Limits of Interference	Standard Limits		Limits of Interference	Standard Limits		Limits of Interference	Standard Limits	
		Hole	Shaft		Hole	Shaft		Hole	Shaft		Hole	Shaft		Hole	Shaft
0–0.12	0.05	+0.25	+0.5	0.2	+0.4	+0.85	...	...	...	0.3	+0.4	+0.95	0.3	+0.6	+1.3
	0.5	−0	+0.3	0.85	−0	+0.6	...	...	...	0.95	−0	+0.7	1.3	−0	+0.9
0.12–0.24	0.1	+0.3	+0.6	0.2	+0.5	+1.0	...	...	...	0.4	+0.5	+1.2	0.5	+0.7	+1.7
	0.6	−0	+0.4	1.0	−0	+0.7	...	...	...	1.2	−0	+0.9	1.7	−0	+1.2
0.24–0.40	0.1	+0.4	+0.75	0.4	+0.6	+1.4	...	...	...	0.6	+0.6	+1.6	0.5	+0.9	+2.0
	0.75	−0	+0.5	1.4	−0	+1.0	...	...	...	1.6	−0	+1.2	2.0	−0	+1.4
0.40–0.56	0.1	+0.4	+0.8	0.5	+0.7	+1.6	...	...	...	0.7	+0.7	+1.8	0.6	+1.0	+2.3
	0.8	−0	+0.5	1.6	−0	+1.2	...	...	...	1.8	−0	+1.4	2.3	−0	+1.6
0.56–0.71	0.2	+0.4	+0.9	0.5	+0.7	+1.6	...	...	...	0.7	+0.7	+1.8	0.8	+1.0	+2.5
	0.9	−0	+0.6	1.6	−0	+1.2	...	...	...	1.8	−0	+1.4	2.5	−0	+1.8
0.71–0.95	0.2	+0.5	+1.1	0.6	+0.8	+1.9	...	...	...	0.8	+0.8	+2.1	1.0	+1.2	+3.0
	1.1	−0	+0.7	1.9	−0	+1.4	...	...	...	2.1	−0	+1.6	3.0	−0	+2.2
0.95–1.19	0.3	+0.5	+1.2	0.6	+0.8	+1.9	0.8	+0.8	+2.1	1.0	+0.8	+2.3	1.3	+1.2	+3.3
	1.2	−0	+0.8	1.9	−0	+1.4	2.1	−0	+1.6	2.3	−0	+1.8	3.3	−0	+2.5
1.19–1.58	0.3	+0.6	+1.3	0.8	+1.0	+2.4	1.0	+1.0	+2.6	1.5	+1.0	+3.1	1.4	+1.6	+4.0
	1.3	−0	+0.9	2.4	−0	+1.8	2.6	−0	+2.0	3.1	−0	+2.5	4.0	−0	+3.0
1.58–1.97	0.4	+0.6	+1.4	0.8	+1.0	+2.4	1.2	+1.0	+2.8	1.8	+1.0	+3.4	2.4	+1.6	+5.0
	1.4	−0	+1.0	2.4	−0	+1.8	2.8	−0	+2.2	3.4	−0	+2.8	5.0	−0	+4.0
1.97–2.56	0.6	+0.7	+1.8	0.8	+1.2	+2.7	1.3	+1.2	+3.2	2.3	+1.2	+4.2	3.2	+1.8	+6.2
	1.8	−0	+1.3	2.7	−0	+2.0	3.2	−0	+2.5	4.2	−0	+3.5	6.2	−0	+5.0

[a]ANSI B4.1.

TABLE A-23. STEEL-WIRE NAILS

American Steel & Wire Company Gage

Size	Length	Common Wire Nails and Brads		Casing Nails		Finishing Nails	
		Gage, Dia.	No. to Pound	Gage, Dia.	No. to Pound	Gage, Dia.	No. to Pound
2d	1.00	15	876	15½	1010	16½	1351
3d	1.25	14	568	14½	635	15½	807
4d	1.50	12½	316	14	473	15	584
5d	1.75	12½	271	14	406	15	500
6d	2.00	11½	181	12½	236	13	309
7d	2.25	11½	161	12½	210	13	238
8d	2.50	10¼	106	11½	145	12½	189
9d	2.75	10¼	96	11½	132	12½	172
10d	3.00	9	69	10½	94	11½	121
12d	3.25	9	64	10½	87	11½	113
16d	3.50	8	49	10	71	11	90
20d	4.00	6	31	9	52	10	62
30d	4.50	5	24	9	46		
40d	5.00	4	18	8	35		
50d	5.50	3	14				
60d	6.00	2	11				

TABLE A-24. AMERICAN NATIONAL STANDARD LARGE RIVETS

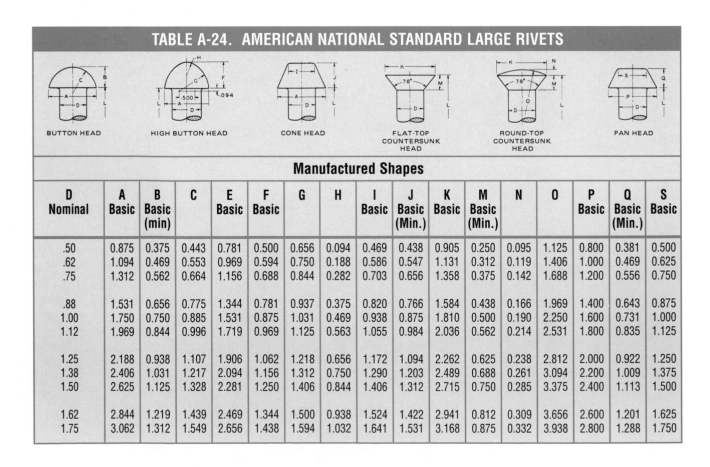

BUTTON HEAD   HIGH BUTTON HEAD   CONE HEAD   FLAT-TOP COUNTERSUNK HEAD   ROUND-TOP COUNTERSUNK HEAD   PAN HEAD

Manufactured Shapes

D Nominal	A Basic	B Basic (min)	C	E Basic	F Basic	G	H	I Basic	J Basic (Min.)	K Basic	M Basic (Min.)	N	O	P Basic	Q Basic (Min.)	S Basic
.50	0.875	0.375	0.443	0.781	0.500	0.656	0.094	0.469	0.438	0.905	0.250	0.095	1.125	0.800	0.381	0.500
.62	1.094	0.469	0.553	0.969	0.594	0.750	0.188	0.586	0.547	1.131	0.312	0.119	1.406	1.000	0.469	0.625
.75	1.312	0.562	0.664	1.156	0.688	0.844	0.282	0.703	0.656	1.358	0.375	0.142	1.688	1.200	0.556	0.750
.88	1.531	0.656	0.775	1.344	0.781	0.937	0.375	0.820	0.766	1.584	0.438	0.166	1.969	1.400	0.643	0.875
1.00	1.750	0.750	0.885	1.531	0.875	1.031	0.469	0.938	0.875	1.810	0.500	0.190	2.250	1.600	0.731	1.000
1.12	1.969	0.844	0.996	1.719	0.969	1.125	0.563	1.055	0.984	2.036	0.562	0.214	2.531	1.800	0.835	1.125
1.25	2.188	0.938	1.107	1.906	1.062	1.218	0.656	1.172	1.094	2.262	0.625	0.238	2.812	2.000	0.922	1.250
1.38	2.406	1.031	1.217	2.094	1.156	1.312	0.750	1.290	1.203	2.489	0.688	0.261	3.094	2.200	1.009	1.375
1.50	2.625	1.125	1.328	2.281	1.250	1.406	0.844	1.406	1.312	2.715	0.750	0.285	3.375	2.400	1.113	1.500
1.62	2.844	1.219	1.439	2.469	1.344	1.500	0.938	1.524	1.422	2.941	0.812	0.309	3.656	2.600	1.201	1.625
1.75	3.062	1.312	1.549	2.656	1.438	1.594	1.032	1.641	1.531	3.168	0.875	0.332	3.938	2.800	1.288	1.750

TABLE A-25. AMERICAN NATIONAL STANDARD SLOTTED-HEAD WOOD SCREWS

Maximum Dimensions

Nominal Size	D	A	B	C	E	F	G	H	I	J	Number Threads per Inch
0	0.060	0.119	0.035	0.023	0.015	0.030	0.056	0.053	0.113	0.039	32
1	0.073	0.146	0.043	0.026	0.019	0.038	0.068	0.061	0.138	0.044	28
2	0.086	0.172	0.051	0.031	0.023	0.045	0.080	0.069	0.162	0.048	26
3	0.099	0.199	0.059	0.035	0.027	0.052	0.092	0.078	0.187	0.053	24
4	0.112	0.225	0.067	0.039	0.030	0.059	0.104	0.086	0.211	0.058	22
5	0.125	0.252	0.075	0.043	0.034	0.067	0.116	0.095	0.236	0.063	20
6	0.138	0.279	0.083	0.048	0.038	0.074	0.128	0.103	0.260	0.068	18
7	0.151	0.305	0.091	0.048	0.041	0.081	0.140	0.111	0.285	0.072	16
8	0.164	0.332	0.100	0.054	0.045	0.088	0.152	0.120	0.309	0.077	15
9	0.177	0.358	0.108	0.054	0.049	0.095	0.164	0.128	0.334	0.082	14
10	0.190	0.385	0.116	0.060	0.053	0.103	0.176	0.137	0.359	0.087	13
12	0.216	0.438	0.132	0.067	0.060	0.117	0.200	0.153	0.408	0.096	11
14	0.242	0.491	0.148	0.075	0.068	0.132	0.224	0.170	0.457	0.106	10
16	0.268	0.544	0.164	0.075	0.075	0.146	0.248	0.187	0.506	0.115	9
18	0.294	0.597	0.180	0.084	0.083	0.160	0.272	0.204	0.555	0.125	8
20	0.320	0.650	0.196	0.084	0.090	0.175	0.296	0.220	0.604	0.134	8
24	0.372	0.756	0.228	0.094	0.105	0.204	0.344	0.254	0.702	0.154	7

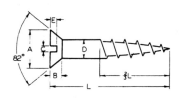
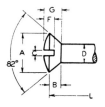

TABLE A-26. AMERICAN WELDING SOCIETY STANDARD WELDING SYMBOLS

Basic Welding Symbols and Their Location Significance

Location Significance	Fillet	Plug or Slot	Spot or Projection	Seam	Back or Backing	Surfacing	Flange Edge	Flange Corner
Arrow Side					Groove weld symbol			
Other Side						Not used		
Both Sides		Not used			Groove weld symbol	Not used	Not used	
No Arrow Side or Other Side Significance	Not used	Not used			Not used	Not used	Not used	Not used

Basic Welding Symbols and Their Location Significance

Location Significance	Square	Groove: V	Groove: Bevel	Groove: U	Groove: J	Flare-V	Flare-Bevel
Arrow Side							
Other Side							
Both Sides							
No Arrow Side or Other Side Significance	Not used	Not used	Not used	Not used	Not used	Not used	Not used

Supplementary Symbols Used with Welding Symbols

Convex Contour Symbol — Convex contour symbol indicates face of weld to be made flush.

Flush Contour Symbol — Flush contour symbol indicates face of weld to be made flush without subsequent finishing.

Field Weld Symbol — Field Weld symbol indicates that weld is to be made at a place other than that of initial construction.

Weld-All-Around Symbol — Weld-all-around symbol indicates that weld extends completely around the joint.

Complete Penetration — Indicates complete penetration regardless of type of weld or joint preparation.

Supplementary Symbols

	Weld-All-Around	Field Weld	Melt-Thru	Backing, Spacer	Contour Flush	Contour Convex	Contour Concave

Typical Welding Symbols

Slot Welding Symbol
Plug Welding Symbol
Backgouging Welding Symbol
Flash or Upset Welding Symbol

Square-Groove Welding Symbol
Chain Intermittent Fillet Welding Symbol
Back or Backing Welding Symbol
Staggered Intermittent Fillet Welding Symbol
Spot Welding Symbol
Seam Welding Symbol
Welding Symbols for Combined Welds

Flare-V and Flare-Bevel-Groove Welding Symbols
Edge- and Corner- Flange Welding Symbols
Surfacing Welding Symbol Indicating Built-up Surface
Single-V Groove Welding Symbol Indicating Root Penetration
Double-Bevel-Groove Welding Symbol
Projection Welding Symbol
Double-Fillet Welding Symbol

Location of Elements of a Welding Symbol

Basic Joints—Identification of Arrow Side and Other Side of Joint

Butt Joint · Corner Joint · Lap Joint · T-Joint · Edge Joint

Process Abbreviations

Where process abbreviations are to be included in the tail of the welding symbol, reference is made to Table A. Designation of Welding and Allied Processes by Letters, of AWS 2.4/79, 7l.

A2.1-79
© 1979 by American Welding Society

AMERICAN WELDING SOCIETY, INC.

TABLE A-27. GEOMETRIC DIMENSIONING SYMBOL SIZES

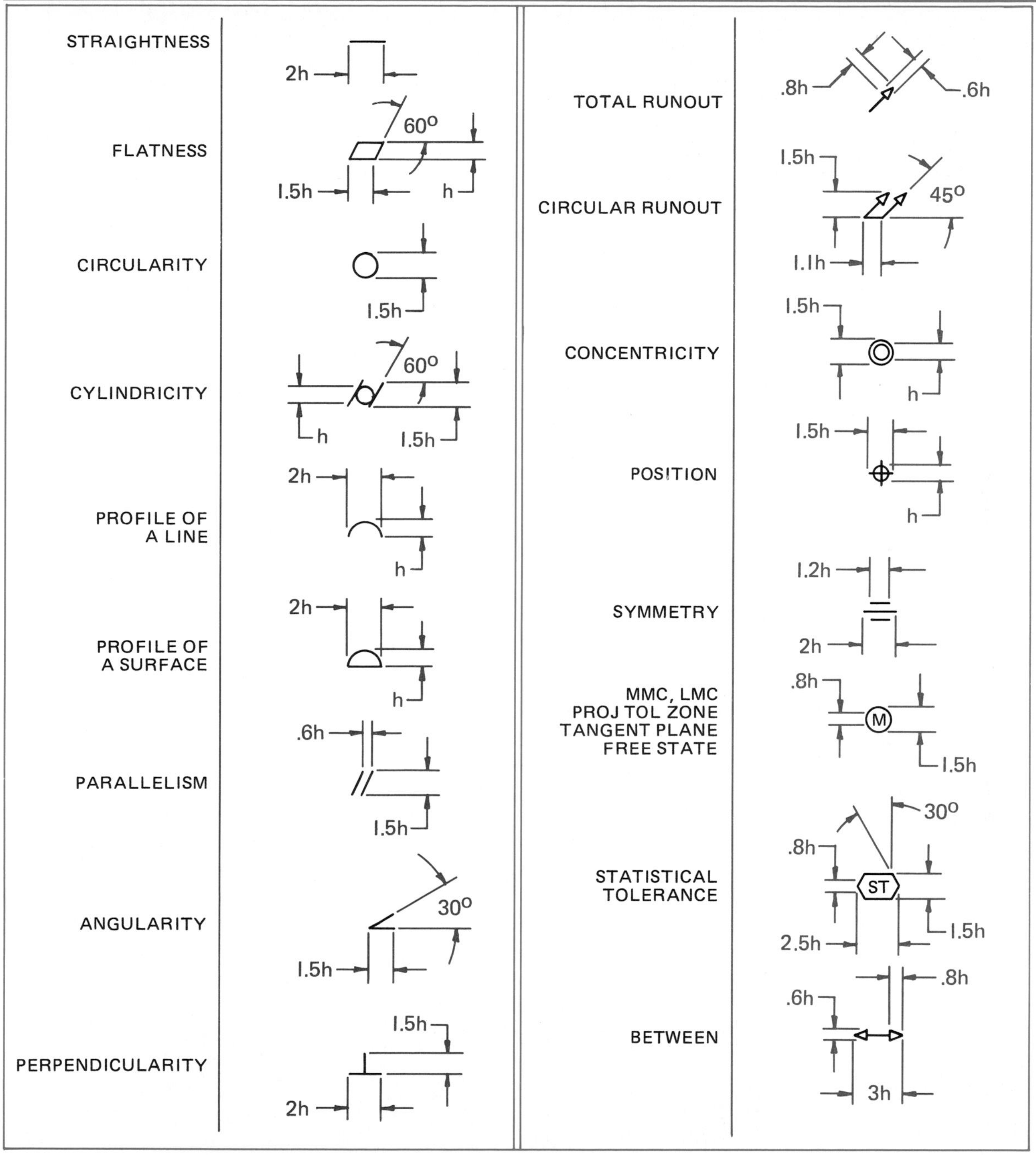

STRAIGHTNESS	
FLATNESS	
CIRCULARITY	
CYLINDRICITY	
PROFILE OF A LINE	
PROFILE OF A SURFACE	
PARALLELISM	
ANGULARITY	
PERPENDICULARITY	

STRAIGHTNESS — 2h

FLATNESS — 60°, 1.5h, h

CIRCULARITY — 1.5h

CYLINDRICITY — 60°, h, 1.5h

PROFILE OF A LINE — 2h, h

PROFILE OF A SURFACE — 2h, h

PARALLELISM — .6h, 1.5h

ANGULARITY — 30°, 1.5h

PERPENDICULARITY — 1.5h, 2h

TOTAL RUNOUT — .8h, .6h

CIRCULAR RUNOUT — 1.5h, 45°, 1.1h

CONCENTRICITY — 1.5h, h

POSITION — 1.5h, h

SYMMETRY — 1.2h, 2h

MMC, LMC PROJ TOL ZONE TANGENT PLANE FREE STATE — .8h, 1.5h

STATISTICAL TOLERANCE — 30°, .8h, 1.5h, 2.5h

BETWEEN — .8h, .6h, 3h

THE METRIC SYSTEM

DEVELOPMENT OF THE METRIC SYSTEM

The metric system of measurement was developed in France in the late 1700s. Within a century, most other countries had adopted it. This helped their trade, because they now manufactured to the same standards and could use each other's goods more easily. The main holdout countries were Great Britain and the United States. Great Britain converted to metrics during the 1960s and 1970s. The United States is now doing the same in many of its industries.

The metric system is a very practical system built around units of 10. It was to be a *natural measurement* based on the Earth's own dimensions. The meter was originally defined as one ten-millionth of the distance from the North Pole to the equator. Measures of volume and weight were related to linear distance. No such unified system had ever been made before. Today, the meter is more precisely defined as a wavelength of the red-orange light of krypton 86.

In 1960, the International Organization of Weights and Measures, representing countries from all over the world, gave the metric system its formal title of Système International d'Unités, which is abbreviated SI. The basic SI units are shown in Fig. B-1.

USING THE METRIC SYSTEM

The most common measurements of the metric system are the *meter*, for length; the *gram*, for mass or weight; and the *liter*, for volume.

The parts and multiples of these units are based on the number 10. This is similar to our money system and our number system, which are also based on 10. It is easier to figure with units of 10 than to divide by 12 (inches) or by 16 (ounces).

Special names are given to certain multiples and divisions of the basic unit. These names become a *prefix* to the name of that unit. The prefixes apply to any of the unit names—meter, gram, or liter. The prefixes and their corresponding amounts are shown in Fig. B-2.

Adding the prefixes to the word *meter* gives the following:
millimeter (mm) = one thousandth of a meter
centimeter (cm) = one hundredth of a meter
decimeter (dm) = one tenth of a meter
dekameter (dam) = ten meters
hectometer (hm) = one hundred meters
kilometer (km) = one thousand meters

Adding the prefixes to the word *gram* gives the following:
milligram (mg) = one thousandth of a gram
centigram (cg) = one hundredth of a gram
decigram (dg) = one tenth of a gram
dekagram (dag) = ten grams
hectogram (hg) = one hundred grams
kilogram (kg) = one thousand grams

Unit	Name	Symbol
Length	Meter	m
Mass	Kilogram	kg
Time	Second	s
Electric current	Ampere	A
Temperature	Kelvin	K
Luminous intensity	Candela	cd

FIG. B-1. The basic SI units.

Prefix	Amount	Fraction	Decimal
Milli	One-thousandth	$\frac{1}{1000}$	0.001
Centi	One-hundredth	$\frac{1}{100}$	0.01
Deci	One-tenth	$\frac{1}{10}$	0.1
Deka	Ten	10	10.0
Hecto	Hundred	100	100.0
Kilo	Thousand	1000	1000.0

FIG. B-2. Metric prefixes and their corresponding amounts.

Length		Volume	
Centimeter	= 0.3937 inch	Cubic centimeter	= 0.0610 cubic inch
Meter	= 3.28 feet	Cubic meter	= 35.3 cubic feet
Meter	= 1.094 yards	Cubic meter	= 1.308 cubic yards
Kilometer	= 0.621 statute mile	Cubic inch	= 16.39 cubic centimeters
Kilometer	= 0.5400 nautical mile	Cubic foot	= 0.0283 cubic meter
Inch	= 2.54 centimeters	Cubic yard	= 0.765 cubic meter
Foot	= 0.3048 meter		
Yard	= 0.9144 meter	**Capacity**	
Statute mile	= 1.61 kilometers	Milliliter	= 0.0338 U.S. fluid ounce
Nautical mile	= 1.852 kilometers	Liter	= 1.057 U.S. liquid quarts
		Liter	= 0.908 U.S. dry quart
Area		U.S. fluid ounce	= 29.57 milliliters
Square centimeter	= 0.155 square inch	U.S. liquid quart	= 0.946 liter
Square meter	= 10.76 square feet	U.S. dry quart	= 1.101 liters
Square meter	= 1.196 square yards		
Hectare	= 2.47 acres	**Mass or weight**	
Square kilometer	= 0.386 square miles	Gram	= 15.43 grains
Square inch	= 6.45 square centimeters	Gram	= 0.0353 avoirdupois ounce
Square foot	= 0.0929 square meter	Kilogram	= 2.205 avoirdupois pounds
Square yard	= 0.836 square meter	Metric ton	= 1.102 short, or net, tons
Acre	= 0.405 hectare	Grain	= 0.0648 gram
Square mile	= 2.59 square kilometers	Avoirdupois ounce	= 28.35 grams
		Avoirdupois pound	= 0.4536 kilogram
		Short, or net, ton	= 0.907 metric ton

FIG. B-3. Metric system equivalents.

Adding the prefixes to the word *liter* gives the following:

milliliter (ml) = one thousandth of a liter

centiliter (cl) = one hundredth of a liter

deciliter (dl) = one tenth of a liter

dekaliter (dal) = ten liters

hectoliter (hl) = one hundred liters

kiloliter (kl) = one thousand liters

As a basis for comparison, approximately U.S. customary and metric equivalents to remember are:

- one liter is a little larger than a quart
- one kilogram is a little greater than two pounds
- one meter is a little longer than a yard
- one kilometer is about five eighths of a mile.

Figure B-3 is a list of metric system equivalents. Figure B-4 shows the actual size as well as the relationship between metric measures of length, volume, and mass.

The meter (m), unit of length, is divided into 100 cm

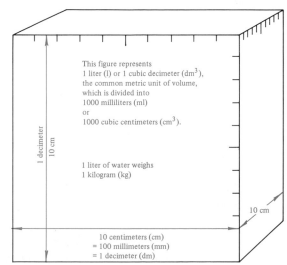

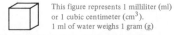

FIG. B-4. Relationship of metric measures of length, volume, and mass. (Courtesy of Metric Association)

THE METRIC SYSTEM IN DRAFTING

In drafting, the primary dimension is length. On drawings, the millimeter is used for dimensions. This is true whether the drawing is prepared with all metric dimensions or is dual-dimensioned, with sizes expressed in both metric and U.S. customary units. In cases where drawings are not dual-dimensioned, a metric equivalent chart (Fig. B-5) may be used to make the conversions. However, not all dimensions are found on the chart. Therefore it is useful to know how to convert mathematically from one system to the other.

Inches to Millimeters

One inch is equal to approximately 25.4 mm. Multiplying an inch dimension by 25.4 will give an answer in millimeters. To simplify the calculation, the inch dimension should be in decimal form.

Example: How many millimeters equal 2.62 in.? Multiply 2.62 by 25.4:

$$
\begin{array}{r}
2.62 \\
\times\ 25.4 \\
\hline
1048 \\
1310 \\
524 \\
\hline
66.548\ \text{mm}
\end{array}
$$

Millimeters to Inches

One millimeter is equal to approximately .0394 in. Multiplying a millimeter dimension by .0394 will give the answer in inches.

Example: How many inches equal 66.5 mm? Multiply 66.5 by .0394:

$$
\begin{array}{r}
66.5 \\
\times\ .0394 \\
\hline
2660 \\
5985 \\
1995 \\
\hline
2.62010\ \text{in., or } 2.62\ \text{in.}
\end{array}
$$

Rounding Off

It is sometimes necessary to round numbers to fewer digits. For example, 25.63220 mm rounded to three decimal places would be 25.632 mm. The same number rounded to two places would be 25.63 mm. The procedure is as follows:

1. When the first number (digit) dropped is less than 5, the last number kept does not change. For example, 15.232 rounded to two decimal places would be 15.23.
2. When the first number dropped is greater than 5, the last number kept should be increased by 1. For example, 6.436 rounded to two decimal places would be 6.44.
3. When the first number dropped is 5 followed by at least one number greater than zero, the last number kept should be increased by 1. For example, 8.4253 rounded to two decimal places would be 8.43.
4. When the first number dropped is 5 followed by zeros, the last number kept is increased by 1 if it is an odd number. If it is an even number, no change is made. For example, 3.23500 rounded to two places would be 3.24. However, 3.22500 would be 3.22.

Angular Dimensions

Angular dimensions do not need to be converted. Angles dimensioned in degrees, minutes, and seconds or in decimal degrees are common to both systems.

mm	in.*	mm	in.*	in.		mm§	in.		mm§
1 = 0.0394		17 = 0.6693		¹⁄₃₂	(0.03125) =	0.794	¹⁷⁄₃₂	(0.53125) =	13.493
2 = 0.0787		18 = 0.7087		¹⁄₁₆	(0.0625) =	1.587	⁹⁄₁₆	(0.5625) =	14.287
3 = 0.1181		19 = 0.7480		³⁄₃₂	(0.9375) =	2.381	¹⁹⁄₃₂	(0.59375) =	15.081
4 = 0.1575		20 = 0.7874		¹⁄₈	(0.1250) =	3.175	⅝	(0.6250) =	15.875
5 = 0.1969		21 = 0.8268		⁵⁄₃₂	(0.15625) =	3.968	²¹⁄₃₂	(0.65625) =	16.668
6 = 0.2362		22 = 0.8662		³⁄₁₆	(0.1875) =	4.762	¹¹⁄₁₆	(0.6875) =	17.462
7 = 0.2756		23 = 0.9055		⁷⁄₃₂	(0.21875) =	5.556	²³⁄₃₂	(0.71875) =	18.256
8 = 0.3150		24 = 0.9449		¼	(0.2500) =	6.349	¾	(0.7500) =	19.050
9 = 0.3543		25 = 0.9843		⁹⁄₃₂	(0.28125) =	7.144	²⁵⁄₃₂	(0.78125) =	19.843
10 = 0.3937		26 = 1.0236		⁵⁄₁₆	(0.3125) =	7.937	¹³⁄₁₆	(0.8125) =	20.637
11 = 0.4331		27 = 1.0630		¹¹⁄₃₂	(0.34375) =	8.731	²⁷⁄₃₂	(0.84375) =	21.431
12 = 0.4724		28 = 1.1024		⅜	(0.3750) =	9.525	⅞	(0.8750) =	22.225
13 = 0.5118		29 = 1.1418		¹³⁄₃₂	(0.40625) =	10.319	²⁹⁄₃₂	(0.90625) =	23.018
14 = 0.5512		30 = 1.1811		⁷⁄₁₆	(0.4375) =	11.112	¹⁵⁄₁₆	(0.9375) =	23.812
15 = 0.5906		31 = 1.2205		¹⁵⁄₃₂	(0.46875) =	11.906	³¹⁄₃₂	(0.96875) =	24.606
16 = 0.6299		32 = 1.2599		½	(0.5000) =	12.699	1	(1.0000) =	25.400

*Calculated to *nearest* fourth decimal place.   §Calculated to *nearest* third decimal place.

FIG. B-5. Inch-millimeter equivalent chart.

ABBREVIATIONS AND SYMBOLS USED ON TECHNICAL DRAWINGS

FIGURE C-1. ABBREVIATIONS AND SYMBOLS USED ON TECHNICAL DRAWINGS

Across Flats ....ACRFLT
American National Standards Institute ....ANSI
And ....&
Angular ....ANLR
Approximate ....APPROX
Assembly ....ASSY
Basic ....BSC
Bill of Material ....B/M
Bolt Circle ....BC
Brass ....BR
Brown and Sharpe Gage ....B&S GA
Bushing ....BUSH
Canada Standards Institute ....CSI
Carbon Steel ....CS
Casting ....CSTG
Cast Iron CI
Center Line ....CL or $\mathbb{C}$
Center to Center ....C to C
Centimeter ....cm
Chamfer ....CHAM
Circularity ....CIR
Cold-Rolled Steel ....CRS
Concentric ....CONC
Counterbore ....⌴ or CBORE
Counterdrill ....CDRILL
Countersink ....∨ or CSK
Cubic Centimeter ....cm³
Cubic Meter ....m³
Datum ....DAT
Degree ....° or DEG
Depth ....DP or ↓
Diameter ....Ø or DIA
Diametral Pitch ....DP
Dimension ....DIM
Drawing ....DWG
Eccentric ....ECC
Equally Spaced ....EQL SP
Figure ....FIG
Finish All Over ....FAO
Flat ....FL
Gage ....GA
Gray Iron ....GI
Head ....HD
Heat Treat ....HT TR
Heavy ....HVY
Hexagon ....HEX
Hydraulic ....HYDR
Inside Diameter ....ID
International Organization for Standardization ...ISO
International Pipe Standard ....IPS
Kilogram ....kg
Kilometer ....km
Left Hand ....LH
Length ....LG
Liter L
Machine Steel ....MST

Machined ....✓
Malleable Iron ....MI
Material ....MATL
Maximum ....MAX
Maximum Material Condition ....Ⓜ or MMC
Meter ....m
Metric Thread ....M
Micrometer ....µm
Millimeter ....mm
Minimum ....MIN
Module ....MDL
Newton ....N
Nominal ....NOM
Not to Scale ....x̲x̲
Number ....NO
On Center ....OC
Outside Diameter ....OD
Parallel ....PAR
Pascal ....Pa
Perpendicular ....PERP
Pitch ....P
Pitch Circle ....PC
Pitch Diameter ....PD
Plate ....PL
Radius ....R
Reference or Reference Dimension ......( ) or REF
Regardless of Feature Size ....Ⓢ

{ISO $\boxed{A}$ ANSI $-A-$}

Revolutions per Minute ....rev/min
Right Hand ....RH
Root Diameter ....RD
Second (Arc) ....(")
Second (Time) ....SEC
Section ....SECT
Slotted ....SLOT
Socket ....SOCK
Spherical ....SPHER
Spotface ....⌴ or SFACE
Square ....□ or SQ
Square Centimeter ....cm²
Square Meter ....m²
Steel ....STL
Straight ....STR
Symmetrical ....╫---╫ or SYM

Taper—Flat ....◁
—Round ....⊏▷
Taper Pipe Thread ....NPT
Thread ....THD
Through ....THRU
Tolerance ....TOL
True Profile ....TP
U.S. Gage ....USG
Watt ....W
Wrought Iron ....WI
Wrought Steel ....WS

TOPOGRAPHIC MAP SYMBOLS

CONTROL DATA AND MONUMENTS

Aerial photograph roll and frame number* 3-20

Horizontal control

Third order or better, permanent mark	Neace △ Neace ⊕
With third order or better elevation	BM △ 45.1 Pike ⊕ BM 45.1
Checked spot elevation	△ 19.5
Coincident with section corner	△ Cactus ⊕ Cactus
Unmonumented*	+

Vertical control

Third order or better, with tablet	BM × 16.3
Third order or better, recoverable mark	× 120.0
Bench mark at found section corner	BM + 18.6
Spot elevation	× 5.3

Boundary monument

With tablet	BM □ 21.6 BM ⊕ 71
Without tablet	□ 171.3
With number and elevation	67 □ 301.1
U.S. mineral or location monument	▲

CONTOURS

Topographic

Intermediate	
Index	
Supplementary	
Depression	
Cut; fill	

Bathymetric

Intermediate	
Index	
Primary	
Index Primary	
Supplementary	

BOUNDARIES

National	
State or territorial	
County or equivalent	
Civil township or equivalent	
Incorporated city or equivalent	
Park, reservation, or monument	
Small park	

*Provisional Edition maps only
Provisional Edition maps were established to expedite completion of the remaining large scale topographic quadrangles of the conterminous United States. They contain essentially the same level of information as the standard series maps. This series can be easily recongnized by the title "Provisional Edition" in the lower right hand corner.

LAND SURVEY SYSTEMS

U.S. Public Land Survey System

Township or range line	
Location doubtful	
Section line	
Location doubtful	
Found section corner; found closing corner	
Witness corner; meander corner	WC + MC

Other land surveys

Township or range line	
Section line	
Land grant or mining claim; monument	
Fence line	

SURFACE FEATURES

Levee	Levee
Sand or mud area, dunes, or shifting sand	Sand
Intricate surface area	Strip mine
Gravel beach or glacial moraine	Gravel
Tailings pond	Tailings Pond

MINES AND CAVES

Quarry or open pit mine	✕
Gravel, sand, clay, or borrow pit	✕
Mine tunnel or cave entrance	⌐
Prospect; mine shaft	X ■
Mine dump	Mine dump
Tailings	Tailings

VEGETATION

Woods	
Scrub	
Orchard	
Vineyard	
Mangrove	Mangrove

GLACIERS AND PERMANENT SNOWFIELDS

Contours and limits	
Form lines	

MARINE SHORELINE

Topographic maps

Approximate mean high water	
Indefinite or unsurveyed	

Topographic-bathymetric maps

Mean high water	
Apparent (edge of vegetation)	

APPENDIX D   TOPOGRAPHIC MAP SYMBOLS CONTINUED

COASTAL FEATURES

Foreshore flat	
Rock or coral reef	
Rock bare or awash	
Group of rocks bare or awash	
Exposed wreck	
Depth curve; sounding	
Breakwater, pier, jetty, or wharf	
Seawall	

BATHYMETRIC FEATURES

Area exposed at mean low tide; sounding datum	
Channel	
Offshore oil or gas: well; platform	
Sunken rock	

RIVERS, LAKES, AND CANALS

Intermittent stream	
Intermittent river	
Disappearing stream	
Perennial stream	
Perennial river	
Small falls; small rapids	
Large falls; large rapids	
Masonry dam	
Dam with lock	
Dam carrying road	
Perennial lake; Intermittent lake or pond	
Dry lake	
Narrow wash	
Wide wash	
Canal, flume, or aqueduct with lock	
Elevated aqueduct, flume, or conduit	
Aqueduct tunnel	
Well or spring; spring or seep	

SUBMERGED AREAS AND BOGS

Marsh or swamp	
Submerged marsh or swamp	
Wooded marsh or swamp	
Submerged wooded marsh or swamp	
Rice field	
Land subject to inundation	

BUILDINGS AND RELATED FEATURES

Building	
School; church	
Built-up Area	
Racetrack	
Airport	
Landing strip	
Well (other than water); windmill	
Tanks	
Covered reservoir	
Gaging station	
Landmark object (feature as labeled)	
Campground; picnic area	
Cemetery: small; large	

ROADS AND RELATED FEATURES

Roads on Provisional edition maps are not classified as primary, secondary, or light duty. They are all symbolized as light duty roads.

Primary highway	
Secondary highway	
Light duty road	
Unimproved road	
Trail	
Dual highway	
Dual highway with median strip	
Road under construction	
Underpass; overpass	
Bridge	
Drawbridge	
Tunnel	

RAILROADS AND RELATED FEATURES

Standard gauge single track; station	
Standard gauge multiple track	
Abandoned	
Under construction	
Narrow gauge single track	
Narrow gauge multiple track	
Railroad in street	
Juxtaposition	
Roundhouse and turntable	

TRANSMISSION LINES AND PIPELINES

Power transmission line: pole; tower	
Telephone line	
Aboveground oil or gas pipeline	
Underground oil or gas pipeline	

GEOLOGICAL TERMINOLOGY

Maps are important in understanding the news, in studying geography and history, in making car trips, and in surveying geology, civil engineering, petroleum engineering, and space exploration.

Brief definitions for some of the terms used in map drafting are listed in this appendix. For more complete descriptions, refer to books on surveying and geology.

bearing The direction of a line as shown by its angle with a north-south line (meridian).

meridian North-south line.

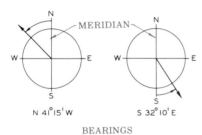

BEARINGS

grid A series of uniformly spaced horizontal and perpendicular lines used to locate points by coordinates, or to enlarge or reduce a figure.

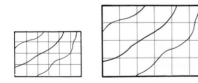

profile A section of the earth on a vertical plane, showing the intersection of the surface of the earth and the plane.

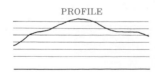

PROFILE

cut Earth to be removed to prepare for construction, such as a desired level or slope for a road.

grade A particular level or slope, as a downgrade.

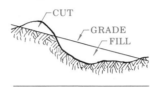

contour A line of constant level showing where a level plane cuts through the surface of the earth.

block diagram A pictorial drawing (generally isometric) of a block of earth, showing profiles and contours.

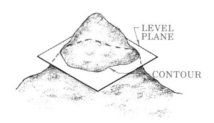

fault A break in the earth's crust with a movement of one side of the break parallel to the line of the break.

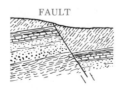

FAULT

stratum A layer of rock, earth, sand, and the like, horizontal or inclined, arranged in flat form clearly different from the matter next to it.

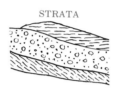

STRATA

fold A bend in a layer or layers of rock brought about by forces acting upon the rock after it has been formed.

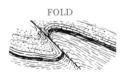

FOLD

dip The angle that an inclined stratum or similar geological feature makes with a horizontal plane. The direction of the dip is perpendicular to the strike.

strike The direction at the surface of the intersection of a stratum with a horizontal plane.

GLOSSARY

(v) = verb

(n) = noun

(adj) = adjective

A/E/C (adj) Architectural, engineering, and construction

acme thread (n) A thread whose depth equals one half of its pitch.

acute angle (n) An angle that measures less than 90°.

addendum (n) The distance a gear tooth extends above the pitch circle.

aeronautics (n) The science of designing, building, and operating all types of flying vehicles.

aerospace industry (n) An industry that includes the design and manufacture of commercial airplanes, leisure aircraft, corporate jets, helicopters, and various spacecraft, as well as many similar products.

airbrush (n) A miniature spray gun used primarily to render illustrations and to retouch photographs.

airfoils (n) The movable parts that help control the aircraft, including ailerons, rudders, stabilizers, flaps, and tabs.

aligned system (n) A system of dimensioning in which the dimensions are placed in line with the dimension lines.

allowance (n) The smallest space permitted between mating parts.

alloys (n) Mixtures of more than one metal or composite material.

alphabet of lines (n) The various lines and line symbols that have specific meanings when used on technical drawings.

alternating current (AC) (n) A form of electrical current in which the electrons flow in one direction for a specific period, and then in the opposite direction during a similar period.

ammeter (n) An instrument that measures electric current in amperes.

angle (n) Two straight lines that meet at one end.

arc (n) Part of a true circle.

arc welding (n) A form of fusion welding in which an electric arc forms between the work (part to be welded) and an electrode.

assembly drawing (n) A drawing, usually three-dimensional, in which all the parts of a machine are shown together in their relative positions.

assembly working drawing (n) An assembly drawing that gives complete information about a part and can be used as a working drawing.

atom (n) The smallest part of an element that still displays the properties of that element.

attributes (n) In CAD, custom information that can be assigned to a block of entities.

auxiliary plane (n) A plane that is parallel to an inclined surface on an object.

auxiliary section (n) A section that results from a cutting plane that is not parallel to any of the normal views.

auxiliary view (n) A projection on an auxiliary plane that is parallel to an inclined surface. It is a view that looks directly at the inclined surface in a direction perpendicular to it, providing a clear, undistorted image of its shape and size.

axis (n) (plural *axes*) A straight, imaginary line that follows the width, height, or depth of an object. The most common names for these axes are: X axis (for width), Y axis (for height or length), and Z axis (for depth).

axis of revolution (n) An imaginary axis, perpendicular to one of the three principal planes, about which the object is revolved in a drawing.

axonometric projection (n) Any projection that uses three axes at angles to show three sides of an object.

azimuth (n) A measurement that defines the direction of a line off due north. The azimuth is always measured off the north-south line in the horizontal plane.

balloon framing (n) A type of framing in which the studs are two stories high; a false girt inserted into the stud wall carries the second-floor joists.

bar chart (n) A type of chart in which bars are used to show relative amounts or values.

basic hole system (n) A system of specifying dimensions and tolerances of close-fitting cylindrical parts in which the design size of the hole is the basic size and the allowance is applied to the shaft.

basic shaft system (n) A system of specifying dimensions and tolerances of close-fitting cylindrical parts in which the design size of the shaft is the basic size and the allowance is applied to the hole.

bearing (n) The angle a line makes with a north-south line in the top view.

bevel (n) A surface slanted to another surface. Called a *miter* when the angle is 45°.

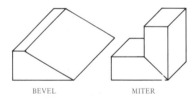

BEVEL MITER

bevel gears (n) Gears used to transfer motion when two gear shafts intersect.

bilateral tolerance (n) A tolerance that allows variations in both directions from a design size. Bilateral variations are usually given with locating dimensions and with any other dimensions that can vary in either direction.

bill of materials (n) A list of the names of parts, material, number required, part numbers,

and so forth needed to manufacture a product.

bisect (v) Divide into two equal parts.

block diagram (n) (1) In mapping, a three-dimensional projection of a map developed using an isometric view. (2) In electrical drafting, a diagram made up of squares and similar symbols joined by single lines; the diagram shows how the components or stages of an electrical or electronic circuit are working.

bolt circle (n) A circular centerline locating the centers of holes arranged in a circle.

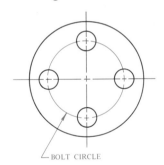

BOLT CIRCLE

boss (v) A raised circular surface as used on a casting or forging.

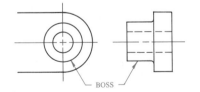

BOSS

branching menu (n) In CAD and other computer software, a menu that branches out from a main or high-level menu.

broken-out section (n) A section that shows an object as it would look if a portion of it were removed by a cutting plane and then "broken off" to reveal the cut surface and insides.

bulkheads (n) The shaped walls that make up the fuselage structure in an airplane.

burr (n) A rough or jagged edge caused by cutting or punching.

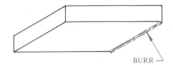

BURR

cabinet oblique (adj) An oblique drawing in which the receding lines are drawn at one-half size.

CAD (n) Computer-aided drafting *or* computer-aided design, depending on the context.

cam (n) A machine part mounted on a turning shaft, used to change rotary (turning) motion into back-and-forth motion.

camera-ready copy (n) A paper master copy of a document from which the printer photographs and creates prints.

cartographers (n) People who have been trained to gather the necessary information and prepare maps.

cartography (n) The field of mapmaking.

cavalier oblique (adj) An oblique drawing in which the receding lines are drawn at their full length.

chamfer (v) To bevel an edge that has been beveled.

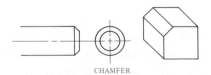

CHAMFER

chord (n) A straight line between any two points on a circle.

circular pitch (n) The distance from a point on one gear tooth to the same point on the next tooth measured along the pitch circle.

circumference (n) The distance around the perimeter of a circle.

circumscribed (adj) A term for a regular polygon that completely surrounds a circle so that each side of the polygon is tangent to the circle.

closed catalog system (n) A filing system in which one person is responsible for keeping all drawings safe.

coincide (v) In a three-dimensional drawing, two lines at different elevations that appear as a single line in a particular view.

combination drawing (n) A two-part detail drawing that can serve two purposes, such as dimensions and notes both for forging and for machining.

comparison bar chart (n) A bar chart in which each item has two or more bars that compare the item in different years, colors, etc.

compass (n) An instrument that allows drafters to draw circles and arcs.

composites (n) Materials such as carbon fiber that are made artificially from metals, plastics, etc.

composition (n) In lettering, the arrangement, style, and size of words and lines on a drawing.

compound bar chart (n) A bar chart in which the total length of each bar is made up of more than one part. Each part represents a time period or other property of the item.

computer literacy (n) An understanding of how to operate a computer.

computer-aided design and drafting (CADD) (n) A method of designing new products and creating the working drawings using computer software meant for this purpose.

computer-aided manufacturing (CAM) (n) A method of manufacturing in which robots and other computerized machines cut, move, drill, and stock production parts.

computer-integrated manufacturing (CIM) (n) Manufacturing systems in which computers are used to manufacture products directly from CAD drawings.

concentric circles (n) circles, usually of different sizes, that share the same center point.

conceptual design (n) A rough design idea; the first attempt to put a new product "on paper."

concurrent engineering (n) An engineering system in which a complete information database is available to every department at the same time.

conductors (n) Materials with small resistance to the flow of electrons.

contour distance (n) The vertical distance between contour lines on a contour map.

contours (n) Lines of constant level on a map.

conventional break (n) A break that allows a uniform part of a very long object to be removed, making it easier to draw and understand.

conversion chart (n) A chart that provides a convenient way to convert from one system or scale to another.

counterbore (v) To make a hole deeper and wider. **(n)** A counterbored hole.

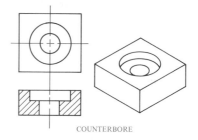

COUNTERBORE

countersink (v) To drill a cone shape at the end of a hole. **(n)** A counter sunk hole.

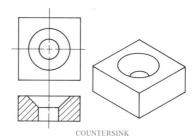

COUNTERSINK

crease lines (n) The fold or bend lines on a pattern development.

cross section (n) In architectural drafting, a full section that cuts across the entire structure to aid in interpreting the relationships of the important spaces.

crosshatching (n) Evenly spaced lines that show a cut surface on a section. Also called *section lining*.

crossing window (n) In AutoCAD, a selection window that selects all objects inside or partly inside the window.

crown (n) The contour of the face of a belt pulley, rounded or angular, used to keep the belt in place. The belt tends to climb to the highest point.

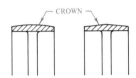

current (n) A steady stream of electrons flowing through a wire or metal surface.

curve (n) In a chart, a line which may be straight, curved, broken, or stepped.

cutting plane (n) The plane through an object at which a section is created.

cutting-plane line (n) A line on a normal view that shows where the cutting plane passes through an object.

cycle (n) In electronics, one complete AC alternation.

cycloidal curve (n) The path of a curve formed by a point on a rolling circle.

data cartridges (n) Cartridges that look similar to audio tapes, but can be used to store large amounts of data, including drawing files, for backup purposes.

datum (n) A point, line, or surface that is assumed to be exact, and from which other dimensions can be computed or located.

dead load (n) In architecture, the weight of the building or structure.

dedendum (n) The distance a gear tooth extends below the pitch circle.

descriptive geometry (n) A method used by designers to solve spacial and geometric problems.

design drafting technician (n) A member of the engineering design team who has the skills to combine drafting and design.

design drawings (n) Drawings that help designers visualize various design options; preliminary drawings based on the conceptual design.

design engineer (n) An engineer who applies math, science, and technology principles to solve problems for production and construction.

designer (n) A person with drafting and design experience and training who works on product designs.

desktop publishing (n) The process of producing typeset pages, which include text, illustrations, and photographs arranged in an attractive format that is ready for printing.

detail drawing (n) A drawing for a single part that includes all the dimensions, notes, and information needed to make the part.

detail section (n) In architectural drafting, a section drawn at a much larger scale

than the elevation, in which a vertical plane cuts through walls to show construction details.

detailed representation (n) A method of drawing screw threads that approximates the actual look of the threads.

development (n) A full-size layout of an object made on a single flat plane.

dialog box (n) In CAD, a menu that appears on the screen as a result of certain commands, allowing the drafter to make a selection or enter further information.

diameter (n) A straight line describing the distance from any one point on a circle through the center of the circle to a corresponding point on the other side.

diametral pitch (n) The number of gear teeth per inch of pitch diameter.

diazo (n) A reproduction method in which prints have dark lines on a white background.

die stamping (v) pressing a flat sheet into shape under heavy pressure.

dihedral angle (n) The angle formed when two planes intersect. A dihedral angle can only be measured perpendicular to the line of intersection.

dimension line (n) A thin line in a dimension that shows where a measurement begins and where it ends, or that shows the size of an angle.

dimensioning (n) Another name for the size description of an object or part.

dimetric projection (n) An axonometric projection in which the angles between only two of the axes are equal.

dip (n) The slope of a line of surface exposure of a contact between two formations on a geological surface map.

direct current (n) A form of electrical current in which the electrons flow in one direction only.

displacement diagram (n) A diagram that shows the shape of a cam and the motion the cam will produce through one 360° revolution.

dock (v) In AutoCAD, to attach a toolbar to the top or sides of the drawing area.

double thread (n) Two helical ridges side by side; a thread in which the lead is twice the pitch.

drafter (n) A person who has been trained in drafting technology and can create accurate drawings of products for use in manufacturing, quality control, and other aspects of product development.

drafting film (n) A film that is sometimes used instead of vellum for technical drawings.

drafting technology (n) The tools and techniques used by designers, drafters, and engineers to develop ideas for new products into usable technical drawings.

dual dimensioning system (n) A dimensioning system in which the size description is given in both decimal inches and millimeters.

elbow (n) A joint in a pipe or duct; a place at which two pieces meet at an angle other than 180°.

electrostatic reproduction (n) A process in which electrostatic force (static electricity) is used to reproduce original drawings; also called *xerography.*

elements (n) The imaginary flat surfaces of a cylinder that is considered to be composed of a many-sided prism.

elevation (n) In architectural drafting, a drawing of a facade of a structure.

ellipse (n) A regular oval that has two centers of equal radius.

engineer (n) A person who has at least a four-year degree in an engineering specialty such as mechanical, electrical, architectural, or nuclear engineering.

entities (n) In CAD, the individual geometric elements such as lines and circles from which a drawing is developed; also called *objects.*

entity attributes (n) In CAD, the color, linetype, and line width of an entity.

erasing shield (n) A thin, flat plastic or metal guard that allows a drafter to erase a small area without accidentally erasing nearby lines.

extension lines (n) Thin lines in a dimension that extend the lines or edges of the views to the dimension line.

FAO (n) A notation on a drawing that stands for "finish all over."

fault (n) In geology, the line along which a layer of earth broke.

fillet (n) The rounded-in corner between two surfaces.

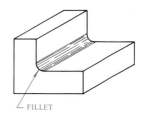

FILLET

fillet weld (n) A weld that is similar in shape to a groove weld, but sits on top of the work material instead of inside a groove.

filtering (v) In CAD, a method of entity selection that selects only those entities that match specified criteria or attributes.

finish mark (n) A surface-texture symbol which shows that a surface is to be machined, or finished.

finite element analysis (FEA) (n) A process in which computer programs are used to assign materials to computer models of structural members and then test them under various loads.

first-angle projection (n) A form of projection used in European countries, in which the front view is located above the top view.

fixed-wing aircraft (n) Aircraft such as airplanes, in which two or more nonmoving wings are fastened to the fuselage.

flange (n) A rim extension, as at the end of a pipe.

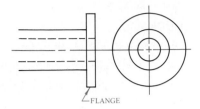

flash welding (n) A kind of resistance welding done by placing the parts to be welded in very light contact or by leaving a very small air gap and allowing an electric current to flash, or arc, which melts the ends of the parts together.

flowchart (n) A chart that shows sequential information.

flyout (n) In AutoCAD, a related group of icons that appears when you hold down the mouse button over an icon in a toolbar.

follower (n) The machine part in contact with a cam that receives the back-and-forth motion.

foreshortened (adj) Appearing shorter than its true length.

formations (n) Natural geological features.

frequency (n) In electronics, the number of times a cycle is repeated in 60 seconds.

frisket paper (n) A special paper used to protect areas of a drawing while using an airbrush on other areas.

front view (n) In a drawing, the view that shows the exact shape and size of the front of an object.

frontal line (n) In descriptive geometry, a line constructed parallel to a reference line in the horizontal plane and projected into the vertical plane to show an inclined line in its true length.

full section (n) A sectional view that shows an object as if it were cut completely apart from one end or side to the other.

fuselage (n) The central body of an airplane, which holds the pilot, passenger, and cargo compartments.

fusion welding (n) Any welding process in which the welder applies heat to create the weld.

Gantt chart (n) An organizational chart that breaks an operation or project into discrete steps and assigns a certain amount of time for each step.

gas and shielded arc welding (n) A welding process that combines arc welding and gas welding.

gas welding (n) A welding process in which a combination of gases, usually oxygen and acetylene, creates the heat for the welding.

gasket (n) A thin piece of material placed between two surfaces to produce a tight joint.

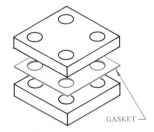

gear (n) A device that transmits motion using a series of teeth.

geology (n) A science that deals with the makeup and structure of the earth's surface and interior depths.

geometric construction (n) A very accurate drawing made up of individual lines and points drawn in proper relationship to one another; usually made with instruments.

geometric dimensioning and tolerancing (n) The specification of permissible variations for the material, size, shape, and geometric characteristics of an object or part, in the form of tolerance or limits.

geometry (n) The study of the size and shape of things; the relationship among the elements of a drawing.

Gothic lettering (n) A single-stroke lettering style; the most commonly used style for lettering drawings.

grade (n) In civil engineering, the incline of a highway or land, measured as a percentage.

groove weld (n) A weld located in a groove or notch in the work material.

guidelines (n) Lightly sketched lines on a drawing used to keep lettering straight and uniform.

half section (n) A section in which one quarter of the original object has been cut away; one half of a full section.

hanging indent (n) In desktop publishing, a format in

which the first line of a paragraph "hangs" out to the left, and the rest of the paragraph is indented; commonly used for numbered or bulleted lists.

hard copy (n) In CAD, the final paper copy of a drawing made by printing or plotting.

harmonic motion (n) A type of motion used by cams when a smooth starting and stopping movement is necessary.

helix (n) The curving path that a point would follow if it were to travel in an even spiral around a cylinder and parallel to (in line with) the axis of that cylinder.

hemming (v) A way of stiffening metal edges by folding them over once (single-hemmed edge) or twice (double-hemmed edge).

horizontal plane (n) The side-to-side plane of an orthographic projection.

hydraulics (n) Fluid under pressure.

hypotenuse (n) The side of a right triangle that is opposite the right angle.

icon (n) In AutoCAD, a small picture representing a CAD command or function that you can activate by clicking it.

igneous (adj) A type of rocks made of the basic crystalline materials that make up the earth's crustal ring. This rock was once molten. It has cooled, but it has not been eroded, and its makeup as not changed.

implementation (n) In drafting, the process of draw-

ing an object that has been visualized.

inclined (adj) Slanted; neither horizontal nor vertical.

inclined line (n) A line that is slanted in one of the three main reference planes but is parallel to one other reference plane.

inclined plane (n) A plane that is perpendicular to one reference plane and inclined to the other two.

indent (n) In desktop publishing, an extra wide margin on the left side of the first line of a paragraph.

india ink (n) A completely opaque ink used for technical drawings.

inscribed (adj) A term for a regular polygon that is completely contained within a circle so that each vertex of the polygon touches the circle.

insulators (n) Materials through which electrons do not flow easily.

intelligent attributes (n) In CAD, attributes that describe the function of an entity.

interconnection diagram (n) A type of connection or wiring diagram that shows only connections outside a component.

intermediate (n) A copy of a drawing used in place of the original; intermediates allow drafters to change a drawing without spending time tracing and redrawing.

intersect (v) Cross. The point at which two lines, arcs, or circles cross is called the *point of intersection.*

involute curve (n) The shape of the teeth on involute gears that lets the teeth mesh smoothly.

irregular curve (n) A drafting aid that provides templates for various noncircular curves. Also called a French curve. Irregular curves are available in many different shapes.

isometric axes (n) Three axes in two-dimensional space, equally spaced 120° apart.

isometric drawing (n) A drawing in which the object is aligned with isometric axes; similar to an isometric sketch, except an isometric drawing is made using instruments.

isometric lines (n) Lines on an isometric drawing that are parallel to the isometric axes.

isometric plane (n) A plane that is parallel to one of the faces of the isometric cube.

isometric sketch (n) A sketch that shows height, width, and depth of an object by using three axes set at 120° angles to one another.

kerf (n) A slot or groove made by a cutting tool.

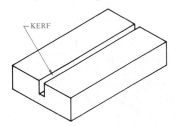

kerning (n) In desktop publishing, the horizontal space between letters.

keyway or keyset (n) A groove or slot in a shaft or hub into which a key fits.

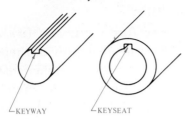

KEYWAY   KEYSEAT

laminating (v) Cutting material into thin slabs and then gluing them together.

leader (n) A thin line drawn from a note or dimension to the place where it applies. It begins with a short horizontal line, then slants at an angle and ends with an arrowhead at the point of application.

leading (n) In desktop publishing, the vertical space between lines.

left-hand thread (n) A thread that screws in when it is turned counterclockwise.

lettering (v) Printing clear, concise words on a drawing to help people understand the drawing.

level line (n) In descriptive geometry, a horizontal line constructed in the vertical reference plane.

lift (n) The upward push on an aircraft that results from the air being pushed downward by the wings.

line (n) The path between two points.

line graphs (n) Charts in which a line is used to show trends or how something changes over time.

live load (n) In architectural drafting, any weight a building must bear that is not part of the weight of the building itself (such as snow or furniture).

lug (n) An "ear" forming a part of, and extending from a part.

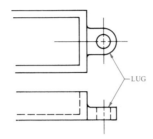

LUG

macro (n) In word processing, a short script that accomplishes a task that will be needed over and over again.

main menu (n) A default menu that appears when you open a CAD program.

major diameter (n) (1) In an ellipse, the longer of the two axes. (2) In detailed representation of a screw thread, the crest diameter.

masking (v) In CAD, a method of entity selection that selects only those entities that match specified criteria or attributes.

measuring lines (n) Vertical construction lines in a radial-line development.

media (n) (singular: *medium*) Materials such as film and paper on which drafters create drawings.

metamorphic (adj) Sedimentary rocks that have been deeply buried, heated to high temperatures, and subjected to great pressure. The combined heat and pressure recompose the rocks so that they can no longer be identified as sedimentary.

microfilm (n) A storage medium on which a large drawing is reduced to a very small size and recorded on film.

minor diameter (n) (1) In an ellipse, the shorter of the two axes. (2) In detailed representation of a screw thread, the root diameter.

model space (n) In Auto-CAD, the mode in which a drawing is created.

multiview drawing (n) An exact representation of a three-dimensional object in one plane, using two or more views.

neck (v) To cut a groove around a cylindrical part, generally at a change in diameter.

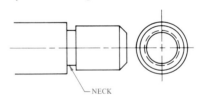

NECK

negative cylinder (n) In a drawing of a three-dimensional object, a hole. It consists of "nothing," yet it has size (the dimensions of the hole).

nominal size (n) The dimensions of lumber before the wood has been surfaced, or prepared for use.

nomograms (n) Engineering charts that show the solutions to problems containing three or more variables.

nonisometric lines (n) Lines on an isometric drawing that do not run parallel to the isometric axes.

normal line (n) In descriptive geometry, a line that is perpendicular to one of the three reference planes.

normal oblique (n) An oblique drawing in which the receding lines are drawn at three-quarter size.

normal plane (n) In descriptive geometry, a plane that is perpendicular to two of the reference planes and parallel to the third.

normal views (n) The front, top, and right-side views of a multiview drawing.

object snaps (n) Functions in a CAD program that allow the drafter to "snap to," or pick accurately, the exact midpoint, endpoint, center, or other special point on an entity.

objects (n) In CAD, the lines, arcs, circles, and polygons used to define the shape of a part; also called *entities*.

oblique line (n) In descriptive geometry, a line that appears inclined in all three reference planes.

oblique plane (n) In descriptive geometry, a plane that appears inclined in all three reference planes.

oblique sketch (n) A sketch in which the height and width are shown in their true size (with axes at 90° to each other), but in which the depth can be drawn at any angle.

obtuse angle (n) An angle that measures more than 90°.

offset section (n) A section in which the cutting plane is not a straight line but shifts at one or more places to show a detail or miss a part.

ogee curve (n) A reverse curve that looks something like an S.

omission chart (n) A chart in which part of the chart is left out; used when the vertical scale needs to be much larger than the horizontal scale to show the information adequately.

opaque (adj) Dark; does not allow light to pass through.

open file system (n) A filing system in which original drawings are always available to anyone in a department.

operations map (n) A map that shows the relationship between the land's physical features and the operation that is to be performed.

orthographic projection (n) The method of representing the exact form of an object in two or more views on planes, usually at right angles to each other, by lines drawn perpendicular from the object to the planes.

outline view (n) A view that shows the shape of an object's flat surface or surfaces.

overlay (n) A piece of tracing paper placed on top of a sketch or drawing so that you can see the drawing through it. Overlays allow you to trace the desired parts of the drawing underneath quickly.

paper space (n) In AutoCAD, the mode in which a drawing is laid out in one or more views to present a final working drawing.

paperless environment (n) A system in which computerized documents and drawing files are used exclusively.

parallel (adj) Side-by-side. By definition, parallel lines are exactly the same distance apart; every point on the first line is exactly the same distance from its corresponding point on the second line. Parallel lines never intersect.

parallel circuits (n) Circuits in which current can flow through more than one path simultaneously.

parallel-line development (n) A way of making a pattern by drawing the edges of an object as parallel lines.

partial auxiliary view (n) An auxiliary view in which some elements have been omitted for clarity.

pattern (n) A flat surface development that can be folded, rolled, or otherwise formed into the required shape.

perpendicular (adj) At right angles to (90°).

perspective drawing (n) A three-dimensional representation of an object as it looks to the eye from a particular point.

PERT chart (n) (Project Evaluation and Review Technique) An organizational chart used in the PERT system, designed by the U.S. Navy. Various types of PERT charts define the tasks required for a project, assign probabilities to production factors such as efficiency and productivity, and then plot them against project activities and operations.

phantom section (n) A hidden section used to show, in one view, the inside and outside of an object that is not completely symmetrical.

photo retouching (n) A process used to change details on a photograph.

photodrafting (n) A drafting process in which a photograph is overlaid with line work to create an enhanced, annotated picture of a product.

photogrammetry (n) A method of mapmaking in which photographs taken by satellites are used to make three-dimensional measurements.

pictograph (n) A pictorial graphic chart similar to a bar chart, except that pictures or symbols are used instead of bars.

pictorial drawing (n) A drawing that shows an object as it appears to the human eye or in a photograph.

picture plane (n) In oblique and perspective drawing, the plane on which the view is drawn.

pie chart (n) A circle chart in which the circle equals 360°. Various sectors of the "pie" represent percentages of the whole.

piercing point (n) The point at which a line intersects a plane.

pin-bar drafting (n) An overlay system that shortens the time it takes to produce a set of drawings by allowing drafters in different trade disciplines (plumbing, electrical, and mechanical) to work on different aspects of the project at the same time.

pinion (n) A small spur gear used with involute gears and rack-and-pinion gears.

pitch (n) (1) The distance from a point on one thread form to the corresponding point on the next thread form. (2) In architecture, the slope of a roof.

pitch diameter (n) The point of tangency of two meshing gears.

pixels (n) Picture elements on a computer screen.

plane (n) In geometry, a flat surface that can be defined by two nonparallel lines or by any two-dimensional polygon.

plank and beam framing (n) A type of framing that uses heavier posts and beams than other systems, allowing ceilings to be higher and more open, with fewer supporting members.

plat (n) A map used to show the boundaries of a piece of land and to identify it.

plug weld (n) A weld that fits into a small hole in the work material.

point (n) A symbol that describes a location in space.

point of origin (n) In CAD, an insertion point that is defined when you create an entity.

polygon (n) A closed figure made up of straight line segments. A *regular* polygon is one in which all the sides are the same length and all the angles are equal. An *irregular* polygon is one in which the sides and angles have different values.

pop-up window (n) In CAD, a menu that appears on the screen as a result of certain commands, allowing the drafter to make a selection or enter further information.

pressure angle (n) The angle of the teeth of a gear in relation to the pitch circle.

prestressed concrete (n) Concrete in which the reinforcing bars are stretched before the concrete is poured over them.

primary auxiliary views (n) Auxiliary views developed directly from the normal views.

production drawings (n) The final drawings used to manufacture a product.

profile line (n) A line constructed parallel to the profile reference plane that shows true length.

profile plane (n) The side view of an orthographic projection.

progressive chart (n) A bar chart in which the horizontal or vertical scale provides a range over which the bars may vary; therefore, not all of the bars begin at the same place.

projection weld (n) A weld in which one part has a boss projection.

proportion (n) Size relative to the size of other objects in the same drawing.

prototype (n) A functional model of a product that allows engineers to test and analyze it before it goes into full production.

prototype drawing (n) In CAD, a drawing in which a drafter sets up commonly used parameters such as units and layers. This drawing is used as a template for other drawings that use those parameters.

protractor (**n**) A drafting instrument that measures the angles of arcs and circles.

pull-down menu (**n**) In CAD, a menu that appears when you pick the name of the menu; names are usually displayed across the top of the screen.

quadrants (**n**) Quarters of a circle; the 0°, 90°, 180°, and 270° points of a circle.

radial-line development (**n**) A development in which the measuring lines radiate from a single point, such as in the development of a cylinder or cone.

radius (**n**) (plural *radii*) The distance from the center of a circle to its edge.

rapid prototyping (**n**) A process in which a solid model created by a computer is used directly to create a prototype of the product made of polymer.

reference assembly drawing (**n**) A special assembly drawing that identifies parts to be assembled.

regular polygon (**n**) A polygon in which all the angles are equal and all the sides are of equal length.

reinforced concrete (**n**) Concrete in which steel bars have been embedded.

removed section (**n**) A section that has been moved from its normal place to somewhere else on the drawing sheet.

reproduction (**n**) The process of making copies of a drawing.

resistance (**n**) In electronics, the amount of difficulty electrons have flowing through a material.

resistance welding (**n**) Any welding process in which a combination of heat and pressure are used to create the weld.

revolution (**n**) A drawing in which the object has been revolved, or turned, to place its primary features parallel to the reference planes.

rib (**n**) A thin, flat part of an object used to strengthen or brace another part of the object.

right angle (**n**) An angle that measures exactly 90°.

right-hand thread (**n**) A thread that screws in when it is turned clockwise.

right-side view (**n**) In a drawing, the view that shows the exact shape and size of the right side of an object.

root diameter (**n**) The diameter of a gear at the base of the teeth.

round (**n**) The rounded-over corner of two surfaces.

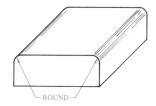

scales (**n**) Drafting instruments used to lay off and mark distances on a drawing.

schedules (**n**) In architectural drafting, lists that define and describe details shown by symbols on the actual drawings.

schematic diagram (**n**) In electronics, an elementary diagram that shows the ways a circuit is connected and what it does.

schematic representation (**n**) A method of drawing threads in which the threads are shown as symbols rather than as they really look.

scratchboard drawing (**n**) A form of line rendering in which scratchboard is coated with india ink, the image is drawn on the inked surface of the scratchboard, and then scratches are made through the ink to expose the lines or surfaces.

screw-thread series (**n**) A method of grouping and organizing thread diameters and pitches in the Unified screw-thread system.

seam weld (**n**) A weld formed along the seam of two adjacent parts.

secondary auxiliary view (**n**) A view that has been projected from a primary auxiliary view; used to find the true size and shape of an inclined surface that lies on an oblique plane.

section lining (**n**) Thin, evenly spaced lines that show a cut surface on a section. Also called *crosshatching*.

sectional view (**n**) A view that shows an object as if part of it were cut away to expose the insides. Also called a *section*.

sedimentary (**adj**) A type of rocks that have been deposited, usually in water, in layers of different thicknesses.

semiconductors (n) Materials that act like conductors at high temperatures but like insulators at low temperatures.

series circuits (n) Circuits in which current can flow through only one path at a time.

service bureaus (n) Companies that can take completed desktop publishing files and create film for printing.

simplified representation (n) A method of drawing screw threads in which the threads are represented by drawing the crest and root lines as dashed lines. Individual threads are not shown.

single thread (n) A single ridge in the form of a helix; a thread in which the lead is equal to the pitch.

single-line diagram (n) A diagram that shows the course of an electric circuit and the parts of the circuit using single lines and graphic symbols.

site plan (n) An architectural drawing that shows the lot and locates the house on it, as well as driveways, sidewalks, utility easements, and other pertinent information required by the building inspector.

skew lines (n) Two lines that are not parallel and do not intersect in three-dimensional space.

slope (n) In descriptive geometry, the angle of a line, in degrees, from the horizontal plane.

slot weld (n) A weld that fits into a slot-shaped hole in the work material.

solid model (n) In CAD, a three-dimensional model that has mass properties.

spars (n) On an airplane, the beams that run the length of the wings and are attached to the ribs that form the shape of the wings.

specifications (n) In construction and manufacturing, a detailed list of facts concerning materials and measurements.

spherical (adj) Perfectly round in three dimensions.

spot weld (n) A weld in which current and pressure are confined to a small area between electrodes.

spur gear (n) A small, toothed wheel that transfers rotary motion from one shaft to another.

step chart (n) A special type of line graph which shows data that remains constant during regular or irregular intervals.

strata graph (n) A shaded-surface graph that compares two different items, such as use of different kinds of materials, over a period of time.

stratum (n) (plural: *strata*) A distinct layer in the earth's surface.

stretchout (n) A full-size layout of an object made on a single flat plane.

stretchout line (n) A line on a development that shows the full length of the pattern when it is unfolded.

strike (n) In geology, the direction of contact between two formations.

structure (n) Anything that has been constructed, such as buildings, dams, and bridges.

successive revolutions (n) The process of revolving an object around an axis perpendicular to one plane, and then revolving it again around an axis perpendicular to a different plane.

surface development (n) A full-size layout of an object made on a single flat plane.

symbol library (n) In CAD, a drawing that contains a large number of related symbols that can be inserted into other drawings.

symmetrical (adj) Mirrored around an axis; for example, an object in which one side is an exact mirror image of the other.

syntax (n) The structure of a CAD program.

T-square (n) A drafting instrument that consists of a head, which fits against a true edge of the drafting table, and an arm, which provides a true edge on the drawing surface.

tabulated drawing (n) A drawing in which dimensions are identified by letters; a table placed on the drawing tells what each dimension is for different sizes of the part.

tag (n) In desktop publishing, a label assigned to a paragraph that controls all the formatting for the paragraph.

tangent (adj) An arc, line, or circle that touches another arc, line, or circle at one point only.

technical drawings (n) A set of accurate drawings that describe the exact size and shape of an object or product.

technical illustration (n) A pictorial drawing that provides technical information by visual methods.

technical promotional drawings (n) Technical illustrations used to promote a product on the market.

template (n) (1) A drafting aid that consists of a piece of metal or plastic with shapes cut out, allowing drafters to draw common shapes (such as ellipses) quickly. (2) In desktop publishing, a blank page that contains all the formatting for a page in a document.

texture (n) The surface quality of an object.

thermit welding (n) A welding process in which a charge of finely divided aluminum and iron oxide is ignited by a small amount of ignition powder. The charge burns rapidly to produce a very high temperature, which melts the metal. The metal then flows into molds and fuses mating parts.

third-angle projection (n) A form of projection used in the United States and Canada, in which the top view is located above the front view.

title block (n) An area on a drawing that contains information about the drawing, the company, the drafter and so on.

tolerance (n) The amount by which a machined part is allowed to vary from absolute measurements.

toner (n) A negatively charged powder that adheres to the charged surfaces of a positively charged plate in electrostatic reproduction.

toolbars (n) In AutoCAD, groups of related icons that can be displayed on the screen.

top view (n) In a drawing, the view that shows the exact shape and size of the top of an object.

topographic maps (n) Maps that present complete pictorial descriptions of the areas shown, including boundaries, natural features, structures, vegetation, and relief (elevations and depressions).

tracking (n) In desktop publishing, the amount of space between words.

transition pieces (n) Pieces used to connect pipes or openings of different shapes, sizes, or positions.

transparent (adj) Something that does not block light completely and therefore can be seen through.

triangulation (n) A method used for making approximate developments of surfaces that cannot be developed exactly.

trimetric projection (n) An axonometric projection in which none of the angles between the axes are equal.

triple thread (n) Three ridges side by side; a thread in which the lead is three times the pitch.

truss (n) A configuration of structural elements that adds strength to a structure.

type (n) In desktop publishing, the letters and other characters as they appear on a page.

type family (n) A group of typefaces that have the same general appearance.

typesetting (n) The process of arranging text, illustrations, and photographs in an attractive format that is ready for printing.

unidirectional system (n) A system of dimensioning in which all the dimensions read from the bottom of the sheet, no matter where they appear on the drawing.

uniform motion (n) "Straight-line" movement in which time and distance are directly proportional.

uniformly accelerated and decelerated motion (n) The motion of cams that do not operate at a steady speed but at steadily increasing and decreasing speeds.

unilateral tolerance (n) A tolerance that allows variation on only one side of the design size of a part.

user coordinate system (UCS) (n) In CAD, a custom coordinate system specified by the drafter.

vanishing point (n) A point in a perspective drawing at which objects theoretically appear so small that they can no longer be seen by the human eye.

vellum (n) Paper that has been treated with resins to make it transparent.

vertex (n) The point at which two lines or sides come together in an angle or polygon.

vertical plane (n) The up-and-down, or vertical, view of an orthographic projection.

visualization (n) The ability to see clearly in the mind's eye what a machine, device, or other object looks like.

voltage (n) Electrical pressure.

voltmeter (n) An instrument that measures electromotive force (pressure) in volts.

wash rendering (n) A form of watercolor rendering done with watercolor and watercolor brushes that is commonly used for rendering architectural drawings and product advertisements.

web (n) A thin, flat part of an object used to strengthen or brace another part of the object.

welding (n) A way of fastening metal pieces together using heat or a combination of heat and pressure.

western framing (n) A method of framing in which each floor is framed separately. The first floor is a platform built on top of the foundation wall; one-story studs are used to develop and support the framework for the second story and the load-bearing interior walls. Also called *platform framing*.

wireframe model (n) In CAD, a model that exists in three dimensions but has no surfaces and no mass properties.

wiring (v) A process that involves reinforcing open ends or edges of articles by enclosing a wire in the edge.

wiring diagram (n) A diagram that shows how the components of a circuit are connected.

working drawing (n) A drawing that gives all the information needed to complete a part or a complete machine or structure.

worm gear (n) A gear that is similar to a screw and can have single or multiple threads.

INDEX